Bayesian Reliability Analysis

Probability and Mathematical Statistics (Continued)

SCHEFFE • The Analysis of Variance
SEBER • Linear Regression Analysis
SEN • Sequential Nonparametrics: Invariance Principles and Statistical Inference
SERFLING • Approximation Theorems of Mathematical Statistics
TJUR • Probability Based on Radon Measures
WILLIAMS • Diffusions, Markov Processes, and Martingales, Volume I: Foundations
ZACKS • Theory of Statistical Inference

Applied Probability and Statistics

ANDERSON, AUQUIER, HAUCK, OAKES, VANDAELE, and WEISBERG • Statistical Methods for Comparative Studies
ARTHANARI and DODGE • Mathematical Programming in Statistics
BAILEY • The Elements of Stochastic Processes with Applications to the Natural Sciences
BAILEY • Mathematics, Statistics and Systems for Health
BARNETT • Interpreting Multivariate Data
BARNETT and LEWIS • Outliers in Statistical Data
BARTHOLOMEW • Stochastic Models for Social Processes, *Third Edition*
BARTHOLOMEW and FORBES • Statistical Techniques for Manpower Planning
BECK and ARNOLD • Parameter Estimation in Engineering and Science
BELSLEY, KUH, and WELSCH • Regression Diagnostics: Identifying Influential Data and Sources of Collinearity
BENNETT and FRANKLIN • Statistical Analysis in Chemistry and the Chemical Industry
BHAT • Elements of Applied Stochastic Processes
BLOOMFIELD • Fourier Analysis of Time Series: An Introduction
BOX • R. A. Fisher, The Life of a Scientist
BOX and DRAPER • Evolutionary Operation: A Statistical Method for Process Improvement
BOX, HUNTER, and HUNTER • Statistics for Experimenters: An Introduction to Design, Data Analysis, and Model Building
BROWN and HOLLANDER • Statistics: A Biomedical Introduction
BROWNLEE • Statistical Theory and Methodology in Science and Engineering, *Second Edition*
BURY • Statistical Models in Applied Science
CHAMBERS • Computational Methods for Data Analysis
CHATTERJEE and PRICE • Regression Analysis by Example
CHERNOFF and MOSES • Elementary Decision Theory
CHOW • Analysis and Control of Dynamic Economic Systems
CHOW • Econometric Analysis by Control Methods
CLELLAND, BROWN, and deCANI • Basic Statistics with Business Applications, *Second Edition*
COCHRAN • Sampling Techniques, *Third Edition*
COCHRAN and COX • Experimental Designs, *Second Edition*
CONOVER • Practical Nonparametric Statistics, *Second Edition*
CORNELL • Experiments with Mixtures: Designs, Models and The Analysis of Mixture Data
COX • Planning of Experiments
DANIEL • Biostatistics: A Foundation for Analysis in the Health Sciences, *Second Edition*
DANIEL • Applications of Statistics to Industrial Experimentation

continued on back

Bayesian Reliability Analysis

HARRY F. MARTZ

and

RAY A. WALLER
Los Alamos National Laboratory

JOHN WILEY & SONS
New York Chichester Brisbane Toronto Singapore

Library of Congress Cataloging in Publication Data:

Martz, Harry F. (Harry Franklin) 1942–
Bayesian reliability analysis.

(Wiley series in probability and mathematical statistics. Applied probability and statistics section)
Includes bibliographical references and index.
1. Reliability (Engineering)—Statistical methods.
2. Bayesian statistical decision theory. I. Waller, Ray A. II. Title. III. Series.

TS173.M38 620′.00452 81-21878
ISBN 0-471-86425-0 AACR2

Printed in the United States of America

10 9 8 7 6 5 4 3 2 1

To Joe, Jeff, and Angela
and my parents

H.F.M.

To Carolyn, Lance, and Jay

R.A.W.

Preface

Performance evaluation of complex devices produced by technological advances is often accomplished by reliability analyses. Recent emphasis on increased productivity has heightened interest in developing reliability techniques that utilize *all* available pertinent information in a cost-effective manner. The past two decades witnessed the growth and implementation of a large body of *Bayesian reliability analysis* methods which meet this challenge. Although there exist numerous books that present the basic ideas of classical reliability analyses, this book focuses attention on Bayesian techniques.

Chapters 2, 3, and 4 provide basic materials in probability, statistics, and classical reliability. The basic notions for understanding, performing, and interpreting Bayesian reliability analyses are given in Chapter 5. Techniques for selecting and evaluating prior distributions are given in Chapter 6 in which the primary emphasis is on noninformative and natural conjugate prior distributions, although other classes of prior distributions are also often considered. Chapter 7 discusses Bayesian estimation methods based on attribute life test data for binomial, Pascal, and Poisson sampling models. Chapters 8 and 9 cover Bayesian estimation procedures for continuous lifetime models including exponential, Weibull, normal, log normal, inverse Gaussian and gamma distributions. Special topics in Bayesian reliability demonstration testing, system reliability assessment, and availability of maintained systems are covered in Chapters 10, 11, and 12, respectively. Finally, Chapter 13 addresses empirical Bayes reliability estimation.

We have included the chapters on probability, statistics, and classical reliability in order to provide a completely self-contained book. The topics in Chapters 1–7 are dependent in a sequential fashion. As some persons may have had previous training in the elementary topics of Chapters 2, 3, and 4, the book is appropriate for several different courses. First, Chapters

1–9 followed by selected topics from Chapters 10–13 could serve as a text for a one-year senior-level course. Second, for students with prior training in probability, statistics, and classical reliability, the material in Chapters 5–9 supplemented by selected topics from Chapters 10–13 could provide a one-semester graduate-level course in Bayesian reliability analysis.

The book is also a comprehensive reference and survey of many recent Bayesian reliability procedures previously available only in research journals or technical reports. For this reason the book is useful to reliability managers and engineers who need convenient access to such methods.

Over 200 examples are used throughout the book to illustrate the methods presented. Many of the examples have a real data component and are based on reliability data pertinent to the United States commercial nuclear power reactor industry.

Each chapter includes exercises to further illustrate and emphasize the text material. Over 250 exercises, representing a range of difficulty from simple to challenging, are provided for Chapters 2–13.

We acknowledge our debt to the many researchers who are responsible for the methods presented in this text. In appreciation of their efforts we have carefully cited each contributor. In the event that we have neglected a merited citation, incorrectly given credit, or incorrectly interpreted the results, we offer our apologies. Further, we alone are responsible for errors of either omission or commission.

Partial support for the efforts resulting in this book was provided by the Nuclear Reactor Programs of the U.S. Department of Energy. We gratefully acknowledge the support and encouragement received from F. X. Gavigan and J. H. Carlson. We also acknowledge the support provided by the U.S. Office of Naval Research and extend our appreciation to Mr. Seymour Selig for his support and encouragement. We are also indebted to Dr. Danny Dyer, University of Texas Arlington, and Dr. Thomas McWilliams, University of Denver, for their assistance in drafting early portions of the book. We express our gratitude to Dr. Benjamin S. Duran, Texas Tech University, for contributing some of the exercises and reviewing the original draft of the book. We extend our appreciation to our co-authors and colleagues, Ms. Myrle Johnson and Dr. Michael S. Waterman, for generating many of the statistical tables that appear in the appendices and for their permission to reprint various tables and figures from the article "Gamma Prior Distribution Selection for Bayesian Analysis of Failure Rate and Reliability."* We wish to thank the Illustration Services Group at the Los Alamos National

Laboratory for preparing the illustrations, Ms. Elizabeth Courtney for painstakingly typing the manuscript, and Mss. Cheryl Markham and Sue Martin for their typing support. Finally, we are grateful to the Literary Executor of the late Sir Ronald A. Fisher, F.R.S., to Dr. Frank Yates, F.R.S., and to Longman Group Ltd. London, for permission to adapt Table III of their book *Statistical Tables for Biological, Agricultural and Medical Research* (6th edition, 1974).

H. F. MARTZ
R. A. WALLER

Los Alamos, New Mexico
February 1982

Contents

Bayesian Reliability Analysis

CHAPTER 1

Introduction

The primary purpose of this book is to aggregate and systematically introduce both the numerous and varied Bayesian techniques for analysis of reliability data. Our intent is to introduce the basic notions that constitute a Bayesian reliability analysis, present the many analytic methods, illustrate the application of Bayesian procedures, and catalog the relevant Bayesian reliability literature. Whereas numerous reliability textbooks present one chapter on Bayesian methods, this book is completely dedicated to the Bayesian approach.

1.1. BAYESIAN METHODS IN RELIABILITY

Modern products have become increasingly more sophisticated in light of rapid technological advances. Very detailed and complex equipment has been researched, designed, developed, and implemented for space exploration, military applications, and commercial uses. In general, each piece of equipment is composed of numerous elementary components and/or subsystems that work as a unit either to achieve specific objectives or to perform a variety of functions. As a consequence, increasing attention has been focused on the evaluation of whether a given device successfully performs its intended function. Studies of operating data for pieces of equipment are used in these evaluations.

Reliability analyses evaluate the performance of equipment. In particular, we use the term *reliability* to mean the probability that a piece of equipment (component, subsystem, or system) successfully performs its intended function for a given period of time under specified (design) conditions. For example, we can let a variable, T, denote the time-to-failure of a given 100 W light bulb. Then the reliability, say $R(t)$, of the bulb as a function of operating time, t, can be written symbolically as follows:

$$R(t) = \Pr(T > t) \tag{1.1}$$

where $\Pr(T > t)$ denotes the probability that the variable T exceeds t. In other words, the probability that the bulb does not fail before time t is the reliability of the bulb at time t. Classical reliability analysis methods as presented in such books as Barlow and Proschan (1975) and Mann, Schafer, and Singpurwalla (1974) use operating data from the equipment of interest for evaluating performance. In contrast, Bayesian methods as presented in this book permit the combining of operating data with any other relevant information available for reliability studies. Some possible sources of supplemental information are engineering design and test data; operating data in different environments; engineering judgments and personal experience; and operating experiences with similar equipment. The particular methods for incorporating such related data and the philosophical relationships between classical and Bayesian methods will be addressed in later chapters of the book, primarily in Chapter 5.

1.2. PURPOSE AND SCOPE

Research activities in recent years have produced many technical papers presenting Bayesian methods for analyzing reliability data. Most of these methods are assembled in this book along with appropriate references for convenient referral. The book also includes necessary background material to cover basic probability concepts, fundamental statistics results, and classical reliability ideas so that the book is self-contained. References are provided for supplemental reading by those seeking additional material on selected topics.

The substance of this book is divided into twelve chapters. Chapters 2, 3, and 4 present necessary background material for completeness. Chapter 2 gives elementary probability concepts including the presentation of introductory applications of Bayes' theorem, a basic theorem underlying the Bayesian approach and for which the approach is named. Chapter 3 covers the basics of descriptive statistics and statistical inference. The fundamental ideas and relevant distributions for classical reliability analyses are presented in Chapter 4.

Chapter 5 introduces the basics of Bayesian inference for reliability studies. The use of Bayes' theorem in analyzing reliability data is introduced and illustrated. Chapter 6 addresses the problem of selecting prior distributions for Bayesian analyses. Noninformative, natural conjugate, and ad hoc procedures for determining prior distributions are presented with emphasis on the first two categories. Particular emphasis is placed on the selection of the widely applied beta, gamma, and negative-log gamma distributions as prior models in Bayesian reliability analyses. In all cases, the particular emphasis is on the evaluation and justification of the prior used in any analysis.

Chapter 7 concentrates on the use of attribute life test data for estimating relevant quantities in equipment performance studies. Emphasis is on the estimation of survival probabilities, failure rates, and reliabilities for binomial, Pascal, and Poisson sampling under selected prior model assumptions.

Chapters 8 and 9 present procedures for analyses using the standard continuous lifetime reliability distributions. Chapter 8 is exclusively devoted to the widely used exponential distribution. Estimation procedures for the Weibull, normal, log normal, inverse Gaussian, and gamma distributions are presented in Chapter 9. For each distribution, the results for estimating the distribution parameters and relevant reliability quantities are given.

Chapters 10, 11, and 12 address more specialized topics. Reliability demonstration testing from a Bayesian perspective is provided in Chapter 10. Concepts of and methods for evaluating whole-system performance through system reliability assessment are presented in Chapter 11. The emphasis is on system evaluation using component data for series, parallel, and r-out-of-k systems. Chapter 12 presents Bayesian techniques for estimating the availability of maintained systems for selected failure/repair models.

Chapter 13 covers the subject of empirical Bayes methods for binomial, Poisson, exponential, Weibull, gamma, and log normal distributions. The discussions include parameter estimation and inference about reliability functions.

Exercises are included at the end of each chapter to provide the reader with an opportunity to experience application of the methods presented.

1.3. NOTATION AND CONVENTIONS

In assembling the literature on Bayesian reliability we have selected particular notation and followed various conventions in an attempt to standardize the presentation. To the extent possible we have tried to honor notational conventions existing in both the reliability and statistics literature. The final notation, is, however, a compromise that we hope is clearly defined and acceptable to the reader. While we also recognize that many of the results in this book apply to the analysis of biological survival data, the notation for the parameters and functions of interest is consistent with that commonly used in reliability engineering.

To start, we distinguish between the symbol denoting a random variable (the function) and the number representing the realization of the random variable. In particular, we use upper case letters to denote the random variable and lower case letters to represent observed values. For example, T may denote the time-to-failure for a given light bulb whereas $t=1053$ h would indicate that a bulb failed (burned out) after 1053 h. There are, however, some exceptions to this convention, particularly in later chapters of the book, and these will be noted.

To indicate the probability of any event, say A, we use the symbolism $\Pr(A)$. The event A may be represented in a variety of ways. For example, $A=\{t|t>1000\}$ shows that A is the set of all survival times, t, that exceed 1000 h. The probability of event A could then be represented in any one of the following equivalent symbolic ways:

$$\Pr(A)=\Pr(T>1000)=\Pr\{t|t>1000\}$$

Appendix A provides an extensive listing of notation symbols used throughout the book. For example $\mathfrak{N}(\mu,\sigma^2)$ denotes a normal distribution with mean μ and variance σ^2. Most of the commonly used reliability distributions are represented in a similar way. Both sampling and marginal distributions are generically represented by $f(\cdot)$ while prior and posterior distributions are represented as $g(\cdot)$. Other important notational conventions are indicated in Appendix A as a guide for the reader. Definitions of various acronyms used throughout the book are referenced in the index.

Equations are sequentially numbered within each chapter, for example, (2.5) references the fifth equation in Chapter 2. Tables and Figures in the text are also numbered sequentially within chapters and referenced by Table 4.1 and Figure 4.5. Tables in the appendices are numbered sequentially within each appendix and denoted by the corresponding letter of the appendix and a number, for example, Table A2 indicates the second table in Appendix A.

REFERENCES

Barlow, R. E. and Proschan, F. (1975). *Statistical Theory of Reliability and Life Testing*, Holt, Rinehart & Winston, New York.

Mann, N. R., Schafer, R. E., and Singpurwalla, N. D. (1974). *Methods for Statistical Analysis of Reliability and Life Data*, Wiley, New York.

CHAPTER 2

Elements of Probability

In this presentation we use the term *item* to denote some system or some component of a system being investigated. Our primary interest throughout the book is related to the reliability of the item being studied. Since reliabilities are special probabilities, we use this chapter to present some basic concepts and results for operating with probabilities.

2.1. EXPERIMENTS, RANDOM VARIABLES, SAMPLE SPACES, AND SET THEORY

We consider any repeatable process for which the outcome (result) is uncertain to be an *experiment*. Two examples of experiments are as follows:

Example 2.1. Measurement of the lifetime (time-to-failure) for a given item.

Example 2.2. Testing of an item to be classified as either "defective" or "nondefective." ■

We do not expect all repetitions or trials of a defined experiment to yield the same result because of the uncertainty associated with the process; thus we identify the set of all possible outcomes for an experiment as the *sample space*, denoted by S, of the experiment. The sample space of the experiment may include points in a continuum or distinct discrete points. For example, the lifetime for a light bulb may be "any positive real number" whereas a switch may be either "off" or "on." As indicated in Example 2.2, the sample space of an experiment may not necessarily be a set of real numbers. Yet for our studies we are concerned with the analysis of real numbers, therefore, we define the concept *random variable* (r.v.) as any real-valued function defined on the sample space of an experiment. The set of values assumed by the r.v. is called the sample space of the r.v.

An r.v. is said to be *discrete* if its sample space is finite or countably infinite. That is, the sample space of X is $S=\{x_1, x_2, \ldots, x_k\}$ where k may be either finite or infinite. If an r.v. X has a sample space which is a continuum of values, then X is said to be a *continuous* r.v. An r.v. that takes on values in an interval (or intervals) of values is a continuous r.v. In statistical studies it is common to use upper case letters to denote the r.v. (i.e., the function) and lower case letters to denote specific values of the r.v. Exceptions to this convention in later chapters are noted. The two following examples illustrate the preceding remarks.

Example 2.3. Consider an experiment in which 60 W light bulbs from a designated supplier are being tested. Let T denote the measured lifetime in hours. Then T is a continuous r.v., $t=932$ h would indicate that a particular test bulb survived for 932 h, and the sample space for T is the set of all positive real numbers (i.e., $S=\{t|0<t\}$, read S is "the set of all values t such that t is greater than zero").

Example 2.4. Consider an experiment of classifying light switches from a designated supplier as either "defective" or "nondefective." Each outcome of the experiment is either defective or nondefective and therefore, the sample space of the experiment is the set of two possibilities, that is, $S=\{\text{defective, nondefective}\}$. To define a r.v. we must associate a real-valued function, say X, with the two points in the sample space. One such r.v. is

$$X=0 \text{ if "defective" is observed and}$$

$$X=1 \text{ if "nondefective" is observed.}$$

The sample space of the discrete r.v. X is $S=\{0, 1\}$. ■

Remarks. As illustrated by the foregoing examples, some experiments yield real-valued outcomes such that the sample space of the experiment and a r.v. are coincident (e.g., T and the experiment in Example 2.3). Other experiments may yield qualitative outcomes for which a functional association with real numbers must be defined to denote a r.v. (e.g., X in Example 2.4).

Since sample spaces are sets, we can use the results of set theory to manipulate interesting subsets of sample spaces. The subsets of particular interest to us are those of sample spaces of r.v.'s. In particular, we define an *event* to be any subset of a sample space of a r.v. For example, we could consider an event, $E=\{t|0<t<100\}$, to denote the subset of light bulbs that operate originally ($t>0$) but fail prior to 100 h. E is then a subset of $S=\{t|0<t\}$ and we write $E \subset S$ (read E is a subset or subevent of S). The individual outcomes (points) in a sample space are known as *elementary* events.

We call the sample space S the *universal* set. We call S a *certain* event because each trial of the experiment must yield a value of the r.v. which is in S, the sample space. Similarly, the *empty* or *null* set, $\varnothing$, is a subset (and thus an event) of S. We call $\varnothing$ an *impossible* event because we cannot conduct a trial of the experiment without observing some value for the defined r.v.; that is, the number of outcomes favorable to $\varnothing$ is 0.

We define two set operations for use in studying events. First, the *intersection* (denoted by $\cap$) of two sets (events), A and B, is written $A \cap B$ (or AB) and denotes those values common to both A *and* B. Second, the *union* (denoted by $\cup$) of two sets, A and B, is written $A \cup B$ and denotes those values that occur in either A alone, B alone, *or* both A and B. Examples are as follows:

Example 2.5. Let $A=\{t|100<t<1000\}$ and $B=\{t|0<t<200\}$

Then
$$A\cap B=\{t|100<t<200\},$$
$$A\cup B=\{t|0<t<1000\}.$$

Example 2.6. Let $A=\{0,1,2,\ldots,100\}$, $B=\{50,51,\ldots,150\}$, and $C=\{101,102,\ldots,200\}$.

Then
$$A\cap B=\{50,51,\ldots,100\},$$
$$A\cup B=\{0,1,2,\ldots,150\},$$
$$A\cap C=\varnothing,$$
$$B\cap C=\{101,102,\ldots,150\},$$
$$A\cap B\cap C=\varnothing,$$
$$A\cup B\cup C=\{0,1,2,\ldots,200\},$$
$$A\cup C=\{0,1,2,\ldots,200\}.$$
■

2.2. PROBABILITY

Associated with any event E ($E \subset S$) of a sample space S is a number denoted by $\Pr(E)$ and called the probability of E. As each event represents a particular set of outcomes for an experiment, the values of $\Pr(E)$ are deduced from the probabilities or likelihoods associated with the outcomes of an experiment. Those numbers are often defined as being representative of the *long-run relative frequency* for event E. To illustrate this, suppose the r.v. X denotes the number of defective light bulbs in inspected samples of 100 bulbs. Now suppose we inspect n samples of 100 bulbs. The relative frequency of samples with 0 defective bulbs is

$$f(0)=\frac{n_0}{n},$$

where n_0 is the number of samples with 0 defective bulbs. The long-run relative frequency of 0 defectives is the $\lim_{n\to\infty} f(0)$. When that limit exists we call it the probability of 0 defectives and write

$$\Pr(0)=\lim_{n\to\infty} f(0).$$

It should be noted that the limit involved here is not the usual point-wise limit. What is implied is that the observed frequencies $f(0)$ stabilize about some value, $\Pr(0)$, as n increases without bound. These ideas are discussed further in Section 3.7. The relative frequency definition (or interpretation) of probability is stated as follows:

Definition 2.1. If an experiment is repeated a large number of times n then the observed relative frequency of occurrence n_E/n of the event E will tend to stabilize at some unknown constant $\Pr(E)$, which is called the *probability* of E.

Another definition of probability is the so-called classical definition:

Definition 2.2. If an experiment can result in n equally likely and mutually exclusive outcomes and if n_E of these outcomes possess attribute E, then the *probability* of E is the ratio n_E/n.

That the outcomes are equally likely cannot be proved but must be assumed. The frequency definition is more useful than the classical definition as the true probability structure associated with the sample space is not known. Both definitions, however, suggest the following axiomatic (mathematical) definition.

Definition 2.3. Let S be a sample space. If $\Pr(\cdot)$ is a function defined on events E of S satisfying

(i) $\Pr(E) \geqslant 0$, for every event E,
(ii) $\Pr(E_1 \cup E_2 \cup \cdots) = \Pr(E_1) + \Pr(E_2) + \cdots$,
where the events $E_1, E_2, \ldots,$ are such that no two have a point in common, and
(iii) $\Pr(S) = 1$,

then Pr is called a *probability function.*

From the axioms in Definition 2.3 it is easy to show that the probability of any impossible event (empty set) is zero, that is, $\Pr(\varnothing) = 0$.

Next we present some rules for calculating the probability of the union of two events.

Rule 2.1. In general, for any two events E_1 and E_2,

$$\Pr(E_1 \cup E_2) = \Pr(E_1) + \Pr(E_2) - \Pr(E_1 \cap E_2).$$

One special result of the above rule is that the probability of the complement of E, say $\bar{E}$, is given by

Rule 2.2. $\Pr(\bar{E}) = 1 - \Pr(E)$.

That result follows because $E \cap \bar{E} = \varnothing$ and $E \cup \bar{E} = S$ by definition. That is, $1 = \Pr(S) = \Pr(E \cup \bar{E}) = \Pr(E) + \Pr(\bar{E})$ yields the indicated result.

Calculation of the probability of two or more events occurring simultaneously requires the concept of *statistical independence*. Two events E_1 and E_2 are said to be statistically independent if the occurrence or nonoccurrence of one event, say E_1, has no effect on the probability of occurrence (or nonoccurrence) for the other event. In symbols *we define* $\Pr(E_2 | E_1)$ *to be the conditional probability that event* E_2 *occurs given that* E_1 *has occurred.*

We now give the formal definition of conditional probability, leading up to the definition of statistical independence:

Definition 2.4. Let E_1 and E_2 be two events on a sample space S, on which is defined a probability function $\Pr(\cdot)$. The *conditional probability* of the event E_2, given the event E_1, is defined by

$$\Pr(E_2 | E_1) = \Pr(E_2 \cap E_1) / \Pr(E_1) \text{ if } \Pr(E_1) > 0,$$

and if $\Pr(E_1) = 0$, then $\Pr(E_2 | E_1)$ is undefined.

Now, E_2 is said to be *independent* of E_1 if $\Pr(E_2|E_1)=\Pr(E_2)$. If E_2 is independent of E_1 then it easily follows that E_1 is independent of E_2. These ideas, together with Definition 2.4 suggest the definition of independence of two events.

Definition 2.5. The events E_2 and E_1, defined on the same space S, are said to be *independent* if

$$\Pr(E_2 \cap E_1) = \Pr(E_2)\Pr(E_1).$$

The foregoing ideas are illustrated in the following example:

Example 2.7. Consider a process using bolts from two independent suppliers. Suppose that the probabilities of defective bolts are $\Pr(D_1)=0.05$ and $\Pr(D_2)=0.07$ for suppliers #1 and #2, respectively. If one bolt is selected at random for each supplier, what is the probability that *both* are defective? Because we assume that the supplier processes are independent

$$\Pr(D_1 D_2) = \Pr(D_1)\Pr(D_2) = (0.05)\ (0.07) = 0.0035.$$

What is the probability that one or both bolts are defective?

$$\Pr(D_1 \cup D_2) = \Pr(D_1) + \Pr(D_2) - \Pr(D_1 D_2)$$

$$= 0.05 + 0.07 - 0.0035 = 0.1165.$$

That is, there is an 11.65% chance that at least one (i.e., one or both) of two selected bolts will be defective. Finally, what is the probability $\Pr(D_2|D_1)$? By the above formulas and the assumed independence

$$\Pr(D_2|D_1) = \frac{\Pr(D_1 D_2)}{\Pr(D_1)} = \frac{\Pr(D_1)\Pr(D_2)}{\Pr(D_1)} = \Pr(D_2).$$

In numbers,

$$\Pr(D_2|D_1) = \frac{0.0035}{0.05} = 0.07 = \Pr(D_2). \qquad \blacksquare$$

At the beginning of this section we defined probability as a limiting or long-run process for a ratio of counts, that is,

$$\Pr(E) = \frac{\text{number of ways the event } E \text{ can occur}}{\text{number of trials of the experiment}} = \frac{\#(E)}{n}.$$

can be the positions *DE*, *PE*, and *QE* be filled from the six candidates?

$$_6P_3 = \frac{6!}{(6-3)!} = \frac{6\cdot5\cdot4\cdot3\cdot2\cdot1}{3\cdot2\cdot1} = 120.$$

To illustrate that permutations are a special case of the general counting rule, note that there are six choices for the *DE*. After the *DE* is selected there are five choices for the *PE*, and after the *DE* and *PE* are selected, there are four choices for the *QE*. Then the joint selection of *DE*, *PE*, and *QE* can be accomplished in $6\cdot5\cdot4=120$ ways. ■

A second specialization of the general counting rule, Rule 2.3, pertains to the number of samples (different unordered groups or *combinations*) of size r that can be selected from n distinct items. This counting method gives the number of *combinations of r items that can be selected from n distinct items without replacement*, and is denoted by $\binom{n}{r}$. It easily follows that

$$\binom{n}{r} = \frac{n!}{r!(n-r)!}, \qquad r=0,1,2,\ldots,n, \tag{2.1}$$

since

$$_nP_r = \binom{n}{r} r!.$$

Example 2.10. A package of 20 new light bulbs has been received by ACE Engineering Company. How many samples of size 4 can be selected from the 20 bulbs?

From (2.1) the number of samples is

$$\binom{20}{4} = \frac{20!}{4!(20-4)!} = \frac{20\cdot19\cdot18\cdot17\cdot16!}{4\cdot3\cdot2\cdot1\cdot16!} = 4845. \quad ■$$

To illustrate the use of combinations, suppose two of the 20 bulbs are defective. What is the probability that the selected sample of four bulbs contains exactly one defective bulb? From the classical definition of probability we have

$$\Pr(1 \text{ defective}) = \frac{\text{number of samples of 4 bulbs containing 1 defective}}{\text{number of samples of 4 bulbs}}.$$

In the next section, we present counting techniques to provide as obtaining the elements of the ratio defining Pr(E). In Section 5.1 notion of probability, subjective probability, will be introduced. or personal probability plays a major role in Bayesian reliability

2.3. COUNTING TECHNIQUES

The purpose of this section is to provide methods for counting of ways in which a given event can occur. That information is t for use in the probability formulas of the preceding section. We a general rule that is useful in developing formulas for specifi

Rule 2.3 (General Counting Rule). If an event E_1 can occ of m different ways, and if for each of those different ways in occur, an event E_2 can occur in n different ways, the joint e can occur in mn different ways.

This rule is easily extended to the case of k events E_1, E_2,

---

Example 2.8. Suppose three components, C_1, C_2 and C assemble a given tool. If component C_1 is available from suppliers, component C_2 is available from two different component C_3 is available from five different suppliers, hov there for assembling a tool? The three components can l assembled in $3 \cdot 2 \cdot 5 = 30$ different ways.

A first specialization of the general counting rule, Rule number of ways in which distinct items can be sequentiall the number of arrangements of n distinct items (when r method of counting yields the number of *permutations o sets of r items* and is denoted by ${}_nP_r$. It is easily shown t

$$ {}_nP_r = n!/(n-r)!, \qquad r = 0, 1, 2, \ldots, n $$

where $a! = a(a-1)\cdots 2 \cdot 1$ denotes a factorial and $0!$ is

---

Example 2.9. Suppose there are six equally well available to serve as Design Engineer (DE), Productior Quality Engineer (QE) for the ABC Company. In how

From Example 2.10, the denominator is 4845. The numerator is found by partitioning the 20 bulbs into 2 defectives and 18 nondefectives and noting that we must select 1 bulb from the 2 defectives and 3 bulbs from the 18 nondefectives. Thus, the number of samples containing one defective is $\binom{2}{1}\binom{18}{3}$ and

$$\Pr(1\ D) = \frac{\binom{2}{1}\binom{18}{3}}{\binom{20}{4}} = \frac{2(816)}{4845} = \frac{1632}{4845} = 0.337.$$

A third generalization of the general counting rule, Rule 2.3, involves repeated experiments. In particular, it applies to composite experiments that are repetitions of a simpler experiment. For example, consider an experiment where one light bulb is selected from a recent shipment of bulbs and is classified as either D (defective) or G (nondefective). Now suppose this experiment is repeated five times. How many different sequences of five Ds and Gs can be produced by the composite experiment? In general, we consider the experiment of interest to be k repetitions of a simpler experiment each of which can yield any one of n results. The composite experiment then has a sample space of n^k outcomes.

Example 2.11. Returning to the question posed above, we have a simple experiment with two outcomes (D or G) repeated five times. Thus, there are $2^5 = 32$ different outcomes (ordered sequences of five Ds and Gs). These outcomes are

(5-D, 0-G)	$DDDDD$	1
(4-D, 1-G)	$DDDDG$, $DDDGD$, $DDGDD$, $DGDDD$, $GDDDD$	5
(3-D, 2-G)	$DDDGG$, $DDGDG$, $DGDDG$, $GDDDG$, $GDDGD$	10
	$GDGDD$, $GGDDD$, $DDGGD$, $DGDGD$, $DGGDD$	
(2-D, 3-G)	$DDGGG$, $DGDGG$, $DGGDG$, $DGGGD$, $GDGGD$	10
	$GGDGD$, $GGGDD$, $GGDDG$, $GGDGD$, $GDDGG$	
(1-D, 4-G)	$DGGGG$, $GDGGG$, $GGDGG$, $GGGDG$, $GGGGD$	5
(0-D, 5-G)	$GGGGG$	1
		32

■

This counting technique will play an important role in the calculation of probabilities associated with the binomial model discussed in the next section.

In this and the preceding sections we have defined the probability of an event and have provided some tools for calculating the desired numbers. With that as a basis, we present some selected probability distributions (models) that are basic to our study of reliability.

2.4. DISCRETE PROBABILITY DISTRIBUTIONS (MODELS)

Several probability models play dominant roles in reliability analysis. In this and Section 2.5 we use selected models to illustrate the basic properties of probability distributions. The remaining necessary models are introduced throughout the book as needed for various analyses. This section discusses discrete probability distributions, whereas Section 2.5 presents selected continuous distributions.

A valid probability model for an experiment (random variable) is a function which assigns a total probability of one to the set of all points in the sample space. Thus, for a discrete r.v. with sample space $S=\{x_1, x_2,\ldots,x_k\}$ where k may be either finite or infinite, the function $\Pr(\cdot)$ is a valid probability model if

$$\Pr(x_i)\geqslant 0, \qquad i=1,2,\ldots,k$$

and

$$\sum_{i=1}^{k} \Pr(x_i)=1.$$

As is implied in the definition of $\Pr(x)$ there are at most a countable infinity of values that have a positive probability. In stating the function $\Pr(\cdot)$, generally only those values of x for which $\Pr(x)$ is positive are given. The probability model of a discrete r.v. may be referred to as a probability function, probability distribution, probability mass function, or a discrete probability density function.

We now consider some examples of probability functions.

2.4.1 Hypergeometric Model

The hypergeometric distribution is introduced with an example.

Example 2.12. Consider again the situation in Example 2.10 where 4 bulbs are selected from 20 bulbs (18 good, 2 defective). A probability distribution for the number of defectives in the sample is as follows:

Let $X = \#$ (defectives) in the sample. Then the possible values of X are $S=\{0,1,2\}$. Let $\Pr(x)=\Pr(X=x)$, then a probability distribution for X is given in the table below

x	$\Pr(x)$
0	$\binom{2}{0}\binom{18}{4}\Big/\binom{20}{4}=\frac{3060}{4845}=0.632$
1	$\binom{2}{1}\binom{18}{3}\Big/\binom{20}{4}=\frac{1632}{4845}=0.337$
2	$\binom{2}{2}\binom{18}{2}\Big/\binom{20}{4}=\frac{153}{4845}=0.032$
Total	1[1.001 (round off)]

■

Example 2.12 presents a special case of a general set of probability distributions known as *hypergeometric models*. In general, the following describes the two-category hypergeometric model:

N = number of units in a population,
D = number of defective units in the population,
n = number of units selected,
x = number of defective units selected.

[We note that $(N-D)$=number of nondefective units in the population and that $(n-x)$=number of nondefective units in the selected sample.] We suppose the n units are selected *without replacement*. The possible values of x are integers between $\max(0, n+D-N)$ and $\min(D, n)$. (In practice, n and D are usually small relative to N, so the lower limit on x is 0.) The probability model is

$$\Pr(x)=\frac{\binom{D}{x}\binom{N-D}{n-x}}{\binom{N}{n}}, \qquad x=\max(0, n+D-N),\ldots,\min(D,n). \tag{2.2}$$

The symbol $\binom{n}{m}$ takes care of the limits on the values of x, since $\binom{n}{m}=0$ if $m>n$ or if $m<0$. Since $\Pr(x)$ is defined to be the ratio of nonnegative numbers, it follows that $\Pr(x)\geqslant 0$ for all x. Further, it can be shown that $\Sigma_{x\in S}\Pr(x)=1$, where S denotes the sample space of X and $\in$ denotes "is contained". Therefore, $\Pr(x)$ is a valid probability function. For the previous example, $N=20$, $D=2$, and $n=4$. Thus the sample space of X is $S=\{0,1,2\}$.

2.4.2 Binomial Distribution

As a second example of a discrete probability model, we consider the following situation. Suppose n light bulbs are placed on test for a specified period of time, for example, 100 hours. At the end of the test each bulb is classified as having either survived the test (success) or failed the test (failure). Of interest is the probability that exactly x bulbs survive the test. If we assume that the probability of survival of any randomly selected bulb is p and if we assume that the bulbs fail independently of each other, then the family of *binomial distributions* provides models for this situation. The probability function

$$\Pr(x)=\begin{cases}\binom{n}{x}p^x q^{n-x}, & x=0,1,2,\ldots,n \\ 0, & \text{otherwise,}\end{cases} \tag{2.3}$$

where $q=1-p=$probability a randomly selected bulb fails, defines a binomial model with parameters p and n.

Example 2.13. We assume that $p=0.9$ (i.e., the probability that a bulb survives a life test of 500 hours is 0.9) and that bulbs fail independently. What is the binomial probability model for X, the number of survivors in a test of four bulbs? The following table gives the distribution of X, according to (2.3).

x	$\Pr(x)$
0	$\binom{4}{0}(0.9)^0(0.1)^4=0.0001$
1	$\binom{4}{1}(0.9)^1(0.1)^3=0.0036$
2	$\binom{4}{2}(0.9)^2(0.1)^2=0.0486$
3	$\binom{4}{3}(0.9)^3(0.1)^1=0.2916$
4	$\binom{4}{4}(0.9)^4(0.1)^0=0.6561$
Total	1.0000

■

2.4.3 Poisson Distribution

A distribution that occurs quite frequently as a model for many random phenomena is the Poisson distribution. The values of a Poisson r.v. X are the nonnegative integers, that is the sample space of X is $S=\{0,1,2,\ldots\}$. Many random phenomena require a count of some sort. For example, one might consider the number of electronic components that fail per unit time, the number of radioactive particles emitted per unit time, or the number of telephone calls coming into a telephone switchboard per unit time. These are just a few examples of such random phenomena.

The probability function of a Poisson r.v. is

$$\Pr(x)=\begin{cases} \dfrac{\rho^x e^{-\rho}}{x!}, & \rho>0,\ x=0,1,2,\ldots \\ 0, & \text{otherwise.} \end{cases} \tag{2.4}$$

Let $X(t)$ denote the number of events (components failing, telephone calls arriving, etc.) that occur during the time interval $(0,t)$. The r.v. $X(t)$, for $t\geqslant 0$, is often assumed to have a Poisson distribution with parameter $\rho=\lambda t$. In this case $X(t)$, $t\geqslant 0$, is referred to as a *Poisson process* with *intensity* $\lambda>0$. The probability function of $X(t)$ is

$$\Pr(x)=\frac{(\lambda t)^x e^{-\lambda t}}{x!}, \qquad \lambda>0,\ x=0,1,2,\ldots \tag{2.5}$$

As will be seen later the expected (average) number of events occurring in the interval 0 to t is λt.

Example 2.14. Suppose that components necessary for the operation of a machine fail according to a Poisson process with intensity $\lambda=1$ (each week); that is, at the rate of one every week. What is the probability that (a) One component fails during a four week period? (b) No components fail during a four week period? (c) If one spare component is available, what is the probability that the machine remains in operation for a given six week period, assuming there is perfect switching?

(a) The probability that one component fails during a four week period is, by (2.5),

$$\Pr[X(4)=1]=\frac{[1(4)]^1 e^{-1(4)}}{1!}=0.0733,$$

since $\lambda=1$ and $t=4$.

(b) $\Pr[X(4)=0]=e^{-4}=0.0183$.

(c) The probability that the machine remains in operation for a six week period, if there is a spare component, is

$$\begin{aligned}\Pr[X(6)=0 \text{ or } 1] &= \Pr[X(6)=0]+\Pr[X(6)=1] \\ &= e^{-6}+6e^{-6} \\ &= 0.0174.\end{aligned}$$

■

If the r.v. X has a binomial distribution with a small value of p (say $p \leqslant 0.1$), then the distribution of X can be approximated by a Poisson distribution with parameter $\rho = np$. The approximation works quite well when p is small and n is large, while np is of moderate size (see Exercise 48, Chapter 2). The approximation is obtained by substituting np for ρ in (2.4).

Example 2.15. Suppose that a computer uses 4000 chips and that each chip has a probability $p=0.001$ of failing (per unit time). Assume that the chips operate independently of each other. What is the probability that there are (a) no failures? (b) at least two failures?

(a) Let X denote the number of failures per unit time. Then $\rho = np = 4000\,(0.001) = 4$ and, by (2.4),

$$\Pr(X=0)=\frac{\rho^0 e^{-\rho}}{0!}=e^{-4}=0.0183.$$

(b)

$$\begin{aligned}\Pr(X \geqslant 2) &= 1-\Pr(X \leqslant 1) \\ &= 1-0.0916 \\ &= 0.9084.\end{aligned}$$

■

Other discrete probability models that are appropriate for reliability analysis are introduced as needed. The book by Johnson and Kotz (1969) is an excellent reference regarding discrete models.

2.5. CONTINUOUS PROBABILITY DISTRIBUTIONS (MODELS)

The continuous analog to the foregoing discussion of discrete models concerns *probability density functions* (pdf) for continuous r.v.'s associated with experiments of interest. Any function, $f(x)$, such that

$$f(x) \geqslant 0, \qquad -\infty < x < \infty$$

and

$$\int_{-\infty}^{\infty} f(x)\,dx = 1$$

is a valid (mathematically) pdf. That is, any nonnegative function which allocates unit mass to the points on the real line is a candidate model for a pdf. In stating the form of a pdf it is generally acceptable to state the pdf only for those values of x for which the pdf is positive. Other names that are used for pdf include density function, continuous density function, frequency function, integrating density function, or probability distribution.

2.5.1 Exponential Distribution

A widely used pdf for modeling a time-to-failure variable, say T, is the family of *exponential models* defined by

$$\begin{aligned} f(t) &= \lambda e^{-\lambda t}, \qquad t > 0 \\ &= 0, \qquad t \leqslant 0. \end{aligned} \tag{2.6}$$

This family of models will be discussed in more detail in Section 4.2.1. Note that the requirements for a valid model are met because $f(t) \geqslant 0$ for all real t and $\int_{-\infty}^{\infty} f(t)\,dt = 1$. The function $f(t)$ assigns densities to positive values of t. To calculate probabilities associated with the variable T, we use intervals. For example, $\Pr(1 \leqslant T \leqslant 5) = \int_1^5 f(t)\,dt = e^{-\lambda} - e^{-5\lambda}$. One characteristic of the exponential family of models is that the likelihood (density) of failure decreases as time increases. Other more flexible univariate models for time-to-failures are gamma, Weibull, and log normal distributions. These, in addition to other probability models, are described in Section 4.2 as statistical models in reliability. Johnson and Kotz (1970) provide detailed discussions of these models.

2.5.2 Beta Distribution

A second example of a continuous pdf is the beta distribution that may serve as a model for a reliability variable, say $R = R(t) = \Pr(T \geqslant t)$, which is just the probability that a system lasts at least t units of time. That is, if R is the reliability of a system (or component) of interest, a potential model for the variable R is the beta model given by

$$g(r)=\frac{\Gamma(n_0)}{\Gamma(x_0)\Gamma(n_0-x_0)}r^{x_0-1}(1-r)^{n_0-x_0-1}, \qquad 0\leqslant r\leqslant 1$$

$$=0, \quad \text{otherwise}, \tag{2.7}$$

where $\Gamma(n)=\int_0^\infty x^{n-1}e^{-x}\,dx$, the gamma function. If n is a positive integer, $\Gamma(n)=(n-1)!$

The beta distribution is used later in Bayesian analyses. It is a very versatile family of distributions that can exhibit shapes varying from decreasing to unimodal right-skewed to symmetric to uniform to U-shaped to unimodal left-skewed to increasing. These and other properties are discussed further in Chapters 4, 6, and 7, and illustrated in Figures 4.9 and 4.10.

2.6. CUMULATIVE DISTRIBUTION FUNCTION

The preceding sections present some selected discrete probability distributions and continuous probability density functions that are used in Bayesian reliability analyses. These functions provide either point probabilities (for discrete r.v.'s) or point densities (for continuous r.v.'s). A related function useful in both probability and reliability analyses is the *cumulative distribution function* (cdf), $F(\cdot)$. The cdf is defined as follows:

$$F(x)=\Pr(X\leqslant x)=\sum_{t\leqslant x}\Pr(t), \qquad \text{for a discrete r.v. } X$$

$$=\int_{-\infty}^{x} f(t)\,dt, \qquad \text{for a continuous r.v. } X.$$

Example 2.16. For the binomial model with parameter p, the cdf is

$$F(x)=\begin{cases}0, & x<0\\ \sum_{t=0}^{x}\binom{n}{t}p^t(1-p)^{n-t}, & x=0,1,2,\ldots,n\\ 1, & x\geqslant n.\end{cases}$$

Example 2.17. For the exponential model with parameter λ, the cdf is

$$F(x)=\begin{cases}0, & x\leqslant 0\\ \int_0^x \lambda e^{-\lambda t}\,dt=1-e^{-\lambda x}, & x>0.\end{cases}$$

■

2.7. SAMPLING

In Sections 2.4 and 2.5 we introduced some examples of probability distributions that can be used to model observed phenomena. That is to say the utility of any probability model depends on how well it describes (or models) a set of observations. In this section we relate some of the preceding models to sampling situations. First, we illustrate further the concepts of joint, marginal, and conditional probabilities as they relate to sampling either with or without replacement. The illustration is based on two examples.

Example 2.18. Suppose that a bin contains ten transistors of which two are defective. Two transistors are selected either simultaneously or sequentially (without replacement) and classified as being either D (defective) or $\bar{D}$ (not defective). The possibilities can be summarized along with the respective probabilities in the following two way table.

		Second Selection		
		D_2	$\bar{D}_2$	
First Selection	D_1	2/90	16/90	0.2
	$\bar{D}_1$	16/90	56/90	0.8
		0.2	0.8	1.0

The numbers in the table are derived as follows:

$\Pr(D_1D_2)=\Pr(D_1)\Pr(D_2\,|\,D_1)=2/10\cdot 1/9=2/90,$
$\Pr(D_1\bar{D}_2)=\Pr(D_1)\Pr(\bar{D}_2\,|\,D_1)=2/10\cdot 8/9=16/90,$
$\Pr(\bar{D}_1D_2)=\Pr(\bar{D}_1)\Pr(D_2\,|\,\bar{D}_1)=8/10\cdot 2/9=16/90,$
$\Pr(\bar{D}_1\bar{D}_2)=\Pr(\bar{D}_1)\Pr(\bar{D}_2\,|\,\bar{D}_1)=8/10\cdot 7/9=56/90,$
$\Pr(D_1)=0.2\ (=2/90+16/90)$, $\Pr(\bar{D}_1)=0.8\ (=16/90+56/90)$,
$\Pr(D_2)=2/90+16/90=0.2$, and $\Pr(\bar{D}_2)=16/90+56/90=0.8$. ■

Example 2.18 illustrates the names joint, marginal, and conditional probabilities in the following manner. First, the joint probabilities $\Pr(D_1 D_2)$, $\Pr(D_1 \bar{D}_2)$, $\Pr(\bar{D}_1 D_2)$, and $\Pr(\bar{D}_1 \bar{D}_2)$ appear as interior table values with each joint event represented as the intersection of a row and column of the table. Second, the marginal probabilities $\Pr(D_1)$, $\Pr(D_2)$, $\Pr(\bar{D}_1)$, and $\Pr(\bar{D}_2)$ appear along with margins of the table. Finally, the conditional probabilities $\Pr(D_2 | D_1)$, $\Pr(\bar{D}_2 | D_1)$, $\Pr(D_2 | \bar{D}_1)$, and $\Pr(\bar{D}_2 | \bar{D}_1)$ are each the ratio of a joint probability divided by a marginal probability. In particular, $\Pr(\bar{D}_2 | D_1) = \Pr(D_1 \bar{D}_2) / \Pr(D_1) = \frac{16}{90} \div 0.2 = \frac{8}{9}$. (In this simple example the $\frac{8}{9}$ follows directly from the fact that after one defective transistor is selected, there are nine transistors of which eight are not defective.)

The events in the Example 2.18 are dependent. This property can be established by noting that the probability for any joint event is not equal to the product of the two marginal events involved. For example

$$\Pr(D_1 D_2) = \tfrac{2}{90} \neq (0.2)(0.2) = \Pr(D_1)\Pr(D_2).$$

The dependency is induced by sampling without replacement. That is, the probabilities associated with the second selection depend on the result of the first selection. The next example illustrates what happens when the first transistor is classified and replaced prior to the selection of a second transistor.

Example 2.19. Consider a bin that contains ten transistors of which two are defective. Suppose two transistors are selected with replacement (i.e., the first transistor selected is inspected and replaced prior to the selection of the second transistor) and classified as being either D (defective) or $\bar{D}$ (not defective). The possibilities along with their respective probabilities can be summarized in a table as follows:

		Second Selection		
		D_2	$\bar{D}_2$	
First Selection	D_1	0.04	0.16	0.2
	$\bar{D}_1$	0.16	0.64	0.8
		0.20	0.80	1.0

Since the sampling is being conducted with replacement, the probability of a defective remains at 0.2 for both selections. Also $\Pr(\bar{D}) = 0.8$ for each selection. Therefore

$$\begin{aligned}
\Pr(D_1 D_2) &= (0.2)(0.2) = 0.04,\\
\Pr(D_1 \bar{D}_2) &= (0.2)(0.8) = 0.16,\\
\Pr(\bar{D}_1 D_2) &= (0.8)(0.2) = 0.16,\\
\Pr(\bar{D}_1 \bar{D}_2) &= (0.8)(0.8) = 0.64.
\end{aligned}$$

We note that the two selections are independent because the joint probabilities are equal to the product of the marginal probabilities. ■

In terms of the models of Section 2.4, the sampling without replacement situation is described by the family of hypergeometric distributions, whereas the sampling with replacement situation is described by the family of binomial distributions. Further, the hypergeometric distribution models the dependency induced by sampling without replacement, whereas the binomial distribution models the independence of sampling with replacement.

2.8. EMPIRICAL DISTRIBUTIONS

Another type of sampling can be illustrated by the following situation. Suppose fifty 100 W light bulbs are placed on a life test to observe times-to-failure and that the resulting data (measured in 1000 h) are grouped in classes as given in the following table. The intervals together with the observed (empirical) frequencies form what is termed an empirical distribution.

Class Intervals (1000 h)	Frequency
0–1	20
1–2	15
2–3	3
3–4	5
4–5	3
5–6	3
6–7	0
7–8	1
Total	50

A question of interest is as follows: what pdf describes the time-to-failure variable, say T, for which the 50 observed lifetimes are a sample? Or more specifically, is the r.v. T being sampled (observed) adequately represented by an exponential pdf with $\lambda=0.5$ ($1/\lambda=2000$ h = mean time-to-failure)? Formal answers to the preceding questions are provided by goodness-of-fit tests presented in Section 3.8. Those tests are based on comparisons between observed frequencies of events and expected frequencies associated with a particular assumed model. Therefore, we conclude this section by illustrating the calculations necessary for determining the expected frequencies associated with an exponential model with $\lambda=0.5$. First we calculate the probabilities associated with the data intervals in the above table. We use $F(x)$ for an exponential model (see Example 2.17) with $\lambda=0.5$ to write

$$\Pr(a \leqslant X \leqslant b) = F(b) - F(a) = (1-e^{-0.5b}) - (1-e^{-0.5a})$$

$$= e^{-0.5a} - e^{-0.5b}, \quad \text{for any } a \geqslant 0 \text{ and } b \geqslant 0.$$

Thus

$$p_1 = \Pr(0 \leqslant X \leqslant 1) = 1 - e^{-0.5(1)} = 0.39347,$$

$$p_2 = \Pr(1 \leqslant X \leqslant 2) = e^{-0.5(1)} - e^{-0.5(2)} = 0.23865,$$

$$p_3 = \Pr(2 \leqslant X \leqslant 3) = 0.14475 \cdots p_8 = \Pr(7 \leqslant X \leqslant 8) = 0.01188,$$

$$p_9 = \Pr(8 \leqslant X) = 0.01832.$$

By multiplying the probability of the ith interval (p_i) by the total frequency (50) we can calculate the expected frequency for each cell to allocate the sample of 50 times-to-failure according to an exponential distribution with $\lambda=0.5$. The results are summarized in the following table.

Class Intervals	Class Probability	Expected Frequency	Observed Frequency
0–1	0.39347	19.674	20
1–2	0.23865	11.933	15
2–3	0.14475	7.238	3
3–4	0.08779	4.390	5
4–5	0.05325	2.663	3
5–6	0.03230	1.615	3
6–7	0.01959	0.980	0
7–8	0.01188	0.594	1
8–	0.01832	0.916	0
Total	1.00000	50.003	50

A casual comparison of the expected and observed frequencies in the table above reveals that similar patterns are exhibited and that the largest discrepancy is approximately four. To determine whether the departure from an exponential model with $\lambda=0.5$ is statistically significant, we would need to apply a formal goodness-of-fit test such as discussed in Section 3.8.

2.9. JOINT, MARGINAL, AND CONDITIONAL DISTRIBUTIONS

Statistical methods are based very heavily on the idea of selecting a sample of size n from a probability distribution (discrete or continuous). A *sample* of n observations from a distribution modeled by $f(x)$ is denoted by

$$(X_1=x_1, X_2=x_2,\ldots,X_n=x_n)=(x_1,x_2,\ldots,x_n).$$

The values $x_1, x_2,\ldots,x_n$ are merely values (realizations) of the r.v. X which has the distribution $f(x)$.

If X_1, X_2 are two r.v.'s, then we may speak about the *joint distribution* of the two r.v.'s in terms of their joint pdf $f(x_1, x_2)$, where (x_1, x_2) is a point in a two-dimensional Euclidean space. If X is an r.v. with sample space S and X_1 and X_2 are two r.v.'s corresponding to selecting two values from $f(x)$, the pdf of X, then the sample space of (X_1, X_2) is the Cartesian product $S^{(2)}=S\times S=\{(x_1,x_2)|x_1\in S, x_2\in S\}$. The pdf $f(x_1, x_2)$ then has domain $S^{(2)}$. These ideas are easily extended from two r.v.'s to n r.v.'s.

Let $X_1, X_2,\ldots,X_n$ denote n r.v.'s. The *joint* pdf of these n r.v.'s is a function $f(x_1,x_2,\ldots,x_n)$ that satisfies the following conditions:

$$\begin{aligned}
&\text{(i)}\quad f(x_1,x_2,\ldots,x_n)\geqslant 0, \qquad -\infty<x_i<\infty, \qquad i=1,2,\ldots,n. &(2.8)\\
&\text{(ii)}\quad \textstyle\int_{-\infty}^{\infty}\int_{-\infty}^{\infty}\cdots\int_{-\infty}^{\infty} f(x_1,x_2,\ldots,x_n)\,dx_1\,dx_2\cdots dx_n=1.
\end{aligned}$$

Any function that satisfies conditions (i) and (ii) may be a joint pdf. Events related to (2.8) are n-dimensional rectangles, or sets of n-dimensional rectangles, and their probabilities are obtained by integrating (2.8) over the n-dimensional regions. For example, the probability of the rectangle $a_i< x_i<b_i, i=1,2,\ldots,n$, is

$$\int_{a_n}^{b_n}\cdots\int_{a_1}^{b_1} f(x_1,x_2,\ldots,x_n)\,dx_1\,dx_2\cdots dx_n.$$

The *marginal* pdf of X_i is defined by

$$f_i(x_i)=\int_{-\infty}^{\infty}\int_{-\infty}^{\infty}\cdots\int_{-\infty}^{\infty} f(x_1,x_2,\ldots,x_n)\,dx_1\,dx_2\cdots dx_{i-1}dx_{i+1}\cdots dx_n.$$

That is, the joint pdf (2.8) is integrated over $(n-1)$ dimensions, leaving a function of x_i.

The *conditional* pdf of X_1 given X_2 is denoted by $g(x_1|x_2)$ and defined by

$$g(x_1|x_2)=\frac{f(x_1,x_2)}{f_2(x_2)}, \qquad f_2(x_2)\neq 0. \tag{2.9}$$

The conditional pdf is a bona fide pdf and satisfies the appropriate conditions. In sampling a conditional pdf (2.9), the outcomes of X_1 are only those that can occur with a fixed value of $X_2=x_2$. The idea of conditional distributions extends easily to more than two r.v.'s.

Two r.v.'s X_1 and X_2 are said to be *independent* if their pdf factors into the product of the two individual pdf's, that is,

$$f(x_1,x_2)=f_1(x_1)f_2(x_2).$$

Similarly, $X_1, X_2,\ldots,X_n$ are independent r.v.'s if

$$f(x_1,x_2,\ldots,x_n)=f_1(x_1)f_2(x_2)\cdots f_n(x_n). \tag{2.10}$$

In sampling from a distribution (or population), it is generally assumed that the n observations are taken at random, in which case it is assumed that $X_1, X_2,\ldots,X_n$ are independent. Thus we have the following definition:

Definition 2.6. A sample $X_1, X_2,\ldots,X_n$ taken from a distribution $f(x)$ is a *random sample* if the joint pdf, denoted by h, of the observations is

$$h(x_1,x_2,\ldots,x_n)=f(x_1)f(x_2)\cdots f(x_n). \tag{2.11}$$

The r.v.'s satisfying (2.11) are said to be *independent and identically distributed* (i.i.d).

The foregoing ideas have been discussed in terms of continuous r.v.'s. The development for discrete r.v.'s is analogous and is not included.

We now consider an example to illustrate the ideas presented in this section.

Example 2.20. Suppose the r.v.'s X_1 and X_2 have the pdf

$$f(x_1,x_2)=\begin{cases} c(x_1^2+x_2), & 0<x_1<1, \quad 0<x_2<1\\ 0, & \text{otherwise.}\end{cases}$$

If f is to be a pdf, then

$$\int_0^1 \int_0^1 c(x_1^2 + x_2)\, dx_1\, dx_2 = 1,$$

$$c\frac{(x_2 + 1/3)^2}{2}\Bigg|_0^1 = 1,$$

or

$$c = \frac{6}{5}.$$

The marginal pdf's are

$$f_1(x_1) = \frac{6}{5}\int_0^1 (x_1^2 + x_2)\, dx_2$$

$$= \frac{3}{5}(1 + 2x_1^2), \qquad 0 < x_1 < 1$$

and

$$f_2(x_2) = \frac{6}{5}\int_0^1 (x_1^2 + x_2)\, dx_1$$

$$= \frac{6}{5}\left(x_2 + \frac{1}{3}\right), \qquad 0 < x_2 < 1.$$

The conditional pdf's are

$$g(x_1 | x_2) = \frac{f(x_1, x_2)}{f_2(x_2)} = \frac{x_1^2 + x_2}{x_2 + 1/3}, \qquad 0 < x_1 < 1, \qquad 0 < x_2 < 1$$

and

$$g(x_2 | x_1) = \frac{2(x_1^2 + x_2)}{(1 + 2x_1^2)}, \qquad 0 < x_1 < 1, \qquad 0 < x_2 < 1.$$

Since

$$f(x_1, x_2) = \frac{6}{5}(x_1^2 + x_2) \neq f_1(x_1) f_2(x_2) = \frac{3}{5}(1 + 2x_1^2) \cdot \frac{6}{5}\left(x_2 + \frac{1}{3}\right),$$

the r.v.'s X_1 and X_2 are *not* independent.

The unconditional probability that $0 < X_2 < 1/2$ is

$$\Pr\left(0 < X_2 < \frac{1}{2}\right) = \int_0^{1/2} \frac{6}{5}\left(x_2 + \frac{1}{3}\right) dx_2$$

$$= \frac{7}{20}$$

and

$$\Pr\left(0 < X_2 < \frac{1}{2} \middle| X_1 = \frac{1}{2}\right) = \int_0^{1/2} \frac{2(1/4 + x_2)}{(1 + 1/2)} dx_2$$

$$= \frac{1}{3}. \qquad \blacksquare$$

2.10. BAYES' THEOREM AND APPLICATIONS

As stated earlier, the focus of this book is on Bayesian reliability techniques. Therefore, Bayes' theorem [Bayes (1958)] is basic to our development. This section presents Bayes' theorem, discusses its relation to conditional probability, and provides some examples. A thorough discussion of Bayes' theorem and its role in Bayesian reliability is given in Section 5.2.3.

Bayes' Theorem

If $A_1, A_2, \ldots, A_n$ are a sequence of disjoint events and if B is any other event (subset of the union of A_i) such that $\Pr(B) > 0$, then

$$\Pr(A_i | B) = \frac{\Pr(B | A_i)\Pr(A_i)}{\Pr(B)}, \tag{2.12}$$

where

$$\Pr(B) = \sum_{j=1}^{n} \Pr(B | A_j)\Pr(A_j).$$

The result in (2.12) follows immediately from Definition 2.4 as follows:

$$\Pr(A_i | B) = \frac{\Pr(BA_i)}{\Pr(B)} = \frac{\Pr(B | A_i)\Pr(A_i)}{\Pr(B)}.$$

Example 2.21. Suppose the supply of transistors is produced by three systems, S_1, S_2, and S_3. Further, suppose that S_1 produces 20%, S_2 produces 30%, and S_3 produces 50% of the supply, and that the defective (D) rates for the systems are 0.01, 0.02, and 0.05 respectively for S_1, S_2, and S_3. In terms of probabilities we have $\Pr(S_1)=0.20$, $\Pr(S_2)=0.30$, $\Pr(S_3)=0.50$, $\Pr(D|S_1)=0.01$, $\Pr(D|S_2)=0.02$, and $\Pr(D|S_3)=0.05$. If a transistor is randomly selected from the supply and found to be defective, what are $\Pr(S_1|D)$, $\Pr(S_2|D)$, and $\Pr(S_3|D)$? First we find

$$\begin{aligned}\Pr(D) &= \Pr(D|S_1)\Pr(S_1)+\Pr(D|S_2)\Pr(S_2)+\Pr(D|S_3)\Pr(S_3)\\ &= 0.01(0.20)+0.02(0.30)+0.05(0.50)\\ &= 0.033.\end{aligned}$$

Then, by Bayes' theorem

$$\Pr(S_i|D)=\Pr(D|S_i)\Pr(S_i)/\Pr(D), \qquad i=1,2,3.$$

Therefore

$$\begin{aligned}\Pr(S_1|D)&=0.01(0.20)/0.033=0.061,\\ \Pr(S_2|D)&=0.02(0.30)/0.033=0.182,\\ \Pr(S_3|D)&=0.05(0.50)/0.033=0.757.\end{aligned}$$ ■

In the jargon of Bayesian statistics, the $\Pr(A_i)$ terms in Bayes' theorem are called *prior* (or a priori) probabilities for the events A_i and the $\Pr(A_i|B)$ terms are called *posterior* (or a posteriori) probabilities for the events A_i. For the preceding example, the terms $\Pr(S_1)$, $\Pr(S_2)$, and $\Pr(S_3)$ are the prior probabilities that a randomly selected transistor was produced by systems S_1, S_2, and S_3, respectively. Also, $\Pr(S_1|D)$, $\Pr(S_2|D)$, and $\Pr(S_3|D)$ are the posterior probabilities that a randomly selected defective transistor was produced by systems S_1, S_2, and S_3, respectively. Notice for example that system S_3 produces 50% [$\Pr(S_3)=0.50$] of all transistors, but that system S_3 produces 75% [$\Pr(S_3|D)=0.757$] of all defective transistors.

Whereas the previous statement of Bayes' theorem is stated for disjoint discrete events and discrete probability models, it is of interest to state the analogous result for pdf's as follows.

If X is a continuous r.v. whose pdf depends on a variable θ so that the conditional pdf of X, given θ, is given by $f(x|\theta)$, and if the prior pdf of Θ is denoted by $g(\theta)$, then for every x such that $f(x)>0$ exists, the posterior pdf of Θ, given $X=x$ is

$$g(\theta|x)=\frac{f(x|\theta)g(\theta)}{f(x)} \tag{2.13}$$

where

$$f(x)=\int f(x|\theta)g(\theta)\,d\theta$$

is the marginal pdf of X.

In the continuous form of Bayes' theorem, we again observe the use of prior and posterior pdf's to denote the beliefs concerning the likelihood of various values of a designated r.v. (Θ) prior to and posterior to the observing of a value of another r.v. (X).

Example 2.22. Assume that the time-to-failure (T) for a given type light bulb has an exponential distribution with parameter λ, that is, $f(t|\lambda)=\lambda e^{-\lambda t}$, $t>0$, $\lambda>0$. Further assume that Λ is a variable with prior pdf

$$g(\lambda)=2e^{-2\lambda}, \qquad \lambda>0.$$

By Bayes' theorem, the posterior pdf of Λ given $T=t$ is

$$\begin{aligned}g(\lambda|t)&=(\lambda e^{-\lambda t})(2e^{-2\lambda})\Big/\int_0^{\infty}2\lambda e^{-\lambda(t+2)}\,d\lambda\\ &=2\lambda e^{-\lambda(t+2)}\Big/2\left(\frac{1}{t+2}\right)^2\\ &=\lambda(t+2)^2e^{-\lambda(t+2)}, \qquad t>0, \qquad \lambda>0.\end{aligned}$$

■

In this chapter we have introduced some basic probability ideas, presented selected probability models, indicated some ideas of sampling, and introduced Bayes' theorem. The next chapter provides a similar basis for statistical concepts.

EXERCISES

2.1. Let $S=\{t|0\leqslant t\leqslant 1000\}$ be the sample space of an experiment. Let $A=\{t|10<t\leqslant 100\}$, $B=\{t|50<t<150\}$, and $C=\{t|125\leqslant t\leqslant 175\}$

be three events. Find the following events:

(a) $A \cap B$	(g) $A \cap B \cap C$	(m) $\overline{A}$
(b) $A \cup B$	(h) $A \cup B \cup C$	(n) $\overline{B}$
(c) $A \cap C$	(i) $A \cup (B \cap C)$	(o) $\overline{A \cup B}$
(d) $B \cap C$	(j) $(A \cup B) \cap (A \cup C)$	(p) $\overline{A \cap B}$
(e) $A \cup C$	(k) $A \cap (B \cup C)$	(q) $\overline{A} \cap \overline{B}$
(f) $B \cup C$	(l) $(A \cap B) \cup (A \cap C)$	(r) $\overline{A} \cup \overline{B}$

2.2. Let $A = \{t \mid t$ is an even integer$\}$ and $B = \{t \mid t^2 - 7 + 12 = 0\}$. Find $A \cap B$ and $A \cup B$.

2.3. Let $A = \{x \mid x^2 - 2x - 24 = 0\}$ and $B = \{x \mid x^2 \geqslant 15\}$. Determine if $A \subset B$

2.4. Let $S = \{-2, 0, \pi, \sqrt{2}, 1, 3\}$. Let $A = \{-2, 0, 1\}$, $B = \{0, \pi, \sqrt{2}, 3\}$, and $C = \{\pi, \sqrt{2}, 3\}$. Find

(a) $A \cup B$	(e) $\overline{A} \cup B$	(i) $\overline{A \cap B}$
(b) $A \cap B$	(f) $\overline{A} \cap C$	(j) $\overline{A} \cap \overline{B}$
(c) $A \cap C$	(g) $(A \cup B) \cap C$	(k) $\overline{A} \cup \overline{B}$
(d) $A \cap \overline{B}$	(h) $\overline{A \cup B}$	(l) $(A \cap C) \cup (B \cap C)$

2.5. Suppose four companies are being considered as fastener suppliers for a process. Let N denote that a fastener is nondefective and D that it is defective. Describe the sample space where each outcome describes the defective-nondefective nature of each of four fasteners.

(a) List all the outcomes in S.

(b) List the outcomes in the event, A, that there are two defectives.

(c) List the outcomes in the event, B, that there are at least two defectives.

(d) List the outcomes in the event, C, that there are at most three defectives.

Find (e) $B \cap C$, (f) $\overline{B}$, (g) $A \cap B$, (h) $\overline{C}$, (i) $(B \cap \overline{A}) \cap C$.

2.6. In Exercise 2.5 let X denote the number of defectives in the sample of four fasteners.

(a) Find the sample space which gives the values of X.
List the outcomes which correspond to

(b) $X = 0$, (c) $X = 1$ (d) $X = 2$, (e) $X = 3$, (f) $X = 4$.

2.7. Let X denote the number of boys in a family of four children.

(a) List the values in the sample space of X.

(b) List the elementary events corresponding to each value of X.

(c) Suggest a probability model for the r.v. X.

2.8. Toss a coin 100 times and graph $f(H) = \#H/n$ versus n, where n is the number of tosses, for $n = 5, 10, 15, \ldots, 100$.

(a) Does $f(H)$ approach $1/2$ as n increases?

(b) Does the graph tend to crisscross the line $f(H) = 1/2$ or remain on one side consistently?

(c) What would you conclude if the graph approached some value other than $1/2$?

2.9. (a) If a fair die is rolled, list the values in the sample space of X, the number of spots showing up.

(b) Find the probability distribution, $f(x)$, of X.

(c) Find the probability $\Pr(1 \leq X < 5)$.

(d) Find the probability that the outcome is even.

2.10. (a) If two fair dice are rolled, describe the 36 outcomes in the bivariate sample space. Find the sample space of X, the total number showing up on both dice.

(b) Find the probability distribution of X.

(c) What value of X has the largest probability? Give that probability.

(d) Find $\Pr(6 \leq X < 9)$.

2.11. (a) Find the probability distribution for the number of boys, X, in a family of four children.

(b) Find the probability that both sexes are equally represented among the four children.

2.12. The same die is rolled twice. What is the probability of getting a 7 by rolling, in order, a 2 followed by a 5? How does this compare with the probability of rolling a 7 by any combination of two dice which are rolled together?

2.13. A card is drawn at random from an ordinary deck of 52 cards. Find the probability that it is (a) a king, (b) a queen of clubs, (c) a four of diamonds or a five of hearts, (d) a spade, (e) any suit except clubs, (f) a six or a club, (g) neither a three nor a diamond.

2.14. If the probability that Jack will have an accident during the next year is 0.7, and if the probability that Jane will have an accident during the next year is 0.8, what is the probability that both individuals will have an accident during the next year? What is the probability that Jack or Jane will have an accident during the next year? (Assume that the individuals have accidents independently.)

2.15. Let A and B be two events such that $\Pr(A)=0.3$, $\Pr(B)=0.8$, and $\Pr(AB)=0.2$. Find each of the following:

(a) $\Pr(\bar{A})$ (e) $\Pr(A\bar{B})$
(b) $\Pr(\bar{B})$ (f) $\Pr(\bar{A}B)$
(c) $\Pr(\overline{AB})$ (g) $\Pr(A|B)$
(d) $\Pr(A \cup B)$ (h) $\Pr(\bar{B}|A)$

2.16. If $\Pr(A)=0.6$ and $\Pr(B)=0.2$, and if A and B are independent events, what is the probability that the following occur?

(a) A and B occur.
(b) Either A or B occurs.
(c) Either A or $\bar{B}$ occurs.
(d) At least one of the events, A, B, occurs.

2.17. If A_1 and A_2 are independent events, show that $\Pr(A_1 \cup A_2)= \Pr(A_1)+\Pr(A_2)\Pr(\bar{A}_1)$ whenever A_1 and A_2 are not mutually exclusive.

2.18. In tossing a fair coin three times, what is the probability of two heads occurring, given that at least one head has occurred?

2.19. A coin is tossed until a head occurs. What is the probability that

(a) The first head occurs on the fifth toss?
(b) More than three tosses are required?

2.20. The probability of a mass spectrometer operating satisfactorily for a month is 0.7. The probability of an oscilloscope performing adequately during the same period is 0.9. What is the probability of both instruments needing a service call this month? Of one needing a service call?

2.21. A series system of components functions if and only if all its components function. If the components function independently of each other, what is the probability that the system is in operation if it consists of k components and each component functions with a probability of p? Compute the probability if $k=5$ and $p=0.9$.

2.22. A parallel system of components functions if and only if at least one component functions. If the components function independently of each other, what is the probability that the system is in operation if it consists of k components and each component functions with a probability of p? Compute the probability if $k=5$ and $p=0.9$.

2.23. In a class of 30 students, in how many ways can a president, a vice-president, and a secretary be chosen?

2.24. In a class of 30 students, in how many ways can a committee of three be chosen?

2.25. In how many ways can six people sit on a bench if there is only room for three?

2.26. In how many ways can eight books be arranged on a shelf if
(a) Any arrangement is allowed.
(b) Four particular books must always stand together.
(c) Two particular books must occupy the ends?

2.27. In how many ways can seven people be arranged in a row? In how many ways can n people be arranged in a row?

2.28. In how many ways can seven people be arranged in a circle? In how many ways can n people be arranged in a circle?

2.29. From six statisticians and four reliability engineers a committee consisting of three statisticians and two reliability engineers is to be formed. In how many ways can a committee be formed if
(a) No restrictions are imposed.
(b) Two specific statisticians must be on the committee.
(c) One specific reliability engineer cannot be on the committee.

2.30. Suppose a coin is tossed 20 times. Find the probability of obtaining
(a) 10 heads.
(b) At least 2 heads.
(c) At least 9 but no more than 12 heads.
(d) 20 heads.
(e) At most 19 heads.

2.31. A k-out-of-n system of components functions if and only if at least k of its n components function. If the components function independently of each other and if each component has a probability p of functioning, what is the probability that the system is functioning? Compute the probability that the system is functioning if $k=14$, $n=20$, and $p=0.90$.

2.32 If the probability is $p=0.9$ that a component survives a life test of 600 h, find the probability distribution of X, the number of survivors in a test of 15 components. Find the probability that 10 components survive the test, given that at least 5 survive.

2.33. Suppose a lot of 40 electronic tubes contains 8 defectives and 32 nondefectives. If $n=5$ tubes are chosen at random from the lot find the distribution of X, the number of defectives in the sample. Find the probability that the sample contains four defectives, given it contains at least three.

2.34. Suppose a process produces electronic components and that 20% are defective. Find the distribution of X, the number of defectives in a sample of size 5. Find the probability that the sample contains four defectives, given it contains at least three. Discuss this exercise in relation to Exercise 2.33.

2.35. Suppose the number of incoming calls, X (per minute), at a switchboard has a Poisson distribution with parameter $\lambda=5$. Find the probability that there are no calls in one minute; between two and eight calls in one minute (including two and eight). Find the probability that there are four calls, given there are between two and eight calls.

2.36. If 1% of electronic components manufactured by a certain process are defective, find the probability that in a sample of 1000 components, the number of defectives is

(a) 0, (e) 1, (h) 3,
(b) 5, (f) 8, (i) 10.
(c) 12, (g) Between 6 and 11 (inclusive),
(d) At least 9,

Use the Poisson approximation to the binomial.

2.37. In Exercise 2.36 find the probability that there are at least 8 defectives in a sample of 100 components, given there are no more than 12 defectives.

2.38. Suppose the probability of an engine failure on a flight between two cities is 0.006. Find approximately the probability of

(a) No engine failure in 1000 flights.
(b) At least one failure in 1000 flights.
(c) At least two failures in 1000 flights.

Use the Poisson approximation to the binomial.

2.39. If X has a Poisson distribution with mean λ, find the probability that $X=4$, given that $X>0$.

2.40. If X has the exponential pdf $f(x;\lambda)=\lambda e^{-\lambda t}$, $t>0$, and $f(x;\lambda)=0$, $t\leqslant 0$, find $\Pr(X\geqslant a+b \mid X\geqslant a)$ where $a, b>0$. If X denotes the life length of a certain component undergoing a life test, what does this say about the life of a component that is not new?

2.41. Suppose T has an exponential pdf with parameter λ, where λ has an exponential pdf with parameter a. Find the posterior pdf of λ, that is, the pdf $g(\lambda \mid \mathrm{t})$.

2.42. Suppose a component used in a system is produced by three companies, C_1, C_2, and C_3. Further, suppose that C_1 produces 10%, C_2 produces 20%, and C_3 produces 70% of the supply, and finally suppose the defective (D) rates are 0.02, 0.02, and 0.03, respectively, for C_1, C_2, and C_3. If a component is randomly selected from the supply and found to be defective, what are $\Pr(C_1|D)$, $\Pr(C_2|D)$, and $\Pr(C_3|D)$?

2.43. A box contains eight red and six white balls. A second box contains ten red and four white balls. A ball is drawn at random from the first box and placed into the second box. Then a ball is drawn at random from the second box.

(a) Find the probability that the ball drawn from the second box is red.

(b) If the ball drawn from the second box is red, what is the probability that a white ball was transferred from the first box to the second?

2.44. Prove that if $\Pr(E)>\Pr(F)$ then $\Pr(E|F)>\Pr(F|E)$.

2.45. If A and B are independent, prove that

(a) A and $\bar{B}$ are independent.

(b) $\bar{A}$ and B are independent.

(c) $\bar{A}$ and $\bar{B}$ are independent.

2.46. If A, B, C are independent, prove that

(a) A and $B \cap C$,

(b) A and $B \cup C$,

(c) $-A$ and $B\bar{C}$, are independent.

2.47. Show that $\Sigma_{j=0}^{n} \binom{n}{j} = 2^n$.

2.48. If X has a binomial distribution with p small and n large so that $\rho = np =$ constant, show that the binomial distribution can be approximated with the Poisson distribution with parameter $\rho = np$.

REFERENCES

Bayes, T. (1958). Essay Towards Solving a Problem in the Doctrine of Chances, *Biometrika*, Vol. 45, pp. 298–315.

Johnson, N. L. and Kotz, S. (1969). *Discrete Distributions*, Houghton Mifflin, Boston.

Johnson, N. L. and Kotz, S. (1970). *Continuous Univariate Distributions − 1 and − 2*, Houghton Mifflin, Boston.

CHAPTER 3

Elements of Statistics

The purpose of this chapter is to introduce selected basic concepts of statistics as a foundation for use throughout the remainder of the text. The chapter presents some characteristics of models, demonstrates the transformation of r.v.'s, introduces the ideas of statistical inference, illustrates sequences of r.v.'s, and discusses goodness-of-fit techniques. Ideas presented here are basic concepts, most of which are expanded and specialized in subsequent chapters.

3.1. EXPECTATION OF RANDOM VARIABLES

In the preceding chapter we defined an r.v. as any real-valued function defined on a sample space for an experiment. We also presented some selected probability distributions in Sections 2.4 and 2.5. In this section we use those results to develop the idea of the expected value (expectation) of an r.v.

Definition 3.1. If $\Pr(x)$ is a probability model for a discrete r.v. X with sample space $\{x_1, x_2, \ldots, x_k\}$ and $g(x)$ is a real-valued function of X, then the *expected value* of $g(X)$ is defined to be

$$E[g(X)] = \sum_{i=1}^{k} g(x_i)\Pr(x_i), \tag{3.1}$$

whenever the sum on the right converges absolutely. Notice that k may be infinite in some cases.

Example 3.1. Suppose X is a binomial r.v. with parameter p and sample size n. Find $E(X)$ and $E[X(X-1)]$.

By definition, when $g(X)=X$

$$E(X)=\sum_{x=0}^{n} x\binom{n}{x}p^x q^{n-x}$$

$$=\sum_{x=1}^{n} x\frac{n!}{x!(n-x)!}p^x q^{n-x}$$

$$=np\sum_{x=1}^{n}\frac{(n-1)!}{(x-1)!(n-x)!}p^{x-1}q^{(n-1)-(x-1)}$$

$$=np\sum_{y=0}^{n-1}\binom{n-1}{y}p^y q^{(n-1)-y}, \qquad \text{where } y=x-1$$

$$=np(q+p)^{n-1}, \qquad \text{by binomial theorem}$$

$$=np.$$

Similarly, if $g(X)=X(X-1)$, then

$$E[X(X-1)]=\sum_{x=0}^{n} x(x-1)\binom{n}{x}p^x q^{n-x}$$

$$=\sum_{x=2}^{n} x(x-1)\frac{n!}{x!(n-x)!}p^x q^{n-x}$$

$$=n(n-1)p^2\sum_{y=0}^{n-2}\binom{n-2}{y}p^y q^{(n-2)-y}$$

$$=n(n-1)p^2.$$

■

In like fashion, we can define the expected value of a function of a continuous r.v. as follows:

Definition 3.2. If $f(x)$ is a pdf for a continuous r.v. X and if $g(x)$ is a real-valued function of X, then the *expected value* of $g(X)$ is defined to be

$$E[g(X)]=\int_{-\infty}^{\infty} g(x)f(x)\,dx, \tag{3.2}$$

whenever the integral converges.

Example 3.2. Suppose X is an exponential r.v. with parameter λ. Find $E(X)$ and $E(X^2)$.

By definition, with $g(X) = X$,

$$E(X) = \int_0^\infty x\lambda e^{-\lambda x}\,dx$$

$$= -xe^{-\lambda x}\Big|_0^\infty + \int_0^\infty e^{-\lambda x}\,dx$$

$$= 0 - \frac{e^{-\lambda x}}{\lambda}\Big|_0^\infty$$

$$= \frac{1}{\lambda}.$$

Similarly, for $g(X) = X^2$, integration by parts gives

$$E(X^2) = \int_0^\infty x^2\lambda e^{-\lambda x}\,dx = \frac{2}{\lambda^2}. \qquad \blacksquare$$

Some expectations are given special names because they are widely used. First, a common measure of location (central value) for an r.v. X is the *mean* (center of gravity) defined as

$$\mu = E(X). \tag{3.3}$$

Second, a measure of dispersion (or variation) of the r.v. about its mean is the *variance*, denoted by σ^2 or $\text{Var}(X)$. The variance is the second moment about the mean (moment of inertia) and is defined as

$$\text{Var}(X) = \sigma^2 = E(X-\mu)^2 = E(X^2) - \mu^2. \tag{3.4}$$

Third, the *ordinary moments* of an r.v. X are defined as

$$\mu'_r = E(X^r), \qquad r = 1, 2, \ldots. \tag{3.5}$$

Fourth, the *central moments* (moments about the mean) of an r.v. X are defined as

$$\mu_r = E(X-\mu)^r, \qquad r = 2, 3, \ldots. \tag{3.6}$$

From the definition of μ it follows that $\mu_1 = 0$. Finally, the *factorial moments* of a variable X are defined as

$$\mu'_{(r)} = E[X(X-1)(X-2)\cdots(X-r+1)], \qquad r = 1, 2, \cdots. \quad (3.7)$$

The preceding moments are characteristics of distributions of r.v.'s in that a distribution can be characterized by its moments. Frequently, we use the first few (two, three, or four) moments to describe a distribution. Two commonly stated characteristics for a distribution are its mean (μ) and variance (σ^2).

Example 3.3. Find μ'_1, μ_2, and $\mu'_{(2)}$ for a binomial distribution with parameter p. From Example 3.1, we have

$$\mu'_1 = \mu = E(X) = np \text{ (mean)} \qquad \text{and} \qquad \mu'_{(2)} = E[X(X-1)] = n(n-1)p^2.$$

Now

$$E[X(X-1)] = E(X^2 - X) = E(X^2) - E(X).$$

Therefore,

$$\begin{aligned} E(X^2) &= E[X(X-1)] + E(X) \\ &= n(n-1)p^2 + np \\ &= n^2p^2 - np^2 + np. \end{aligned}$$

Thus, the variance μ_2 is given by

$$\begin{aligned} \mu_2 = \sigma^2 &= E(X^2) - \mu^2 \\ &= n^2p^2 - np^2 + np - (np)^2 \\ &= np(1-p). \end{aligned}$$

Example 3.4. Find the mean and variance of an exponential distribution with parameter λ. From Example 3.2 we have that the mean is

$$\mu = E(X) = \frac{1}{\lambda} \text{ and that } E(X^2) = \frac{2}{\lambda^2}.$$

Example 3.2. Suppose X is an exponential r.v. with parameter λ. Find $E(X)$ and $E(X^2)$.

By definition, with $g(X)=X$,

$$\begin{aligned}
E(X) &= \int_0^\infty x\lambda e^{-\lambda x}\,dx \\
&= -xe^{-\lambda x}\Big|_0^\infty + \int_0^\infty e^{-\lambda x}\,dx \\
&= 0 - \frac{e^{-\lambda x}}{\lambda}\Big|_0^\infty \\
&= \frac{1}{\lambda}.
\end{aligned}$$

Similarly, for $g(X)=X^2$, integration by parts gives

$$E(X^2)=\int_0^\infty x^2\lambda e^{-\lambda x}\,dx=\frac{2}{\lambda^2}.$$ ■

Some expectations are given special names because they are widely used. First, a common measure of location (central value) for an r.v. X is the *mean* (center of gravity) defined as

$$\mu = E(X). \tag{3.3}$$

Second, a measure of dispersion (or variation) of the r.v. about its mean is the *variance*, denoted by σ^2 or $\mathrm{Var}(X)$. The variance is the second moment about the mean (moment of inertia) and is defined as

$$\mathrm{Var}(X)=\sigma^2=E(X-\mu)^2=E(X^2)-\mu^2. \tag{3.4}$$

Third, the *ordinary moments* of an r.v. X are defined as

$$\mu_r' = E(X^r), \qquad r=1,2,\ldots. \tag{3.5}$$

Fourth, the *central moments* (moments about the mean) of an r.v. X are defined as

$$\mu_r = E(X-\mu)^r, \qquad r=2,3,\ldots. \tag{3.6}$$

From the definition of μ it follows that $\mu_1 = 0$. Finally, the *factorial moments* of a variable X are defined as

$$\mu'_{(r)} = E[X(X-1)(X-2)\cdots(X-r+1)], \qquad r = 1, 2, \cdots. \qquad (3.7)$$

The preceding moments are characteristics of distributions of r.v.'s in that a distribution can be characterized by its moments. Frequently, we use the first few (two, three, or four) moments to describe a distribution. Two commonly stated characteristics for a distribution are its mean (μ) and variance (σ^2).

Example 3.3. Find μ'_1, μ_2, and $\mu'_{(2)}$ for a binomial distribution with parameter p. From Example 3.1, we have

$$\mu'_1 = \mu = E(X) = np \text{ (mean)} \qquad \text{and} \qquad \mu'_{(2)} = E[X(X-1)] = n(n-1)p^2.$$

Now

$$E[X(X-1)] = E(X^2 - X) = E(X^2) - E(X).$$

Therefore,

$$E(X^2) = E[X(X-1)] + E(X)$$

$$= n(n-1)p^2 + np$$

$$= n^2p^2 - np^2 + np.$$

Thus, the variance μ_2 is given by

$$\mu_2 = \sigma^2 = E(X^2) - \mu^2$$

$$= n^2p^2 - np^2 + np - (np)^2$$

$$= np(1-p).$$

Example 3.4. Find the mean and variance of an exponential distribution with parameter λ. From Example 3.2 we have that the mean is

$$\mu = E(X) = \frac{1}{\lambda} \text{ and that } E(X^2) = \frac{2}{\lambda^2}.$$

Thus, the variance is

$$\sigma^2 = E(X^2) - \mu^2 = \frac{2}{\lambda^2} - \left(\frac{1}{\lambda}\right)^2 = \frac{1}{\lambda^2}. \qquad \blacksquare$$

The idea of expectation is easily extended to multivariate (joint) distributions as discussed in Section 2.9. The expectation of a function g of continuous r.v.'s $X_1, X_2, \ldots, X_n$ is

$$E[g(X_1, X_2, \ldots, X_n)] = \int_{-\infty}^{\infty} \int_{-\infty}^{\infty} \cdots \int_{-\infty}^{\infty} g(x_1, x_2, \ldots, x_n)$$

$$\times f(x_1, x_2, \ldots, x_n)\, dx_1\, dx_2 \cdots dx_n,$$

where f is the joint pdf of the X_i. For the case of discrete r.v.'s $X_1, X_2, \ldots, X_n$, the n-fold integral is replaced with an n-fold summation.

The *covariance* of two r.v.'s X_1 and X_2 is defined to be

$$\operatorname{Cov}(X_1, X_2) = E\{[X_1 - E(X_1)][X_2 - E(X_2)]\}$$

and it is easily shown that

$$\operatorname{Cov}(X_1, X_2) = E(X_1 X_2) - E(X_1)E(X_2). \tag{3.8}$$

The *correlation* between X_1 and X_2, denoted by $\rho(X_1, X_2)$, is defined as

$$\rho(X_1, X_2) = \frac{\operatorname{Cov}(X_1, X_2)}{[\operatorname{Var}(X_1)\operatorname{Var}(X_2)]^{1/2}}.$$

In applications of statistical theory one often works with sums of observations that arise in the sampling process. If $X_1, X_2, \ldots, X_n$ are r.v.'s, then it is easily shown, for constants $a_1, a_2, \ldots, a_n$, that

(i) $E\left(\sum_{i=1}^{n} a_i X_i\right) = \sum_{i=1}^{n} a_i E(X_i)$,

(ii) $\operatorname{Var}\left(\sum_{i=1}^{n} a_i X_i\right) = \sum_{i=1}^{n} a_i^2 \operatorname{Var}(X_i) + \sum_{\substack{i=1 \\ i \neq j}}^{n} \sum_{j=1}^{n} a_i a_j \operatorname{Cov}(X_i, X_j)$.

If the X_i are independent, then (ii) becomes

$$\text{(iii)}\quad \operatorname{Var}\left(\sum_{i=1}^{n} a_i X_i\right) = \sum_{i=1}^{n} a_i^2 \operatorname{Var}(X_i).$$

Example 3.5. Find the covariance of X_1 and X_2 if their joint pdf is

$$f(x_1, x_2) = \frac{6}{5}(x_1^2 + x_2), \qquad 0 < x_1 < 1, 0 < x_2 < 1.$$

We need to find $E(X_1 X_2)$, $E(X_1)$, $E(X_2)$, and then use (3.8). We have

$$E(X_1) = \frac{6}{5}\int_0^1 \int_0^1 x_1(x_1^2 + x_2)\,dx_1\,dx_2 = \frac{3}{5},$$

$$E(X_2) = \frac{6}{5}\int_0^1 \int_0^1 x_2(x_1^2 + x_2)\,dx_1\,dx_2 = \frac{3}{5},$$

and

$$E(X_1 X_2) = \frac{6}{5}\int_0^1 \int_0^1 x_1 x_2(x_1^2 + x_2)\,dx_1\,dx_2 = \frac{7}{20}.$$

Therefore

$$\operatorname{Cov}(X_1, X_2) = \frac{7}{20} - \left(\frac{3}{5}\right)\left(\frac{3}{5}\right) = \frac{7}{20} - \frac{9}{25} = -\frac{1}{100}. \qquad \blacksquare$$

3.2. CHEBYSHEV INEQUALITY

One reason that the mean and variance are said to characterize a distribution is illustrated by the *Chebyshev Inequality*. First, we define the *standard deviation* of a distribution as the positive square root of the variance, that is, $\sigma = \sqrt{\sigma^2}$. Second, the Chebyshev Inequality states that

$$\Pr(|X - \mu| < k\sigma) \geq 1 - \frac{1}{k^2}, \qquad k > 0, \tag{3.9}$$

whenever $\sigma < \infty$.

The practical application of (3.9) is that for any distribution at least 75% of all values of the variable are within two standard deviations (2σ) of the mean. Further, at least 89% (8/9) of all values are within three standard deviations (3σ) of the mean. It is in this sense that the mean and variance (standard deviation) can be said to describe a distribution. We note that the bounds provided by the Chebyshev Inequality are valid for any distribution; however, for many widely used distributions the bounds are very loose.

Example 3.6. Find $\Pr(|X-\mu|<k\sigma)$ for $k=1,2,3$ for (a) a binomial distribution with $n=16$ and $p=0.5$, (b) an exponential distribution with $\lambda=1/2$.

(a) From Example 3.3, the mean and variance of a binomial distribution with $n=16$ and $p=0.5$ are 8 and 4, respectively. Therefore, $\mu=8$ and $\sigma=2$. Thus, for $k=1$ we seek

$$\Pr(|X-8|<2)=\Pr(7\leqslant X\leqslant 9)$$

$$=\sum_{k=7}^{9}\binom{16}{k}(0.5)^{16}=0.5456,$$

for $k=2$ we want

$$\Pr(|X-8|<4)=\Pr(5\leqslant X\leqslant 11)=0.9232,$$

and for $k=3$ we need

$$\Pr(|X-8|<6)=\Pr(3\leqslant X\leqslant 13)=0.9958.$$

These probabilities can be obtained from Table B6, Appendix B.

(b) From Example 3.4, the mean and variance of an exponential distribution with $\lambda=1/2$ are 2 and 4, respectively. Therefore, $\mu=2$ and $\sigma=2$. Thus, for $k=1$ we want

$$\Pr(|X-2|<2)=\Pr(0<X<4)$$

$$=\int_0^4 \frac{1}{2}e^{-(1/2)t}\,dt=1-e^{-2}=0.8647,$$

for $k=2$

$$\Pr(|X-2|<4)=\Pr(0<X<6) \qquad (\text{since } X>0)$$

$$=1-e^{-3}=0.9502,$$

and for $k=3$

$$\begin{aligned}\Pr(|X-2|<6) &= \Pr(0<X<8)\\ &= 1-e^{-4}=0.9817.\end{aligned}$$

■

The preceding examples illustrate that for two specific distributions, Chebyshev bounds are quite loose. Table 3.1 summarizes the results from Example 3.6, the corresponding Chebyshev bounds, and some guideline bounds which hold for most of the distributions discussed in this text.

Table 3.1. Probabilities Associated with $\Pr(|X-\mu|<k\sigma)$

k	Chebyshev Bound	Exact Binomial Probability	Exact Exponential Probability	Practical Guidelines
1	0	0.5456	0.8647	0.50–0.75
2	0.75	0.9232	0.9502	0.75–0.95
3	0.89	0.9958	0.9817	0.90–1.00

The guidelines given in Table 3.1 help us to visualize the range of values likely to occur when sampling a distribution with known values of μ and σ. Similar results apply to empirical distributions when μ and σ are replaced by the sample mean (or average) and the sample standard deviation, respectively. These and other statistics associated with a sample are discussed in the next section.

3.3. DESCRIPTIVE STATISTICS

The ideas presented in Section 3.2 pertain to characteristics of populations or distributions. Those ideas can be developed in terms of the observations that comprise a random sample chosen from a distribution. In Section 2.8 the idea of an empirical distribution, based on a sample, was presented as it relates to the sampled distribution.

Let $X_1, X_2, \ldots, X_n$ denote a random sample from a distribution $f(x)$. The *sample mean*, denoted by $\bar{X}$, is defined as

$$\bar{X}=\sum_{i=1}^{n}\frac{X_i}{n}$$

and the *sample variance*, denoted by S^2, is defined as

$$S^2 = \frac{\Sigma(X_i - \bar{X})^2}{n-1},$$

the denominator $n-1$ being used so as to make S^2 "unbiased" for estimating the population variance σ^2. The *sample moments* are defined by

$$m'_j = \sum_{i=1}^{n} \frac{X_i^j}{n}, \qquad j = 1, 2, \ldots.$$

Similar ideas can be developed in the case of samples from multivariate distributions.

A function of observable r.v.'s that can be computed once the data are collected is called a *statistic*. One use of statistics is to estimate the parameters of a distribution. For example, the sample mean or average, $\bar{X}$, estimates the distribution mean, μ, and the sample variance, S^2, estimates the distribution variance, σ^2. Section 3.6 provides a more complete discussion of statistical estimation.

3.4. MOMENT GENERATING FUNCTIONS

A tool that is quite useful in statistics is the *moment generating function*. Let the r.v. X have a distribution $f(x)$, continuous or discrete. The moment generating function of the r.v. X is defined by $M(t; x) = E(e^{Xt})$ if M exists in an interval $-h < t < h$. Thus

$$M(t; x) = \begin{cases} \int_{-\infty}^{\infty} e^{tx} f(x)\, dx, & \text{if } X \text{ is continuous} \\ \sum_i e^{tx_i} f(x_i), & \text{if } X \text{ is discrete.} \end{cases}$$

By expanding e^{tX} it follows that

$$M(t; x) = E\left(1 + tX + \frac{t^2 X^2}{2!} + \cdots\right)$$

$$= 1 + tE(X) + \frac{t^2}{2!} E(X^2) + \cdots + \frac{t^j}{j!} E(X^j) + \cdots$$

and $E(X^j)$ is the coefficient of $t^j/j!$. Thus the moments can be found from a series expansion of $M(t; x)$.

Also, the moments can be found by differentiating $M(t; x)$ with respect to t and setting $t=0$. More specifically, the jth derivative of $M(t; x)$, if X is continuous, is

$$\frac{d^j M(t; x)}{dt^j} = \int_{-\infty}^{\infty} x^j e^{tx} f(x)\, dx$$

and

$$\left.\frac{d^j M(t; x)}{dt^j}\right|_{t=0} = \int_{-\infty}^{\infty} x^j f(x)\, dx = E(X^j).$$

An important property of moment generating functions is that if $M(t; x)$ exists, then it uniquely determines a distribution. For additional properties and results the reader is referred to books on mathematical statistics.

Example 3.7. Determine the moment generating function of $f(x) = xe^{-x}$ and find the first two moments by differentiation.

The moment generating function is

$$M(t; x) = \int_0^{\infty} e^{tx} x e^{-x}\, dx$$

$$= \int_0^{\infty} x e^{-x(1-t)}\, dx$$

$$= (1-t)^{-2}.$$

Thus

$$E(X) = -2(1-t)^{-3}(-1)|_{t=0} = 2$$

and

$$E(X^2) = 2(-3)(1-t)^{-4}(-1)|_{t=0} = 6.$$

The variance would then be $\text{Var}(X) = 6 - 2^2 = 2$. ■

3.5. TRANSFORMATIONS OF RANDOM VARIABLES

In working with statistics we frequently work with transformations of our data rather than the actual data. Some common transformations are "square roots," "logarithm," and "linear." The purposes of these transformations are to stabilize the variance, to provide better fitting distributions, and to calculate distributions of statistics. In this section we illustrate some results of transformations.

A common transformation of data is the linear one defined by

$$Y = A + BX. \tag{3.10}$$

Direct application of (3.1), (3.3), and (3.4) gives

$$E(Y) = A + BE(X)$$

and (3.11)

$$\operatorname{Var}(Y) = B^2 \operatorname{Var}(Y).$$

Thus, the mean of a linear function of X is the same linear function of the mean of X. However, the variance of X is unchanged by the location factor A but changed by the square of the scale factor B.

Two transformations used for stabilizing the variance and adjusting the distribution of r.v.'s are the "square root" and the "logarithm." Recall that for the binomial distribution in Example 3.3, the mean is np and the variance is npq. Therefore, the mean and variance both increase as n increases. For this case, a square root transformation of the data can be used to stabilize the variance. Data distributions that are highly skewed to the right (i.e., distributions having relatively large values with low probability) can be adjusted to be more symmetric in appearance by taking logarithms.

In Bayesian analyses we are frequently interested in some transformation of a parameter of interest.

As an example, we suppose that T, a failure time r.v., is exponential with parameter λ. In this case, the reliability of the item [see Section 4.1.1] is

$$R(t) = \Pr(T > t) = \int_t^\infty \lambda e^{-\lambda x}\, dx = e^{-\lambda t}.$$

Further suppose that the parameter λ is an r.v. having an exponential distribution with parameter λ_0. One can then determine the expected value of $R(t)$ with respect to the distribution on Λ.

Example 3.8. If $R(t)$ is the reliability function for an exponential r.v. having parameter λ, and Λ has an exponential distribution with parameter 0.5, find $E[R(t)]$. By (3.2),

$$E[R(t)] = \int_0^\infty R(t) f(\lambda)\, d\lambda = \int_0^\infty e^{-\lambda t} 0.5 e^{-0.5\lambda}\, d\lambda$$

$$= 0.5 \int_0^\infty e^{-(t+0.5)\lambda}\, d\lambda = \frac{0.5}{t+0.5}. \qquad \blacksquare$$

3.5.1 Distribution of Functions

The problem of determining the distribution of functions of r.v.'s is a very common problem in statistics. We present here the simplest case.

Suppose the r.v. X has the pdf $f(x)$ and assume that $y = u(x)$ is a one-to-one function of X. Then the pdf of Y, $g(y)$, is

$$g(y) = f[u^{-1}(y)] \left| \frac{dx}{dy} \right|,$$

where $x = u^{-1}(y)$.

Example 3.9. Let T have an exponential pdf with parameter λ. Find the pdf of $Y = e^{-\lambda T}$.

We have

$$y = u(t) = e^{-\lambda t} \qquad \text{or} \qquad t = -\frac{\ln y}{\lambda}.$$

Thus, $dt/dy = -1/\lambda y$ and the pdf of $Y = e^{-\lambda T}$ is

$$g(y) = \lambda y \left| -\frac{1}{\lambda y} \right| = 1, \qquad 0 < y < 1. \qquad \blacksquare$$

For the case of more general transformations $y = u(x)$ and those based on multidimensional r.v.'s, the reader is referred to Hogg and Craig (1978).

3.5.2 Derived Distributions

A distribution that models a population that is sampled is known as a *sampling distribution*. These basic sampling distributions can be used to derive other distributions which are called *derived distributions*. Three commonly used derived distributions are the t-distribution, the χ^2-distribution, and the $\mathcal{F}$-distribution.

The t-distribution with k degrees of freedom is

$$f(t)=\frac{[(k-1)/2]!}{\sqrt{n\pi}\,[(k-2)/2]!}\left(1+\frac{t^2}{k}\right)^{-(k+1)/2}, \qquad -\infty<t<\infty,\; k=1,2,\ldots, \tag{3.12}$$

the χ^2-distribution with n degrees of freedom is

$$f(u)=\frac{u^{n/2-1}e^{-u/2}}{\Gamma\left(\frac{n}{2}\right)2^{n/2}}, \qquad 0<u<\infty,\; n=1,2,\ldots, \tag{3.13}$$

and the $\mathcal{F}$-distribution with n_1 numerator and n_2 denominator degrees of freedom is

$$f(x)=\frac{\Gamma\left(\frac{n_1+n_2}{2}\right)}{\Gamma\left(\frac{n_1}{2}\right)\Gamma\left(\frac{n_2}{2}\right)}\left(\frac{n_1}{n_2}\right)^{n_1/2}x^{n_1/2-1}\left(1+\frac{n_1}{n_2}x\right)^{-(n_1+n_2)/2},$$

$$0<x<\infty,\; n_1,n_2=1,2,\ldots. \tag{3.14}$$

The t, χ^2, and $\mathcal{F}$ distributions are referred to by the expressions $t(k)$, $\chi^2(n)$, and $\mathcal{F}(n_1, n_2)$, respectively.

These three distributions can be derived by considering the appropriate statistics based on a sample of n independent observations from a normal distribution with pdf

$$f(x)=\frac{1}{\sqrt{2\pi}\,\sigma}\exp\left\{-\frac{1}{2}\left(\frac{x-\mu}{\sigma}\right)^2\right\}, \qquad -\infty<x<\infty.$$

The normal distribution, denoted by $\mathcal{N}(\mu, \sigma^2)$, is discussed in Section 4.2.3 as a reliability distribution. For details on the derivation of these three derived distributions the reader is referred to any basic mathematical statistics textbook [see e.g., Hogg and Craig (1978), Chapter 4].

3.6. STATISTICAL INFERENCE

Section 3.1 presents moments and functions of moments of distributions of r.v.'s with the purpose of *describing* (characterizing) the distributions. In this section we direct our attention to a different set of questions. In particular, we are interested in using sample data to answer questions about the distribution of an r.v. Some examples are as follows:

1. If the failure time T of a safety system component in a nuclear power plant is assumed to be an exponential r.v., what is the best estimate of λ, the failure rate of the component? Or how can we use a sample of failure times, say $t_1, t_2, \ldots, t_n$, to estimate λ?
2. If the number, say X, of 60 W light bulbs that survive when 10 bulbs are placed on life test for 1000 h is assumed to be a binomial r.v. with $n = 10$ and unknown parameter p, how can we use sample data to estimate p? (The parameter p denotes the probability that a randomly selected bulb will survive a life test of 1000 h.)
3. How can we use failure time data in question 1 above to test whether $\lambda = 10^{-4}$ failures per year?
4. How can we use life test data from question 2 above to test whether $p = 0.95$?

Questions 1–4 above illustrate the types of information provided by statistical inference. In general, *statistical inference* techniques are designed to use sample data to make (infer) statements about the distribution (model or population) from which the sample observations were obtained. As we are inferring from a *specific* sample to a *general* distribution in our inference statements, we are using inductive reasoning. That is, we are considering a *part* (sample) to talk about the *whole* (model, distribution, or population). Thus, this reasoning cannot be accomplished with certainty. Therefore, each inference technique will involve some uncertainty. It is the purpose of this section to describe general statistical inference procedures and to explain how the necessary uncertainty is incorporated in the statistical analyses.

Two broad categories of inference methods are (1) the estimation of distributions (models) and their characteristics, and (2) the tests of statistical hypotheses. We now address these two topics.

3.6.1 Statistical Point Estimation

Let $f(x; \theta)$ denote a distribution of the r.v. X where θ represents an unknown parameter. In this section we present techniques for estimating the parameter θ with a single value; that is, methods for obtaining *point estimates* of θ.

Suppose that experimental results $X_1, X_2, \ldots, X_n$ are available and that we desire *one number* to estimate the unknown value θ of a parameter. There are many methods for accomplishing the desired result and the reader can find complete discussions in Hogg and Craig (1978) or Kendall and Stuart (1973). Waller (1979) presents an intuitive discussion of estimation. Here we briefly present and illustrate some commonly used methods of point estimation.

Let $X_1, X_2, \ldots, X_n$ denote a random sample from $f(x; \theta)$. The *likelihood function* for the random sample is the joint pdf of $X_1, X_2, \ldots, X_n$, which is

$$L(\theta; x_1, x_2, \ldots, x_n) = \prod_{i=1}^{n} f(x_i; \theta),$$

when considered as a function of θ. The *maximum-likelihood* (ML) *estimate* of θ is defined as the value $\hat{\theta}$ such that $L(\hat{\theta}; x_1, x_2, \ldots, x_n) \geqslant L(\theta; x_1, x_2, \ldots, x_n)$ for every value of θ. That is, the ML estimate of θ is the value $\hat{\theta}$ that maximizes the likelihood function. The ML estimate is a function of the observed random sample $x_1, x_2, \ldots, x_n$. When $\hat{\theta}$ is considered to be a function of the random sample $X_1, X_2, \ldots, X_n$, then $\hat{\theta}$ is an r.v., and is called the *ML estimator* of θ.

Example 3.10. Let $T_1, T_2, \ldots, T_n$ be n failure times that are exponentially distributed with failure rate λ. Find the ML estimator for λ. The likelihood function is

$$L = L(\lambda; t_1, t_2, \ldots, t_n) = \prod_{i=1}^{n} \lambda e^{-\lambda t_i} = \lambda^n e^{-\lambda \Sigma t_i}.$$

It is convenient to use the logarithm of L to find its maximum. We have

$$\ln L = n \ln \lambda - \lambda \sum t_i$$

and

$$\frac{\partial \ln L}{\partial \lambda} = \frac{n}{\lambda} - \sum t_i = 0$$

implies $\hat{\lambda} = n / \Sigma t_i = 1/\bar{t}$ is the ML estimate for λ. The ML estimator of λ is thus $\hat{\lambda} = 1/\bar{T}$. ■

Another method of estimation is the *method of moments*, which consists of equating the distribution moments $\mu_j' = E(X^j)$ to the sample moments $m_j' = \Sigma X_i^j/n$, for $j=1,2,\ldots,k$, if the pdf $f(x;\theta_1,\theta_2,\ldots,\theta_k)$ contains k parameters. The k equations in k unknowns

$$E(X^j) = \sum X_i^j/n, \qquad j=1,2,\ldots,k$$

or

$$\mu_j' = m_j', \qquad j=1,2,\ldots,k,$$

are solved for the k unknowns $\theta_1,\theta_2,\ldots,\theta_k$ and the solutions $\hat{\theta}_1,\hat{\theta}_2,\ldots,\hat{\theta}_k$ are called the estimators obtained by the method of moments.

Example 3.11. Let X have a $\mathfrak{N}(\mu,\sigma^2)$ distribution [see Section 4.2.3]. Find the estimators for μ and σ^2 by the method of moments.

The first two moments of the normal distribution are

$$\mu_1' = \mu \qquad \text{and} \qquad \mu_2' = E(X^2) = \sigma^2 + \mu^2.$$

Thus

$$\mu = \frac{\Sigma X_i}{n} = \overline{X},$$

$$\sigma^2 + \mu^2 = \frac{\Sigma X_i^2}{n},$$

and the solution is $\hat{\mu} = \overline{X}$ and $\hat{\sigma}^2 = \dfrac{\Sigma X_i^2}{n} - \overline{X}^2 = \dfrac{\Sigma(X_i - \overline{X})^2}{n}$. ■

A third method of estimation is *least squares*. In this method the r.v.'s are assumed to have the form

$$X_i = h_i(\theta_1,\theta_2,\ldots,\theta_k) + e_i, \qquad i=1,2,\ldots,n,$$

where h_i is a known function of the unknown parameters and the e_i are r.v.'s. The least squares estimators of the θ_is are the values of the θ_is that minimize the sum of squares

$$\sum_{i=1}^{n} e_i^2 = \sum_{i=1}^{n} \left[X_i - h_i(\theta_1,\theta_2,\ldots,\theta_k)\right]^2.$$

There are numerous properties for estimators that are desirable. Two of these are unbiasedness and sufficiency.

Definition 3.3. An estimator $\hat{\theta}$ is *unbiased* for a parameter θ if $E(\hat{\theta})=\theta$.

In sampling from a normal distribution the sample mean $\overline{X}$ and variance S^2 are unbiased for μ and σ^2; however, the method of moments variance estimator, $\hat{\sigma}^2$, is biased.

Definition 3.4. Let $X_1, X_2,\ldots,X_n$ be a random sample from $f(x;\theta_1, \theta_2,\ldots,\theta_k)$. Let $\hat{\theta}_1, \hat{\theta}_2,\ldots,\hat{\theta}_m$, be functions (statistics) of the X_is. The statistics $\hat{\theta}_1, \hat{\theta}_2,\ldots,\hat{\theta}_m$, are said to be a set of *jointly sufficient statistics* if the conditional pdf of the X_is given the statistics $\hat{\theta}_i$s, $g(x_i, x_2,\ldots,x_n \mid \hat{\theta}_1, \hat{\theta}_2,\ldots,\hat{\theta}_m)$, is independent of the parameters.

If it is assumed that the range of the r.v. X does not depend on θ, then there is a simple criterion for determining a set of sufficient statistics for $\theta_1, \theta_2,\ldots,\theta_k$. A necessary and sufficient condition for $\hat{\theta}_1, \hat{\theta}_2,\ldots,\hat{\theta}_m$ to be a set of jointly sufficient statistics is that the pdf of the X_is be factored as

$$\prod_{i=1}^{n} f(x_i;\theta_1,\theta_2,\ldots,\theta_k)=h(\hat{\theta}_1,\hat{\theta}_2,\ldots,\hat{\theta}_m;\theta_1,\theta_2,\ldots,\theta_k)\cdot k(x_1,x_2,\ldots,x_n),$$

where k does not contain the θ_is.

The idea of sufficiency is important since it leads to estimators which are minimum variance, and it also greatly simplifies Bayesian estimation methods [see Section 5.5.2].

There are several other methods of estimation and other desirable properties for estimators. For a treatment of these the reader is referred to the previously mentioned references. The idea of Bayesian estimation has received much attention in recent years, and this topic is introduced in Chapter 5.

As an example of the foregoing ideas, consider a set of n objects placed on a life test for a specified number of hours. At the conclusion of the test, X survivors are observed. We suppose that the assumptions of independence and common probability of survival, say p, are met so that a binomial model is reasonable. Our aim is to use the data, X and n, to estimate the common, but unknown value of p. A common estimator of p is

$$\hat{p}=\frac{X}{n}. \tag{3.15}$$

That is, the sample proportion of survivors estimates the distribution parameter p, the probability of survival. The results in Examples 3.1 and 3.3 show that

$$E(\hat{p}) = p \qquad \text{and} \qquad \operatorname{Var}(\hat{p}) = \frac{p(1-p)}{n}. \tag{3.16}$$

Thus, $\hat{p}$ is unbiased for p and its variance decreases as the sample size n increases. It can be shown that $\hat{p}$ is a sufficient statistic and a ML estimator for p.

Example 3.12. A life test of 1000 h for a sample of 100 60 W light bulbs produced 93 survivors. Find a point estimate of p, the unknown probability of surviving the test.

The point estimate is

$$\hat{p} = \tfrac{93}{100} = 0.93.$$

■

As a second illustration of point estimation, we look at the exponential distribution with failure rate λ. To do so, consider a component that is placed on life test. We suppose n items are tested and that the failure times $t_1, t_2, \ldots, t_n$ are recorded. The purpose of the exercise is then to use the t_i values to estimate λ. It was shown in Example 3.10 that the ML estimator of λ is

$$\hat{\lambda} = \frac{n}{\sum_{i=1}^{n} T_i}. \tag{3.17}$$

In other words, the failure rate estimate is the number of failures n divided by the observed total time on test Σt_i.

It can be shown that

$$E(\hat{\lambda}) = \frac{n}{n-1}\lambda \qquad \text{and} \qquad \operatorname{Var}(\hat{\lambda}) = \frac{n^2}{(n-1)^2(n-2)}\lambda^2. \tag{3.18}$$

Example 3.13. Ten 60 W light bulbs are placed on life test and the individual failure times t_i, are recorded. The $\Sigma_1^{10} t_i = 9083$ h. Find $\hat{\lambda}$.

$$\hat{\lambda} = \tfrac{10}{9083} = 0.0011$$

Note. If $\lambda = 0.001$ (i.e., expect 1 failure/1000 h), then

$$E(\hat{\lambda}) = \tfrac{10}{9}(0.001) = 0.0011 \qquad \text{and}$$

$$\mathrm{Var}(\hat{\lambda}) = \frac{100}{(81)(8)}(0.001)^2 = 1.54 \times 10^{-7}. \qquad \blacksquare$$

3.6.2 Interval Estimation

We turn now to the problem of interval estimation. Two types of techniques are considered. First, we discuss confidence intervals (CI). Second, probability intervals (PI) as they relate to Bayesian inference are introduced, to be further considered in Section 5.5.3.

For confidence intervals the idea is to find two functions of the data, say $L(\cdot)$ and $U(\cdot)$, so that prior to observing the sample there is a known probability that the parameter (or function of parameters) of interest, say Ω, is contained in the interval $[L(\cdot), U(\cdot)]$. That is,

$$\Pr[L(\cdot) \leqslant \Omega \leqslant U(\cdot)] = 1 - \gamma, \qquad 0 < \gamma < 1,$$

is a valid statement based on the sampling distribution of the data. After the sample is observed, the functions $L(\cdot)$ and $U(\cdot)$ yield two numbers, say L and U. We call the interval (L, U) a *two-sided confidence interval* (TCI) of level $(1-\gamma)$ or a $100(1-\gamma)\%$ confidence interval. Such an interval will be referred to as a $100(1-\gamma)\%$ TCI. The reasoning is as follows: Prior to obtaining the sample, the probability that the data obtained will yield an interval (L, U) containing the unknown *point* Ω, is $(1-\gamma)$. However, once the data are in hand, the values of L and U are fixed and the interval (L, U) either *contains* the point or *does not contain* the point Ω. Thus, we speak of our *confidence* in the interval (L, U) as being $100(1-\gamma)\%$ and the (L, U) a $(1-\gamma)$ level confidence interval. *Upper one-sided confidence intervals* (UCI) or *lower one-sided confidence intervals* (LCI) which specify only an upper or a lower limit, respectively, are similarly defined.

In contrast to the above classical confidence interval where the parameter is assumed to be an unknown constant, the Bayesian views Ω as an r.v. that has a known prior probability distribution, say $g(\omega)$. Then any two values of the *variable*, Ω, say ω_1 and ω_2 such that

$$\Pr(\omega_1 \leqslant \Omega \leqslant \omega_2) = \int_{\omega_1}^{\omega_2} g(\omega)\, d\omega = 1 - \gamma$$

is a $(1-\gamma)$ level *two-sided Bayes probability interval* (TBPI) for Ω, which we denote by $100(1-\gamma)\%$ TBPI. If data are combined with the prior distribution by Bayes' theorem to provide a posterior distribution, say $g(\omega|\cdot)$, then we would use ω_1 and ω_2 such that

$$\Pr(\omega_1 \leqslant \Omega \leqslant \omega_2|\cdot) = \int_{\omega_1}^{\omega_2} g(\omega|\cdot)\, d\omega = 1-\gamma$$

to determine a $100(1-\gamma)\%$ TBPI for Ω. One-sided Bayesian probability intervals are similarly defined in Section 5.5.3.

The concepts of confidence intervals and probability intervals are now illustrated for two models, binomial and exponential. For the binomial data giving X successes in n tests, it can be shown that a $100(1-\gamma)\%$ TCI estimate of p is given by the following formulas:

$$L(x) = \frac{x}{x+(n-x+1)\mathcal{F}_{1-\gamma/2}(2n-2x+2, 2x)},$$

$$U(x) = \frac{(x+1)\mathcal{F}_{1-\gamma/2}(2x+2, 2n-2x)}{(n-x)+(x+1)\mathcal{F}_{1-\gamma/2}(2x+2, 2n-2x)}, \qquad (3.19)$$

where $\mathcal{F}_{1-\gamma/2}(n_1, n_2)$ is the $100(1-\gamma/2)$th percentile of the $\mathcal{F}(n_1, n_2)$ distribution which is defined in (3.14). Table B4 in Appendix B gives values of $\mathcal{F}_\gamma(n_1, n_2)$ for selected values of γ, n_1, and n_2.

Example 3.14. Example 3.12 shows that 93 successes in 100 tests of 60 W light bulbs yield a point estimate $\hat{p} = 0.93$. Find a 95% TCI estimate for p, the probability that any selected bulb successfully survives the test.

From (3.19)

$$L(x) = \frac{93}{93+8\mathcal{F}_{0.975}(16, 186)} = \frac{93}{93+8(1.91)} = 0.86,$$

$$U(x) = \frac{94\mathcal{F}_{0.975}(188, 14)}{7+94\mathcal{F}_{0.975}(188, 14)} = \frac{94(2.54)}{7+94(2.54)} = 0.97.$$

Therefore, we have

$$95\%\ \text{TCI}: 0.86 \leqslant p \leqslant 0.97.$$

■

Properties of the $\chi^2(n)$ distribution show that a $100(1-\gamma)\%$ TCI estimate for λ, the parameter of an exponential model, is given by

$$100(1-\gamma)\%\ \text{TCI:} \frac{\chi^2_{\gamma/2}(2n)}{2t} \leqslant \lambda \leqslant \frac{\chi^2_{1-\gamma/2}(2n)}{2t}, \tag{3.20}$$

when $t = \sum_{i=1}^{n} t_i$ is the total time on test and $\chi^2_\gamma(n)$ is the 100γth percentile of a $\chi^2(n)$ distribution, which is given in (3.13). Values of $\chi^2_\gamma(n)$ are given in Table B3 in Appendix B for selected values of γ and n.

Example 3.15. Example 3.13 shows that 10 light bulbs burn for a total time on test of 9083 h yielding a point estimate $\hat{\lambda} = 0.0011$.

Find a 95% TCI for λ. From (3.20) we have

$$\frac{\chi^2_{0.025}(20)}{2(9083)} \leqslant \lambda \leqslant \frac{\chi^2_{0.975}(20)}{2(9083)}.$$

Table B3 gives

$$\chi^2_{0.025}(20) = 9.591 \qquad \text{and} \qquad \chi^2_{0.975}(20) = 34.170.$$

Therefore

$$95\%\ \text{TCI: } 0.00053 \leqslant \lambda \leqslant 0.0019.$$

■

If the probability of success p is assumed to follow a beta distribution [see Section 4.2.8] with parameters x_0 and n_0, then the prior $100(1-\gamma)\%$ TBPI estimate of p is given by

$$100(1-\gamma)\%\ \text{TBPI:}\ \frac{x_0}{x_0 + (n_0 - x_0)\mathcal{F}_{1-\gamma/2}(2n_0 - 2x_0, 2x_0)} \leqslant p$$

$$\leqslant \frac{x_0 \mathcal{F}_{1-\gamma/2}(2x_0, 2n_0 - 2x_0)}{(n_0 - x_0) + x_0 \mathcal{F}_{1-\gamma/2}(2x_0, 2n_0 - 2x_0)} \tag{3.21}$$

[see Section 7.2.4]. Further, if a binomial sample of n items yields x successful test completions, then the posterior $100(1-\gamma)\%$ TBPI estimate of

p is given by

$$100(1-\gamma)\%\ \text{TBPI:}$$

$$\frac{x+x_0}{x+x_0+(n+n_0-x-x_0)\mathcal{F}_{1-\gamma/2}(2n+2n_0-2x-2x_0,2x+2x_0)} \leqslant p$$

$$\leqslant \frac{(x+x_0)\mathcal{F}_{1-\gamma/2}(2x+2x_0,2n+2n_0-2x-2x_0)}{n+n_0-x-x_0+(x+x_0)\mathcal{F}_{1-\gamma/2}(2x+2x_0,2n+2n_0-2x-2x_0)} \tag{3.22}$$

[see Section 7.2.4].

Example 3.16. Suppose that p, the probability that a 60 W light bulb survives a life test of 1000 h, is a beta r.v. with parameters $n_0 = 50$ and $x_0 = 48$. Further suppose a test of 100 bulbs yields 93 survivors of a life test of 1000 h. (See Example 3.12.) Compute a prior 95% TBPI estimate of p and a posterior 95% TBPI estimate of p.

Prior Probability Interval

Using (3.21) we have a prior probability interval,

$$95\%\ \text{TBPI:}\ \frac{48}{48+2\mathcal{F}_{.975}(4,96)} \leqslant p \leqslant \frac{48\mathcal{F}_{.975}(96,4)}{2+48\mathcal{F}_{.975}(96,4)},$$

$$95\%\ \text{TBPI:}\ \frac{48}{48+2(2.94)} \leqslant p \leqslant \frac{48(8.33)}{2+48(8.33)},$$

$$95\%\ \text{TBPI:}\ 0.891 \leqslant p \leqslant 0.995.$$

Posterior Probability Interval

Using (3.22) we have a posterior probability interval,

$$95\%\ \text{TBPI:}\ \frac{141}{141+9\mathcal{F}_{.975}(18,282)} \leqslant p \leqslant \frac{141\mathcal{F}_{.975}(282,18)}{9+141\mathcal{F}_{.975}(282,18)},$$

$$95\%\ \text{TBPI:}\ \frac{141}{141+9(1.81)} \leqslant p \leqslant \frac{141(2.21)}{9+141(2.21)},$$

$$95\%\ \text{TBPI:}\ 0.896 \leqslant p \leqslant 0.972.$$

■

3.6.3 Tests of Statistical Hypotheses

In this section we consider techniques for testing hypotheses about parameters. Whereas the previous section was concerned about using data to say something about the magnitude of a parameter, our concern now is to use the data to determine whether a given statement about a parameter is acceptable. To begin we define the *null hypothesis*, denoted by H_0, as the statement being tested. For example, we could write H_0: $p=0.99$ to indicate that we wish to test whether the true probability that a specified brand of 60 W light bulbs will pass a 1000 h life test is 0.99. Or we could use H_0: $\lambda=0.001$ to state our belief that the failure rate of 60 W light bulbs is 0.001.

Now when we state a null hypothesis, it can either be *true* or *false*. Further, our testing procedure can either *reject* or *not reject* (accept) the hypothesis. Thus, in any testing situation, there are four distinct events that can occur. To discuss these possibilities and the philosophy of testing hypotheses, we introduce the following ideas and terms. Using the conditional probability notation of Section 2.2, we write

$$\begin{aligned}
\Pr(H_0 \text{ is accepted} | H_0 \text{ is true}) &= 1-\gamma \\
&= \text{Level of confidence}, \\
\Pr(H_0 \text{ is rejected} | H_0 \text{ is true}) &= \gamma \\
&= \Pr(\text{Type I error}), \\
&= \text{Level of significance}, \\
\Pr(H_0 \text{ is accepted} | H_0 \text{ is false}) &= \delta \\
&= \Pr(\text{Type II error}),
\end{aligned}$$

and

$$\begin{aligned}
\Pr(H_0 \text{ is rejected} | H_0 \text{ is false}) &= 1-\delta \\
&= \text{Power of the test}.
\end{aligned}$$

The four possibilities are summarized in Table 3.2.

Of the four possibilities, two are desirable and two are undesirable. Clearly we would prefer to have $\delta=0$ and $\gamma=0$. However, those idealistic preferences cannot generally be achieved in practice. Note that the double line in Table 3.2 is to emphasize the fact that for any given problem, either

Table 3.2. Possibilities for Testing Hypothesis H_0

Decision \ True Situation	H_0 is true	H_0 is false
Accept H_0	$\Pr(\text{accept } H_0 \mid H_0 \text{ is true}) = 1-\gamma$ Level of confidence	$\Pr(\text{accept } H_0 \mid H_0 \text{ is false}) = \delta$ Pr(Type II error)
Reject H_0	$\Pr(\text{reject } H_0 \mid H_0 \text{ is true}) = \gamma$ Pr(Type I error) Level of significance	$\Pr(\text{reject } H_0 \mid H_0 \text{ is false}) = 1-\delta$ Power

H_0 is true or H_0 is false, but never both. Therefore, we are either working on the right side or left side of the table, but we are uncertain as to which. The approach is to fix γ, the level of significance desired, and to choose a decision rule which minimizes δ for the selected γ. Waller (1979) provides illustrations and an applied view of the testing procedure. Theoretical discussions are given in Hogg and Craig (1978), and Kendall and Stuart (1973).

In summary, the steps of a test of a statistical hypothesis H_0 are as follows:

1. State the null hypothesis, H_0: $\Omega=\Omega_0$, and an appropriate alternative hypothesis (usually either H_1: $\Omega\neq\Omega_0$ or H_1: $\Omega>\Omega_0$ or H_1: $\Omega<\Omega_0$).
2. Select γ and define the test (i.e., determine the rejection region R).
3. Observe the data.
4. Calculate the value of the test statistic, say t.
5. If $t\in R$ reject H_0; otherwise, do not reject H_0 (accept H_0).
6. State any appropriate conclusions.

To illustrate the preceding ideas, let us consider the situation where a manufacturer claims that their 60 W bulbs are such that the probability is p_0 that any randomly selected bulb will survive a test of 1000 h. The null and alternative hypotheses are

$$H_0: p=p_0 \qquad \text{and} \qquad H_1: p\neq p_0.$$

A common test statistic for testing H_0 is

$$Y=\frac{(X-np_0)^2}{np_0(1-p_0)}, \tag{3.23}$$

which has approximately a $\chi^2(1)$ distribution when H_0 is true. Thus the rejection region is

$$R=\{y \mid y>\chi^2_{1-\gamma}(1)\},$$

where $\chi^2_{1-\gamma}(1)$ is obtained from Table B3 in Appendix B.

Example 3.17. A manufacturer of 60 W bulbs claims that the survival probability for 1000 h life tests is 0.99. If a sample of 100 bulbs shows 93 survivors, is the claim acceptable at $\gamma=0.05$? Thus

$$H_0: \quad p=0.99,$$

$$H_1: \quad p\neq 0.99, \qquad 1-\gamma=0.95,$$

$$\text{Reject } H_0 \text{ if } y>\chi^2_{0.95}(1)=3.84.$$

Now 100 (0.99)=99 and $x=93$, therefore

$$y=\frac{(93-99)^2}{100(0.99)(0.01)}=\frac{36}{0.99}>3.84.$$

We reject H_0 and conclude that the true value of $p<0.99$. [That conclusion follows because we expected to find 99 survivors under the null hypothesis and found only 93 (too few).] ■

As a second illustration, suppose the light bulb manufacturer claims the failure rate of the 60 W bulbs is λ_0. We then wish to test the hypothesis

$$H_0: \lambda=\lambda_0 \qquad \text{versus} \qquad H_1: \lambda\neq\lambda_0.$$

A common test statistic based on exponential life test data for n failures and total time on test $\mathfrak{T}$ is

$$Y=2\mathfrak{T}\lambda, \tag{3.24}$$

which follows a $\chi^2(2n)$ distribution [see Section 4.5.1].

Example 3.18. A manufacturer of pumps used in pressurized water reactors (PWRs) claims that $\lambda\leqslant 1.0\times 10^{-6}$ f/h, assuming an exponential failure time model. Recent operating data show six failures in 3,504,000 h of operation. Use these data to test H_0: $\lambda \leqslant 1.0 \times 10^{-6}$ against H_1: $\lambda > 1.0 \times 10^{-6}$ with $\gamma = 0.05$. Now

$$H_0: \lambda \leqslant 1.0 \times 10^{-6}, \qquad \gamma = 0.05,$$

$$H_1: \lambda > 1.0 \times 10^{-6},$$

$$\text{Reject } H_0 \text{ if } y = 2t\lambda_0 > \chi^2_{0.95}(12) = 21.026,$$

$$y = 2(3.504 \times 10^6)(1.0 \times 10^{-6}) = 7.008.$$

Since $y = 7.008 < 21.026$, we accept H_0 and conclude that $\lambda \leqslant 1.0 \times 10^{-6}$ f/h.

■

3.6.4 Neyman–Pearson Fundamental Lemma

In hypothesis testing, for a fixed value of the level of significance γ, it is desirable to determine the "best test" against a specific simple alternative. That is, if one is testing H_0: $\theta = \theta_0$ against H_1: $\theta = \theta_1$, where θ is some parameter of interest, then, for fixed γ, it is desirable to determine the critical region for which the power is maximized. That this can in fact be done is demonstrated by the *Neyman–Pearson fundamental lemma* which follows:

Let $X_1, X_2, \ldots, X_n$ denote a sample of size n from $f(x; \theta)$ and let $f(x_1, x_2, \ldots, x_n | H_0)$ be the joint pdf of the sample when H_0 is true. Likewise, $f(x_1, x_2, \ldots, x_n | H_1)$ is the pdf of the sample when H_1 is true. The best region, that is, the critical region that maximizes the power for fixed γ and n, is the set of all $(x_1, x_2, \ldots, x_n)$ such that

$$\frac{f(x_1, x_2, \ldots, x_n | H_0)}{f(x_1, x_2, \ldots, x_n | H_1)} \leqslant c_\gamma, \tag{3.25}$$

when H_0 is true. That is, H_0 is rejected if (3.25) holds.

The ratio in (3.25) is called a *likelihood ratio* as it is just a ratio of two likelihood functions. A very significant feature of (3.25) is that the test related to it is usually determined by a single statistic (function). In such cases it is not necessary to determine c_γ. This is best illustrated by means of an example.

Example 3.19. Suppose that it is desired to test H_0: $\lambda = \lambda_0$ against H_1: $\lambda = \lambda_1$ $(\lambda_1 > \lambda_0)$, where λ is the failure rate of an exponential distribution.

Let $T_1, T_2, \ldots, T_n$ denote the n failure times. The pdf of the sample is $f(t_1, t_2, \ldots, t_n; \lambda) = \lambda^n e^{-\lambda \Sigma t_i}$. Thus, the inequality (3.25) becomes

$$\frac{\lambda_0^n e^{-\lambda_0 \Sigma t_i}}{\lambda_1^n e^{-\lambda_1 \Sigma t_i}} \leqslant c_\gamma.$$

Solving this for Σt_i we have

$$\sum t_i \leqslant \frac{[\ln c_\gamma - n \ln(\lambda_0/\lambda_1)]}{(\lambda_1 - \lambda_0)}$$

or

$$\sum t_i \leqslant k \text{ (say)}.$$

Thus, the test statistic is $\mathcal{T} = \Sigma T_i$ and k is determined so that $\Pr(\mathcal{T} \leqslant k \mid H_0) = \gamma$. The critical region is then the set of all values of $\mathcal{T} \leqslant k$. It is known that $2\lambda_0 \Sigma T_i = 2\lambda_0 \mathcal{T} = Y$, where $\mathcal{T}$ denotes the total time on test, has a $\chi^2(2n)$ distribution. The critical region is then $y \leqslant \chi_\gamma^2(2n)$. The critical region in this example is valid as long as $\lambda_1 > \lambda_0$. The test in this case is called a *uniformly most powerful* test, since it is most powerful for each $\lambda_1 > \lambda_0$. ■

For a development of the theory of hypothesis testing the reader is referred to Hogg and Craig (1978) and Kendall and Stuart (1973).

3.7. SEQUENCES OF RANDOM VARIABLES

In the subsequent chapters we will have occasion to apply selected theoretical results that relate to sequences of r.v.'s. The purpose of this section is to summarize some selected results for later reference.

A widely occurring model is the $\mathcal{N}(\mu, \sigma^2)$ distribution. The two parameters are $\mu = E(X)$ and $\sigma = \sqrt{\operatorname{Var}(X)}$. Any r.v. X having a $\mathcal{N}(\mu, \sigma^2)$ distribution can be transformed to the standard normal variable by

$$Z = \frac{X - \mu}{\sigma}.$$

The standard normal r.v. has mean 0, variance 1, and pdf

$$f(z) = \frac{1}{\sqrt{2\pi}} \exp\left(-\frac{z^2}{2}\right), \qquad -\infty < z < \infty.$$

It is denoted by $\mathfrak{N}(0,1)$. The normal distribution has been studied extensively and the reader is referred to Johnson and Kotz (1970) for details.

A part of the importance of the $\mathfrak{N}(\mu, \sigma^2)$ distribution is that it occurs as the limiting distribution for sample sums and sample means. The result is called a *central limit theorem* and can be stated as follows:

Theorem 3.1.

If $X_1, X_2, \ldots, X_n$ are independent and identically distributed r.v.'s such that $E(X_i)=\mu$ and $\text{Var}(X_i)=\sigma^2<\infty$, $i=1,2,\ldots,n$, then for every fixed y, as n tends to ∞,

$$\Pr\left(\frac{\Sigma X_i - n\mu}{\sigma\sqrt{n}} < y\right) \to \int_{-\infty}^{y} \frac{1}{\sqrt{2\pi}} e^{-z^2/2}\,dz.$$

This theorem is a special case of a very general theorem that was proved by Lindberg and can be found in Feller (1970). A proof of Theorem 3.1 is provided by Hogg and Craig (1978) and a detailed motivation is given by Waller (1979).

The implication of the result is that, regardless of what family of distributions is being sampled, as long as its mean μ and variance σ^2 are finite the sum of n observations tends toward a $\mathfrak{N}(n\mu, n\sigma^2)$ distribution. The importance of this result is that when the distribution of an r.v. is unknown, we can obtain approximate probability results for the sum of n observations from the extensive and readily available tables of the $\mathfrak{N}(0,1)$ distribution. Since $\bar{X}=\Sigma X_i/n$ is a linear transformation of ΣX_i, it follows that the central limit theorem applies. Thus, we can say that the limiting distribution of $\bar{X}$, for any distribution having $\mu<\infty$ and $\sigma^2<\infty$, is such that $\sqrt{n}(\bar{X}-\mu)/\sigma \to \mathfrak{N}(0,1)$. The next example illustrates the procedure.

Example 3.20. Suppose the failure time T for a component has mean $\mu=1000$ h and variance 2500 h^2, but the distribution of T is unspecified. What is the probability that a random sample of 100 components yields a mean survival time of less than 990 h?

The central limit theorem states that for $\bar{T}=\Sigma_{i=1}^{100} T_i/100$

$$\Pr(\bar{T} \leqslant 990 \text{ h}) \simeq \Pr\left(Z=\frac{\bar{T}-\mu}{\sigma/\sqrt{n}} \leqslant \frac{990-1000}{50/\sqrt{100}} = -2\right),$$

where $\simeq$ means "approximately equal to." From Table B1 in Appendix B we find

$$\Pr(\bar{T} \leqslant 990 \text{ h}) \simeq .023. \qquad \blacksquare$$

The central limit theorem is an example of convergence in distribution or law. The implication is that, as n becomes large, the $\mathcal{N}(\mu, \sigma^2)$ distribution can be used to approximate the unknown distribution of a sample sum or a sample mean.

Another type of convergence is *convergence in probability*. Here we are looking for a limit value to which the function of interest converges as n increases.

Definition 3.5. If for every $\varepsilon > 0$,

$$\lim_{n \to \infty} \Pr(|T_n - \Omega| < \varepsilon) = 1,$$

we say that T_n *converges to* $\Omega (T_n \to \Omega)$ *in probability*.

An example is that the sample mean, say $\bar{T}$, converges in probability to $E(T)$.

A third type of convergence is called *almost sure* convergence and is defined as follows.

Definition 3.6. If for every $\varepsilon > 0$,

$$\Pr\left(\left|\lim_{n \to \infty} T_n - \Omega\right| < \varepsilon\right) = 1,$$

then $T_n \to \Omega$ *almost surely* (a.s.).

Almost sure convergence implies that the set of points where the sequence T_n *does not* converge to Ω has probability zero. Thus, almost sure convergence is also called *convergence with probability one*.

We sometimes refer to convergence in probability as the "weak law of large numbers," and to almost sure convergence as the "strong law of large numbers." Furthermore, almost sure convergence implies convergence in probability. That is, if T_n converges almost surely, then T_n converges in probability. However, the converse of this statement is not true.

Example 3.21. Show that the observed proportion of components that fail a life test converges in probability to p, the probability that a randomly selected component fails the test.

Assume that the number of components X in a sample of size n that fail the test has a binomial distribution with mean $\mu = np$ and variance $\sigma^2 = np(1-p)$. Then by (3.16), $\hat{p} = X/n$ has mean p and variance $p(1-p)/n$. Thus, by the Chebyshev inequality (3.9), replacing $k\sigma$ with ε

$$\Pr(|\hat{p} - p| < \varepsilon) \geq 1 - \frac{\sigma^2}{\varepsilon^2} = 1 - \frac{p(1-p)}{n\varepsilon^2}.$$

Now, the limit of the right-hand side as $n \to \infty$ is 1, and since a probability cannot exceed 1, it follows that

$$\lim_{n \to \infty} \Pr(|\hat{p} - p| < \varepsilon) = 1.$$

Thus, $\hat{p} \to p$ in probability. ■

3.8. GOODNESS-OF-FIT PROCEDURES

In Chapter 2 we introduced some selected probability models as candidates for representing various observed data. Chapter 4 presents additional models for use in modeling empirical reliability data. In this section we present two standard procedures for use in checking whether an assumed model is adequate to represent a set of data. Some specialized graphic methods for checking the adequacy of a model are presented in Section 4.3.

The two procedures described here are the chi-square and Kolmogorov goodness-of-fit methods. In general, both methods are based on a comparison of how well the empirical (sample) data agree with a theoretical data set that is determined by the assumed model.

3.8.1 Chi-Square Goodness-of-Fit Tests

The name *chi-square* results from the test statistic used for the statistical test having an approximate $\chi^2(n)$ distribution.

We suppose that a random sample of n observations, $X_1, X_2, \ldots, X_n$, is split into k classes to form an empirical distribution. The assumed model, say $f_0(x)$, is then used to calculate the probability that a random observation falls into each class. We denote the probabilities by $p_1, p_2, \ldots, p_k$. We should be careful to choose nonoverlapping class intervals which cover the range of values from $f_0(x)$. In that way we are assured that $\Sigma_{i=1}^{k} p_i = 1$. The "expected frequency", say E_i, of the ith class is given by $E_i = np_i$. If we let

O_i denote the "observed frequency" of the ith class, the objective is to compare the O_i with the E_i for the k classes. The test procedure is as follows:

STEP 1. H_0: The variable X is distributed as $f_0(x)$.
H_1: The variable X is *not* distributed as $f_0(x)$.

STEP 2. Select γ, the level of significance of the test.

STEP 3. Specify the rejection region

$$R = \{ y \mid y > \chi^2_{1-\gamma}(k-1-m) \},$$

where m is the number of parameters in $f_0(x)$ estimated from the data being tested.

STEP 4. Calculate the test statistic

$$y = \sum_{i=1}^{k} \frac{(O_i - E_i)^2}{E_i} = \sum_{i=1}^{k} \frac{O_i^2}{E_i} - n.$$

STEP 5. If $y \in R$, reject H_0; otherwise, accept H_0.

STEP 6. Either conclude that $f_0(x)$ provides an adequate fit for the data (accept H_0) or conclude that the fit is *not* adequate (reject H_0).

Example 3.22. To illustrate the chi-square goodness-of-fit test, we refer to the light bulb data given in Section 2.8. We will conduct a test of

H_0: The failure time T is an exponential r.v. with $\lambda = 0.5$

H_1: The failure time T is *not* an exponential r.v. with $\lambda = 0.5$,

with $\gamma = 0.05$. For reasons discussed following the example, we summarize the information in the tables of Section 2.8.

Class Intervals	Observed Frequencies	Expected Frequencies
0–1	20	19.674
1–2	15	11.933
2–3	3	7.238
3–4	5	4.390
over 4	7	6.768
Total	50	50.003

We have

$$H_0\text{: The pdf of } T \text{ is } f_0(t) = 0.5e^{-0.5t},$$

$$H_1\text{: The pdf of } T \text{ is } \textit{not } f_0(t),$$

$$\gamma = 0.05.$$

Since λ is specified as 0.5, no parameters are being estimated by the data. Thus, $m = 0$ and the rejection region is

$$R = \{y \mid y > \chi^2_{0.95}(4) = 9.488\}.$$

$$y = \frac{(20-19.674)^2}{19.674} + \frac{(15-11.933)^2}{11.933} + \frac{(3-7.238)^2}{7.238} + \frac{(5-4.390)^2}{4.390} + \frac{(7-6.768)^2}{6.768} = 3.37.$$

Since $y = 3.37 < 9.488$ we accept H_0 and conclude that an exponential model with $\lambda = 0.5$ provides an adequate fit for the light bulb failure time data. ■

We conclude this section with some remarks and caution concerning the chi-square procedure. First, notice that we must group the data into classes in order to obtain observed frequencies for comparison with calculated expected frequencies. Second, any class that has one or more observed values but has a small expected frequency can make a major contribution to the value of y. For example, suppose a class has an expected frequency of 0.1 and an observed value of 1. The contribution of that class to y is $(1-0.1)^2/0.1 = 8.1$. To prevent one random observation from having an undue effect on the test of the model, we combine classes so that the expected frequencies of the final classes used are greater than or equal to five. That rule of thumb provides good results in practice. However, the grouping of tail observations into combined classes means that the chi-square procedure is not very effective in detecting model discrepancies for extreme values of a variable. Third, in Example 3.22, we specified λ to be 0.5. Thus, if we had rejected H_0, the rejection could have been due to either the wrong family of models (exponential) or the wrong parameter value (0.5) or both. If however, we had used the data to estimate λ by $\hat{\lambda} = n/\Sigma t_i$, then we could have tested the hypothesis that T is distributed as an exponential variable with parameter $\hat{\lambda}$. Rejection in that case would indicate that an exponential model would not provide an adequate model for the data. For a more comprehensive discussion of the chi-square goodness-of-fit procedure see Conover (1980).

3.8.2 Kolmogorov Procedure

The preceding chi-square procedure required substantial amounts of data to provide expected class frequencies of approximately five or more. The Kolmogorov procedure uses each individual data point and thus is equally effective for either small or large samples. The sample (empirical) cdf is compared to a hypothesized (theoretical) cdf for a Kolmogorov test of fit. We suppose that we observe an ordered sample of data $t_1 \leqslant t_2 \leqslant t_3 \leqslant \cdots \leqslant t_n$. The sample cdf is defined by

$$S_n(t_j) = \frac{j}{n}; \qquad j = 1, 2, \ldots, n,$$

$$S_n(t) = 0 \text{ for } t < t_1 \qquad \text{and} \qquad S_n(t) = 1 \text{ for } t \geqslant t_n.$$

With that brief introduction, we can now define the test procedure as follows:

STEP 1. H_0: The r.v. T has cdf $F(t)$
H_1: T does *not* have cdf $F(t)$.

STEP 2. Select γ, the level of significance of the test.

STEP 3. Specify the rejection region

$$R = \{ D_{\max} \mid D_{\max} > D_n^{(\gamma)} \},$$

where $D_n^{(\gamma)}$ is obtained from Table B5 in Appendix B.

STEP 4. Calculate the statistic:

$$D_{\max} = \max_i \{ |F(t_i) - S_n(t_i)|, |F(t_i) - S_n(t_{i-1})| \}.$$

STEP 5. If $D_{\max} > D_n^{(\gamma)}$, reject H_0 and conclude that $F(t)$ does not describe the data; otherwise, accept H_0 and conclude that $F(t)$ describes the data.

Example 3.23. Are the failure times 8, 20, 34, 46, 63, 86, 111, 141, 186, and 266 h adequately modeled by an exponential model with parameter $\lambda = 0.01$ f/h?

We have

$$H_0: \quad \text{The r.v. } T \text{ has cdf } F(t) = 1 - e^{-0.01t},$$

$$H_1: \quad T \text{ does } \textit{not} \text{ have cdf } F(t),$$

$$\gamma = 0.05,$$

$$R = \{D_{\max} \mid D_{\max} > D_{10}^{(0.05)} = 0.410\}.$$

The table below is a convenient method for calculating $D_{\max}$ for a small sample.

i	t_i	$S_n(t_i)$	$S_n(t_{i-1})$	$F(t_i)$	$\lvert F(t_i) - S_n(t_i) \rvert$	$\lvert F(t_i) - S_n(t_{i-1}) \rvert$
1	8	0.10	0.00	0.077	0.023	0.077
2	20	0.20	0.10	0.181	0.019	0.081
3	34	0.30	0.20	0.288	0.012	0.088
4	46	0.40	0.30	0.369	0.031	0.064
5	63	0.50	0.40	0.467	0.033	0.067
6	86	0.60	0.50	0.577	0.023	0.072
7	111	0.70	0.60	0.670	0.030	0.070
8	141	0.80	0.70	0.756	0.044	0.056
9	186	0.90	0.80	0.844	0.056	0.044
10	266	1.00	0.90	0.930	0.070	0.030

$$D_{\max} = 0.088(\lvert F(t_3) - S_n(t_2) \rvert = 0.088).$$

Since $D_{\max} = 0.088 < 0.410$, we accept H_0 and conclude that an exponential model with $\lambda = 0.01$ provides an adequate fit for the 10 observations. ■

This section presents two goodness-of-fit procedures. There are other procedures with special properties that meet situations where parameters are unknown or two samples are tested to determine whether they can be modeled by the same distribution function. Conover (1980) provides a good description of the various procedures.

EXERCISES

3.1. Determine the constant c so that $f(x)=cxe^{-x}$, $x>0$, and $f(x)=0$, otherwise, is a pdf. Find the mean and variance of X.

3.2. Find the mean and variance of X, where X has a Poisson distribution.

3.3. Find the moment generating function of the Poisson distribution and use it to find the mean and variance.

3.4. Find the mean and variance of the hypergeometric distribution.

3.5. Find the moment generating function of the binomial distribution and use it to find the mean and variance.

3.6. Find the constant c so that $f(x, y)=cxy$, $0\leqslant x, y\leqslant 1$, and $f(x, y)=0$, otherwise, is a joint pdf of X and Y. Are X and Y independent? Find $E(X)$, $E(Y)$, $E(XY)$, $\mathrm{Var}(X)$, $\mathrm{Var}(Y)$, $\mathrm{Cov}(X, Y)$.

3.7. Find the constant c so that $f(x, y)=c(x+y)$, $0\leqslant x, y\leqslant 1$, and $f(x, y)=0$, otherwise, is a joint pdf of X and Y. Are X and Y independent? Find $E(X)$, $E(Y)$, $E(XY)$, $\mathrm{Var}(X)$, $\mathrm{Var}(Y)$, $\mathrm{Cov}(X, Y)$, and the correlation coefficient of X and Y.

3.8. If $f(x, y)=1$, $0\leqslant x, y\leqslant 1$, and $f(x, y)=0$, otherwise, find $\Pr(|X-Y|\leqslant\varepsilon)$ for $\varepsilon>0$.

3.9. If $f(x, y)=1/x$, $0\leqslant y<x$, $0<x\leqslant 1$, and $f(x, y)=0$, otherwise, find $\Pr(|X-Y|\leqslant\varepsilon)$ for $\varepsilon>0$.

3.10. For the pdf $f(x, y)=c(x+y)$ in Exercise 3.7, find the conditional pdf of Y given $X=x$; that is, $g(y|X=x)$. Find $\Pr(0\leqslant Y\leqslant\frac{1}{2}|X=\frac{1}{2})$, $\Pr(0\leqslant Y\leqslant\frac{1}{2}|\frac{1}{2}\leqslant X\leqslant 1)$.

3.11. Let T have the $t(k)$ distribution. Find the mean and variance of T.

3.12. Find the moment generating function of the exponential pdf $f(t)=\lambda e^{-\lambda t}$, $t>0$, and use it to find the mean and variance.

3.13. Let $f(x)=3e^{-3x}$, $x>0$, and $f(x)=0$, otherwise.

(a) Find $\Pr(|X-\mu|>1)$.

(b) Use Chebyshev's inequality to obtain an upper bound on $\Pr(|X-\mu|>1)$ and compare it to the result in (a).

(c) Find $\Pr(|X-\mu|>h\sigma)$ and compare it to the upper bound in Chebyshev's inequality $\Pr(|X-\mu|>h\sigma)\leqslant 1/h^2$.

3.14. Use Chebyshev's inequality to determine how many times a fair coin should be tossed so that the probability will be at least 0.90 that the observed frequency of heads will be between 0.4 and 0.6.

3.15. An r.v. Y has mean 5 and variance 3. Use Chebyshev's inequality to find an upper bound for

(a) $\Pr(|Y-5| \geqslant 3)$.

(b) $\Pr(|Y-5| \geqslant 1)$.

3.16. Let X have the uniform distribution $f(x)=1/(\theta_2-\theta_1)$, $\theta_1 \leqslant x \leqslant \theta_2$, and $f(x)=0$, otherwise. Find

(a) $E(X)$

(b) $\mathrm{Var}(X)$

(c) The moment generating function of X.

3.17. Let X have the gamma distribution $f(x)=x^{\alpha-1}e^{-x/\beta}/\beta^{\alpha}\Gamma(\alpha)$, $x>0$, $\alpha>0$, $\beta>0$, and $f(x)=0$, otherwise. Find

(a) $E(X)$

(b) $\mathrm{Var}(X)$

(c) The moment generating function of X.

3.18. Let R have the beta distribution defined by (2.7). Find $E(R)$ and $\mathrm{Var}(R)$.

3.19. Let X have the normal distribution $f(x)=(\sqrt{2\pi}\,\sigma)^{-1}e^{-(x-\mu)^2/(2\sigma^2)}$ $-\infty<x<\infty$, $-\infty<\mu<\infty$, $\sigma>0$. Find the mean and variance of X. Find the moment generating function of X.

3.20. If $f(x, y)$ denotes any probability model, discrete or continuous, show that $E(X+Y)=E(X)+E(Y)$.

3.21. If X and Y are independent r.v.'s with probability model $f(x, y)$ show that $\mathrm{Var}(X+Y)=\mathrm{Var}(X)+\mathrm{Var}(Y)$.

3.22. Let $X_1, X_2, \ldots, X_n$ denote a random sample from a population having the pdf $f(x)=(1/\theta)e^{-x/\theta}$, $0<x<\infty$, $0<\theta<\infty$, and zero elsewhere. Show that $\bar{X}$ is unbiased for θ and find its variance.

3.23. Show that the statistic $\Sigma X_i^2/n$, based on a random sample of size n from a normal distribution with mean zero and variance σ^2, $0<\sigma^2<\infty$, is unbiased for σ^2 and find its variance.

3.24. Let $\hat{\theta}_1$ and $\hat{\theta}_2$ be two independent unbiased statistics for θ. If the variance of $\hat{\theta}_1$ is three times as large as the variance of $\hat{\theta}_2$, find the constants c_1 and c_2 so that $L=c_1\hat{\theta}_1+c_2\hat{\theta}_2$ is unbiased for θ with the smallest possible variance for such a linear combination.

3.25. Let $X_1, X_2, \ldots, X_n$ denote a random sample from the pdf $f(x;\theta)=\theta x^{\theta-1}$, $0<x<1$, $\theta>0$, and zero elsewhere. Find a sufficient statistic for the parameter θ.

3.26. Find a sufficient statistic for λ in sampling from a Poisson distribution with parameter λ.

3.27. Find a sufficient statistic for θ in sampling from the pdf $f(x;\theta)=(1/\theta)e^{-x/\theta}$, $x>0$, $\theta>0$, and zero elsewhere. Find the mean and variance of the statistic. Find a function of the sufficient statistic that is unbiased for θ and find its variance. What can you say about this variance relative to the variance of any other unbiased statistic for θ?

3.28. Find joint sufficient statistics for the mean μ and variance σ^2 of a normal distribution based on a random sample of size n. Determine the unbiased statistics for μ and σ^2 that have the smallest possible variances and find these variances.

3.29. Find the unbiased estimator for θ that has the least variance in sampling from the pdf $f(x;\theta)=\theta^x(1-\theta)^{1-x}$, $x=0,1$, $0<\theta<1$, and zero elsewhere. Find the variance of the estimator.

3.30. Let $X_1, X_2, \ldots, X_n$ denote a random sample from the dichotomous population with the pdf given in Exercise 3.29. Find the unbiased statistic for $\text{Var}(\Sigma X_i/n)$ that has minimum variance. Also, find the minimum variance unbiased statistic for $\text{Var}(\Sigma X_i)$. Compare the two results.

3.31. Let $X_1, X_2, \ldots, X_n$ denote a random sample from each of the following pdf's:

(a) $f(x;\theta)=(1/\theta)e^{-x/\theta}$, $x>0$, $\theta>0$, and zero elsewhere.

(b) $f(x;\theta)=1/\theta$, $0<x<\theta$, and zero elsewhere.

(c) $f(x;\theta)=\theta^x(1-\theta)^{1-x}$, $x=0,1$, $0<\theta<1$, and zero elsewhere.

(d) $f(x;\theta)=\theta^x e^{-\theta}/x!$, $x=0,1,2,\ldots$, $\theta>0$, and zero elsewhere.

(e) $f(x;\theta)=\theta x^{\theta-1}$, $0<x<1$, $\theta>0$, and zero elsewhere.

(f) $f(x;\theta)=e^{-(x-\theta)}$, $\theta\leqslant x<\infty$, $-\infty<\theta<\infty$, and zero elsewhere.

For each of the pdf's find the ML estimator $\hat{\theta}$ for θ.

3.32. Find the ML statistics for the mean μ and variance σ^2 of a normal distribution, based on a sample of size n. Write these ML statistics as functions of the joint sufficient statistics for μ and σ^2.

3.33. A sample of 90 voters chosen at random in a given city showed that 48 favored a particular candidate. Find (a) 90%, (b) 95%, (c) 99% confidence limits for the proportion of all voters who favor this candidate.

3.34. Let $\overline{X}_n$ denote the sample mean of a random sample of size n from a normal distribution with mean 0 and variance σ^2. Show that the limiting distribution function of $\overline{X}_n$ is degenerate.

3.35. Let X_n have the pdf $f_n(x)=1$ at $x=3+1/n$ and zero elsewhere. Find $\lim_{n\to\infty} f_n(x)$. Find the limiting distribution function of X_n.

3.36. Let X_n have the pdf $g_n(x)=(5-x/n)^{n-1}/5^n$, $0<x<5n$, and zero elsewhere. Find the limiting distribution function of X_n.

3.37. Let $\overline{X}_n$ denote the mean of a random sample of size n from a normal distribution with mean μ and variance σ^2. Show that $\overline{X}_n$ converges in probability to μ.

3.38. A sample of 50 digits using a random number generator yielded the following:

Digit	0	1	2	3	4	5	6	7	8	9
Frequency	4	8	8	4	10	3	2	2	4	5

Is there any reason to doubt the digits are uniformly distributed?

3.39. The frequency distribution of 100 grades in an elementary statistics course is given below.

Score Interval	Frequency
45–55	7
55–65	18
65–75	35
75–85	28
85–95	12

Use these data to test whether a normal distribution with $\mu=70$ and $\sigma=10$ is an appropriate model for the score being analyzed.

3.40. If X has a negative-log gamma $\mathcal{NLG}(\alpha,\theta)$ distribution (see Table A1 in Appendix A) show that $Y=(-\ln X)$ has a gamma $\mathcal{G}_1(\alpha,\theta)$ distribution.

3.41. If $M(t; x)$ is the moment generating function of X, find the moment generating function of $Y=aX+b$.

3.42. If $X_1, X_2,\ldots,X_n$ is a random sample from $f(x)$ and X has the moment generating function $M(t; x)$, find the moment generating function of $\overline{X}=\Sigma X_i/n$.

3.43. Derive the confidence limits for the parameter of an exponential distribution given in (3.20). (Hint: Use the relationship between the exponential and χ^2 pdf's).

REFERENCES

Conover, W. J. (1980). *Practical Nonparametric Statistics* (2nd Ed.), Wiley, New York.

Feller, W. (1970). *An Introduction to Probability Theory and its Applications, Vol. II* (2nd Ed.), Wiley, New York.

Hogg, R. V. and Craig, A. T. (1978). *Introduction to Mathematical Statistics*, Macmillan, New York.

Johnson, N. L. and Kotz, S. (1970). *Continuous Univariate Distributions–1*, Houghton-Mifflin, New York.

Kendall, M. G. and Stuart, A. (1973). *The Advanced Theory of Statistics, Vol. 2*, Hafner, New York.

Waller, R. A. (1979). *Statistics: An Introduction to Numerical Reasoning*, Holden-Day, San Francisco.

CHAPTER 4

Elements of Reliability

This chapter is concerned with the introduction of some of the fundamental notions of reliability that provide a foundation for the remainder of the book. The presentation adheres to the classical frequency approach to reliability, without regard for the Bayesian point of view, which is introduced in Chapter 5.

Any reliability analysis of a system must be based on precisely defined concepts in order to make comparisons between systems and to provide a logical basis for improving the reliability of a system. Section 4.1 presents the basic definitions of the terms commonly used in reliability. Terms such as reliability function, mean time to failure, hazard rate, and reliable life are defined and various examples are used to reinforce the meaning of these concepts. The estimation of such quantities is the subject of much of the remainder of the book and a clear understanding of the meaning of such terms is of fundamental importance.

Section 4.2 examines various distribution models useful in both classical and Bayesian reliability analyses. The models are described and various properties of each model, such as the moments and general shape characteristics, are discussed. Various reliability measures associated with each distribution are also presented as appropriate.

Section 4.3 is largely concerned with the answer to the question, "How might an analyst determine if a given set of data agrees with a prescribed distribution?" Various properties of the hazard rate function are exploited that can be used to identify certain candidate distributions. Total time on test plots, as well as certain other probability plots, are also discussed. Such techniques have been found to be useful in providing evidence for or against candidate distributions.

Once a distribution has been selected, the reliability measure of interest must then be estimated from the available data. Statistical methods play a crucial role in estimating such measures. *Life testing* is the name given to a variety of testing methods commonly used to obtain the necessary statistical data. Several of the most commonly used life testing methods are discussed in Section 4.4. Estimation of various reliability measures for the exponential model is considered in Section 4.5 which also introduces the basic use of statistics in reliability estimation.

Section 4.6 introduces the notion of reliability demonstration testing, which is analogous to testing a statistical hypothesis. Results are presented for the exponential distribution.

The main item of concern in many reliability analyses is the attainment of a measure of the reliability of a system, which is a logical configuration of components/subsystems. Such reliability measures for a variety of systems are presented in Section 4.7.

Repair is an important consideration in most complex systems, and Section 4.8 discusses the influence of repair on overall system performance.

Throughout this chapter and the remainder of the book, for convenience the class of minimum variance unbiased estimators will be referred to as MVU estimators. As in Chapters 2 and 3, the class of maximum likelihood estimators will be denoted as ML estimators.

4.1. MEASURES OF RELIABILITY

Recall from Chapter 1 that the reliability of a device is defined as the probability that the device will perform its intended function for at least a specified period of time under specified environmental conditions. Since reliability is a quantitative measure involving probability, the treatment of this subject is based essentially on the material introduced in Chapters 2 and 3. Thus, the rules of probability introduced in Chapter 2 can be used to calculate the reliability of a complex system, provided that the reliabilities of the individual components are known. Estimates of the component reliabilities are frequently obtained from either statistical life tests, as discussed in Section 4.4, or from field use data.

Different measures of reliability are necessary, as different devices may have different objectives. The use of a certain device actually determines the kind of reliability measure that is most meaningful and most useful. For example, the reliability measure associated with nuclear power reactor components is frequently taken to be the failure rate, since failure of a reactor is of primary concern. On the other hand, a power supply for a deep space probe must function without failure for the entire mission duration and so the probability of survival for the mission, the reliability, is the important measure of reliability. In addition, at different times during its operating life a device may be required to have a different probability of successfully performing its required function under the stated conditions.

In reliability, the term *failure* means that the device is incapable of performing its required function. We consider only the case in which the device either is capable of performing its function or not, and exclude the case where varying degrees of capability are considered. It should be kept in mind, however, that generally the term *capable* must be precisely defined.

4.1.1 Reliability Function

Let the random variable T denote the failure time (or time to failure) of some device of interest under stated environmental conditions and let $f(t)$ denote the pdf of T. In general, $f(t)$ will change as the environmental conditions change. The probability of failure as a function of time can be defined by

$$\Pr(T \leqslant t) = \int_0^t f(\tau)\,d\tau = F(t), \qquad t \geqslant 0, \tag{4.1}$$

where $F(t)$ is the probability that the device will fail by time t. Occasionally, time is replaced by another more pertinent measure of interest such as cycles, stress, and so on. In such cases we speak of cycle to failure, stress to failure, and so on. According to Section 2.6, $F(t)$ is the failure cdf, which is also sometimes referred to as the *unreliability function*. If we define reliability as the probability of success, that is, the probability that the device will perform its intended function for at least a period of time t, then we can write

$$R(t) = \Pr(T \geqslant t) = \int_t^{\infty} f(\tau)\,d\tau = 1 - F(t), \tag{4.2}$$

where $R(t)$ is the *reliability function*. Thus for a given failure time distribution, the reliability function can be found directly. We defer examples to Section 4.2 which illustrates this for several different distributions.

Often it is of interest to consider the situation in which only the success or failure of a device is recorded and not its time to failure. In these situations the probability of success is used as a measure of the reliability of the device. This approach to reliability, independent of time, generates what is known as *attribute* test data which are discussed in Chapter 7.

4.1.2 Mean Time to Failure

The *mean time to failure* (MTTF) is the expected time during which the component will perform successfully and, according to (3.2),

$$E(T) = \int_0^{\infty} t f(t)\,dt. \tag{4.3}$$

$E(T)$ is often referred to also as the *expected life*. If $\lim_{t\to\infty} tR(t)=0$, then another convenient method for determining the MTTF is given by

$$E(T)=\int_0^{\infty} R(t)\,dt, \tag{4.4}$$

which may be shown to be true by integration by parts.

If the device under consideration is renewed through maintenance and repair, $E(T)$ is also known as the *mean time between failures* (MTBF). Much of renewal theory assumes that repaired systems are as good as new from a failure standpoint. In general, perfect renewal is not possible, and in such cases, it is appropriate to consider terms such as the mean time to the jth failure.

We can also define the mean time to failure of a device that has survived to time t; namely,

$$M(t)=\frac{1}{R(t)}\int_0^{\infty} \tau f(t+\tau)\,d\tau, \tag{4.5}$$

which is known as the *mean residual life*. Thus

$$M(0)=E(T).$$

4.1.3 Failure and Hazard Rates

The failure process is usually quite complex and it is often difficult to understand the physics of the underlying process. It is even more difficult to mathematically describe such a process. Consequently, a failure distribution is used to provide a statistical summary account of the length of life of a device and the choice of distribution is largely an art. If one attempts to use observed failure times (often referred to as *life test data*) to distinguish among various asymmetrical distributions, one is still faced with the problem. The reason is that asymmetric distributions are importantly different in their tail regions and actual observations are sparse, particularly in the right-hand tail region, due to small sample sizes.

It has been recognized that these difficulties can be overcome by appealing to a concept that permits different distributions to be distinguished on the basis of a physical consideration. Such a concept is expressed as a hazard rate. A closely related concept is that of failure rate, which we now discuss.

The rate at which failures occur in a certain time interval $[t_1, t_2]$ is called the *failure rate* during that interval. It is defined as the probability that a failure per unit time occurs in the interval, given that a failure has not occurred prior to the beginning of the interval t_1. Mathematically, the interval failure rate is

$$FR(t_1, t_2) = \left[\frac{R(t_1) - R(t_2)}{R(t_1)}\right]\left[\frac{1}{t_2 - t_1}\right], \tag{4.6}$$

where the first bracketed factor is the conditional probability of failure during $[t_1, t_2]$, given survival to time t_1, and the second factor is a dimensional characteristic used to express the conditional probability on a per-unit time basis. Although the "rate" in the above definition is expressed as failures per unit time, frequently the "time" units might be "cycles", "stress", "distance", and so on, as previously discussed.

The *hazard rate* (or hazard rate function or, simply, hazard function) is defined as the limit of the failure rate as the length of the interval $[t_1, t_2]$ approaches zero. Thus, it is the instantaneous failure rate. The hazard rate $h(t)$ is defined as

$$h(t) = \lim_{\Delta t \to 0} \frac{R(t) - R(t + \Delta t)}{\Delta t\, R(t)} = \frac{1}{R(t)}\left[-\frac{dR(t)}{dt}\right]$$

$$= -\frac{d \ln R(t)}{dt} = \frac{f(t)}{R(t)}, \tag{4.7}$$

since $-dR(t)/dt = f(t)$, the failure time pdf. It is noted that $h(t)\,dt$ represents the probability that a device which has survived to time t will fail in the small interval of time t to $t + dt$, and thus $h(t)$ is the rate of change of the conditional probability of failure given survival to time t. Also worth noting is that $f(t)$ is the rate of change of the ordinary (unconditional) probability of failure. The importance of the hazard rate is that it indicates the change in the failure rate over the lifetime of a population of devices. If $h(t)$ is increasing (decreasing) in $t \geqslant 0$, then $f(t)$ is said to be an *increasing failure rate* (*IFR*) [*decreasing failure rate* (*DFR*)] distribution.

A typical hazard rate generally has the so-called *bathtub* shape shown in Figure 4.1. Three distinct failure regions are indicated. The first, called the *initial failure* region, is characterized by a decreasing failure rate. It represents early failures due to material or manufacturing defects. Good quality control and burn-in product testing usually eliminate many substandard devices, and thus avoid this high initial failure rate. The second one, called the *chance* or *random failure* region, is characterized by a near constant failure rate. It represents chance failures caused by sudden stresses, unusually severe and unpredictable operating conditions, and so on. To eliminate these would require a device that is overdesigned for the vast majority of situations. The third portion, called the *wearout failure* region, is typified by

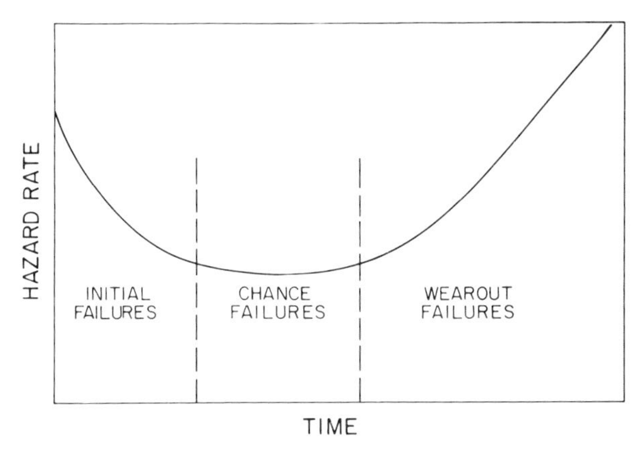

Figure 4.1. A typical (bathtub) hazard rate curve.

an increasing failure rate, resulting from equipment deterioration, accumulated shocks, fatigue, and the like.

As stated earlier, it may be more convenient to select a distribution on the basis of the shape characteristics of the hazard rate rather than the shape of the pdf. The characteristic shapes for some well-known distributional models are presented in the next section.

There is an important mathematical relationship between the hazard rate and the reliability function. It is assumed that $R(0)=1$. Integrating both sides of (4.7) yields

$$\int_0^t h(x)\,dx = -\int_0^t \frac{dR(x)/dx}{R(x)}\,dx = -\ln R(x)|_0^t$$

$$= -\ln R(t) + \ln R(0) = -\ln R(t),$$

and thus

$$R(t) = \exp\left[-\int_0^t h(x)\,dx\right]. \qquad (4.8)$$

From (4.7) and (4.8) it also follows that

$$f(t) = h(t)\exp\left[-\int_0^t h(x)\,dx\right], \qquad (4.9)$$

and thus $f(t)$, $R(t)$, and $h(t)$ are all related and any one uniquely determines the other two.

Example 4.1. Suppose that $h(t)$ is a linearly increasing function given by

$$h(t)=\beta t, \qquad \beta, t>0,$$

where β governs the rate of increase of the hazard rate. Then

$$\int_0^t \beta x\, dx=\tfrac{1}{2}\beta t^2,$$

from which, by (4.9),

$$f(t)=\beta t \exp\left[-\tfrac{1}{2}\beta t^2\right],$$

and

$$R(t)=\exp\left[-\tfrac{1}{2}\beta t^2\right].$$

This distribution of T is known as a *Rayleigh distribution.* ■

Barlow and Proschan (1975) define

$$H(t)=\frac{1}{t}\int_0^t h(\tau)\, d\tau,$$

as the *failure (hazard) rate average*. Thus $H(t)=-(1/t)\ln R(t)$. If $H(t)$ is increasing (decreasing) in $t \geqslant 0$, then $f(t)$ is said to be an *increasing failure rate average* (IFRA) [*decreasing failure rate average* (DFRA)] distribution.

4.1.4 Reliable Life

It is sometimes necessary to consider the time t_R for which the reliability will be R. This time is known as the *reliable life* and is also a measure of the reliability of a device. In Example 4.1, we find that the reliable life is given by

$$t_R=(-2\ln R/\beta)^{1/2}.$$

The reliable life may also be thought of as the time t_R for which $100R\%$ of the population will survive. Determination of t_R is equivalent to computing the $100(1-R)$th percentile of the failure time distribution. If we are interested in the estimation of t_R based on a sample of data from the underlying failure time distribution, an LCI estimate on the reliable life is known as a *tolerance interval* estimate. Either TCI or LCI estimates on t_R are often of interest. More will be said about such estimates in Section 4.5.

Finally, Example 8.1 further illustrates several situations in which reliability, MTTF, hazard rate, and reliable life would be the preferred reliability measure of interest.

4.1.5 System Effectiveness

Thus far, we have been concerned only with the reliability of a device or system. However, to a consumer the term "reliability" usually has a broader meaning than that provided by the reliability function. For example, if a consumer must choose between a system that is highly reliable but difficult to repair and one that is less reliable but easier to repair, he may choose the less reliable system. Clearly, other measures may be necessary in order to appropriately describe the consumer's concept of reliability. Let us now examine some of these other measures several of which will be used later in the book, particularly in Chapter 12.

Table 4.1 contains the definitions of some concepts that relate to the overall effectiveness of a system. These definitions allude to various time categories which are defined in Table 4.2. *Availability* is a measure of the ratio of the operating time of the system to the operating time plus total downtime excluding only any free downtime. Thus, it considers both reliability and maintainability. On the other hand, *intrinsic availability* is a more restrictive measure than availability, since only the active repair time portion of downtime is considered. Thus, the availability will always be less than the intrinsic availability, and the decrease due to the administrative and logistics time connected with the downtime cycle. In other words, the intrinsic availability measures the upper limit on availability that can be achieved if the administrative and logistics time can be eliminated from the downtime cycle.

Let us illustrate these concepts with a simple example. Table 4.3 gives the time (in h) a certain system spent in various successive time categories for two uptime and one downtime cycle. Although we consider only a few cycles here, in practice many more cycles of data would be needed in order to accurately assess the system effectiveness.

The total uptime in both cycles is 1098 h, and the operating time in each cycle is 582 h and 465 h, respectively. These represent two samples from the failure time distribution.

The total downtime in one complete cycle is 192 h. The active repair time is 8 h, which is one sample from the repairability distribution.

Table 4.1. System Effectiveness Concepts

Concept	Definition
System Effectiveness	The probability that the system can successfully meet an operational demand within a given time period when operated under specified conditions.
Operational Readiness	The probability that, at any point in time, the system is either operating or can operate satisfactorily when operated under specified conditions.
Availability	The probability that the system is operating satisfactorily at any point in time when operating under specified conditions, where the time categories considered include all but free time.
Intrinsic Availability	The probability that the system is operating satisfactorily at any point in time when operated under specified conditions, where the time categories considered are only operating time and active repair time.
Maintainability	The probability that a failed system can be restored to operating conditions in a specified interval of downtime, which excludes only any free downtime.
Repairability	The probability that a failed system can be restored to operating condition in a specified interval of active repair time.
Serviceability	The ease with which a system can be repaired.

The intrinsic availability A_I is estimated as

$$A_I = \frac{\text{Operating Time}}{\text{Operating Time} + \text{Active Repair Time}} = \frac{1047}{1055} = 99\%,$$

while the availability A is estimated to be

$$A = \frac{\text{Operating Time}}{\text{Operating Time} + \text{Down Time (Excluding Free Time)}} = \frac{1047}{1215} = 86\%.$$

Thus, there is a potential for 13% improvement in availability which can be realized by elimination of 160 h of administrative and logistic time.

Finally, the operational readiness of the system is estimated to be

$$\text{Operational Readiness} = \frac{\text{Total Up Time}}{\text{Total Up time} + \text{Total Down Time}} = \frac{1098}{1290} = 85\%,$$

Table 4.2. **Definitions of Time Categories Used to Quantify System Effectiveness Concepts**

Concept	Definition
Uptime	The total time during which the system is in acceptable operating condition.
Operating time	The portion of uptime that the system is operating in an acceptable manner.
Free Time	The portion of uptime or downtime during which the operational use of the system is not required.
Downtime	The total time during which the system is not in acceptable operating condition.
Active Repair Time	The portion of downtime during which repairmen are actively repairing the system, which includes set-up time, fault diagnosis time, fault correction time, and final check-out time.
Logistic Time	The portion of downtime spent waiting for replacement parts.
Administrative Time	The portion of downtime not included under active repair time and logistic time, which includes the time spent preparing orders for repair, waiting for repairmen, etc.

Table 4.3. System Effectiveness Example Data

Activity No.	System Status	Time	Time Category	Time Units (h)
1	Available	Uptime-Cycle 1	Operating Time	338
2			Free Time	26
3			Operating Time	244
4			Free Time	10
5	Unavailable	Downtime-Cycle 1	Free Time	24
6			Administrative Time	36
7			Logistic Time	120
8			Active Repair Time	8
9			Administrative Time	4
10	Available	Uptime-Cycle 2	Operating Time	280
11			Free Time	15
12			Operating Time	185

which means that the system is ready and able to perform its function 85% of the time.

In general, the availability and maintainability functions are time dependent as is clear from their definitions. The above static example did not take explicit account of this time dependency, as the data represented just a few samples from the corresponding distributions. It is possible to explicitly mathematically incorporate the time dependency into the availability and maintainability expressions. This leads to the so-called *transient* and *steady state availability* expressions. Such analysis requires the use of certain distributional models to be presented, and so we defer discussion until Section 4.8.

4.2. STATISTICAL MODELS USED IN RELIABILITY

In the following section, we shall examine some of the common failure time distributions and their related hazard rates. We will also introduce some distributions that are particularly useful in Bayesian reliability analysis. The first six subsections contain the most commonly used failure time and repairability/maintainability distributions, whereas the last two describe distributions that are commonly used as prior models in Bayesian reliability analysis. The gamma distribution discussed in Section 4.2.6 is used as both a failure time and prior distribution. Some other distributions, less widely used in Bayesian reliability analysis, are introduced as needed.

4.2.1 The Exponential Distribution

The exponential distribution is perhaps as widely used in reliability as the normal distribution is in other areas of statistics. There are many reasons for this that have been put forth in arguments by authors such as Barlow and Proschan (1965), Davis (1952), Epstein (1958), and others.

Most complex system reliability models assume that only random component failures need be considered; thus, interest is focused only on the chance failure region in Figure 4.1. When all initial failures have been removed by burn-in, and the time to occurrence of wearout failures is very great, then this assumption is reasonable. Frequently, electronic parts typify this situation. Bayesian analysis of the exponential distribution is considered in Chapter 8 and we also make extensive use of this distribution in Chapters 10–12.

The exponential distribution is inherently associated with the Poisson process. Suppose that random "shocks" to a device occur according to the postulates of a Poisson process. Thus, the random number of shocks $X(t)$

occurring in a time interval of length t is described by the Poisson distribution

$$\Pr[X(t)=n]=\frac{e^{-\lambda t}(\lambda t)^n}{n!}, \qquad n=0,1,2,\ldots, \quad \lambda, t>0,$$

where λ is the rate at which the shocks occur. Further suppose that the device fails immediately upon receiving a single shock and will not fail otherwise. Let the random variable T denote the failure time of the device. Thus,

$$\begin{aligned} R(t;\lambda) &= \Pr[\text{the device survives at least to time } t] \\ &= \Pr[\text{no shocks occur in } (0,t)] \\ &= \Pr[X(t)=0] = e^{-\lambda t}, \end{aligned}$$

where we have written $R(t;\lambda)$ to denote the explicit functional dependency of $R(t)$ on λ, a convention that we shall adopt for the remainder of this chapter. Also,

$$f(t;\lambda) = -\frac{dR(t;\lambda)}{dt} = \lambda e^{-\lambda t}, \qquad t \geqslant 0, \quad \lambda > 0, \tag{4.10}$$

which is the *exponential distribution* introduced in Chapter 2 with parameter λ, and denoted simply as the $\mathcal{E}(\lambda)$ distribution. According to (4.7), the hazard rate is

$$h(t;\lambda) = \frac{\lambda e^{-\lambda t}}{e^{-\lambda t}} = \lambda, \tag{4.11}$$

a constant, which is often referred to simply as the $\mathcal{E}(\lambda)$ failure rate or, the failure rate. The $\mathcal{E}(\lambda)$ failure time distribution is thus uniquely associated with a constant failure rate model. Figure 4.2 shows a typical $\mathcal{E}(\lambda)$ pdf, reliability, and hazard function.

The MTTF and variance of T are

$$E(T;\lambda) = \frac{1}{\lambda}, \tag{4.12}$$

and

$$\operatorname{Var}(T;\lambda) = \frac{1}{\lambda^2}, \tag{4.13}$$

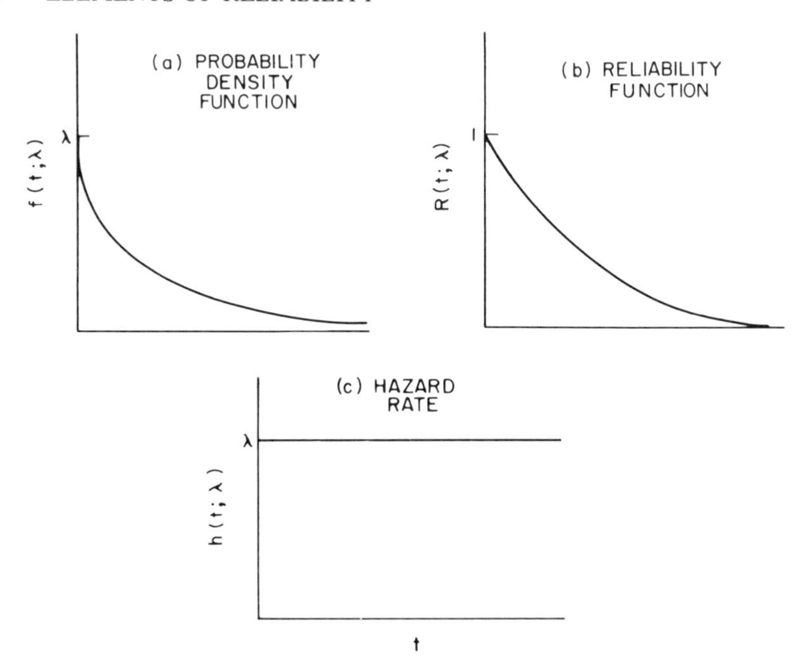

Figure 4.2. The probability density function, reliability function, and hazard rate for the exponential distribution.

respectively, and the reliable life t_R is obtained by solving $R = \exp(-\lambda t)$ for t; namely,

$$t_R = -\frac{1}{\lambda} \ln R. \tag{4.14}$$

In particular, the median life $t_{0.50}$ is $-\ln(0.5)/\lambda$.

The constant failure rate λ can be interpreted to mean that the failure process has no memory. Using the laws of conditional probability

$$\begin{aligned} \Pr(t \leqslant T \leqslant t + \Delta t | T > t) &= \frac{R(t; \lambda) - R(t + \Delta t; \lambda)}{R(t; \lambda)} \\ &= \frac{e^{-\lambda t} - e^{-\lambda(t + \Delta t)}}{e^{-\lambda t}} = 1 - e^{-\lambda \Delta t}, \end{aligned} \tag{4.15}$$

which is independent of t. Thus, if the device is still functioning at time t, it is as good as new and the remaining life has the same $\mathcal{E}(\lambda)$ distribution given in (4.10).

A somewhat more general form for the $\mathcal{E}(\lambda)$ distribution is the so-called *two-parameter exponential distribution* given by

$$f(t;\lambda,\mu)=\lambda e^{-\lambda(t-\mu)}, \qquad t\geqslant\mu, \tag{4.16}$$

where μ is referred to as the *threshold parameter* and physically represents a period of guaranteed life. Such a parameter may be used to describe such things as a warranty period, a period with no initial failures such as occurs with bearing failures in reciprocating engines, and so on. The two-parameter exponential distribution will be referred to as the $\mathcal{E}(\lambda,\mu)$ distribution.

4.2.2 The Weibull Distribution

The Weibull distribution is widely used in reliability. It was first presented by Weibull (1939) and its use in reliability was discussed by Weibull (1951). It generalizes the use of the $\mathcal{E}(\lambda,\mu)$ distribution to include nonconstant hazard rates. In particular, the Weibull distribution encompasses both increasing and decreasing hazard rates, and has successfully been used to describe both initial failures [Von Alven (1964)] as well as wearout failures [Lieblein and Zelen (1956)]. When a system is composed of a number of components and failure is due to the most severe defect of a large number of possible defects, the Weibull distribution is often especially appropriate.

Suppose we consider the hazard rate to be the power function of t given by

$$h(t;\alpha,\beta,\theta)=\frac{\beta}{\alpha}\left(\frac{t-\theta}{\alpha}\right)^{\beta-1}, \qquad \beta,\alpha>0, \qquad 0<\theta\leqslant t<\infty. \tag{4.17}$$

It is easily shown that $h(\cdot)$ is decreasing (increasing) in $t-\theta$ if $\beta<1$ ($\beta>1$), and is constant if $\beta=1$. The $\mathcal{E}(\alpha^{-1},\theta)$ distribution is thus a special case of the Weibull distribution when $\beta=1$. When $\beta=2$, the hazard rate is a linearly increasing function of $t-\theta$, and the resulting distribution is known as a *Raleigh distribution*. Since β controls the shape of the distribution, it is called the *shape* parameter, while α and θ are usually referred to as the *scale* and *location* parameters, respectively. Again, θ corresponds to a period of guaranteed life that is not present in many applications; thus $\theta=0$ in such instances. It is noted that the Weibull distribution has many different forms, depending on the way in which the parameters are defined in (4.17). Although all forms are essentially the same in regard to the important distributional properties, there isn't a "standard" form for the Weibull distribution.

Using (4.9) and (4.8), the corresponding Weibull pdf and reliability functions become

$$f(t;\alpha,\beta,\theta)=\frac{\beta}{\alpha}\left(\frac{t-\theta}{\alpha}\right)^{\beta-1}\exp\left[-\left(\frac{t-\theta}{\alpha}\right)^{\beta}\right],\qquad t\geqslant\theta,\quad (4.18)$$

and

$$R(t;\alpha,\beta,\theta)=\exp\left[-\left(\frac{t-\theta}{\alpha}\right)^{\beta}\right],\quad (4.19)$$

respectively. The three-parameter Weibull distribution will be subsequently referred to as the $\mathcal{W}(\alpha,\beta,\theta)$ distribution. Some typical Weibull density, reliability, and hazard functions are shown in Figure 4.3 for the case where $\theta=0$ and $\alpha=1$. When $\theta=0$, the $\mathcal{W}(\alpha,\beta,\theta)$ distribution will be denoted simply by $\mathcal{W}(\alpha,\beta)$.

The MTTF and variance of (4.18) are

$$E(T;\alpha,\beta,\theta)=\theta+\alpha\Gamma\left(\frac{\beta+1}{\beta}\right),\quad (4.20)$$

$$\mathrm{Var}(T;\alpha,\beta,\theta)=\alpha^2\left[\Gamma\left(\frac{\beta+2}{\beta}\right)-\Gamma^2\left(\frac{\beta+1}{\beta}\right)\right],\quad (4.21)$$

respectively, The reliable life t_R is also found from (4.19) to be

$$t_R=\theta+\alpha(-\ln R)^{1/\beta},\quad (4.22)$$

and, in particular, the median of (4.18) is $\theta+\alpha(0.69315)^{1/\beta}$. It is possible to reparameterize (4.18) thus yielding different forms of the Weibull distribution. Some of these are considered in Section 9.1.

4.2.3 The Normal Distribution

The normal distribution has the well-known bell shape and is symmetrical about its mean value. Although the support of the normal distribution is $(-\infty,\infty)$, by taking the mean μ to be a sufficiently large positive value and the standard deviation σ to be sufficiently small relative to μ, the probability content below zero becomes sufficiently small and can be ignored. If such is not the case, then the normal distribution can be truncated below zero and rescaled accordingly [see Johnson and Kotz (1970)]. Bayesian estimation in the normal distribution is considered in Chapter 9. Recall from Section 3.5.2

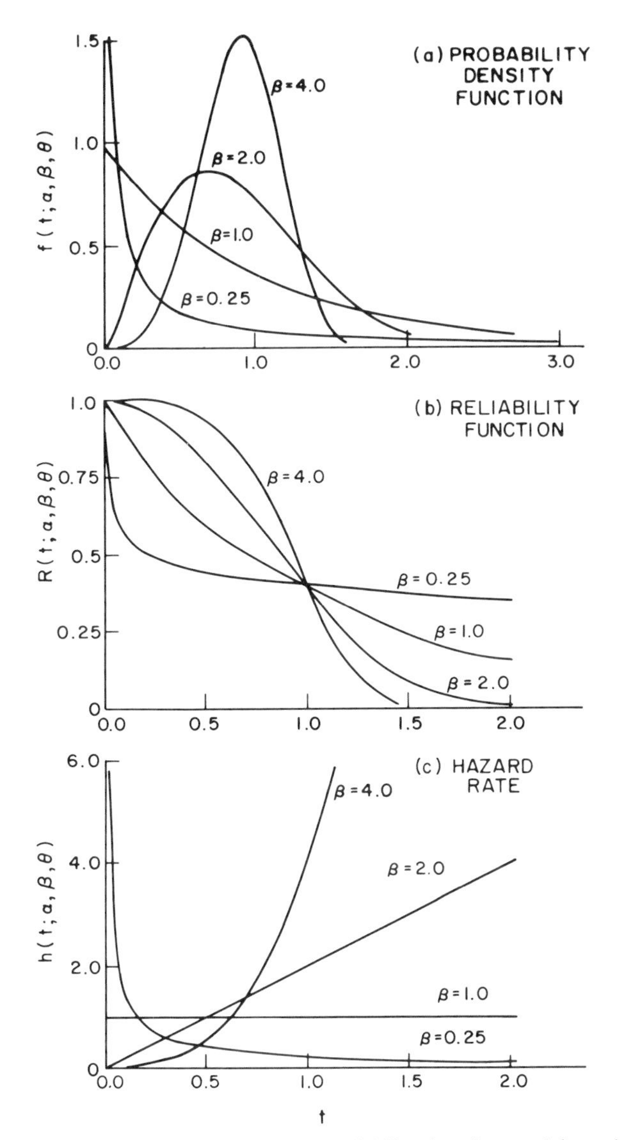

Figure 4.3. The probability density function, reliability function, and hazard rate for the Weibull distribution.

that the normal pdf is given by

$$f(t;\mu,\sigma^2)=\frac{1}{\sigma\sqrt{2\pi}}\exp\left[-\frac{1}{2\sigma^2}(t-\mu)^2\right],\qquad \begin{array}{l}-\infty<t<\infty,\\ 0<\mu<\infty,\\ \sigma^2>0,\end{array}\tag{4.23}$$

where μ is the MTTF and σ^2 is the variance of the failure time. Recall that this distribution is denoted as the $\mathcal{N}(\mu,\sigma^2)$ distribution. Now (4.23) can be expressed in terms of the standard normal pdf $\phi(z)$ given by

$$\phi(z)=\frac{1}{\sqrt{2\pi}}\exp\left(-\frac{z^2}{2}\right),\qquad -\infty<z<\infty,\tag{4.24}$$

which is well-tabulated. Thus,

$$f(t;\mu,\sigma^2)=\phi\left(\frac{t-\mu}{\sigma}\right)\Big/\sigma.\tag{4.25}$$

Let us also consider the standard normal cdf given by

$$\Phi(x)=\int_{-\infty}^{x}\frac{1}{\sqrt{2\pi}}\exp\left(-\frac{t^2}{2}\right)dt,\tag{4.26}$$

which is tabulated in Table B1 of Appendix B. Thus, in terms of $\Phi(x)$, the reliability function becomes

$$R(t;\mu,\sigma^2)=1-\Phi\left(\frac{t-\mu}{\sigma}\right),\tag{4.27}$$

which cannot be expressed in closed form.

The hazard rate for a normally distributed r.v. can be shown to be a monotonically increasing function of t [see Kapur and Lamberson (1977)] and according to (4.7) is given by

$$h(t;\mu,\sigma^2)=\frac{\phi\left(\frac{t-\mu}{\sigma}\right)}{\sigma-\sigma\Phi\left(\frac{t-\mu}{\sigma}\right)}.\tag{4.28}$$

Such a function is appropriate for the wearout region in Figure 4.1. The $\mathcal{N}(\mu,\sigma^2)$ distribution has also been used to model stress failure, where the r.v. is stress rather than time. Some typical $\mathcal{N}(\mu,\sigma^2)$ pdf, reliability, and hazard functions are shown in Figure 4.4.

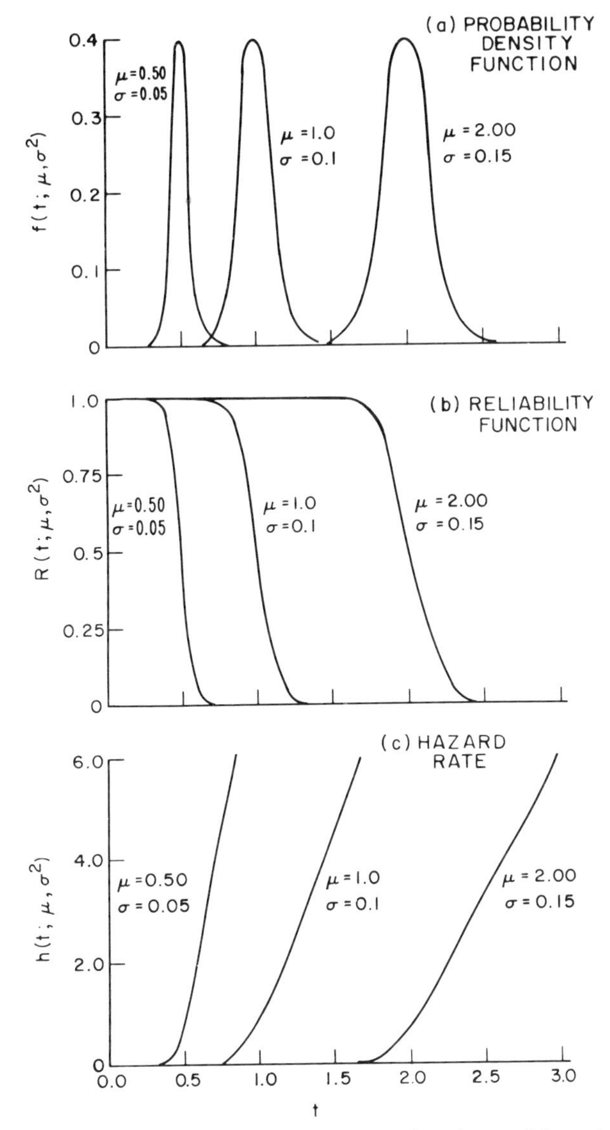

Figure 4.4. The probability density function, reliability function, and hazard rate for the normal distribution.

The MTTF and variance of (4.25) are, of course,

$$E(T;\mu,\sigma^2)=\mu, \qquad \operatorname{Var}(T;\mu,\sigma^2)=\sigma^2. \tag{4.29}$$

The reliable life is given by

$$t_R=\sigma\Phi^{-1}(1-R)+\mu, \tag{4.30}$$

where $\Phi^{-1}(1-R)$ denotes the $100(1-R)$th percentile of the standard normal distribution. For example, if $R=0.99$, then $\Phi^{-1}(0.01)=1$st percentile$=-2.33$, as $\Phi(-2.33)=0.01$ from Table B1 of Appendix B.

4.2.4 The Log Normal Distribution

The use of the log normal distribution in reliability has become increasingly widespread. Early in the 1960s, Goldthwaite (1961) justified its use as a failure time distribution. Mann, Schafer, and Singpurwalla (1974) derive the log normal as a model for failure due to fatigue cracks. The log normal distribution has also emerged as a suitable distribution in maintainability analysis. The general properties of the log normal distribution may be found in Aitchison and Brown (1957). Bayesian analysis of the log normal distribution is undertaken in Chapter 9.

An r.v. T, such as failure or down time, is said to have a *log normal distribution* if $X=\ln T$ is normally distributed. By means of a simple logarithmic transformation of variable it can easily be shown that the log normal pdf is

$$f(t;\xi,\sigma^2)=\frac{1}{\sigma t\sqrt{2\pi}}\exp\left[-\frac{1}{2\sigma^2}(\ln t-\xi)^2\right], \qquad \begin{array}{l} 0\leq t<\infty, \\ -\infty<\xi<\infty, \\ \sigma^2>0, \end{array} \tag{4.31}$$

where $\xi=E(\ln T)$ and $\sigma^2=\operatorname{Var}(\ln T)$. The log normal distribution will be denoted as the $\mathcal{LN}(\xi,\sigma^2)$ distribution. Using (4.24) we can express (4.31) in terms of ϕ as

$$f(t;\xi,\sigma^2)=\frac{\phi\left(\dfrac{\ln t-\xi}{\sigma}\right)}{\sigma t}, \tag{4.32}$$

and thus $\mathcal{LN}(\xi,\sigma^2)$ probabilities can be found by using tables of the standard normal cdf. The mean and variance of T can be shown to be

$$E(T;\xi,\sigma^2)=\exp\left(\xi+\frac{\sigma^2}{2}\right), \tag{4.33}$$

and

$$\operatorname{Var}(T;\xi,\sigma^2)=\left[e^{2\xi+\sigma^2}\right]\left[e^{\sigma^2}-1\right], \tag{4.34}$$

respectively. In addition, the mode of (4.31) is $\exp(\xi-\sigma^2)$ and the median is $\exp(\xi)$.

Because of its logarithmic relationship with the $\mathcal{N}(\mu,\sigma^2)$ distribution, its reliability and hazard function involve ϕ and Φ, as for the $\mathcal{N}(\mu,\sigma^2)$ distribution. The reliability and hazard functions are given by

$$R(t;\xi,\sigma^2)=1-\Phi\left(\frac{\ln t-\xi}{\sigma}\right), \tag{4.35}$$

and

$$h(t;\xi,\sigma^2)=\frac{\phi\left(\dfrac{\ln t-\xi}{\sigma}\right)}{\sigma t-\sigma t\Phi\left(\dfrac{\ln t-\xi}{\sigma}\right)}, \tag{4.36}$$

respectively. Goldthwaite (1961) shows that the $\mathcal{LN}(\xi,\sigma^2)$ hazard rate initially increases over time and then decreases, approaching zero for sufficiently large values of time. Thus, the $\mathcal{LN}(\xi,\sigma^2)$ distribution is useful for describing those situations in which early failures or occurrences dominate the distribution. The $\mathcal{LN}(\xi,\sigma^2)$ pdf, reliability, and hazard functions are shown in Figure 4.5 for various choices of ξ and σ. The reliable life t_R may also be expressed as

$$t_R=\exp\left[\sigma\Phi^{-1}(1-R)+\xi\right], \tag{4.37}$$

where Φ^{-1} is defined in conjunction with (4.30).

4.2.5 The Inverse Gaussian Distribution

Inverse Gaussian distributions represent another general class of statistical distributions. The name derives from the observation that an inverse relationship exists between the cumulant generating functions of these distributions and those of normal (Gaussian) distributions.

Recently, the inverse Gaussian distribution has been found to be a useful model for those situations whenever early failures or occurrences dominate the lifetime distribution. In the past, such cases suggested the use of the $\mathcal{LN}(\xi,\sigma^2)$ distribution due to its characteristic initially increasing and then decreasing hazard rate. As will be seen, the hazard rate of the inverse

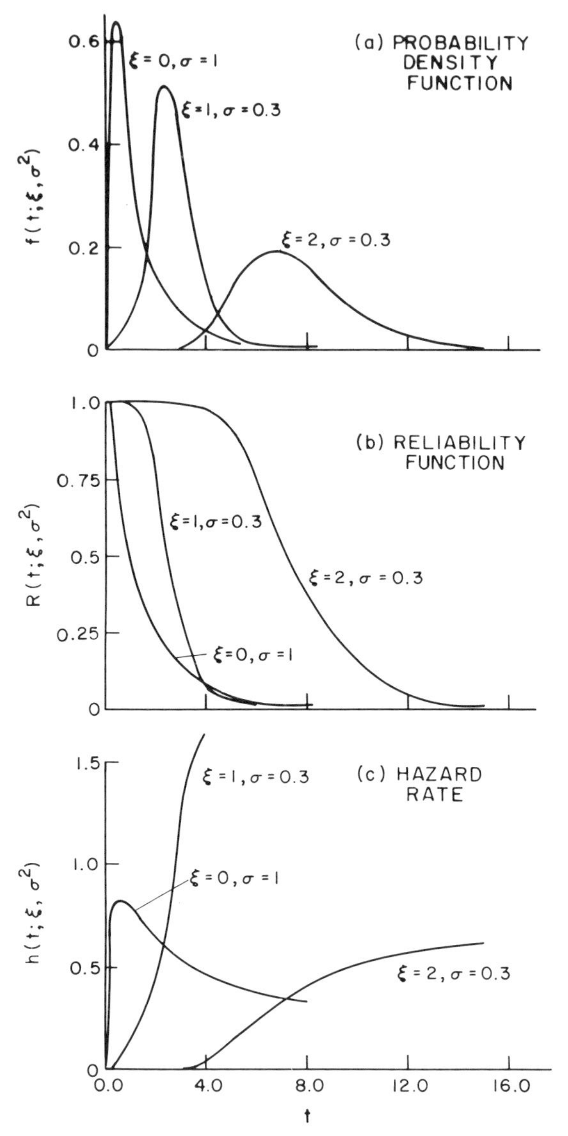

Figure 4.5. The probability density function, reliability function, and hazard rate for the log normal distribution.

Gaussian distribution exhibits similar behavior to the $\mathcal{LN}(\xi, \sigma^2)$ hazard rate. However, as pointed out by Chhikara and Folks (1977), there are several advantages of the inverse Gaussian distribution as a failure time or repairable/maintainability model. First, there is less difficulty justifying the inverse Gaussian model on physical grounds, since it arises as the first passage time distribution in a Brownian motion process [Cox and Miller (1965)]. Second, it addresses a wider class of failure time distributions. Lastly, the small sample statistical properties and inference procedures are well developed and often parallel those of the normal distribution [Chhikara and Folks (1974, 1976, 1977), Johnson and Kotz (1970), and Tweedie (1957)].

Padgett (1979) has also developed a lower confidence interval estimator for the inverse Gaussian reliability function. Bayesian estimation in this model is considered in Chapter 9.

The pdf of an *inverse Gaussian distribution* with parameters μ and λ is

$$f(t;\mu,\lambda)=\left(\frac{\lambda}{2\pi t^3}\right)^{1/2}\exp\left[-\frac{\lambda}{2\mu^2 t}(t-\mu)^2\right], \qquad t,\mu,\lambda,>0, \quad (4.38)$$

where λ is a shape parameter. This will be referred to as the $\mathcal{IN}(\mu,\lambda)$ distribution. The mean and variance of (4.38) are

$$E(T;\mu,\lambda)=\mu, \quad (4.39)$$

and

$$\operatorname{Var}(T;\mu,\lambda)=\frac{\mu^3}{\lambda}, \quad (4.40)$$

respectively, and thus μ is not a location parameter in the usual sense.

Chhikara and Folks (1977) give the reliability and hazard functions in terms of the standard normal cdf as

$$R(t;\mu,\lambda)=\Phi\left[\left(\frac{\lambda}{t}\right)^{1/2}\left(1-\frac{t}{\mu}\right)\right]-e^{2\lambda/\mu}\Phi\left[-\left(\frac{\lambda}{t}\right)^{1/2}\left(1+\frac{t}{\mu}\right)\right], \quad (4.41)$$

and

$$h(t;\mu,\lambda)=\frac{(\lambda/2\pi t^3)^{1/2}\exp\left[-\lambda(t-\mu)^2/2\mu^2 t\right]}{\Phi\left[(\lambda/t)^{1/2}(1-t/\mu)\right]-e^{2\lambda/\mu}\Phi\left[-(\lambda/t)^{1/2}(1+t/\mu)\right]}, \quad (4.42)$$

respectively. The $\mathcal{IN}(\mu, \lambda)$ distribution is unimodel with mode

$$t_m = -\frac{3\mu^2}{2\lambda} + \mu\left(1 + \frac{9\mu^2}{4\lambda^2}\right)^{1/2}. \tag{4.43}$$

Chhikara and Folks (1977) show that $h(\cdot)$ is increasing for $t < t_m$, decreasing for $t > 2\lambda/3$, and attains its maximum value at the solution to the equation

$$\frac{\lambda}{2\mu^2} + \frac{3}{2t} - \frac{\lambda}{2t^2} = 0. \tag{4.44}$$

They also show that $h(\cdot)$ approaches $\lambda/2\mu^2$ as $t \to \infty$. The pdf, reliability, and hazard functions for several values of λ, when $\mu = 1$, are shown in Figure 4.6.

4.2.6 The Gamma Distribution

The gamma distribution is a natural extension of the $\mathcal{E}(\lambda)$ distribution and is sometimes used as a failure time model [Gupta and Groll (1961)]. It is also often used as a prior distribution in Bayesian reliability analysis [see Chapters 7–9].

The gamma distribution can be derived by considering the time to the occurrence of the nth event in a Poisson process [Mann, Schafer, and Singpurwalla (1974)]. For example, if the time T_i between successive failures of a system has an $\mathcal{E}(\lambda)$ distribution, then $T = T_1 + \cdots + T_n$, the cumulative time to the nth failure, follows a gamma distribution with scale parameter λ and shape parameter n. The gamma distribution is thus said to be the n-fold convolution of an $\mathcal{E}(\lambda)$ distribution [Feller (1965)].

There is an alternative way in which the gamma distribution may arise as a failure time distribution. Consider a situation in which a system operates in an environment where shocks of common magnitude occur according to a Poisson process. Further, suppose that the system fails upon receipt of exactly n shocks and not before. The system failure time T, which denotes the random time of occurrence of the nth shock, follows a gamma distribution with parameters λ and n.

The pdf of a *gamma distribution* with parameters α and β is given by

$$f(t; \alpha, \beta) = \frac{1}{\beta^\alpha \Gamma(\alpha)} t^{\alpha-1} \exp(-t/\beta), \qquad \alpha, \beta, t > 0, \tag{4.45}$$

where $\Gamma(\alpha)$ denotes the gamma function evaluated at α. The mean and

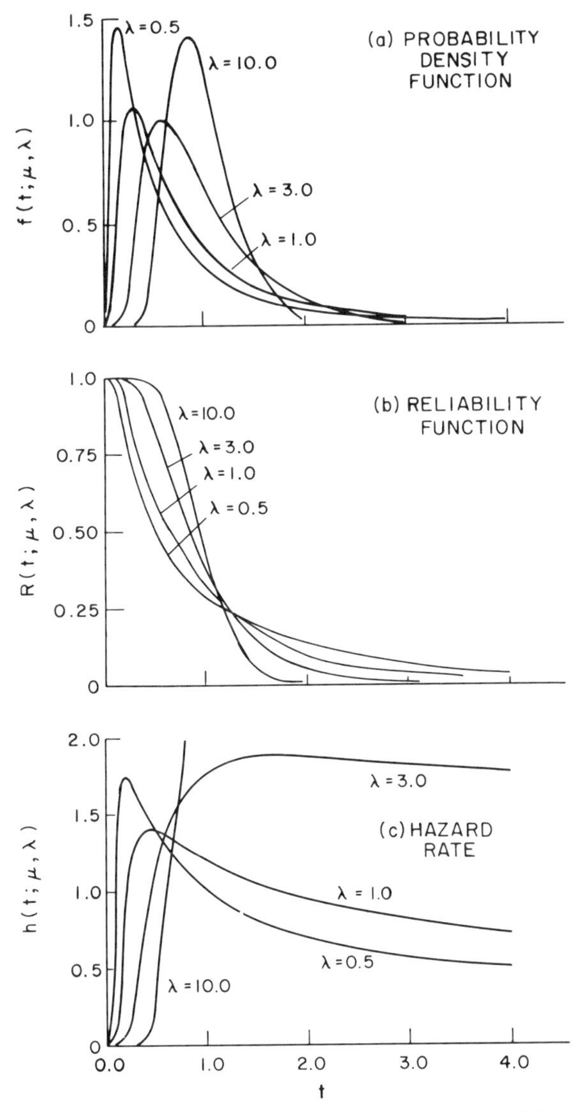

Figure 4.6. The probability density function, reliability function, and hazard rate for the inverse Gaussian distribution.

variance of T are given in (4.51) and (4.52), respectively. The gamma distribution will be denoted by $\mathcal{G}(\alpha, \beta)$. The parameters α and β are referred to as the *shape* and *scale* parameters of the $\mathcal{G}(\alpha, \beta)$ distribution. Sometimes the scale parameter is defined as β^{-1}, and both forms will be considered in later chapters. For integer values of α, the $\mathcal{G}(\alpha, \beta)$ distribution is also known as the *Erlangian distribution*. If $\alpha=1$, it is observed that the $\mathcal{G}(\alpha, \beta)$ distribution reduces to the $\mathcal{E}(\beta^{-1})$ distribution. Also, for the special case where $\alpha=n/2$ and $\beta=2$, the $\mathcal{G}(\alpha, \beta)$ distribution is known as a *chi-square distribution* with n degrees of freedom, which we represent as $\chi^2(n)$. The pdf of the $\chi^2(n)$ distribution is given in Section 3.5.2.

The reliability and hazard functions are not available in closed form unless α happens to be an integer; however, they may be expressed in terms of the *standard incomplete gamma function* $\Gamma(a, z)$ defined by

$$\Gamma(a, z)=\int_0^z y^{a-1}\exp(-y)\,dy, \qquad a>0, \tag{4.46}$$

which is widely tabulated and for which good computer subroutines have been written. Of course, $\Gamma(\alpha, \infty)=\Gamma(\alpha)$, the usual (complete) gamma function. In terms of $\Gamma(a, z)$, the reliability function for the $\mathcal{G}(\alpha, \beta)$ distribution is given by

$$R(t; \alpha, \beta)=\frac{\Gamma(\alpha)-\Gamma(\alpha, t/\beta)}{\Gamma(\alpha)}, \tag{4.47}$$

which, if α is an integer, becomes

$$R(t; \alpha, \beta)=\sum_{k=0}^{\alpha-1}\frac{(t/\beta)^k\exp(-t/\beta)}{k!}. \tag{4.48}$$

The hazard rate is likewise given by

$$h(t; \alpha, \beta)=\frac{t^{\alpha-1}\exp(-t/\beta)}{\beta^{\alpha}[\Gamma(\alpha)-\Gamma(\alpha, t/\beta)]} \tag{4.49}$$

for any $\alpha>0$ and, if α is an integer, has the closed form

$$h(t; \alpha, \beta)=\frac{t^{\alpha-1}}{\beta^{\alpha}\Gamma(\alpha)\sum_{k=0}^{\alpha-1}(t/\beta)^k/k!}. \tag{4.50}$$

It is apparent from (4.49) that $h(\cdot)$ is a decreasing function of t for $\alpha<1$, a constant for $\alpha=1$, and increasing for $\alpha>1$. The $\mathcal{G}(\alpha,\beta)$ distribution is thus appropriate for use in all three regions in Figure 4.1. Figure 4.7 illustrates the $\mathcal{G}(\alpha,\beta)$ pdf, reliability, and hazard functions for several choices of α and β. Further discussion on the various shapes of $\mathcal{G}(\alpha,\beta)$ distributions may be found in Section 7.4.3.

The mean and variance of (4.45) are

$$E(T;\alpha,\beta)=\alpha\beta, \tag{4.51}$$

and

$$\mathrm{Var}(T;\alpha,\beta)=\alpha\beta^2, \tag{4.52}$$

respectively. The reliable life t_R may be expressed as

$$t_R=\beta\Gamma^{-1}[\alpha,\Gamma(\alpha)(1-R)], \tag{4.53}$$

where $\Gamma^{-1}(\alpha,u)$ denotes that value of z such that $\Gamma(\alpha,z)=u$.

4.2.7 The Inverted Gamma Distribution

The inverted gamma distribution is often used as a prior distribution when estimating the MTTF of an $\mathcal{E}(\lambda)$ distribution using Bayesian methods. If Λ is a random variable having a $\mathcal{G}(\alpha,\beta^{-1})$ distribution, then $\Theta=1/\Lambda$ has an *inverted gamma distribution* with pdf

$$g(\theta;\alpha,\beta)=\frac{\beta^\alpha}{\Gamma(\alpha)}\left(\frac{1}{\theta}\right)^{\alpha+1}\exp\left(-\frac{\beta}{\theta}\right),\qquad \alpha,\beta,\theta>0, \tag{4.54}$$

which is denoted by $\mathcal{IG}(\alpha,\beta)$. As in the case of a $\mathcal{G}(\alpha,\beta)$ distribution, α and β are the shape and scale parameters, respectively. The mean and variance of (4.54) are

$$E(\Theta;\alpha,\beta)=\frac{\beta}{\alpha-1},\qquad \alpha>1, \tag{4.55}$$

and

$$\mathrm{Var}(\Theta;\alpha,\beta)=\frac{\beta^2}{(\alpha-1)^2(\alpha-2)},\qquad \alpha>2, \tag{4.56}$$

respectively. It is observed that, if $\alpha\leq 1$, the mean (and higher moments as

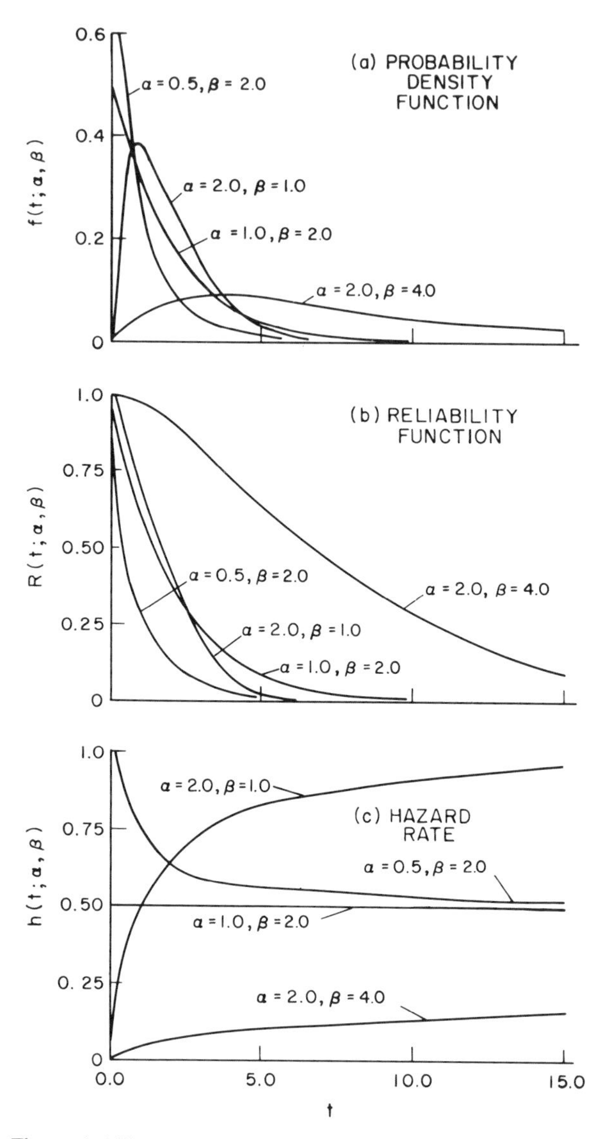

Figure 4.7. The probability density function, reliability function, and hazard rate for the gamma distribution.

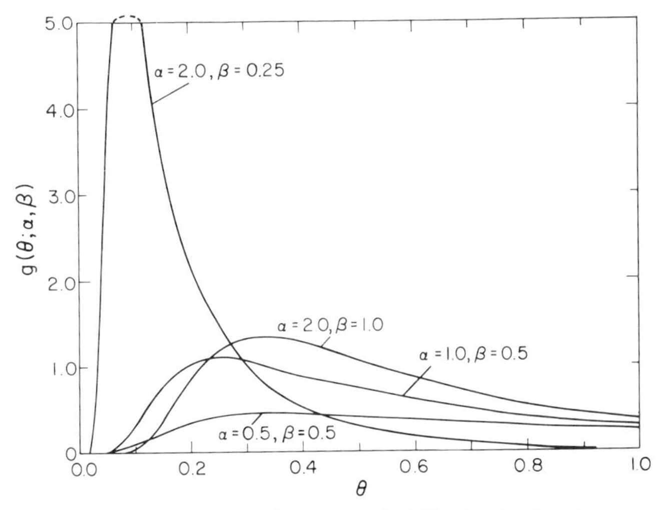

Figure 4.8. The inverted gamma probability density function.

well) does not exist, whereas if $1<\alpha\leqslant 2$, the variance does not exist. In general it can be shown that, for k greater than the largest integer smaller than α, the kth and larger order moments do not exist. Nonetheless, the distribution is well behaved in other respects and this shortcoming has little practical consequence in Bayesian estimation [see Section 8.3.3].

Since the $\mathcal{IG}(\alpha, \beta)$ distribution is not used as a failure time model, there is little practical significance in the reliability and hazard functions and so these are not presented. Figure 4.8 shows several $\mathcal{IG}(\alpha, \beta)$ pdfs for selected choices of α and β.

4.2.8 The Beta Distribution

The beta distribution is widely used in Bayesian reliability as a prior distribution on the random survival probability in a binomial distribution. This will be considered in Section 7.2.4. The pdf of a *beta distribution* with parameters x_0 and n_0 was introduced in Section 2.5.2 and is given by

$$g(y; x_0, n_0)=\frac{\Gamma(n_0)}{\Gamma(x_0)\Gamma(n_0-x_0)}y^{x_0-1}(1-y)^{n_0-x_0-1}, \qquad 0\leqslant y\leqslant 1, \quad n_0>x_0>0, \tag{4.57}$$

which will be referred to as a $\mathcal{B}(x_0, n_0)$ distribution. The mean and

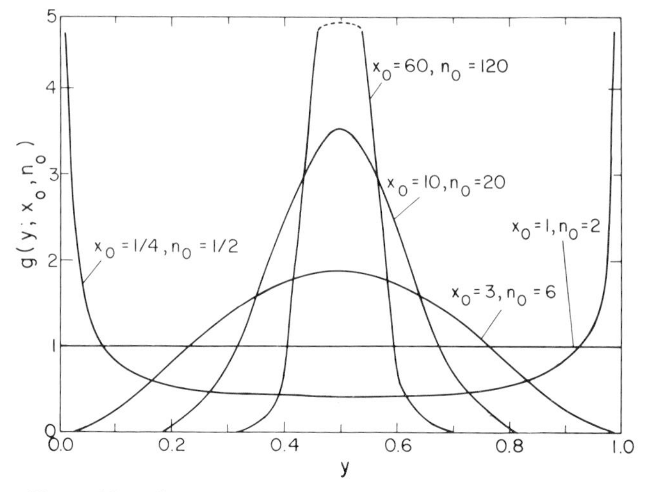

Figure 4.9. The beta probability density function with mean 0.5.

variance of (4.57) are

$$E(Y; x_0, n_0) = \frac{x_0}{n_0}, \tag{4.58}$$

and

$$\operatorname{Var}(Y; x_0, n_0) = \frac{x_0(n_0 - x_0)}{n_0^2(n_0 + 1)}, \tag{4.59}$$

respectively. A reparameterized alternate form of the $\mathcal{B}(x_0, n_0)$ distribution

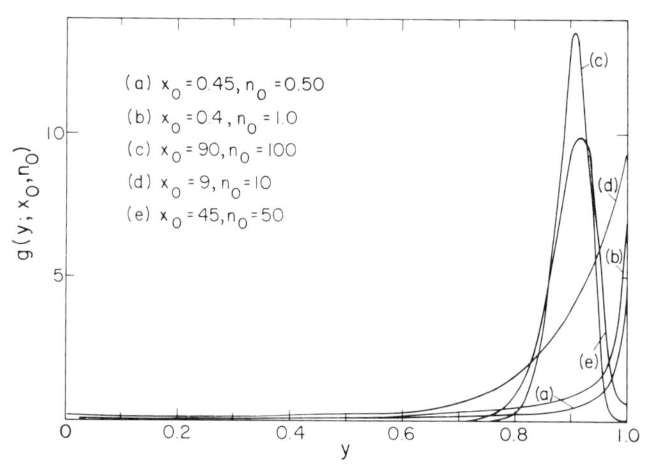

Figure 4.10. The beta probability density function with mean 0.9.

is sometimes considered in which $p = x_0$ and $q = n_0 - x_0$ are the two parameters and this form will be denoted as the $\mathcal{B}_1(p, q)$ distribution.

The $\mathcal{B}(x_0, n_0)$ distribution can assume a variety of both symmetrical and asymmetrical shapes. For a mean of 0.5, the general shapes range from uniform to bell-shaped to U-shaped. Some of these are shown in Figure 4.9. Figure 4.10 illustrates some asymmetrical $\mathcal{B}(x_0, n_0)$ distributions with a mean of 0.9. Further discussion on the various shapes of $\mathcal{B}(x_0, n_0)$ distributions may be found in Section 7.2.4.

4.3. MODEL IDENTIFICATION AND SELECTION

In this section we take up an important question that arises in the analysis of reliability data; namely, "Which model should be used to describe a given set of data"? Although this is generally a difficult question to answer, there are several techniques available which can assist the analyst in making such a decision. Some of these are now considered.

4.3.1 Nonparametric Hazard Rate Estimation

We now discuss some nonparametric procedures (i.e., those procedures that do not require a distribution to be specified) that can be used to estimate the pdf, reliability, and hazard functions from either a small sample of ungrouped data or a large sample of data that have been grouped into intervals. The shape of the hazard rate estimate is often used to suggest an appropriate distribution model for the data.

Let us first consider the case of a small sample of ungrouped failure data.

Ungrouped Failure Data. Let $t_1 \leqslant t_2 \leqslant \cdots \leqslant t_n$ represent n ordered observed failure times, where n is the sample size. The nonparametric estimate of the hazard rate $h(t_i)$ is given by

$$\hat{h}(t_i) = \frac{1}{(n - i + 0.625)(t_{i+1} - t_i)}, \qquad i = 1, 2, \ldots, n-1. \tag{4.60}$$

The unknown reliability and density functions are likewise each estimated by means of

$$\hat{R}(t_i) = \frac{n - i + 0.625}{n + 0.25}, \qquad i = 1, 2, \ldots, n, \tag{4.61}$$

and

$$\hat{f}(t_i) = \frac{1}{(n + 0.25)(t_{i+1} - t_i)}, \qquad i = 1, 2, \ldots, n-1, \tag{4.62}$$

respectively.

Although these estimators are in common use, they are by no means the only estimators that can be employed. The estimator in (4.61) is originally due to Blom (1958) and is known to produce good empirical results. The other direct estimators of $R(t_i)$, such as $(n-i+1)/(n+1)$, $(n-i+0.7)/(n+0.4)$, $(n-i+0.5)/n$, and $(n-i)/n$, are based on various properties of the rank distribution, which is the joint distribution of an ordered sample of size n from the underlying population. Kimball (1960) has done a limited study comparing some of the contenders. Although his determinations are limited, he concludes that the estimator given by (4.61) is close to being optimal and recommends its use. An excellent comparison of several nonparametric hazard rate estimators is given by Barlow and Van Zwet (1969).

Example 4.2. The following ordered failure times were observed for a heat exchanger used in the alkylation unit of a gasoline refinery. The failure mode was "leaking in one or more sections of a four-in-a-bank unit." The observed failure times were 0.41, 0.58, 0.75, 0.83, 1.00, 1.08, 1.17, 1.25, and 1.35 years. The following table contains the necessary calculations for the estimated hazard rate, reliability function, and pdf.

i	t_i	$t_{i+1}-t_i$	$\hat{h}(t_i)$	$\hat{R}(t_i)$	$\hat{f}(t_i)$
1	0.41	0.17	$\frac{1}{(8.625)(0.17)}=0.68$	$\frac{8.625}{9.25}=0.93$	$\frac{1}{(9.25)(0.17)}=0.64$
2	0.58	0.17	$\frac{1}{(7.625)(0.17)}=0.72$	$\frac{7.625}{9.25}=0.82$	$\frac{1}{(9.25)(0.17)}=0.64$
3	0.75	0.08	$\frac{1}{(6.625)(0.08)}=1.89$	$\frac{6.625}{9.25}=0.72$	$\frac{1}{(9.25)(0.08)}=1.35$
4	0.83	0.17	$\frac{1}{(5.625)(0.17)}=1.05$	$\frac{5.625}{9.25}=0.61$	$\frac{1}{(9.25)(0.17)}=0.64$
5	1.00	0.08	$\frac{1}{(4.625)(0.08)}=2.70$	$\frac{4.625}{9.25}=0.50$	$\frac{1}{(9.25)(0.08)}=1.35$
6	1.08	0.09	$\frac{1}{(3.625)(0.09)}=3.07$	$\frac{3.625}{9.25}=0.39$	$\frac{1}{(9.25)(0.09)}=1.20$
7	1.17	0.08	$\frac{1}{(2.625)(0.08)}=4.76$	$\frac{2.625}{9.25}=0.28$	$\frac{1}{(9.25)(0.08)}=1.35$
8	1.25	0.10	$\frac{1}{(1.625)(0.10)}=6.15$	$\frac{1.625}{9.25}=0.18$	$\frac{1}{(9.25)(0.10)}=1.08$
9	1.35	—	—	$\frac{0.625}{9.25}=0.07$	—

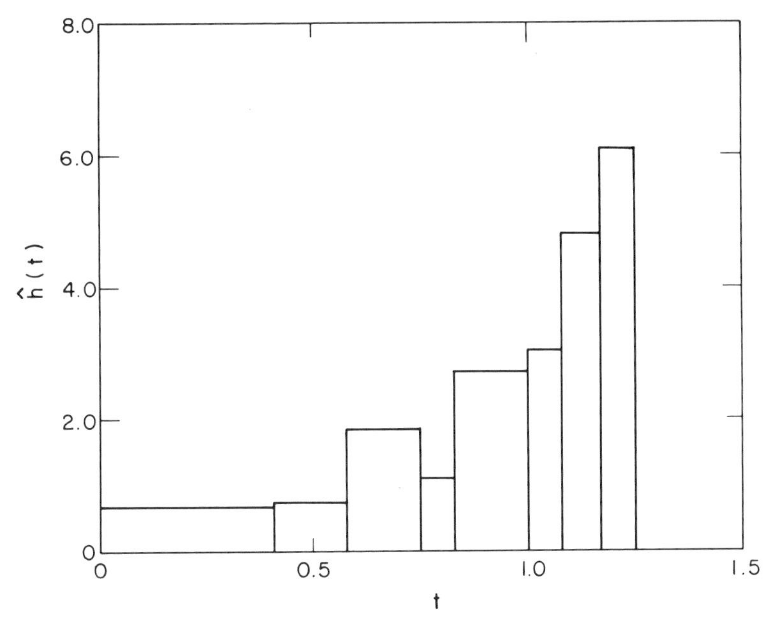

Figure 4.11. A plot of the hazard rate in Example 4.2.

Figure 4.11 gives a plot of $\hat{h}(t)$ and from this figure it is clear that the hazard rate is increasing, which suggests that a distribution such as the $\mathcal{W}(\alpha, \beta)$ is appropriate for the theoretical failure time distribution.

However, in analyzing small samples of data such as this, extreme caution must be used because a single outlying observation can have considerable influence. ■

Grouped Failure Data. Frequently, for larger samples of failure data, the data are grouped into intervals and only the number of failures in each interval is reported. Such data are referred to as *grouped failure data.*

Let $n(t)$ denote the observed number of units surviving at time t and let n represent the sample size. The nonparametric estimate of the reliability function becomes

$$\hat{R}(t) = \frac{\text{No. of Survivors at time } t}{\text{Sample Size}} = \frac{n(t)}{n}, \qquad t \geqslant 0, \tag{4.63}$$

where t is usually taken to be the upper endpoint of each interval. The pdf is similarly estimated by

$$\hat{f}(t) = \frac{\text{No. of Failures in } (t, t+\Delta t)}{\text{Sample Size} \times \text{Interval Width}} = \frac{n(t) - n(t+\Delta t)}{n \cdot \Delta t}, \qquad t \geqslant 0, \tag{4.64}$$

where Δt is the width of the interval at time t. Finally, the hazard rate is estimated by appealing directly to (4.6) as

$$\hat{h}(t) = \frac{\text{No. of Failures in } (t, t+\Delta t)}{\text{No. of Survivors at } t \times \text{Interval Width}} = \frac{n(t) - n(t+\Delta t)}{n(t)\Delta t}, \qquad t \geqslant 0. \tag{4.65}$$

Example 4.3. At a large government laboratory resistance measurements on various component parts are made with a certain product tester. The battery packs on such testers have been a continuing source of trouble and the following data are the failure times (in months) for a random sample of 50 battery packs.

Interval	Number of Failures
$0 \leqslant t < 3$	21
$3 \leqslant t < 6$	10
$6 \leqslant t < 9$	7
$9 \leqslant t < 12$	9
$12 \leqslant t < 15$	2
$15 \leqslant t < 18$	1
TOTAL	50

The following table contains the results for all six intervals

t	$t+\Delta t$	$n(t) - n(t+\Delta t)$	$\hat{R}(t)$	$\hat{f}(t)$	$\hat{h}(t)$
0	3	$50-29=21$	$\frac{50}{50}=1.00$	$\frac{21}{50(3)}=0.14$ f/month	$\frac{21}{50(3)}=0.14$ f/month
3	6	$29-19=10$	$\frac{29}{50}=0.58$	$\frac{10}{50(3)}=0.07$	$\frac{10}{29(3)}=0.11$
6	9	$19-12=7$	$\frac{19}{50}=0.38$	$\frac{7}{50(3)}=0.05$	$\frac{7}{19(3)}=0.12$
9	12	$12-3=9$	$\frac{12}{50}=0.24$	$\frac{9}{50(3)}=0.06$	$\frac{9}{12(3)}=0.25$
12	15	$3-1=2$	$\frac{3}{50}=0.06$	$\frac{2}{50(3)}=0.01$	$\frac{2}{3(3)}=0.22$
15	18	$1-0=1$	$\frac{1}{50}=0.02$	$\frac{1}{50(3)}=0.01$	$\frac{1}{1(3)}=0.33$
18	—	—	$\frac{0}{50}=0$	—	—

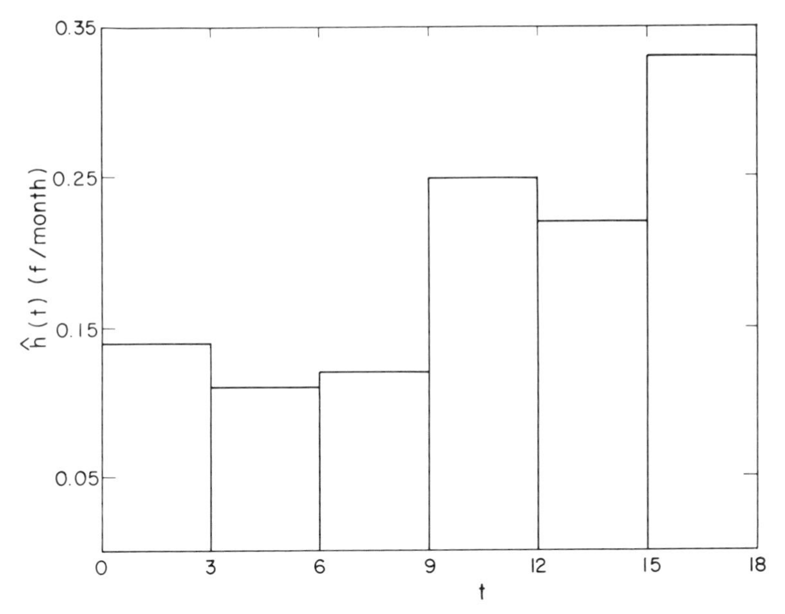

Figure 4.12. A plot of the hazard rate in Example 4.3.

The hazard rate is plotted in Figure 4.12. From this plot it appears that the $\mathcal{E}(\lambda)$ distribution would be a reasonable choice of model. In the plot of the hazard rate the large increase in the last interval is an inevitable result of the discretized calculation method. ■

4.3.2 Total Time On Test Plots

Barlow and Campo (1975) and Barlow (1978) discuss the use of total time on test plots for failure data analysis. The theory of total time on test plots and their properties are outlined in Barlow, Bartholomew, Bremner, and Brunk (1972). Bergman (1977) and Barlow and Davis (1977) discuss their use in determining optimal age replacement intervals.

Here we concentrate on the use of total time on test plots as a means for determining whether the data represent an increasing failure rate (IFR), decreasing failure rate (DFR), or constant failure rate over time. For convenience, we also restrict our discussion to the case where all n failure time observations are complete, that is, a failure time is observed for each observation and no observations are incomplete. An incomplete observation x_i is one in which it is known only that the device failed sometime *after* x_i

hours. Incomplete as well as grouped failure data may also be analyzed by use of this technique [Barlow and Campo (1975)].

If $n(\tau)$ units survive to time τ, then

$$\int_0^t n(\tau)\,d\tau$$

is the *total time on test* to age t. If units are observed to fail at (ordered) times

$$t_1 \leqslant t_2 \leqslant \cdots \leqslant t_n,$$

then

$$\mathfrak{t}(t_i) = \int_0^{t_i} n(\tau)\,d\tau = nt_1 + (n-1)(t_2 - t_1) + \cdots + (n-i+1)(t_i - t_{i-1}) \tag{4.66}$$

is the observed total time on test to time t_i. The ratio

$$0 \leqslant \frac{\mathfrak{t}(t_i)}{\mathfrak{t}(t_n)} \leqslant 1 \tag{4.67}$$

is called the (observed) *scaled total time on test* at time t_i. A *total time on test plot* is a plot of $\mathfrak{t}(t_i)/\mathfrak{t}(t_n)$ versus i/n for $i = 1, 2, \ldots, n$ and provides a great deal of information about the underlying hazard rate.

The corresponding total time on test transform [Barlow and Campo (1975)] of the $\mathcal{E}(\lambda)$ distribution plots as a straight line. Thus, the total time on test plot of $\mathcal{E}(\lambda)$ failure data is a 45° line. If $h(t)$ is *increasing* in t, then the total time on test plot is *concave* (i.e., above the 45° line); if $h(t)$ is *decreasing* in t, then the corresponding plot is *convex* (i.e., below the 45° line). Thus, the curvature of the total time on test plot gives an indication of the slope of the hazard rate and may suggest an appropriate distributional model.

There are several advantages of total time on test plots over other forms of plotting data such as hazard plots [Nelson (1969)] and probability plots which are introduced in the next section. Very incomplete sets of data can be analyzed and there is a sound theoretical basis for the analysis. All plots are graphs on the unit square and thus different distributions may be compared on the same scale. The plots are scale invariant and no special probability paper is needed, as in the case of probability plotting to be discussed. Finally, the interpretation of the plots is simple and straightforward.

Example 4.4. Let us again consider the heat exchanger failure data in Example 4.2. The necessary calculations for constructing a total time on test plot of these data are given in the following table.

i	t_i	$t_i - t_{i-1}$	$n-i+1$	$(n-i+1)(t_i - t_{i-1})$	$\mathfrak{t}(t_i)$	$\mathfrak{t}(t_i)/\mathfrak{t}(t_n)$
1	0.41	0.41	9	3.69	3.69	0.44
2	0.58	0.17	8	1.36	5.05	0.60
3	0.75	0.17	7	1.19	6.24	0.74
4	0.83	0.08	6	0.48	6.72	0.80
5	1.00	0.17	5	0.85	7.57	0.90
6	1.08	0.08	4	0.32	7.89	0.94
7	1.17	0.09	3	0.27	8.16	0.97
8	1.25	0.08	2	0.16	8.32	0.99
9	1.35	0.10	1	0.10	8.42	1.00

The total time on test plot is given in Figure 4.13. The concavity of the plot is an indication that the hazard rate is increasing over time, which supports the conclusion in Example 4.2. ■

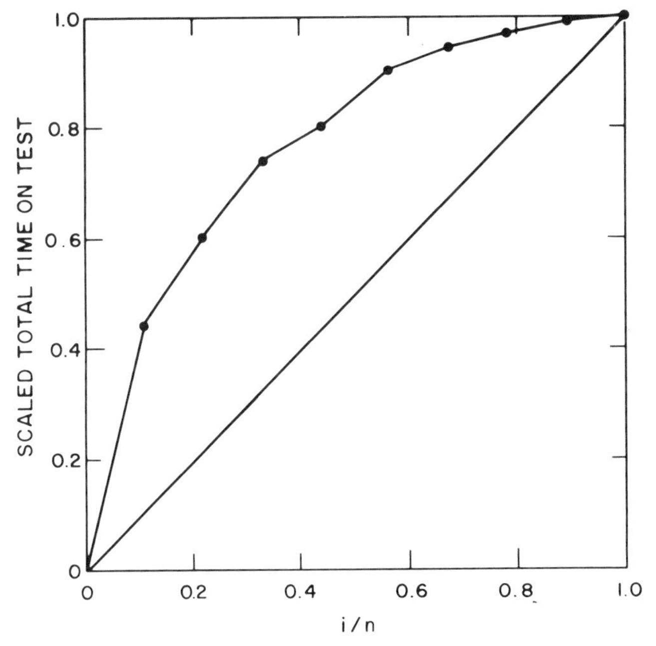

Figure 4.13. A scaled total time on test plot of the heat exchanger data in Example 4.4.

4.3.3 Probability Plotting

Probability plotting is a graphical technique whereby observed data and their corresponding empirical cumulative frequencies (or some transformation of these) are plotted on suitably constructed graph paper. Hahn and Shapiro (1967) and Ang and Tang (1975) are good references for the general engineering applications of probability plots, while Mann, Schafer, and Singpurwalla (1974) consider their specific use in failure data analysis.

Usually, a probability paper is constructed using a transformed probability scale in such a way that data from the underlying distribution plot as a straight line. Thus, the linearity, or lack thereof, of the plotted data can be used as a basis for determining the degree to which the distribution of the data conforms to that of the probability paper. If the plotted data are reasonably linear, the underlying parameter values in the model can also be obtained as functions of values of specified properties of the fitted line. Although there is probability paper for many distributions, we will consider only the use of $\mathcal{E}(\lambda)$, $\mathcal{W}(\alpha, \beta)$, $\mathcal{N}(\mu, \sigma^2)$, and $\mathcal{LN}(\xi, \sigma^2)$ probability plots.

Probability plots are thus often used as graphical goodness-of-fit tests, which were discussed in Chapter 3. Although statistical goodness-of-fit tests, such as the chi-square and Kolmogorov tests, are generally more powerful procedures, probability plots are visually important and are often used in the preliminary or exploratory phases of failure data analysis due to their simplicity.

Exponential Probability Plot. Consider a complete set of (ordered) failure data $t_1 \leqslant t_2 \cdots \leqslant t_n$. To determine how well the $\mathcal{E}(\lambda)$ distribution describes this data, plot the n values

$$\left\{t_i, \frac{n+0.25}{n-i+0.625}\right\}, \qquad i=1,2,\ldots,n, \tag{4.68}$$

on semi-log paper, in which the arithmetic scale is used for t_i and the logarithmic scale is used for $(n+0.25)/(n-i+0.625)$. If the data plot as a straight line, the $\mathcal{E}(\lambda)$ distribution is appropriate and the slope of the fitted line provides a graphical estimate of the failure rate λ. The graphical estimator of λ is a rather crude estimate, so analytic estimators are presented in Section 4.5.

It is also mentioned that there exist numerous more powerful statistical goodness-of-fit tests for testing $\mathcal{E}(\lambda)$ fit, and Mann, Schafer, and Singpurwalla (1974) refer to and discuss many of these tests.

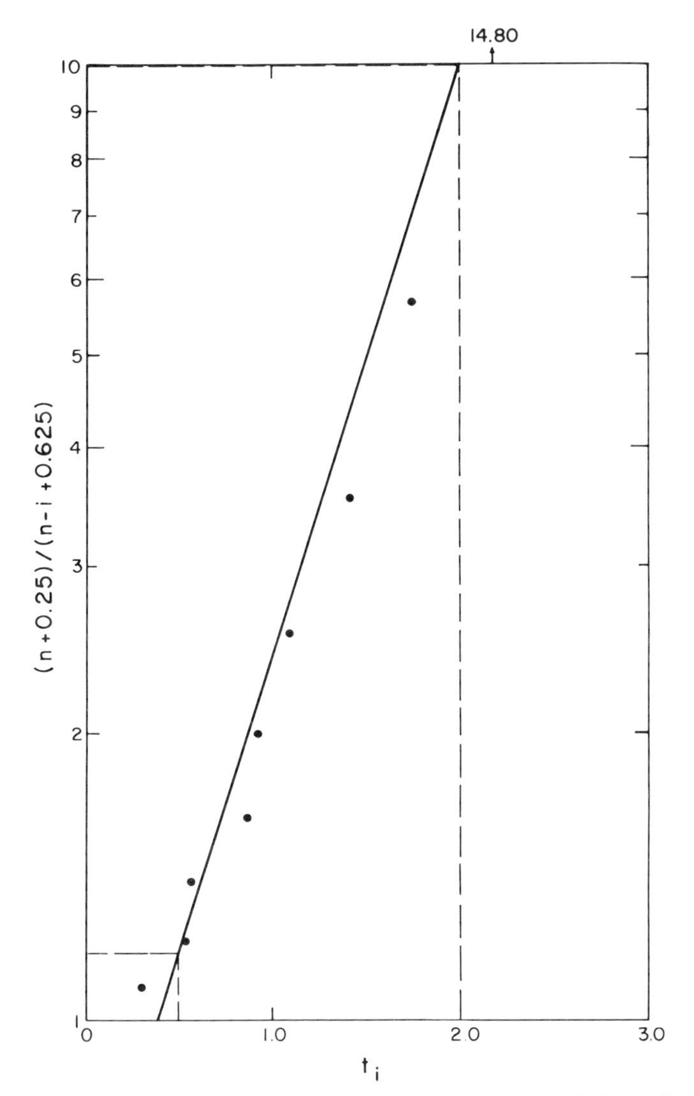

Figure 4.14. An exponential plot for the oscilloscope failure data in Example 4.5.

Example 4.5. Failure times of 0.30, 0.55, 0.56, 0.86, 0.93, 1.15, 1.42, 1.75, and 2.16 months were reported for an oscilloscope in use in the electronics shop of a large laboratory. It is desired to determine if the $\mathcal{E}(\lambda)$ distribution reasonably describes this data and, if so, to estimate the failure rate. Figure 4.14 shows an $\mathcal{E}(\lambda)$ plot of this data. The plotted values are quite linear and thus the $\mathcal{E}(\lambda)$ distribution adequately describes this data. The failure rate is graphically estimated to be

$$\hat{\lambda} = \frac{\ln(10.0) - \ln(1.18)}{2.0 - 0.5} = 1.4 \text{ f/month.} \qquad \blacksquare$$

Weibull Probability Plot. Special $\mathcal{W}(\alpha, \beta)$ probability paper is available[a] for use in constructing $\mathcal{W}(\alpha, \beta)$ probability plots. On the special $\mathcal{W}(\alpha, \beta)$ paper we plot

$$\left\{ t_i, \frac{i - 0.375}{n + 0.25}(100) \right\}, \qquad i = 1, 2, \ldots, n, \tag{4.69}$$

where t_i is plotted on the logarithmic scale and $100(i - 0.375)/(n + 0.25)$ is plotted on the axis labeled PERCENT FAILURE. Kapur and Lamberson (1977) discuss the differences in choice of plotting positions, and (4.69) is chosen on the basis of the discussion following (4.62). The degree to which the plotted data follow a straight line determines the degree of conformance of the data to a $\mathcal{W}(\alpha, \beta)$ distribution.

If the data reasonably plot as a straight line, the fitted line may be used in conjunction with appropriately transformed scales on the paper for graphically estimating the shape parameter β and scale parameter α. The estimate of β is given by the intersection on the extreme left-hand axis of a line drawn parallel to the fitted curve through the circle $\oplus$ labeled ORIGIN at the top of the paper. The estimate of α is given by the value along the abscissa of the intersection of a horizontal line drawn through 0.0 on the right-hand vertical axis and its intersection with the fitted line.

Example 4.6. The following failure times of 49, 73, 103, 140, 162, 164, 181, 196, 232, 248, 288, 290, 309, 377, 388, 464, and 500 h are suspected of having a $\mathcal{W}(\alpha, \beta)$ distribution. Is this a reasonable conjecture? If so, graphically estimate the shape and scale parameter values.

Figure 4.15 shows a $\mathcal{W}(\alpha, \beta)$ probability plot on paper similar to the commercial variety. The data follow a fairly straight line, and thus the conjecture appears to be a reasonable one. The graphical estimate of β is observed to be approximately 1.9 and the estimate of α is approximately 280 h. Again, the reader is reminded that such graphical estimators are generally inferior to those estimators having an analytic statistical foundation, such as

[a] Technical and Engineering Aids for Management (TEAM), Box 25, Tamworth, New Hampshire 03886.

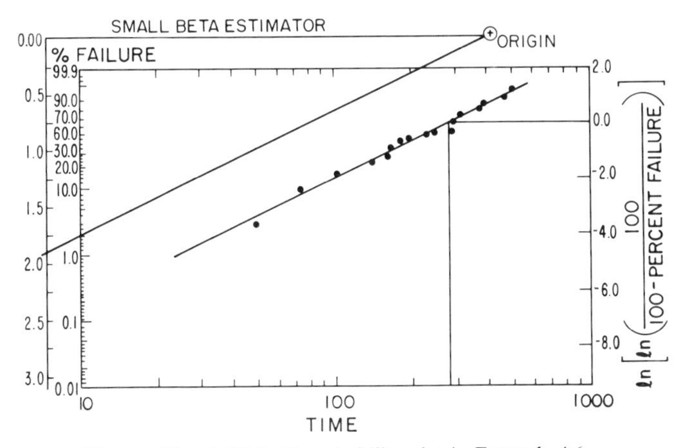

Figure 4.15. A Weibull probability plot in Example 4.6.

the class of MVU or ML estimators. Mann, Schafer, and Singpurwalla (1974) provide an excellent discussion of alternative $\mathcal{W}(\alpha, \beta)$ parameter estimation methods and results. ■

Normal Probability Plot. $\mathcal{N}(\mu, \sigma^2)$ probability paper is constructed on the basis of the standard normal cdf. One axis is an arithmetic scale, whereas the other shows the cumulative probabilities Φ, defined by (4.26), expressed as a percentage. Such papers are also available commercially.

Based on an ordered sample of n complete observations $t_1 \leqslant t_2 \leqslant \cdots \leqslant t_n$, the points

$$\left\{t_i, \frac{i-0.375}{n+0.25}(100)\right\}, \qquad i=1,2,\ldots,n, \tag{4.70}$$

are plotted on the $\mathcal{N}(\mu, \sigma^2)$ paper. Kimball (1960) shows that the plotting position $100(i-0.375)/(n+0.25)$, accredited to Blom (1958), is generally superior to either of the two more commonly used positions $100(i-0.5)/n$ or $100(i)/(n+1)$. A straight line fit indicates that the data are $\mathcal{N}(\mu, \sigma^2)$ distributed.

The mean μ and standard deviation σ of the underlying $\mathcal{N}(\mu, \sigma^2)$ population may also be graphically estimated. The value of t on this line corresponding to $\Phi(t)=50\%$ (i.e., the fiftieth percentile) is the graphical estimate of μ. The slope of the fitted line is the graphical estimate of σ; thus, $\hat{\sigma} \cong t_{0.84} - \hat{\mu}$, where $t_{0.84}$ is the value of t on this line such that $\Phi(t_{0.84})=84\%$ (i.e., the 84th percentile).

Example 4.7. The failure data for 22 bushing failures of a 115 kV power generator are given in the following table along with the corresponding plotting positions.

i	t_i (years)	$100(i-0.375)/(n+0.25)$	i	t_i (years)	$100(i-0.375)/(n+0.25)$
1	8.00	2.81	12	13.91	52.25
2	8.83	7.30	13	14.08	56.74
3	9.50	11.80	14	14.75	61.24
4	10.75	16.29	15	15.00	65.73
5	11.75	20.79	16	15.75	70.22
6	11.83	25.28	17	16.50	74.72
7	11.92	29.78	18	17.60	79.21
8	12.67	34.27	19	17.83	83.71
9	12.83	38.76	20	18.83	88.20
10	13.08	43.26	21	19.17	92.70
11	13.50	47.75	22	20.08	97.19

The data are plotted on $\mathfrak{N}(\mu, \sigma^2)$ probability paper in Figure 4.16. The

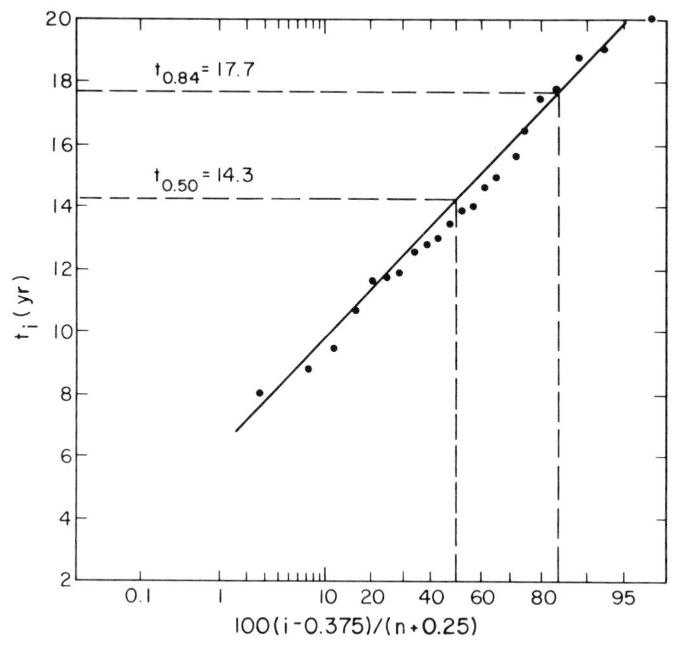

Figure 4.16. A normal probability plot of the bushing failure data in Example 4.7.

plotted data are fairly linear and thus the $\mathcal{N}(\mu, \sigma^2)$ distribution satisfactorily describes these data. The mean μ and standard deviation σ are graphically estimated to be $\hat{\mu} = 14.3$ years and

$$\hat{\sigma} = 17.7 - 14.3 = 3.4 \text{ years},$$

respectively. ■

Log Normal Probability Plot. Recall that, in the case of $\mathcal{LN}(\xi, \sigma^2)$ r.v., the logarithms of a set of data are $\mathcal{N}(\mu, \sigma^2)$ distributed. Thus, $\mathcal{LN}(\xi, \sigma^2)$ probability paper can be obtained from $\mathcal{N}(\mu, \sigma^2)$ probability paper by simply replacing the arithmetic scale with a logarithmic scale. The plotted points are the same as (4.70), where the t_i values are now plotted on a logarithmic scale.

In the case of a linear fit, the $\mathcal{LN}(\xi, \sigma^2)$ parameters ξ and σ are graphically estimated from the fitted line by

$$\hat{\xi} = \ln t_{0.50}, \qquad \hat{\sigma} = \ln\left(\frac{t_{0.84}}{t_{0.50}}\right) = \ln t_{0.84} - \hat{\xi}, \tag{4.71}$$

where $t_{0.50}$ and $t_{0.84}$ are the 50th and 84th percentile values as before.

Example 4.8. The WASH 1400 "Reactor Safety Study" (1975) reported the following distinct failure rate estimates for high quality pipes exceeding 3″ in diameter; namely, 1×10^{-10}, 2×10^{-10}, 2×10^{-9}, 3×10^{-9}, 6×10^{-9}, 1×10^{-8}, 7×10^{-8}, 1×10^{-6}, and 5×10^{-6} f/h. The estimates were reported by various failure data reporting sources. The failure mode was pipe rupture per section. Is the $\mathcal{LN}(\xi, \sigma^2)$ distribution an adequate model for these data? If so, what are the approximate values of ξ and σ?

The data are plotted on $\mathcal{LN}(\xi, \sigma^2)$ paper in Figure 4.17, and the fit is fairly linear. Consequently, the $\mathcal{LN}(\xi, \sigma^2)$ distribution seems to adequately describe this data. The two $\mathcal{LN}(\xi, \sigma^2)$ parameters ξ and σ are graphically estimated from the fitted line to be

$$\hat{\xi} = \ln(1.5\times10^{-8}) = -18.02,$$

and

$$\hat{\sigma} = \ln(7\times10^{-7}) + 18.02 = -14.17 + 18.02 = 3.85.$$ ■

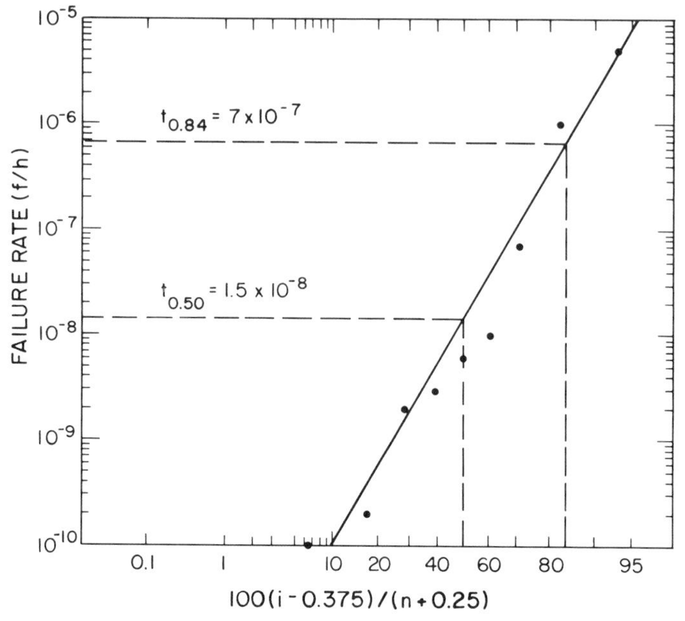

Figure 4.17. A log normal probability plot of the pipe failure rate data in Example 4.8.

4.4. LIFE TESTING

Most reliability estimation methods assume the existence of failure data obtained from *life testing*. Depending on the purpose of such testing, a random sample of n devices from a hypothesized population of such devices is placed on test under specified environmental conditions, and failure times of some or all of the units are observed. If each device that fails is immediately replaced by a new one, the resulting life test is called a *test with replacement*; otherwise, the life test is said to be a *test without replacement*. Frequently, the MTTF of the devices is so large that it is not practical, or economically feasible, to test each device to failure. In such cases, the life test is often *truncated* or *censored*. A *time-truncated life test* is one which is terminated after a fixed period of time has elapsed, whereas an *item-censored life test* is one which is terminated after a prespecified number of failures have occurred. Time-truncated life tests are often referred to as *Type I censored life tests*, whereas item-censored life tests are often called *Type II censored life tests*. In any of these life tests there may also be *withdrawals*, that is, some devices may have been withdrawn from the life test prior to failure in which case only the survival time is observed.

In order to induce failures of very-high-reliability devices, special testing methods known as *accelerated life tests* are often used. In an accelerated life test, the units are tested under environmental conditions that are far more severe than those normally encountered in practice. Either existing or proposed mathematical relationships are then used to extrapolate the observed results to the use conditions [Mann, Schafer, and Singpurwalla (1974), Chapter 9].

Life Test Experiments. Epstein (1958, 1960) considered several possible life test experiments, of which the most commonly used are:

1. Testing is terminated after a prespecified number of failures have occurred; failures are replaced (Type II/item-censored testing with replacement).
2. Testing is terminated after a prespecified number of failures have occurred; failures are not replaced (Type II/item-censored testing without replacement).
3. Testing is terminated after a prespecified time has elapsed; failures are replaced (Type I/time-truncated testing with replacement).
4. Testing is terminated after a prespecified time has elapsed; failures are not replaced (Type I/time-truncated testing without replacement).

In cases 1 and 2, the number of failures is a fixed constant and time is the r.v., whereas in cases 3 and 4 the opposite is true. It is also noted that case 3 frequently denotes the situation encountered when reliability data are collected and reported for field operational devices. In such cases the data are often reported for a certain period of time, such as one year, and the reporting period represents the test duration. Failures during the period are usually repaired or replaced as they occur.

As seen above, a life test can be terminated at a particular time, or after a particular number of failures occur, or, in fact, all items can be tested to failure. In test planning, it is important to remember that the accuracy of the resulting statistical estimation will be determined by the number of failures obtained. Also, the more items placed on test, the quicker one will obtain a preselected number of failures. However, one must balance the economic advantage of a shorter test duration with the economic disadvantage of placing more items on test.

Let $T_1 \leqslant T_2 \cdots \leqslant T_r$ denote the r.v.'s corresponding to the observed sample of r ordered failure times. If the number of failures is an r.v., it is denoted by R. Let n represent the number of items placed on test and let t_0

denote the test termination time. Let $\mathfrak{T}$ represent the total test time accumulated on all items including those that failed and those that did not fail prior to test termination. Corresponding to each of the four cases above we have

$$\begin{aligned}
&\textbf{1.}\quad \mathfrak{T} = nT_r,\ (r \text{ specified; } T_r \text{ random}),\\
&\textbf{2.}\quad \mathfrak{T} = \Sigma_{i=1}^{r} T_i + (n-r)T_r,\ r \leqslant n\ (r \text{ specified; } T_r \text{ random}),\\
&\textbf{3.}\quad \mathfrak{T} = nt_0,\ (t_0 \text{ specified; } R \text{ random}),\\
&\textbf{4.}\quad \mathfrak{T} = \Sigma_{i=1}^{R} T_i + (n-R)t_0,\ R \leqslant n\ (t_0 \text{ specified; } R \text{ random}).
\end{aligned} \tag{4.72}$$

In cases 1 and 3, n is often the number of test facilities available for the life test, in which case r may even exceed n.

4.5. THE EXPONENTIAL MODEL IN LIFE TESTING

The $\mathcal{E}(\lambda)$ distribution is the most widely used distribution in reliability. Consequently, many specialized reliability estimation methods assume an underlying $\mathcal{E}(\lambda)$ distribution. Let us now assume that each item in the population has an $\mathcal{E}(\lambda)$ failure time distribution with failure rate λ and that all failure times are statistically independent.

It is known that $\mathfrak{T}$ is sufficient for estimating the failure rate λ of an $\mathcal{E}(\lambda)$ distribution in cases 1 and 2 above, whereas $\mathfrak{T}$ and R are joint sufficient statistics for λ in case 4. In case 3, R is a single sufficient statistic for λ [Bain (1978)]. Thus, the estimation of λ, MTTF, reliability, and reliable life will be a function of the total time on test $\mathfrak{T}$ and/or R. This is a particularly important result in that only the total time on test and the number of failures are required, and *not* the actual failure times themselves, for reliability analyses based on the $\mathcal{E}(\lambda)$ model. Consequently, many reliability database systems such as NPRDS[b], GIDEP[c], and so on, report only the accumulated operating time and number of failures for successive periods of time for those systems or components monitored.

If n items are tested and the test is terminated at the time of the rth failure, the expected time to test completion is

$$E(T_r) = \frac{1}{\lambda} \sum_{i=1}^{r} \frac{1}{n-i+1}, \tag{4.73}$$

[b] Nuclear Plant Reliability Data System (NPRDS), The Edison Electric Institute, Southwest Research Institute, San Antonio, Texas 78284.

[c] Government-Industry Data Exchange Program (GIDEP), GIDEP Operations Center, Corona, California 91720.

and the variance is

$$\operatorname{Var}(T_r)=\frac{1}{\lambda^2}\sum_{i=1}^{r}\frac{1}{(n-i+1)^2}. \tag{4.74}$$

Thus, the savings in test time can be compared by examining the ratio $E(T_r, n>r)/E(T_r, n=r)$. For example, if 20 items are placed on test without replacement and the test is terminated at the eighth failure, the test will take only 18% as long, on the average, as one in which eight items are placed on test and all are allowed to fail.

4.5.1 Estimation in the One-Parameter Exponential Distribution

The development of the ML estimator for λ in the one parameter $\mathcal{E}(\lambda)$ distribution (4.10) is rather straightforward. It turns out that

$$\hat{\lambda}=\frac{r}{\mathcal{T}}, \tag{4.75}$$

is the ML estimator for λ for all cases considered above. Epstein (1960) shows that the quantity $2r\lambda/\hat{\lambda}=2\lambda\mathcal{T}$ has a $\chi^2(2r)$ distribution in the case of Type II/item-censored testing. In the case of Type I/time-truncated testing with replacement, the sufficient statistic R (the random number of failures) follows a Poisson distribution with parameter $(nt_0\lambda)$ [Bain (1978)] and the relationship between the Poisson and $\chi^2(n)$ distributions permits *approximate* inferences to be based on the $\chi^2(n)$ distribution. In case 4, Bartholomew (1963) gives the exact cumulative distribution of $1/\hat{\lambda}$ as a weighted sum of $\chi^2(n)$ integrals that, for the case where $R(t_0)\geqslant 0.5$, may be approximated by a $\mathcal{N}(\mu,\sigma^2)$ distribution. As the approximation is quite cumbersome, case 4 will not be considered further.

The ML estimators for the MTTF, reliability, and reliable life are

$$\widehat{\text{MTTF}}=\hat{\theta}=\frac{1}{\hat{\lambda}}=\frac{\mathcal{T}}{r}, \tag{4.76}$$

$$\hat{R}(t_0)=\exp(-\hat{\lambda}t_0)=\exp\left(-\frac{rt_0}{\mathcal{T}}\right), \tag{4.77}$$

and

$$\hat{t}_R=-\frac{\ln R}{\hat{\lambda}}=-\frac{\mathcal{T}\ln R}{r}. \tag{4.78}$$

Table 4.4. Confidence Limits on λ and MTTF

Limits	Type II/Item-Censored Testing (with or without replacement)	Type I/Time-Truncated Testing (with replacement)
$100(1-\gamma)\%$ TCI on λ:	$\left[\dfrac{\chi^2_{\gamma/2}(2r)}{2t};\dfrac{\chi^2_{1-\gamma/2}(2r)}{2t}\right]$	$\left[\dfrac{\chi^2_{\gamma/2}(2r)}{2t};\dfrac{\chi^2_{1-\gamma/2}(2r+2)}{2t}\right]$
$100(1-\gamma)\%$ UCI on λ:	$\left[0;\dfrac{\chi^2_{1-\gamma}(2r)}{2t}\right]$	$\left[0;\dfrac{\chi^2_{1-\gamma}(2r+2)}{2t}\right]$
$100(1-\gamma)\%$ TCI on MTTF:	$\left[\dfrac{2t}{\chi^2_{1-\gamma/2}(2r)};\dfrac{2t}{\chi^2_{\gamma/2}(2r)}\right]$	$\left[\dfrac{2t}{\chi^2_{1-\gamma/2}(2r+2)};\dfrac{2t}{\chi^2_{\gamma/2}(2r)}\right]$
$100(1-\gamma)\%$ LCI on MTTF:	$\left[\dfrac{2t}{\chi^2_{1-\gamma}(2r)};\infty\right]$	$\left[\dfrac{2t}{\chi^2_{1-\gamma}(2r+2)};\infty\right]$

Table 4.4 summarizes the $100(1-\gamma)\%$ confidence limits on λ and MTTF, where $\chi^2_\gamma(n)$ represents the $100(\gamma)$th percentile of a $\chi^2(n)$ distribution.

In the case of Type II/item-censored testing the limits are exact, although they are only approximate $100(1-\gamma)\%$ limits in the case of Type I/time-truncated testing with replacement.

Since $R(t_0)=\exp(-\lambda t_0)$ is a monotonically decreasing function of λ for a given t_0, the results in Table 4.4 may be used to construct $100(1-\gamma)\%$ confidence limits on $R(t_0)$. The $100(1-\gamma)\%$ TCI on $R(t_0)$ is given by

$$100(1-\gamma)\%\text{ TCI on } R(t_0):\quad \left[\exp(-\hat{\lambda}_U t_0);\exp(-\hat{\lambda}_L t_0)\right], \tag{4.79}$$

and a $100(1-\gamma)\%$ LCI becomes

$$100(1-\gamma)\%\text{ LCI on } R(t_0):\quad \left[\exp(-\hat{\lambda}_U t_0);1\right], \tag{4.80}$$

where the values $\hat{\lambda}_L$ and $\hat{\lambda}_U$ are the corresponding lower and upper confidence limits on λ given in Table 4.4.

Similarly, we can construct $100(1-\gamma)\%$ confidence limits on the reliable life t_r. Since $t_R=-\ln R/\lambda$ is also a monotonically decreasing function of λ for a given R, the $100(1-\gamma)\%$ TCI on t_R is

$$100(1-\gamma)\%\text{ TCI on } t_R:\quad \left[-\frac{\ln R}{\hat{\lambda}_U};-\frac{\ln R}{\hat{\lambda}_L}\right], \tag{4.81}$$

whereas the $100(1-\gamma)\%$ LCI is

$$100(1-\gamma)\%\text{ LCI on } t_R:\quad \left[-\frac{\ln R}{\hat{\lambda}_U};\infty\right]. \tag{4.82}$$

Example 4.9. The WASH 1400 "Reactor Safety Study" (1975) reported that in 1972 there were 50 instrument failures out of a total of 5613 such instruments in operation in commercial nuclear power reactors. We wish to estimate the failure rate, a 95% TCI on λ, $R(8760\text{ h})$, and a 95% TCI on $R(8760)$. Finally, we wish to estimate the reliable life for $R=0.80$ with both a point and a 95% TCI estimate.

Due to the way in which the data are reported, the testing scheme can be considered to be a Type I/time-truncated test with replacement. Thus, $t=5613(8760)=49{,}169{,}880$ h and $r=50$. Using (4.75) and the results presented in Table 4.1,

$$\hat{\lambda}=\frac{50}{49{,}169{,}880}=1.0\times 10^{-6}\text{ f/h},$$

$$95\%\text{ TCI on }\lambda:\quad \left[\frac{\chi^2_{0.025}(100)}{2t};\frac{\chi^2_{0.975}(100)}{2t}\right]=\left[\frac{74.22}{98{,}339{,}760};\frac{129.56}{98{,}339{,}760}\right]$$

$$=[7.5\times 10^{-7};1.3\times 10^{-6}].$$

Thus, $\hat{\lambda}_L=7.5\times 10^{-7}$ f/h and $\hat{\lambda}_U=1.3\times 10^{-6}$ f/h. Using (4.77) and (4.79),

$$\hat{R}(8760)=\exp[-1.0\times 10^{-6}(8760)]=0.991$$

and

95% TCI on $R(8760)$:

$$[\exp\{-1.3\times 10^{-6}(8760)\};\exp\{-7.5\times 10^{-7}(8760)\}]=[0.989;0.993].$$

Thus, $\hat{R}_L=0.989$ and $\hat{R}_U=0.993$.

Using (4.78) and (4.81),

$$\hat{t}_{0.80}=-\frac{\ln(0.80)}{1.0\times 10^{-6}}=223{,}144\text{ h}=25\text{ years}.$$

Thus, it is estimated that 80% of all such instruments will survive 25 years. Also 95% TCI on $t_{0.80}$: $\left[-\frac{\ln(0.80)}{1.3\times 10^{-6}};-\frac{\ln(0.80)}{7.5\times 10^{-7}}\right]=[171{,}649;\ 297{,}525]$. Hence, $\hat{t}_{0.80,L}=171{,}649$ h and $\hat{t}_{0.80,U}=297{,}525$ h. ■

4.5.2 Estimation in the Two-Parameter Exponential Distribution

There are a variety of estimators for use in estimating λ and μ in the $\mathcal{E}(\lambda,\mu)$ distribution given in (4.16) [Mann, Schafer, and Singpurwalla (1974)]. Some of the more common estimators are the ML estimators, the MVU estimators, and the best invariant estimators [Mann (1969)].

Type II/Item-Censored Testing. In the case of Type II/item-censored testing without replacement, the MVU estimators of λ and μ are

$$\hat{\lambda}' = \frac{r-1}{\sum_{i=1}^{r} T_i + (n-r)T_r - nT_1} \tag{4.83}$$

and

$$\hat{\mu} = T_1 - \frac{1}{\hat{\lambda}' n}, \tag{4.84}$$

where the prime on λ is used to distinguish it from the estimator in the $\mathcal{E}(\lambda)$ distribution.

The reliability function is estimated by

$$\hat{R}(t_0) = \exp\left[-\hat{\lambda}'(t_0 - \hat{\mu})\right]. \tag{4.85}$$

Now $100(1-\gamma)\%$ TCIs for λ and μ are

$$100(1-\gamma)\% \text{ TCI on } \lambda: \left[\frac{\hat{\lambda}'\chi^2_{\gamma/2}(2r-2)}{2r-2}; \frac{\hat{\lambda}'\chi^2_{1-\gamma/2}(2r-2)}{2r-2}\right], \tag{4.86}$$

and

$$100(1-\gamma)\% \text{ TCI on } \mu: \left[t_1 - \frac{\mathcal{F}_{1-\gamma}(2,2r-2)}{\hat{\lambda}' n}; t_1\right], \tag{4.87}$$

respectively, where $\mathcal{F}_{1-\gamma}(n_1,n_2)$ is the $100(1-\gamma)$th percentile of the $\mathcal{F}(n_1,n_2)$ distribution with n_1 numerator and n_2 denominator degrees of freedom. The pdf for the $\mathcal{F}(n_1,n_2)$ distribution was given in Section 3.5.2.

In the case of Type II/item-censored testing with replacement, the MVU estimators of λ and μ are

$$\hat{\lambda}' = \frac{r-1}{n(T_r - T_1)} \tag{4.88}$$

and

$$\hat{\mu} = T_1 - \frac{1}{\hat{\lambda}' n}, \tag{4.89}$$

respectively. Confidence interval estimators for λ and μ are again given by (4.86) and (4.87), in which (4.88) is used in lieu of (4.83).

Example 4.10. The following data represent the cycles to failure of a reciprocating pump when tested under conditions similar to those encountered in practice. Fifteen pumps were tested until the time of the eighth failure. The observed cycles to failure were 237,217; 346,704; 430,057; 447,737; 507,746; 575,877; 1,399,019; and 1,502,329. We wish to estimate both λ and μ, as well as the reliability for 1M cycles, and to compute 95% TCIs on λ and μ. Using (4.83), (4.84), and (4.85)

$$\hat{\lambda}' = \frac{7}{5{,}446{,}686 + 7(1{,}502{,}329) - 15(237{,}217)} = 5.643 \times 10^{-7} \text{ f/cycle},$$

$$\hat{\mu} = 237{,}217 - \frac{1}{(5.643 \times 10^{-7})15} = 119{,}000 \text{ cycles},$$

$$\hat{R}(1.0 \times 10^{6}) = \exp\left[-5.643 \times 10^{-7}(1.0 \times 10^{6} - 119{,}000)\right] = 0.61.$$

From (4.86) and (4.87) we have

$$95\% \text{ TCI on } \lambda: \left[\frac{(5.643 \times 10^{-7})\chi^2_{0.025}(14)}{14}; \frac{(5.643 \times 10^{-7})\chi^2_{0.975}(14)}{14}\right]$$

$$= \left[\frac{(5.643 \times 10^{-7})(5.63)}{14}; \frac{(5.643 \times 10^{-7})(26.12)}{14}\right]$$

$$= [2.27 \times 10^{-7};\ 1.05 \times 10^{-6}].$$

Thus, $\hat{\lambda}'_L = 2.27 \times 10^{-7}$ f/cycle and $\hat{\lambda}'_U = 1.05 \times 10^{-6}$ f/cycle.

$$95\% \text{ TCI on } \mu: \left[237{,}217 - \frac{\mathcal{F}_{0.95}(2, 14)}{(5.643 \times 10^{-7})(15)}; 237{,}217\right]$$

$$= \left[237{,}217 - \frac{3.74}{(5.643 \times 10^{-7})(15)}; 237{,}217\right].$$

In this case, since the lower limit is negative, we must modify the limits to

$$95\% \text{ TCI on } \mu: \ [0; 237{,}217].$$

Since the interval includes zero, it is not a particularly informative interval. ■

Type I/Time-Truncated Testing. For Type I/time-truncated testing without replacement, the ML estimators of λ and μ are

$$\hat{\lambda}' = \frac{R}{\sum_{i=1}^{R} T_i + (n-R)t_0 - nT_1}, \qquad R \neq 0, \tag{4.90}$$

and

$$\hat{\mu} = T_1, \tag{4.91}$$

respectively.

In the case of Type I/time-truncated testing with replacement, the ML estimators of λ and μ are

$$\hat{\lambda}' = \frac{R}{n(t_0 - T_1)}, \qquad R \neq 0, \tag{4.92}$$

and

$$\hat{\mu} = T_1, \tag{4.93}$$

respectively. In either case above one cannot obtain confidence intervals for μ that are invariant under translations of the data.

4.5.3 Hypothesis Testing

Hypothesis testing was introduced in Chapter 3. Here we are concerned with hypotheses associated with the $\mathcal{E}(\lambda, \mu)$ distribution. Recall that γ is the probability of rejecting the null hypothesis H_0 when H_0 is true, which is known as the level of significance (or size) of the test. Thus $1-\gamma$ is the level of confidence of the test.

As before, let $T_1 \leqslant T_2 \leqslant \cdots \leqslant T_r$ be a sequence of r i.i.d $\mathcal{E}(\lambda, \mu)$ r.v.'s. Let $\theta = 1/\lambda$ denote the MTTF of an $\mathcal{E}(\lambda)$ distribution.

Table 4.5. Hypotheses Concerning the Exponential Distribution

Hypothesis Number	Null Hypothesis H_0	Alternative Hypothesis H_1
1	$\mu = 0$	$\mu > 0$
2*	$\theta \geqslant \theta_0, \theta_0$ specified	$\theta < \theta_0, \theta_0$ specified
3**	$\theta \leqslant \theta_0, \theta_0$ specified	$\theta > \theta_0, \theta_0$ specified
4***	$\theta = \theta_0, \theta_0$ specified	$\theta \neq \theta_0, \theta_0$ specified
5	$\lambda_1 = \lambda_2$	$\lambda_1 \neq \lambda_2$

* Equivalent to testing H_0: $\lambda \leqslant \lambda_0$, λ_0 specified, versus H_1: $\lambda > \lambda_0$, where $\lambda_0 = 1/\theta_0$.
** Equivalent to testing H_0: $\lambda \geqslant \lambda_0$, λ_0 specified, versus H_1: $\lambda < \lambda_0$, where $\lambda_0 = 1/\theta_0$.
*** Equivalent to testing H_0: $\lambda = \lambda_0$, λ_0 specified, versus H_1: $\lambda \neq \lambda_0$, where $\lambda_0 = 1/\theta_0$.

Table 4.5 presents a variety of hypotheses concerning the $\mathcal{E}(\lambda, \mu)$ distribution. The first hypothesis concerns the threshold parameter μ and is useful for determining if $\mathcal{E}(\lambda, \mu)$ should be used instead of $\mathcal{E}(\lambda)$. The second, third, and fourth hypotheses concern the value of the MTTF in the $\mathcal{E}(\lambda)$ distribution. The fifth hypothesis concerns whether the underlying failure rate of two $\mathcal{E}(\lambda, \mu)$ populations has the same value. Such an hypothesis is often used to compare two designs, two processing methods, two vendors, and so on.

Table 4.6 summarizes the hypothesis test procedure for testing each of the five hypotheses in Table 4.5 at the γ level of significance. Thus the tests are all of size γ. The expressions in the column headed Calculate are called the *test statistics* and the regions identified in the last column are called the

Table 4.6. Hypothesis Test Procedures for the Exponential Distribution

Hypothesis Number	Calculate	Reject H_0 if
1*	$\mathcal{F}_0 = nt_1\hat{\lambda}'$, where $\hat{\lambda}'$ is given by (4.83)	$\mathcal{F}_0 > \mathcal{F}_{1-\gamma}(2, 2r-2)$
2**	$\chi_0^2 = 2t/\theta_0$	$\chi_0^2 < \chi_\gamma^2(2r)$
3**		$\chi_0^2 > \chi_{1-\gamma}^2(2r)$
4**		$\chi_0^2 > \chi_{1-\gamma/2}^2(2r)$ or $\chi_0^2 < \chi_{\gamma/2}^2(2r)$
5*	$\mathcal{F}_0 = \hat{\lambda}'_1/\hat{\lambda}'_2$, where $\hat{\lambda}'$ is given by (4.83) and the subscript denotes the population to which the sample refers	$\mathcal{F}_0 > \mathcal{F}_{1-\gamma/2}(2r_2-2, 2r_1-2)$ or $\mathcal{F}_0 < 1/\mathcal{F}_{1-\gamma/2}(2r_1-2, 2r_2-2)$

* Assumes Type II/item-censored testing without replacement.
** Assumes Type II/item-censored testing either with or without replacement.

rejection regions. The details surrounding the development of these tests may be found in such books as Kapur and Lamberson (1977), Mann, Schafer, and Singpurwalla (1974), and Bain (1978).

Example 4.11. In our calculations for Example 4.10 we found $\hat{\lambda}'=5.643\times 10^{-7}$ f/cycle and $\hat{\mu}=119{,}000$ cycles. Let us test the hypothesis $H_0:\mu=0$ at the five percent level of significance.

This situation is described by the first hypothesis in Table 4.5 and from Table 4.6 we have

$$\mathcal{F}_0=15(237{,}217)(5.643\times 10^{-7})=2.01.$$

The critical $\mathcal{F}(2,14)$ value is $\mathcal{F}_{0.95}(2,14)=3.74$. As $2.01<3.74$, we cannot conclude that the threshold parameter μ differs from zero. This conclusion was obvious from the confidence limits on μ in Example 4.10 and the equivalence between confidence limits and hypothesis testing should be recognized.

Example 4.12. In Example 4.9, suppose we wish to test the hypothesis

$$H_0: \lambda \geqslant 1.0\times 10^{-6}\text{f/h}$$
$$H_1: \lambda < 1.0\times 10^{-6}\text{f/h},$$

at the 10 percent level of significance.

This situation corresponds to the third hypothesis in Table 4.5. Thus,

$$\chi_0^2=2t\lambda_0=2(49{,}169{,}880)(1.0\times 10^{-6})=98.34.$$

The critical $\chi^2(100)$ value is $\chi^2_{0.90}(100)=118.50$. Since $98.34<118.50$ we cannot reject H_0.

Example 4.13. Martz (1975) presented failure data on the outages of two 115 kV transmission circuits from a common generating station in New Mexico. Nine outages, considered to be failures, were reported for circuit No. 1 and twelve failures were reported for circuit No. 2. The failure times, in mile-years, were recorded as

Circuit No. 1		Circuit No. 2	
3.06	58.98	1.08	27.66
4.35	96.20	4.72	30.07
7.88	222.50	5.27	37.60
20.76	345.86	6.05	41.24
53.16		6.35	101.77
		6.89	146.15

Assuming an $\mathcal{E}(\lambda, \mu)$ distribution for each population, it is desired to test the hypothesis

$$H_0: \lambda_1 = \lambda_2$$

$$H_1: \lambda_1 \neq \lambda_2,$$

at the five percent level of significance.

This situation corresponds to the fifth hypothesis in Table 4.5. From (4.83) we find that

$$\hat{\lambda}'_1 = \frac{8}{812.75 - 9(3.06)} = \frac{8}{785.21} = 0.010$$

and

$$\hat{\lambda}'_2 = \frac{11}{414.85 - 12(1.08)} = \frac{11}{401.89} = 0.027.$$

Thus, according to Table 4.6,

$$\mathcal{F}_0 = \frac{0.010}{0.027} = 0.370.$$

The critical $\mathcal{F}$ values are $\mathcal{F}_{0.975}(22,16) = 2.65$ and $1/\mathcal{F}_{0.975}(16,22) = 1/2.48 = 0.40$. Since $\mathcal{F}_0 < 0.40$, we reject H_0 and conclude that the two circuits have different failure (outage) rates at the five percent level of significance. ■

Statistical estimation and testing procedures for the $\mathcal{W}(\alpha, \beta, \theta)$, $\mathcal{G}(\alpha, \beta)$, extreme value, and other failure time distributions have also been worked out. Bain (1978) and Mann, Schafer, and Singpurwalla (1974) give the results for these distributions and some others as well.

4.6. RELIABILITY DEMONSTRATION TESTING IN THE EXPONENTIAL DISTRIBUTION

In the preceding section, the hypothesis testing procedures all assumed that such things as the sample size n, the truncation time t_0, and the number of failures r were known prior to the hypothesis test. When n and r or t_0 must be determined in advance to satisfy certain criteria (risks), an added degree of difficulty is introduced in the general hypothesis testing situation. When the question of how much data to collect is addressed in reliability, the hypothesis test is often called a *reliability demonstration test*. The design of life tests that carry a probabilistic guarantee that certain risks will not be exceeded is the subject of this section. Once n and r or t_0 have been determined, the data are collected, and the corresponding hypothesis is then tested by computing the value of an appropriate test statistic and determining if it falls in the proper rejection region.

Typically, such demonstration tests arise when it is required to demonstrate that the MTTF of a certain device exceeds a contract specified MTTF with a specified confidence coefficient. A demonstration life test must be designed that satisfies the requirement. If the test is subsequently passed, it can be claimed that the requirement has been met.

Recall from Chapter 3, that when testing a simple null hypothesis $H_0: \theta = \theta_0$ versus a simple alternative hypothesis $H_1: \theta = \theta_1$,

$$\gamma = \Pr(\text{Rejecting } H_0 \text{ when } H_0 \text{ is true})$$

and

$$\delta = \Pr(\text{Accepting } H_0 \text{ when } H_1 \text{ is true}).$$

Recall that the quantity $1 - \delta$ is called the *power* of the test. In reliability demonstration testing, γ and δ have different names: γ is called the *producer's risk* and δ is called the *consumer's risk*, names borrowed from similar conventions in quality control applications.

In this section we shall consider only tests for the MTTF (or, equivalently, for the failure rate λ) of the $\mathcal{E}(\lambda)$ distribution. However, the same procedures can also be used to test hypotheses about the reliability and reliable life of an $\mathcal{E}(\lambda)$ distribution with only minor modifications. Section 4.6.1 considers the case where only the survival or nonsurvival of each test unit for the test duration is observed. The case where the actual failure times are observed is considered in Section 4.6.2. Corresponding Bayesian reliability demonstration testing procedures are presented in Chapter 10.

4.6.1 Tests Based on Survival-Nonsurvival

When testing the MTTF of an $\mathcal{E}(\lambda)$ distribution, it is often economical to observe only the survival or nonsurvival of each test unit for a prespecified test duration t_0. This avoids the necessity of monitoring individual failure times. However, tests based on the observed failure times, which will be discussed in Section 4.6.2, are superior in the sense that the sample size required will be smaller for fixed (γ, δ). Thus, the opportunity exists for comparing the test convenience on the one hand and the test setup costs on the other.

Suppose that we wish to test the hypothesis

$$H_0: \theta = \theta_0$$
$$H_1: \theta = \theta_1, \theta_1 < \theta_0,$$

where θ_0, the *specified MTTF*, and θ_1, the *minimum acceptable MTTF*, are both specified.

The test procedure is to place each of the n items on life test for a prespecified length of time t_0, the test duration. Let the r.v. X represent the total number of failures in the test. Thus, X has a binomial distribution [Section 2.4.2] with parameter $p = 1 - \exp(-t_0/\theta)$, and the situation is known as *binomial sampling* [see Section 7.1.1]. The procedure for deciding between H_0 and H_1 is as follows:

$$\text{If } x \leqslant c, \text{ accept } H_0$$
$$\text{If } x > c, \text{ reject } H_0, \tag{4.94}$$

where c is known as the *acceptance number* in the test. It is desired to determine the pair of values (n, c) *before* the test is conducted that satisfy the two risk criteria given by

$$\gamma = \sum_{x=c+1}^{n} \binom{n}{x} p_0^x (1 - p_0)^{n-x} \tag{4.95}$$

and

$$\delta = \sum_{x=0}^{c} \binom{n}{x} p_1^x (1 - p_1)^{n-x}, \tag{4.96}$$

where γ and δ are specified and

$$p_0 = 1 - e^{-t_0/\theta_0}, \qquad p_1 = 1 - e^{-t_0/\theta_1}. \tag{4.97}$$

Thus,

$$\gamma = \Pr(\text{Rejecting } H_0 \text{ when } \theta = \theta_0)$$

and

$$\delta = \Pr(\text{Accepting } H_0 \text{ when } \theta = \theta_1)$$

are fixed.

The desired values of (n, c) can be found from tables of the binomial distribution or by use of a binomial nomograph developed by Larson (1966) and reproduced in Table B6 in Appendix B. The simple nomograph procedure is as follows:

1. Compute the values of p_0 and p_1.
2. Connect p_0 on the left-hand scale to $(1-\gamma)$ on the right-hand scale with a straight line.
3. Connect p_1 on the left-hand scale to δ on the right-hand scale with a straight line.
4. Read (n, c) at the intersection of the two straight lines, where (n, c) are appropriately rounded to nearest integers.

It is also noted that as t_0 increases, p increases and thus n tends to decrease, indicative of the tradeoff between test duration and sample size.

Example 4.14. Suppose that it is desired to test

$$H_0: \theta = \theta_0 = 1000 \text{ h}$$

$$H_1: \theta = \theta_1 = 500 \text{ h},$$

with $\gamma = \delta = 0.10$ and $t_0 = 100$ h. Then

$$p_0 = 1 - e^{-100/1000} = 0.095$$

$$p_1 = 1 - e^{-100/500} = 0.18.$$

Connecting the points (0.095,0.90) on the nomograph in Appendix B with a straight line and the points (0.18,0.10) with a straight line, yields the desired values ($n = 110, c = 14$) at the intersection. Thus, 110 items would each be tested for 100 h. If more than 14 failures occurred, H_0 would be rejected; otherwise, H_0 would be accepted. ■

Tests can also be developed based on a fixed number of failures, known as Pascal sampling [see Section 7.1.2], in which case the sample size is an r.v., or a sequential testing strategy. The details may be found in Mann, Schafer, and Singpurwalla (1974).

4.6.2 Tests Based on Observed Failure Times

In the preceding section, the only information available from the test was that each test unit either survived or failed the duration of the test. Now let us assume that n identical $\mathcal{E}(\lambda)$ distributed items are tested and the actual lifetime of each unit that fails during the test duration is recorded. All four types of the life tests introduced in Section 4.4 will be considered, and selection of the type of test is largely an engineering decision based on costs, time available for the test, and other considerations.

Let $T_1 \leqslant T_2 \leqslant \cdots \leqslant T_r$ denote the ordered failure times in a sample of size n. As before, we shall denote the observed number of failures by r, the test termination time (in the case of a Type I/time-truncated test) by t_0, and the total time on test by t. Epstein and Sobel (1953) and Epstein (1954, 1960) are credited with most of the results of this section.

In all four cases, we are interested in testing the simple hypothesis

$$H_0: \theta = \theta_0$$

versus

$$H_1: \theta = \theta_1, \quad \theta_1 < \theta_0,$$

where θ_0, θ_1 and

$$\gamma = \Pr(\text{Rejecting } H_0 \text{ when } \theta = \theta_0)$$

and

$$\delta = \Pr(\text{Accepting } H_0 \text{ when } \theta = \theta_1)$$

are specified. The ratio θ_1/θ_0 is called the *discrimination ratio*.

Table 4.7 describes the necessary criteria for designing and implementing each of the four life testing procedures. Table 4.8 summarizes the procedure for determining the necessary parameters for each life test described in Table 4.7. The expression t_c in Tables 4.7 and 4.8 is used to denote the critical value of the total time on test t, which is the criterion value for determining whether the test has been passed.

Two remarks are in order. First, the value of n does not in any way influence the values of r and t_c for the Type II tests in Table 4.8. However, n does affect the expected length of time to observe r failures, according to (4.73), and consequently the test duration is influenced by the choice of n; namely, the larger the value of n, the smaller the expected test duration. Secondly, all tests for which (γ, δ) and θ_1/θ_0 are the same yield the same r value. Thus, it is necessary to index plans for determining r only on the consumer and producer risk and the discrimination ratio.

The following examples illustrate the use of these test procedures.

Table 4.7. Design and Implementation of Exponential Life Tests Based on Observed Failure Time Data

Type of Test	Test Procedure	Test Parameters: Specified	Test Parameters: To Be Determined	Criterion
Type I/time-truncated with replacement	Place n items on life test with replacement until termination at $\min(t_r, t_0)$	$(\gamma, \delta, \theta_0, \theta_1)$; t_0	(r, n)	If $t_r \geqslant t_0$, accept H_0; If $t_r < t_0$, reject H_0 (applies to both Type I tests)
Type I/time-truncated without replacement	Place n items on life test without replacement until termination at $\min(t_r, t_0)$	$(\gamma, \delta, \theta_0, \theta_1)$; t_0	(r, n)	
Type II/item-censored with replacement	Place n items on life test with replacement until termination at $\min(t_r, \mathrm{t}_c/n)$	$(\gamma, \delta, \theta_0, \theta_1,)$; n	(r, t_c)	If $t_r \geqslant \mathrm{t}_c/n$, accept H_0; If $t_r < \mathrm{t}_c/n$, reject H_0
Type II/item-censored without replacement	Place n items on life test without replacement until termination at time t_r	$(\gamma, \delta, \theta_0, \theta_1)$; n	(r, t_c)	If $\mathrm{t} \geqslant \mathrm{t}_c$, accept H_0; If $\mathrm{t} < \mathrm{t}_c$, reject H_0, where $\mathrm{t} = \sum_{i=1}^{r} t_i + (n-r)t_r$

Table 4.8. Determination of Life Test Parameters

Type of Test	Test Parameter to be Determined	Procedure	Test Parameter to be Determined	Procedure
Type I/time-truncated with replacement	r	Smallest r such that $\frac{\chi^2_{\gamma}(2r)}{\chi^2_{1-\delta}(2r)} = \frac{\theta_1}{\theta_0}$	n	$n^* = \left[\!\left[\frac{\theta_0 \chi^2_{\gamma}(2r)}{2t_0} \right]\!\right]$
Type I/time-truncated without replacement	r	(same as above)	n	Hold γ fixed and determine n from the cumulative binomial nomograph [see Example 4.16]*
Type II/item-censored with replacement	r	(same as above)	t_c	$t_c^* = \frac{\theta_0 \chi^2_{\gamma}(2r)}{2}$
Type II/item-censored without replacement	r	(same as above)	t_c	(same as above)

*This procedure guarantees γ exactly, whereas δ may be approximate as a result of the necessity of considering only integer values of r, the number of failures at the time of truncation.

Example 4.15. Suppose that it is desired to test the hypothesis

$$H_0: \theta = 500 \text{ h}$$

$$H_1: \theta = 200 \text{ h}$$

by means of a Type I/time-truncated life test with replacement in which $\gamma = \delta = 10$ percent. Determine the required demonstration test plan and discuss its implementation. Now $\theta_1/\theta_0 = 200 \text{ h}/500 \text{ h} = 0.40$. From Table B3 in Appendix B, one obtains

$$\frac{\chi^2_{0.10}(16)}{\chi^2_{0.90}(16)} = \frac{9.31}{23.5} = 0.396 < 0.40,$$

$$\frac{\chi^2_{0.10}(18)}{\chi^2_{0.90}(18)} = \frac{10.9}{26.0} = 0.419 > 0.40.$$

Since $2r = 16$ is too small, and since $2r = 17$ results in a noninteger value of r, we take $2r = 18$, or $r = 9$. According to the procedure in Table 4.8,

$$n = \left[\!\left[\frac{500(10.9)}{2t_0} \right]\!\right] = \left[\!\left[\frac{2725}{t_0} \right]\!\right].$$

Thus, if $t_0 = 100$ h, $n = [\![27.25]\!] = 27$; if $t_0 = 500$ h, $n = [\![5.45]\!] = 5$; and so on.

For example, if $t_0 = 100$ h, then 27 items would initially be placed on life test with failures to be replaced as they occur. If the time at which the ninth failure occurs is less than 100 h from the start of the test, the test is then terminated and H_0 is rejected. The probability that H_0 has been falsely rejected is 10 percent. On the other hand, if the ninth failure has not occurred prior to 100 h from the start of the test, the test is also terminated and H_0 is accepted. The probability that H_0 has been falsely accepted is also approximately 10 percent.

Example 4.16. Again let us consider Example 4.15 in which a Type I/time-truncated test without replacement is desired.

As in Example 4.15, the same procedure for determining r yields $r = 9$. To determine n from the cumulative binomial nomograph in Appendix B, we connect $p_0 = 1 - \exp(-t_0/\theta_0)$ on the left-hand scale to $(1-\gamma)$ on the right-hand scale with a straight line. The desired value of n is then read at the intersection of this line and the line corresponding to $c = r$. Suppose we choose $t_0 = 100$ h. Thus, by connecting the points $p_0 = 1 - \exp(-100/500) = 0.18$ and $1-\gamma = 0.90$ with a straight line, the intersection of this line and the line labeled $c = 9$ occurs at approximately $n = 37$. As testing is without replacement, 10 more units are required to be initially placed on life test than for the case in which failed units are to be replaced as they occur.

Example 4.17. Again let us consider Example 4.15 in which a Type II/item-censored test with replacement is desired.

As before, $r = 9$. Now

$$t_c = \frac{500(10.9)}{2} = 2725 \text{ h}.$$

Thus, if $n = 25$, then 25 items would be placed on life test with replacement. The test would be passed if no more than eight failures occurred prior to $2725/25 = 109$ h after the start of the test; otherwise, the test would be failed and terminated at the time of the ninth failure. The expected test time to the occurrence of the rth failure is given by

$$E(T_r) = \frac{r\theta}{n}$$

[Mann, Schafer, and Singpurwalla (1974)]. In this case, if H_0 is true, the expected test time is

$$E(T_9) = \frac{9(500)}{25} = 180 \text{ h},$$

and the test most likely would be terminated in acceptance of H_0 at 109 h. On the other hand, if H_1 is true, $E(T_9) = 9(200)/25 = 72$ h and the test would likely terminate in rejection of H_0, on the average, 72 h after the start of the test. ■

Similar reliability demonstration test procedures for the $\mathfrak{W}(\alpha, \beta)$ distribution have been considered by Fertig and Mann (1980).

4.7. SYSTEM RELIABILITY

One of the most important uses of reliability concerns the analysis of the reliability of a system. A system is defined here as a given configuration of subsystems and/or components whose proper functioning over a stated interval of time determines whether the system will perform as designed.

A functional block diagram for a system is frequently used to represent the effect of subsystem failure on system performance. A subsystem will be taken to mean either a particular low-level grouping of components or a single component, depending on the manner in which the system is subdivided. The particular structure of a block diagram for a given system will depend on the definition of reliability for that system. The block diagram is constructed to determine operational success and does not necessarily show physical subsystem configuration. For example, consider a simple system of two diodes, physically arranged in a parallel configuration. If either diode short circuits, the system will fail; thus, the operational success of the system regarding the short circuit failure mode is a series configuration. On the other hand, both diodes must fail open for the system to be unsuccessful regarding the open failure mode; thus, the system configuration is a parallel arrangement for this failure mode.

The following notation will be used:

T_i = failure time of the ith subsystem,
T = failure time of the system,
$f_i(t)$ = failure time pdf of the ith subsystem,
$f_s(t)$ = failure time pdf of the system,
$F_i(t)$ = failure time cdf of the ith subsystem,
$F_s(t)$ = failure time cdf of the system,
$R_i(t)$ = reliability of the ith subsystem,
$R_s(t)$ = reliability of the system,
$h_i(t)$ = hazard rate of the ith subsystem,
$h_s(t)$ = hazard rate of the system,
μ_i = MTTF of the ith subsystem,
μ_s = MTTF of the system,
t = time.

We also limit our consideration to the case where all subsystems function statistically independently. Recall from Chapter 2 that r.v.'s T_1 and T_2 are said to be statistically independent if $f(t_1|t_2) = f(t_1)$. In other words, if knowledge of the failure time of one subsystem does not affect the statistical distribution of the failure time of the remaining subsystem, then the two subsystems are said to be statistically independent. If there are more than two subsystems present, the mutual statistical independence of all possible pairs of subsystems can be used to define statistical independence of all the subsystems. Barlow and Proschan (1975) consider the case where the subsystems are *not* statistically independent.

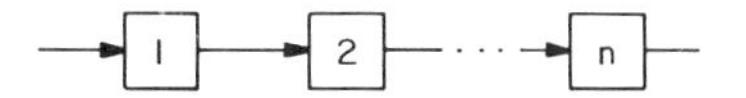

Figure 4.18. A series system block diagram.

4.7.1 Series System

The series configuration is the most commonly encountered situation, and is also the simplest to analyze. A functional block diagram for a simple series system is shown in Figure 4.18. The reliability model is thus

$$\begin{aligned} R_s(t) &= \Pr(T>t) = \Pr(T_1>t)\cdot\Pr(T_2>t)\cdots\Pr(T_n>t) \\ &= \prod_{i=1}^{n} R_i(t), \end{aligned} \tag{4.98}$$

which is known as the *product rule* in reliability. It is observed that the system reliability will always be less than or equal to the least reliable component. Thus

$$R_s(t) \leq \min_i \{R_i(t)\}. \tag{4.99}$$

If equality holds in (4.99), we have what is known as the *chain model* or *weakest link model* [Kapur and Lamberson (1977)].

The hazard rate for a series system is also a convenient expression. Since $h(t) = -d\ln R(t)/dt$, we have from (4.98) that

$$\begin{aligned} h_s(t) &= \frac{-d\ln R_s(t)}{dt} = \frac{-d\ln \prod_{i=1}^{n} R_i(t)}{dt} \\ &= \sum_{i=1}^{n} \frac{-d\ln R_i(t)}{dt} = \sum_{i=1}^{n} h_i(t), \end{aligned} \tag{4.100}$$

and thus the system hazard rate is the sum of the subsystem hazard rates. From (4.100) it is apparent that if each subsystem is IFR (DFR) then the system will also be IFR (DFR). Also, if each subsystem is IFRA (DFRA), then the system will also be IFRA (DFRA).

In the case of subsystems having $\mathcal{E}(\lambda_i)$ distributions, the system failure rate will be

$$\lambda_s = \sum_{i=1}^{n} \lambda_i, \tag{4.101}$$

the sum of the subsystem failure rates, which is also a constant. Also

$$R_s(t)=\prod_{i=1}^{n} e^{-\lambda_i t}=\exp\left(-t\sum_{i=1}^{n}\lambda_i\right) \tag{4.102}$$

and thus, the system failure time likewise has an $\mathcal{E}(\lambda_s)$ distribution. The system MTTF is, of course, given by

$$\mu_s=1\Big/\sum_{i=1}^{n}\lambda_i=1\Big/\sum_{i=1}^{n}(1/\mu_i). \tag{4.103}$$

Example 4.18. A space power reactor (SPAR) system,[d] consisting of a reactor, radiation shield, converter, radiator, and controls, is assumed to be functionally arranged in series, as any subsystem failure will result in overall system failure. Assuming an $\mathcal{E}(\lambda)$ model, preliminary failure rate estimates are thought to be

Subsystem	Failure Rate (f/h)
Reactor	3×10^{-7}
Shield	1×10^{-8}
Converter	2×10^{-7}
Radiator	2×10^{-7}
Controls	3×10^{-7}

The system failure time is $\mathcal{E}(\lambda_s)$ distributed with failure rate $\lambda_s=10.1\times10^{-7}$ f/h. The system reliability for a seven year mission is thus R_s (61,320)= $\exp[-7(8760)(10.1\times10^{-7})]=0.94$, and $\mu_s=990{,}099$ h. ■

4.7.2 Parallel System

A parallel system is shown in Figure 4.19. The system functions if k or more of the subsystems function ($k\leq n$). If $k=1$ the system is sometimes called a *purely parallel system*, whereas if $k>1$, the system is often called a *k-out-of-n system*.

[d] *Selection of Power Plant Elements for Future Reactor Space Electric Power Systems*, Los Alamos Scientific Laboratory Report LA-7858 (September 1979).

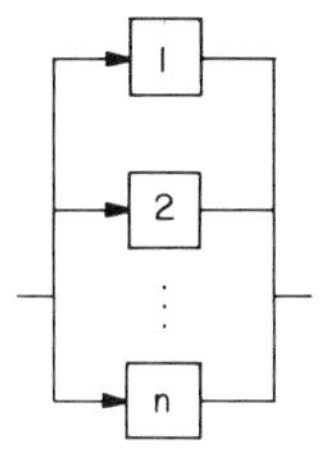

Figure 4.19. A parallel system block diagram.

First consider a purely parallel system ($k=1$). The system will fail if and only if each component fails. Thus,

$$\begin{aligned}\Pr(T \leqslant t) = F_s(t) &= \Pr(T_1 \leqslant t) \cdot \Pr(T_2 \leqslant t) \cdots \Pr(T_n \leqslant t), \\ &= \prod_{i=1}^{n} F_i(t) = \prod_{i=1}^{n} [1 - R_i(t)],\end{aligned}$$

or

$$R_s(t) = 1 - F_s(t) = 1 - \prod_{i=1}^{n} [1 - R_i(t)]. \tag{4.104}$$

The system hazard rate can be obtained by use of (4.7), but the result is fairly complicated. The system MTTF is also difficult to obtain in general.

In the case of subsystems having $\mathcal{E}(\lambda_i)$ distributions

$$R_s(t) = 1 - \prod_{i=1}^{n} (1 - e^{-\lambda_i t}), \tag{4.105}$$

and thus, T does *not* have an $\mathcal{E}(\lambda)$ distribution. It can be shown by use of the binomial expansion that

$$\begin{aligned}\mu_s = {} & \left(\frac{1}{\lambda_1} + \frac{1}{\lambda_2} + \cdots + \frac{1}{\lambda_n}\right) - \left(\frac{1}{\lambda_1 + \lambda_2} + \frac{1}{\lambda_1 + \lambda_3} + \cdots + \frac{1}{\lambda_i + \lambda_j}\right) \\ & + \left(\frac{1}{\lambda_1 + \lambda_2 + \lambda_3} + \cdots + \frac{1}{\lambda_i + \lambda_j + \lambda_k}\right) - \cdots + (-1)^{n+1} \frac{1}{\sum_{i=1}^{n} \lambda_i}.\end{aligned} \tag{4.106}$$

In the special case where all components are identical with failure rate λ (as in the case of an *active redundant system*), then

$$R_s(t)=1-(1-e^{-\lambda t})^n \tag{4.107}$$

and

$$\mu_s=\mu\left(1+\frac{1}{2}+\cdots+\frac{1}{n}\right), \tag{4.108}$$

where $\mu=1/\lambda$. Thus, if an active redundant system consists of two identical parallel subsystems rather than one, the MTTF exceeds that of the single subsystem by 50 percent, rather than doubling it. In general, a rather severe law of diminishing returns for increasing the number of parallel subsystems is apparent in (4.108).

Barlow and Proschan (1975) show that parallel systems of IFR (DFR) components need *not* be IFR (DFR); however, parallel systems of IFR components will be IFRA, whereas parallel systems of DFR components need *not* be DFRA. For example, the system hazard rate for two $\mathcal{E}(\lambda_i)$ subsystems in parallel is given by

$$h(t)=\frac{\lambda_1 e^{-\lambda_1 t}+\lambda_2 e^{-\lambda_2 t}-(\lambda_1+\lambda_2)e^{-(\lambda_1+\lambda_2)t}}{e^{-\lambda_1 t}+e^{-\lambda_2 t}-e^{-(\lambda_1+\lambda_2)t}}, \tag{4.109}$$

which is increasing on $[0, t_0)$ and decreasing on (t_0, ∞), where t_0 depends on the values of λ_1 and λ_2 [Barlow and Proschan (1975)]. Thus, although both subsystems are both IFR and DFR, the system hazard rate is neither IFR nor DFR.

Now let us consider a k-out-of-n system. A general formulation is inconvenient to work with, and is unnecessary in most cases. Thus, let us consider only the case in which all components are identically distributed with common reliability function $R_i(t)=R(t)$, $i=1,2,\ldots,n$. The system will function if *any* subset of k subsystems functions. The binomial distribution is appropriate so that

$$\begin{aligned} R_s(t) &= \sum_{x=k}^{n}\binom{n}{x}[R(t)]^x[1-R(t)]^{n-x} \\ &=1-\sum_{x=0}^{k-1}\binom{n}{x}[R(t)]^x[1-R(t)]^{n-x}. \end{aligned} \tag{4.110}$$

Example 4.19. Again let us consider the SPAR system described in Example 4.18. The reactor core consists of 90 fuel element paths in parallel, in which each element has a projected seven year mission reliability of 0.9799. If at least 86 fuel elements are required to produce the necessary heat for conversion to electricity, then the reliability of the reactor core is

$$R_s(61{,}320)=\sum_{x=86}^{90}\binom{90}{x}(0.9799)^x(0.0201)^{90-x}=0.965. \qquad \blacksquare$$

4.7.3 Standby Redundant Systems

A standby redundant system is a type of parallel system in which only a single subsystem is in operation at a time, the remaining subsystems being subsequently brought into operation upon failure of the operating subsystem. The spare tire on an automobile is a simple example of a standby system. A standby redundant configuration is depicted in Figure 4.20. The switch S may be either an electronic or mechanical piece of equipment, or it may be an operator whose task it is to replace a failed subsystem. The subsystem ordering $1,2,\ldots,n$ is assumed to be the order in which the subsystems are called into operation. Further, the reliability function for a standby system that has n subsystems will be denoted by $R_s^n(t)$.

In a standby system, the system reliability is extremely dependent on the reliability of the switch. A poor switch can yield a system which is worse than a single subsystem. Thus, let us examine both the perfect and imperfect switching cases.

Perfect Switching. The assumption is now made that the switch is completely failure free. First, consider the case of a two-subsystem system ($n=2$). Now

$$\begin{aligned} R_s^2(t)&=\Pr(T_1>t)+\Pr(T_1\leq t \text{ and } T_2>t-T_1)\\ &=R_1(t)+\int_0^t f_1(t_1)R_2(t-t_1)\,dt_1. \end{aligned} \qquad (4.111)$$

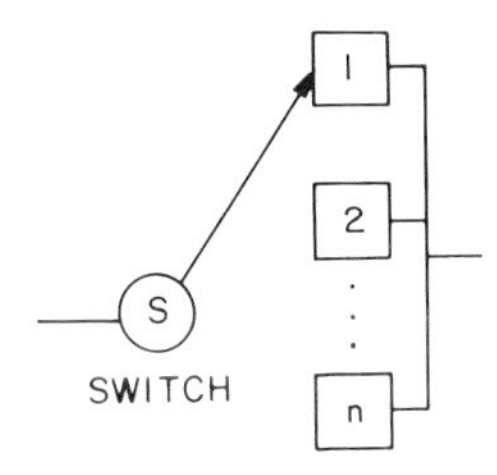

Figure 4.20. A standby system block diagram.

Thus, for specified subsystem pdf's, (4.111) gives the system reliability for a two-subsystem standby system.

Now consider the case of a three-subsystem standby system. We have

$$\begin{aligned}
R_s^3(t) &= \Pr(T_1 > t) + \Pr(T_1 \leqslant t \text{ and } T_2 > t - T_1) \\
&\quad + \Pr(T_1 < t \text{ and } T_2 < t - T_1 \text{ and } T_3 > t - T_1 - T_2) \\
&= R_1(t) + \int_0^t f_1(t_1) R_2(t - t_1)\, dt_1 \\
&\quad + \int_0^t f_1(t_1) \int_0^{t-t_1} f_2(t_2) R_3(t - t_1 - t_2)\, dt_2 dt_1. \qquad (4.112)
\end{aligned}$$

Thus,

$$R_s^3(t) = R_s^2(t) + \int_0^t f_1(t_1) \int_0^{t-t_1} f_2(t_2) R_3(t - t_1 - t_2)\, dt_2 dt_1.$$

Similarly, the reliability for a four-subsystem standby system is given by

$$\begin{aligned}
R_s^4(t) &= R_s^3(t) + \int_0^t f_1(t_1) \int_0^{t-t_1} f_2(t_2) \\
&\quad \times \int_0^{t-t_1-t_2} f_3(t_3) R_4(t - t_1 - t_2 - t_3)\, dt_3 dt_2 dt_1. \qquad (4.113)
\end{aligned}$$

In the case of perfect switching, a standby system possesses the interesting and useful property that $T = T_1 + T_2 + \cdots + T_n$. Thus, $f_s(t)$ is the *convolution* of $f_i(t)$, $i = 1, 2, \ldots, n$ [Wilks (1962)]. For example, if each subsystem has an identical $\mathcal{E}(\lambda)$ distribution, then $f_s(t)$ is a $\mathcal{G}(n, \lambda^{-1})$ distribution. Thus, in this special case,

$$\begin{aligned}
R_s^n(t) &= 1 - \int_0^t \frac{\lambda^n}{\Gamma(n)} x^{n-1} e^{-\lambda x}\, dx \\
&= e^{-\lambda t} \sum_{i=0}^{n-1} \frac{(\lambda t)^i}{i!}. \qquad (4.114)
\end{aligned}$$

Example 4.20. Suppose that two identical $\mathcal{E}(\lambda)$ distributed subsystems with $\lambda = 3 \times 10^{-7}$ f/h are either to be placed in an active parallel or a standby configuration. For $t = 7$ years, what is the reliability gain of a standby over an active parallel arrangement?

From (4.107), the reliability of an active parallel configuration is

$$R_s(61{,}320) = 1 - \left[1 - e^{-(3\times 10^{-7})(61{,}320)}\right]^2 = 0.99967,$$

and, from (4.114), the reliability of a standby arrangement is

$$R_s^2(61{,}320) = e^{-(3\times 10^{-7})(61{,}320)}\left[1 + \frac{(3\times 10^{-7})(61{,}320)}{1}\right] = 0.99983.$$

Thus, a standby arrangement yields a 1.6 percent gain in reliability. ■

Imperfect Switching. Now let us consider the case where the switch can fail in either a static or a dynamic mode. First, consider the static failure mode. In the static mode the switch may simply fail to operate when called upon, and let p represent the probability that the switch performs when required. For a two-subsystem standby system we have

$$R_s^2(t)' = R_1(t) + p\int_0^t f_1(t_1)R_2(t-t_1)\,dt_1, \tag{4.115}$$

and for a three-subsystem standby system

$$R_s^3(t)' = R_s^2(t)' + p^2\int_0^t f_1(t_1)\int_0^{t-t_1} f_2(t_2)R_3(t-t_1-t_2)\,dt_2dt_1. \tag{4.116}$$

Thus, the previous results are easily modified to handle this form of static switch failure.

If the switch is a complex piece of equipment, the dynamic failure mode may be appropriate. In the dynamic case, the switch is assumed to have an $\mathcal{E}(\lambda_s)$ failure time distribution. Thus, the switch may fail before it is needed. In the case of a two-subsystem standby system, we have

$$R_s^2(t)'' = R_1(t) + \int_0^t f_1(t_1)e^{-\lambda_s t_1}R_2(t-t_1)\,dt_1. \tag{4.117}$$

In the case of a three-subsystem standby system we find that

$$R_s^3(t)'' = R_s^2(t)'' + \int_0^t f_1(t_1)\int_0^{t-t_1} f_2(t_2)R_3(t-t_1-t_2)e^{-\lambda_s(t_1+t_2)}dt_2dt_1, \tag{4.118}$$

and if each subsystem has a constant failure rate λ, the expression reduces to

$$R_s^3(t)'' = e^{-\lambda t}\left[1+\frac{\lambda}{\lambda_s}(1-e^{-\lambda t})\right] + e^{-\lambda t}(\lambda/\lambda_s)^2$$
$$\times\left[1-e^{-\lambda_s t}-\lambda_s t e^{-\lambda_s t}\right]. \tag{4.119}$$

4.7.4 Shared Load Systems

A shared load system is a type of parallel system in which the subsystems equally share the load. Upon subsystem failure, the surviving subsystems must sustain an increased load. Thus, the failure rate of surviving subsystems increases as successive subsystems fail. A simple example is the supports of a load bearing structure, such as a bridge.

We consider only the case of two identical subsystems and define

$f_f(t)$ = subsystem failure time pdf under full load (one subsystem has failed)

$f_h(t)$ = subsystem failure time pdf under half load (neither subsystem has failed),

with corresponding reliability functions $R_f(t)$ and $R_h(t)$. It may be shown that [Kapur and Lamberson (1977)]

$$R_s(t) = [R_h(t)]^2 + 2\int_0^t f_h(t_1)R_h(t_1)R_f(t-t_1)\,dt_1. \tag{4.120}$$

In the special case of constant failure rates λ_f and λ_h, respectively, (4.120) reduces to

$$R_s(t) = e^{-2\lambda_h t} + \frac{2\lambda_h}{(2\lambda_h-\lambda_f)}\left[e^{-\lambda_f t}-e^{-2\lambda_h t}\right], \tag{4.121}$$

which is the system reliability for a two-subsystem shared load system with constant failure rates.

4.7.5 Complex Systems

The reliability configuration of many practical systems is such that neither a pure series nor a pure parallel (nor some hybrid combination) reliability model is appropriate. For example, consider the system shown in Figure 4.21*a*. The failure of subsystem 4 eliminates several paths. Thus, we do not have either a pure series or a pure parallel arrangement.

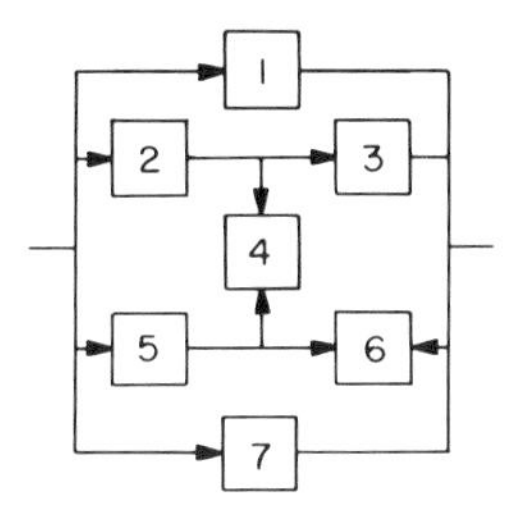

(a) RELIABILITY BLOCK DIAGRAM

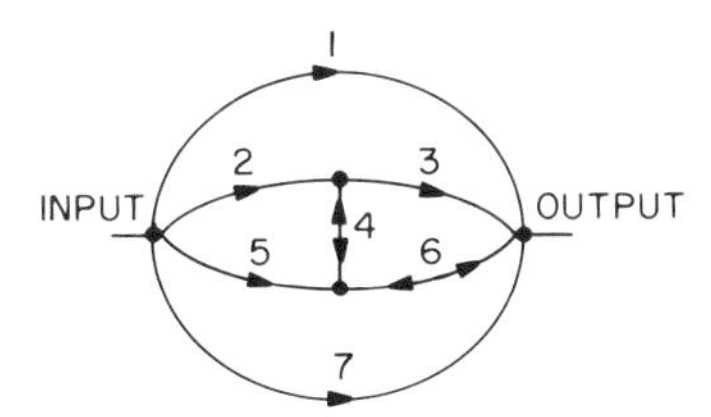

(b) RELIABILITY GRAPH

Figure 4.21. A complex system configuration.

Several methods exist for the analysis of complex systems. Shooman (1968) describes the *inspection method*, the *event-space method*, the *path-tracing method*, and the method of *decomposition*. Most of these methods can be successfully used if the number of subsystems is fairly small. When the number of subsystems exceeds eight or nine, more efficient methods are often necessary.

A very efficient method for computing the reliability of a system of independent subsystems is based on notions of graph theory. The method is based on the notions of a *path set* and a *cut set*. The reliability block diagram is first represented as a reliability graph in which the branches represent the n subsystems. Such a graph is shown in Figure 4.21*b*. The nodes of the graph tie the branches together and form the structure.

A *path set* (or tie set) is a set of branches that forms a connection between input and output when traversed in the direction of the arrows. A path set is thus a "path" through the graph. A *minimal path set* (or minimal tie set) is a path set containing a minimal number of branches such that no node is traversed more than once in tracing out a path set.

Example 4.21. In Figure 4.21*b*, the path sets are $T_1=\{1\}$, $T_2=\{7\}$, $T_3=\{5, 6\}$, $T_4=\{2, 3\}$, $T_5=\{3, 4, 5\}$, $T_6=\{2, 4, 6\}$, $T_7=\{1, 3, 4, 6\}$, $T_8=\{2, 3, 4, 6\}$, $T_9=\{3, 4, 6, 7\}$, and $T_{10}=\{3, 4, 5, 6\}$, whereas the minimal path sets are T_1, T_2, T_3, T_4, T_5, and T_6. ■

A *cut set* is a set of branches that interrupts *all* connections between input and output if removed from the graph. A *minimal cut set* is a minimal set of branches that interrupts all connections between input and output. Physically, a minimal cut set is a minimal set of subsystems whose failure causes the system to fail.

Example 4.22. In Figure 4.21*b*, some cut sets are $C_1=\{1, 3, 6, 7\}$, $C_2=\{1, 2, 5, 7\}$, $C_3=\{1, 3, 4, 5, 7\}$, $C_4=\{1, 2, 4, 6, 7\}$, $C_5=\{1, 2, 3, 4, 5, 6, 7\}$, $C_6=\{1, 2, 5, 6, 7\}$, and so on. The minimal cut sets are C_1, C_2, C_3, and C_4. ■

If a system has i minimal path sets denoted by $T_1, T_2, \ldots, T_i$, then the system reliability is given by

$$R_s(t)=\Pr(T_1 \cup T_2 \cup \cdots \cup T_i), \tag{4.122}$$

where T_i now denotes the "event" that the subsystems in the ith minimal path set all survive for a length of time t. In a large system, the expansion of (4.122) may be formidable, as, in general, the minimal path sets will not be disjoint. For example, in Example 4.21 the minimal path sets T_4 and T_5 are not disjoint as they have subsystem 3 in common. However, a useful upper bound on the system reliability may be obtained by assuming the disjoint property; namely,

$$R_s(t) \leqslant \Pr(T_1)+\Pr(T_2)+ \cdots +\Pr(T_i). \tag{4.123}$$

This bound is known to be a good approximation to $R_s(t)$ in the low reliability region [Shooman (1968)].

Example 4.23. Suppose that each subsystem in Figure 4.21 has an $\mathcal{E}(\lambda_i)$ failure time distribution, $i=1,2,\ldots,7$. Then

$$R_s(t) \leqslant e^{-\lambda_1 t}+e^{-\lambda_7 t}+e^{-(\lambda_5+\lambda_6)t}+e^{-(\lambda_2+\lambda_3)t}$$
$$+e^{-(\lambda_3+\lambda_4+\lambda_5)t}+e^{-(\lambda_2+\lambda_4+\lambda_6)t}.$$ ■

In the same spirit, if the system has j minimal cut sets denoted by $C_1, C_2, \ldots, C_j$, then the system reliability is also given by

$$R_s(t)=1-\Pr\left(\bar{C}_1 \cup \bar{C}_2 \cup \cdots \cup \bar{C}_j\right), \tag{4.124}$$

where C_j now denotes the "event" that the subsystems comprising the jth minimal cut set all fail prior to time t. Again, the expansion of (4.124) is often formidable and we resort to the useful lower bound on system reliability given by

$$R_s(t) \geqslant 1 - \left[\Pr(\bar{C}_1) + \Pr(\bar{C}_2) + \cdots + \Pr(\bar{C}_j)\right], \tag{4.125}$$

which is known to produce a sharp bound in the high reliability region.

Example 4.24. Continuing with Example 4.23, we find that

$$\begin{aligned} R_s(t) \geqslant 1 - &\left[(1-e^{-\lambda_1 t})(1-e^{-\lambda_3 t})(1-e^{-\lambda_6 t})(1-e^{-\lambda_7 t})\right. \\ &+ (1-e^{-\lambda_1 t})(1-e^{-\lambda_2 t})(1-e^{-\lambda_5 t})(1-e^{-\lambda_7 t}) \\ &+ (1-e^{-\lambda_1 t})(1-e^{-\lambda_3 t})(1-e^{-\lambda_4 t})(1-e^{-\lambda_5 t})(1-e^{-\lambda_7 t}) \\ &\left. + (1-e^{-\lambda_1 t})(1-e^{-\lambda_2 t})(1-e^{-\lambda_4 t})(1-e^{-\lambda_6 t})(1-e^{-\lambda_7 t})\right]. \end{aligned}$$

Example 4.25. Suppose that each subsystem in Figure 4.21*b* is identical with probability p of survival for a fixed duration t_0. By expanding (4.122), the system reliability is given by

$$R_s = 2p + p^2 - 2p^3 - 7p^4 + 14p^5 - 9p^6 + 2p^7.$$

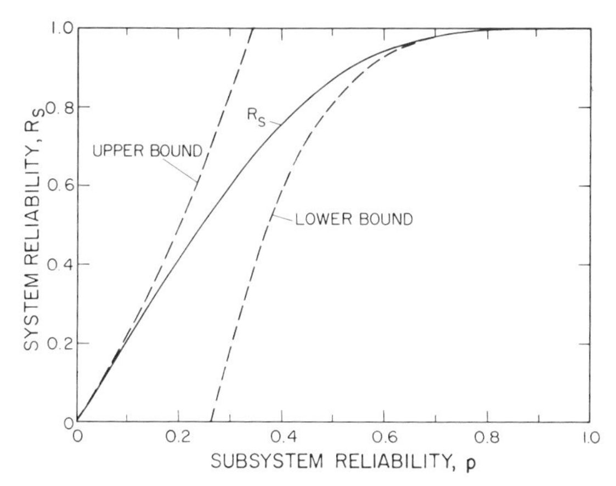

Figure 4.22. A comparison of the system reliability function, upper bound, and lower bound of the complex system in Figure 4.21.

From (4.123) and (4.125), the bounds on R_s become

$$1-2(1-p)^4-2(1-p)^5 \leqslant R_s \leqslant 2p+2p^2+2p^3.$$

In Figure 4.22, we have plotted R_s, as well as the bounds on R_s, as a function of subsystem reliability p. The respective bounds are observed to be good approximations in the high and low reliability regions. ■

Monte Carlo simulation is another efficient method which is often used to determine the reliability of a complex system. Shooman (1968) describes the use of this method.

4.8. REPAIRABLE SYSTEMS

In the preceding section we considered only the probability that a system would operate without failure for a given length of time. If the average repair cost is a fraction of the initial procurement cost, one usually considers system repair. In such systems there are measures of system effectiveness, other than reliability, such as maintainability, repairability, and availability, that are equally important [see Section 4.1.5]. Of course, in some systems, such as those involving life support, space exploration, or surveillance, any failure is catastrophic, and repair is of no avail. Bayesian estimation of the availability of repairable systems is considered in Chapter 12.

Recall from Section 4.1.5 that the availability of a system is the probability that it is operating satisfactorily at time t when operated under specified conditions. The availability function $A(t)$ contains no information on how many (if any) failure-repair cycles have occurred prior to time t. If a system has no repair capability, then $A(t)=R(t)$. However, in the case of repair, since the reliability function requires failure-free operation to time t, in general $R(t) \leqslant A(t)$.

Consider a series system. If we allow repair, then $R(t)$ does not change, but $A(t)$ becomes greater than $R(t)$. The situation changes for any system having more than one path set, (i.e., for those systems in which there is some sort of redundancy present). In such cases, repair can benefit both $R(t)$ and $A(t)$, as certain failed subsystems may be repaired while the system continues to operate successfully.

4.8.1 Markov Models

Markov models are important mathematical tools that are particularly useful in the reliability analysis of repairable systems. However, they have also been used to model dependent failures as well as to study the behavior of simple nonrepairable systems [Shooman (1968)]. The basic concepts of a Markov model are the notions of the "state" of a system and "transitions" between such states. The system is said to occupy a certain *state* whenever it satisfies the conditions that define the state. The dynamical changes in the state of the system are referred to as state *transitions*.

Markov models require the use of the so-called *Markov property*: given the present state of the system, the future behavior of the system is independent of the past behavior. In other words, if the present state of the system is known, the future state transition probabilities are independent of the past states of the system. Although at first glance this seems to be a somewhat restrictive property; nevertheless, in the analysis of repairable systems this property represents an excellent approximation to reality.

There are two basic classes of Markov models. The first is the class of so-called discrete time models that are commonly known as *Markov chains*. The second is the class of continuous time models often referred to as *Markov processes*. In the analysis of repairable systems, our primary concern is with Markov process models as events, such as the failure or repair of a system, can occur at any point in time.

Any Markov model is defined by a set of probabilities p_{ij} which define the probability of transition from any state i to any state j. In the case of a Markov process, p_{ij} is the probability that the system will undergo a transition from state i to state j in the infinitesimal time period t to $t+\Delta t$. The probabilities p_{ij} are often arrayed in a matrix for convenience in subsequent manipulations. For example, if $p_i(t)$, $i=1,2,\ldots,n$, represents the unconditional probability that the system is in state i at time t, then we can write

$$p_j(t+\Delta t)=p_{1j}p_1(t)+p_{2j}p_2(t)+\cdots+p_{nj}p_n(t) \tag{4.126}$$

and the coefficients in this expression are simply the elements in the jth column of the transition matrix $p=\{p_{ij}\}$. If the p_{ij} terms are all independent of time and depend only on constants and Δt, the Markov process is said to be *homogeneous*; otherwise, the process is said to be *nonhomogeneous*. Because of the Markov property, we always obtain a set of first-order differential equations that must be solved for the set of unconditional probabilities that the system is in each of the various possible states at any time t. These probabilities are then related to the desired reliability measure of interest, such as the reliability or availability function. For an homogeneous process, the resulting set of differential equations has constant coefficients, and the solutions are of the form $\exp(-rt)$ or $t^n\exp(-rt)$.

Under certain conditions, the limiting unconditional probabilities that a homogeneous system is in each state do not depend on the initial starting conditions. Such a process is said to be *ergodic*. A finite-state homogeneous process is ergodic if every state can be reached from any other state (not necessarily in one transition) with positive probability. If it is impossible to reach any outside state from some particular state, the latter state is called an *absorbing state*. For example, the failure state of a series system for which the reliability at time t is desired is an absorbing state, as any subsequent repairs will have an impact only on the availability and not the reliability function.

A general introduction to engineering applications of Markov processes is given by Papoulis (1965). Shooman (1968) contains an excellent discussion of the use of Markov process models in reliability analysis.

4.8.2 System-Level Repair

We limit our discussion to the system-level consideration in which the system is thought of as having only a single repairable subsystem. Further, suppose that there are only two possible states: S_1, the subsystem is operational, and S_2, the subsystem has failed. The states of the system at time $t=0$ are called the initial states. For convenience, we shall henceforth refer to states S_1 and S_2 simply as states 1 and 2, respectively.

The transition probabilities must obey the following rules:

1. The probability of transition in the time interval $(t, t+\Delta t)$ from state S_1 to state S_2 is given by $h(t)\Delta t$, where $h(t)$ is the hazard rate of the subsystem.
2. The probability of transition in the time interval $(t, t+\Delta t)$ from state S_2 to state S_1 is given by $w(t)\Delta t$, where $w(t)$ is the repair rate of the subsystem.
3. The probabilities of more than one transition in the time interval $(t, t+\Delta t)$ are infinitesimals of a higher order in Δt and can be neglected.

The transition probability $h(t)\Delta t$ is the probability of subsystem failure in $(t, t+\Delta t)$ given that it was operational at time t. The transition probability $w(t)\Delta t$ is the probability that a failed subsystem will be repaired in $(t, t+\Delta t)$ given that it was in a failed state at time t. The state transition matrix for this simple system is given in Table 4.9. Note that the rows sum to unity. From (4.126) and Table 4.9 we can write

$$p_1(t+\Delta t)=[1-h(t)\Delta t]\, p_1(t)+[w(t)\Delta t]\, p_2(t) \tag{4.127}$$

and

$$p_2(t+\Delta t)=[h(t)\Delta t]\, p_1(t)+[1-w(t)\Delta t]\, p_2(t), \tag{4.128}$$

Table 4.9. State Transition Matrix for a Single Repairable Subsystem

Initial States	Final States	
	$S_1(t+\Delta t)$	$S_2(t+\Delta t)$
$S_1(t)$	$1-h(t)\Delta t$	$h(t)\Delta t$
$S_2(t)$	$w(t)\Delta t$	$1-w(t)\Delta t$

where $p_i(t)$ denotes the unconditional probability that the subsystem is in state i, $i=1,2$, at time t. Rearranging (4.127) and (4.128) yields

$$\frac{p_1(t+\Delta t)-p_1(t)}{\Delta t}=-h(t)p_1(t)+w(t)p_2(t)$$

and

$$\frac{p_2(t+\Delta t)-p_2(t)}{\Delta t}=h(t)p_1(t)-w(t)p_2(t).$$

Taking the limit as $\Delta t \to 0$ now yields

$$\frac{dp_1(t)}{dt}=-h(t)p_1(t)+w(t)p_2(t) \tag{4.129}$$

and

$$\frac{dp_2(t)}{dt}=h(t)p_1(t)-w(t)p_2(t). \tag{4.130}$$

Equations (4.129) and (4.130) can be simultaneous solved in conjunction with the appropriate initial conditions for $p_1(t)$ and $p_2(t)$. The most common initial condition is that the subsystem is operational at time $t=0$, that is, $p_1(0)=1$ and thus $p_2(0)=0$. Equations (4.129) and (4.130) are simple first-order linear differential equations that are easily solved by use of the classical theory involving integrating factors. Solving this system we obtain the following solution in integral form:

$$p_1(t)=\exp\left\{-\int_0^t[h(x)+w(x)]\,dx\right\}\left(\int_0^t\exp\left\{\int_0^x[h(u)+w(u)]\,du\right\}w(x)\,dx+1\right) \tag{4.131}$$

and

$$p_2(t)=1-p_1(t), \tag{4.132}$$

as $p_1(t)+p_2(t)=1$. It is noted that, if $w(t)=0$ (repair is impossible), then $p_1(t)$ reduces to

$$p_1(t)=R(t)=\exp\left\{-\int_0^t h(x)\,dx\right\},$$

which is the same as the general reliability expression given in (4.8).

If $h(t)$ and $w(t)$ are constants, the expression for $p_1(t)$ is fairly straightforward. Thus, let us now suppose that the subsystem failure time (or, more precisely, the time *between* failures) has an $\mathcal{E}(\lambda)$ distribution and that repair time follows an $\mathcal{E}(\mu)$ distribution. Thus,

$$\begin{aligned} p_1(t) &= \exp\left\{-\int_0^t[\lambda+\mu]\,dx\right\}\left(\int_0^t \exp\left\{\int_0^x[\lambda+\mu]\,du\right\}\mu\,dx+1\right) \\ &= \exp\{-\lambda t-\mu t\}\left(\frac{\mu}{\lambda+\mu}\{\exp[(\lambda+\mu)t]-1\}+1\right) \\ &= \frac{\mu}{\lambda+\mu}+\frac{\lambda}{\lambda+\mu}\exp[-(\lambda+\mu)t]. \end{aligned} \tag{4.133}$$

If $\mu=0$, then $p_1(t)=R(t)=\exp(-\lambda t)$, the familiar $\mathcal{E}(\lambda)$ reliability function. If $\mu>0$, then by definition the availability function is the probability that the system is in state 1 at any time t and thus $p_1(t)=A(t)$.

An important difference between $A(t)$ and $R(t)$ is their limiting or steady-state behavior. As $t\to\infty$, $R(t)\to 0$; however,

$$A=\lim_{t\to\infty}A(t)=\frac{\mu}{(\lambda+\mu)}=\frac{\text{MTBF}}{\text{MTBF}+\text{MTTR}}, \tag{4.134}$$

where $\text{MTTR}=1/\mu$, the mean time to repair. Thus, in the limit, the availability function reaches the steady-state value given in (4.134), which is known as the *steady-state availability*. It is also noted that the steady-state availability is the fraction of the average cycle length that the system is operational. This recognition has led to the use of the expression for steady-state availability given by $A=\text{MTBF}/(\text{MTBF}+\text{MTTR})$ in those situations in which failure and/or repair time have other than $\mathcal{E}(\cdot)$ distributions.

Shooman (1968) considers a variety of systems composed of two and three subsystems; however, the basic procedure is the same as illustrated above, although the mathematical calculations become much more tedious.

EXERCISES

4.1. Find the hazard rate $h(t)$ of the distribution given by

$$f(t)=\begin{cases} te^{-\beta t}\beta^2, & t>0, \beta>0 \\ 0, & \text{otherwise.} \end{cases}$$

Determine the reliable life t_R corresponding to the reliability R. Compute $t_{0.90}$ if $\beta=3$.

4.2. Consider the hazard rate $h(t)=k(t/\beta)^k$, $k>1, \beta>0$.
   (a) Determine the pdf corresponding to this hazard rate.
   (b) Determine the reliable life corresponding to $h(t)$.
   (c) Compute $t_{0.80}$ if $k=2$ and $\beta=4$.

4.3. Consider the hazard rate $h(t)=\exp[-t/\beta]/\beta$, $-\infty<t<\infty, \beta>0$.
   (a) Find the pdf corresponding to $h(t)$.
   (b) Find the reliable life t_R.
   (c) Compute $t_{0.90}$ if $\beta=6$.

4.4. Consider $h(t)=\alpha e^{-\beta t}$, $\alpha>0, \beta>0, t>0$.
   (a) Find the corresponding pdf.
   (b) Find t_R.
   (c) Compute $t_{0.70}$ if $\alpha=2$ and $\beta=4$.

4.5. Consider $h(t)=\beta/(\alpha+t)$, $\alpha>0$, $\beta>0$, $t>0$.
   (a) Find the corresponding cdf.
   (b) Find t_R.
   (c) Compute $t_{0.60}$ if $\alpha=2$ and $\beta=8$.

4.6. Consider the following data concerning the status of a given system under study:

Activity No.	System Status	Time	Time Category	Time Units (h)
1	Available	Uptime-Cycle 1	Operating time	278
2			Free time	33
3			Operating time	364
4			Free time	14
5	Unavailable	Downtime-Cycle 1	Free time	10
6			Logistic time	140
7			Active repair time	9
8			Administrative time	6
9	Available	Uptime-Cycle 2	Operating time	240
10			Free time	22

(a) Compute the intrinsic availability A_I.
(b) Compute the availability A.
(c) Compute the operational readiness.

4.7. Consider the "folded-over" normal pdf.

$$f(t;\mu,\sigma)=\begin{cases}2\phi\left(\dfrac{t-\mu}{\sigma}\right), & t\geqslant\mu\\ 0, & \text{otherwise.}\end{cases}$$

(a) Determine the reliability function $R(t;\,\mu,\sigma)$.
(b) Determine the hazard rate $h(t;\,\mu,\sigma)$.
(c) Determine the MTTF and Var(T).
(d) Determine the reliable life t_R.

4.8. Show that the $\mathcal{IN}(\mu,\lambda)$ distribution has a unique mode given by (4.43).

4.9. Verify that the mean and variance of the $\mathcal{IG}(\alpha,\beta)$ distribution are given by (4.55) and (4.56), respectively.

4.10. Derive the reliability function $R(t;\,\alpha,\beta)$ and the hazard function $h(t;\,\alpha,\beta)$ of the $\mathcal{G}(\alpha,\beta)$ distribution.

4.11. Consider the failure times (in units of 1000 h) of a well pump: 1.5, 2.7, 3.9, 4.1, 4.6, 5.6, 5.8, 6.8, 8.3, 9.4. Compute the estimates of the hazard rate, $\hat{h}(t_i)$; reliability function, $\hat{R}(t_i)$; and pdf, $f(\hat{t}_i)$, according to (4.60), (4.61), and (4.62), respectively. Summarize your computations in tabular form as in Example 4.2. Also, plot the hazard rate and discuss its behavior according to the plot.

4.12. Consider the following grouped data on failure times of diaphragm valves in the chemical and volume control systems of two nuclear reactor plants of the same design. The failure times are in units of 1000 h.

Interval	Number of Failures
$0\leqslant t<3$	2
$3\leqslant t<6$	6
$6\leqslant t<9$	20
$9\leqslant t<12$	9
$12\leqslant t<15$	7
$15\leqslant t<18$	6

Compute the estimates of the reliability function, $\hat{R}(t)$; the pdf, $\hat{f}(t)$;

and the hazard rate $\hat{h}(t)$, according to (4.63), (4.64), and (4.65), respectively. Summarize the computations in tabular form as in Example 4.3. Also, plot the hazard rate.

4.13. Consider the data in Exercise 4.11. Tabulate the calculations required for a total time on test plot. Give a graph of the total time on test plot. From the plot discuss the behavior of the hazard rate over time.

4.14. Consider the failure times (in units of 1000 h) of a certain device used in an electric generator: 5.05, 0.42, 3.54, 1.35, 7.01, 1.64, 2.28, 3.22. Plot the pairs $(t_i, (n+0.25)/(n-i+0.625))$, $i=1,2,\ldots,8$, on semi-log paper where the arithmetic scale is used for t_i and the logarithmic scale is used for $(n+0.25)/(n-i+0.625)$. Determine from the plot whether the $\mathcal{E}(\lambda)$ distribution is appropriate for the data. If so, estimate the failure rate λ graphically.

4.15. Consider the failure times of a computer component in hours: 216, 405, 2048, 2460, 4827, 4096, 4565, 1, 195, 1191. Determine from a probability plot whether it is reasonable to assume that the data came from a $\mathcal{W}(\alpha, \beta)$ distribution. If so, estimate the shape and scale parameters graphically.

4.16. The following failure times (in months) are believed to have come from an $\mathcal{N}(\mu, \sigma^2)$ distribution: 8.274, 5.952, 6.591, 9.704, 4.848, 5.424, 8.877, 10.666, 7.684, 4.187, 9.010, 6.894, 6.793, 3.660, 8.281, 8.438, 9.792, 6.801, 3.388, 5.357. Tabulate the data as in Example 4.7 and plot these data on $\mathcal{N}(\mu, \sigma^2)$ probability paper. Determine from the plot whether the $\mathcal{N}(\mu, \sigma^2)$ assumption is reasonable. If it is, graphically estimate the mean μ and variance σ^2.

4.17. The following failure times (in months) are believed to have come from an $\mathcal{LN}(\xi, \sigma^2)$ distribution: 0.83, 41.68, 22.88, 12.18, 167.34, 13.46, 148.41, 6.96, 53.52, 3.13. Plot the data on $\mathcal{LN}(\xi, \sigma^2)$ paper and determine whether the $\mathcal{LN}(\xi, \sigma^2)$ assumption is reasonable. If it is, graphically estimate the parameters ξ and σ^2.

4.18. The following data are the first 10 failure times, in hours, of 268 diaphragm valves in the chemical and volume control systems of two nuclear reactor plants of the same design. The remaining valves continued to operate normally for longer than 16833.4 h. The data are 2827.6, 3916.0, 4416.2, 6271.1, 7211.9, 8186.5, 12203.7, 13088.9, 16144.6, 16833.4. Assume that the $\mathcal{E}(\lambda)$ distribution is a reasonable model.

(a) Estimate the failure rate λ.

(b) Construct a 95% TCI on λ.

(c) Estimate the reliability at $t=15000$ h.

(d) Construct a 95% TCI on $R(15000)$.
(e) Estimate the reliable life for $R=0.85$.
(f) Construct a 95% TCI on $t_{0.85}$.

4.19. Consider the following failure times (in months) of a special type of pump used in a nuclear reactor: 15.1, 10.7, 8.8, 11.3, 12.6, 14.4, 8.7. Twelve pumps were placed on test until seven failed. Assume the data came from an $\mathcal{E}(\mu, \lambda)$ distribution.

(a) Estimate the parameters λ and μ.
(b) Estimate the reliability for $t=12$ months.
(c) Construct a 95% TCI for λ.
(d) Construct a 95% TCI for μ.

4.20. Consider the data in Exercise 4.18. Test the hypothesis H_0: $\lambda=3\times10^{-6}$ against the alternative H_1: $\lambda\neq3\times10^{-6}$ at the 5% level of significance.

4.21. Consider the data in Exercise 4.19. Test the hypothesis H_0: $\mu=0$ against H_1: $\mu\neq0$ at the 10% level of significance.

4.22. Consider the two sets of failure times (in months) corresponding to two mechanical devices: Data Set 1: 0.80, 4.82, 5.99, 1.99, 2.10. Data Set 2: 0.60, 0.74, 1.27, 1.35, 1.69, 1.60. Assume the sets came from two $\mathcal{E}(\lambda_i)$ distributions with possibly different failure rates. Test the hypothesis H_0: $\lambda_1=\lambda_2$ against H_1: $\lambda_1\neq\lambda_2$ at the 5% level of significance.

4.23. In obtaining failure data according to an $\mathcal{E}(\lambda)$ distribution, suppose it is desired to test H_0: $\theta=\theta_0=200$ h against H_1: $\theta=\theta_1=50$ h with $\gamma=\delta=0.10$ and $t_0=20$ h, where t_0 is the duration of the life test. Determine the number n of items that should be put on test and the minimum number c of failures that should be observed in order for H_0 to be rejected. Determine n and c if $\theta_0=500$, $\theta_1=50$, and $t_0=20$.

4.24. Suppose it is desired to test H_0: $\theta=200$ h against H_1: $\theta=50$ h by means of a Type I/time-truncated life test *with* replacement in which $\gamma=\delta=0.10$. Determine the required demonstration plan and determine its implementation. Consider $t_0=20$ h.

4.25. Solve Exercise 4.24 if the test is a Type I/time-truncated life test *without* replacement.

4.26. Solve Exercise 4.24 if the test is a Type II/item-censored test with replacement. Consider $n=10$.

4.27. Verify (4.4).

4.28. Verify that the MTTF and variance of (4.18) are given by (4.20) and (4.21), respectively.

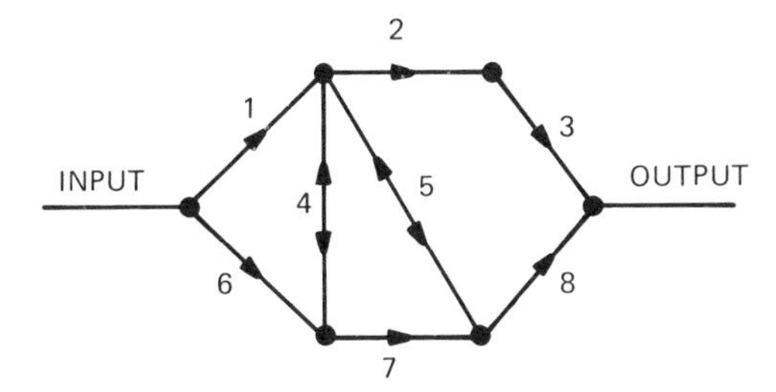

Figure 4.23. A reliability graph for Exercise 4.31.

4.29. Consider a purely parallel system of two independent $\mathcal{E}(\lambda_i)$ components. Find the pdf for the system failure time.

4.30. Consider a purely parallel system of five identical components. Let R denote the reliability of each component for a mission time t. If the system reliability must be 0.999, how poor can each component be?

4.31. Consider the reliability graph given in Figure 4.23.

(a) Find all the path sets for this graph.

(b) Find all the minimal path sets for this graph.

(c) Find all the cut sets for this graph.

(d) Find all the minimum cut sets for this graph.

(e) If each component has reliability 0.9, compute a lower bound on the reliability of the system.

4.32. An airborne communications system has a reliability requirement of 0.99 for 8 h. The best available "off the shelf" communications system has a demonstrated MTTF of 300 h. How could the requirement be met (assume an $\mathcal{E}(\lambda)$ failure time distribution)?

4.33. Consider the system of three identical $\mathcal{E}(\lambda)$ components given in Figure 4.24. Assume that all components dissipate substantial heat when operating. If one component is operating, the component failure rate is λ; if two components are operating, the component failure rate is 2λ; and if three components are operating, the component failure rate is 3λ. Determine the reliability function for the system.

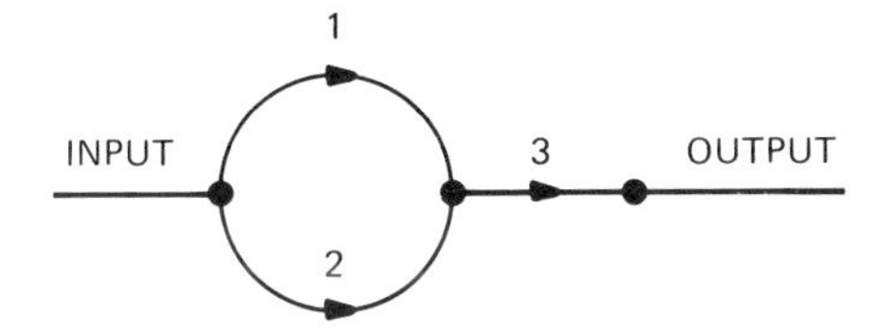

Figure 4.24. The system configuration for Exercise 4.33.

REFERENCES

Aitchison, J. and Brown, J. A. C. (1957). *The Lognormal Distribution*, Cambridge University Press, London.

Ang, A. and Tang, W. (1975). *Probability Concepts in Engineering Planning and Design, Vol. I — Basic Principles*, Wiley, New York.

Bain, L. J. (1978). *Statistical Analysis of Reliability and Life-Testing Models*, Statistics: Textbooks and Monographs, Vol. 24, Marcel Dekker, New York.

Barlow, R. E. (1978). Analysis of Retrospective Failure Data Using Computer Graphics, *Proceedings of the 1978 Annual Reliability and Maintainability Symposium*, pp. 113–116.

Barlow, R. E., Bartholomew, D. J., Bremner, J. M. and Brunk, H. B. (1972), *Statistical Inference Under Order Restrictions*, Wiley, New York.

Barlow, R. E. and Campo, R. A. (1975). Total Time on Test Processes and Applications to Failure Data Analysis, *Reliability and Fault Tree Analysis*, edited by Barlow, Fussell, and Singpurwalla, SIAM, Philadelphia, PA, pp. 451–481.

Barlow, R. E. and Davis, B. (1977). Analysis of Time Between Failures for Repairable Components, *Nuclear Systems Reliability Engineering and Risk Assessment*, edited by Fussell and Singpurwalla, SIAM, Philadelphia, PA, pp. 543–561.

Barlow, R. E. and Proschan, F. (1965). *Mathematical Theory of Reliability*, Wiley, New York.

Barlow, R. E. and Proschan, F. (1975). *Statistical Theory of Reliability and Life Testing (Probability Models)*, Holt, Rinehart, & Winston, New York.

Barlow, R. E. and Van Zwet, W. R. (1969). Comparison of Several Nonparametric Estimators of the Failure Rate Function, *Proceedings of the NATO Conference on Reliability at Turino, Italy*, Israel Publishing House.

Bartholomew, D. J. (1963). The Sampling Distribution of an Estimate Arising in Life Testing, *Technometrics*, Vol. 5, pp. 361–374.

Bergman, B. (1977). Some Graphical Methods for Maintenance Planning, *Proceedings of the 1977 Annual Reliability and Maintainability Symposium*, pp. 467–471.

Blom, G. (1958). *Statistical Estimates and Transformed Beta Variables*, Wiley, New York.

Chhikara, R. S. and Folks, J. L. (1974). Estimations of the Inverse Gaussian Distribution Function, *Journal of the American Statistical Association*, Vol. 69, pp. 250–254.

Chhikara, R. S. and Folks, J. L. (1976). Optimum Test Procedures for the Mean of First Passage Time Distribution in Brownian Motion with Positive Drift (Inverse Gaussian Distribution), *Technometrics*, Vol. 18, pp. 189–193.

Chhikara, R. S. and Folks, J. L. (1977). The Inverse Gaussian Distribution as a Lifetime Model, *Technometrics*, Vol. 19, pp. 461–468.

Cox, D. R. and Miller, H. D. (1965). *The Theory of Stochastic Processes*, Methuen, London.

Davis, D. J. (1952). An Analysis of Some Failure Data, *Journal of the American Statistical Association*, Vol. 47, pp. 113–150.

Epstein, B. (1954). Truncated Life Tests in the Exponential Case, *Annals of Mathematical Statistics*, Vol. 25, pp. 555–564.

Epstein, B. (1958). The Exponential Distribution and Its Role in Life Testing, *Industrial Quality Control*, Vol. 15, pp. 5–9.

Epstein, B. (1960). Estimation from Life Test Data, *Technometrics*, Vol. 2, pp. 447–454.

Epstein, B. and Sobel, M. (1953). Life Testing I, *Journal of the American Statistical Association*, Vol. 48, pp. 486–502.

Feller, W. (1965). *An Introduction to Probability Theory and its Applications* (2nd Ed.), Wiley, New York.

Fertig, K. W. and Mann, N. R. (1980). Life-Test Sampling Plans for Two-Parameter Weibull Populations, *Technometrics*, Vol. 22, pp. 165–177.

Goldthwaite, L. (1961). Failure Rate Study for the Lognormal Lifetime Model, *Proceedings of the Seventh National Symposium on Reliability and Quality Control*, pp. 208–213.

Gupta, S. and Groll, P. (1961). Gamma Distribution in Acceptance Sampling Based on Life Tests, *Journal of the American Statistical Association*, Vol. 56, pp. 943–970.

Hahn, G. and Shapiro, S. (1967). *Statistical Models in Engineering*, Wiley, New York.

Johnson, N. L. and Kotz, S. (1970). *Continuous Univariate Distributions — 1*, Houghton Mifflin, Boston.

Kapur, K. C. and Lamberson, L. R. (1977). *Reliability in Engineering Design*, Wiley, New York.

Kimball, B. F. (1960). On the Choice of Plotting Positions on Probability Paper, *Journal of the American Statistical Association*, Vol. 55, pp. 546–560.

Larson, H. R. (1966). A Nomograph of the Cumulative Binomial Distribution, *Industrial Quality Control*, Vol. 23, pp. 270–278.

Lieblein, J. and Zelen, M. (1956). Statistical Investigations of the Fatigue Life of Deep-Groove Ball Bearings, *Journal of Research, National Bureau of Standards*, Vol. 57, pp. 273–316.

Mann, N. R. (1969). Optimum Estimators for Linear Functions of Location and Scale Parameters, *Annals of Mathematical Statistics*, Vol. 40, pp. 2149–2155.

Mann, N. R., Schafer, R. E., and Singpurwalla, N. D., (1974). *Methods for Statistical Analysis of Reliability and Life Data*, Wiley, New York.

Martz, H. F. (1975). Pooling Life-Test Data by Means of the Empirical Bayes Method, *IEEE Transactions on Reliability*, Vol. R-24, pp. 27–30.

Nelson, W. (1969). Hazard Plotting for Incomplete Failure Data, *Journal of Quality Technology*, Vol. 1, pp. 27–52.

Padgett, W. J. (1979). Confidence Bounds on Reliability for the Inverse Gaussian Model, *IEEE Transactions on Reliability*, Vol. R-28, pp. 165–168.

Papoulis, A. (1965). *Probability, Random Variables, and Stochastic Processes*, McGraw-Hill, New York.

Reactor Safety Study (1975). WASH-1400 (NUREG-75/014), U.S. Nuclear Regulatory Commission.

Shooman, M. L. (1968). *Probabilistic Reliability — An Engineering Approach*, McGraw-Hill, New York.

Tweedie, M. C. K. (1957). Statistical Properties of Inverse Gaussian Distribution, *Annals of Mathematical Statistics*, Vol. 28, pp. 362–377.

Von Alven, W. H. (Ed.) (1964). *Reliability Engineering (ARINC Research Corporation)*, Prentice-Hall, Englewood Cliffs, NJ.

Weibull, W. (1939). A Statistical Theory of Strength of Materials, *Ingeniors Vetenskaps Akademien Handlingar*, No. 151; The Phenomenon of Rupture in Solids, *Ibid.*, No. 153.

Weibull, W. (1951). A Statistical Distribution Function of Wide Applicability, *Journal of Applied Mechanics*, Vol. 18, pp. 293–297.

Wilks, S. S. (1962). *Mathematical Statistics*, Wiley, New York.

CHAPTER 5

Bayesian Inference in Reliability

Rev. Thomas Bayes' (1702–1761) famous paper was published in 1763 and provides the basis for the method known as "Bayesian Statistical Inference." Due to its fundamental importance the paper has been republished [Bayes (1958)]. Since that time interest has alternated between periods of acceptance and rejection of the method as a basis for statistical inference. Often subtle and unsuspected difficulties with alternate methods of statistical inference have been largely responsible for the continued resurgence of the Bayesian method of reasoning. The method is currently riding a high tide of popularity in virtually all areas of statistical application.

Several factors contribute to the recent resurgence. First, the work of several authors, notably DeFinetti (1937), Good (1950), Jeffreys (1961), Lindley (1965), Ramsey (1931), and Savage (1954), has provided a philosophical basis for the method. Second, other theories of inference are based on rather restrictive assumptions that provide convenient mathematical solutions. The realization that such solutions address only a limited set of problems likely to be encountered in practice has necessitated a larger and less restrictive framework for statistical inference. For example, it hardly makes sense to develop a sampling plan that provides high assurance against a specified level of quality if, in fact, that level of quality is unlikely to occur. Third, the advent of high speed computers permits a wider class of data analyses to be considered as compared to the narrow class of analyses based on restrictive special assumptions. Fourth, the interest in more efficient methods of data analysis has kindled interest in Bayesian methods that incorporate various subjective and objective data sources into the analysis.

The extent to which such gains have been recognized is pointed out by Box and Tiao (1973) on p. 2, ..."Bayesian inference alone seems to offer the possibility of sufficient flexibility to allow reaction to scientific complexity free from impediment from purely technical limitation..." Kendall and Stuart (1961) on p. 153, in discussing Bayesian and confidence intervals, comment that ..."The principal argument in favor of confidence intervals, however, is that they can be derived in terms of a frequency theory of probability without any assumptions concerning prior distributions such as are essential to the Bayes approach. This, in our opinion, is undeniable. But it is fair to ask whether they achieve this economy of basic assumptions without losing something which the Bayes theory possesses. Our view is that they do, in fact, lose something on occasion, and that this something may be important for the purposes of estimation..." The effect of this loss in reliability estimation is particularly keen as illustrated in numerous examples throughout the remainder of this book.

5.1. FOUNDATIONS OF BAYESIAN STATISTICAL INFERENCE

5.1.1 Subjective Probability

The cornerstone of the foundation of Bayesian inference is the notion of subjective probability. Such a notion contrasts with the well-known frequency notion of probability which was presented in Chapter 2. Recall that the frequency concept has its axiomatic origins in the properties of events and the success ratios of those events in a repeated series of trials. The axioms of probability, and in particular the probability of an event, are suggested by the limiting values of success ratios. Examples of frequency interpretations of probability are provided by games of chance, such as coin tossing, die rolling, and card drawing. The practical success of such an interpretation is attested to by such sciences as actuarial science, the science of genetics, and the science of radioactive decay. The important distinguishing feature is the notion that the probability of an event can be empirically established by conducting a sufficiently large series of repeated trials in which the event can occur.

Subjective probability deals not only with events but with propositions as well. A *proposition* is considered to be a collection of events that *cannot* be interpreted as an imagined series of repetitions. For example, it is highly artificial to imagine repetitions of a series of trials in which the proposition "nuclear power plant X will suffer a core meltdown" either does or does not occur. In statistical jargon the word *hypothesis* is used instead of proposition because we are usually interested in a proposition whose truth is in question. As evidence increases relevant to the hypothesis we then change our *degree of belief* in the hypothesis. Here the degree of belief in a proposition A, $\Pr(A)$, represents one's strength of conviction that A is true. Who among us does not allude to the use of subjective probability in everyday life when we hear and use phrases like: "He will probably not marry her because...," "I'll probably get that job...," "I'll bet that I pass this test... ." In none of these examples are natural repetitions conceivable. Subjective probability thus refers to the degree of belief in a proposition. At one extreme, if A is believed to be true, $\Pr(A)=1$; at the other extreme, if A is believed to be false, $\Pr(A)=0$. Points in the interval $(0,1)$ express intermediate beliefs between truth and falsehood. Table 5.1 summarizes the distinguishing features of the frequency and subjective notions of probability.

It has been shown by Lindley (1965), Savage (1954), and others as well, that degree of belief does in fact possess real world meaning and obeys the axioms of probability suggested by frequency notions. Savage further goes on to consider a method for numerically assessing degree of belief based on two sets of elements that he calls "states" and "consequences." The attitude we adopt in this book is that these axioms are reasonable, that the deductions from them described in detail by Savage are correct, and as a consequence we may use probabilities in the subjective sense of degree of belief.

One important additional point that should be noted is that the subjective probabilities assigned to a particular hypothesis may indeed be quite subjective. In other words, the probabilities assigned by one individual may be quite different from those that would be assigned by some other individual. This is the reason that subjective probability is sometimes referred to as "personal probability" by Morgan (1968), Savage (1954), and others. It is important to realize that subjectivity is certainly not peculiar to Bayesian inference. As an investigator is seldom certain about the true

Table 5.1. A Summary of the Frequency and Subjective Notions of Probability

	Notion of Probability	
	Frequency	Subjective
Real World Meaning	Yes	Yes
Deals with Events	Yes	Yes
Deals with Propositions	No	Yes
Capable of Experimental Verification	Yes	No*
Axiomatic Foundation	Yes	Yes
Numerical Quantification	Yes	Yes
Personal Differences	No	Yes

*Except under controlled conditions.

nature of the process that generates the observed events, assumptions must be made about the underlying process. Thus, there is nearly always some question about the validity of these assumptions. For example, in reliability analyses it is frequently assumed that data are generated according to an underlying Poisson process even when, as is often the case, there is little or no evidence to support such an assumption.

It is therefore reasonable to conclude that subjectivity enters into nearly all statistical analysis and that such analysis is an art as well as a science. The important and distinguishing difference is the explicit manner in which Bayesian inference utilizes the subjective elements in the analysis.

5.1.2 Sampling Theory versus Bayesian Inference

There are distinctive differences between the sampling theory and Bayesian methods of inference. To illustrate these differences, suppose we are interested in studying the useful life of a certain population of Si-Ge thermoelectric converter elements under specified use conditions. Further suppose that we tentatively assume that the observed lives of these elements are distributed independently and exponentially with mean life θ. The joint probability distribution of a projected sample of n observations $\underset{\sim}{Y}'=(Y_1,\ldots,Y_n)$ would then be

$$f(\underset{\sim}{y}|\theta)=\frac{1}{\theta^n}\exp\left(-\frac{1}{\theta}\sum_{i=1}^{n} y_i\right), \qquad 0<y_i<\infty, \tag{5.1}$$

and we are interested in making inferences about θ given the n data values.

Inferences Based on Sampling Theory. In the sampling theory approach, the unknown mean life θ is assumed to be a fixed constant. A point estimator $\hat{\theta}(\underset{\sim}{Y})$, which is a function of the data set $\underset{\sim}{Y}$, is then chosen according to some principle, such as ML, minimum variance, least squares, or the method of moments. For example, both the ML and method of moments estimators of θ are

$$\hat{\theta}=\hat{\theta}(\underset{\sim}{Y})=\sum_{i=1}^{n}\frac{Y_i}{n}. \tag{5.2}$$

By imagining the set of all possible hypothetical data vectors $\underset{\sim}{Y}_1, \underset{\sim}{Y}_2,\ldots$ that could be generated by (5.1), for a given value of θ, we obtain the corresponding "sampling distribution" of $\hat{\theta}(\underset{\sim}{Y})$. For example, the sampling distribution of $U=2n\hat{\theta}/\theta$ is a $\chi^2(2n)$ distribution with pdf given by

$$f(u)=\frac{1}{2^n\Gamma(n)}u^{n-1}\exp\left(-\frac{u}{2}\right), \qquad 0<u<\infty. \tag{5.3}$$

A confidence interval estimator of θ may be calculated in order to provide some idea how far away the calculated value $\hat{\theta}(\tilde{y})$ is from the true value θ which generated the n observations. For example, the $100(1-\gamma)\%$ TCI estimator for θ is given by

$$100(1-\gamma)\% \text{ TCI for } \theta: \left[\frac{2n\hat{\theta}}{\chi^2_{1-\gamma/2}(2n)};\quad \frac{2n\hat{\theta}}{\chi^2_{\gamma/2}(2n)}\right]. \tag{5.4}$$

Recall from Section 3.6 that, in repeated sampling in which such a confidence interval is computed from each sample, the computed confidence intervals would include the true value θ a proportion $1-\gamma$ of the time. A confidence interval *cannot* be interpreted as a probability statement about θ since θ is not an r.v. This is particularly unfortunate in reliability analysis as it is often desired to combine confidence intervals in various ways. For instance, it may be desired to compute a confidence interval on the MTTF of a radioisotope thermoelectric generator system from confidence intervals on its components, such as its thermoelectric elements. Since "confidence" is not probability regarding θ, the methods for accomplishing this are not as well defined as in the case of probability.

Sampling theory inferences are examples of inductive reasoning. For instance, the upper endpoint of the TCI in (5.4) is the largest value that θ could have been without the observed estimator $\hat{\theta}(\tilde{Y})$ having a probability less than $\gamma/2$, according to (5.3), of occurring. A similar inductive statement may be given for the lower endpoint.

The sampling theory method of inference is depicted in Figure 5.1. The process begins with the postulating of a sampling model worthy of being

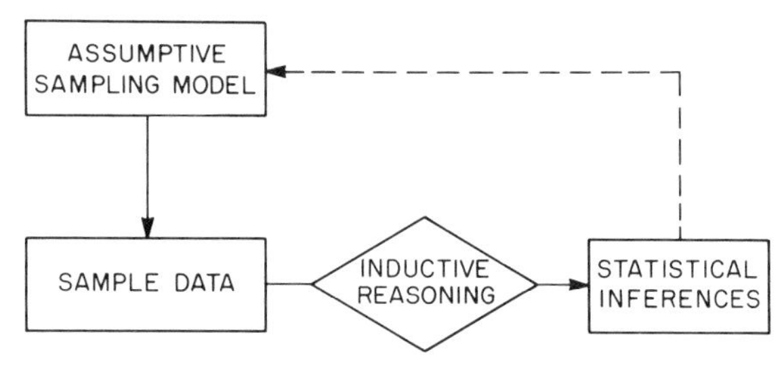

Figure 5.1. Inferences based on sampling theory.

tentatively considered. Inductive reasoning is used in conjunction with the sample observations to produce inferences about the unknown parameters in the assumed model.

Bayesian Inference. The Bayesian method of reasoning is much more direct, that is, deductive. To achieve this direct approach, the mean life θ is assumed to be an r.v. with an *a priori* or *prior* pdf $g(\theta)$. This distribution expresses the state of knowledge or ignorance about θ before the sample data are analyzed. It may be argued that it is reasonable to consider a parameter as the value of an r.v. as most parameters of real interest seldom become known. The a priori uncertainty is modeled by appealing to the use of a prior distribution.

Given the prior distribution, the probability model $f(\underset{\sim}{y}|\theta)$ and the data $\underset{\sim}{y}$, Bayes' theorem is used to calculate the so-called *posterior* pdf $g(\theta|\underset{\sim}{y})$ of Θ, given the data $\underset{\sim}{y}$. For example, suppose that the prior distribution of Θ is taken to be uniform on the range θ_1 to θ_2, where $0<\theta_1<\theta_2<\infty$. That is

$$g(\theta)=\frac{1}{\theta_2-\theta_1}, \qquad 0<\theta_1\leqslant\theta\leqslant\theta_2<\infty, \tag{5.5}$$

which we denote by $\mathcal{U}(\theta_1,\theta_2)$. This says, in effect, that a priori it is believed that θ is equally likely to be any value within the specified interval (θ_1,θ_2). Using Bayes' theorem [see Section 5.2.3], in conjunction with (5.1) and (5.5), the resulting posterior distribution is calculated to be

$$g(\theta|\underset{\sim}{y})=\frac{\left(\sum y_i\right)^{n-1}\exp\left(-\sum y_i/\theta\right)}{\theta^n\left[\Gamma\left(n-1,\sum y_i/\theta_1\right)-\Gamma\left(n-1,\sum y_i/\theta_2\right)\right]}, \qquad \begin{array}{l}\theta_1\leqslant\theta\leqslant\theta_2,\\ n>1,\end{array} \tag{5.6}$$

where the summation on y_i is from 1 to n and where $\Gamma(a,z)$ is the standard incomplete gamma function defined in (4.46).

Deductive arguments are used with the posterior distribution to make Bayesian inferences about θ. For example, a point estimator for θ is the mean of the posterior distribution (5.6) given by

$$E(\Theta|\underset{\sim}{y})=\sum y_i\left[\frac{\Gamma\left(n-2,\sum y_i/\theta_1\right)-\Gamma\left(n-2,\sum y_i/\theta_2\right)}{\Gamma\left(n-1,\sum y_i/\theta_1\right)-\Gamma\left(n-1,\sum y_i/\theta_2\right)}\right], \qquad n>2. \tag{5.7}$$

Bayesian interval estimators for θ are also calculated from the posterior distribution as described in Section 5.5.3.

The prior distribution in a Bayesian analysis usually embodies a subjective notion of probability since the frequencies of the values of a parameter such as θ rarely are known. It is the distribution of degree of belief about θ before the observational data y are obtained. As a consequence, the prior probabilities usually do not admit a direct limiting frequency interpretation and are not usually subject to experimental confirmation. However, in some situations the prior does possess a frequency interpretation. In these cases observed data may often be used to estimate the prior distribution [see Sections 7.6 and 7.7 and Chapter 13]. This had led to a general questioning of the relevance of the prior distribution and controversy regarding the desirability of a Bayesian approach. Whether a prior distribution does, in fact, exist in a manner amenable for quantification in a Bayesian analysis depends on the nature of the problem and can be meaningfully discussed only within this context. For reasons to be discussed, the Bayesian method of inference is particularly well suited for use in reliability analysis.

A distinctive feature of Bayesian inference is that it takes explicit account of prior information in the analysis. This contrasts with the approach based on sampling theory in which this information is considered only in an informal manner, if at all. The Bayesian and sampling theory approaches tend to yield the same results only under rather restrictive conditions and, even then, there is usually a difference in the interpretation of the results. For example, a confidence interval is not a probability statement about θ while a Bayesian interval is.

Figure 5.2 depicts the Bayesian method of inference. The process begins with a postulated sampling model worthy of being tentatively considered. A prior probability distribution is also postulated for those unknown parameters in the assumed sampling model for which Bayesian inferences are desired. The sample data and the prior distribution are then combined by use of Bayes' theorem. Deductive reasoning is then used in conjunction with

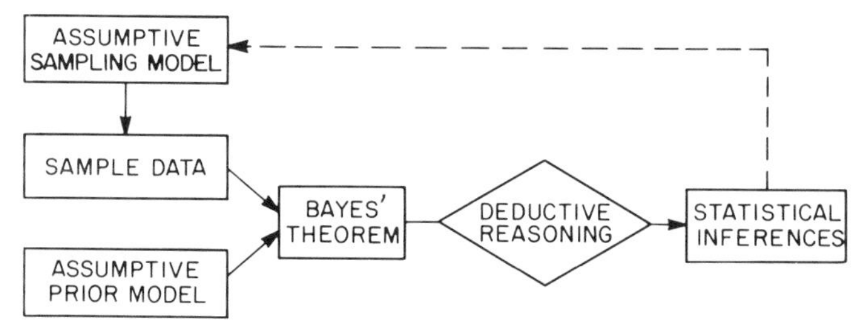

Figure 5.2. Bayesian inference.

the resulting posterior distribution to produce the desired inferences about the parameters of the assumed sampling model.

There are two further distinctive differences between the sampling theory and Bayesian approaches. The statistical inferences based on sampling theory are usually more restrictive than Bayes' due to the exclusive use of sample data. The Bayes' use of relevant past experience, which is quantified by the prior distribution, produces more informative inferences in those cases where the prior distribution accurately reflects the variation in the parameter. The degree to which more informative inferences occur otherwise depends upon the quality of the assessments embodied in the prior distribution.

The second distinction is that the Bayesian method usually requires less sample data to achieve the same quality of inferences than the method based on sampling theory. In many cases this is the practical motivation for using Bayesian methods and represents the practical advantage in the use of prior information. This is an especially important consideration in those areas of application where sample data may be either expensive or difficult to obtain, such as reliability. Table 5.2 provides a comparison of the

Table 5.2. A Summary of Certain Characteristics of the Sampling Theory and Bayesian Methods of Statistical Inference

Characteristic	Sampling Theory	Bayesian
Parameter(s) of Interest	Unknown constant(s)	Random variable(s)
Prior Distribution	Does not exist	Exists and explicitly assumed
Sampling Model	Assumed	Assumed
Posterior Distribution	Does not exist	Explicitly derived
Method of Reasoning	Inductive	Deductive
Type of Interval Estimate	Confidence interval	Probability interval
Role of Past Experience	Not applicable	Applicable
Purpose of Sampling Experiment	Supply the data for making inferences	Confirm or deny expected performance as predicted from past experience
Quality of Inferences	More restrictive than Bayes' because of exclusive use of sample data	Depends on ability to quantitatively relate past experience to the sample data
Quantity of Sample Data	Bayes' approach usually requires less because it utilizes relevant past data	

sampling theory and Bayesian methods of inference for several important characteristics.

5.2. BAYESIAN INFERENCE IN RELIABILITY

5.2.1 Problems with Sampling Theory in Reliability

Over the years reliability estimation methods based on sampling theory have been found to be extremely useful for a wide variety of problems. There are, however, many instances in which the classical methods have been found to be less than satisfactory. Several of these are now considered.

Increasing insistance on cost-effectiveness in reliability testing programs has had a decreasing effect on the case for consideration of sampling theory methods. If one were to consider only the use of sampling theory methods, one would be extremely limited, because of cost and time constraints, to a very small number of samples. Such a limited sample size would result in either a very low level of confidence in the reliability estimate or imprecise estimates. Funds and/or time usually are not available for obtaining a sample size compatible with the degree of precision required in the reliability estimates. To illustrate this fact, consider an example credited to Grohowski, Hausman, and Lamberson (1976). Suppose that, in testing to determine the reliability of a redesigned automobile air conditioning system, there are only sufficient time and funds available for testing a single vehicle for 36,000 miles. If such a test resulted in two failures, then, based on the use of an exponential model, a 90 percent two-sided confidence interval estimate for the system reliability at 12,000 miles (warranty ends) is found to be (12.2%, 88.8%). Such an estimate is so imprecise as to be practically useless.

In a similar manner sampling theory methods are especially inappropriate for reliability analysis based on scarce data. If reliability hardware is intrinsically highly reliable, then failure data will be scarce. For example, from the small database extracted from a recent NRC report,[a] no manual valve failures were observed to occur in $t = 7.9 \times 10^6$ h for a population of 16 boiling water reactor (BWR) commercial nuclear power plants. Consequently, it is not possible to construct a two-sided confidence interval estimate on the constant failure rate, assuming an exponential failure time model, and the point estimate would be zero, an overly optimistic result. As another example, consider estimating the frequency of core meltdowns in

[a] Nuclear Regulatory Commission (1978). *ECCS Valve Failure Rate Analysis* (draft), Washington, D.C.

commercial nuclear power plants operating in the United States. Based on sample data consisting of zero observed meltdowns in many reactor years of commercial operation, we are faced with the same situation as in the preceding example. In situations such as these, the methods based on sampling theory are frequently discarded in favor of more useful methods, such as the Bayesian approach.

It is well known that most engineering designs are evolutionary rather than revolutionary processes in which current equipment is modified to suit the new requirements. In such instances it would appear that the known facts regarding the reliability of the current hardware could possibly be utilized in an attempt to improve the quality of the reliability estimates of the new design. The sampling theory methods are inappropriate for incorporating such related information. For example, suppose that the warranty data on the previous model year air conditioning design yielded an observed system reliability of 78%. The fact that the new design is an evolution of the old design, which had an observed reliability of 78%, is an important consideration that cannot be taken into account using methods based on sampling theory.

Sampling theory methods can incorporate additional information of a somewhat different sort about a reliability parameter; however, it is often mathematically cumbersome to do so. If the parameter of interest is *known* to lie within a specified range it is difficult for sampling methods to consider this information. For example, suppose we are interested in estimating the mean life θ of a certain automobile system, such as steering, brakes, suspension, drivetrain, electrical, and so on, based on the assumption of an exponential model. Generally, it is known that such automotive systems exhibit a mean life between 10^4 and 10^5 miles per failure. Based on sample test data, sampling theory can do no more than compute a confidence interval estimate according to (5.4). Such estimates are still true in the required proportion of cases, but the statements take no account of our prior knowledge about the range of θ and may occasionally be idle. It may be true, but would be absurd, to assert that $10^3 \leqslant \theta \leqslant 10^6$, with a certain degree of confidence, if we already know that $10^4 \leqslant \theta \leqslant 10^5$. If we truncate such interval estimates in accord with our prior information, we then lose precision, and our confidence interval estimates are then made with confidence *at least* $1-\gamma$.

Over the years, various papers such as Bonis (1966, 1975) and Schafer (1973), have pointed out various shortcomings of reliability demonstration programs based on sampling theory methods. Such things as overdesign merely to pass demonstration tests, higher safety margins than needed, reduced confidence levels such as 60% on components, ridiculously large numbers of tests for infeasibly long periods of time, and high producer and consumer risks have been highlighted. An example will serve to illustrate this point. Suppose the mean life θ of an $\mathcal{E}(\lambda)$ distribution is an r.v. and that a reliability demonstration test is used with θ_1 (minimum acceptable MTTF) and $\delta = 10\%$ [see Section 4.6]. Suppose further that $\Pr(\Theta \leqslant \theta_1) = 0.0001$. In this case one is "paying" (in sample size) to provide 10% protection for an event that is quite rare, when in fact a larger δ would more appropriately yield a smaller sample size.

For reasons such as these, clauses in military and government reliability contracts permit other methods, such as Bayesian procedures, with the onus of justification put on the producer.

5.2.2 Advantages of Bayesian Inference in Reliability

There are several benefits in using Bayesian methods in reliability. First of all, it is important to recognize that all statistical inferential theories, whether sampling theory, Bayesian, likelihood, or otherwise, require some degree of subjectivity in their use. Sampling theory requires assumptions concerning such things as a sampling model, confidence coefficient, which estimator to use, and so on. For example, a sampling theory analysis of (5.1) proceeds as if it were believed a priori that the data were *exactly* $\mathcal{E}(\lambda)$ distributed, that each observation had *exactly* the same mean life θ, and that each observation was distributed *exactly* independently of every other sample observation. The Bayesian method provides a satisfactory way of explicitly introducing and organizing assumptions regarding prior knowledge or ignorance. These assumptions lead via Bayes' theorem to posterior inferences, that is, inference obtained once the data have been incorporated into the analysis, about the reliability parameter(s) of interest. Bayes' theorem provides a simple, error-free truism for incorporating the prior information. The engineering judgment and prior knowledge are brought out into the open and is there for everyone to see instead of being quietly hidden. The engineer usually appreciates this opportunity to divulge such prior information in a formalized way.

Is it reasonable to assume that such prior information is usually available in reliability? Engineers are always using their prior knowledge and judgment in some way. If they did not, few production lines would ever have anything rolling off them. Evans (1969) clearly understands this as indicated by his statement that "...if it is the mean time between failures (MTBF) of an exponential distribution that must be evaluated from some tests, he (the engineer) undoubtedly has some idea of what this value will turn out to be. If he does not, he is about to get fired..."

Subjectively held notions of reliability are fundamental to evolutionary system design, which is basic to reliability engineering. The fact that such notions are personal and that each engineer may not agree on the quantification of this subjective knowledge should not be used to discredit the Bayesian argument. Different engineers rarely come up with the same answer to any complex problem, yet engineers do proceed to solve problems adequately, even though the solutions may be neither unique nor superb.

The fact that different choices of a prior distribution may lead to different inferences has worried some statisticians. We feel, however, that explicit dependency on the choice of what is subjectively believed to be true is an advantage of the Bayesian method rather than the opposite. If a group of engineers hold the same degree of belief, then such agreement serves to reassure the reliability analyst that the resulting inferences are probably correct. On the other hand, if there is basic disagreement among the engineers, a real conflict exists. The reliability analyst would, in this case, either have to ignore the judgment of one or more of the engineers, or arrange that further data be obtained to resolve the conflict. In either case, the Bayesian analysis shows to what extent different results may or may not be obtained according to the differences in the prior opinions held. Such a result serves to enhance the value of a Bayesian reliability analysis rather than discredit it.

There are two important practical benefits of a Bayesian analysis. One is the increased quality of the inferences, provided the prior information accurately reflects the true variation in the parameter(s). The other is the reduction in testing requirements (test time and/or sample size) that often occurs in Bayesian reliability demonstration test programs. Both of these are the result of formally including additional information, in the form of the prior distribution, in the analysis.

There is yet another important advantage of a Bayesian analysis; namely, inferences that are unacceptable *must* come from incorrect assumptions and not from inadequacies of the method used to provide the inferences. In this regard, the Bayesian procedure rectifies many shortcomings of the sampling theory method as discussed in the preceding section. Inferences based on deductive arguments are more direct than those based on inductive arguments.

In performing system reliability analysis based on component data, another important advantage of Bayesian inference becomes apparent. The rules for manipulating probability statements on components into corresponding statements on system reliability are well-known, whereas equivalent rules for manipulation of confidence statements are not. The use of these rules is illustrated in Chapters 11 and 12 where systems of components are examined using probabilistic methods. Thus, Bayesian methods have more appeal when inferences on system reliability are desired, as Bayesian statements embody probability notions rather than confidence.

Let us now illustrate one important advantage of Bayesian inference in reliability. Returning to the automobile air conditioning example introduced in the preceding section, Grohowski, Hausman, and Lamberson (1976) constructed a Bayesian interval estimate for system reliability at $t=12{,}000$ miles to be (66%,84%) with a point estimate of 75%. Compared to the previous model year observed value of 78%, the Bayesian results indicate that the system on the average is not quite as good as the old, but is a figure that is believable. On the other hand, the classical estimate of 51% is so low as to be absurdly pessimistic. That the Bayesian method produces believable results goes a long way in convincing the engineer of the utility of Bayesian methods.

5.2.3 Bayes' Theorem with Subjective Probabilities

Bayes' theorem, which was introduced in Section 2.10, is the fundamental tool used to arrive at Bayesian inferences. Before proceeding to a discussion of this important theorem, let us introduce the following notation. Let

$\underset{\sim}{\theta}'=(\theta_1,\ldots,\theta_k)=k\times 1$ vector of parameters,

$\underset{\sim}{x}'=(x_1,x_2,\ldots,x_n)=n\times 1$ vector of statistically independent observations of the r.v. X (the *sample data*),

$g(\underset{\sim}{\theta})=$ the joint prior probability distribution of $\underset{\sim}{\Theta}$ (the *prior model*),

$f(x_i|\underset{\sim}{\theta})=$ the conditional probability distribution of X_i given $\underset{\sim}{\theta}$ (the *sampling model*),

$f(\underset{\sim}{x},\underset{\sim}{\theta})=$ the joint probability distribution of $\underset{\sim}{X}$ and $\underset{\sim}{\Theta}$,

$f(\underset{\sim}{x}|\underset{\sim}{\theta})=\prod_{i=1}^{n} f(x_i|\underset{\sim}{\theta})=$ the joint conditional probability distribution of $\underset{\sim}{X}$ given $\underset{\sim}{\theta}$,

$f(\underset{\sim}{x})=$ the marginal probability distribution of $\underset{\sim}{X}$,

$g(\underset{\sim}{\theta}|\underset{\sim}{x})=$ the joint posterior probability distribution of $\underset{\sim}{\Theta}$ given $\underset{\sim}{x}$ (the *posterior model*).

In the remainder of the book we sometimes refer to the prior and posterior distribution simply as the "prior" and "posterior," respectively. Also, by "probability distribution" we will mean either a pdf in the case of a continuous r.v., or a probability mass function, in the case of a discrete r.v. Occasionally we will also have need to refer to the cdf associated with an r.v. The prior model is assumed to represent the totality of subjective information available concerning the parameter vector $\underset{\sim}{\theta}$ prior to the observation of the sample data $\underset{\sim}{x}$. Thus, it is not functionally dependent upon $\underset{\sim}{x}$. On the other hand, the sampling model depends on the values of the k parameters in $\underset{\sim}{\theta}$ and is thus a conditional probability distribution. The posterior model tells us what is known about $\underset{\sim}{\theta}$ given knowledge of the data $\underset{\sim}{x}$. It is essentially an updated version of our prior knowledge about $\underset{\sim}{\theta}$ in light of knowledge of the sample data—hence, the name *posterior model*. It is intuitive that the posterior model should represent a modification of the subjective knowledge about $\underset{\sim}{\theta}$ expressed by the prior model in light of the observed sample data. If the sample data support our subjective opinion about $\underset{\sim}{\theta}$, then the posterior model should reflect increased confidence in the subjective notions embodied in the prior model. On the other hand, if the sample data do not support the subjective information, the posterior model should reflect a weighted consideration of both assessments. This is precisely the situation that results in using Bayes' theorem to calculate the posterior model.

Bayes' theorem states that the posterior model is related to the prior and sampling models according to

$$g(\underset{\sim}{\theta}|\underset{\sim}{x})=\frac{\left[\prod_{i=1}^{n} f(x_i|\underset{\sim}{\theta})\right]g(\underset{\sim}{\theta})}{f(\underset{\sim}{x})}=\frac{f(\underset{\sim}{x}|\underset{\sim}{\theta})g(\underset{\sim}{\theta})}{f(\underset{\sim}{x})}, \tag{5.8}$$

where the marginal distribution $f(\underset{\sim}{x})$ may be obtained according to

$$f(\underset{\sim}{x})=\begin{cases}\int f(\underset{\sim}{x}|\underset{\sim}{\theta})g(\underset{\sim}{\theta})\,d\underset{\sim}{\theta}, & \underset{\sim}{\theta} \text{ continuous}\\ \sum f(\underset{\sim}{x}|\underset{\sim}{\theta})g(\underset{\sim}{\theta}), & \underset{\sim}{\theta} \text{ discrete,}\end{cases} \tag{5.9}$$

where the summation or integration is taken over the admissible range of $\underset{\sim}{\theta}$. Given the sample data $\underset{\sim}{x}$, $f(\underset{\sim}{x}|\underset{\sim}{\theta})$ may be regarded as a function, not of $\underset{\sim}{x}$, but of $\underset{\sim}{\theta}$. When so regarded, Fisher (1922) refers to this as the *likelihood function* of $\underset{\sim}{\theta}$ given $\underset{\sim}{x}$, which is usually written as $L(\underset{\sim}{\theta}|\underset{\sim}{x})$ to insure its distinct interpretation apart from $f(\underset{\sim}{x}|\underset{\sim}{\theta})$. The likelihood function is important in Bayes' theorem and is the function through which the sample data $\underset{\sim}{x}$ modify prior knowledge of $\underset{\sim}{\theta}$. It can be regarded as the function that represents the information about $\underset{\sim}{\theta}$ contained in the sample data. It is observed in (5.8) that multiplication of the likelihood function by a constant has no effect on the posterior distribution, as the constant multiplier cancels in both the numerator and denominator. In this regard, the likelihood function is said to be defined up to a multiplicative constant.

Normalizing the Likelihood Function. By choosing the right multiplicative constant the likelihood function may be "normalized" in such a way that either its hypervolume or supremum is unity. When the integral $\int L(\underset{\sim}{\theta}|\underset{\sim}{x})\,d\underset{\sim}{\theta}$, taken over the admissible range of $\underset{\sim}{\theta}$, is finite, the normalized likelihood function

$$\ell(\underset{\sim}{\theta}|\underset{\sim}{x}) = \frac{L(\underset{\sim}{\theta}|\underset{\sim}{x})}{\int L(\underset{\sim}{\theta}|\underset{\sim}{x})\, d\underset{\sim}{\theta}} \tag{5.10}$$

is referred to as the *standardized likelihood function*. In a similar way, the normalized likelihood function given by

$$R(\underset{\sim}{\theta}|\underset{\sim}{x}) = \frac{L(\underset{\sim}{\theta}|\underset{\sim}{x})}{\underset{\underset{\sim}{\theta}}{\text{Sup}}\, L(\underset{\sim}{\theta}|\underset{\sim}{x})} \tag{5.11}$$

is known as the *relative likelihood function*, provided that the denominator is finite. Developments of such authors as Barnard, Jenkins, and Winsten (1962), Sprott and Kalbfleisch (1965), and others, have made extensive use of the relative likelihood function as a basis for a likelihood theory of statistical inference. As only the relative value of the likelihood function is important in Bayes' theorem, it is sometimes convenient to consider such normalized likelihood functions in making Bayesian inferences.

The Fundamental Relationship. If we regard $f(\underset{\sim}{x}|\underset{\sim}{\theta})$ in Bayes' theorem as the likelihood function $L(\underset{\sim}{\theta}|\underset{\sim}{x})$, then we may write Bayes' theorem as

$$g(\underset{\sim}{\theta}|\underset{\sim}{x}) \propto g(\underset{\sim}{\theta}) L(\underset{\sim}{\theta}|\underset{\sim}{x}), \tag{5.12}$$

which says that the posterior distribution is proportional to the product of the prior distribution and the likelihood function. The constant of proportionality, necessary to insure that the posterior distribution integrates or sums to one, is the integral or sum of the product, which is the marginal distribution of $\underset{\sim}{X}$ given by (5.9). Thus, in words we have the fundamental relationship given by

$$\begin{aligned} \text{Posterior Distribution} &\propto \text{Prior Distribution} \times \text{Likelihood (Function)} \\ &= \frac{\text{Prior Distribution} \times \text{Likelihood (Function)}}{\text{Marginal Distribution}}. \end{aligned}$$

Bayes' Theorem with Sequential Sample Data. Bayes' theorem provides a mathematical framework for processing new sample data as such data becomes sequentially available over time. The theorem provides a mechanism for continually updating our knowledge about $\underset{\sim}{\theta}$ as more sample data become available.

To illustrate how this is done, suppose that we have an initial set of sample data $\underset{\sim}{x}_1$, as well as a prior distribution for $\underset{\sim}{\theta}$. Bayes' theorem says that

$$g(\underset{\sim}{\theta}|\underset{\sim}{x}_1) \propto g(\underset{\sim}{\theta}) L(\underset{\sim}{\theta}|\underset{\sim}{x}_1). \tag{5.13}$$

Suppose now that we later obtain a second set of sample data $\underset{\sim}{x}_2$ which is statistically independent of the first sample. Then

$$\begin{aligned} g(\underset{\sim}{\theta}|\underset{\sim}{x}_1, \underset{\sim}{x}_2) &\propto g(\underset{\sim}{\theta}) L(\underset{\sim}{\theta}|\underset{\sim}{x}_1) L(\underset{\sim}{\theta}|\underset{\sim}{x}_2) \\ &\propto g(\underset{\sim}{\theta}|\underset{\sim}{x}_1) L(\underset{\sim}{\theta}|\underset{\sim}{x}_2), \end{aligned} \tag{5.14}$$

as the combined likelihood may be factored as a product. It is observed that this expression is of the same form as (5.13) except that the posterior distribution of Θ given $\underset{\sim}{x}_1$ assumes the role of the prior distribution $g(\underset{\sim}{\theta})$. This process may be repeated. In fact, each sample observation can be processed separately if desired. For example, if we have n independent observations that are to be sequentially processed one at a time, then at the ith processing stage we have

$$g(\underset{\sim}{\theta}|x_1, \ldots, x_i) \propto g(\underset{\sim}{\theta}|x_1, \ldots, x_{i-1}) L(\underset{\sim}{\theta}|x_i), \qquad i = 2, \ldots, n \tag{5.15}$$

and where at the first processing stage

$$g(\underset{\sim}{\theta}|x_1) \propto g(\underset{\sim}{\theta}) L(\underset{\sim}{\theta}|x_1).$$

This shows how knowledge about $\underset{\sim}{\theta}$ can be continually updated as new sample data become available. It is also true that "batches" of data may be sequentially processed where each batch may consist of one or more new observations. This basic fact concerning Bayes' theorem forms the basis of real-time processing of sample data by use of such well-known mathematical devices as the Kalman filter [Bucy and Kalman (1961)].

Example 5.1. Suppose two design engineers A and B are given the task of redesigning an industrial engine that is to have a mean life θ of at least 3000 h. Engineer A has had considerable experience in the design of similar engines and can make a moderately good guess of the success of the effort. On the other hand, B has had less experience and is far less certain of the outcome of the task. With the help of a reliability analyst, both *A* and *B* have been separately encouraged to quantify their degree of belief in the success of their task according to the following prior distributions.

θ (h)	$g_A(\theta)$	$g_B(\theta)$
0–1000	0.01	0.15
1000–2000	0.04	0.15
2000–3000	0.20	0.20
3000–4000	0.50	0.20
4000–5000	0.15	0.15
>5000	0.10	0.15
Sum	1.00	1.00

It is observed that *A* believes a priori that the probability is 0.75 that the design will be successful, while *B* believes that the probability is 0.50. Engineer *A* further believes that there is only an a priori 0.05 probability that the design effort will grossly fail (a mean life less than 2000 h), whereas *B* believes that the a priori probability that this will happen is 0.30. In other words, the past experience of *A* leads him to a more optimistic view concerning the success of the task, while *B* has a more pessimistic attitude. These prior distributions are shown in Figure 5.3*a* and *b*.

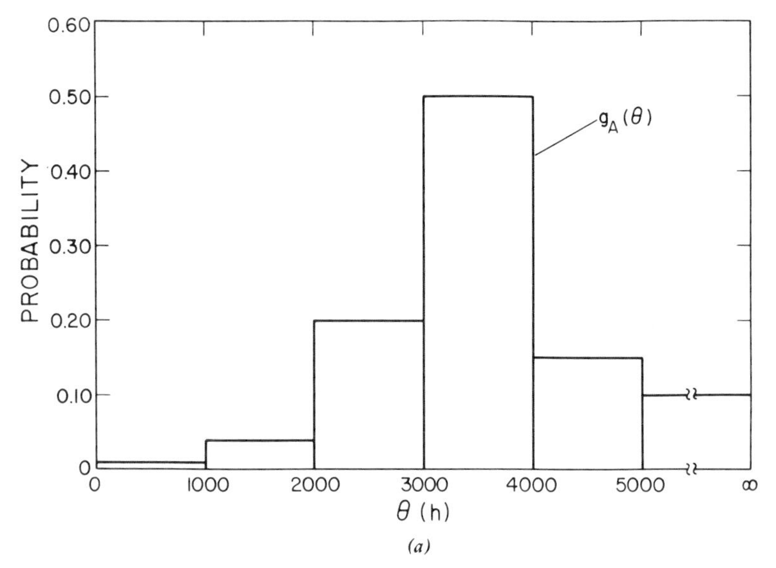

(*a*)

Figure 5.3. The prior and posterior distributions for Engineers *A* and *B*.

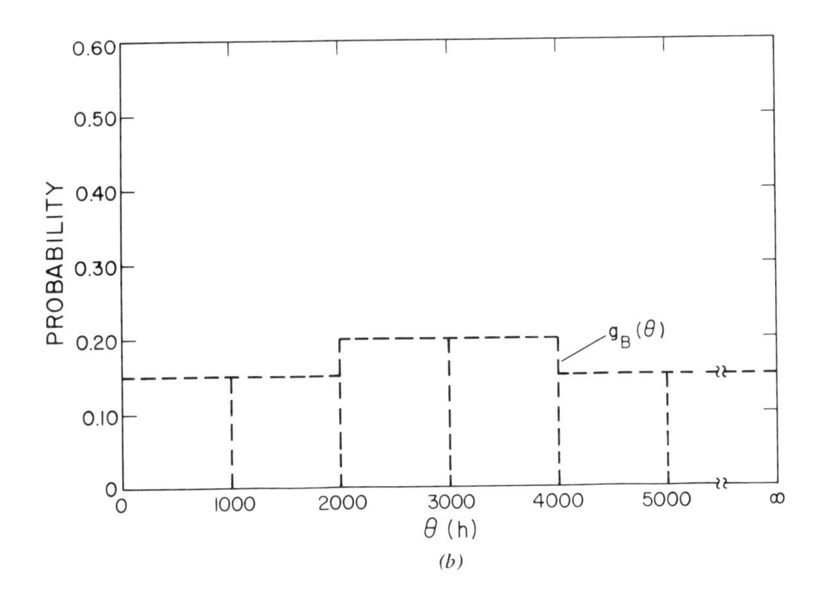

(b)

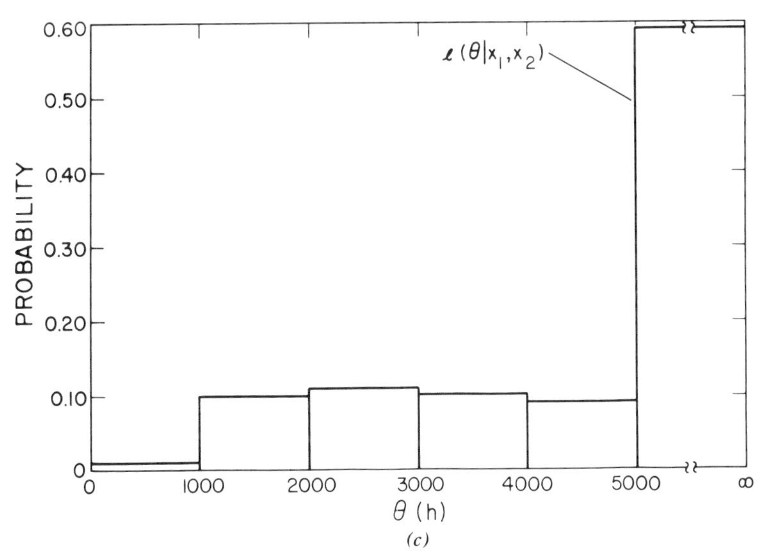

(c)

Figure 5.3. *(Continued)*

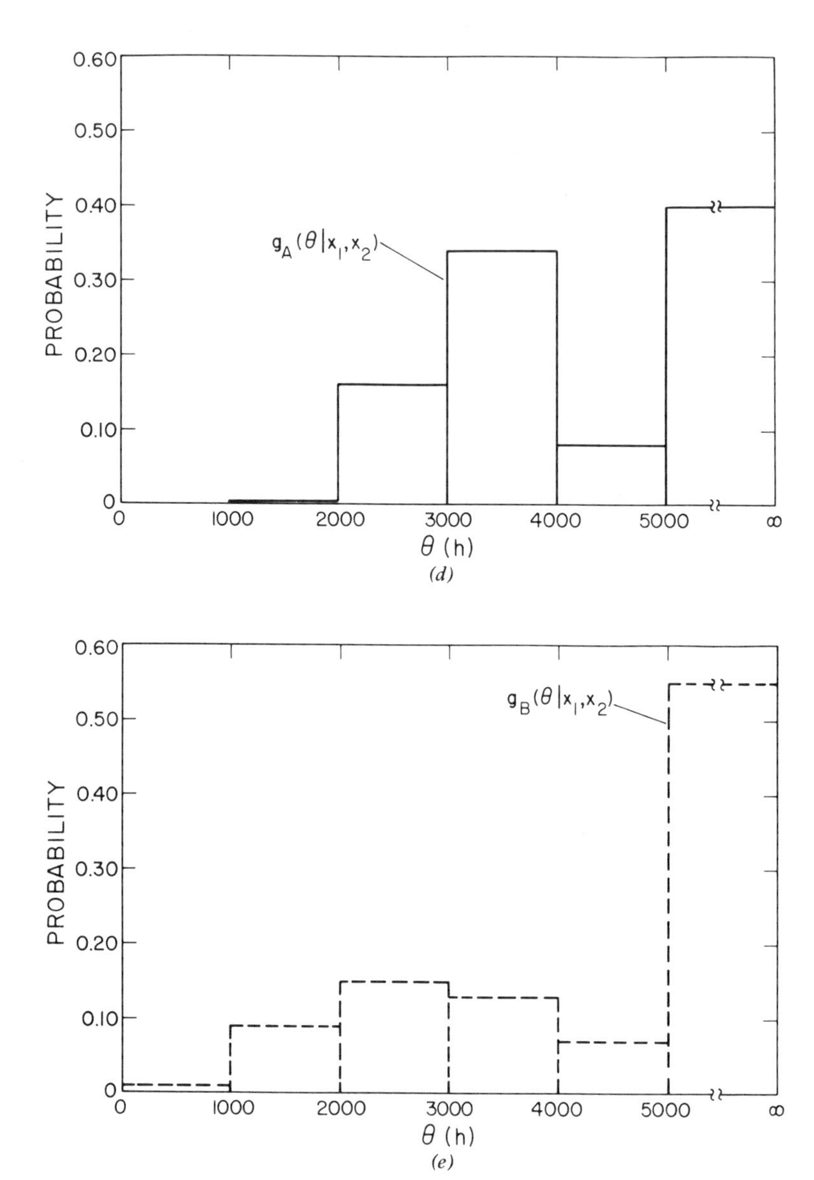

Figure 5.3. (*Continued*)

Following the redesign effort the reliability analyst proposes that two prototype engines be tested until both fail. Assuming an $\mathcal{E}(\lambda)$ failure time distribution with mean life θ, the standardized likelihood function of θ is

$$\ell(\theta|x_1, x_2) = \frac{(x_1+x_2)}{\theta^2}\exp\left[-\frac{(x_1+x_2)}{\theta}\right], \qquad 0<\theta<\infty.$$

Now suppose that the test has been conducted and that $x_1 = 2000$ and $x_2 = 2500$ h are observed. The standardized likelihood that θ is between a and b, where $a<b$, is

$$\ell(a<\theta<b|2000, 2500) = \int_a^b \frac{4500}{\theta^2}\exp\left(-\frac{4500}{\theta}\right) d\theta$$

$$= \exp\left(-\frac{4500}{b}\right) - \exp\left(-\frac{4500}{a}\right).$$

Using this we can easily compute the values of the standardized likelihood function to be

θ (h)	$\ell(\theta\|2000, 2500)$
0–1000	0.01
1000–2000	0.10
2000–3000	0.12
3000–4000	0.10
4000–5000	0.08
>5000	0.59
Sum	1.00

The standardized likelihood function is shown in Figure 5.3*c*. Note that the likelihood is 77% that θ exceeds the requirement of 3000 h, in spite of the fact that neither engine survived 3000 h. This is due to the skewed L-shape of the $\mathcal{E}(\lambda)$ distribution and associated property which says that, in such a model, 63% of the failures are expected to occur prior to the mean life θ. Consequently, it is anticipated that the mean life should exceed 5000 h, as indicated by the likelihood of 59%, given $x_1 = 2000$ and $x_2 = 2500$.

Now let us compute the posterior distribution for A in light of the test results. According to Bayes' theorem

$$g_A(0<\theta<1000|2000, 2500) = \frac{\ell(0<\theta<1000|2000, 2500)\, g_A(0<\theta<1000)}{f_A(2000, 2500)}$$

where $f_A(2000, 2500)$ denotes the standardized marginal distribution for A evaluated at $x_1 = 2000$ and $x_2 = 2500$. It may be calculated according to

$$f_A(2000,2500) = \int_0^\infty \frac{4500}{\theta^2} \exp\left(-\frac{4500}{\theta}\right) g_A(\theta)\, d\theta$$

$$= 0.01 \int_0^{1000} \frac{4500}{\theta^2} \exp\left(-\frac{4500}{\theta}\right) d\theta$$

$$+ 0.04 \int_{1000}^{2000} \frac{4500}{\theta^2} \exp\left(-\frac{4500}{\theta}\right) d\theta$$

$$+ \cdots + 0.10 \int_{5000}^{\infty} \frac{4500}{\theta^2} \exp\left(-\frac{4500}{\theta}\right) d\theta$$

$$= 0.01 \exp\left[\left(-\frac{4500}{1000}\right)\right] + 0.04\left[\exp\left(-\frac{4500}{2000}\right) - \exp\left(-\frac{4500}{1000}\right)\right]$$

$$+ \cdots + 0.10\left[1 - \exp\left(-\frac{4500}{5000}\right)\right]$$

$$= 0.15.$$

Thus

$$g_A(0 < \theta < 1000 | 2000, 2500) = \frac{(0.01111)(0.01)}{0.15} = 7.4 \times 10^{-4}.$$

Similarly, we compute the remainder of the posterior distribution for A to be

θ (h)	$g_A(\theta \| 2000, 2500)$
0–1000	7.4×10^{-4}
1000–2000	2.5×10^{-2}
2000–3000	0.16
3000–4000	0.34
4000–5000	0.08
>5000	0.40
Sum	1.00

This posterior distribution is shown in Figure 5.3*d*. It is noted that, in light of the observed data, the posterior distribution of A indicates that the engineer now believes there is a 0.82 probability that θ exceeds the requirement, upwards from 0.75 prior to the data. By comparing Figure 5.3*a* and *d*, it is observed that *A*'s strong belief that $3000<\theta<4000$ has been reduced, whereas the belief that $\theta>5000$ has increased in view of the test results. We see that, in this case, neither the prior nor the likelihood "dominates" the posterior, but that both are fairly equally weighted in the analysis. This is not always the case as will now be shown.

The posterior distribution for *B* is similarly computed by use of Bayes' theorem; for example,

$$g_B(0<\theta<1000|2000,2500)=\frac{\ell(0<\theta<1000|2000,2500)\,g_B(0<\theta<1000)}{f_B(2000,2500)},$$

where the standardized marginal distribution for *B* evaluated at $x_1=2000$ and $x_2=2500$ is calculated according to

$$f_B(2000,2500)=\int_0^{\infty}\frac{4500}{\theta^2}\exp\left(-\frac{4500}{\theta}\right)g_B(\theta)\,d\theta$$

$$=0.15\int_0^{2000}\frac{4500}{\theta^2}\exp\left(-\frac{4500}{\theta}\right)d\theta+0.20\int_{2000}^{4000}\frac{4500}{\theta^2}\exp\left(-\frac{4500}{\theta}\right)d\theta$$

$$+0.15\int_{4000}^{\infty}\frac{4500}{\theta^2}\exp\left(-\frac{4500}{\theta}\right)d\theta$$

$$=0.16.$$

Thus, the posterior distribution for *B* is computed to be

θ (h)	$g_B(\theta\|2000,2500)$
0–1000	0.01
1000–2000	0.09
2000–3000	0.15
3000–4000	0.13
4000–5000	0.07
>5000	0.55
Sum	1.00

This distribution is shown in Figure 5.3*e*. In view of the observed data, *B* now believes that there is a 0.75 probability that $\theta>3000$, as opposed to the prior probability of 0.50. The likelihood is thus observed to have a fairly influential effect on *B*'s posterior distribution, as clearly seen by observation of Figure 5.3*b*, *c*, and *e*. In fact, apart from a slight rise in the center of the posterior distribution due to the slight influence of the prior, the posterior resembles the likelihood. In such a case we say that the prior is "dominated" by the likelihood. ■

Dominant Likelihood. In general, the sharpness or flatness of the prior distribution relative to the sharpness or flatness of the likelihood determines whether the prior dominates the likelihood. In general, if the prior is flat relative to the likelihood, then the likelihood dominates the prior; otherwise, the opposite is true. Box and Tiao (1973) point out that, often in the case of analysis of scientific data, the likelihood dominates the prior and rarely is the opposite the case. However, in reliability applications the prior distribution often dominates the likelihood as sample data are scarce and extensive engineering analysis is often employed in reliability and risk assessments. If the results of such analyses are used to construct a prior distribution, then such a prior will often be sharp compared to the likelihood based on the sparse data.

For certain types of prior distributions the likelihood will dominate the prior. If the prior distribution is approximately uniform over a certain range and the likelihood is appreciable over this same range, then the likelihood will dominate the prior in such cases. This can be seen by observation of Bayes' theorem in which a constant is substituted for the prior distribution. The resulting posterior distribution is then approximately equal to the standardized likelihood function. This situation is somewhat reflected in Figure 5.3*b* and *e*, since the prior in *b* is almost uniform.

In many cases, the likelihood will increasingly tend to dominate the prior as the number of observations increases. For large sample sizes the posterior distribution will be approximately numerically equal to the standardized likelihood and the difference between Bayesian inferences and inferences based on the likelihood function will be insignificant. In practical problems a moderate sample size will yield a dominant likelihood if the sample results are consistent with the prior distribution. If the sample results are inconsistent with the prior assumptions, then the likelihood will usually not dominate the prior, and a weighted combination of both will be reflected in the posterior distribution. Such situations are examined as they arise in the development of the Bayesian estimators in Chapters 7–9.

5.3. PERFORMING A BAYESIAN RELIABILITY ANALYSIS

A variety of considerations must be addressed once the decision has been made to use a Bayesian approach for making inferences in reliability.

Various difficulties must be overcome and procedures followed in producing defensible inferences.

Before examining some of these issues, it is important to understand what we mean by the term "Bayesian reliability analysis," and so we state the following:

Definition 5.1. A *Bayesian reliability analysis* consists of the use of statistical methods in reliability problems that involve parameter estimation in which one or more of the parameters is considered to be an r.v. with a nondegenerate prior probability distribution which expresses the analyst's prior degree of belief about the parameters.

5.3.1 Problems and Considerations in a Bayesian Reliability Analysis

Most of the difficulties in performing a Bayesian reliability analysis concern the identification, selection, and justification of the prior distribution. Such relevant questions as: "What sort of prior distribution should I use?", "What sources of data are available for selecting a prior model?", "How should I quantify the subjective information?", "What procedure should I use to fit a prior distribution to the subjective data?", must be addressed.

If multiple sources of relevant data are available for use in the analysis, there is even a more fundamental issue that must be settled. It must be decided which data are to be used in fitting the prior distribution and which data are to be used in the likelihood function. This is not always an easy task. Traditionally, the softer and more subjective data sources have been alloted to the prior, whereas the harder and more objective sample test data have been used in the likelihood. For example, suppose we are interested in Bayesian inferences about the frequency of core meltdown in commercial United States nuclear power reactors. Risk assessments have produced estimates of this event which are available to us. These are usually somewhat subjective estimates based on analysis. In addition, there are historical operating data from the population of commercial power reactors, both within and outside the United States. There are also various historical data sources on noncommercial, such as military, power reactors.

Which of these data sources should be used to fit a prior distribution is not clear cut and judgment must be exercised. It may even be decided to use certain subjective data sources in the likelihood portion of Bayes' theorem, as there is nothing inherent in Bayes' theorem that prohibits this usage. Apostolakis and Mosleh (1979) consider Bayesian estimates of reactor core melt frequency in which subjective data are used in the likelihood, and the more objective data sources are used to fit a prior distribution.

The problem of prior identification concerns such things as whether to use a continuous or discrete prior distribution, a noninformative or informative prior [see Section 6.1], or a prior distribution that is mathematically convenient, such as the family of conjugate distributions [see Section 6.2]. There is a variety of considerations in making such decisions and they concern such things as mathematical simplicity and convenience, flexibility of the family, degree of softness or hardness of the subjective data sources, and so on. These topics are further discussed in Chapter 6.

Prior selection considers the fitting of a specific prior distribution based on the available subjective data sources, once the prior has been identified. It deals mainly with the estimation procedure to be used in determining the values of the parameters of the prior distribution (called the *prior parameters*), and by which the subjective data are to be solicited. For example, in the case of information subjectively held by a group of design engineers, are questionnaires, personal interviews, or some other method to be used in quantifying such information? The quality of the subjective information may depend on what information is sought, how the questions are posed, and so on. For example, what information is received when a specified percentile value of a failure rate is requested? Is it reasonable to believe that a specified percentile (e.g., the 90th), can be accurately known and supplied by a design engineer? Are responses consistent? To illustrate this, suppose a respondent supplies three (e.g., 5th, 50th, 95th) percentiles to be used in fitting a two-parameter prior distribution. Would the three fitted distributions determined by the three different pairs of percentiles [(5,50), (5,95), (50,95)] be in close agreement? If not, how are differences to be reconciled? Are 5th percentiles less precisely known by design engineers than, say, 50th percentiles (medians)? Which two percentiles are best for fitting the prior distribution? The answers to questions such as these reflect on the ability to select a prior distribution that accurately assesses the quality of the subjective knowledge to be used in the Bayesian analysis. In many cases the precise answers to questions such as these are unknown and judgment must be used by the analyst in deciding how the prior is to be fitted. Such questions are considered in Chapter 6.

Little justification is all too often given in a Bayesian reliability analysis for the prior distribution selected. Because the prior distribution is based on degree of belief does not remove the analyst's responsibility to adequately defend the basis for its selection. Otherwise, the resulting inferences can often be criticized on the basis that the prior distribution has been biasedly "chosen" to give "selfserving" results that do not reflect the true random variation inherent in the parameter. This is frequently a major criticism of Bayesian inferences.

To prevent such criticism, thorough attention should be directed at adequate justification and detailed description of the prior distribution selected. This should include a clear description of the data sources used to fit the prior, the method used to fit the prior to these data, a preposterior analysis of the fitted prior, and the mathematical implications of the chosen prior [see e.g., Section 6.7].

5.3.2 Preposterior Analysis

Preposterior analysis is a procedure for analyzing a prior distribution, prior to the attainment of test data, based on its impact on the posterior distribution that will result when either confirming or contradictory sample data are obtained. It consists of the following steps:

1. Select a prior distribution tentatively considered to be satisfactory and for which preposterior analysis is desired as an indication of the suitability of the selection.
2. Using the amount of sample test data ultimately expected, consider an ensemble of likely and unlikely hypothetical sample test data results.
3. For the tentative prior distribution and each set of hypothetical test data results, compute the resultant posterior distribution using Bayes' theorem.
4. Study the set of resulting posterior distributions to determine whether they seem reasonable in light of the hypothetical test data.
5. If they are reasonable, the prior distribution becomes a strong candidate for use. If they are not, readjust the prior and reconduct the preposterior analysis until the prior is satisfactory.

Such a preposterior analysis (so named because it examines possible test results using Bayesian methods before they have actually occurred, and while it is therefore still possible to "adjust" the prior) is a valuable aid in performing a Bayesian reliability analysis. A hypothetical example will serve to illustrate its use in practice.

Example 5.2. Suppose that an accelerated life test is proposed to sharpen the degree of belief about the reliability of a certain piece of equipment. Three units are to be tested and only the success or failure of each item is considered. The possible reliability values and the tentative prior distribution have been chosen by an engineer to be:

Reliability	0.1	0.5	0.8	0.95	0.99
Prior	0.1%	0.1%	0.8%	95%	4%

Here the reliability is the probability that a unit will survive the test. We desire a preposterior analysis of this prior distribution.

It is anticipated that all three units will survive the test; however, it is possible, although unlikely, that either zero or one unit may survive the test. Thus, we consider the ensemble of three hypothetical results {3 successes, 0 failures}, {1 success, 2 failures}, and {0 successes, 3 failures} in the analysis.

The likelihood of each of these results and the resulting posterior distributions are computed to be

Reliability Prior	0.1 0.1%	0.5 0.1%	0.8 0.8%	0.95 95%	0.99 4%
Likelihood (3-S, 0-F)	0.001	0.125	0.512	0.857	0.970
Likelihood (1-S, 2-F)	0.243	0.375	0.096	0.007	0.0003
Likelihood (0-S, 3-F)	0.729	0.125	0.008	0.0001	0.000001
Posterior (3-S, 0-F)	0.000001	0.00015	0.0048	0.950	0.045
Posterior (1-S, 2-F)	0.030	0.047	0.095	0.826	0.002
Posterior (0-S, 3-F)	0.719	0.123	0.063	0.094	0.00004

By observation of the posterior distributions, it may seem unreasonable to the engineer that if one success and two failures occur, the posterior degree of belief for a reliability value of 0.95 should be as large as 83%. It also appears to be somewhat unreasonable that, if zero successes occur, this posterior degree of belief should be as large as 9%. This posterior distribution does not agree with the engineer's subjective notions of what should occur; so the tentative prior has to change.

The engineer now considers a different tentative prior. The following table shows this prior, the likelihoods, and the resulting posterior distributions.

Reliability Prior	0.1 1 %	0.5 1 %	0.8 1.5%	0.95 95%	0.99 1.5%
Likelihood (3-S, 0-F)	0.001	0.125	0.512	0.857	0.970
Likelihood (1-S, 2-F)	0.243	0.375	0.096	0.007	0.0003
Likelihood (0-S, 3-F)	0.729	0.125	0.008	0.0001	0.000001
Posterior (3-S, 0-F)	0.00001	0.002	0.009	0.972	0.017
Posterior (1-S, 2-F)	0.170	0.263	0.101	0.466	0.0003
Posterior (0-S, 3-F)	0.833	0.143	0.013	0.011	0.000002

Although there is still a 47% posterior degree of belief for a reliability value of 0.95, given one success and two failures are observed, this may be much more reasonable than before. Also, the former 9% peak at 0.95 for zero successes has now been reduced to 1%, a result more believable to the engineer. This prior thus represents a "better" prior than the original prior distribution, although additional analysis should be performed before its actual adoption. ■

5.3.3 Essential Elements in a Bayesian Reliability Analysis

There are several elements in a good Bayesian reliability analysis; namely,

1. A detailed justification and analysis of the prior distribution selected, with a clear understanding of the mathematical implications of the prior,
2. A thorough documentation of the data sources used in identifying and selecting the prior,
3. A preposterior analysis of the prior distribution with hypothetical test results,
4. A clearly defined posterior distribution on the parameter(s) of interest,
5. An analysis of the sensitivity of the Bayesian inferences to the prior model selected.

Now 1–3 have been discussed in Sections 5.3.1–5.3.2. In regard to 4, it is often insufficient to provide only a mathematical statement of the posterior distribution, as most of the features of the distribution are not readily apparent in such a form. A deeper appreciation is provided by giving several percentile values, as well as certain measures of location, such as the mean and mode (provided they exist) of the posterior. A plot of the posterior is also quite informative. For example, to state that a posterior distribution for reactor core meltdown is a $\mathcal{G}(2, 1.43 \times 10^{-4})$ distribution is not particularly informative. However, by stating that the 5th, 50th, and 95th percentiles are 5.1×10^{-5}, 2.4×10^{-4}, and 6.8×10^{-4} meltdowns per reactor-year, and that the mean and mode are 2.9×10^{-4} and 1.4×10^{-4}, respectively, is somewhat more informative to the practitioner.

Martz and Waller (1979) illustrate 5 by means of an example related to Bayesian reliability demonstration testing [see Chapter 10].

5.4. BAYESIAN DECISION THEORY

Statistical decision theory is a field of statistics concerned with the development of techniques for making decisions in situations in which stochastic components play a crucial role. Important applications exist in business decision making, management science, operations research, as well as the areas of statistical inference concerned with parameter estimation and testing of hypotheses.

The field of statistical decision theory was originally developed by Wald (1950), following the development of game theory by Von Neumann and Morgenstern (1947). Some introductory and advanced texts on statistical decision theory are Blackwell and Girshick (1954), Chernoff and Moses (1959), De Groot (1970), Ferguson (1967), Lindley (1970), Raiffa and Schlaifer (1961), and Schlaifer (1959). The specialized areas of statistical decision theory known as Bayesian decision theory and Bayesian statistics are considered in such texts as Box and Tiao (1973), Morgan (1968), Schmitt (1969), and Winkler (1972).

Statistical decision theory concerns the situation in which a decision maker has to make a choice from a given set of available actions $(a_1, \ldots, a_p)$ and where the loss of a given action depends upon a state of nature θ which is unknown. In Bayesian decision theory, Θ is assumed to have a prior distribution. The decision maker combines prior knowledge of θ and stochastic information provided by an experiment in the form of a posterior distribution of θ and then chooses the action that minimizes the expected loss over the posterior distribution.

5.4.1 Basic Elements

The basic elements of statistical decision theory are

1. A space $\Omega_\theta = \{\underset{\sim}{\theta}\}$, which may be vector-valued, of the possible states of nature,
2. An action space $A = \{a\}$ of the possible courses of action,
3. A loss function $L(\underset{\sim}{\theta}, a)$ representing the loss incurred when action a is taken and the state of nature is $\underset{\sim}{\theta}$.

Example 5.3. Consider a decision whether to accept or reject a lot of components as having met a certain MTTF θ reliability requirement. Here there are only two actions, $a_1 =$ accept and $a_2 =$ reject. The state of nature is the true underlying value of θ which is unknown, and we consider only two values of θ. Suppose that $\theta_1 = 1000$ h is the requirement and $\theta_2 = 250$ h is the minimum acceptable MTTF. Based on an analysis of the economic consequences of each decision for each state of nature, the economic net loss per item is given by

	State of Nature	
Action	$\theta_1 = 1000$	$\theta_2 = 250$
Accept Lot (a_1)	\$ 0	\$15
Reject Lot (a_2)	\$10	\$ 5

These entries can be considered to be the values of a monetary loss function $L(\theta, a)$ where $L(\theta_1, a_1) = \$0$, $L(\theta_2, a_1) = \$15$, $L(\theta_1, a_2) = \$10$, and $L(\theta_2, a_2) = \$5$. ■

A sampling experiment is often conducted in order to provide some data to assist in making a decision about which action should be taken. When this is the case, we add the following elements to the basic structure:

4. An observable r.v. $\underset{\sim}{X}$, which may be vector-valued, defined on a sample space $\mathscr{X} = \{\underset{\sim}{x}\}$, such that when $\underset{\sim}{\theta}$ is the true state of nature, $\underset{\sim}{X}$ has probability distribution $f(\underset{\sim}{x} \mid \underset{\sim}{\theta})$,
5. A decision space $D = \{\delta(\underset{\sim}{x})\}$ of possible decision functions defined on $\mathscr{X}$ that maps $\mathscr{X}$ into the action space A.

Example 5.4. Continuing with Example 5.3, suppose we decide to place six randomly selected items on lifetest. Assuming an $\mathcal{E}(\lambda)$ failure time distribution and i.i.d. lifetimes, the joint distribution of $\underset{\sim}{X}'=(X_1, X_2,\ldots,X_6)$ is given by

$$f(\underset{\sim}{x}|\theta)=\frac{1}{\theta^6}\exp\left(-\frac{\mathrm{t}}{\theta}\right),$$

where $\mathrm{t}=\Sigma x_i=$observed total time on test. Two possible decision functions are

5.4.1. If $\mathrm{t}>2613$ h, accept the lot; otherwise, reject the lot. Thus, letting δ_1 represent the decision function, we have

$$\delta_1(\underset{\sim}{x})=\begin{cases} a_1, & \text{if } \mathrm{t}>2613 \text{ h} \\ a_2, & \text{if } \mathrm{t}\leqslant 2613 \text{ h}. \end{cases}$$

Note. The choice of 2613 h will become clear later.

5.4.2. If $\min(x_1, x_2,\ldots,x_6)>1000$ h, accept the lot; otherwise, reject the lot. Thus, we have

$$\delta_2(\underset{\sim}{x})=\begin{cases} a_1, & \text{if } \min(x_1, x_2,\ldots,x_6)>1000 \text{ h} \\ a_2, & \text{if } \min(x_1, x_2,\ldots,x_6)\leqslant 1000 \text{ h}. \end{cases}$$ ■

Finally, for a Bayesian decision theoretic structure, we further assume that

6. A prior probability distribution g is defined on the space $\Omega_{\underset{\sim}{\theta}}$.

Example 5.5. In Example 5.3, suppose on the basis of lots previously produced, it is estimated that the prior probability that $\theta=1000$ h is 0.9 and the prior probability that $\theta=250$ h is 0.1. ■

5.4.2 Risk

For a given decision function δ the loss function may be written as

$$L\{\underset{\sim}{\theta}, \delta(\underset{\sim}{x})\}, \tag{5.16}$$

since our action a depends on the particular sample data $\underset{\sim}{x}$ that we observe. Thus, we see that the loss is an r.v. and depends on the sample outcome. Therefore, let us define the *risk* to be the expected value of the loss function. That is, the risk $R(\underset{\sim}{\theta}, \delta)$ is a function of $\underset{\sim}{\theta}$, δ, and the loss function L such that

$$R(\underset{\sim}{\theta}, \delta) = E[L\{\underset{\sim}{\theta}, \delta(\underset{\sim}{X})\} | \underset{\sim}{\theta}] = \int_{\mathfrak{X}} L\{\underset{\sim}{\theta}, \delta(\underset{\sim}{x})\} f(\underset{\sim}{x} | \underset{\sim}{\theta})\, d\underset{\sim}{x}. \tag{5.17}$$

Obviously, a good decision function would be one that minimizes the risk for all values of $\underset{\sim}{\theta}$ in Ω_{θ}. Unfortunately, in most realistic problems, there does not exist a single decision function that minimizes the risk for all possible values of $\underset{\sim}{\theta}$. As the value of $\underset{\sim}{\theta}$ is unknown, this limits the usefulness of risk as a criterion for selecting a decision function. However, risk can be used as a guide. For example, if δ_1 and δ_2 are two decision functions we can compare the risk $R(\underset{\sim}{\theta}, \delta_1)$ with $R(\underset{\sim}{\theta}, \delta_2)$ for specified values of $\underset{\sim}{\theta}$. If $R(\underset{\sim}{\theta}, \delta_1) \leq R(\underset{\sim}{\theta}, \delta_2)$, for all values of $\underset{\sim}{\theta}$, and $R(\underset{\sim}{\theta}, \delta_1) < R(\underset{\sim}{\theta}, \delta_2)$, for at least one value of $\underset{\sim}{\theta}$, then δ_2 is said to be an *inadmissible* decision function [see Ferguson, (1967)]. On the other hand, if $R(\underset{\sim}{\theta}, \delta_1) < R(\underset{\sim}{\theta}, \delta_2)$ for some value of $\underset{\sim}{\theta}$, then δ_1 is a "better" decision function than δ_2 for that value of $\underset{\sim}{\theta}$.

Example 5.6. Let us again consider the preceding example. First consider the decision δ_1 from Example 5.4. Now

$$R(\theta_1, \delta_1) = \int_{\text{all } \underset{\sim}{x} \ni t > 2613} \$0\, f(\underset{\sim}{x} | \theta_1)\, d\underset{\sim}{x} + \int_{\text{all } \underset{\sim}{x} \ni t \leq 2613} \$10\, f(\underset{\sim}{x} | \theta_1)\, d\underset{\sim}{x}$$

$$= \$0 \Pr(\mathfrak{T} > 2613 | \theta_1) + \$10 \Pr(\mathfrak{T} \leq 2613 | \theta_1).$$

Now it may be shown that the conditional distribution of $\mathfrak{T}$ given θ is a $\mathcal{G}(6, \theta)$ distribution. Thus

$$R(\theta_1, \delta_1) = \$10 \int_0^{2613} \frac{1}{\theta_1 \Gamma(6)} \left(\frac{t}{\theta_1}\right)^5 \exp\left(-\frac{t}{\theta_1}\right) dt$$

$$= \frac{\$10\, \Gamma(6, 2613/\theta_1)}{\Gamma(6)},$$

where $\Gamma(a, x)$ is the incomplete gamma function. Hence $R(1000, \delta_1) = \$10 \times \Gamma(6, 2.61)/5! = \$10(5.97)/120 = \$0.50$. Similarly,

$$R(\theta_2, \delta_1) = \int_{\text{all } \underset{\sim}{x} \ni t > 2613} \$15\, f(\underset{\sim}{x} | \theta_2)\, d\underset{\sim}{x} + \int_{\text{all } \underset{\sim}{x} \ni t \leqslant 2613} \$5\, f(\underset{\sim}{x} | \theta_2)\, d\underset{\sim}{x}$$

$$= \$15 - \frac{\$10\, \Gamma(6, 10.45)}{\Gamma(6)} = \$15 - \frac{\$10\, (113.78)}{120} = \$5.52.$$

Now consider δ_2. Letting $W = \min(X_1, X_2, \ldots, X_6)$, it is easily shown [see Galambos (1978), p. 7] that the conditional distribution of W given θ is an $\mathcal{E}(6/\theta)$ distribution. Thus

$$R(\theta_1, \delta_2) = \$0 \Pr(W > 1000 | \theta_1) + \$10 \Pr(W \leqslant 1000 | \theta_1)$$

$$= \$10 \left[1 - \exp(-6000/\theta_1)\right].$$

Hence $R(1000, \delta_2) = \$10\,[1 - \exp(-6)] = \$9.98$. Similarly,

$$R(\theta_2, \delta_2) = \$15 \Pr(W > 1000 | \theta_2) + \$5 \Pr(W \leqslant 1000 | \theta_2)$$

$$= \$5 + \$10 \exp(-24) = \$5.00.$$

Thus, upon comparing the risks of δ_1 and δ_2, we see that δ_1 is "better" than δ_2 for $\theta = 1000$ h; however, δ_2 is "better" than δ_1 for $\theta = 250$ h. ■

5.4.3 Bayes Risk

Let us now define the *Bayes risk* of a decision function δ as the expected value of the risk $R(\underset{\sim}{\theta}, \delta)$ with respect to the prior distribution g on $\Omega_{\underset{\sim}{\theta}}$; namely,

$$r(g, \delta) = E[R(\underset{\sim}{\Theta}, \delta)] = \int_{\Omega_{\underset{\sim}{\theta}}} \int_{\mathcal{X}} L\{\underset{\sim}{\theta}, \delta(\underset{\sim}{x})\} f(\underset{\sim}{x} | \underset{\sim}{\theta}) g(\underset{\sim}{\theta})\, d\underset{\sim}{x}\, d\underset{\sim}{\theta}. \quad (5.18)$$

The Bayes risk is of value as it sets up a linear ordering on the decision space D (i.e., the decision maker prefers decision function δ_1 to δ_2 if it has smaller Bayes risk).

Example 5.7. For the prior distribution in Example 5.5, the Bayes risk associated with δ_1 and δ_2 in the preceding example become

$$r(g,\delta_1)=0.9(\$0.50)+0.1(\$5.52)=\$1.00$$

and

$$r(g,\delta_2)=0.9(\$9.98)+0.1(\$5.00)=\$9.48.$$

Thus, for this prior distribution, the decision maker would prefer decision function δ_1 as it has a smaller Bayes risk. ■

5.4.4 Bayes Decision Functions

Since the Bayes risk sets up a linear ordering on the decision space D, the search begins for the decision function δ which minimizes the Bayes risk for a specified prior distribution g. If such a decision function, say $\delta_g(\underset{\sim}{x})$, exists it is known as the *Bayes decision function* and its associated Bayes risk

$$r(g)=r(g,\delta_g)=\min_{\delta\in D} r(g,\delta) \tag{5.19}$$

is known as the *minimum Bayes risk*.

Note. Some authors, such as De Groot (1970), refer to (5.18) as the "risk," rather than the "Bayes risk," and refer to (5.19) as the "Bayes risk" instead of the "minimum Bayes risk." Ferguson (1967) and others adopt the same terminology that is used here.

Construction of Bayes Decision Functions. Provided that the order of integration can be reversed, which we shall assume here, the Bayes risk becomes

$$r(g,\delta)=\int_{\mathscr{X}} f(\underset{\sim}{x})\left\{\int_{\Omega_{\underset{\sim}{\theta}}} L\{\underset{\sim}{\theta},\delta(\underset{\sim}{x})\}g(\underset{\sim}{\theta}|\underset{\sim}{x})\,d\underset{\sim}{\theta}\right\}d\underset{\sim}{x}, \tag{5.20}$$

where, according to (5.8), $g(\underset{\sim}{\theta}|\underset{\sim}{x})$ is the joint posterior probability distribution of $\underset{\sim}{\Theta}$ given $\underset{\sim}{x}$. Therefore, to minimize the Bayes risk, a decision function $\delta(\underset{\sim}{x})$ should be chosen such that the inner integral is a minimum. The inner integral is simply the conditional expectation $E[L\{\underset{\sim}{\Theta},\delta(\underset{\sim}{x})\}|\underset{\sim}{x}]$ of the loss with respect to the joint posterior distribution of Θ given $\underset{\sim}{x}$, or simply, the *posterior risk*. Thus, the Bayes decision function with respect to a given g can be found *without* computing the value of the minimum Bayes risk. If we let $\phi(\delta, \underset{\sim}{x})$ denote the posterior risk and let

$$\phi(\underset{\sim}{x}) = \phi(\delta_g, \underset{\sim}{x}) = \min_{\delta \in D} \phi(\delta, \underset{\sim}{x}) = \min_{\delta \in D} E[L\{\Theta, \delta(\underset{\sim}{x})\} | \underset{\sim}{x}], \quad (5.21)$$

then the minimum Bayes risk becomes

$$r(g) = \int_{\mathfrak{X}} \phi(\underset{\sim}{x}) f(\underset{\sim}{x})\, d\underset{\sim}{x} = E[\phi(\underset{\sim}{X})], \quad (5.22)$$

which is the expectation of $\phi(\underset{\sim}{X})$ using the joint marginal distribution of $\underset{\sim}{X}$.

Example 5.8. Let us continue with the preceding example. The statistic $\mathfrak{T} = \Sigma_{i=1}^{6} X_i$ in Example 5.4 is a sufficient statistic for θ [see Example 5.12]. Therefore, let us consider the class of decision functions $D = \{\delta_t\}$, based on $\mathfrak{T}$, defined by

$$\delta_t(\mathrm{t}) = \begin{cases} a_1, & \text{if } \mathrm{t} > t \text{ h} \\ a_2, & \text{if } \mathrm{t} \leq t \text{ h}, \end{cases}$$

where t is the observed value of $\mathfrak{T}$, and determine the Bayes decision function within this class. Rather than use (5.20), it is more straightforward to minimize (5.18) directly. According to (5.18)

$$r(g, \delta_t) = (0.9)[\$0 \Pr(\mathfrak{T} > t | 1000) + \$10 \Pr(\mathfrak{T} \leq t | 1000)]$$
$$+ (0.1)[\$15 \Pr(\mathfrak{T} > t | 250) + \$5 \Pr(\mathfrak{T} \leq t | 250)].$$

Since $f(\mathrm{t} | \theta)$ is a $\mathcal{G}(6, \theta)$ distribution [see Example 5.6], upon simplification we have

$$r(g, \delta_t) = \$1.50 + \frac{\$9.00\Gamma(6, t/1000)}{\Gamma(6)} - \frac{\$1.00\Gamma(6, t/250)}{\Gamma(6)}.$$

Figure 5.4 shows a plot of $r(g, \delta_t)$ as a function of t. It is observed that the Bayes decision function within this class occurs for $t = 2040$ h. That is, the

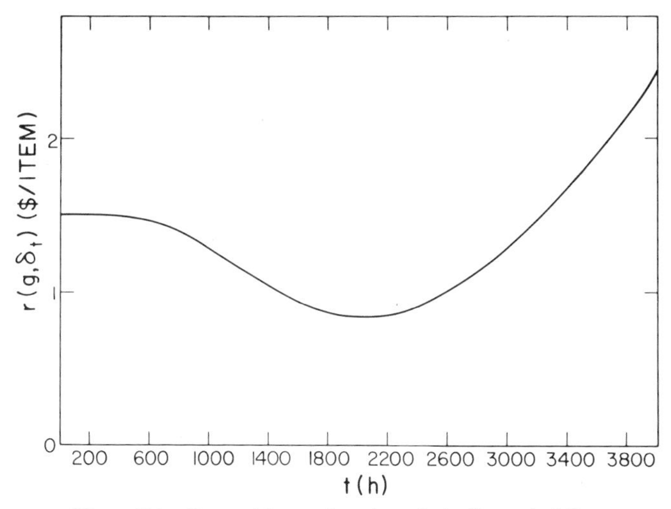

Figure 5.4. Bayes risk as a function of t in Example 5.8.

Bayes risk is minimized if we use the Bayes decision function δ_{2040}; namely,

$$\delta_{2040}(t) = \begin{cases} \text{accept the lot,} & \text{if } t = \sum_{i=1}^{6} x_i > 2040 \text{ h} \\ \text{reject the lot,} & \text{if } t = \sum_{i=1}^{6} x_i \leq 2040 \text{ h.} \end{cases}$$

If we use this rule, the associated Bayes risk, the minimum Bayes risk, will be \$0.84 per item; that is, $r(g) = r(g, \delta_{2040}) = \$0.84$.

Note. If economic considerations are ignored, the classical decision function for testing the hypothesis $\theta = 1000$ h with $\gamma = 5\%$ against the alternative hypothesis $\theta = 250$ h with $\delta = 10\%$ turns out to be $n = 6$ and $T_c = 2613$ h [see Section 4.6.2]. That is, six items are tested and the lot is accepted if $t = \sum_{i=1}^{6} x_i > 2613$ h. This is precisely the decision function δ_1 defined in Example 5.4. If the economic considerations and the prior distribution given in Examples 5.3 and 5.5, respectively, are imposed, the Bayes risk of this classical decision function was shown in Example 5.7 to be \$1.00 per item, which is \$0.16 per item more than the minimum Bayes risk. ■

The examples in this section have illustrated the use of Bayesian methods for testing the simple statistical hypothesis H_0: $\theta = \theta_1 = 1000$ h versus H_1: $\theta = \theta_2 = 250$ h when a loss function is specified [see Example 5.3]. The method involved selecting a decision function that minimized the Bayes risk. In the absence of such an explicit loss function one could consider the *posterior odds ratio*, which according to Bayes' theorem becomes

$$\frac{\Pr(H_0|\underset{\sim}{x})}{\Pr(H_1|\underset{\sim}{x})} = \frac{\Pr(H_0)}{\Pr(H_1)} \times \frac{L(H_0|\underset{\sim}{x})}{L(H_1|\underset{\sim}{x})}.$$

That is, the posterior odds for H_0 versus H_1, are the prior odds multiplied by the likelihood ratio. We would then consider rejecting H_0 in favor of H_1 for sufficiently small values of the posterior odds ratio. However, this approach does not lend itself to a simple computation of the critical region as, in order to determine this region, the distribution of the posterior probabilities is required. In Chapter 10, Bayesian hypothesis testing will again be considered from another slightly different viewpoint in which critical regions will be determined based on the specification of maximum allowable error probabilities. However, with the exception of Chapter 10, the primary focus of the remainder of the book is on Bayesian estimation rather than hypothesis testing.

5.5. BAYESIAN ESTIMATION THEORY

A statistical estimation problem is a special kind of statistical decision problem in which the decision made by the analyst is the estimate of the value of an unknown parameter $\underset{\sim}{\theta}$, which may be vector-valued.

5.5.1 Bayesian Point Estimation

In the case of Bayesian point estimation, the action space A consists of the possible point "estimates" of the parameter $\underset{\sim}{\theta}$ and thus is a subset of the parameter space Ω_{θ}. The decision space D consists of the possible "estimators" for $\underset{\sim}{\theta}$. Thus, our decision function is

$$\underset{\sim}{\hat{\theta}} = \delta(\underset{\sim}{x}), \tag{5.23}$$

where $\underset{\sim}{x}$ is the observed value of $\underset{\sim}{X}$. The function $\underset{\sim}{\hat{\theta}}$ is called a *point estimator* for $\underset{\sim}{\theta}$. Once X has been observed, the function $\delta(\underset{\sim}{x})$ can be evaluated and the resulting "action" $\underset{\sim}{\hat{\theta}}$ (to act as if $\underset{\sim}{\hat{\theta}}$ were the true value of $\underset{\sim}{\theta}$) is referred to as a *point estimate* of $\underset{\sim}{\theta}$. In other words, the estimates (the actions) are the values of an estimator (a decision function).

The loss incurred in estimating $\underset{\sim}{\theta}$ by $\hat{\underset{\sim}{\theta}}$ should reflect the discrepancy between the value of $\underset{\sim}{\theta}$ and the estimate $\hat{\underset{\sim}{\theta}}$ and should also be zero if and only if $\hat{\underset{\sim}{\theta}} = \underset{\sim}{\theta}$. For this reason the loss function L in an estimation problem is often assumed to be of the form

$$L(\underset{\sim}{\theta}, \hat{\underset{\sim}{\theta}}) = h(\underset{\sim}{\theta})\gamma(\underset{\sim}{\theta} - \hat{\underset{\sim}{\theta}}), \tag{5.24}$$

where γ is a nonnegative function of the error $\underset{\sim}{\theta} - \hat{\underset{\sim}{\theta}}$ such that $\gamma(\underset{\sim}{0}) = 0$ and h is a nonnegative weighting function that reflects the relative seriousness of a given error for different values of $\underset{\sim}{\theta}$. In determining Bayes decision functions based on this loss function, the function h can be considered as a component of the prior distribution $g(\underset{\sim}{\theta})$. For this reason, it is frequently assumed that the function h in (5.24) is a constant.

In some estimation problems the analyst is only required to estimate some of the components of $\underset{\sim}{\theta}$, the others either having known or assumed values. In this case the components of $\underset{\sim}{\theta}$ that are not required to be estimated are called *nuisance parameters*. For example, if only the $\mathcal{W}(\alpha, \beta)$ scale parameter α is to be estimated, the shape parameter β would be a nuisance parameter.

When the parameter θ is one-dimensional, the loss function in an estimation problem can often be expressed as

$$L(\theta, \hat{\theta}) = a|\theta - \hat{\theta}|^b, \tag{5.25}$$

where $a > 0$ and $b > 0$. If $b = 2$, the loss function is quadratic and is called a *squared-error loss function*; if $b = 1$, the loss function is proportional to the absolute value of the estimation error and is called an *absolute-error loss function*.

Squared-Error Loss Function. The squared-error loss function lends itself to mathematical manipulation. It also represents a second-order approximation to the more general loss function $\gamma(\theta - \hat{\theta})$, when γ can be differentiated at least twice. For these reasons it is widely used in statistical estimation.

If the squared-error loss function

$$L(\theta, \hat{\theta}) = a(\theta - \hat{\theta})^2 \tag{5.26}$$

is used for each component of $\underset{\sim}{\theta}$, then the risk according to (5.17) becomes

$$R(\theta, \hat{\theta}) = aE\left[(\theta - \hat{\theta})^2 | \theta\right] \tag{5.27}$$

and $E[(\theta-\hat{\theta})^2|\theta]$ is called the *mean-squared error*. Following the discussion in Section 5.4.2, an estimator that minimizes the mean-squared error for all values of θ usually does not exist. This is evident, as for a specific value of θ, θ_0 say, the estimator $\hat{\theta}=\theta_0$ has a mean-squared error which is zero for this value of θ. However, this is not usually the case when we consider Bayes risk as the criterion.

When the loss function is specified by (5.26), according to (5.20) the Bayes estimator, for any specified prior distribution $g(\theta)$, will be that estimator that minimizes the posterior risk given by

$$E\left[a(\Theta-\hat{\theta})^2|\underset{\sim}{x}\right]=\int_{\Omega_\theta} a(\theta-\hat{\theta})^2 g(\theta|\underset{\sim}{x})\,d\theta, \tag{5.28}$$

provided that this expectation exists, which we shall assume. After adding and subtracting $E(\Theta|\underset{\sim}{x})$ and simplifying, we have

$$E\left[a(\Theta-\hat{\theta})^2|\underset{\sim}{x}\right]=a\left[\hat{\theta}-E(\Theta|\underset{\sim}{x})\right]^2+a\operatorname{Var}(\Theta|\underset{\sim}{x}), \tag{5.29}$$

which is clearly minimized when

$$\hat{\theta}=E(\Theta|\underset{\sim}{x})=\int_{\Omega_\theta}\theta g(\theta|\underset{\sim}{x})\,d\theta. \tag{5.30}$$

Thus, for a squared-error loss function, the Bayes estimator is simply the posterior mean of Θ given $\underset{\sim}{x}$. The posterior risk associated with this Bayes estimator, the minimum posterior risk, is $a\operatorname{Var}(\Theta|\underset{\sim}{x})$, provided it exists, which will be assumed. That is, according to (5.21) and (5.29),

$$\phi(\underset{\sim}{x})=\min_{\hat{\theta}} E\left[a(\Theta-\hat{\theta})^2|\underset{\sim}{x}\right]=a\operatorname{Var}(\Theta|\underset{\sim}{x}). \tag{5.31}$$

According to (5.22), the minimum Bayes risk associated with $\hat{\theta}=E(\Theta|\underset{\sim}{x})$ is

$$r(g)=aE_{\underset{\sim}{x}}\left[\operatorname{Var}_{\theta|\underset{\sim}{x}}(\Theta|\underset{\sim}{X})\right], \tag{5.32}$$

which is, of course, the Bayes risk of the Bayes estimator. It is remarked here that, by way of notation, when we are taking an expectation of either a conditional mean or variance we use the r.v. $\underset{\sim}{X}$ rather than its observed value $\underset{\sim}{x}$ in the inner expression.

In practice, it is often a difficult task to compute $r(g)$ in all but the simplest applications. There are at least two methods that can be used to compute $r(g)$. In many applications it will be easiest to take the expectation of $\operatorname{Var}(\Theta|\underset{\sim}{x})$ directly with respect to the marginal distribution of $\underset{\sim}{X}$, once $f(\underset{\sim}{x})$ and $\operatorname{Var}(\Theta|\underset{\sim}{x})$ have been determined. In other applications it will be helpful to first recognize the fact that $r(g)$ can be expressed as

$$r(g) = aE_{\theta}\left\{E_{\underset{\sim}{x}|\theta}\left[\operatorname{Var}(\Theta|\underset{\sim}{X})|\Theta\right]\right\}, \tag{5.33}$$

using the laws of conditional expectation. That is, the expectation of $\operatorname{Var}(\Theta|\underset{\sim}{x})$ is first taken with respect to the sampling distribution $f(\underset{\sim}{x}|\theta)$. The expectation of this result, as a function of θ, is then taken with respect to the prior distribution of Θ. Both of these methods are appropriately used in the remaining chapters of this book.

The Bayes estimator $\hat{\theta} = E(\Theta|\underset{\sim}{x})$ is an r.v., since $\underset{\sim}{X}$ is an r.v. The mean of the Bayes estimator, provided it exists, is

$$E(\hat{\theta}) = E_{\underset{\sim}{x}}\left[E_{\theta|\underset{\sim}{x}}(\Theta|\underset{\sim}{X})\right] = E(\Theta), \tag{5.34}$$

and the mean of the Bayes estimator is the same as the prior mean under squared-error loss. In general, an estimator $\hat{\theta}$ is said to be *Bayes unbiased* if $E(\hat{\theta}) = E(\Theta)$. Thus the posterior mean is Bayes unbiased. The variance of the Bayes estimator, provided it exists, is given by

$$\operatorname{Var}(\hat{\theta}) = \operatorname{Var}_{\underset{\sim}{x}}\left[E_{\theta|\underset{\sim}{x}}(\Theta|\underset{\sim}{X})\right] = \operatorname{Var}(\Theta) - E_{\underset{\sim}{x}}\left[\operatorname{Var}_{\theta|\underset{\sim}{x}}(\Theta|\underset{\sim}{X})\right]. \tag{5.35}$$

Table 5.3 summarizes the important properties of the Bayes estimator for a squared-error loss function.

Table 5.3. Some Properties of the Bayes Estimator for a Squared-Error Loss Function $L(\theta, \hat{\theta}) = a(\theta - \hat{\theta})^2, \quad a > 0$

Property	Value	Explanation	R.V.?
Risk	$a\int_{\mathscr{X}}[\theta - E(\Theta\|\underset{\sim}{x})]^2 f(\underset{\sim}{x}\|\theta)\,d\underset{\sim}{x}$	—	Yes (function of Θ)
Posterior Risk	$a\operatorname{Var}(\Theta\|\underset{\sim}{x})$	Proportional to the posterior variance	Yes (function of $\underset{\sim}{X}$)
Bayes Risk	$aE_{\underset{\sim}{x}}[\operatorname{Var}_{\theta\|x}(\Theta\|\underset{\sim}{X})]$	Proportional to the expectation of the posterior variance	No
Mean	$E(\Theta)$	Prior mean	No
Variance	$\operatorname{Var}(\Theta) - E_{\underset{\sim}{x}}[\operatorname{Var}_{\theta\|x}(\Theta\|\underset{\sim}{X})]$	Difference between the prior variance and the expectation of the posterior variance	No

Example 5.9. Diesel engines are used to provide emergency power for commercial power reactors. Suppose we are interested in estimating the frequency that such engines fail to start upon demand. For diesels less than 500 kw, 16 failures were observed in 382 trials.[a] Assuming independent trials and a constant probability of failure to start p in each trial, the binomial distribution is the appropriate sampling model. Further suppose that the prior distribution of P is a $\mathcal{U}(0,1)$ distribution.

For the case of a binomial sampling model in which $x=16$ and $n=382$ and a $\mathcal{U}(0,1)$ prior distribution, the posterior distribution of P is a $\mathcal{B}(17,384)$ distribution [see Section 7.2.1]. The Bayes estimate of p for a squared-error loss function with $a=1$ is given by the posterior mean, which is $E(P|x=16)=17/384$. Furthermore, the posterior risk is the posterior variance given in (7.11) which is computed to be $(17)(367)/[(384)^2(385)]$.

The Bayes risk may be calculated by recognizing that the variance of the posterior distribution is

$$\operatorname{Var}(P|x)=\frac{382x-x^2+383}{(384)^2(385)}.$$

Also,

$$E(X)=E_p\left[E_{x|p}(X|P)\right]=382E(P)=191$$

and

$$E(X^2)=E_p\left[E_{x|p}(X^2|P)\right]=382E(P)+(381)(382)E(P^2)=48705,$$

since $E(P)=1/2$ and $E(P^2)=1/3$. Thus, according to (5.32), the Bayes risk is given by

$$r(g)=\frac{382E(X)-E(X^2)+383}{(384)^2(385)}=\frac{(382)(191)-48705+383}{(384)^2(385)}$$

$$=\frac{1}{2304}.$$

Finally, the mean and variance of the Bayes estimator are $1/2$ and $1/12-1/2304=191/2304$, respectively. ■

[a]AEC Operations Report, 00E-ES-002, June 1974.

Quadratic Loss Function. Suppose that we are interested in estimating the parameter vector $\underset{\sim}{\theta}'=(\theta_1,\theta_2,\ldots,\theta_p)$, where $p\geqslant 2$. A generalization of the squared-error loss function in (5.26) is the *quadratic loss function* L given by

$$L(\underset{\sim}{\theta},\hat{\underset{\sim}{\theta}})=(\underset{\sim}{\theta}-\hat{\underset{\sim}{\theta}})'A(\underset{\sim}{\theta}-\hat{\underset{\sim}{\theta}}), \tag{5.36}$$

where A is a symmetric $p\times p$ nonnegative definite matrix. If A is positive definite, then any nonzero error vector $(\underset{\sim}{\theta}-\hat{\underset{\sim}{\theta}})$ leads to a positive loss. However, if A is positive semidefinite, there will be nonzero error vectors that yield zero loss. Such vectors exist if the analyst is interested in estimating only some of the components of $\underset{\sim}{\theta}$, the others being regarded as nuisance parameters.

We shall suppose that the mean vector $E(\Theta|\underset{\sim}{x})$ and covariance matrix $\text{Cov}(\Theta|\underset{\sim}{x})$ of the posterior distribution of Θ given $\underset{\sim}{x}$ exist. According to (5.20) the Bayes estimator is that estimator that minimizes the posterior risk given by

$$\begin{aligned}E[(\underset{\sim}{\Theta}-\hat{\underset{\sim}{\theta}})'A(\underset{\sim}{\Theta}-\hat{\underset{\sim}{\theta}})|\underset{\sim}{x}]&=E\{[\underset{\sim}{\Theta}-E(\underset{\sim}{\Theta}|\underset{\sim}{x})]'A[\underset{\sim}{\Theta}-E(\underset{\sim}{\Theta}|\underset{\sim}{x})]|\underset{\sim}{x}\}\\&\quad+[E(\underset{\sim}{\Theta}|\underset{\sim}{x})-\hat{\underset{\sim}{\theta}}]'A[E(\underset{\sim}{\Theta}|\underset{\sim}{x})-\hat{\underset{\sim}{\theta}}]\\&=\text{tr}[A\,\text{Cov}(\underset{\sim}{\Theta}|\underset{\sim}{x})]+[E(\underset{\sim}{\Theta}|\underset{\sim}{x})-\hat{\underset{\sim}{\theta}}]'A[E(\underset{\sim}{\Theta}|\underset{\sim}{x})-\hat{\underset{\sim}{\theta}}].\end{aligned} \tag{5.37}$$

This is clearly minimized when

$$\hat{\underset{\sim}{\theta}}=E(\underset{\sim}{\Theta}|\underset{\sim}{x})=\int_{\Omega_{\underset{\sim}{\theta}}}\underset{\sim}{\theta}g(\underset{\sim}{\theta}|\underset{\sim}{x})\,d\underset{\sim}{\theta}, \tag{5.38}$$

and thus the posterior mean $E(\Theta|\underset{\sim}{x})$ is a Bayes estimator for $\underset{\sim}{\theta}$. Furthermore, if A is positive definite, $\hat{\underset{\sim}{\theta}}$ is the unique Bayes estimator, whereas if A is only positive semidefinite, there will be more than one value $\hat{\underset{\sim}{\theta}}$ that minimizes (5.37). Table 5.4 summarizes the important properties of the Bayes estimator for a quadratic loss function.

Absolute-Error Loss Function. The absolute-error loss function

$$L(\theta,\hat{\theta})=a|\theta-\hat{\theta}|,\qquad a>0, \tag{5.39}$$

assumes that the loss is proportional to the absolute value of the estimation

Table 5.4. Some Properties of the Bayes Estimator for a Quadratic Loss Function $L(\underset{\sim}{\theta}, \hat{\underset{\sim}{\theta}}) = (\underset{\sim}{\theta} - \hat{\underset{\sim}{\theta}})' A (\underset{\sim}{\theta} - \hat{\underset{\sim}{\theta}})$, A is Symmetric Nonnegative Definite

Property	Value	Explanation	R.V.?
Risk	$\int_{\mathscr{X}} [\underset{\sim}{\theta} - E(\underset{\sim}{\Theta} \mid \underset{\sim}{x})]' A [\underset{\sim}{\theta} - E(\underset{\sim}{\Theta} \mid \underset{\sim}{x})] \times f(\underset{\sim}{x} \mid \underset{\sim}{\theta})\, d\underset{\sim}{x}$	—	Yes (function of $\underset{\sim}{\Theta}$)
Posterior Risk	$\mathrm{tr}[A\, \mathrm{Cov}(\underset{\sim}{\Theta} \mid \underset{\sim}{x})]$	Trace of the product of A and the posterior covariance matrix	Yes (function of $\underset{\sim}{X}$)
Bayes Risk	$\mathrm{tr}\{A E_{\underset{\sim}{x}}[\mathrm{Cov}_{\underset{\sim}{\theta} \mid \underset{\sim}{x}}(\underset{\sim}{\Theta} \mid \underset{\sim}{X})]\}$	Trace of the product of A and the expectation of the posterior covariance matrix	No
Mean	$E(\underset{\sim}{\Theta})$	Prior mean vector	No
Covariance	$\mathrm{Cov}(\underset{\sim}{\Theta}) - E_{\underset{\sim}{x}}[\mathrm{Cov}_{\underset{\sim}{\theta} \mid \underset{\sim}{x}}(\underset{\sim}{\Theta} \mid \underset{\sim}{X})]$	Difference between the prior covariance matrix and the expectation of the posterior covariance matrix	No

error. According to (5.21), the Bayes estimator will be that estimator that minimizes the posterior risk given by $E(a|\Theta-\hat{\theta}||\underset{\sim}{x})$. Chernoff and Moses (1959, p. 319) show that the value of $\hat{\theta}$ that minimizes this posterior risk is any median of the posterior distribution of Θ given $\underset{\sim}{x}$, where a number $\hat{\theta}$ is a *median* of the distribution of Θ given $\underset{\sim}{x}$ if $\Pr(\Theta \geqslant \hat{\theta}|\underset{\sim}{x}) \geqslant 0.5$ and $\Pr(\Theta \leqslant \hat{\theta}|\underset{\sim}{x}) \geqslant 0.5$. Although the median may not be unique, it will usually be unique for those posterior distributions considered in the remaining chapters of this book. According to (5.22), the Bayes risk becomes

$$r(g) = aE_{\underset{\sim}{x}}\left[E_{\theta|\underset{\sim}{x}}\left(|\Theta-\hat{\theta}||\underset{\sim}{X}\right)\right], \tag{5.40}$$

where $\hat{\theta}$ is a median of the posterior distribution of Θ given $\underset{\sim}{x}$.

Example 5.10. Let us again consider Example 5.9. The median of the posterior $\mathcal{B}(17,384)$ distribution may be computed by means of (7.18) in which $x=16$, $n=382$, $x_0=1$, $n_0=2$, and $\gamma=0.50$. Thus the median is $17/[17+367\mathcal{F}_{0.50}(734,34)]$. From Table B4 in Appendix B, $\mathcal{F}_{0.50}(734,34)=1.02$. Hence, the Bayes estimate for an absolute-error loss function is the median of the $\mathcal{B}(17,384)$ distribution which is $17/391.34=0.0434$. ■

Other Estimators and Loss Functions. Another Bayes estimator for either a univariate or vector parameter is the value of the parameter that maximizes the posterior distribution of $\underset{\sim}{\Theta}$ given $\underset{\sim}{x}$. That is, the Bayes estimator $\hat{\underset{\sim}{\theta}}$ is that value of $\underset{\sim}{\theta}$ that satisfies the relation

$$g(\hat{\underset{\sim}{\theta}}|\underset{\sim}{x}) = \max_{\underset{\sim}{\theta}} g(\underset{\sim}{\theta}|\underset{\sim}{x}). \tag{5.41}$$

If such a value $\hat{\underset{\sim}{\theta}}$ exists for any value of $\underset{\sim}{x}$, then it is referred to as a *generalized ML estimator* for $\underset{\sim}{\theta}$ [De Groot (1970)]. Although such an estimator may not be a Bayes estimator for any standard loss function, it is a reasonable estimator, as it measures the location of the posterior distribution analogous to the posterior mean and median.

Example 5.11. The generalized ML estimator for p in Example 5.9 is the mode of the posterior $\mathcal{B}(17,384)$ distribution which is x/n. Thus, for a $\mathcal{U}(0,1)$ prior distribution on P, the generalized ML estimator is the same as the classical ML estimator, and the estimate is $16/382$. ■

Canfield (1970) considers an asymmetric loss function in which the loss incurred in underestimation of reliability is less than the loss incurred in over estimation by the same amount. He then uses the loss function to derive a Bayes estimator for the reliability in an $\mathcal{E}(\lambda)$ failure model.

Harris (1977) suggests a loss function that depends on how well one estimates $(1-R)^{-1}$, since $R=0.99$ implies, on the average, one failure out of 100, while $R=0.999$ implies, on the average, one failure out of 1000. Thus, $R=0.999$ is 10 times as good as $R=0.99$.

Higgins and Tsokos (1978) compare five different classes of loss functions for estimating the failure intensity λ of a Poisson failure model that is assumed to have a $\mathcal{G}(\alpha,\beta)$ prior distribution. In addition to the usual squared-error loss, they consider a squared relative error loss and its generalization, a linear weighted loss, the loss function of Harris (1977), and an exponential loss function. They conclude that the squared-error loss function is *not* robust with respect to the other four loss functions, and that Bayes estimates of λ are quite sensitive to changes in the loss function.

Britney and Winkler (1968) study symmetric loss functions for symmetric probability and also consider a class of asymmetric loss functions applied to the $\mathfrak{N}(\mu,\sigma^2)$ distribution. El-Sayyad (1967) considers a class of weighted squared-error loss functions for the $\mathcal{E}(\lambda)$ distribution.

5.5.2 Bayesian Estimation Based on a Sufficient Statistic

Recall that $\underset{\sim}{X}$ represents an observable r.v. from the conditional probability distribution $f(\underset{\sim}{x}|\underset{\sim}{\theta})=\prod_{i=1}^{n} f(x_i|\underset{\sim}{\theta})$. If there exists a set of jointly sufficient statistics $\underset{\sim}{Z}'=(Z_1, Z_2,\ldots,Z_q)$ for estimating $\underset{\sim}{\theta}'=(\theta_1,\theta_2,\ldots,\theta_p)$, then, according to the Neyman factorization criterion [Mood, Graybill, and Boes (1974)], $f(\underset{\sim}{x}|\underset{\sim}{\theta})$ factors into

$$f(\underset{\sim}{x}|\underset{\sim}{\theta})=f(\underset{\sim}{z}|\underset{\sim}{\theta})h(\underset{\sim}{x}), \tag{5.42}$$

where $f(\underset{\sim}{z}|\underset{\sim}{\theta})$ is the conditional probability distribution of $\underset{\sim}{Z}$ given $\underset{\sim}{\theta}$ and $h(\underset{\sim}{x})$ is some nonnegative function of $\underset{\sim}{x}$ not involving $\underset{\sim}{\theta}$.

Example 5.12. In Example 5.4, the joint distribution of $\underset{\sim}{X}'=(X_1, X_2,\ldots,X_6)$ was given as

$$f(\underset{\sim}{x}|\theta)=\frac{1}{\theta^6}\exp\left(-\frac{t}{\theta}\right), \qquad t=\sum_{i=1}^{6} x_i,$$

which may be factored as

$$f(\underset{\sim}{x}|\theta) = f(\mathrm{t}|\theta)h(\underset{\sim}{x}),$$

where

$$f(\mathrm{t}|\theta) = \frac{1}{\theta\Gamma(6)}\left(\frac{\mathrm{t}}{\theta}\right)^5 \exp\left(-\frac{\mathrm{t}}{\theta}\right), \qquad h(\underset{\sim}{x}) = \frac{\Gamma(6)}{\mathrm{t}^5}.$$

Thus, according to (5.42), $\mathfrak{T}$ is a sufficient statistic for estimating θ. ■

If $g(\underset{\sim}{\theta}|\underset{\sim}{x}) = g(\underset{\sim}{\theta}|\underset{\sim}{z})$, then the sufficient statistic $\underset{\sim}{Z}$ can be used in lieu of $\underset{\sim}{X}$ in constructing Bayes estimators. That is, the sampling distribution of the sufficient statistic can be used in place of the sampling model of the observable r.v. $\underset{\sim}{X}$ in computing the posterior distribution. Frequently, the use of a sufficient statistic will reduce the dimensionality, and thus the complexity, of Bayesian estimation problems. For example, in the preceding example, the posterior distribution of Θ would be conditional on the value of $\mathfrak{T}$, which is univariate, rather than the value of $\underset{\sim}{X}$, which is multivariate.

It is easily shown that $g(\underset{\sim}{\theta}|\underset{\sim}{x}) = g(\underset{\sim}{\theta}|\underset{\sim}{z})$. According to Bayes' theorem

$$g(\underset{\sim}{\theta}|\underset{\sim}{x}) = \frac{f(\underset{\sim}{x}|\underset{\sim}{\theta})g(\underset{\sim}{\theta})}{\int_{\underset{\sim}{\Omega_\theta}} f(\underset{\sim}{x}|\underset{\sim}{\theta})g(\underset{\sim}{\theta})\,d\underset{\sim}{\theta}} = \frac{f(\underset{\sim}{z}|\underset{\sim}{\theta})h(\underset{\sim}{x})g(\underset{\sim}{\theta})}{\int_{\underset{\sim}{\Omega_\theta}} f(\underset{\sim}{z}|\underset{\sim}{\theta})h(\underset{\sim}{x})g(\underset{\sim}{\theta})\,d\underset{\sim}{\theta}}$$

$$= \frac{f(\underset{\sim}{z}|\underset{\sim}{\theta})g(\underset{\sim}{\theta})}{\int_{\underset{\sim}{\Omega_\theta}} f(\underset{\sim}{z}|\underset{\sim}{\theta})g(\underset{\sim}{\theta})\,d\underset{\sim}{\theta}} = g(\underset{\sim}{\theta}|\underset{\sim}{z}), \tag{5.43}$$

since $h(\underset{\sim}{x})$ is not a function of $\underset{\sim}{\theta}$ and thus can be cancelled in both the numerator and denominator of this expression. In particular, for a quadratic loss function, the Bayes estimator for $\underset{\sim}{\theta}$ becomes

$$\hat{\underset{\sim}{\theta}} = E(\Theta|\underset{\sim}{x}) = E(\Theta|\underset{\sim}{z}). \tag{5.44}$$

5.5.3 Bayesian Interval Estimation

The Bayesian approach to interval estimation is much more direct than the classical approach based on confidence intervals. We restrict our attention to the case of a single parameter θ for which an interval estimate is desired. The case of a parameter vector $\underset{\sim}{\theta}$ will be considered in the next subsection. Once the posterior distribution of Θ given $\underset{\sim}{x}$ has been obtained, a *symmetric* $100(1-\gamma)\%$ *two-sided Bayes probability interval* [$100(1-\gamma)\%$ TBPI] estimate of θ is easily obtained by solving the two equations

$$\int_{-\infty}^{\theta_*} g(\theta \mid \underset{\sim}{x})\, d\theta = \frac{\gamma}{2} \tag{5.45}$$

and

$$\int_{\theta^*}^{\infty} g(\theta \mid \underset{\sim}{x})\, d\theta = \frac{\gamma}{2} \tag{5.46}$$

for the lower limit θ_* and the upper limit θ^*, so that $\Pr(\theta_* \leqslant \Theta \leqslant \theta^* \mid \underset{\sim}{x}) = 1-\gamma$. The interval (θ_*, θ^*) is referred to, simply, as a "$100(1-\gamma)\%$ TBPI estimate" of θ. Bayes interval estimates are sometimes called "probability interval estimates" as the coverage is an explicit statement of the probability that Θ lies in the given interval.

Furthermore, a $100(1-\gamma)\%$ *lower one-sided Bayes probability interval* [$100(1-\gamma)\%$ LBPI] estimate of θ may be obtained by solving (5.45) for θ_* with $\gamma/2$ replaced by γ. Similarly, a $100(1-\gamma)\%$ *upper one-sided Bayes probability interval* [$100(1-\gamma)\%$ UBPI] estimate of θ may be obtained by solving (5.46) for θ^* with $\gamma/2$ replaced by γ. It is also noted that γ is frequently chosen to be some small quantity such as 0.01, 0.05, 0.10, or 0.20, and in many reliability applications, $\gamma = 0.05$ or 0.10. As (5.45) and (5.46) are valid probability statements, they may be combined with other statements according to the rules of probability. As discussed in Section 5.2.2, this represents a distinct advantage over confidence interval estimates which cannot be so easily combined.

The TBPI given above is referred to here as a *symmetric interval* since γ is equally divided between the two tails of the posterior distribution. Such intervals are not necessarily the shortest interval, as every point included may not have higher probability density than every point excluded. Box and Tiao (1973) define a *highest posterior density interval* as a Bayes interval such that, for a given probability content, it is the shortest interval among the class of Bayes estimates. A natural consequence is that the probability density of every point inside such an interval is at least as large as that of any point outside it. However, such an interval is often difficult to obtain. For this reason, the use of a symmetric Bayes interval, which is considerably easier to obtain, is considered in the remainder of the book. In most practical reliability problems, the difference between the two intervals is slight.

Example 5.13. Let us compute a 90% TBPI estimate of the probability p that the diesel engines considered in Example 5.9 will fail to start. Solving the pair of equations given by

$$\int_0^{\theta_*} \frac{\Gamma(384)}{\Gamma(17)\Gamma(367)} p^{16}(1-p)^{366}\, dp = 0.05$$

and

$$\int_{\theta^*}^1 \frac{\Gamma(384)}{\Gamma(17)\Gamma(367)} p^{16}(1-p)^{366}\, dp = 0.05$$

for θ_* and θ^* yields the required interval. By means of (7.18) and (7.20), the solution is computed to be $\theta_* = 0.029$ and $\theta^* = 0.062$. Thus, the required 90% TBPI estimate is (0.029,0.062). Figure 5.5 gives a plot of the posterior pdf given in Example 5.9. The 90% TBPI estimate, as well as the mode, median, and mean, are also indicated. ■

5.5.4 Bayesian Regions

Suppose now that we are interested in determining a region R in the parameter space $\Omega_{\underset{\sim}{\theta}}$, where $\underset{\sim}{\theta}$ is a parameter vector, for which the probability

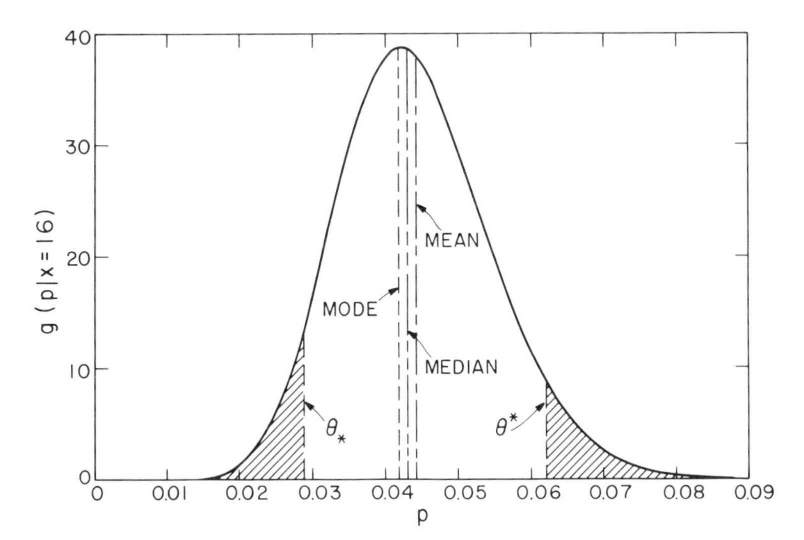

Figure 5.5. The posterior probability distribution in Example 5.9.

content of the posterior distribution is $1-\gamma$. Following Box and Tiao (1973), such a region is called a *Bayesian highest posterior density* $100(1-\gamma)\%$ *region* if

$$\int_R g(\underset{\sim}{\theta}|\underset{\sim}{x})\,d\underset{\sim}{\theta}=1-\gamma \tag{5.47}$$

and

$$g(\underset{\sim}{\theta}_1|\underset{\sim}{x}) \geqslant g(\underset{\sim}{\theta}_2|\underset{\sim}{x}), \qquad \text{for every } \underset{\sim}{\theta}_1 \in R \text{ and } \underset{\sim}{\theta}_2 \notin R. \tag{5.48}$$

However, in many practical problems it is not an easy task to determine R. One way of accomplishing this will now be described. In some situations, depending upon the form of $g(\underset{\sim}{\theta}|\underset{\sim}{x})$, there exists some convenient function of $\underset{\sim}{\theta}$, say $Q(\underset{\sim}{\theta})$, for which $g(\underset{\sim}{\theta}|\underset{\sim}{x})$ is a monotonic function and for which we know the univariate posterior distribution. The inequality $Q(\underset{\sim}{\theta}) \leqslant Q_{1-\gamma}$ therefore defines the required region R in the parameter space Ω_{θ}, where $Q_{1-\gamma}$ is the $100(1-\gamma)$th percentile of the posterior distribution of $Q(\underset{\sim}{\theta})$ and $(1-\gamma)$ is the posterior probability content of R. For example, if $g(\underset{\sim}{\theta}|\underset{\sim}{x})$ is a multivariate normal distribution with mean vector $\underset{\sim}{u}(\underset{\sim}{x})$ and known covariance matrix Σ, then $g(\underset{\sim}{\theta}|\underset{\sim}{x})$ is a monotonic decreasing function of the quadratic form $Q(\underset{\sim}{\theta})=[\underset{\sim}{\theta}-\underset{\sim}{u}(\underset{\sim}{x})]'\Sigma^{-1}[\underset{\sim}{\theta}-\underset{\sim}{u}(\underset{\sim}{x})]$ that, according to Anderson (1958), has a posterior $\chi^2(p)$ distribution, where p is the dimension of $\underset{\sim}{\theta}$. Therefore, the inequality $Q(\underset{\sim}{\theta}) \leqslant \chi^2_{1-\gamma}(p)$ defines the required region with content $1-\gamma$.

Example 5.14. Consider two dependent components whose lifetimes $\underset{\sim}{X}'=(X_1, X_2)$ follow an approximate bivariate normal distribution with unknown mean vector $\underset{\sim}{\theta}'=(\theta_1, \theta_2)$ and known covariance matrix

$$A=\begin{bmatrix} 100 & 120 \\ 120 & 150 \end{bmatrix}.$$

Further suppose that $\underset{\sim}{\Theta}$ has a bivariate normal prior distribution with mean vector $\underset{\sim}{\eta}'=(100, 150)$ and covariance matrix

$$B=\begin{bmatrix} 64 & 48 \\ 48 & 144 \end{bmatrix}.$$

If $\underset{\sim}{\bar{x}}'=(90, 130)$ represents the sample mean vector from a simple random sample of size five observed lifetimes, then the posterior distribution of Θ given $\underset{\sim}{\bar{x}}$ is also a bivariate normal distribution [see De Groot (1970), p. 175] with mean vector

$$\hat{\underset{\sim}{\theta}} = \begin{bmatrix} 0.814\bar{x}_1 - 0.061\bar{x}_2 + 27.735 \\ -0.214\bar{x}_1 + 0.916\bar{x}_2 + 33.970 \end{bmatrix} = \begin{bmatrix} 93.065 \\ 133.790 \end{bmatrix}$$

and covariance matrix

$$C = (5A^{-1} + B^{-1})^{-1} = \begin{bmatrix} 5824/393 & 20880/1179 \\ 20880/1179 & 2928/131 \end{bmatrix}.$$

$\hat{\underset{\sim}{\theta}}$ is, of course, the Bayesian point estimate of $\underset{\sim}{\theta}$ for a quadratic loss function. Thus, the quadratic form

$$\begin{aligned} Q(\underset{\sim}{\theta}) &= (\underset{\sim}{\theta} - \hat{\underset{\sim}{\theta}})' C^{-1} (\underset{\sim}{\theta} - \hat{\underset{\sim}{\theta}}) \\ &= 1.3(\theta_1 - 93.1)^2 - 2.0(\theta_1 - 93.1)(\theta_2 - 133.8) + 0.8(\theta_2 - 133.8)^2 \end{aligned}$$

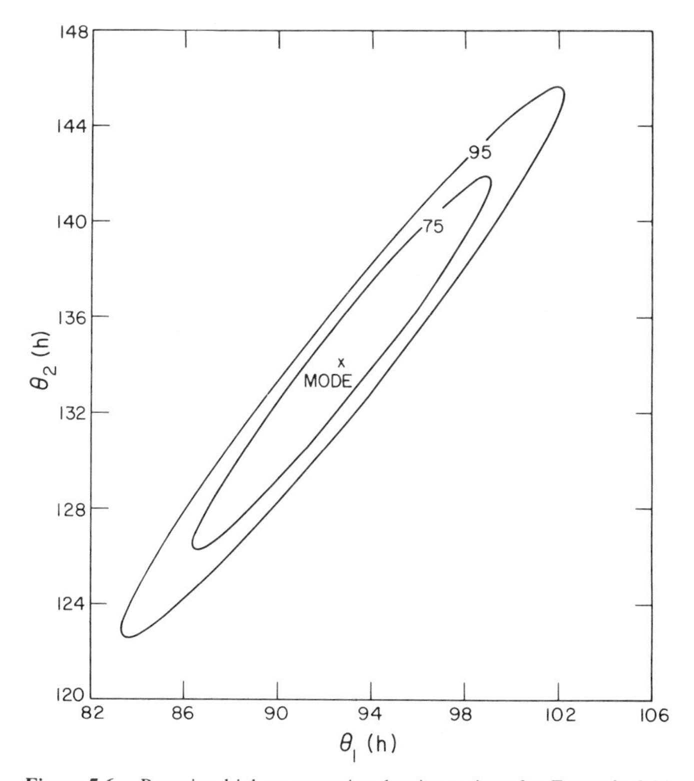

Figure 5.6. Bayesian highest posterior density regions for Example 5.14.

has a posterior $\chi^2(2)$ distribution. Bayesian regions with coverage probability of 75% and 95% are found by setting $Q(\underset{\sim}{\theta})$ equal in turn to 2.77 and 5.99 which are, respectively, the 75th and 95th percentiles of the $\chi^2(2)$ distribution from Table B3 in Appendix B. These regions are shown in Figure 5.6. ■

EXERCISES

5.1. Suppose that the ten values 0.80, 4.82, 5.99, 1.99, 2.10, 0.33, 0.89, 1.20, 1.83, 0.58 come from an $\mathcal{E}(\lambda)$ sampling distribution. Assume Θ has a $\mathcal{G}(2, 0.5)$ prior distribution.

(a) Find the posterior pdf $g(\theta \mid \underset{\sim}{t})$ explicitly.

(b) Compare the prior $g(\theta)$ with the posterior $g(\theta \mid \underset{\sim}{t})$ by graphing both on a common set of axes.

(c) Which of the two pdf's, the prior or the posterior, has a larger variance?

(d) What type of data (i.e., Σt_i) would result in about the same variance for the prior and posterior pdf's?

5.2. Let $X_1, X_2, \ldots, X_n$ denote a random sample from the pdf $f(x \mid \theta) = \theta^x(1-\theta)^{1-x}$, $x = 0, 1$, and $f(x \mid \theta) = 0$, otherwise. Assume that Θ has a uniform prior, that is, $g(\theta) = 1$, $0 < \theta < 1$ and $g(\theta) = 0$, otherwise.

(a) Find the posterior Bayes estimator of θ for a squared-error loss function.

(b) Is the estimator in (a) unbiased?

(c) Compare the posterior Bayes estimator with the classical ML estimator.

(d) Find the posterior Bayes estimator of the parameter $\theta(1-\theta)$ assuming a squared-error loss function.

(e) Is the estimator in (d) unbiased?

(f) Compare the posterior Bayes estimator with the classical estimator of $\theta(1-\theta)$.

5.3. Let $T_1, T_2, \ldots, T_n$ denote a random sample from the pdf $f(x \mid \theta) = e^{-(x-\theta)}$, $x \geq \theta$, $-\infty < \theta < \infty$, and $f(x \mid \theta) = 0$, otherwise. Assume the prior pdf of Θ to be $g(\theta) = e^{-\theta}$, $\theta \geq 0$, and $g(\theta) = 0$, otherwise. Find the posterior Bayes estimator of θ for a squared-error loss function.

5.4. Let $X_1, X_2, \ldots, X_n$ denote a random sample from a $\mathcal{N}(\mu, 1)$ distribution. Assume that μ has a $\mathcal{N}(\theta, \tau^2)$ prior distribution.

(a) Find the posterior pdf of μ.

(b) Assume a squared-error loss function and find the Bayes estimator of μ.

(c) Find the posterior risk of the estimator in (b).

(d) Find the Bayes risk of the estimator in (b).

(e) Find the mean and variance of $E(\mu \mid \underset{\sim}{x})$.

(f) Use the posterior pdf of μ to find $\Pr(LL \leqslant \mu \leqslant UL \mid \underset{\sim}{x}) = 0.95$, where LL and UL are symmetric about $E(\mu \mid \underset{\sim}{x})$.

(g) Compare the probability interval for μ in (f) with the classical confidence interval for μ.

(h) Show that the two intervals compared in (g) are "about" the same if the variance $1/n$ of $\overline{X}$ is much smaller than the variance τ^2 of μ.

(i) Find the Bayes estimator of μ for an absolute-error loss function.

5.5. A company has received several complaints about a specific type of defect on a washing machine that it manufactures. The company wishes to determine the proportion θ of washers sold during the previous year that possessed the defect, and sends questionnaires to purchasers of the washers. It receives 100 replies to the questionnaire. Assume that a purchaser's response is independent of whether the machine has a defect. Let θ have a $\mathcal{B}(1, 10)$ prior distribution.

(a) Determine the posterior distribution for θ if there were 30 defects reported in the 100 returns.

(b) Estimate the proportion of washers that contain the defect.

5.6. Suppose the lifetime, in whole years t, of an electronic component has the geometric distribution $f(t \mid \theta) = \theta(1-\theta)^{t-1}$, $t = 1, 2, \ldots$, $0 < \theta < 1$, and $f(t \mid \theta) = 0$, otherwise.

(a) Show that $E(T \mid \theta) = 1/\theta$.

(b) Let $T_1, T_2, \ldots, T_n$ denote the lifetimes of n electronic components and assume a $\mathcal{B}(x_0, n_0)$ prior distribution for Θ. Find the posterior distribution of Θ.

(c) Let $x_0 > 2$, $n_0 - x_0 > 0$ and define $\theta^* = 1/\theta$, $\theta_0^* = E(\Theta^*)$, $\sigma_0^2 = \operatorname{Var}(\Theta^*)$. Show that

$$\theta_0^* = \frac{(n_0 - 1)}{(x_0 - 1)}$$

and

$$\sigma_0^2 = \frac{(n_0-1)(n_0-x_0)}{(x_0-1)^2(x_0-2)}.$$

(d) Show, using (c) that

$$x_0 = 2 + \frac{\theta_0^*(\theta_0^*-1)}{\sigma_0^2};\ n_0 - x_0 = \frac{(\theta_0^*-1)(\sigma_0^2+\theta_0^{*2}-\theta_0^*)}{\sigma_0^2}.$$

(e) Suppose five components lasted 6, 3, 1, 8, 5 years and the prior distribution of mean life $\Theta^* = 1/\Theta$ has mean $\theta_0^* = 7$ and standard deviation $\sigma_0 = 1$. Assume also that $1/\Theta^*$ has a $\mathcal{B}(x_0, n_0)$ prior distribution. Find the posterior mean and variance of Θ^*.

(f) Find $\lim_{\sigma_0^2\to\infty} E(\Theta^*|\underset{\sim}{t})$ and $\lim_{\sigma_0^2\to\infty} \mathrm{Var}(\Theta^*|\underset{\sim}{t})$.

(g) Find these limits when $\theta_0^* = 7$ and the data in (e).

5.7. Let Θ have a $\mathcal{B}(x_0, n_0)$ distribution with mean $\mu = E(\Theta)$ and variance $\sigma^2 = \mathrm{Var}(\Theta)$. Find the quantity $E[\Theta/(1-\Theta)]$. Give all the restrictions required for the result to be valid.

5.8. Suppose X has the distribution $f(x|\theta) = \theta^x(1-\theta)^{1-x}$, $x = 0, 1$, $0 < \theta < 1$, and zero, elsewhere. Show that the transformation $\tau(\theta) = \sin^{-1}\sqrt{\theta}$ leaves the standardized likelihood function $\ell(\tau|\underset{\sim}{x})$ very nearly data translated; that is, the likelihood curves are nearly identical in shape and differ only in location [see Section 6.1].

5.9. Suppose X has a $\mathfrak{N}(\mu, \sigma^2)$ distribution with known mean μ and unknown standard deviation σ.

(a) Show that the standardized likelihood $\ell(\sigma|\underset{\sim}{x}, \mu)$ is not data translated.

(b) Find a transformation on σ, $\tau(\sigma)$, such that the likelihood $\ell(\tau|\underset{\sim}{x}, \mu)$ is exactly data translated.

5.10. Let $X_1, X_2, \ldots, X_n$ denote a random sample from a $\mathfrak{N}(\mu, \sigma^2)$ distribution with mean μ and known variance σ^2. Assume μ has a $\mathfrak{N}(\mu_0, \sigma_0^2)$ prior distribution. Show that μ has a normal posterior distribution with mean

$$E(\mu|\underset{\sim}{x}) = \frac{\mu_0\sigma_{\bar{x}}^2 + \bar{x}\sigma_0^2}{\sigma_0^2 + \sigma_{\bar{x}}^2} \qquad \text{and variance}$$

$$\mathrm{Var}(\mu|\underset{\sim}{x}) = \frac{\sigma_0^2\sigma_{\bar{x}}^2}{\sigma_0^2 + \sigma_{\bar{x}}^2}, \qquad \text{where } \sigma_{\bar{x}} = \frac{\sigma^2}{n} \text{ and } \bar{x} = \frac{\sum_{i=1}^{n} x_i}{n}.$$

(a) Discuss the posterior distribution when σ_0 is small relative to $\sigma_{\bar{x}}$. When would such a prior on μ be used for such a value of σ_0?

(b) Discuss the posterior distribution when σ_0 is large relative to $\sigma_{\bar{x}}$. What does the use of a prior with such a value of σ_0 imply about μ?

(c) Find the limiting posterior variance as $n \to \infty$. What does this imply about the posterior distribution of μ when n is large?

(d) Show that the influence of the prior distribution on the posterior disappears as $n \to \infty$.

5.11. Suppose that the r.v. X has a continuous pdf $f(x|\theta)$ and further suppose that the parameter θ has a discrete prior $g(\theta)$, $\theta = a_1, a_2, \ldots$, and $g(\theta) = 0$, otherwise. Also, let $X_1, X_2, \ldots, X_n$ denote a random sample from $f(x|\theta)$.

(a) Describe the likelihood function of θ.

(b) Describe the standardized likelihood function of θ.

(c) Find the standardized likelihood function when $\theta = a_k$.

(d) Find the posterior probability that $\Theta = a_k$.

(e) If $g(\theta)$ is a discrete uniform pdf, that is, $g(\theta) = c$, $\theta = a_1, a_2, \ldots$, show that the posterior pdf of Θ is given in terms of the likelihood function only.

5.12. Let X have a continuous pdf $f(x|\theta)$ and let Θ have a continuous prior $g(\theta)$. Further suppose that the distribution of Θ is unknown. However, it is known that $\Pr(\Theta \leqslant c) = p$ and $\Pr(\Theta > c) = 1 - p$, for a given constant c and known p. The r.v. X corresponds to the lifetime of a well pump and it is desirable that its mean lifetime exceed c h. Assume that the mean of X is θ. Let $X_1, X_2, \ldots, X_n$ denote a random sample of lifetimes of n well pumps. It is desired to use this data to make a posterior assessment about the mean life θ.

(a) Find the posterior distribution of Θ, in terms of the likelihood function, using p in its denominator; that is, in the marginal pdf of $\underset{\sim}{X}$.

(b) Find the posterior distribution of Θ, in terms of the standardized likelihood function, and show that it is the same as in (a).

(c) Use the posterior distribution in (b) to determine the posterior probability $\Pr(\Theta > c|\underset{\sim}{x})$.

(d) Suppose $c = 1000$ h, $n = 2$, $f(x|\theta) = (1/\theta)e^{-x/\theta}$, $x \geqslant 0$, $x_1 = 1200$, $x_2 = 1400$, and $p = 0.30$. Find $\Pr(\Theta > 1000|x_1 = 1200, x_2 = 1400)$ and interpret the result.

(e) Suppose $p = 0.60$ in (d). Find $\Pr(\Theta > 1000|x_1 = 1200, x_2 = 1400)$ and interpret the result.

5.13. Let T have a continuous pdf $f(t|\theta), 0<t<\infty$, and let Θ have a continuous prior $g(\theta)$ which is unknown. However, it is known that $\Pr(0 \leqslant \Theta \leqslant a_1) = p_1$, $\Pr(a_1 \leqslant \Theta \leqslant a_2) = p_2, \ldots, \Pr(a_{k-1} \leqslant \Theta < \infty) = p_k$, where $a_1, a_2, \ldots, a_{k-1}$ are given constants and $p_1, p_2, \ldots, p_k$ are known. The r.v. T can be taken to be the lifetime, in hours, of a certain piece of equipment. Let $T_1, T_2, \ldots, T_n$ denote the lifetimes of n pieces of equipment. Assuming that the mean life time is θ, it is desired to make a posterior assessment about θ.

(a) Find the posterior distribution of Θ, in terms of the likelihood function, using $p_1, p_2, \ldots, p_k$ in its denominator; (i.e., in the marginal pdf of T).

(b) Find the posterior distribution of Θ, in terms of the standardized likelihood function, and show that it is the same as in (a).

(c) Use the posterior distribution in (b) to determine the posterior probability $\Pr(a_i \leqslant \Theta \leqslant a_{i+1} | \underset{\sim}{t})$.

(d) Suppose $k=6$ and a_i and p_i are as follows:

i	$[a_{i-1}, a_i]$	p_i
1	[0, 1000]	0.05
2	[1000, 2000]	0.09
3	[2000, 3000]	0.20
4	[3000, 4000]	0.40
5	[4000, 5000]	0.15
6	[5000, ∞)	0.11

Assume further that $n=2$ and $f(x|\theta) = (1/\theta)e^{-x/\theta}$, $x \geqslant 0$. Find $\Pr(\theta > 3000 | x_1 = 2000, x_2 = 3000)$ and interpret the result. Does the likelihood dominate the prior?

(e) Suppose $p_1 = 0.15$, $p_2 = 0.15$, $p_3 = 0.25$, $p_4 = 0.20$, $p_5 = 0.15$, $p_6 = 0.10$ in (d). Compute $\Pr(\Theta > 3000 | x_1 = 2000, x_2 = 3000)$ and interpret the result. Does the likelihood dominate the prior?

5.14. Summarize the solution to Example 5.1 using the results in Exercise 5.13 (b), (c).

5.15. Suppose the reliability of a mechanical device is under study. Not much is known about the prior distribution of the reliability of the device. It is thus desired to conduct a preposterior analysis to determine if a candidate for a prior distribution is adequate. Four devices are put under test for a given length of time and only the success or failure of each item is to be recorded. Let R denote the reliability, that is the probability that the item will survive the test.

Following are five reliability values and the corresponding prior probabilities:

Reliability (R)	0.1	0.4	0.7	0.9	0.95
Prior	0.1%	0.2%	0.6%	95%	4.1%

It is expected that at least three of the four items will survive the test and highly unlikely that at most one unit will survive the test. In view of this information the four hypothetical results (i) {4 successes (S), 0 failures (F)}, (ii) {3-S, 1-F}, (iii) {1-S, 3-F}, (iv) {0-S, 4-F}, are to be considered in a preposterior analysis.

(a) Compute the likelihood for each of the four hypothetical results (i)–(iv) given the five reliability values.

(b) Compute the marginal probability for each of the four hypothetical results.

(c) Compute the posterior probabilities of the five reliability values, given each of the four hypothetical results.

(d) Present the final results in tabular form, as in Example 5.2.

(e) Make appropriate conclusions, based on the preposterior analysis, regarding the appropriateness of the prior distribution. Does the suggested prior agree with the subjective notions of what should occur?

5.16. Determine a prior distribution of reliability that agrees better with the subjective notions than the one used in Exercise 5.15.

5.17. Suppose that the lifetime of a device follows an $\mathcal{E}_1(\theta)$ distribution, where Θ also has an $\mathcal{E}_1(\theta_0)$ prior distribution.

(a) It is anticipated that the average lifetime of two devices put on test will be at least 2000 h. If the parameter of the prior is $\theta_0 = 10^4$ find the posterior probability that Θ is at least 2000 h. Discuss the appropriateness (or sensitivity) of the Bayesian inferences to the prior.

(b) Determine a prior (that is, a value of θ_0), for which the inference is more compatible with the choice of the prior.

5.18. Suppose a lot of electronic components is being considered for acceptance or rejection, depending on whether it meets a reliability criterion. There are two actions available, a_1 (accept) and a_2 (reject). The state of nature is the true value of the mean life θ which is unknown. Only two values of θ are considered. An economic analysis of the decisions and states of nature involved yield the following net loss/unit.

		State of Nature	
		$\theta_1 = 2000$	$\theta_2 = 500$
Action	Accept Lot (a_1)	\$0	\$20
	Reject Lot (a_2)	\$15	\$5

(a) List the four values of the loss function $L(\theta, a)$.

(b) A random sample of ten electronic components are put on test and yield the lifetimes $x_1, x_2, \ldots, x_{10}$. Assume X has an $\mathcal{E}_1(\theta)$ distribution with mean θ. Describe the joint pdf of $X_1, X_2, \ldots, X_{10}$, letting $\mathcal{T} = \Sigma X_i$ be the total time on test.

(c) Let a decision function δ_1 be

$$\delta_1(\underset{\sim}{x}) = \begin{cases} a_1, & \text{if t} > 900 \text{ h} \\ a_2, & \text{if t} \leq 900 \text{ h.} \end{cases}$$

Determine the risk $R(\theta, \delta_1)$ for $\theta = \theta_1$ and θ_2.

(d) Let Θ have the prior distribution $g(\theta_1) = g(2000) = 0.8$ and $g(\theta_2) = g(500) = 0.2$. Find the Bayes risk of $\delta_1(\underset{\sim}{x})$.

(e) Let $D = \{\delta_t\}$ be a class of decision functions defined by

$$\delta_t(\text{t}) = \begin{cases} a_1, & \text{if t} > t \text{ h} \\ a_2, & \text{if t} \leq t \text{ h.} \end{cases}$$

Find the Bayes decision function over this class.

5.19. Let a decision function δ_2 be defined by

$$\delta_2(\underset{\sim}{x}) = \begin{cases} a_1, & \text{if } \min(x_1, \ldots, x_{10}) > 2000 \text{ h} \\ a_2, & \text{if } \min(x_1, \ldots, x_{10}) \leq 2000 \text{ h.} \end{cases}$$

Determine the risk and Bayes risk of $\delta_2(\underset{\sim}{x})$ using the same assumptions as in Exercise 5.18.

5.20. Let $X_1, X_2, \ldots, X_n$ denote a random sample from an $\mathcal{E}_1(\theta)$ distribution. Show that the pdf of the r.v. $W = \min(X_1, X_2, \ldots, X_n)$ is $\mathcal{E}_1(\theta/n)$. [Hint: Find the cdf of W, $\Pr(W \leqslant w|\theta) = 1 - \Pr(W \geqslant w|\theta) = 1 - \prod_{i=1}^{n} \Pr(X_i \geqslant w|\theta)$].

5.21. Let the decision space be $D = \{\delta|\delta \in (0, \infty)\}$ and the action space be $A = \{a|a \in (0, \infty)\}$. The problem is to estimate the parameter θ based on a value of the r.v. X which has the pdf $f(x|\theta) = 1/\theta$, $0 < x < \theta$, and zero otherwise. Suppose Θ has the prior $g(\theta) = \theta e^{-\theta}$, $\theta > 0$, and zero otherwise.

(a) Find the posterior pdf of Θ.

(b) Assume the loss function $L(\theta, a) = c(\theta - a)^2$. Find $E(\Theta|X = x)$.

(c) Determine the Bayes decision rule.

(d) Verify that the decision function in (c) is $\delta(x) = E(\Theta|X = x)$.

5.22. Suppose $X_1, X_2, \ldots, X_n$ is a random sample from the pdf $f(x|\theta) = 1/\theta$, $0 < x < \theta$ and let Θ have the prior $g(\theta) = 1$, $0 < \theta < 1$.

(a) Find the posterior pdf of Θ.

(b) Let the loss function be $L(\theta, \hat{\theta}) = c(1 - \hat{\theta}/\theta)^2$. Find the posterior expected loss given $X = x$.

5.23. Verify the posterior pdf $g(\theta|\underset{\sim}{y})$ in (5.6).

5.24. Verify the computations in Example 5.2.

5.25. Work through Exercise 5.18 using the following decision functions:

(a) $$\delta_2(\underset{\sim}{x}) = \begin{cases} a_1, & \text{if } \dfrac{\max(x_1, \ldots, x_{10}) - \min(x_1, \ldots, x_{10})}{2} > 2000 \text{ h} \\ a_2, & \text{otherwise;} \end{cases}$$

(b) $$\delta_3(\underset{\sim}{x}) = \begin{cases} a_1, & \text{if } \dfrac{x_{(5)} + x_{(6)}}{2} > 2000 \text{ h} \\ a_2, & \text{if } \dfrac{x_{(5)} + x_{(6)}}{2} \leqslant 2000 \text{ h}, \end{cases}$$

where $X_{(i)}$ denotes the ith smallest X among $X_1, X_2, \ldots, X_n$.

REFERENCES

Anderson, T. W. (1958). *Introduction to Multivariate Statistical Analysis*, Wiley, New York.

Apostolakis, G. and Mosleh, A. (1979). Expert Opinion and Statistical Evidence: An Application to Reactor Core Melt Frequency, *Nuclear Science and Engineering*, Vol. 70, pp. 135–149.

Barnard, G. A., Jenkins, G. M., and Winsten, C. B. (1962). Likelihood Inference and Time Series, *Journal of the Royal Statistical Society*, Series A, Vol. 125, pp. 321–372.

Bayes, T. (1958). Essay Towards Solving a Problem in the Doctrine of Chances, *Biometrika*, Vol. 45, pp. 293–315.

Blackwell, D. and Girshick, M. A. (1954). *Theory of Games and Statistical Decisions*, Wiley, New York.

Bonis, A. J. (1966). Bayesian Reliability Demonstration Plans, *Annals of Reliability and Maintainability*, Vol. 5, pp. 861–873.

Bonis, A. J. (1975). Why Bayes is Better, *Proceedings 1975 Annual Reliability and Maintainability Symposium*, pp. 340–348.

Box, G. E. P. and Tiao, G. C. (1973). *Bayesian Inference in Statistical Analysis*, Addison-Wesley, Reading, MA.

Britney, R. R. and Winkler, R. L. (1968). Bayesian Point Estimation Under Various Loss Function, *Proceedings of the Business and Economic Statistics Section, American Statistical Association*, pp. 356–364.

Bucy, R. S. and Kalman, R. E. (1961). New Results in Linear Filtering and Prediction Theory, *Transactions of ASME, Series D: Journal of Basic Engineering*, Vol. 83, pp. 95–108.

Canfield, R. V. (1970). A Bayesian Approach to Reliability Estimation Using a Loss Function, *IEEE Transactions on Reliability*, Vol. R-19, pp. 13–16.

Chernoff, H. and Moses, L. (1959). *Elementary Decision Theory*, Wiley, New York.

De Finetti, B. (1937). La Prévision: Ses Lois Logiques, ses Sources Subjective, *Annales de l'Institute Henri Poincaré*, Vol. 7, No. 1. English Translation in *Studies in Subjective Probability*, H. E. Kyburg, Jr. and Howard E. Smokler (Eds.), 1964, Wiley, New York, pp. 93–158.

De Groot, M. H. (1970). *Optional Statistical Decisions*, McGraw-Hill, New York.

El-Sayyad, G. M. (1967). Estimation of the Parameter of an Exponential Distribution, *Journal of the Royal Statistical Society, Series B*, Vol. 29, pp. 525–532.

Evans, R. A. (1969). Prior Knowledge, Engineers Versus Statisticians, *IEEE Transactions on Reliability*, Vol. R-18, p. 143.

Ferguson, T. S. (1967). *Mathematical Statistics, A Decision Theoretic Approach*, Academic, New York.

Fisher, R. A. (1922). On the Mathematical Foundations of Theoretical Statistics, *Phil. Trans. Roy. Soc., Series A*, Vol. 222, pp. 308–358.

Galambos, J. (1978). *The Asymptotic Theory of Extreme Order Statistics*, Wiley, New York.

Good, I. J. (1950). *Probability and the Weighting of Evidence*, Charles Griffin and Co., London.

Grohowski, G., Hausman, W. C., and Lamberson, L. R. (1976). A Bayesian Statistical Inference Approach to Automotive Reliability Estimation, *Journal of Quality Technology*, Vol. 8, pp. 197–208.

Harris, B. (1977). A Survey of Statistical Methods in Systems Reliability Using Bernoulli Sampling of Components, *The Theory and Applications of Reliability*, Vol. II, Academic, New York, pp. 275–297.

Higgins, J. J. and Tsokos, C. P. (1978). *A Study of the Effect of the Loss Function on Bayes Estimates of Failure Intensity, MTBF, and Reliability*, Unpublished Manuscript, University of South Florida, Tampa, FL.

Jeffreys, H. (1961). *Theory of Probability* (3rd ed.), Clarendon Press, Oxford.

Kendall, M. G. and Stuart, A. (1961). *The Advanced Theory of Statistics, Vol. 2: Inference and Relationship*, Hafner, New York.

Lindley, D. V. (1965). *Introduction to Probability and Statistics from a Bayesian Viewpoint, Part 1, Probability and Part 2, Inference*, University Press, Cambridge.

Lindley, D. V. (1970). *Making Decisions*, Wiley-Interscience, New York.

Martz, H. F. and Waller, R. A. (1979). A Bayesian Zero-Failure (BAZE) Reliability Demonstration Testing Procedure, *Journal of Quality Technology*, Vol. II, pp. 128–138.

Mood, A., Graybill, F., and Boes, D. (1974). *Introduction to the Theory of Statistics* (3rd ed.), McGraw-Hill, New York.

Morgan, B. W. (1968). *An Introduction to Bayesian Statistical Decision Processes*, Prentice-Hall, Englewood Cliffs, NJ.

Raiffa, H. and Schlaifer, R. (1961). *Applied Statistical Decision Theory*, Harvard University Press, Cambridge, MA.

Ramsey, F. P. (1931). *The Foundation of Mathematics and Other Logical Essays*, Routledge and Kegan Paul, London.

Savage, L. J. (1954). *The Foundations of Statistics*, Wiley, New York.

Schafer, R. E. (1973). The Role of Bayesian Methods in Reliability, *Proceedings 1973 Annual Reliability and Maintainability Symposium*, pp. 290–291.

Schlaifer, R. (1959). *Probability and Statistics for Business Decisions*, McGraw-Hill, New York.

Schmitt, S. A. (1969). *Measuring Uncertainty: An Elementary Introduction to Bayesian Statistics*, Addison-Wesley, New York.

Sprott, D. A. and Kalbfleisch, J. G. (1965). Use of the Likelihood Function in Inference, *Psychological Bulletin*, Vol. 64, pp. 15–22.

Von Neumann, J. and Morgenstern O. (1947). *Theory of Games and Economic Behavior* (2nd ed.) Princeton University Press, Princeton, NJ.

Wald, A. (1950). *Statistical Decision Functions*, Wiley, New York.

Winkler, R. L. (1972). *Introduction to Bayesian Inference & Decision*, Holt, Rinehart & Winston, New York.

CHAPTER 6

The Prior Distribution

The fundamentals of Bayesian estimation were presented in Chapter 5 with particular emphasis on reliability analysis. Those developments imply that a fundamental part of any Bayesian analysis is the prior distribution. Therefore, we devote this chapter to a discussion of prior distributions and explore methods for selecting, fitting, and evaluating prior distributions.

As stated previously, the prior distribution $g(\theta)$ represents all that is known or assumed about the parameter θ (either scalar or vector) prior to the observation of empirical data. Thus, the information summarized by the prior distribution may be either objective or subjective or both. Two examples of objective input to the prior distribution are operational data and observational data from a previous comparable experiment. Subjective information may include an engineer's quantification of personal experiences and judgments, a statement of one's degree of belief regarding the parameter, design information, and personal opinions.

Prior distributions may be categorized in different ways. One common classification is a dichotomy that separates "proper" and "improper" priors. A *proper prior* is one that allocates positive weights that total one to the possible values of the parameter. Thus a proper prior is a weight function that satisfies the definition of a probability mass function or a pdf, whichever the case may be. An *improper prior* is any weight function that sums (or integrates) over the possible values of the parameter to a value other than one, say p. If p is finite, then an improper prior can induce a proper prior by normalizing the function. However, when p is infinite (or otherwise doesn't exist) over the range of possible parameter values, an improper prior either remains improper and plays the role of a weighting function or an ad hoc method of determining a proper prior is used [Box and Tiao (1973)].

Other classifications of priors, either by properties or by distributional forms, are discussed in this chapter. Section 6.1 presents noninformative priors and Section 6.2 discusses the important class of conjugate priors. A collection of less frequently used types of priors is discussed in Section 6.3. Section 6.4 treats the problem of evaluating priors, while Sections 6.5 and 6.6 present techniques for selecting particular prior models from the beta and gamma families of distributions respectively. Section 6.7 considers the negative-log gamma prior distribution.

6.1. NONINFORMATIVE PRIOR DISTRIBUTIONS

One general class of prior distributions is called *noninformative priors*. They are also called priors of ignorance. Neither name produces a necessarily positive image so it is important to clarify the intent of the titles. Box and Tiao (1973, pp. 25–60) provide a thorough discussion of noninformative priors for one or more parameters. In this section we introduce the general philosophy of noninformative priors and provide some relevant examples. Specific noninformative priors are considered in subsequent chapters.

Rather than a state of complete ignorance, the noninformative prior refers to the case when relatively little (or very limited) information is available a priori. In other words a priori information about the parameter is not considered substantial relative to the information expected to be provided by the sample of empirical data. Further, it frequently means that there exists a set of parameter values that the experimenter believes to be equally likely choices for the parameter. That is, the experimenter is indifferent to a selected range of parameter values. The goal here is to select a prior model that adequately represents the state of prior knowledge about the parameter. One way of expressing our indifference is to select a prior distribution that is locally uniform, that is, a prior that is approximately uniformly distributed over the interval of interest. However, there are other ways of defining noninformative priors, several of which are discussed by Jeffreys (1961). The specifics of a noninformative prior are summarized in the following definition.

Definition 6.1. [Box and Tiao (1973), p. 32] If $\phi(\theta)$ is a one-to-one transformation of θ, we shall say that a prior distribution of Θ that is locally proportional to $|d\phi/d\theta|$ is *noninformative* for the parameter θ if, in terms of ϕ, the likelihood curve is *data translated*; that is, the data $\underset{\sim}{x}$ only serve to change the location of the likelihood $L(\phi|\underset{\sim}{x})$.

It can be shown [Box and Tiao (1973), pp. 25–46] that any data translated likelihood is expressible as

$$L(\theta|\underset{\sim}{x}) = h[\phi(\theta) - f(\underset{\sim}{x})], \tag{6.1}$$

where $h(\cdot)$ is known and independent of $\underset{\sim}{x}$ and $f(\cdot)$ is a function of $\underset{\sim}{x}$ only. Thus, one way of determining noninformative prior distributions is to exploit the property in (6.1). Further, when (6.1) does not hold exactly, we can obtain approximate noninformative priors by transformations that approximate (6.1). As will be noted later, the derivation of noninformative priors is mathematically very closely associated with variance stabilizing transformations [Bartlett (1937)] and Fisher's information [Fisher (1922, 1925)].

Box and Tiao (1973, pp. 36–38) provide details to show that, if a single sufficient statistic exists for estimating a parameter θ, then an approximate noninformative prior can be obtained as follows:

STEP 1. Let $L(\theta|\underset{\sim}{x}) = \ln \prod_{i=1}^{n} f(x_i|\theta)$ denote the log-likelihood of the sample.

STEP 2. Let $J(\hat{\theta}) = \left(-\frac{1}{n}\frac{\partial^2 L}{\partial \theta^2} \right)_{\theta=\hat{\theta}}$,

where $\hat{\theta}$ is the ML estimator of θ.

STEP 3. The approximate noninformative prior for Θ is given by

$$g(\theta) \propto J^{1/2}(\theta). \tag{6.2}$$

A part of the above development shows that the transformation providing data translation is given by

$$\phi(\theta) \propto \int J^{1/2}(t)\,dt\big|_{t=\theta}. \tag{6.3}$$

Also, $J(\theta)$ in Step 2 is a special case of *Fisher's information* defined as $-E_{\underset{\sim}{x}|\theta}(\partial^2 L/\partial\theta^2)$ and, in general, an approximate noninformative prior is taken proportional to the square root of Fisher's information. This is known as *Jeffreys' Rule* [Jeffreys (1961)].

Example 6.1. Let X_i follow an $\mathfrak{N}(\mu, \sigma^2)$ distribution, $i = 1, 2, \ldots, n$. Find a noninformative prior for μ. Find $\phi(\mu)$. Let $\underset{\sim}{x}' = (x_1, x_2, \ldots, x_n)$ and let $\underset{\sim}{\theta}' = (\mu, \sigma)$.

$$L(\underset{\sim}{\theta}|\underset{\sim}{x}) = -\frac{n}{2}\ln(2\pi) - n\ln\sigma - \frac{1}{2\sigma^2}\Sigma(x_i - \mu)^2,$$

$$\frac{\partial L}{\partial \mu} = \frac{2}{2\sigma^2}\sum_{i=1}^{n}(x_i - \mu),$$

$$\frac{\partial^2 L}{\partial \mu^2} = -n,$$

$$\frac{\partial L}{\partial \mu} = 0 \text{ gives } \hat{\mu} = \frac{\Sigma X_i}{n} = \bar{X}(\text{ML estimator}),$$

$$J(\hat{\mu}) = \left[-\frac{1}{n}(-n) \right]_{\hat{\mu} = \bar{x}} = 1.$$

Therefore, the noninformative prior for μ is a uniform distribution

$$g(\mu) = c.$$

Further, $\phi(\mu) = \int dt|_{t=\mu} = \mu$ is the transformation providing data translated likelihood.

Example 6.2. Use the information in Example 6.1 to find a noninformative prior for σ. Using $L(\underset{\sim}{\theta}|\underset{\sim}{x})$ from Example 6.1,

$$\frac{\partial L}{\partial \sigma} = -\frac{n}{\sigma} + \frac{1}{\sigma^3}\Sigma(x_i - \hat{\mu})^2,$$

$$\frac{\partial^2 L}{\partial \sigma^2} = \frac{n}{\sigma^2} - \frac{3}{\sigma^4}\Sigma(x_i - \hat{\mu})^2.$$

Therefore, the ML estimator for σ is

$$\hat{\sigma} = \left[\frac{\Sigma(X_i - \hat{\mu})^2}{n} \right]^{1/2}.$$

Thus,

$$J(\hat{\sigma}) = \left(-\frac{1}{n}\left[\frac{n}{\hat{\sigma}^2} - \frac{3}{\hat{\sigma}^4} \sum (X_i - \hat{\mu})^2 \right] \right)$$

$$= \left(-\frac{1}{\hat{\sigma}^2} + \frac{3}{\hat{\sigma}^2} \right) = \frac{2}{\hat{\sigma}^2}.$$

Therefore,

$$g(\sigma) \propto \frac{1}{\sigma}$$

is a noninformative prior for σ. Further,

$$\phi(\sigma)=\int \frac{dt}{t}\Big|_{t=\sigma}=\ln \sigma$$

provides a data translated likelihood. ■

The noninformative prior distributions in Examples 6.1 and 6.2 are improper priors for μ on the real line and σ on the positive half of the real line because they integrate to ∞. That is not a serious problem in practice in that the prior can either be used as a weight function assigning equal weight over an infinite interval, or be normalized to assign a total weight of one to any finite interval of interest. In particular the prior $g(\mu)=c$, in Example 6.1, assigns equal weight $cd\mu$ to all values of μ between $-\infty$ and $+\infty$. Similarly, the prior $g(\sigma)=1/\sigma$, in Example 6.2, assigns equal weight to all values of $\ln \sigma$ for σ between 0 and $+\infty$. In most cases such improper prior distributions yield proper posterior distributions as the posterior distribution is uniquely defined up to a normalizing constant on $g(\theta)$ [see e.g., Exercise 6.4].

6.2. CONJUGATE PRIOR DISTRIBUTIONS

Raiffa and Schlaifer (1961, pp. 43–58) present a formal development of conjugate prior distributions. Intuitively, a *conjugate prior distribution*, say $g(\theta)$, for a given sampling distribution, say $f(x|\theta)$, is such that the posterior distribution $g(\theta|x)$ and the prior $g(\theta)$ are members of the *same* family of distributions. In developing the fundamentals of conjugate prior distributions, Raiffa and Schlaifer (1961, p. 44) seek a family of priors to satisfy the following criteria:

1. Analytic tractibility: (a) Easy to determine posterior from the sample and prior distributions. (b) Easy to obtain expectations of utility functions of interest. (c) Prior and posterior should be members of same family of distributions (closed).
2. Flexible and rich: a wide variety of information and beliefs should be describable by the family of priors.
3. Easily interpreted: The parameters should be such that an experimenter can readily relate his or her beliefs to characteristics of the prior distribution.

In this section we present and illustrate the methodology for obtaining conjugate prior distributions. The methodology relies on the concepts of sufficient statistics and likelihood functions introduced in Chapters 3 and 5. In particular, we say that $\underset{\sim}{T}$ is a sufficient statistic of dimension k, if for every sample of size n, $\underset{\sim}{T}$ is a k-dimensional vector function of the sample $\underset{\sim}{x}$. Recall that we refer to the product of the sample density functions expressed as a function of the parameter θ given the sample values as the likelihood function. That is

$$g(\theta \mid \underset{\sim}{x}) = \prod_{i=1}^{n} f(x_i \mid \theta) \tag{6.4}$$

is the likelihood function of θ, where we now write $g(\theta \mid \underset{\sim}{x})$ rather than the usual $L(\theta \mid \underset{\sim}{x})$.

If $\underset{\sim}{T}(\underset{\sim}{x})$ is a sufficient statistic (either scalar or vector), then according to (5.42), (6.4) can be rewritten as

$$g(\theta \mid \underset{\sim}{x}) = h_1(\underset{\sim}{x}) h_2[\underset{\sim}{T}(\underset{\sim}{x}), \theta]. \tag{6.5}$$

Let $g(\theta \mid \underset{\sim}{\alpha})$ represent a family of pdf's where $\underset{\sim}{\alpha}$ denotes the parameter(s) that index specific members of the family. If for every pair of pdf's $g(\theta \mid \underset{\sim}{\alpha}_1)$ and $g(\theta \mid \underset{\sim}{\alpha}_2)$, there is another pdf $g(\theta \mid \underset{\sim}{\alpha}_3)$ such that

$$g(\theta \mid \underset{\sim}{\alpha}_3) \propto g(\theta \mid \underset{\sim}{\alpha}_1) g(\theta \mid \underset{\sim}{\alpha}_2), \tag{6.6}$$

we say the family is *closed under multiplication*.

The foregoing results allow us to construct conjugate prior distributions by determining a family of pdf's that satisfies two criteria:

1. The likelihood function of θ for any given sample must be proportional to a member of the family.
2. The family must be closed under multiplication.

Both De Groot (1970, p. 163) and Raiffa and Schlaifer (1961, pp. 47–48) provide proof that when a sufficient statistic exists, a family of conjugate prior distributions exists. We now illustrate the procedure by an example.

Example 6.3. Consider n trials in a binomial experiment where the probability of success is p. Find a family of conjugate prior distributions for P.

In Chapter 2, the binomial distribution was given as

$$\Pr(X = x) = \binom{n}{x} p^x (1-p)^{n-x}, \quad x = 0, 1, \ldots, n,$$

where X = number of successes in n trials. It can be shown that X is a sufficient statistic for p for this sampling distribution by the Neyman factorization criterion given in (5.42) or (6.5). Here

$$h_1(x)=\binom{n}{x} \text{ and } h_2(x,p)=p^x(1-p)^{n-x}.$$

Therefore,

$$g(p|x)\propto p^x(1-p)^{n-x}.$$

Thus $g(p|x)$ is proportional to a $\mathcal{B}_1(\cdot,\cdot)$ distribution.

Consider a $\mathcal{B}_1(\alpha_1,\beta_1)$ and $\mathcal{B}_1(\alpha_2,\beta_2)$ distribution on Θ. Then

$$\mathcal{B}_1(\alpha_1,\beta_1)\cdot\mathcal{B}_1(\alpha_2,\beta_2)\propto\theta^{\alpha_1-1}(1-\theta)^{\beta_1-1}\theta^{\alpha_2-1}(1-\theta)^{\beta_2-1}$$

$$=\theta^{\alpha_1+\alpha_2-2}(1-\theta)^{\beta_1+\beta_2-2}.$$

We recognize that the product of the two $\mathcal{B}_1(\alpha_i,\beta_i)$ distributions is yet another $\mathcal{B}_1(\alpha_1+\alpha_2-1,\ \beta_1+\beta_2-1)$ distribution. Therefore, the beta family is closed under multiplication and the beta family provides a family of conjugate prior distributions for the binomial parameter p. ■

Simple conjugate prior distributions are widely used in Bayesian analysis and numerous illustrations of conjugate priors are provided in the following chapters. De Groot (1970, pp. 164–179) and Raiffa and Schlaifer (1961, pp. 53–58) give conjugate prior distributions for selected sampling models.

6.3. OTHER PRIOR DISTRIBUTIONS

The noninformative and conjugate priors discussed in Sections 6.1 and 6.2 are widely used in practice. However, any number of other weighting functions can be used as prior distributions. The purpose of this section is to indicate a few ad hoc procedures that have been proposed for determining prior models. We present only a selection of ideas available in the literature and refer the reader to the references below for further reading.

MacFarland (1971, 1972) proposes the use of a discrete prior for both classical and sequential decision making. Here the experimenter selects a finite set of distinct points as possible prior values for the parameter. A total probability of one is then allocated to these points. The selected discrete values can be thought of as "middle" values of cells or intervals to approximate a continuous prior. The assigned probabilities are then associated with the respective cells. The discrete form of Bayes' theorem given in (5.8) could then be used for posterior analysis. A note of caution is needed here as we alternate between probabilities of discrete points and the required densities to provide the same probability to cells containing the assumed points. Particular attention needs to be given to graphing the information, (i.e., whether the vertical scale is "probability" or "density").

The discrete method can be useful in that it provides a convenient way for the analyst to talk to the engineer regarding the quantification of knowledge of a particular piece of equipment. That is, a selected set of points can be discussed and relative weights of importance assigned in a straightforward manner. Those relative weights can be normalized to a total weight of one.

Example 6.4. MacFarland (1972) gives the following data[a] to illustrate the foregoing method, where reliability is the parameter of interest

Cell Number (i)	Cell Value (r_i)	Cell Range	Cell Probability
1	0.999999	0.9999945–1.000000	0.02
2	0.99999	0.999945 –0.9999945	0.03
3	0.9999	0.9997 –0.999945	0.05
4	0.9995	0.99925 –0.9997	0.15
5	0.999	0.997 –0.99925	0.25
6	0.995	0.9925 –0.997	0.20
7	0.99	0.97 –0.9925	0.15
8	0.95	0.925 –0.97	0.11
9	0.9	0.85 –0.925	0.03
10	0.8	0.75 –0.85	0.002
11	0.7	0.65 –0.75	0.001
12	0.6	0.55 –0.65	0.001
13	0.5	0.45 –0.55	0.001
14	0.4	0.35 –0.45	0.001
15	0.3	0.25 –0.35	0.001
16	0.2	0.15 –0.25	0.001
17	0.1	0.05 –0.15	0.001
18	0.0	0.00 –0.05	0.001
		Total	1.000

[a]© 1972 IEEE. Adapted, with permission, from Bayes' Equation, Reliability and Multiple Hypothesis Testing, by W. J. MacFarland, appearing in *IEEE Transactions on Reliability*, Vol. R-21, pp. 136–147, August 1972.

Remarks.

1. Eighteen arbitrary cell values are allocated positive probabilities.
2. The cell intervals have varying lengths. Thus, graphing uniform densities for each cell would require plotting $g(r)$, where $g(r_i)\cdot$ cell length = cell probability is solved for $g(r_i)$ for the 18 values of r_i.
3. MacFarland (1972) treats the prior as discrete for 18 values and plots the probabilities showing that 0.999 is the most likely value. ■

As a second example of other prior formulations, suppose that an informed engineer believes that the most likely value of the reliability of a safety system is 0.999. Further, the engineer believes that the probability is 0.60 that the reliability is greater than or equal to 0.999. One way of representing this information is to use a triangle prior as illustrated in Figure 6.1. The technique uses two triangles joined at $r = 0.999$. The triangle over the interval 0.999–1.000 has area 0.60, whereas the triangle over the interval 0.000–0.999 has area 0.40.

Use of triangles having specified probabilities over intervals is one way of translating subjective information into a prior model for a parameter of interest. This approach could be extended to use linear functions to join a selected number of nodes. The resulting lines could be joined to give a piecewise continuous (but not necessarily differentiable) pdf. Such a technique is used by Martz and Lian (1977).

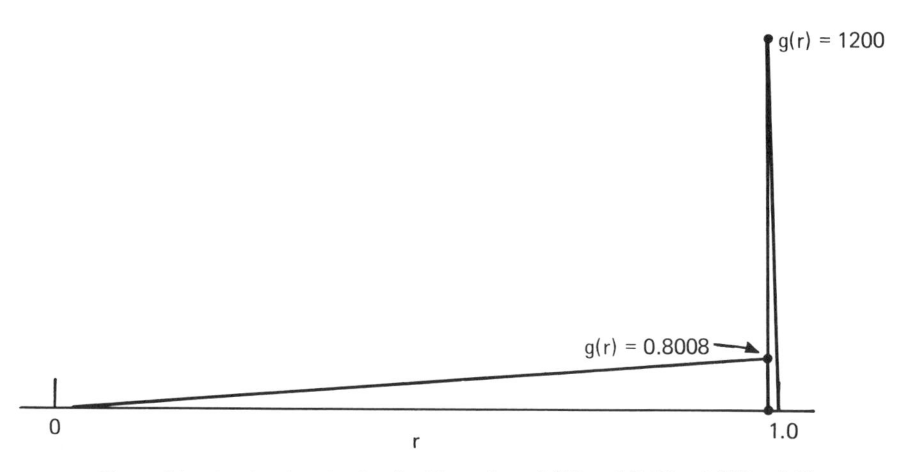

Figure 6.1. A triangle prior for R with mode at 0.999 and $\Pr(R > 0.999) = 0.60$.

In an attempt to remove some of the subjectivity innate in the selection of prior models, several articles have addressed the problem of obtaining *one* prior distribution for given information. That is, how can we be sure that all users of some specified information will use the same prior distribution? Deeley, Tierney and Zimmer (1970), Jaynes (1968), and Shimi and Tsokos (1977) present a method based on maximum entropy and transformation groups to address the preceding question. This method assumes that there are n distinct possible values of X where n is at most countably infinite and that the relevant information can be summarized in the form of constraints on the probabilities $p_j = \Pr(X = x_j)$, $j = 1, 2, \ldots, n$. For example, the information could be the expected values of some functions, say $k_1(x), \ldots, k_m(x)$, $m < n$. That is, we could be given the values $A_1, \ldots, A_m$ such that

$$\sum_{j=1}^{n} p_j k_i(x_j) = A_i, \qquad i = 1, 2, \ldots, m.$$

The *method of maximum entropy* is then to find p_j, $j = 1, 2, \ldots, n$, to maximize

$$H = -\sum_{j=1}^{n} p_j \ln p_j \tag{6.7}$$

subject to $p_j \geqslant 0$ for all j, $\Sigma_{j=1}^{n} p_j = 1$, and

$$\sum_{j=1}^{n} p_j k_i(x_j) = A_i \qquad \text{for } i = 1, 2, \ldots, n.$$

After Jaynes (1968) solves the above optimization problem, he then chooses the prior model so any changes in the parameters do not change the entropy. This is accomplished through transformation group techniques.

Examples of the maximum entropy method are given by Deeley, Tierney, and Zimmer (1970) for binomial and $\mathcal{E}(\lambda)$ sampling models and by Jaynes (1968) for binomial and $\mathfrak{N}(\mu, \sigma^2)$ models. Further, Jaynes (1968) presents the development for continuous priors that is analogous to the discrete case in (6.7).

Measures of information, such as entropy, can also be used to assess the amount of information about a parameter θ that is contained in the sample relative to the prior distribution $g(\theta)$. Goel and De Groot (1981) consider the information at various levels of a hierarchical model by defining an appropriate measure of information. Such an approach can also be used to determine prior distributions by relating the amount of information a prior contains to an equivalent sample size. These methods should find increasing application in determining prior distributions that contain a precise quantity of prior "information" without additional unwanted information being inadvertently introduced.

In this section we have highlighted some of the ideas relating to the choice of a credible prior that represents the available relevant information. The main point is that the prior model must be formulated by the decision maker working closely with the individual who must quantify the available information for use in Bayes' theorem. The types of information pooled together in the agreed-on prior model may be some or all of engineering design, engineering judgment, operational experience, previous empirical results, and experience with similar hardware. The close liaison between the decision maker and hardware expert regarding specification of the prior distribution is only the beginning. Once the prior is specified, it should be carefully evaluated so that its implications and consequences are completely understood.

6.4. EVALUATION OF PRIOR MODELS

The preceding sections have presented some useful methods for selecting prior distributions. From those discussions it should be evident that the selection of priors is a subjective, and frequently ad hoc, exercise that should be examined from various points of view. The purpose of this section is to suggest some methods for evaluating a selected prior distribution. The presentation is not meant to be exhaustive, rather it is hoped that the suggestions will serve as a stimulus for further investigations on prior evaluation.

It is important that we understand the consequences of selecting a certain prior distribution so that we can defend its use in a Bayesian analysis. To accomplish this we need to address such questions as: What is the prior to represent? How is the prior to be identified? What are the characteristics of the prior? How does the prior behave in the presence of experimental (observed) data?

Representativeness is a basic requirement for any prior distribution. The knowledgable researcher or engineer (hereafter referred to as the *expert*) and the decision maker must work closely together in determining and evaluating the prior distribution. It is the decision maker's role to accurately and most appropriately assess the expert's prior degree of belief about the parameter in question. It is also the decision maker's role to explain what the prior distribution represents, whereas the expert must provide objective and quantitative statements of his or her degree of belief.

Prior identification is a cooperative effort between the expert and the decision maker. The quantification of subjectively held notions regarding degree of belief has been the subject of numerous investigative efforts both by psychologists interested in such cognitive processes as well as statisticians.

The broad subject area involving quantification of subjectively held degree of belief information may be loosely subdivided into two major areas. The first area represents the philosophy and practice of *probability encoding*, which is the process of extracting and quantifying individual judgment about uncertain quantities. The distinguishing characteristic of this area is that only a single expert is considered. The second major area concerns the techniques and practice by which subjective probability assessments may be derived as a consensus of a group or panel of experts. In this case more than one expert is involved and thus more than one subjectively held degree of belief notion must be pooled to arrive at a consensus subjective probability statement. This is the so-called *consensus problem*. Hogarth (1975) presents a comprehensive bibliography and Hampton, et al. (1973) provide a good introduction to both areas.

6.4.1 Encoding Subjective Probabilities

It is well-known that various biases may accompany the subjective probability values held by an expert. Large and systematic departures from proper calibration were reported in the experiments performed by Alpert and Raiffa (1969). They found that experienced adults often overestimate the degree of certainty of their estimates and thus are too sure of their information. For example, they observed that 98 percent interval estimates were frequently found to actually correspond to about 70 percent intervals. Tversky and Kahneman (1974) attribute such bias in part to "anchoring." Dalkey and Brown (1971) also observe that underestimation of a probability of interest is common and observe that the 65th percentile is often a better estimate of the true probability than either the mean or the geometric mean.

It is also well-known that the manner chosen by the decision maker to encode the information held by the expert is crucial and may affect the precision of the information [Du Charme and Donnell (1973), Seaver, et al. (1978) and Winkler (1967)]. Spetzler and Staël von Holstein (1975) describe and recommend a structured interview process and a number of techniques to reduce biases in the quantification of judgment. Hampton, et al. (1973) and Staël von Holstein (1970, 1971) provide comprehensive reviews of the literature on encoding techniques.

Another aspect of the encoding problem concerns the reconciliation of incoherent probability assessments supplied by an expert. Roughly, an expert is said to be *calibrated* if the expert's subjective probabilities agree with those obtained from a long sequence of independent events of the proper type. Lindley, et al. (1979) provide a general model for reconciling incoherent assessments and an excellent survey of the state-of-the-art in the calibration area is provided by Lichtenstein, et al. (1977).

Saaty, in a series of papers [(1972, 1974, 1977, 1978)] and a book [Saaty (1980)], develops an eigenvalue method that may be used to encode subjective probabilities [Yager (1979)]. The method Saaty proposes is based on pairwise (binary) evaluation of the likelihood of a series of events pertaining to the quantity of interest. For example, consider an expert in the reliability of a certain valve used in nuclear power reactor applications. Suppose that a decision maker must assess the prior distribution of the constant valve failure rate Λ subjectively held by the expert. A series of three pertinent events might be $E_1:\{\Lambda < 10^{-7}\}$, $E_2:\{10^{-7} \leq \Lambda \leq 10^{-5}\}$, and $E_3:\{\Lambda > 10^{-5}\}$. The Saaty method is to compare the relative subjective likelihood of the pairs E_1 to E_2, E_1 to E_3, and E_2 to E_3 on an integer scale of 1–9 which measures the strength of the preference. The matrix of such preferences is then used to arrive at the expert's subjective probabilities of each of the three events. There is an important feature of the Saaty method that distinguishes it from other calibration procedures. The method Saaty proposes does *not* require consistency among the pairwise comparisons that, in many instances, may be intransitive. As it is a fact of life that such comparisons supplied by an expert are rarely consistent, this is an important practical advantage over those methods that require consistent expert assessments of the probability of events such as E_1, E_2, and E_3. Saaty's method also provides a measure of consistency of the expert.

6.4.2 Consensus Prior Distributions

Suppose now that a decision maker consults a number of experts and the subjective (prior) probability distribution of a quantity of interest has been encoded for each expert. In this case, the decision maker must somehow combine or pool the distributions assessed by the experts to form a single consensus prior distribution for use in a Bayesian analysis.

Several methods for obtaining such a consensus prior have been suggested in the literature. One of earliest methods is the "parimutuel" method credited to Eisenberg and Gale (1959). Brown and Helmer (1964) and Helmer (1963) investigate the consensus of experts' point estimates. Winkler (1968) and Winkler and Cummings (1972) discuss several methods, some involving feedback and/or group discussion, for combining subjective distributions. Among the methods discussed are weighted averages, both with equal weights and weights proportional to ranking or self-rating, and the use of natural conjugate prior distributions that are combined via Bayes' theorem to arrive at a consensus distribution. In the case where the consensus prior is a mixture of natural conjugate distributions, an application of Bayes' theorem results in another mixture of natural conjugate distributions.

De Groot (1974) presents a Markov chain technique for determining a consensus prior that is a mixture of individual expert priors. Stone (1961) refers to such a mixture as an "opinion pool". Morris (1974, 1977) develops a Bayesian structure in which the consensus problem may be logically formulated and conceptually solved. Morris (1977) accomplishes this by defining a "joint calibration function" that is used to measure each expert's probability assessment ability and to encode the degree of dependence among the experts. He then defines a consensus prior that is equal to the normalized product of the joint calibration function and the individual expert priors. Mirkin (1979) also discusses several methods that can be used in arriving at a consensus decision among a panel of experts based on their evaluations of a set of objects or events. His method is quite similar to Saaty's eigenvalue approach in the sense that experts need not be consistent in their evaluations and differences need not be reconciled.

Breiphol (1978) and Martz and Waller (1978) consider specialized methods for arriving at a consensus prior based on expert opinion regarding component or system reliability.

It should also be mentioned that the consensus problem falls within the framework of what has come to be known as *group decision theory*, an excellent account of which is given by Dalkey (1977) and Mirkin (1979).

Returning to the general question of prior identification, Fearn (1979), discusses an iterative interactive computer code that uses percentiles to select an inverted χ^2 prior distribution having designated properties. Section 6.5 provides a method using a combination of moments and percentiles for specifying $\mathcal{B}(x_0, n_0)$ prior distributions. Section 6.6 discusses a percentile technique for translating subjective degree of belief into $\mathcal{G}(\alpha_0, \beta_0)$ or $\mathcal{G}_1(\alpha_0, \gamma_0)$ prior distributions. Section 6.7 likewise considers $\mathcal{NLG}(\alpha_0, \gamma_0^{-1})$ priors.

A prior can be characterized in many different ways. First, we can describe a distribution by its parameters as was done in Chapter 3. Second, we can use one or more moments to describe such characteristics as the prior's location, variance, skewness, and so on. Third, we can use percentiles to describe tail values, middle values, and dispersion. Dialogue between the expert(s) and the decision maker regarding "most likely" values and dispersion information can be used to identify and/or refine the prior distribution. Fourth, the prior can be characterized by general shape characteristics such as the number of modes, whether it is symmetric or skewed, whether it is increasing (decreasing) over a certain range of interest, and so on. Free-hand sketches, graphs, and plots may also be useful in motivating and refining discussions between the expert(s) and the decision maker.

6.5. SELECTING BETA PRIOR DISTRIBUTIONS

The $\mathcal{B}(x_0, n_0)$ family of distributions is frequently used as a prior model for the survival (or failure) probability of a component in binomial sampling [Section 7.2.4]. $\mathcal{B}(x_0, n_0)$ prior distributions are also used in the Pascal sampling model [Section 7.3.1]. The purpose of this section is to discuss methods for selecting $\mathcal{B}(x_0, n_0)$ priors having specified properties. In so doing, we can represent available subjective and/or objective information by selecting a particular $\mathcal{B}(x_0, n_0)$ prior having the properties designated by knowledgeable professionals.

To illustrate one methodology for selecting a $\mathcal{B}(x_0, n_0)$ prior distribution, we assume that R is the reliability of an item of interest and that we desire a prior distribution for R. The development which follows is taken from Waterman, Martz, and Waller (1976).

To use Bayesian methods in practice, the experimenter must specify values for x_0 and n_0. The choices are often subjective in nature and are an attempt to convert "beliefs" and "judgments" into numbers. One method for doing this is presented by Weiler (1965). The technique proposed by Weiler is valid for the following situations: (1) when the expected reliability $E(R)$ and the probability p such that $\Pr(R > 2E(R)) = p$, are known, and (2) when the probabilities, say p_a and p_b, such that, given R_a and R_b, $\Pr(R > R_a) = p_a$ and $\Pr(R < R_b) = p_b$, are known. The methodology proposed here is similar to that of Weiler in spirit, but it provides more explicit use of an experimenter's subjective information. In addition, the tables provided in Appendix C cover a range of values of reliability likely to be encountered in practice.

We begin by defining

1. The prior mean reliability $R_1[R_1 = E(R)]$,
2. The prior 95th percentile $R_2[\Pr(R > R_2) = 0.05]$,
3. The prior 5th percentile $R_3[\Pr(R < R_3) = 0.05]$.

Thus R_2 is that value of reliability such that, before the test is performed, there is only a 5% chance that the reliability R will exceed R_2. Similarly, R_3

is that value of reliability such that, before the test results are obtained, there is only a 5% chance that the reliability R will be less than R_3. Usually, any two of the above quantities are sufficient to determine uniquely the parameters x_0 and n_0. This technique is usually subjective in practice because the user subjectively selects pairs of values for either (R_1, R_2) or (R_1, R_3). What follows is a methodology that permits reliability experts to use judgments, beliefs, and experiences in their assessment of $\mathcal{B}(x_0, n_0)$ parameters for a reliability prior distribution.

Example 6.5. Table C1 in Appendix C presents an easy to use table of values of prior parameters x_0 and n_0 corresponding to an extensive set of values of R_1 and R_2. For example, suppose that a $\mathcal{B}(x_0, n_0)$ distribution is desired for which the prior mean reliability is believed to be 0.99. Suppose further that it is believed there is only a 5% chance that the true prior reliability exceeds 0.9999. Thus $R_1 = 0.99$ and $R_2 = 0.9999$. From Table C1, this prior belief is consistent with specifying the corresponding $\mathcal{B}(x_0, n_0)$ parameters to be $x_0 = 60.06226$ and $n_0 = 60.66895$.

Example 6.6. Table C2 in Appendix C presents a table of values of prior parameters x_0 and n_0 corresponding to an extensive set of values of R_1 and R_3. For example, suppose that the prior mean reliability is believed to be 0.995 and that it is believed there is only a 5% chance that R is less than 0.980. Thus $R_1 = 0.995$ and $R_3 = 0.98$. From Table C2 this prior belief is consistent with a $\mathcal{B}(x_0, n_0)$ distribution having parameters $x_0 = 88.26738$ and $n_0 = 88.71094$. ■

The basis for constructing Tables C1 and C2 of Appendix C may be found in Waterman, Martz, and Waller (1976). The preceding examples and the work by Weiler (1965) provide methods for translating information about means and percentiles into $\mathcal{B}(x_0, n_0)$ prior distributions. Another method of selecting a $\mathcal{B}(x_0, n_0)$ prior is given in Columbo and Constantini (1980). Their work uses the concept of a relevance quotient to select integer valued parameters. The relevance quotient is a ratio of sequential binomial probabilities.

6.6. SELECTING GAMMA PRIOR DISTRIBUTIONS

The material in this section is based on technical papers by Martz and Waller (1979), and Waller, Johnson, Waterman, and Martz (1977a, 1977b).

As discussed in Chapter 4, it is frequently assumed that a failure time r.v. follows an $\mathcal{E}(\lambda)$ distribution. For Bayesian analysis in this model, the family of $\mathcal{G}_1(\alpha_0, \beta_0)$ distributions provides conjugate prior models for Λ to be used in Bayesian reliability/failure rate analyses. The selection of a particular prior distribution is accomplished by identifying values for the prior shape parameter α_0 and prior scale parameter β_0. The mean and variance of Λ are given by (α_0/β_0) and (α_0/β_0^2), respectively.

A simple method for determining values for α_0 and β_0 as presented in Martz and Waller (1979) is now described. The method requires an expert to provide upper and lower percentile values. Once these are given, a simple graphical or table look-up, in addition to a few simple calculations, yields the corresponding values of α_0 and β_0.

The expert must provide two values of the failure rate λ, referred to as the *lower prior limit* (LL) and *upper prior limit* (UL), such that

$$\Pr(\lambda < LL) = \Pr(\lambda > UL) = \frac{(1.0 - p_0)}{2}, \tag{6.8}$$

where p_0 is required to be equal to one of the values 0.95, 0.90, or 0.80, and where $LL < UL$. That is, LL and UL are specified such that there is an equal $50(1.0 - p_0)\%$ chance that the true (unknown) failure rate is either less than LL or greater than UL, respectively. Thus LL and UL are the respective $50(1.0 - p_0)$th and $50(1.0 + p_0)$th percentiles of the prior $\mathcal{G}_1(\alpha_0, \beta_0)$ distribution. For example, suppose that an expert's best prior judgement or belief is that $\Pr(\lambda < 1.0 \times 10^{-7}\ \text{f/h}) = 5\%$ and that $\Pr(\lambda > 1.0 \times 10^{-5}\ \text{f/h}) = 5\%$. Thus $(1.0 - p_0)/2 = 0.05$, $p_0 = 0.90$, $LL = 1.0 \times 10^{-7}$ f/h, and $UL = 1.0 \times 10^{-5}$ f/h. The quantity $100p_0\%$ is the *prior assurance* that the interval (LL, UL) contains the failure rate of interest. As 5% error probabilities are becoming increasingly common, it is likely that $p_0 = 0.90$ will normally be used. For example, in the Reactor Safety Study (1975), 90% prior assurance was considered. If the prior assurance is free to be selected, we recommend that $p_0 = 0.90$ be used. In this case, LL and UL become the lower and upper prior 5% probability bounds on Λ. The procedure can be justified as follows: It is required to find α_0 and β_0 which satisfies

$$\int_{LL}^{UL} \frac{\beta_0^{\alpha_0}}{\Gamma(\alpha_0)} x^{\alpha_0 - 1} e^{-\beta_0 x}\, dx = p_0, \tag{6.9}$$

where

$$\int_0^{LL} \frac{\beta_0^{\alpha_0}}{\Gamma(\alpha_0)} x^{\alpha_0 - 1} e^{-\beta_0 x}\, dx = \int_{UL}^{\infty} \frac{\beta_0^{\alpha_0}}{\Gamma(\alpha_0)} x^{\alpha_0 - 1} e^{-\beta_0 x}\, dx = \frac{(1 - p_0)}{2}.$$

Letting $y = x/LL$ in (6.9), we have

$$\int_1^{UL/LL} \frac{(\beta_0 LL)^{\alpha_0}}{\Gamma(\alpha_0)} y^{\alpha_0 - 1} e^{-(\beta_0 LL)y}\, dy = p_0. \tag{6.10}$$

Since β_0 is a scale parameter, set $\beta_0 LL = 1$, and solve (6.10) for the shape parameter α_0. Thus α_0 depends only upon the value of UL/LL, or equivalently, $\log(UL/LL)$. Once α_0 has been numerically determined, we can solve (6.10) for a temporary value of β_0, say b_0, corresponding to a temporary lower limit of, say, 1.0×10^{-6} f/h. Since β_0 is a scale parameter, we know that

$$\beta_0 LL = b_0(1.0 \times 10^{-6})$$

from which

$$\beta_0 = b_0 \frac{(1.0 \times 10^{-6} \text{ f/h})}{LL}.$$

The steps of the procedure are as follows:

STEP 1. Specify the values of LL, UL, and $p_0 = 0.80$, 0.90, or 0.95 that represent in totality your best judgment and belief about the failure rate λ of interest. These values are selected in accordance with (6.8).

STEP 2. Compute the value of $\log_{10}(UL/LL)$.

STEP 3. For the value of p_0 chosen in Step 1 and the value of $\log_{10}(UL/LL)$ calculated in Step 2, read the required value of shape parameter α_0 from Figure C1 in Appendix C.

STEP 4. For the value of p_0 from Step 1 and for the value of α_0 found in Step 3, read the value of b_0 from Figure C2.

Note. Table C3 (in Appendix C) may be used in lieu of Figure C2 to obtain b_0, depending on which is more convenient.

STEP 5. For the value of LL from Step 1 and the value of b_0 from Step 4, calculate the required value of the scale parameter β_0 according to

$$\beta_0 = \frac{b_0(1.0 \times 10^{-6} \text{ f/h})}{LL}.$$

Example 6.7. For a certain component of interest, suppose it is believed that the failure rate Λ is such that $\Pr(\Lambda < 1.0 \times 10^{-7} \text{ f/h}) = 5\%$ and $\Pr(\Lambda >$

1.0×10^{-5} f/h)=5%. It is required to identify the particular $\mathcal{G}_1(\alpha_0,\beta_0)$ distribution which is consistent with this belief.

STEP 1. $LL=1.0\times10^{-7}$ f/h, $UL=1.0\times10^{-5}$ f/h, and $p_0=0.90$.

STEP 2. $\log_{10}(UL/LL)=\log_{10}(10^2)=2.0$.

STEP 3. From Figure C1, for $p_0=0.90$ and $\log_{10}(UL/LL)=2.0$, we find $\alpha_0=0.84$.

STEP 4. For $\alpha_0=0.84$ and $p_0=0.90$, Table C3 yields $b_0=2.6723\times10^4$ h.

STEP 5. The required scale parameter β_0 becomes

$$\beta_0=\frac{(2.6723\times10^4\text{ h})(1.0\times10^{-6}\text{ f/h})}{(1.0\times10^{-7}\text{ f/h})}$$

$$=2.6723\times10^5\text{ h}.$$

■

By means of an incomplete gamma function code the actual tail area probabilities for a $\mathcal{G}_1(0.84, 2.6723\times10^5)$ distribution are 0.05, as desired. However, this is not always the case. Due to numerical and round-off errors, the upper tail area may not be exactly equal to $(1-p_0)/2$. In Step 5, the denominator of the expression for β_0 was LL. This was done to insure that the lower tail area will always be $(1-p_0)/2$, whereas the upper tail area may depart somewhat from the desired value $(1-p_0)/2$. We chose to hold the lower tail area fixed because of the positively skewed nature of the $\mathcal{G}_1(\alpha_0,\beta_0)$ distribution.

One final note concerns the usefulness of Figures C1 and C2 in practice. The effective range of values of α_0 considered in Figures C1 and C2 is between 0.25 and 10.0. Experience with fitting $\mathcal{G}_1(\alpha_0,\beta_0)$ prior distributions to failure rate data indicates that this range should contain nearly all situations likely to be encountered in practice. This range is consonant with ratios of UL to LL roughly between 0.3 and 4.0 orders of magnitude.

The preceding method requires that equal probabilities be excluded from both tails of the $\mathcal{G}_1(\alpha_0,\beta_0)$ distribution. In some cases this requirement may be inappropriate, therefore we proceed with a graphic method presented by Waller, et al. (1977a, 1977b) which selects a $\mathcal{G}(\alpha_0,\beta_0)$ prior for Λ by use of any two percentiles.

In practice there exists a set of values for (α_0,β_0) which satisfies

$$\Pr(\Lambda<\lambda_p)=\int_0^{\lambda_p}g(\lambda)\,d\lambda=p \tag{6.11}$$

for any given value of p where $g(\lambda)$ is defined by (4.45). That is, there is a set of values (α_0, β_0) which satisfies (6.11) for a given pth percentile, λ_p. Thus, we ask that information in the form of two distinct percentiles be provided to generate a pair of simultaneous equations. In particular, we need to specify λ_1 and λ_2 such that for $\lambda_1 < \lambda_2$

$$\Pr(\Lambda < \lambda_1) = p_1 \quad \text{and} \quad \Pr(\Lambda < \lambda_2) = p_2. \tag{6.12}$$

Clearly, the specifications of λ_1 and λ_2 are made with reference to the probabilities, p_1 and p_2, where $p_1 < p_2$. The simultaneous solution of (6.12) will select the pair of values for (α_0, β_0) that determines the $\mathcal{G}(\alpha_0, \beta_0)$ prior that summarizes the available information. A specific outline of the methodology is as follows:

STEP 1. The expert specifies the values for λ_1, λ_2, p_1, and p_2 that best represent the totality of the expert's experiences, judgments and beliefs about the failure rate λ. These values provide (6.12).

STEP 2. A search procedure is used to determine α_0 and β_0 that simultaneously satisfy the conditions of Step 1.

Table C4 and Figures C3–C15 in Appendix C present values of α_0 and β_0 for selected choices of λ_0 and p_0 where

$$\Pr(\Lambda < \lambda_0) = p_0. \tag{6.13}$$

By overlaying transparencies of the two graphs that present the specific choices of λ_1, λ_2, p_1, and p_2 of interest, we can determine graphically the desired values of α_0 and β_0.

Example 6.8. Suppose we are studying the reliability of an item for which the available expert opinion indicates that the failure rate λ is such that

$$\Pr(\Lambda < 1.0 \times 10^{-5}) = 0.05 \text{ and}$$

$$\Pr(\Lambda < 1.0 \times 10^{-3}) = 0.50.$$

By overlaying transparencies of the graphs in Figures C6 and C9 of Appendix C, we determine that $\alpha_0 = 0.505$ and $\beta_0 = 0.004$ provide a $\mathcal{G}(\alpha_0, \beta_0)$ prior distribution that possesses the percentile properties given by the stated conditions. The selected prior is thus given by

$$g(\lambda) = \frac{\lambda^{-0.495} e^{-\lambda/0.004}}{(0.004)^{0.505}\Gamma(0.505)}, \qquad \lambda > 0,$$

and is graphically represented in Figure 6.2.

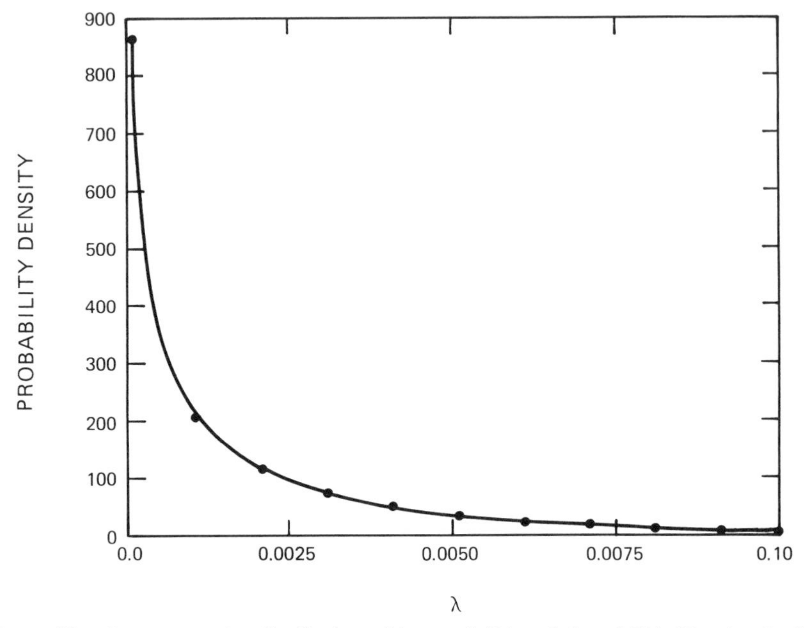

Figure 6.2. A gamma prior distribution with $\alpha_0 = 0.505$ and $\beta_0 = 0.004$. [Reprinted with permission of Society for Industrial and Applied Mathematics from *Nuclear Systems Reliability Engineering and Risk Assessment*, J. B. Fussell and G. R. Burdick, Eds. © 1977 Society for Industrial and Applied Mathematics, Philadelphia. All rights reserved.] ∎

6.7. SELECTING NEGATIVE-LOG GAMMA PRIOR DISTRIBUTIONS

Waller, et al. (1977a, 1977b) present an alternative approach to selecting a $\mathcal{G}(\alpha_0, \beta_0)$ prior by use of percentiles of the reliability function. This procedure would be preferable to the failure rate method when the expert is more knowledgeable about reliability properties than failure rate characteristics of the item under study.

To begin the development we transform the $\mathcal{G}(\alpha_0, \beta_0)$ pdf for Λ into a distribution for the reliability, $R = R(t) = e^{-\lambda t}$. The resulting distribution is

$$g(r) = \frac{(-\ln r)^{\alpha_0 - 1} r^{1/t\beta_0 - 1}}{(t\beta_0)^{\alpha_0}\Gamma(\alpha_0)}, \qquad 0 < r < 1, \tag{6.14}$$

which is known as a *negative-log gamma* distribution and indicated by

$\mathcal{NLG}(\alpha_0, 1/t\beta_0)$. It can be shown that the mean and variance of R are $(1+t\beta_0)^{-\alpha_0}$ and $[(1+2t\beta_0)^{-\alpha_0}-(1+t\beta_0)^{-2\alpha_0}]$, respectively. The density in (6.14) has been used by Mann (1970), Mastran and Singpurwalla (1974) and Springer and Thompson (1965, 1967). Locks (1973) provides a discussion of the $\mathcal{NLG}(\alpha, \theta)$ distribution. Further discussion of the $\mathcal{NLG}(\alpha, \delta)$ distribution may be found in Section 7.5.2 including its use in Bayesian reliability analysis. Thus, the prior distribution induced on the reliability by assuming a conjugate $\mathcal{G}(\alpha_0, \beta_0)$ prior for the failure rate occurs frequently in the literature. Our purpose here is to present a methodology that allows a reliability engineer to use available information as a tool for selecting values of α_0 and β_0 to be used in a Bayesian reliability analysis.

From (6.14) it is clear that there is a different prior on R for each choice of time t. Yet for analysis, it is convenient to reparameterize the density in (6.14) by using $\gamma_0 = t\beta_0$. With that change, the density of interest becomes

$$g(r) = \begin{cases} \dfrac{(-\ln r)^{\alpha_0 - 1} r^{1/\gamma_0 - 1}}{\gamma_0^{\alpha_0}\Gamma(\alpha_0)}, & 0<r<1, \\ 0, & \text{otherwise.} \end{cases} \tag{6.15}$$

The pth percentile of the reliability at time t is $R_p = R_p(t)$. In symbols we write

$$\Pr(R<R_p) = \int_0^{R_p} g(r)\,dr = p.$$

Once the two percentiles are provided with respect to a reference time, say t_0, we can set up two simultaneous equations whose solution provides values for α_0 and γ_0. The desired values of (α_0, β_0) are then given by α_0 and $\beta_0 = \gamma_0/t_0$. An outline of the method is as follows:

STEP 1. The reference time t_0 is fixed.

STEP 2. With respect to t_0, two percentiles R_1 and R_2 for p_1 and p_2, respectively, that best summarize available experiences, judgments, and beliefs about the reliability are specified such that

$$\Pr(R<R_1) = p_1 \quad \text{and} \quad \Pr(R<R_2) = p_2.$$

STEP 3. A search procedure is used to determine α_0 and γ_0 that satisfy the probability statements in Step 2 for the density in (6.15).

STEP 4. Solve for $\beta_0 = \gamma_0/t_0$. The values so determined for (α_0, β_0) are the selected parameters for either the $\mathcal{G}(\alpha_0, \beta_0)$ prior on Λ or the $\mathcal{NLG}(\alpha_0, 1/t_0\beta_0)$ prior on $R(t_0)$. Tables C5–C17 and Figures C16–C28 in

Appendix C give values of α_0 and γ_0 which satisfy

$$\Pr(R < R_0) = p_0 \tag{6.16}$$

for selected values of R_0 and p_0. By overlaying transparencies of the two graphs that contain the values of $R_1(t_0)$, $R_2(t_0)$, p_1, and p_2 used in Step 2 of the procedure, we determine the values of α_0 and γ_0. The value of β_0 is given by γ_0 / t_0 where t_0 is the reference time provided by the expert in Step 1.

Example 6.9. Suppose a reliability expert believes that the reliability of a motor is such that for $t_0 = 100$ h

$$\Pr[R(100) < 0.99] = 0.50,$$

$$\Pr[R(100) < 0.99999] = 0.95.$$

By overlaying transparencies of the graphs in Figures C22 and C25, we

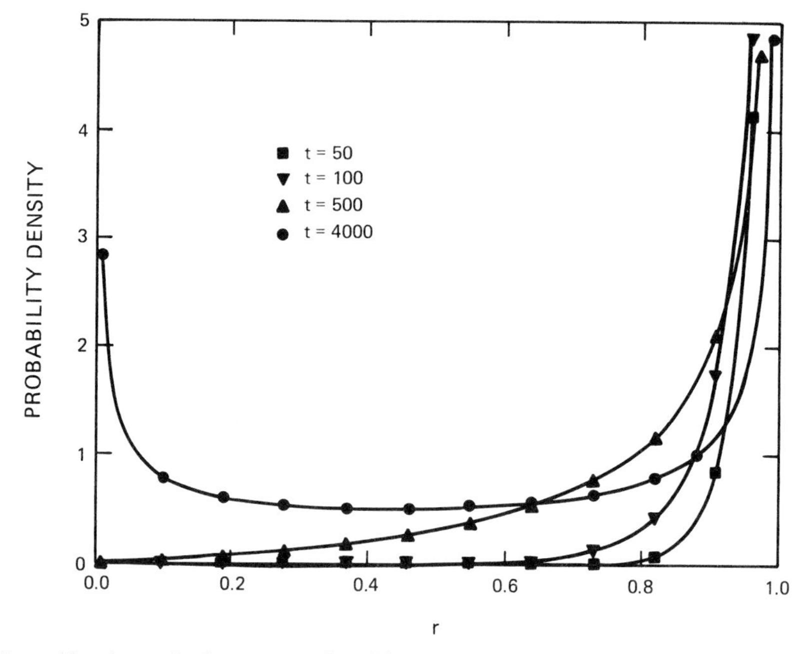

Figure 6.3. A negative-log gamma prior with $\alpha_0 = 0.35$, $\beta_0 = 0.001$, and $t = 50, 100, 500, 4000$. [Reprinted with permission of Society for Industrial and Applied Mathematics from *Nuclear Systems Reliability Engineering and Risk Assessment*, J. B. Fussell and G. R. Burdick, Editors. © 1977 Society for Industrial and Applied Mathematics, Philadelphia. All rights reserved.]

determine that $\alpha_0 = 0.35$ and $\gamma_0 = 0.10$. Therefore,

$$g(r) = \begin{cases} \dfrac{(-\ln r)^{-0.65} r^{1000/t-1}}{(t/1000)^{0.35}\Gamma(0.35)}, & 0 < r < 1, \\ 0, & \text{otherwise.} \end{cases}$$

Recall that we get a different prior for each choice of mission time t. A graph of the selected prior is given for $t = 50$, 100, 500, 4000 h in Figure 6.3. ■

Our interest is to select a pair of prior parameter values (α_0, γ_0) for the $\mathcal{NLG}(\alpha_0, \gamma_0^{-1})$ distribution. However, since $\gamma_0 = t\beta_0$, it follows that γ_0 increases as t increases for a fixed value of β_0. Thus, we have a different prior on the reliability for each choice of t. That property was illustrated in Figure 6.3 for $\alpha_0 = 0.35$, $\beta_0 = 0.001$, and $t = 50$, 100, 500, 4000 h. The range of α_0 frequently encountered in reliability analyses is $0 < \alpha_0 < 1$. Now for $0 < \alpha_0 < 1$ and t small enough so that $0 < \gamma_0 = t\beta_0 \leqslant 1$, the prior distribution $g(r)$ is an increasing function of r. But, when t is such that $\gamma_0 = t\beta_0 > 1$ and $0 < \alpha_0 < 1$, $g(r)$ is U-shaped with its minimum (antimode) at $r = \exp[\gamma_0(\alpha_0 - 1)/(\gamma_0 - 1)]$. Thus, the $\mathcal{NLG}(\alpha_0, \gamma_0^{-1})$ prior on R is concentrated near one in early life (small t). As t increases to large values (late life), the distribution becomes U-shaped. That is, values of r near either zero or one are more likely to occur than other values. Thus, in "old age" the $\mathcal{NLG}(\alpha_0, \gamma_0^{-1})$ prior for $\alpha_0 < 1$ predicts the item will be either quite reliable or quite unreliable. The pdf in Figure 6.3 for $t = 4000$ h illustrates this point. Further analytic properties of the negative-log gamma distribution are given in Table 7.1 in Section 7.5.2.

In the final sections of this chapter we have illustrated techniques for selecting prior distributions from the $\mathcal{B}(x_0, n_0)$, $\mathcal{G}(\alpha_0, \beta_0)$, and $\mathcal{NLG}(\alpha_0, \gamma_0^{-1})$ families. In each case we used *two* bits of information to determine *two* conditions that in turn select a particular member of the *two* parameter family being used to model the totality of information available. In the event that more than two pieces of information are provided, we have over-specified the prior. For example, an engineer could provide three percentiles, say 5th, 50th, and 95th. By overlaying three appropriate transparencies of graphs in Appendix C, we probably would find that the three curves do not intersect at the same point. That indicates that the information is inconsistent for the family of distributions being considered. However, if we were to use the three percentiles in pairs, say (5, 50), (5, 95), and (50, 95) we can select three different prior models. Graphs and properties of the three prior models can then be discussed with the expert to select the prior for analysis. In summary, we emphasize again that the selection of a prior model is a cooperative interactive process where the expert is responsible for providing information and the analyst is responsible for both translating the information into a prior model and communicating (graphically or analytically or both) the properties of the selected model. Subsequent chapters include more examples and illustrations of the basic ideas presented here.

EXERCISES

6.1. Let X denote the number of pump failures per year in a power plant cooling system. Further, suppose X is a Poisson r.v. with distribution

$$f(x|\rho)=\frac{\rho^x e^{-\rho}}{x!}.$$

(a) Find a noninformative prior distribution for ρ.

(b) What is the corresponding transformation ϕ that provides data translation of the likelihood?

6.2. Repeat Exercise 6.1 when X is assumed to be a binomial r.v. such that

$$f(x|p)=\binom{n}{x}p^x(1-p)^{n-x}.$$

6.3. Let T denote the failure time for Brand X chain saws.

(a) Find a noninformative prior for Λ when

$$f(t|\lambda)=\lambda e^{-\lambda t}.$$

(b) What is the corresponding transformation ϕ that provides data translation of the likelihood?

6.4. Assume that $f(x|\mu)=(1/\sqrt{2\pi}\,\sigma)e^{-[(x-\mu)/\sigma]^2/2}$ and that $g(\mu)=d\mu$. Show that $g(\mu)$ is an improper prior distribution, yet the posterior distribution, $g(\mu|x)$, is a valid probability distribution.

6.5. Let $f(x|\lambda)=\lambda e^{-\lambda t}$. Show that

$$g(\lambda)=\frac{1}{\beta^\alpha\Gamma(\alpha)}\lambda^{\alpha-1}\exp\left(-\frac{\lambda}{\beta}\right), \qquad \lambda>0$$

is a conjugate prior distribution for Λ.

6.6. Let

$$f(x|\mu)=\frac{1}{\sqrt{2\pi}}\exp\left\{-\frac{1}{2}(x-\mu)^2\right\}.$$

Show that

$$g(\mu)=\frac{1}{\sqrt{2\pi}\,\sigma_0}\exp\left\{-\frac{1}{2}\left(\frac{\mu-\mu_0}{\sigma_0}\right)^2\right\}, \qquad -\infty<\mu<\infty,$$

is a conjugate prior distribution for μ.

6.7. The distribution of R, the reliability for automobile generators, is assumed to be a $\mathcal{B}(x_0, n_0)$ distribution with a mean of 0.95 and a 95th percentile of 0.99. Find and graph the $\mathcal{B}(x_0, n_0)$ distribution having the two specified properties.

6.8. Find and graph a $\mathcal{B}(x_0, n_0)$ prior distribution for R such that $E(R)=0.95$ and $\Pr(R<0.90)=0.05$.

6.9. Suppose that the failure rate, Λ, for automobile brakes is such that $\Pr(\Lambda<1.0\times10^{-4}\ \text{f/h})=0.05$ and $\Pr(\Lambda>1.0\times10^{-3}\ \text{f/h})=0.05$. Find a $\mathcal{G}_1(\alpha_0, \beta_0)$ prior distribution that is consistent with the given percentile information. Graph the resulting distribution.

6.10. Assume that the failure rate, Λ, for a certain type of television picture tubes is such that $\Pr(\Lambda<1.0\times10^{-5}\ \text{f/h})=0.10$ and $\Pr(\Lambda<1.0\times10^{-3})=0.95$. Find and graph a $\mathcal{G}(\alpha_0, \beta_0)$ prior model to represent the given information.

6.11. Historical records indicate that Brand Y batteries have a reliability, R, such that, for $t_0=1000$ h, $\Pr(R<0.95)=0.05$ and $\Pr(R<0.999)=0.95$. Find an $\mathcal{NLG}(\alpha_0, \gamma_0^{-1})$ prior distribution to represent the available information. Graph the resulting distribution for $t=500$, 1000, and 10000 h.

6.12. Assume that the reliability, R, has the $\mathcal{NLG}(\alpha_0, 1/t\beta_0)$ prior distribution given by (6.14). Show that R has mean $(1+t\beta_0)^{-\alpha_0}$ and variance $[(1+2t\beta_0)^{-\alpha_0}-(1+t\beta_0)^{-2\alpha_0}]$.

6.13. The Reactor Safety Study (1975) gives the 5th and 95th percentiles on the failure rate for prematurely opening relief valves as 3×10^{-6} and 3×10^{-5} f/h, respectively.

(a) Find the $\mathcal{G}_1(\alpha_0, \beta_0)$ prior distribution that is consistent with this prior information.

(b) How well does the median of this fitted prior agree with the assessed prior median of 1×10^{-5} given in the Reactor Safety Study?

(c) Discuss ways and means of reconciling any differences in (b).

REFERENCES

Alpert, M. and Raiffa, H. (1969). *A Progress Report on the Training of Probability Assessors*, Unpublished paper, August 28, 1969.

Bartlett, M. S. (1937). Properties of Sufficiency and Statistical Tests, *Proceedings of the Royal Society of London*, *Series A*, Vol. 160, pp. 268–282.

Box, G. E. P. and Tiao, G. C. (1973). *Bayesian Inference in Statistical Analysis*, Addison–Wesley, Reading, MA.

Breiphol, A. (1978). Group Reliability Predictions, *IEEE Transactions on Reliability*, Vol. R-27, pp. 139–144.

Brown B. and Helmer, O. (1964). *Improving the Reliability of Estimates Obtained from a Consensus of Experts*, P-2986, The Rand Corporation, Santa Monica, CA.

Columbo, A. G. and Constantini, D. (1980). Ground-Hypotheses for Beta Distribution as Bayesian Prior, *IEEE Transactions on Reliability*, Vol. R-29, pp. 17–21.

Dalkey, N. (1977). *Group Decision Theory*, UCLA-ENG-7749, The University of California at Los Angeles, Los Angeles, CA.

Dalkey, N. and Brown, B. (1971). *Comparison of Group Judgment Techniques with Short-range Predictions and Almanac Questions*, R-678-ARPA, The Rand Corporation, Santa Monica, CA.

Deeley, J. J., Tierney, M. S., and Zimmer, W. J. (1970). On the Usefulness of Maximum Entropy Principle in the Bayesian Estimation of Reliability, *IEEE Transactions on Reliability*, Vol. R-19, pp. 110–115.

De Groot, M. H. (1970). *Optimal Statistical Decisions*, McGraw-Hill, New York.

De Groot, M. H. (1974). Reaching a Consensus, *Journal of the American Statistical Association*, Vol. 69, pp. 118–121.

Du Charme, W. M. and Donnell, M. L. (1973). Intrasubject Comparison of Four Response Modes for 'Subjective Probability' Assessment, *Organizational Behavior and Human Performance*, Vol. 10, pp. 108–117.

Eisenberg, E. and Gale, D. (1959). Consensus of Subjective Probabilities: the Parimutuel Method, *Annals of Mathematical Statistics*, Vol. 30, pp. 165–168.

Fearn, T. (1979). Aiding the Choice of Inverse Chi Prior Distributions, *Journal of Statistical Computation and Simulation*, Vol. 10, pp. 25–29.

Fisher, R. A. (1922). On the Mathematical Foundations of Theoretical Statistics, *Philosophical Transactions of the Royal Society of London*, *Series A*, Vol. 222, pp. 309–368.

Fisher, R. A. (1925). Theory of Statistical Estimation, *Proceedings of the Cambridge Philosophical Society*, Vol. 22, pp. 700–725.

Goel, P. K. and De Groot, M. H. (1981). Information About Hyperparameters in Hierarchical Models, *Journal of the American Statistical Association*, Vol. 76, pp. 140–147.

Hampton, J. M., Moore, P. G., Thomas, H. (1973). Subjective Probability and Its Measurement, *Journal of the Royal Statistical Society*, *Series A*, Vol. 136, pp. 21–42.

Helmer, O. (1963). *The Systematic Use of Expert Judgement in Operations Research*, P-2795, The Rand Corporation, Santa Monica, CA.

Hogarth, R. M. (1975). Cognitive Processes and the Assessment of Subjective Probability Distributions, *Journal of the American Statistical Association*, Vol. 70, pp. 271–289.

Jaynes, E. T. (1968). Prior Probabilities, *IEEE Transactions on Systems Science and Cybernetics*, Vol. SSC-4, pp. 227–241.

Jeffreys, H. (1961). *Theory of Probability* (3rd ed.), Clarendon Press, Oxford.

Lichtenstein, S., Fischhoff, B. and Phillips, L. D. (1977). Calibration of Probabilities: the State of the Art, in *Decision Making and Change in Human Affairs*, H. Jungerman and G. de Zeewv (eds.), Reidel, Dordrecht, pp. 275–324.

Lindley, D. V., Tversky, A. and Brown, R. V. (1979). On the Reconciliation of Probability Assessments, *Journal of the Royal Statistical Society, Series A*, Vol. 142, pp. 146–180.

Locks, M. O. (1973). *Reliability, Maintainability and Availability Assessment*, Hayden, Rochelle Park, NJ.

MacFarland, W. J. (1971). Sequential Analysis and Bayesian Demonstration, *Annals of Reliability and Maintainability*, pp. 692–703.

MacFarland, W. J. (1972). Bayes' Equation Reliability and Multiple Hypothesis Testing, *IEEE Transactions on Reliability*, Vol. R-21, pp. 136–147.

Mann, N. R. (1970). Computer-Aided Selection of Prior Distributions for Generating Monte Carlo Confidence Bounds for System Reliability, *Naval Research Logistics Quarterly*, Vol. 17, pp. 41–45.

Martz, H. F. and Lian, M. G. (1977). Bayes and Empirical Bayes Point and Interval Estimation of Reliability for the Weibull Model, *The Theory and Applications of Reliability*, Vol. II, Academic, New York, pp. 203–233.

Martz, H. F. and Waller, R. A. (1978). *An Exploratory Comparison of Methods for Combining Failure Rate Data from Different Data Sources*, LA-7556-MS, Los Alamos Scientific Laboratory, Los Alamos, NM.

Martz, H. F. and Waller, R. A. (1979). A Bayesian Zero-Failure (BAZE) Reliability Demonstration Testing Procedure, *Journal of Quality Technology*, Vol. II, pp. 128–138.

Mastran, D. V. and Singpurwalla, N. D. (1974). *A Bayesian Assessment of Coherent Structures*, Social T-293, George Washington University, School of Engineering and Applied Science Institute for Management Science and Engineering, Washington, DC.

Mirkin, B. G. (1979). *Group Choice*, Winston, Washington, DC.

Morris, P. A. (1974). Decision Analysis Expert Use, *Management Science*, Vol. 20, pp. 1233–1241.

Morris, P. A. (1977). Combining Expert Judgments: A Bayesian Approach, *Management Science*, Vol. 23, pp. 679–693.

Raiffa, H. and Schlaifer, R. (1961). *Applied Statistical Decision Theory*, Harvard University, Boston.

Reactor Safety Study, Appendix III — Failure Data (1975). WASH-1400 (NUREG-75/014), U.S. Nuclear Regulatory Commission.

Saaty, T. L. (1972). *An Eigenvalue Allocation Model in Contingency Planning*, University of Pennsylvania, Philadelphia, PA.

Saaty, T. L. (1974). Measuring the Fuzziness of Sets, *Journal of Cybernetics*, Vol. 4, pp. 53–61.

Saaty, T. L. (1977). A Scaling Method for Priorities in Hierarchical Structures, *Journal of Mathematical Psychology*, Vol. 15, pp. 234–281.

Saaty, T. L. (1978). Exploring the Interface between Hierarchies, Multiple Objects, and Fuzzy Sets, *Fuzzy Sets and Systems*, Vol. 1, pp. 57–68.

Saaty, Thomas L. (1980). *The Analytic Hierarchy Process*, McGraw-Hill, New York.

Seaver, D. A., Winterfeldt, D. V., and Edwards, W. (1978). Eliciting Subjective Probability Distributions on Continuous Variables, *Journal of Organizational Behavior and Human Performance*, Vol. 21, pp. 379–391.

Shimi, I. and Tsokos, C. P. (1977). The Bayesian Nonparametric Approach to Reliability Studies: A Survey of Recent Work, *The Theory and Applications of Reliability*, Vol. II, Academic, New York, pp. 5–47.

Spetzler, C. S. and Staël Von Holstein, C. A. S. (1975). Probability Encoding in Decision Analysis, *Management Science*, Vol. 22, pp. 340–358.

Springer, M. D. and Thompson, W. E. (1965). Bayesian Confidence Limits for Reliability of Redundant Systems when Tests are Terminated at First Failure, *Technometrics*, Vol. 10, pp. 29–36.

Springer, M. D. and Thompson, W. E. (1967). Bayesian Confidence Limits for the Reliability of Cascade Exponential Subsystems, *IEEE Transactions on Reliability*, Vol. R-16, pp. 86–89.

Staël Von Holstein, C. A. (1970). *Assessment and Evaluation of Subjective Probability Distributions*, The Economic Research Institute at the Stockholm School of Economics, Stockholm.

Staël Von Holstein, C. A. (1971). Two Techniques for Assessment of Subjective Probability Distributions—An Experimental Study, *Acta Psychologica*, Vol. 35, pp. 478–494.

Stone, M. (1961). The Opinion Pool, *Annals of Mathematical Statistics*, Vol. 32, pp. 1339–1342.

Tversky, A. and Kahneman, D. (1974), Judgment Under Uncertainty: Heuristics and Biases, *Science*, Vol. 185, pp. 1124–1131.

Waller, R. A., Johnson, M. M., Waterman, M. S. and Martz, H. F. (1977a). Gamma Prior Distribution Selection for Bayesian Analysis of Failure Rate and Reliability, *Nuclear Systems Reliability Engineering and Risk Assessment*, Society for Industrial and Applied Mathematics, Philadelphia, PA, pp. 584–606.

Waller, R. A., Johnson, M. M., Waterman, M. S. and Martz, H. F. (1977b). *Gamma Prior Distribution Selection for Bayesian Analysis of Failure Rate and Reliability*, LA-6879-MS, Los Alamos Scientific Laboratory, Los Alamos, NM.

Waterman, M. S., Martz, H. F., and Waller, R. A. (1976). *Fitting Beta Prior Distributions in Bayesian Reliability Analysis*, LA-6395-MS, Los Alamos Scientific Laboratory, Los Alamos, NM.

Weiler, H. (1965). The Use of Incomplete Beta Functions for Prior Distributions in Binomial Sampling, *Technometrics*, Vol. 7, pp. 335–347.

Winkler, R. L. (1967). The Assessment of Prior Distributions in Bayesian Analysis, *Journal of the American Statistical Association*, Vol. 62, pp. 776–800.

Winkler, R. L. (1968). The Consensus of Subjective Probability Distribution, *Management Science*, Vol. 15, pp. B61–B75.

Winkler, R. L. and Cummings, L. L. (1972). On the Choice of a Consensus Distribution in Bayesian Analysis, *Organizational Behavior and Human Performance*, Vol. 7, pp. 63–76.

Yager, R. R. (1979). An Eigenvalue Method of Obtaining Subjective Probabilities, *Behavioral Science*, Vol. 24, pp. 382–387.

CHAPTER 7

Estimation Methods Based on Attribute Life Test Data

An *attribute life test* is one in which the only sample test data to be considered in the estimation procedure is whether each test unit survives or fails a specified life test. For example, it may be known that six pumps of a certain size and type failed during the past year out of 400 such pumps in use in commercial pressurized water nuclear power reactors. Data of this type are sometimes called *attribute test data* because only the survival/failure, *not* the lifetime, is recorded for each test item. In contrast to this, a test in which the actual lifetimes are recorded is called a *variables life test* and the data are referred to as *variables test data*.

There are two main advantages of an attribute life test over a variables life test. First, it is generally more economical to conduct an attribute test than a variables test, as less test equipment, test monitoring, and data recording capability are required. Second, a probability distribution for the failure time r.v. is sometimes unnecessary, and the results are valid regardless of the underlying failure time model.

There are also two main disadvantages when attribute testing is compared to variables testing. First, if the failure times are monitored and recorded and if the probability distribution of these failure times is known, the resulting reliability estimates are generally superior (e.g., shorter Bayesian interval estimates) to those based on corresponding attribute data. Second, the reliability estimates based on attribute data may or may not be functionally dependent on time. That is, the estimates may pertain only to the reliability for a time period equal to the test duration, and reliability estimates cannot be given for other time periods. It is also important to recognize that variables test data can be converted to attribute test data but not vice versa.

7.1. ATTRIBUTE SAMPLING MODELS

7.1.1 Binomial Sampling

The binomial distribution was discussed in Chapter 2, being the distribution of the number of occurrences of an event in n independent trials in which the event has a constant probability of occurrence p in each trial. The process that generates these independent observations for each trial is called a *Bernoulli process* and has a *Bernoulli probability distribution* defined by

$$f(y \mid p) = p^y(1-p)^{1-y}, \qquad 0<p<1, \quad y=0,1, \tag{7.1}$$

where $Y=0(1)$ is the r.v. used to denote the nonoccurrence (occurrence) of the event in question. In the area of reliability, the "event" is usually the survival of the system/subsystem/component under consideration in a specified test. The n observations can be obtained by renewing a unit after each trial or by observing n identical units during the test. In the first case, the total time required to obtain the n observations will exceed the duration of the test, whereas in the second case, the total time will be equal to the test duration. The parameter p is the probability of a unit surviving the test and is the reliability parameter of interest to be estimated. In the nuclear industry the parameter $(1-p)$ is often referred to as the "failure-on-demand" probability where the "test" is a demand for the unit to function successfully. In those cases where the test is of duration t, we have that $R(t)=p$, where $R(t)$ is the reliability function discussed in Section 4.1.1. The usual point estimator of p is (number of successes)/n, and the number of successes is a sufficient statistic for estimating p.

If the sample data are obtained from a Bernoulli process in such a way that n is fixed in advance, and the number of survivors $X=\Sigma_{i=1}^{n} Y_i$ is left to chance, the situation is known as *binomial sampling*. The probability distribution of the number of survivors is the binomial distribution given by

$$\begin{aligned} f(x \mid p) &= \Pr\{x \text{ survivors will occur in } n \text{ trials} \mid p\} \\ &= \frac{n!}{(n-x)!x!} p^x(1-p)^{n-x}, \qquad x=0,1,\ldots,n, \quad 0<p<1. \end{aligned} \tag{7.2}$$

In adherence to the Bayesian approach, we now assume that P is an r.v. having a prior distribution $g(p)$.

To illustrate binomial sampling and to supply numerical data for use in subsequent examples, let us consider the following example.

Example 7.1. The WASH-1400 "Reactor Safety Study" (1975) reported the following data on the number of pump failures observed in 1972 in eight pressurized water reactors (PWRs) in commercial operation in the United States. The designated failure mode was failure to run normal.

j	PWR	n_j	s_j	x_j	$t_j(h)$	$\frac{x_j}{n_j}$
1	Haddam Neck	50	0	50	438,000	1.00
2	Yankee Rowe	50	0	50	438,000	1.00
3	Indian Point 1	50	1	49	438,000	0.98
4	San Onofre 1	50	1	49	438,000	0.98
5	Ginna	50	0	50	438,000	1.00
6	Point Beach 1	50	0	50	438,000	1.00
7	Robinson 2	50	1	49	438,000	0.98
8	Palisades	50	3	47	438,000	0.94
	Total	400	6	394	5,504,000	7.88

Here s_j is the observed number of failures in n_j pumps, and t_j is the total test time in h. The sampling is clearly binomial with n_j fixed in advance. The r.v. X_j is the number of survivors during the one year test. ■

In most of the examples in Section 7.2 that refer to the above data, for purposes of illustration it will be assumed that the pump survival probability p has the same unknown value for each PWR. Thus plant-to-plant differences in p are *not* assumed to exist and the combined data will be used in the examples. However, in the examples in Section 7.6, for illustration purposes each PWR will be assumed to have an underlying pump survival probability p which is an independent realization from a common prior distribution. In this case plant-to-plant differences *are* assumed to exist and the data for each PWR is used to determine the prior distribution. A similar situation exists in the examples in Section 7.7.

7.1.2 Pascal Sampling

In binomial sampling, the sample size was held fixed at n and X was the r.v. that was left to chance. In planning a life test experiment that produces the sample data, it may turn out that testing costs depend more on s, the number of failures, than on n. This is particularly true if failures cost significantly more than survivors. In such cases, it is advantageous to fix s in advance and to consider N as an r.v. whose value is left to chance. For example, we may decide to sequentially test a series of test units for a specified duration until two failures have been observed. The number of test units required in order to accomplish this will be at least two, and is an r.v. at the outset of the test. Such a sampling scheme is referred to as *Pascal sampling*.

The conditional probability, given p, that n trials will be required to produce s failures is given by the *Pascal probability distribution*

$$f(n \mid p) = \Pr\{n \text{ trials are required to produce } s \text{ failures} \mid p\}$$

$$= \frac{(n-1)!}{(s-1)!(n-s)!} p^{n-s}(1-p)^s, \qquad 0 < p < 1, \tag{7.3}$$

$$n = s, s+1, \ldots, \qquad s = 1, 2, \ldots$$

The factorials arise from the fact that, since the nth trial must yield a failure, there must then be $s-1$ failures in the preceding $n-1$ trials. The factorials thus represent the number of ways that the $s-1$ failures can occur in the $n-1$ trials. It will henceforth also be assumed that P is an r.v. with prior distribution $g(p)$, and it is required to estimate p. The distribution (7.3) is also commonly known as the *negative binomial distribution*.

Example 7.2. LX-13 or Extex is an extrudable high explosive used in a variety of systems. The detonation velocity is an important measure of the reliability of the material. A recent specification required that production lots of Extex have a detonation velocity less than 7300 m/sec, and lots whose detonation velocity exceeded this specification were required to be destroyed. In order to control the number of lots destroyed, it was decided to successively test samples from a sequence of lots until two lots were obtained that did not meet the specification. The testing was conducted and the second "failure" was observed on the twelfth lot tested. Thus, for $s = 2$, n was observed to be equal to 12. ■

7.1.3 Poisson Sampling

Recall from Section 4.2.1 that a Poisson process can be defined as a process that generates a sequence of events for which the times between successive events are independently and identically distributed according to the $\mathcal{E}(\lambda)$ distribution. The parameter λ is called the *intensity* of the process, and in reliability applications for which the "event" is failure of an item of interest, λ is the failure rate. Thus, for an $\mathcal{E}(\lambda)$ failure time distribution, according to the theory of Poisson processes the distribution of the number of failures s occurring in fixed total test time t is given by the Poisson distribution

$$f(s|\lambda)=\Pr\{s \text{ failures occur in total test time t}|\lambda\}$$

$$=\frac{e^{-\lambda \mathrm{t}}(\lambda \mathrm{t})^{s}}{s!}, \qquad \lambda>0, \quad s=0,1,\ldots, \tag{7.4}$$

where failed items are assumed to be replaced. This is known as *Poisson sampling*. Thus, if each of n items is tested for t_0 time units and failed items are replaced, then $\mathrm{t}=nt_0$, which is the total time on test or total test time. We will further assume that the failure rate Λ is an r.v. with prior distribution $g(\lambda)$, and it is of interest to estimate λ.

Example 7.3. The WASH-1400 "Reactor Safety Study," (1975) stated that in 1972 there were a total of 32 valve failures reported for the eight PWRs in commercial operation in the United States. Each plant had an average of 289 valves in operation during the year. Thus $s=32$, $n=8(289)=2312$, and $\mathrm{t}=2312(8760)=20{,}253{,}120$ h. If we assume that valve failure times follow an $\mathcal{E}(\lambda)$ distribution and that failed valves are replaced, then this depicts a Poisson sampling situation. ■

7.2. ESTIMATING THE SURVIVAL PROBABILITY—BINOMIAL SAMPLING

We are interested in obtaining both point and Bayesian interval estimators for the reliability p in the binomial sampling model given in (7.2). For a general prior distribution $g(p)$, the posterior distribution according to (5.8) is given by

$$g(p|x)=\frac{[n!/(n-x)!x!]\,p^{x}(1-p)^{n-x}g(p)}{\int_0^1[n!/(n-x)!x!]\,p^{x}(1-p)^{n-x}g(p)\,dp}$$

$$=\frac{p^{x}(1-p)^{n-x}g(p)}{\int_0^1 p^{x}(1-p)^{n-x}g(p)\,dp}, \qquad 0<p<1, \tag{7.5}$$

where x is the observed number of survivors in n trials. The likelihood is proportional to $p^x(1-p)^{n-x}$ and the denominator of (7.5) is proportional to the marginal distribution $f(x)$ of X, the proportionality constant being the factorial terms that cancel in (7.5). Although (7.5) may be computed (either numerically or in closed form) for *any* prior distribution whose support is a subset of the unit interval, there is a variety of prior distributions that have been found to be particularly appropriate. Some of these are now considered.

7.2.1 Uniform Prior Distribution

The $\mathcal{U}(0, 1)$ prior distribution represents a state of total ignorance that is *not* characteristic of most reliability situations. Nevertheless, it serves as a nice starting point to illustrate the mechanics of the Bayesian approach. Thus

$$g(p) = \begin{cases} 1, & 0 < p < 1 \\ 0, & \text{elsewhere,} \end{cases} \tag{7.6}$$

and, from (7.5), the posterior distribution of P given x becomes

$$g(p|x) = \frac{p^{(x+1)-1}(1-p)^{(n-x+1)-1}}{\int_0^1 p^{(x+1)-1}(1-p)^{(n-x+1)-1}\,dp}, \qquad 0 < p < 1.$$

Now

$$\int_0^1 p^{(x+1)-1}(1-p)^{(n-x+1)-1}\,dp = \frac{\Gamma(x+1)\Gamma(n-x+1)}{\Gamma(n+2)}, \qquad n \geqslant x, \tag{7.7}$$

where $\Gamma(z)$ is the gamma function of z. When z is an integer, as is the case here, $\Gamma(z) = (z-1)!$ Thus

$$g(p|x) = \frac{\Gamma(n+2)}{\Gamma(x+1)\Gamma(n-x+1)} p^{(x+1)-1}(1-p)^{(n-x+1)-1}, \qquad 0 < p < 1, \tag{7.8}$$

which is a $\mathcal{B}(x+1, n+2)$ distribution.

The marginal distribution of X is the discrete uniform distribution given by

$$f(x)=\frac{n!\Gamma(x+1)\Gamma(n-x+1)}{(n-x)!x!\Gamma(n+2)}=\frac{1}{n+1}, \qquad x=0,1,\ldots,n. \qquad (7.9)$$

Example 7.4. In Example 7.1, the posterior distribution of the annual PWR pump reliability p, for a $\mathcal{U}(0,1)$ prior distribution, is

$$g(p|x=394)=\frac{401!}{394!6!}p^{394}(1-p)^6, \qquad 0<p<1,$$

which is plotted in Figure 7.1.

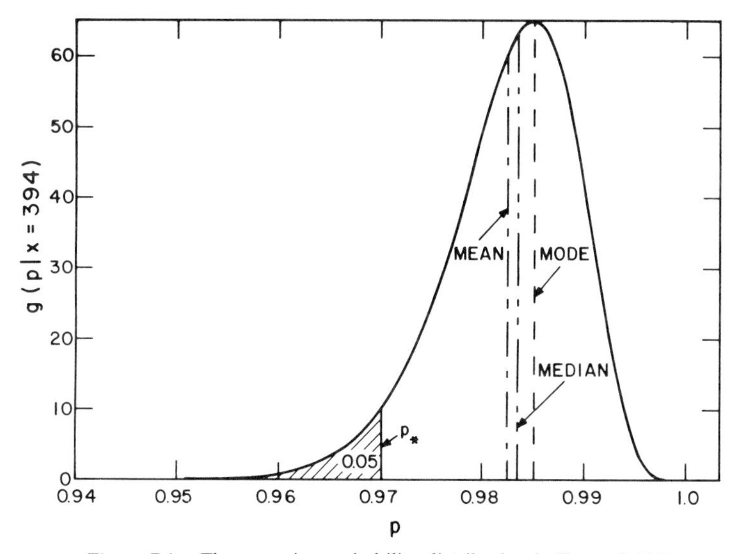

Figure 7.1. The posterior probability distribution in Example 7.4. ■

Point Estimation. The posterior mean of $g(p|x)$ is the mean of the $\mathcal{B}(x+1, n+2)$ distribution given by

$$E(P|x)=\frac{x+1}{n+2}, \qquad (7.10)$$

which is the Bayes point estimator for p under a squared-error loss function.

This estimator is close to the ML estimator, and when $x = n/2$, the two estimators are identically $\frac{1}{2}$. Thus, when x is near $n/2$ the two estimators are approximately equal. Also, for x near zero or n the two estimators are close.

Further, for a squared-error loss function, the posterior risk, $aE\{[P - E(P|x)]^2|x\}$, of the Bayes estimator is proportional to the posterior variance given by

$$\mathrm{Var}(P|x) = \frac{(x+1)(n-x+1)}{(n+2)^2(n+3)}, \tag{7.11}$$

and thus, according to (5.31), the posterior risk of (7.10) is

$$\phi(x) = \frac{a(x+1)(n-x+1)}{(n+2)^2(n+3)}. \tag{7.12}$$

The minimum Bayes risk, $aE_x[\mathrm{Var}_{p|x}(P|X)]$, associated with (7.10) may be found by recognizing that

$$\phi(x) = a\frac{nx - x^2 + (n+1)}{(n+2)^2(n+3)}$$

and using the fact that, according to (7.9),

$$E(X) = \sum_{x=0}^{n} \frac{x}{n+1} = \frac{n}{2}$$

and

$$E(X^2) = \sum_{x=0}^{n} \frac{x^2}{n+1} = \frac{n(2n+1)}{6}.$$

Thus, the minimum Bayes risk of $E(P|x)$ is, according to (5.32), given by

$$r(g) = a\frac{nE(X) - E(X^2) + (n+1)}{(n+2)^2(n+3)} = \frac{a}{6(n+2)}. \tag{7.13}$$

Finally, the mean and variance of $E(P|x)$ are, according to (5.34) and (5.35), given by

$$E_x\left[E_{p|x}(P|X)\right]=\frac{1}{2} \tag{7.14}$$

and

$$\mathrm{Var}_x\left[E_{p|x}(P|X)\right]=\frac{n}{12(n+2)}, \tag{7.15}$$

respectively.

Example 7.5. In Example 7.1, the Bayes point estimate of the annual PWR pump reliability for a squared-error loss function in which $a=1$ is the posterior mean given by $395/402=0.98$, with associated minimum Bayes risk equal to $1/6(402)=4.1\times 10^{-4}$. This estimate is shown in Figure 7.1, along with the posterior median and mode. ■

Interval Estimation. According to (5.45) and (5.46) a symmetric $100(1-\gamma)\%$ TBPI estimate for p is found by solving the two equations given by

$$\Pr(P\leqslant p_*|x)=\int_0^{p_*}\frac{\Gamma(n+2)}{\Gamma(x+1)\Gamma(n-x+1)}p^{(x+1)-1}(1-p)^{(n-x+1)-1}dp=\frac{\gamma}{2} \tag{7.16}$$

and

$$\Pr(p\geqslant p^*|x)=\int_{p^*}^{1}\frac{\Gamma(n+2)}{\Gamma(x+1)\Gamma(n-x+1)}p^{(x+1)-1}(1-p)^{(n-x+1)-1}dp=\frac{\gamma}{2} \tag{7.17}$$

for the lower limit p_* and upper limit p^*. These equations can be most easily solved by considering the change of variable defined by the transformation

$$P=\left[\left(\frac{n-x+1}{x+1}\right)\left(\frac{1}{\mathcal{F}}\right)+1\right]^{-1}.$$

As P has a posterior $\mathcal{B}(x+1,n+2)$ distribution, then $\mathcal{F}$ is known to have a posterior $\mathcal{F}(2x+2, 2n-2x+2)$ distribution. Let us consider (7.16). Now

$$\Pr(P \leqslant p_*|x) = \frac{\gamma}{2} = \Pr\left\{\left[\left(\frac{n-x+1}{x+1}\right)\left(\frac{1}{\mathcal{F}}\right)+1\right]^{-1} \leqslant p_*|x\right\}$$

$$= \Pr\left\{\left(\frac{n-x+1}{x+1}\right)\left(\frac{1}{\mathcal{F}}\right)+1 \geqslant p_*^{-1}|x\right\}$$

$$= \Pr\left\{\left(\frac{n-x+1}{x+1}\right)\left(\frac{1}{\mathcal{F}}\right) \geqslant p_*^{-1}-1|x\right\}$$

$$= \Pr\left\{\mathcal{F} \leqslant \left(\frac{n-x+1}{x+1}\right)\left(\frac{p_*}{1-p_*}\right)|x\right\}.$$

Thus

$$\mathcal{F}_{\gamma/2}(2x+2, 2n-2x+2) = \left(\frac{n-x+1}{x+1}\right)\left(\frac{p_*}{1-p_*}\right)$$

from which

$$p_* = \frac{x+1}{x+1+(n-x+1)\mathcal{F}_{1-\gamma/2}(2n-2x+2, 2x+2)}, \tag{7.18}$$

where we have used the well-known reciprocal $\mathcal{F}$ relationship given by

$$\frac{1}{\mathcal{F}_{\gamma/2}(2x+2, 2n-2x+2)} = \mathcal{F}_{1-\gamma/2}(2n-2x+2, 2x+2). \tag{7.19}$$

Note. $\mathcal{F}_{1-\gamma/2}(2n-2x+2, 2x+2)$ is often referred to as the $100(\gamma/2)$th *percentage point* of an $\mathcal{F}(2n-2x+2, 2x+2)$ distribution, and many $\mathcal{F}$ tables tabulate such percentage points.

The upper limit p^* may be similarly computed as

$$p^* = \frac{(x+1)\mathcal{F}_{1-\gamma/2}(2x+2, 2n-2x+2)}{n-x+1+(x+1)\mathcal{F}_{1-\gamma/2}(2x+2, 2n-2x+2)}, \tag{7.20}$$

and thus it may be claimed that $\Pr(p_* \leqslant P \leqslant p^*|x) = 1-\gamma$.

A $100(1-\gamma)\%$ LBPI estimate for p may be found from (7.18) in which $\gamma/2$ is replaced by γ and it may be claimed that $\Pr(P \geqslant p_*|x) = 1-\gamma$. A $100(1-\gamma)\%$ UBPI estimate may similarly be computed from (7.20).

Example 7.6. Again let us consider Example 7.1. A 95% LBPI estimate of p is given by

$$p_* = \frac{395}{395+7(1.695)} = 0.97,$$

since $\mathcal{F}_{0.95}(14,790)=1.695$. Thus $\Pr(P \geqslant 0.97 | x=394)=0.95$. This limit is also shown in Figure 7.1. ■

It is noted that the posterior median, which is the Bayes point estimator of p for an absolute-error loss function, may be obtained by letting $\gamma=1.0$ in (7.18). This was used to compute the posterior median shown in Figure 7.1.

One may also be concerned with estimation of the failure probability $1-p$. The failure probability can be estimated by letting $p' \equiv 1-p$, and letting s, the observed number of failures in n trials, assume the role of x. That is, in (7.2), we let $p' \equiv 1-p$ and $s \equiv x$. In this case, we are frequently interested in either a TBPI or a UBPI estimate of p. Such a case was illustrated in Examples 5.9 and 5.13.

7.2.2 Noninformative Prior Distribution

The notion of a noninformative prior distribution was presented in Section 6.1. According to (7.2), the likelihood is proportional to $p^x(1-p)^{n-x}$ which is clearly *not* data translated. Recall that a likelihood is said to be data translated if different sets of data affect *only* the location of the likelihood. However, if we consider the well-known asymptotic variance stabilizing transformation of p given by [see Bartlett (1937)]

$$\phi(p)=\sin^{-1}\sqrt{p}\,, \tag{7.21}$$

then the likelihood can be shown to be nearly data translated [see Box and Tiao (1973)]. Thus, using this transformation, a locally uniform prior distribution is nearly noninformative. This in turn implies, according to (6.3), that the corresponding nearly noninformative prior for P is proportional to

$$g(p) \propto \left| \frac{d\phi}{dp} \right| = [p(1-p)]^{-1/2}. \tag{7.22}$$

Thus $g(p)$ is a $\mathcal{B}(0.5,1)$ distribution. For this noninformative prior distribution, the corresponding posterior distribution for P given x becomes

$$g(p|x)=\frac{\Gamma(n+1)}{\Gamma(x+1/2)\Gamma(n-x+1/2)}p^{(x+1/2)-1}(1-p)^{(n-x+1/2)-1}, \qquad 0<p<1, \tag{7.23}$$

which is also a $\mathcal{B}(x+1/2, n+1)$ distribution. By comparing (7.23) with (7.8), it is observed that the only difference in using the noninformative prior for P, rather than the $\mathcal{U}(0,1)$ prior, is to reduce the number of survivors and the number of failures by 0.5. Thus, for moderately sized values of n, the posterior distributions are nearly equivalent.

7.2.3 Truncated Uniform Prior Distribution

Suppose now that P has a $\mathcal{U}(p_0, p_1)$ prior distribution. Thus

$$g(p; p_0, p_1) = \begin{cases} \dfrac{1}{p_1 - p_0}, & 0 \leq p_0 < p < p_1 \leq 1 \\ 0, & \text{elsewhere.} \end{cases} \tag{7.24}$$

The posterior distribution of P given x is

$$g(p|x; p_0, p_1) = \frac{\dfrac{\Gamma(n+2)}{\Gamma(x+1)\Gamma(n-x+1)} p^{(x+1)-1}(1-p)^{(n-x+1)-1}}{\displaystyle\int_{p_0}^{p_1} \frac{\Gamma(n+2)}{\Gamma(x+1)\Gamma(n-x+1)} p^{(x+1)-1}(1-p)^{(n-x+1)-1}\,dp}$$

$$= \frac{\dfrac{\Gamma(n+2)}{\Gamma(x+1)\Gamma(n-x+1)} p^{(x+1)-1}(1-p)^{(n-x+1)-1}}{I(p_1; x+1, n-x+1) - I(p_0; x+1, n-x+1)},$$

$$0 \leq p_0 < p < p_1 \leq 1, \tag{7.25}$$

where $I(z;\ a, b)$ is the standard incomplete beta function ratio defined by

$$I(z; a, b) = \frac{\Gamma(a+b)}{\Gamma(a)\Gamma(b)} \int_0^z t^{a-1}(1-t)^{b-1}\,dt. \tag{7.26}$$

Numerous computer codes and tables, such as those credited to Pearson (1934), exist which can be used to compute values of $I(z;\ a, b)$. In the examples that follow, we have used an excellent computer code written by Amos and Daniel (1969).

The marginal distribution of X is easily computed to be

$$f(x; p_0, p_1) = \frac{I(p_1; x+1, n-x+1) - I(p_0; x+1, n-x+1)}{(n+1)(p_1 - p_0)},$$

$$x = 0, 1, \ldots, n. \tag{7.27}$$

Point Estimation. The posterior mean and variance of (7.25) are

$$E(P|x; p_0, p_1) = \frac{x+1}{n+2} \left\{ \frac{I(p_1; x+2, n-x+1) - I(p_0; x+2, n-x+1)}{I(p_1; x+1, n-x+1) - I(p_0; x+1, n-x+1)} \right\} \tag{7.28}$$

and

$$\mathrm{Var}(P|x; p_0, p_1) = \frac{(x+2)(x+1)}{(n+3)(n+2)} \left\{ \frac{I(p_1; x+3, n-x+1) - I(p_0; x+3, n-x+1)}{I(p_1; x+1, n-x+1) - I(p_0; x+1, n-x+1)} \right\} - \frac{(x+1)^2}{(n+2)^2} \left\{ \frac{I(p_1; x+2, n-x+1) - I(p_0; x+2, n-x+1)}{I(p_1; x+1, n-x+1) - I(p_0; x+1, n-x+1)} \right\}^2, \tag{7.29}$$

respectively.

Example 7.7. Again considering Example 7.1, suppose it is known that the annual PWR pump reliability is no less than 0.98. Thus $p_0 = 0.98$ and

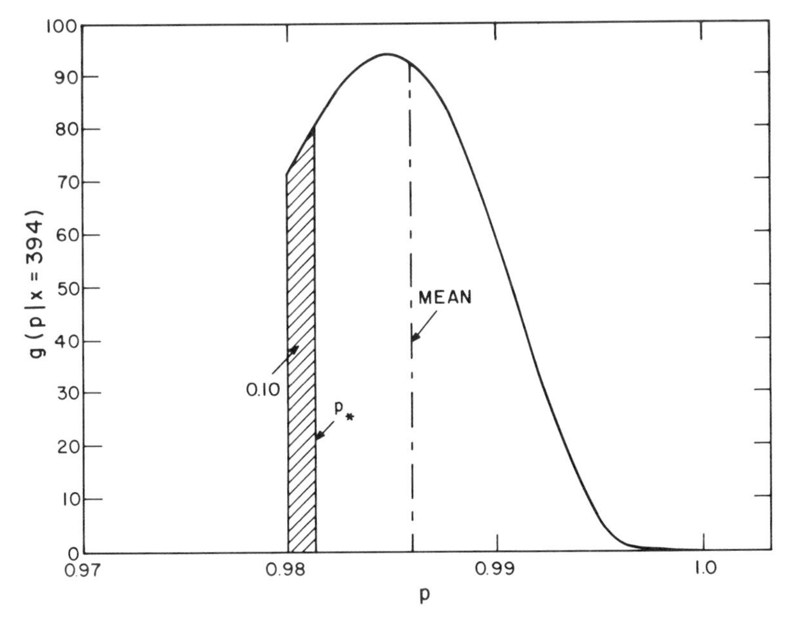

Figure 7.2. The posterior probability distribution in Example 7.7.

$p_1 = 1$. For $x = 394$ and $n = 400$, the posterior distribution of P given $x = 394$, for a $\mathcal{U}(0.98, 1)$ prior distribution, is

$$g(p|x = 394; 0.98, 1) = \frac{[\Gamma(402)/\Gamma(395)\Gamma(7)]\, p^{394}(1-p)^6}{1.0 - 0.30846}, \qquad 0.98 < p < 1,$$

which is plotted in Figure 7.2. The posterior mean is computed to be

$$E(P|x = 394; 0.98, 1) = \frac{395}{402}\left\{\frac{1.0 - 0.30602}{1.0 - 0.30846}\right\} = 0.986,$$

which is also shown in Figure 7.2. ■

Interval Estimation. The symmetric $100(1-\gamma)\%$ TBPI limits for p are given by the solutions to the equations given by

$$I(p_*; x+1, n-x+1) = \left(1 - \frac{\gamma}{2}\right) I(p_0; x+1, n-x+1) + \left(\frac{\gamma}{2}\right) I(p_1; x+1, n-x+1) \qquad (7.30)$$

and

$$I(p^*; x+1, n-x+1) = \left(1 - \frac{\gamma}{2}\right) I(p_1; x+1, n-x+1) + \left(\frac{\gamma}{2}\right) I(p_0; x+1, n-x+1) \qquad (7.31)$$

for p_* and p^*, respectively. A $100(1-\gamma)\%$ LBPI estimate for p_0 may be found by replacing $\gamma/2$ with γ in (7.30). A $100(1-\gamma)\%$ UBPI may be found from (7.31) by replacing $\gamma/2$ with γ.

Example 7.8. In Example 7.7, a lower 90% Bayesian interval estimate of p is given by the solution to

$$I(p_*; 395, 7) = 0.90\, I(0.98; 395, 7) + 0.10 = 0.90(0.30846) + 0.10 = 0.378.$$

The solution is found to be $p_* = 0.9813$, and thus $\Pr(P \leqslant 0.9813 | x = 394; 0.98, 1) = 0.90$. This limit is also shown in Figure 7.2. ■

7.2.4 Beta Prior Distribution

The most widely used prior distribution for P is the $\mathcal{B}(x_0, n_0)$ distribution discussed in Section 4.2.8. Recall that the $\mathcal{U}(0,1)$ distribution is a special $\mathcal{B}(x_0, n_0)$ distribution in which $x_0 = 1$ and $n_0 = 2$. It turns out that the $\mathcal{B}(x_0, n_0)$ distribution is a conjugate prior distribution for the binomial sampling model. That is, the posterior distribution is also a beta distribution. This was illustrated in Section 6.2. The use of the $\mathcal{B}(x_0, n_0)$ distribution has been recommended on the basis of its mathematical tractability as guaranteed by the natural conjugate criterion as well as its versatility. Weiler (1965) shows that the effect of assuming a $\mathcal{B}(x_0, n_0)$ distribution, when in fact the true prior distribution is not of the beta type, is negligible in many practical applications. He shows that rather severe deviations in the beta prior parameter values produce only slight changes in the corresponding posterior distributions.

We consider the family of $\mathcal{B}(x_0, n_0)$ prior distributions given by (4.57). The prior mean of (4.57) is $E(P) = x_0/n_0$, and for this reason the prior parameters x_0 and n_0 are interpreted as *pseudo number of survivors* and *pseudo sample size*, respectively, of a life test of duration t. The modifier "pseudo" should be thought of as meaning "pretended."

Different prior models may be chosen by specifying different values for x_0 and n_0. Various prior models with means of 0.5 are shown in Figure 4.9. Notice that the prior is unimodal and bell-shaped for relatively large values of x_0 and n_0. For small values of x_0 and n_0, the prior tends to become U-shaped with the probability tending to "pile-up" at the extremes of p. The $\mathcal{U}(0,1)$ distribution is also shown.

A similar variety of shapes exist for larger values of the prior mean, although the curves become asymmetric for small values of n_0. Figure 4.10 illustrates some typical $\mathcal{B}(x_0, n_0)$ distributions with means of 0.9. Generally, for a fixed mean, the larger the value of n_0 the more the $\mathcal{B}(x_0, n_0)$ distribution is concentrated around the mean. For example, from Figure 4.10, $x_0 = 0.90$ and $n_0 = 1$ yields a beta distribution with a mean of 0.90 and standard deviation 0.21, whereas $x_0 = 90$ and $n_0 = 100$ yields a beta prior with the same mean but with a standard deviation of 0.03.

We repeat that the most important consideration in selecting a prior model for p is that the model selected must represent the analyst's knowledge and experience concerning p. Each of the curves in Figures 4.9 and 4.10 convey different prior beliefs about p. For example, in Figure 4.10 the model (c) conveys the essential beliefs that (1) the prior mean is 0.90; (2) the "effective" range of p values is 0.8, 0.975; and (3) values of p in the immediate vicinity of 0.9 are quite likely. Thus, in this case, the analyst strongly believes in the relative high reliability of the item in question before obtaining the test results.

Now let us consider model (b) in Figure 4.10. Although the analyst believes the prior mean to be 0.90, uncertainty remains over the closeness of p to this value. This is reflected in the distribution of density over a wider range of p values. Several techniques for determining values of x_0 and n_0 consistent with prior belief in p were considered in Section 6.5. Other methods, based on the objective use of past data, are presented in Section 7.6.

A few remarks are in order about the variety of shapes of beta priors that can occur depending on the magnitude and relationship of x_0 and n_0. If $n_0 - x_0 > 1$ and $x_0 > 1$, then the $\mathcal{B}(x_0, n_0)$ prior distribution has a single mode at $(x_0 - 1)/(n_0 - 2)$. If $n_0 - x_0 < 1$ and $x_0 < 1$, then there is an antimode at $(x_0 - 1)/(n_0 - 2)$, and the distribution is U-shaped. Otherwise, $(n_0 - x_0 - 1)(x_0 - 1) < 1$, and the distribution is J-shaped or L-shaped.

From (7.5), with a $\mathcal{B}(x_0, n_0)$ prior distribution, the posterior distribution of P given x is a $\mathcal{B}(x + x_0, n + n_0)$ distribution with pdf

$$g(p|x; x_0, n_0) = \frac{\Gamma(n+n_0)}{\Gamma(x+x_0)\Gamma(n+n_0-x-x_0)} p^{(x+x_0)-1}(1-p)^{(n+n_0-x-x_0)-1}, \quad 0 < p < 1. \tag{7.32}$$

The parameter $(x + x_0)$, which may be referred to as the *combined number of survivors*, is the sum of the observed and pseudo number of survivors. Similarly, $(n + n_0)$ represents the *combined sample size*.

The marginal distribution of X is given by

$$f(x; x_0, n_0) = \frac{n!\Gamma(n_0)\Gamma(x+x_0)\Gamma(n+n_0-x-x_0)}{x!(n-x)!\Gamma(n+n_0)\Gamma(x_0)\Gamma(n_0-x_0)}, \quad x = 0, 1, \ldots, n, \tag{7.33}$$

which is known as a *beta-binomial* distribution [Raiffa and Schlaifer (1961), p. 237]. The discrete uniform distribution given in (7.9) is, of course, a special case of the beta-binomial distribution when $x_0 = 1$ and $n_0 = 2$. The mean and variance of (7.33) are

$$E(X; x_0, n_0) = nx_0/n_0$$

and

$$\text{Var}(X; x_0, n_0) = \frac{nx_0(n+n_0)(n_0-x_0)}{n_0^2(n_0+1)},$$

respectively.

Example 7.9. Suppose that in Example 7.1 a $\mathcal{B}(x_0, n_0)$ prior distribution with $x_0 = 23$ and $n_0 = 24$ is assigned to the annual pump reliability P. The posterior distribution becomes

$$g(p|x=394;\, 23, 24) = \frac{\Gamma(424)}{\Gamma(417)\Gamma(7)} p^{416}(1-p)^6, \qquad 0<p<1,$$

which is plotted in Figure 7.3 along with the prior distribution.

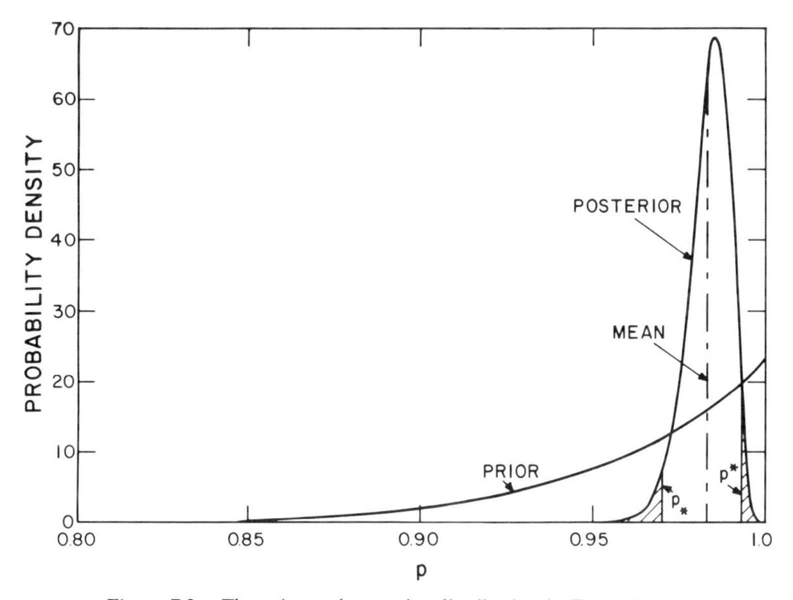

Figure 7.3. The prior and posterior distribution in Example 7.9. ■

Point Estimation. The posterior mean of (7.32) is

$$E(P|x;\, x_0, n_0) = \frac{x + x_0}{n + n_0}, \tag{7.34}$$

which is, of course, the Bayesian point estimator for p under a squared-error loss function. The generalized maximum likelihood estimator for p (when it exists) is the mode of (7.32) given by $(x + x_0 - 1)/(n + n_0 - 2)$. As n approaches infinity, the Bayes estimator approaches x/n, the classical ML estimator. That is, the likelihood tends to dominate the prior as n increases.

The behavior of (7.34) in the case of a strong or weak prior is of interest. For example, suppose that an analyst strongly believes that the reliability of an item is 0.90. Further assume that the analyst expresses this belief by use of a $\mathcal{B}(90, 100)$ prior distribution. Thus, $E(P|90, 100)=90/100=0.90$. Suppose now that subsequent life test is performed in which $x=10$ survivors are observed in $n=10$ trials. The mean of this posterior distribution becomes $E(P|x=10;\ 90, 100)=100/110=0.91$. This slight increase in the mean is reasonable.

However, what happens if $x=0$ survivors are observed? The posterior mean now becomes $E(P|x=0;\ 90, 100)=90/110=0.82$. To most engineers, this is an unreasonably high value in view of 10 observed failures in 10 trials. Such a highly concentrated beta prior is thus likely to be unsatisfactory.

Now suppose that the analyst's best prior estimate was 0.90 but that he or she felt unsure about it. In this case the analyst may choose to formulate the weaker belief by means of a more diffuse beta distribution in which $x_0=0.90$ and $n_0=1$. Thus, $E(P|0.90, 1)=0.90$ as before. Again, if $x=10$ and $n=10$, then the posterior mean is $E(P|x=10;\ 0.90, 1)=10.90/11=0.99$, which is reasonable in light of 10 observed survivors. Also, if $x=0$ and $n=10$, then $E(P|x=0;\ 0.90, 1)=0.90/11=0.08$. This result (unlike the previous case) is more reasonable in the sense that the analyst observes the reliability value of the item eroded away, as it should have been. The use of a diffuse beta prior thus represents a more reasonable choice in this particular case.

For a squared-error loss function, the posterior risk of (7.34) is proportional to the posterior variance given by

$$\operatorname{Var}(P|x;\ x_0, n_0)=\frac{(x+x_0)(n+n_0-x-x_0)}{(n+n_0)^2(n+n_0+1)}. \tag{7.35}$$

The minimum Bayes risk $r(g)$ of (7.34) is computed by recognizing that

$$\operatorname{Var}(P|x;\ x_0, n_0)=\frac{(n+n_0-2x_0)x-x^2+x_0n+x_0n_0-x_0^2}{(n+n_0)^2(n+n_0+1)}. \tag{7.36}$$

Thus

$$r(g)=a\frac{(n+n_0-2x_0)E(X;\ x_0, n_0)-E(X^2;\ x_0, n_0)+x_0n+x_0n_0-x_0^2}{(n+n_0)^2(n+n_0+1)}, \tag{7.37}$$

where a is the constant in (5.32) and where the first two moments of the marginal distribution of X are

$$E(X;\, x_0, n_0) = \frac{n x_0}{n_0} \tag{7.38}$$

and

$$E(X^2;\, x_0, n_0) = \frac{n x_0 (n + n_0)(n_0 - x_0) + n^2 x_0^2 (n_0 + 1)}{n_0^2 (n_0 + 1)}. \tag{7.39}$$

Finally, the mean and variance of (7.34) are given by

$$E_x\left[E_{p|x}(P \mid X;\, x_0, n_0)\right] = E(P;\, x_0, n_0) = \frac{x_0}{n_0} \tag{7.40}$$

and

$$\operatorname{Var}_x\left[E_{p|x}(P \mid X;\, x_0, n_0)\right] = \frac{n}{n + n_0} \operatorname{Var}(P;\, x_0, n_0) = \frac{n x_0 (n_0 - x_0)}{n_0^2 (n + n_0)(n_0 + 1)}, \tag{7.41}$$

respectively.

Example 7.10. In Example 7.9, the Bayes point estimate of the annual PWR pump reliability for a squared-error loss function in which $a = 1$ is

$$E(P \mid x = 394;\, 23,\, 24) = \frac{394 + 23}{400 + 24} = 0.983,$$

with associated minimum Bayes risk

$$r(g) = \frac{378(383.33) - 147{,}486.22 + 9200 + 552 - 529}{(424)^2 (425)} = 8.68 \times 10^{-5}.$$

The mean and variance of $E(P \mid x = 394;\, 23,\, 24)$ are 0.958 and 1.51×10^{-3}, respectively. ■

Interval Estimation. The transformation given in Section 7.2.1 relating the $\mathcal{B}(x_0, n_0)$ to the $\mathcal{F}(n_1, n_2)$ distribution can be used again here to construct

Bayesian interval estimates for p. Using this transformation, a symmetric $100(1-\gamma)\%$ TBPI estimate for p is given by

$$p_* = \frac{x+x_0}{x+x_0+(n+n_0-x-x_0)\mathcal{F}_{1-\gamma/2}(2n+2n_0-2x-2x_0, 2x+2x_0)} \tag{7.42}$$

and

$$p^* = \frac{(x+x_0)\mathcal{F}_{1-\gamma/2}(2x+2x_0, 2n+2n_0-2x-2x_0)}{n+n_0-x-x_0+(x+x_0)\mathcal{F}_{1-\gamma/2}(2x+2x_0, 2n+2n_0-2x-2x_0)}, \tag{7.43}$$

where p_* and p^* are the lower and upper interval endpoints, respectively. It may be claimed that $\Pr\{p_* \leqslant P \leqslant p^* | x;\ x_0, n_0\} = 1-\gamma$. $100(1-\gamma)\%$ LBPI and UBPI estimates for p may be found from (7.42) and (7.43) respectively, in which $\gamma/2$ is replaced by γ.

Example 7.11. In Example 7.9, a 95% TBPI estimate of p is easily computed to be

$$p_* = \frac{417}{417+7(1.870)} = 0.970$$

and

$$p^* = \frac{417(2.487)}{7+417(2.487)} = 0.993,$$

since $\mathcal{F}_{0.975}(14,\ 834) \cong 1.870$ and $\mathcal{F}_{0.975}(834,\ 14) \cong 2.487$ from Table B4 in Appendix B. Thus, $\Pr\{0.970 \leqslant P \leqslant 0.993 | x=394;\ 23,\ 24\} \cong 0.95$. This interval estimate, along with the mean, are shown in Figure 7.3. ■

Truncated Beta Prior Distribution. It is possible to consider a broader class of beta prior distributions that includes (4.57) as a special case. Consider the four-parameter family of beta distributions given by

$$g(p; x_0, n_0, p_0, p_1) = \frac{\Gamma(n_0)}{\Gamma(x_0)\Gamma(n_0 - x_0)(p_1 - p_0)^{n_0}}(p - p_0)^{x_0 - 1}(p_1 - p)^{(n_0 - x_0) - 1},$$

$$0 \leqslant p_0 < p < p_1 \leqslant 1, n_0 > x_0 > 0. \quad (7.44)$$

We shall refer to this distribution as a *truncated beta prior distribution*.

According to (7.5), the posterior distribution of P given x becomes

$$g(p|x; x_0, n_0, p_0, p_1) = \frac{p^x(p - p_0)^{x_0 - 1}(p_1 - p)^{(n_0 - x_0) - 1}(1 - p)^{n - x}}{\int_{p_0}^{p_1} p^x(p - p_0)^{x_0 - 1}(p_1 - p)^{(n_0 - x_0) - 1}(1 - p)^{n - x}\, dp},$$

$$p_0 < p < p_1, \quad (7.45)$$

where the denominator must be numerically evaluated. Once this has been done, the posterior mean and variance can be obtained by numerical integration, along with the desired Bayesian interval estimate. It is further noted that the family in (7.44) is *not* the natural conjugate family except in the special case where $p_0 = 0$ and $p_1 = 1$.

7.2.5 Normal Generated Prior Distribution

Chiu (1974) proposed a prior distribution for $Q = 1 - P$ based on an assumption that a process manufactures batches of items of a normally distributed quality characteristic. Any item whose quality measurement is less than a certain specified value is classified as defective, otherwise it is nondefective. Chiu also assumed that the batch mean μ has an $\mathcal{N}(m, \sigma^2)$ prior distribution, where m and σ^2 are known and are independent of the batch size.

The resulting *normal generated prior distribution* is found to be

$$g(q; m, \sigma^2) = \frac{1}{\sigma}\exp\left\{\frac{[\Phi^{-1}(q)]^2}{2} - \frac{[\Phi^{-1}(q) + m]^2}{2\sigma^2}\right\}, \quad 0 \leqslant q \leqslant 1, \quad (7.46)$$

where $\Phi^{-1}(q)$ is that value of z such that $\Phi(z) = q$ and where $\Phi(z)$ denotes

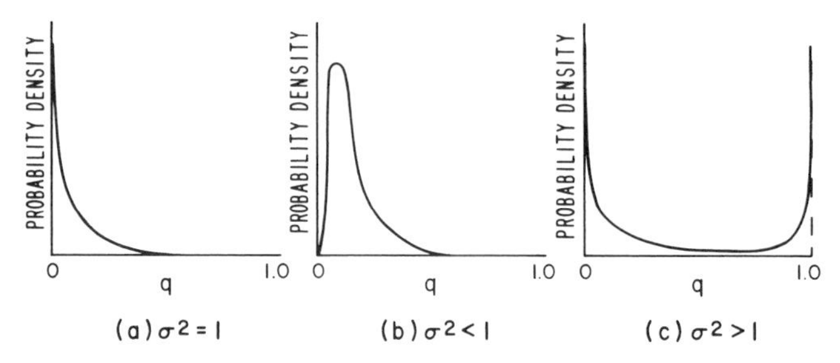

Figure 7.4. The normal generated prior distribution.

the standard normal cdf defined in (4.26). For example, for $q=0.05$ we have $\Phi^{-1}(0.05)=-1.645$. When $\sigma^2=1$, $g(q;\, m, \sigma^2)$ has a unique infinite mode at $q=0$ for positive values of m as shown in Figure 7.4*a*. When $\sigma^2<1$, the curve has a unique finite mode at $q=\Phi(-\mu)$, where $\mu=m/(1-\sigma^2)$ as in Fig. 7.4*b*. When $\sigma^2>1$, the distribution is U-shaped with two infinite modes at $q=0$ and $q=1$ [see Figure 7.4*c*]. The mean and variance of $g(q;\, m, \sigma^2)$ are found to be

$$E(Q;\, m, \sigma^2)=\Phi(-k) \tag{7.47}$$

and

$$\operatorname{Var}(Q;\, m, \sigma^2)=\frac{1}{2\pi\sqrt{1-\rho^2}}\int_{-\infty}^{-k}\int_{-\infty}^{-k}\exp\left[-\frac{y^2-2\rho yz+z^2}{2(1-\rho^2)}\right]dy\,dz-\Phi^2(-k), \tag{7.48}$$

where $k=m/\sqrt{1+\sigma^2}$ and $\rho=\sigma^2/(1+\sigma^2)$. Chiu gives a table of values of $\operatorname{Var}(Q;\, m, \sigma^2)$ as a function of selected values of σ^2 and $\Phi(-k)$. Chiu goes on to show that there may be marked differences between the shape of the normal generated and $\mathcal{B}(x_0, n_0)$ prior distributions for the same mean and variance and that, in certain situations, the normal generated prior may fit actual data better than the $\mathcal{B}(x_0, n_0)$ distribution.

Based on (7.46) the posterior distribution of Q given x is given by

$$g(q|x;\, m, \sigma^2)=\frac{(1-q)^x q^{n-x}\exp\left\{[\Phi^{-1}(q)]^2/2-[\Phi^{-1}(q)+m]^2/(2\sigma^2)\right\}}{\int_0^1(1-q)^x q^{n-x}\exp\left\{[\Phi^{-1}(q)]^2/2-[\Phi^{-1}(q)+m]^2/(2\sigma^2)\right\}dq},\quad 0\leqslant q\leqslant 1, \tag{7.49}$$

and the corresponding posterior mean becomes

$$E(Q|x; m, \sigma^2) =$$

$$\frac{\int_0^1 (1-q)^x q^{n-x+1} \exp\left\{[\Phi^{-1}(q)]^2/2 - [\Phi^{-1}(q)+m]^2/(2\sigma^2)\right\} dq}{\int_0^1 (1-q)^x q^{n-x} \exp\left\{[\Phi^{-1}(q)]^2/2 - [\Phi^{-1}(q)+m]^2/(2\sigma^2)\right\} dq}. \tag{7.50}$$

Numerical integration may be used to calculate (7.49) and to evaluate (7.50). Interval estimates of q may likewise be found by numerical integration of (7.49).

Example 7.12. Let us again consider the data in Example 7.1. We consider the annual pump unreliability $q = 1 - p$. Suppose we assume a prior mean of 0.04 and a prior variance of 0.0016 for Q. Using Table 1 of Chiu, we find that this yields a normal generated prior distribution in which $m = 1.917$ and $\sigma^2 = 0.200$. The prior and posterior distributions in which $x = 394$ and $n = 400$ become

$$g(q;\ 1.917, 0.200) = \frac{1}{0.4472} \exp\left\{\frac{[\Phi^{-1}(q)]^2}{2} - \frac{[\Phi^{-1}(q)+1.917]}{0.3999}\right\}$$

and

$$g(q|394;\ 1.917, 0.200) = \frac{1}{1.721\times 10^{-15}}(1-q)^{394} q^6 \times$$

$$\exp\left\{[\Phi^{-1}(q)]^2/2 - [\Phi^{-1}(q)+1.917]^2/0.3999\right\},$$

respectively, in which the 16-point Legendre-Gauss quadrature formula is used to perform the numerical integration. These distributions are plotted in Figure 7.5. The posterior mean is computed to be

$$E(Q|394;\ 1.917, 0.200) = \frac{3.985\times 10^{-17}}{1.721\times 10^{-15}} = 0.023,$$

which is also shown in Figure 7.5. It is observed that the posterior distribution is much less diffuse than the prior distribution as a consequence

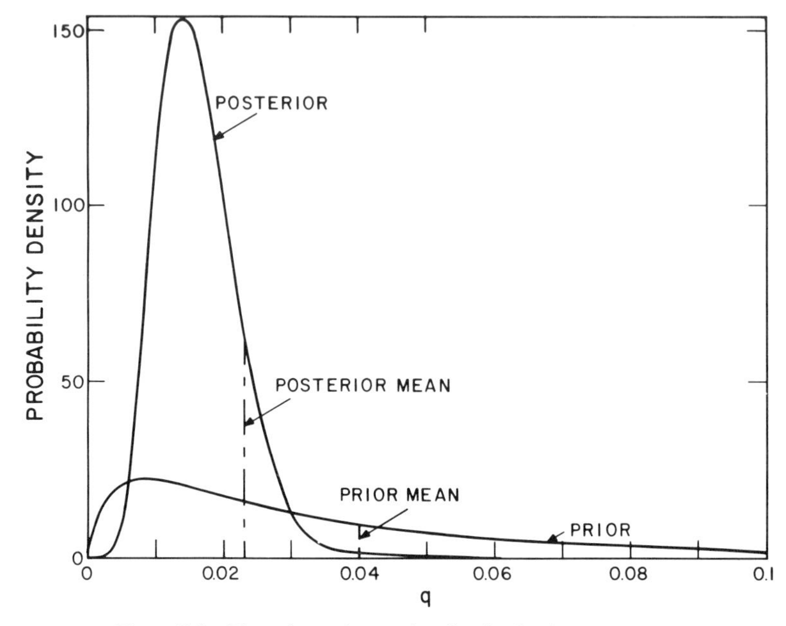

Figure 7.5. The prior and posterior distribution in Example 7.12.

of incorporating the observed data. Although the posterior distribution is positively skewed as in Figure 7.4*b*, there is no requirement that it be a member of the normal generated family of distributions, such as in the case of conjugate distributions. ■

7.2.6 Other Prior Distributions

There are various other distributions that can be used as prior distributions for P. Their suitability depends upon the subjective prior notions which the analyst wishes to express in the prior distribution. Although the following distributions generally require numerical methods for determining such things as the posterior distribution, posterior mean and variance, and interval estimates, nevertheless, this is easily accomplished and is not a serious handicap. Numerical integration is easily and efficiently accomplished by means of such techniques as Gauss quadrature using Legendre polynomials.

If it is assumed that $\ln[P/(1-P)]$ has an $\mathcal{N}(\alpha, \beta^2)$ prior distribution, then one obtains the two-parameter prior distribution given by

$$g(p;\alpha,\beta^2)=\frac{1}{p(1-p)\beta\sqrt{2\pi}}\exp\left[-\frac{1}{2\beta^2}\left\{\ln\left(\frac{p}{1-p}\right)-\alpha\right\}^2\right], \quad (7.51)$$

$$0<p<1, \qquad -\infty<\alpha<\infty, \qquad \beta^2>0.$$

The posterior distribution is easily obtained according to (7.5) which may be evaluated by appropriate numerical methods.

Another family of prior distributions is obtained by assuming that $-\ln(1-P)$ follows a $\mathcal{G}(\alpha,\beta)$ prior distribution. The resulting *log gamma* prior distribution on P becomes

$$g(p;\alpha,\beta)=\frac{1}{\beta^{\alpha}\Gamma(\alpha)}(1-p)^{1/\beta-1}\left[\ln\left(\frac{1}{1-p}\right)\right]^{\alpha-1}, \quad (7.52)$$

$$0<p<1, \qquad \alpha,\beta>0.$$

In a similar way it may be assumed that $-\ln(1-P)$ follows an $\mathcal{LN}(\alpha,\beta^2)$ prior distribution. In this case the resulting *log log normal* prior distribution is given by

$$g(p;\alpha,\beta^2)=\frac{1}{(1-p)\ln[1/(1-p)]\beta\sqrt{2\pi}}\exp\left\{\frac{-1}{2\beta^2}\left[\ln\ln\left(\frac{1}{1-p}\right)-\alpha\right]^2\right\}, \quad (7.53)$$

$$0<p<1, \qquad -\infty<\alpha<\infty, \qquad \beta^2>0.$$

Additional prior distributions may be derived by considering the transformation $X=-\ln(1-P)$ where X is an r.v. defined on the range $(0,\infty)$.

7.3. ESTIMATING THE SURVIVAL PROBABILITY—PASCAL SAMPLING

Pascal sampling was introduced in Section 7.1.2. We are interested in obtaining both Bayesian point and interval estimators for the survival probability p in the Pascal sampling model given in (7.3). For a general prior distribution $g(p)$, the posterior distribution according to (5.8) becomes

$$g(p|n) = \frac{[(n-1)!/(s-1)!(n-s)!]\, p^{n-s}(1-p)^s g(p)}{\int_0^1 [(n-1)!/(s-1)!(n-s)!]\, p^{n-s}(1-p)^s g(p)\, dp}$$

$$= \frac{p^{n-s}(1-p)^s g(p)}{\int_0^1 p^{n-s}(1-p)^s g(p)\, dp}, \qquad 0<p<1, \tag{7.54}$$

where n is the trial at which the prespecified sth failure is observed to occur. It is noted that (7.54) is of the same form as the posterior distribution for the binomial sampling model given in (7.5) where $n-s=x$. Thus, by simply making this change, the estimation results for the binomial model in the preceding section will apply to the Pascal model as well. However, such things as the marginal distribution of n cannot be so directly obtained since the factorial terms differ by a factor of $(n+1)/(s+1)$ in the two sampling models.

Since the uniform distribution is a special case of the $\mathcal{B}(x_0, n_0)$ distribution, we proceed directly to the more general case.

7.3.1 Beta Prior Distribution

The $\mathcal{B}(x_0, n_0)$ prior distribution is the natural conjugate prior distribution for the Pascal sampling model. We again consider the family of $\mathcal{B}(x_0, n_0)$ prior distribution given in (4.57).

The posterior distribution of P given n thus becomes

$$g(p|n; x_0, n_0)$$

$$= \frac{p^{n-s}(1-p)^s[\Gamma(n_0)/\Gamma(x_0)\Gamma(n_0-x_0)]\, p^{x_0-1}(1-p)^{(n_0-x_0)-1}}{\int_0^1 p^{n-s}(1-p)^s[\Gamma(n_0)/\Gamma(x_0)\Gamma(n_0-x_0)]\, p^{x_0-1}(1-p)^{(n_0-x_0)-1}\, dp}$$

$$= \frac{p^{n+x_0-s-1}(1-p)^{n_0-x_0+s-1}}{\int_0^1 p^{n+x_0-s-1}(1-p)^{n_0-x_0+s-1}\, dp}$$

$$= \frac{p^{n+x_0-s-1}(1-p)^{n_0-x_0+s-1}}{\Gamma(n+x_0-s)\Gamma(n_0-x_0+s)/\Gamma(n+n_0)}$$

$$= \frac{\Gamma(n+n_0)}{\Gamma(n+x_0-s)\Gamma(n_0-x_0+s)} p^{(n+x_0-s)-1}(1-p)^{(n_0-x_0+s)-1}, 0<p<1, \tag{7.55}$$

which is a $\mathcal{B}(n+x_0-s, n+n_0)$ distribution.

The marginal distribution of N is given by

$$f(n;x_0,n_0)=\frac{(n-1)!\Gamma(n_0)\Gamma(n+x_0-s)\Gamma(n_0-x_0+s)}{(s-1)!(n-s)!\Gamma(n+n_0)\Gamma(x_0)\Gamma(n_0-x_0)},$$

$$n=s,s+1,\ldots, \tag{7.56}$$

which is known as a *beta-Pascal* distribution [Raiffa and Schlaifer (1961), p. 238], with mean and variance given by

$$E(N;x_0,n_0)=\frac{s(n_0-1)}{n_0-x_0-1}$$

and

$$\text{Var}(N;x_0,n_0)=\frac{sx_0(n_0-x_0+s-1)(n_0-1)}{(n_0-x_0-1)^2(n_0-x_0-2)},$$

respectively.

Point Estimation. The posterior mean of (7.55) is

$$E(P|n;x_0,n_0)=\frac{n+x_0-s}{n+n_0}, \tag{7.57}$$

which is the Bayesian point estimator for p under a squared-error loss function. The generalized maximum likelihood estimator for p (when it exists) is the mode of (7.55) which is $(n+x_0-s-1)/(n+n_0-2)$.

The posterior variance is also given by

$$\text{Var}(P|n;x_0,n_0)=\frac{(n+x_0-s)(n_0-x_0+s)}{(n+n_0)^2(n+n_0+1)}, \tag{7.58}$$

which is proportional to the posterior risk of (7.57). No closed form solution can be found for the minimum Bayes risk $r(g)$ of (7.57) for a squared-error loss function, although in terms of an infinite series $r(g)$ as defined in (5.32) may be expressed as

$$r(g)=\frac{a\Gamma(n_0)\Gamma(n_0-x_0+s+1)}{(s-1)!\Gamma(x_0)\Gamma(n_0-x_0)}\sum_{n=s}^{\infty}\frac{(n-1)!\Gamma(n+x_0-s+1)}{(n+n_0)(n-s)!\Gamma(n+n_0+2)}. \tag{7.59}$$

The mean of (7.57) is

$$E_n\left[E_{p|n}(P|N;x_0;n_0)\right]=\frac{x_0}{n_0}, \tag{7.60}$$

and, again, no closed form solution can be found for the variance of (7.57), although it can be expressed as an infinite series.

Example 7.13. In Example 7.2, suppose we assign a $\mathcal{B}(x_0, n_0)$ prior distribution to the probability P that a lot of Extex meets the specification on detonation velocity. Based on historical data it is decided to consider a $\mathcal{B}(50.4, 63)$ prior distribution (prior mean and standard deviation equal to 0.80 and 0.05, respectively). For $n=12$ and $s=2$, the posterior distribution becomes

$$g(p|n=12;50.4,63)=\frac{\Gamma(75)}{\Gamma(60.4)\Gamma(14.6)}p^{59.4}(1-p)^{13.6}, \qquad 0<p<1,$$

which is plotted in Figure 7.6 along with the prior distribution.

The posterior mean is

$$E(P|n=12;\ 50.4,\ 63)=\frac{12+50.4-2}{12+63}=0.805,$$

which is also shown in Figure 7.6, and the posterior variance becomes

$$\text{Var}\,(P|n=12;\ 50.4,\ 63)=\frac{(12+50.4-2)(63-50.4+2)}{(12+63)^2(12+63+1)}=0.00206.$$

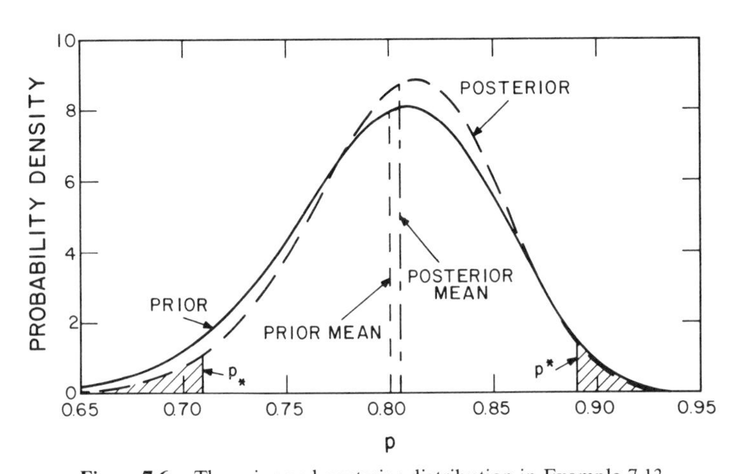

Figure 7.6. The prior and posterior distribution in Example 7.13.

The Bayes risk according to (7.59) turns out to be 0.00217 where the infinite series in (7.59) was expanded to 1000 terms to insure accuracy. The generalized maximum likelihood estimate of p is calculated to be 0.814. ■

Interval Estimation. The Bayesian interval estimates proceed as for the binomial sampling model, where we now replace x by $n-s$, and remember that N is now the r.v. rather than X. An example will serve to illustrate their construction.

Example 7.14. Suppose that in the preceding example we desire a symmetric 95% TBPI estimate of p. According to (7.42) and (7.43) we find that

$$p_* = \frac{12-2+50.4}{12-2+50.4+(63-50.4+2)(1.70)} = 0.709$$

and

$$p^* = \frac{(12-2+50.4)(1.89)}{63-50.4+2+(12-2+50.4)(1.89)} = 0.887,$$

where we have used the fact that $\mathcal{F}_{0.975}(29.2,\ 120.8) \cong 1.70$ and $\mathcal{F}_{0.975}(120.8,\ 29.2) \cong 1.89$. Thus, $\Pr\{0.709 \leq p \leq 0.887 | n=12;\ 50.4,\ 63\} = 0.95$ and this interval estimate is also shown in Figure 7.6. ■

7.3.2 Normal Generated Prior Distribution

The normal generated prior distribution was defined in (7.46). Using this prior distribution, the posterior distribution of $Q=1-P$ in the case of the Pascal sampling distribution given in (7.3) becomes

$$g(q|n;\ m,\sigma^2) =$$

$$\frac{(1-q)^{n-s} q^s \exp\left\{[\Phi^{-1}(q)]^2/2 - [\Phi^{-1}(q)+m]^2/(2\sigma^2)\right\}}{\int_0^1 (1-q)^{n-s} q^s \exp\left\{[\Phi^{-1}(q)]^2/2 - [\Phi^{-1}(q)+m]^2/(2\sigma^2)\right\} dq},$$

$$0 \leq q \leq 1, \quad (7.61)$$

which is the same form as (7.49) in which x is replaced by $n-s$. Thus, the results of Section 7.2.5 apply here as well.

Example 7.15. Again let us consider Example 7.2. We consider $q = 1 - p$, the probability of *not* meeting the specification on detonation velocity. Assuming a prior mean and standard deviation of 0.10 and 0.05, respectively, yields $m = 1.330$ and $\sigma^2 = 0.080$ from Chiu's tables. The prior and posterior distributions given by (7.46) and (7.61), respectively, in which $m = 1.330, \sigma^2 = 0.080$, $n = 12$, and $s = 2$, and where we let $q = 1 - p$, are plotted in Figure 7.7.

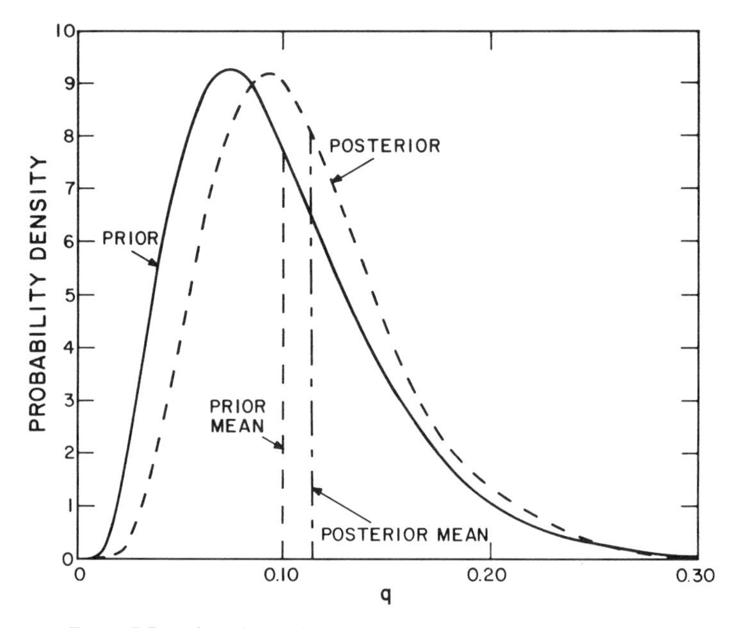

Figure 7.7. The prior and posterior distribution in Example 7.15.

The posterior mean becomes

$$E(Q|12;\,1.330,\,0.080) = \frac{9.86\times10^{-5}}{8.51\times10^{-4}} = 0.116,$$

which is also shown in Figure 7.7 along with the prior mean. ■

7.4. ESTIMATING THE FAILURE RATE—POISSON SAMPLING

The notion of Poisson sampling to estimate a constant failure rate was introduced in Section 7.1.3. We are interested in obtaining both point and

Bayesian interval estimators for the failure rate λ in the Poisson sampling model given in (7.4). Papadopoulos and Rao (1977) develop Bayesian probability intervals for the parameter $\theta = 1/\lambda$ in the Poisson model as well as for the probability of at least t failures in time t. They consider both a $\mathcal{U}(a, b)$ and an $\mathcal{IG}(\nu, \mu)$ prior distribution on Θ. Using Monte Carlo simulation, they observe that the Bayesian bounds are superior to the classical bounds. On the average, they find that the Bayesian bounds on θ have approximately a 50% smaller mean squared error than the corresponding classical bounds and that the Bayesian interval is always shorter, on the average, than the classical interval. Similar conclusions were also observed in estimating the probability of at least t failures in time t and for estimation of the Poisson hazard rate function.

According to (5.8), for any suitable prior distribution $g(\lambda)$, the posterior distribution of Λ given the observed number of failures s in total test time t becomes

$$g(\lambda \mid s) = \frac{\left[e^{-\lambda \mathrm{t}}(\lambda \mathrm{t})^s / s!\right] g(\lambda)}{\int_0^\infty \left[e^{-\lambda \mathrm{t}}(\lambda \mathrm{t})^s / s!\right] g(\lambda)\, d\lambda}$$

$$= \frac{e^{-\lambda \mathrm{t}} \lambda^s g(\lambda)}{\int_0^\infty e^{-\lambda \mathrm{t}} \lambda^s g(\lambda)\, d\lambda}, \qquad 0 < \lambda < \infty. \tag{7.62}$$

For $\mathcal{U}(0, \lambda_1)$, noninformative, $\mathcal{G}_1(\alpha, \beta)$, and a few other less widely used prior distributions on Λ, we derive both point and interval estimators for λ. In Section 7.5 we consider Bayesian estimation of the reliability function $r(t) = \exp(-\lambda t)$, which is the Poisson probability that no failures occur in the time interval $(0, t)$ according to (7.4), where $\mathrm{t} \equiv t$. This is also the reliability for the $\mathcal{E}(\lambda)$ failure time model presented in Section 4.2.1, as the Poisson sampling model implies that the failure times follow an $\mathcal{E}(\lambda)$ distribution and vice versa.

7.4.1 Uniform Prior Distribution

Suppose that the prior distribution of Λ is the $\mathcal{U}(0, \lambda_1)$ distribution given by

$$g(\lambda; \lambda_1) = \begin{cases} \dfrac{1}{\lambda_1}, & 0 < \lambda < \lambda_1 \\ 0, & \text{elsewhere.} \end{cases} \tag{7.63}$$

It follows from (7.62) that the posterior distribution of Λ given s is

$$g(\lambda|s;\lambda_1)=\frac{e^{-\lambda t}\lambda^s}{\int_0^{\lambda_1} e^{-\lambda t}\lambda^s\,d\lambda}, \qquad 0<\lambda<\lambda_1. \tag{7.64}$$

To evaluate the denominator of (7.64) let $y=\lambda t$. Then

$$\int_0^{\lambda_1} e^{-\lambda t}\lambda^s\,d\lambda=\int_0^{\lambda_1 t}\frac{y^s}{t^{s+1}}e^{-y}\,dy=\frac{\Gamma(s+1,\lambda_1 t)}{t^{s+1}},$$

where $\Gamma(a,z)$ is the incomplete gamma function. In the examples that follow, and throughout the book, we have used an excellent computer subroutine written by Amos and Daniel (1972), called INCGAM, to evaluate $\Gamma(a,z)$. The code is quite accurate and uses a confluent hypergeometric series in conjunction with backward recursion on a two-term formula to perform the necessary computations. Therefore, the posterior distribution of Λ given s becomes

$$g(\lambda|s;\lambda_1)=\frac{t^{s+1}\lambda^s e^{-\lambda t}}{\Gamma(s+1,\lambda_1 t)}, \qquad 0<\lambda<\lambda_1. \tag{7.65}$$

The marginal distribution of S is also given by

$$f(s;\lambda_1)=\frac{\Gamma(s+1,\lambda_1 t)}{ts!}, \qquad s=0,1,\ldots. \tag{7.66}$$

Example 7.16. In Example 7.3, suppose we consider a $\mathcal{U}(0,10^{-5}f/h)$ prior distribution for Λ. The posterior distribution thus becomes

$$g(\lambda|s=32;\,10^{-5})=\frac{(20,253,120)^{33}\lambda^{32}e^{-20,253,120\lambda}}{\Gamma(33,\,202.53)}, \qquad 0<\lambda<10^{-5},$$

which is plotted in Figure 7.8 along with the prior distribution.

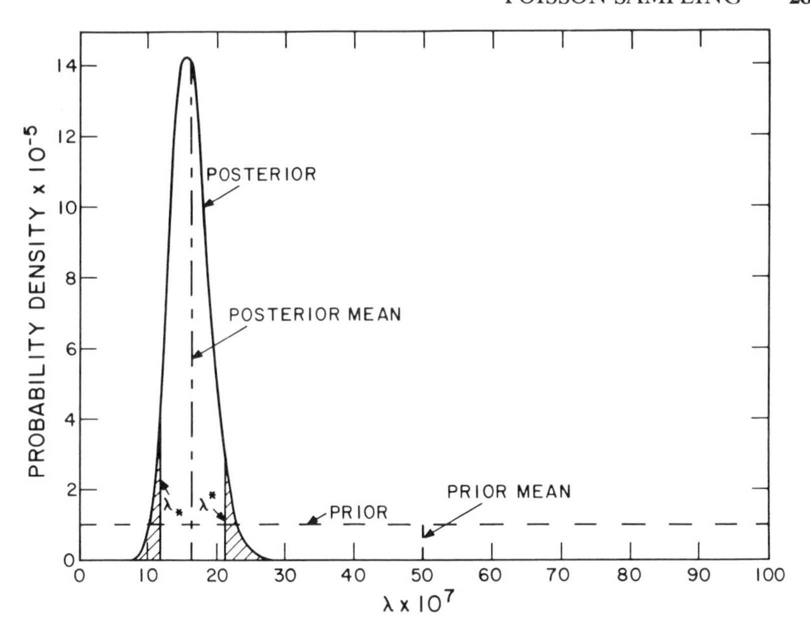

Figure 7.8. The prior and posterior distribution in Example 7.16. ■

Point Estimation. The posterior mean and variance of (7.65) are computed to be

$$E(\Lambda|s;\lambda_1)=\frac{\Gamma(s+2,\lambda_1 t)}{t\Gamma(s+1,\lambda_1 t)} \tag{7.67}$$

and

$$\operatorname{Var}(\Lambda|s;\lambda_1)=\frac{1}{t^2\Gamma(s+1,\lambda_1 t)}\left\{\Gamma(s+3,\lambda_1 t)-\frac{\Gamma^2(s+2,\lambda_1 t)}{\Gamma(s+1,\lambda_1 t)}\right\}, \tag{7.68}$$

respectively. Thus, according to (5.31), the posterior risk of (7.67) is given by

$$\phi(s)=\frac{a}{t^2\Gamma(s+1,\lambda_1 t)}\left\{\Gamma(s+3,\lambda_1 t)-\frac{\Gamma^2(s+2,\lambda_1 t)}{\Gamma(s+1,\lambda_1 t)}\right\}. \tag{7.69}$$

The minimum Bayes risk of (7.67) for a squared-error loss function, which is the expectation of (7.68) with respect to the distribution given in (7.66), becomes

$$r(g)=\frac{a}{t^3}\left\{\sum_{s=0}^{\infty}\frac{\Gamma(s+3,\lambda_1 t)}{s!}-\sum_{s=0}^{\infty}\frac{\Gamma^2(s+2,\lambda_1 t)}{s!\Gamma(s+1,\lambda_1 t)}\right\}, \qquad (7.70)$$

which is not readily expressible in closed form. The generalized maximum likelihood estimator for λ is the mode of (7.65) which is s/t. Finally, the mean and variance of (7.67) are given by

$$E_s\left[E_{\lambda|s}(\Lambda|S;\lambda_1)\right]=\frac{\lambda_1}{2} \qquad (7.71)$$

and

$$\mathrm{Var}_s\left[E_{\lambda|s}(\Lambda|S;\lambda_1)\right]=\frac{\lambda_1^2}{12}-\frac{1}{t^3}\left\{\sum_{s=0}^{\infty}\frac{\Gamma(s+3,\lambda_1 t)}{s!}-\sum_{s=0}^{\infty}\frac{\Gamma^2(s+2,\lambda_1 t)}{s!\Gamma(s+1,\lambda_1 t)}\right\}, \qquad (7.72)$$

respectively.

Example 7.17. In Example 7.16, the Bayes point estimate of the valve failure rate, based on 1972 sample data, for a squared-error loss function in which $a=1$ is

$$E(\Lambda|s=32;\,10^{-5})=\frac{8.6833\times10^{36}}{(20,253,120)(2.6313\times10^{35})}=16.3\times10^{-7},$$

which is shown in Figure 7.8. The posterior variance is also computed to be 8.0451×10^{-14}. The Bayes risk according to (7.70) is computed to be 2.1699×10^{-18}, where the infinite series has been expanded to 500 terms. Thus, according to (7.71) and (7.72), $E(\Lambda|s=32;\,10^{-5})$ is the value of an r.v. having mean 0.5×10^{-5} and variance $8.3333\times10^{-12}-2.1699\times10^{-18}=8.3333\times10^{-12}$. The generalized maximum likelihood estimate of λ is $32/20,253,120=15.8\times10^{-7}$, which is also shown in Figure 7.8. ■

Interval Estimation. From (5.45) and (5.46), a symmetric $100(1-\gamma)\%$ TBPI estimate for λ is obtained by solving the pair of equations given by

$$\Pr(\Lambda \leqslant \lambda_*|s;\lambda_1) = \int_0^{\lambda_*} \frac{t^{s+1}\lambda^s e^{-\lambda t}}{\Gamma(s+1,\lambda_1 t)} d\lambda = \frac{\Gamma(s+1,\lambda_* t)}{\Gamma(s+1,\lambda_1 t)} = \frac{\gamma}{2} \tag{7.73}$$

and

$$\Pr(\Lambda \geqslant \lambda^*|s;\lambda_1) = \int_{\lambda^*}^{\lambda_1} \frac{t^{s+1}\lambda^s e^{-\lambda t}}{\Gamma(s+1,\lambda_1 t)} d\lambda = 1 - \frac{\Gamma(s+1,\lambda^* t)}{\Gamma(s+1,\lambda_1 t)} = \frac{\gamma}{2} \tag{7.74}$$

for the lower and upper limits, λ_* and λ^*, respectively. $100(1-\gamma)\%$ LBPI and UBPI estimates for λ may be found from (7.73) and (7.74), respectively, in which $\gamma/2$ is replaced by γ.

Example 7.18. In Example 7.16, for $\gamma = 0.10$, by direct computer search using INCGAM we find that

$$\Pr(\Lambda \leqslant 11.926 \times 10^{-7}|s=32;10^{-5}) = \frac{\Gamma(33,24.1539)}{\Gamma(33,202.5312)}$$

$$= \frac{1.3163 \times 10^{34}}{2.6313 \times 10^{35}} = 0.05$$

and

$$\Pr(\Lambda \geqslant 21.220 \times 10^{-7}|s=32;10^{-5}) = 1 - \frac{\Gamma(33,42.9771)}{\Gamma(33,202.5312)} = 0.05.$$

Thus,

$$\lambda_* = \frac{24.1539}{20{,}253{,}120} = 1.19 \times 10^{-6} \text{ f/h}$$

and

$$\lambda^* = \frac{42.9771}{20{,}253{,}120} = 2.12 \times 10^{-6} \text{ f/h},$$

and hence $\Pr(1.19 \times 10^{-6} \text{ f/h} \leqslant \Lambda \leqslant 2.12 \times 10^{-6} \text{ f/h}|s=32;10^{-5}) = 0.90$. This interval is also shown in Figure 7.8. ■

7.4.2 Noninformative Prior Distribution

As only a single Poisson sample observation s is observed, the logarithm of the likelihood function is

$$\ln L(\lambda|s) = s\ln(\lambda t) - \lambda t - \ln s!, \tag{7.75}$$

from which

$$\frac{\partial \ln L}{\partial \lambda} = \frac{s}{\lambda} - t, \qquad \frac{\partial^2 \ln L}{\partial \lambda^2} = -\frac{s}{\lambda^2}.$$

Thus, following Section 6.1,

$$J(\hat{\lambda}) = \left(-\frac{\partial^2 \ln L}{\partial \lambda^2}\right)_{\lambda=\hat{\lambda}} = \frac{t}{\hat{\lambda}}, \tag{7.76}$$

where $\hat{\lambda} = S/t (s \neq 0)$ is the ML estimator of λ obtained from $\partial \ln L/\partial\lambda = 0$. According to (6.2), a noninformative prior distribution for Λ is

$$g(\lambda) \propto J^{1/2}(\lambda) \propto \lambda^{-1/2}, \qquad \lambda > 0, \tag{7.77}$$

and $\phi(\lambda) = \lambda^{1/2}$ is the transformation for which the approximately noninformative prior distribution is locally uniform. The fact that $J(\hat{\lambda})$ is a function of the nuisance parameter t does not violate the requirement that $J(\hat{\lambda})$ be a function of $\hat{\lambda}$ alone, as the sample statistic S does not appear in (7.76).

The prior in (7.77) is improper over its entire admissible range; however, the corresponding posterior distribution is calculated to be

$$g(\lambda|s) = \frac{t^{s+1/2}}{\Gamma(s+1/2)}\lambda^{s-1/2}e^{-\lambda t}, \qquad \lambda > 0, \tag{7.78}$$

which is a proper distribution. In fact, it may be shown that, given s, $2\Lambda t$ follows a $\chi^2(2s+1)$ distribution.

The posterior mean and variance of (7.78) are

$$E(\Lambda|s) = \frac{2s+1}{2t} \tag{7.79}$$

and

$$\mathrm{Var}(\Lambda|s) = \frac{2s+1}{2t^2}, \tag{7.80}$$

respectively. For large values of t, the estimator in (7.79) is approximately equal to the classical ML estimator.

A symmetric $100(1-\gamma)\%$ TBPI estimate for λ may be obtained by use of the $\chi^2(2s+1)$ distribution; namely,

$$\Pr(\Lambda \leqslant \lambda_* | s) = \Pr(2\Lambda t \leqslant 2\lambda_* t | s) = \Pr\left[\chi^2(2s+1) \leqslant 2\lambda_* t | s\right] = \frac{\gamma}{2}$$

and

$$\Pr(\Lambda \geqslant \lambda^* | s) = \Pr(2\Lambda t \geqslant 2\lambda^* t | s) = 1 - \Pr\left[\chi^2(2s+1) \leqslant 2\lambda^* t | s\right] = \frac{\gamma}{2}.$$

Thus, the desired interval has lower and upper endpoints given by

$$\lambda_* = \frac{\chi^2_{\gamma/2}(2s+1)}{2t} \tag{7.81}$$

and

$$\lambda^* = \frac{\chi^2_{1-\gamma/2}(2s+1)}{2t}, \tag{7.82}$$

respectively. Corresponding $100(1-\gamma)\%$ LBPI and UBPI estimates may be constructed by appropriate use of (7.81) or (7.82) in which $\gamma/2$ is replaced by γ. The interval estimate given by (7.81) and (7.82) is quite close to the corresponding $100(1-\gamma)\%$ TCI estimate of λ given in Section 4.5.1. Recall that the TCI estimate has either $2s$ or $2s+2$ degrees of freedom, depending on the testing scheme considered. Thus, both procedures yield approximately the same interval estimate for large values of s.

Example 7.19. Again, let us consider Example 7.3. The posterior distribution is given by

$$g(\lambda | s=32) = \frac{(20{,}253{,}120)^{32.5}}{\Gamma(32.5)} \lambda^{31.5} e^{-20{,}253{,}120\lambda}, \qquad \lambda > 0,$$

which is shown in Figure 7.9. The posterior mean and variance are

$$E(\Lambda | s=32) = \frac{65}{40{,}506{,}240} = 1.60 \times 10^{-6} \text{ f/h}$$

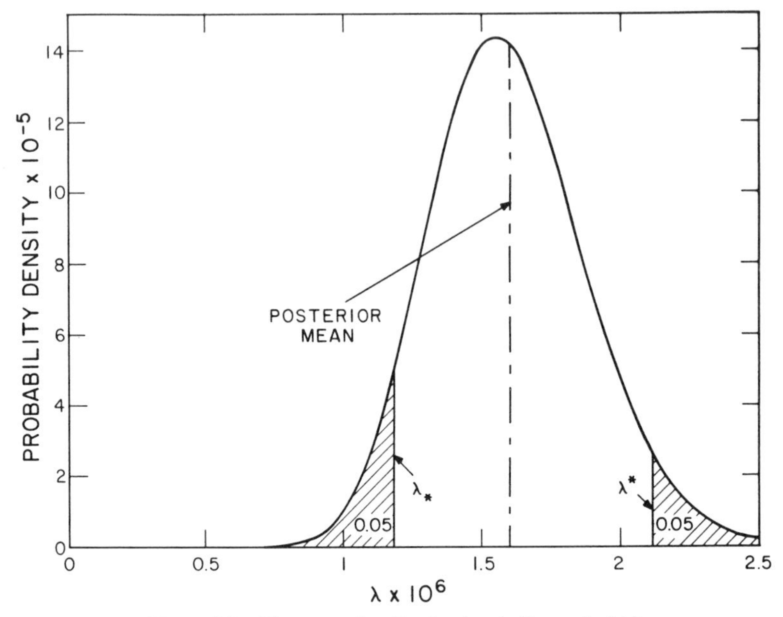

Figure 7.9. The posterior distribution in Example 7.19.

and

$$\operatorname{Var}(\Lambda \mid s=32)=\frac{65}{8.2038\times 10^{14}}=7.9232\times 10^{-14}\ \mathrm{f}^2/\mathrm{h}^2,$$

respectively. A 90% TBPI estimate of λ is given by

$$\lambda_*=\frac{\chi^2_{0.05}(65)}{40{,}506{,}240}=\frac{47.450}{40{,}506{,}240}=1.17\times 10^{-6}\ \mathrm{f/h}$$

and

$$\lambda^*=\frac{\chi^2_{0.95}(65)}{40{,}506{,}240}=\frac{84.821}{40{,}506{,}240}=2.09\times 10^{-6}\ \mathrm{f/h}.$$

Thus $\Pr(1.17\times 10^{-6}\ \mathrm{f/h}\leq \Lambda \leq 2.09\times 10^{-6}\ \mathrm{f/h} \mid s=32)=0.90$, and the interval is also shown in Figure 7.9 along with the posterior mean. ■

7.4.3 Gamma Prior Distribution

Probably the most widely used prior distribution for Λ is the $\mathcal{G}(\alpha, \beta)$ distribution presented in Section 4.2.6. The main reason for this general acceptability is the mathematical tractability resulting from the fact that the $\mathcal{G}(\alpha, \beta)$ distribution is the natural conjugate prior distribution.

Such authors as Apostolakis and Mosleh (1979), Grohowski, Hausman, and Lamberson (1976), and others have found the gamma prior distribution to be sufficiently versatile for practical reliability applications.

Consider a gamma prior distribution with pdf given by

$$g(\lambda; \alpha, \beta) = \frac{\beta^{\alpha}}{\Gamma(\alpha)} \lambda^{\alpha-1} e^{-\beta\lambda}, \qquad \lambda, \alpha, \beta > 0, \tag{7.83}$$

which we shall denote as a $\mathcal{G}_1(\alpha, \beta)$ distribution in order to distinguish it from the reparameterized version given in (4.45) in which the scale parameter is β^{-1} instead of β. Recall that (4.45) is denoted as a $\mathcal{G}(\alpha, \beta)$ distribution. Because the prior mean of Λ in (7.83) is α/β, the shape parameter α can be interpreted as the *pseudo number of failures* in a prior life test of duration β *pseudo time units*. It is further noted that the shape parameter α is a unitless quantity, whereas the scale parameter β in the $\mathcal{G}_1(\alpha, \beta)$ distribution has units of scale, in this case, time, associated with it. Figure 4.7 illustrates four $\mathcal{G}(\alpha, \beta)$ models to indicate the versatility of gamma prior models.

We note again that the dominant factor in selecting a prior model for λ is that the selected model represent the analyst's knowledge and experience concerning λ. That is, the prior model should reflect the analyst's prior beliefs about λ. For example, Figure 4.7 shows that the $\mathcal{G}(1,2)$ distribution places a high density on small values of λ, whereas $\mathcal{G}(2,4)$ distributes the density over a wide range of λ values. Therefore, these two prior models convey different prior beliefs about λ. The flexibility present in the gamma family through the choices of α and β allows the analyst to select the model that best expresses the current state of knowledge about λ. In Section 6.6 we considered the important question of how to select values of α and β consistent with various kinds of prior belief. In Section 7.7 we consider two methods for selecting values of α and β based on the objective use of Poisson sampling data.

The positive skewness of the gamma family can account for the occurrence of less likely but large variations in the failure rate, such as abnormally high failure rates due to batch defects, environmental degradation, and other such anomalies. Since density is given to values that can be skewed toward large values, a protective positive type (and thus conservative) bias is provided in such an analysis.

By differentiating (7.83) with respect to λ, we find that $g(\lambda; \alpha, \beta)$ is a unimodal function of λ with a mode at $\lambda = (\alpha - 1)/\beta$ when $\alpha > 1$. For $\alpha = 1$, we have an $\mathcal{E}(\beta)$ distribution. For $0 < \alpha < 1$, $g(\lambda; \alpha, \beta)$ is L-shaped (decreasing).

From (7.62), the posterior pdf of Λ given s is given by

$$g(\lambda | s; \alpha, \beta) = \frac{e^{-\lambda t} \lambda^s \lambda^{\alpha - 1} e^{-\beta \lambda}}{\int_0^\infty e^{-\lambda t} \lambda^s \lambda^{\alpha - 1} e^{-\beta \lambda}\, d\lambda}$$

$$= \frac{\lambda^{s + \alpha - 1} e^{-(t + \beta)\lambda}}{\int_0^\infty \lambda^{s + \alpha - 1} e^{-(t + \beta)\lambda}\, d\lambda}, \qquad \lambda > 0.$$

Letting $y = \lambda(t + \beta)$, we find that

$$\int_0^\infty \lambda^{s + \alpha - 1} e^{-(t + \beta)\lambda}\, d\lambda = \frac{1}{(t + \beta)^{s + \alpha}} \int_0^\infty y^{s + \alpha - 1} e^{-y}\, dy = \frac{\Gamma(s + \alpha)}{(t + \beta)^{s + \alpha}}. \tag{7.84}$$

Thus, the posterior pdf of Λ given above reduces to

$$g(\lambda | s; \alpha, \beta) = \frac{(t + \beta)^{s + \alpha}}{\Gamma(s + \alpha)} \lambda^{s + \alpha - 1} e^{-(t + \beta)\lambda}, \qquad \lambda > 0, \tag{7.85}$$

which is recognized as a $\mathcal{G}_1(s + \alpha, t + \beta)$ distribution. Analogous to the case of the binomial sampling model and a $\beta(x_0, n_0)$ prior distribution, the parameter $(s + \alpha)$ is referred to as the *combined number of failures*, whereas $(t + \beta)$ is the *combined total test time*. Thus, the effect of the $\mathcal{G}_1(\alpha, \beta)$ prior distribution on λ is to increase the observed number of failures s by α and to increase the observed total test time t by β. This is a clear interpretation of the effect of the prior distribution in the analysis.

The marginal distribution of S is given by

$$f(s; \alpha, \beta) = \frac{\beta^\alpha t^s \Gamma(s + \alpha)}{\Gamma(\alpha) s! (t + \beta)^{s + \alpha}}, \qquad s = 0, 1, 2, \ldots, \tag{7.86}$$

with marginal mean and variance given by

$$E(S; \alpha, \beta) = \frac{\alpha t}{\beta} \tag{7.87}$$

and

$$\operatorname{Var}(S; \alpha, \beta) = \frac{\alpha t(t+\beta)}{\beta^2}, \tag{7.88}$$

respectively.

Example 7.20. Suppose that in Example 7.3 a $\mathcal{G}_1(2.45, 183{,}200)$ prior distribution is assigned to the valve failure rate λ. This is a prior distribution consistent with fifth and ninety-fifty percentile bounds identified in the WASH-1400 "Reactor Safety Study" (1975) as 3×10^{-6} f/h and 3×10^{-5} f/h, respectively. The corresponding prior mean and variance are 1.34×10^{-5} and 7.30×10^{-11}, respectively. The posterior pdf of Λ is given by

$$g(\lambda \mid s=32; 2.45, 183{,}200) = \frac{(20{,}436{,}320)^{34.45}}{\Gamma(34.45)}\lambda^{33.45}e^{-20{,}436{,}320\lambda}, \qquad \lambda > 0,$$

which is plotted in Figure 7.10 along with the prior distribution. It is observed that the likelihood dominates the prior in the analysis due to the extensive operating data reported in 1972.

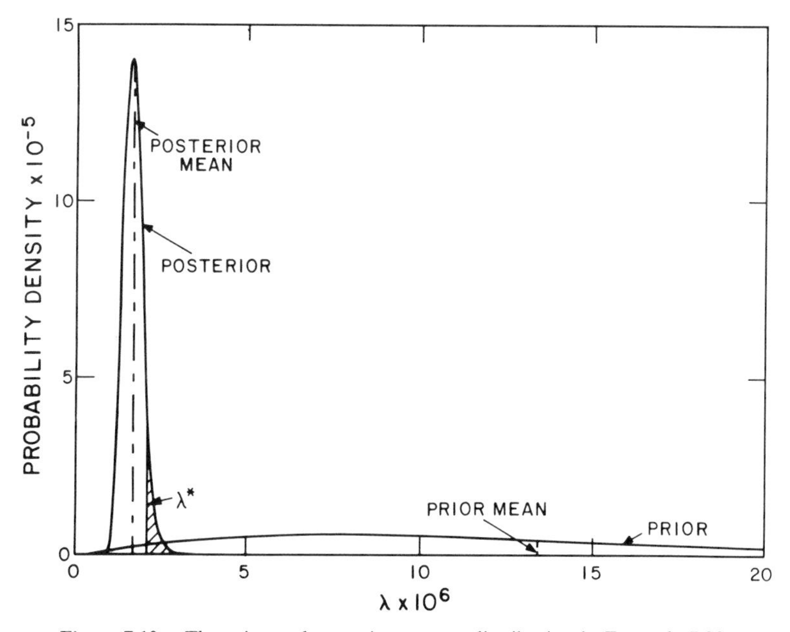

Figure 7.10. The prior and posterior gamma distribution in Example 7.20. ■

Point Estimation. The posterior mean of (7.85) is

$$E(\Lambda|s;\alpha,\beta)=\frac{s+\alpha}{\mathrm{t}+\beta}=\frac{\text{Combined Number of Failures}}{\text{Combined Total Test Time}}, \tag{7.89}$$

which is the Bayesian point estimator for λ under a squared-error loss function. The generalized maximum likelihood estimator for λ exists when $s+\alpha>1$ and is the mode of (7.85) which is $(s+\alpha-1)/(\mathrm{t}+\beta)$. As t approaches infinity, the Bayes estimator approaches the ML estimator S/t and thus, as t increases, the likelihood tends to dominate the prior.

For a squared-error loss function, the posterior risk of (7.89) is proportional to the posterior variance given by

$$\operatorname{Var}(\Lambda|s;\alpha,\beta)=\frac{s+\alpha}{(\mathrm{t}+\beta)^2}. \tag{7.90}$$

The minimum Bayes risk $r(g)$ of (7.89) is proportional to the expectation of (7.90) with respect to the marginal distribution of S given in (7.86). Thus

$$\begin{aligned} r(g)&=aE_s\left[\operatorname{Var}_{\lambda|s}(\Lambda|S;\alpha,\beta)\right]\\ &=a\sum_{s=0}^{\infty}\frac{(s+\alpha)}{(\mathrm{t}+\beta)^2}\,\frac{\beta^{\alpha}\mathrm{t}^{s}\Gamma(s+\alpha)}{\Gamma(\alpha)s!(\mathrm{t}+\beta)^{s+\alpha}}, \end{aligned}$$

which, upon simplification, becomes

$$r(g)=\frac{a}{\mathrm{t}(\mathrm{t}+\beta)}E(S;\alpha,\beta)=\frac{a\alpha}{\beta(\mathrm{t}+\beta)}. \tag{7.91}$$

Finally, the mean and variance of (7.89) are

$$E_s\left[E_{\lambda|s}(\Lambda|S;\alpha,\beta)\right]=\frac{\alpha}{\beta} \tag{7.92}$$

and

$$\operatorname{Var}_s\left[E_{\lambda|s}(\Lambda|S;\alpha,\beta)\right]=\frac{\alpha\mathrm{t}}{\beta^2(\mathrm{t}+\beta)}, \tag{7.93}$$

respectively.

It is of interest to compare the risk of (7.89) to the risk of the usual classical estimator for λ. The ML (and MVU) estimator for λ is $\hat{\lambda}=S/\mathrm{t}$ and its mean-squared error is λ/t. The Bayes risk of $\hat{\lambda}$ is obtained by averaging λ/t over the $\mathcal{G}_1(\alpha, \beta)$ prior distribution and is

$$r(g, \hat{\lambda}) = \frac{\alpha}{\beta t}.$$

Thus, the ratio of the minimum Bayes risk of (7.89) to the Bayes risk of the MVU estimator is

$$\frac{r(g)}{r(g, \hat{\lambda})} = \frac{1}{1 + \beta/t}.$$

This equation never exceeds unity since (7.89) minimizes the Bayes risk. The known comparative Bayes risk advantage of the Bayes estimator over the MVU estimator is clearly indicated in this equation, provided the assumptions underlying both estimators are valid. As the relative magnitude of β, the pseudo test time, to t, the observed test duration, increases, the Bayes estimator will enjoy an increasing risk advantage over the MVU estimator.

Example 7.21. In Example 7.20, the Bayes point estimate of λ for a squared-error loss function in which $a = 1$ is

$$E(\Lambda | s = 32; 2.45, 183{,}200) = \frac{32 + 2.45}{20{,}253{,}120 + 183{,}200} = 1.69 \times 10^{-6} \text{ f/h}$$

with associated minimum Bayes risk

$$r(g) = \frac{2.45}{(183{,}200)(20{,}436{,}320)} = 6.54 \times 10^{-13}.$$

This estimate is shown in Figure 7.10. The generalized ML estimate, using (7.85), for λ is $(33.45)/(20{,}436{,}320) = 1.64 \times 10^{-6}$ f/h. ■

Interval Estimation. Let us consider the transformed r.v. $\Theta = 2\Lambda(t + \beta)$. From (7.85), we find that the conditional distribution of Θ given s has pdf

$$g(\theta | s; \alpha, \beta) = \frac{(t + \beta)^{s+\alpha}}{\Gamma(s + \alpha)} \left(\frac{\theta}{2t + 2\beta} \right)^{s+\alpha-1} e^{-(t+\beta)\frac{\theta}{2(t+\beta)}} \left| \frac{d\lambda}{d\theta} \right|$$

$$= \frac{1}{2^{(2s+2\alpha)/2}\Gamma[(2s + 2\alpha)/2]} \theta^{(2s+2\alpha)/2-1} e^{-\theta/2}, \qquad \theta > 0,$$

(7.94)

which is recognized as a $\chi^2(2s+2\alpha)$ distribution. This observation allows us to use $\chi^2(2s+2\alpha)$ percentile values in calculating Bayesian interval estimates of λ.

From (5.45) and (5.46) a symmetric $100(1-\gamma)\%$ TBPI estimate for λ is given by

$$\begin{aligned}\Pr(\Lambda \leqslant \lambda_* | s; \alpha, \beta) &= \Pr[2\Lambda(t+\beta) \leqslant 2\lambda_*(t+\beta) | s; \alpha, \beta] \\ &= \Pr[\Theta \leqslant 2\lambda_*(t+\beta) | s; \alpha, \beta] = \frac{\gamma}{2}\end{aligned}$$

and

$$\begin{aligned}\Pr(\Lambda \geqslant \lambda^* | s; \alpha, \beta) &= \Pr[2\Lambda(t+\beta) \geqslant 2\lambda^*(t+\beta) | s; \alpha, \beta] \\ &= 1 - \Pr[\Theta \leqslant 2\lambda^*(t+\beta) | s; \alpha, \beta] = \frac{\gamma}{2}.\end{aligned}$$

Since Θ given s has a $\chi^2(2s+2\alpha)$ distribution, it follows that

$$\chi^2_{\gamma/2}(2s+2\alpha) = 2\lambda_*(t+\beta)$$

and

$$\chi^2_{1-\gamma/2}(2s+2\alpha) = 2\lambda^*(t+\beta).$$

Solving for λ_* and λ^*, the desired Bayesian interval estimate becomes

$$\lambda_* = \frac{\chi^2_{\gamma/2}(2s+2\alpha)}{2(t+\beta)} \tag{7.95}$$

and

$$\lambda^* = \frac{\chi^2_{1-\gamma/2}(2s+2\alpha)}{2(t+\beta)}. \tag{7.96}$$

Thus, it may be claimed that $\Pr(\lambda_* \leqslant \Lambda \leqslant \lambda^* | s; \alpha, \beta) = 1-\gamma$. A $100(1-\gamma)\%$ UBPI estimate for λ is also given by (7.96) in which $\gamma/2$ is replaced by γ.

Example 7.22. In Example 7.20, let us construct a 90% UBPI estimate of λ. Now $\chi^2_{0.90}(68.9) = 84.307$, from which we have

$$\lambda^* = \frac{84.307}{2(20{,}436{,}320)} = 2.06 \times 10^{-6} \text{ f/h}.$$

Thus, $\Pr(\Lambda \leq 2.06 \times 10^{-6}\ \text{f/h} \mid s=32; 2.45, 183{,}200) = 0.90$, and this estimate is also shown in Figure 7.10. ■

It is noted here that the Bayes point estimate of λ for an absolute-error loss function is the posterior median which may be computed from (7.96) by letting $\gamma = 1.0$.

Example 7.23. In Example 7.20, the median of the posterior distribution is computed to be

$$\frac{68.235}{2(20{,}436{,}320)} = 1.67 \times 10^{-6}\ \text{f/h},$$

since $\chi^2_{0.50}(68.9) = 68.235$. ■

Sensitivity to a Gamma Prior Distribution. An important consideration in Bayesian reliability analysis concerns the sensitivity of the performance of a Bayes estimator derived under an assumed prior distribution. Canavos (1975) considers the sensitivity of a $\mathcal{G}_1(\alpha, \beta)$ prior distribution for the failure rate in a Poisson sampling model. If the chosen $\mathcal{G}_1(\alpha, \beta)$ prior distribution happens to be the true prior, then the minimum Bayes risk of the Bayesian point estimator under squared-error loss is given by (7.91).

Now let us consider the risk of (7.89) when the $\mathcal{G}_1(\alpha, \beta)$ prior is *not* the true underlying prior distribution. First, let us assume that the true prior is a $\mathcal{U}(0, \lambda_1)$ distribution. Canavos (1975) shows that the risk of (7.89) when the true prior is uniform is

$$r[\mathcal{U}(0, \lambda_1), (s+\alpha)/(t+\beta)] = \frac{2\lambda_1^2\beta^2 + 3\lambda_1(t - 2\alpha\beta) + 6\alpha^2}{6(t+\beta)^2},$$

which is clearly a function of the parameters of both the assumed $\mathcal{G}_1(\alpha, \beta)$ prior and the true $\mathcal{U}(0, \lambda_1)$ prior distribution. Equivalently, the Bayes risk of the MVU estimator with respect to a $\mathcal{U}(0, \lambda_1)$ prior distribution is

$$r[\mathcal{U}(0, \lambda_1), \hat{\lambda}] = \frac{\lambda_1}{2t}.$$

In Figure 7.11 we have plotted the ratio of $r[\mathcal{U}(0, \lambda_1), (s+\alpha)/(t+\beta)]$ to $r[\mathcal{U}(0, \lambda_1), \hat{\lambda}]$ as a function of λ_1 for $\alpha = 2.45$, $\beta = 183{,}200$ and $t = 20{,}253{,}120$ h, which are the parameter values previously considered. In

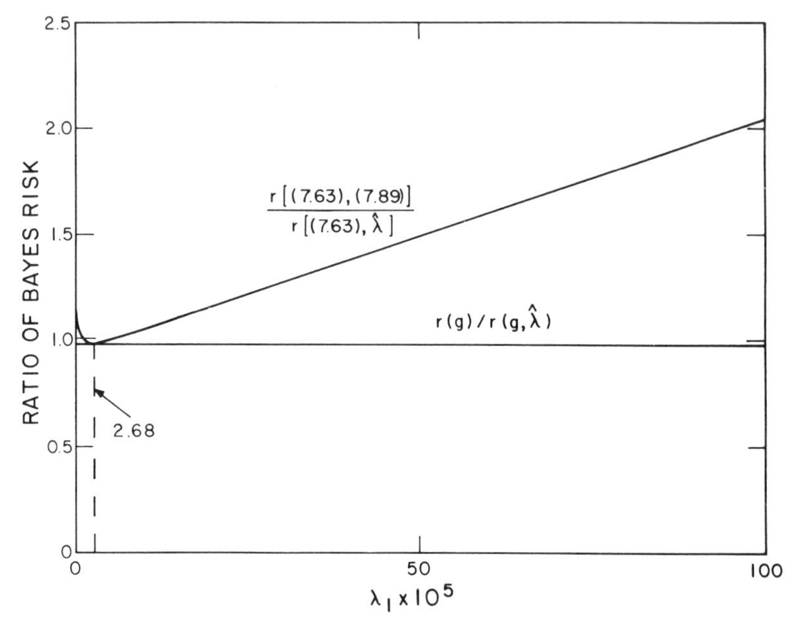

Figure 7.11. The sensitivity of an assumed gamma prior distribution when the true prior distribution is uniform $(0, \lambda_1)$.

Figure 7.11 we also plot the ratio of the minimum Bayes risk for a true $\mathcal{G}_1(\alpha, \beta)$ prior distribution (7.91) to the corresponding Bayes risk of the MVU estimator. The vertical line labeled 2.68 is the value of λ_1 which yields a prior mean of 1.34×10^{-5} f/h, which is the $\mathcal{G}_1(\alpha, \beta)$ prior mean. It is observed that, although the MVU estimator possesses nearly the same Bayes risk as the Bayes estimator due to the large value of t relative to β, the risk of (7.89) is not particularly sensitive to the $\mathcal{G}_1(\alpha, \beta)$ prior assumption, particularly when the means of the assumed and true prior distributions are essentially the same (i.e., in the vicinity of $\lambda_1 = 2.68 \times 10^{-5}$ f/h).

Canavos (1975) obtains similar results for $\mathcal{U}(0, \beta)$, $\mathcal{W}(\gamma\sqrt{2}, 2)$, $\mathcal{N}(\mu, \sigma^2)$ and extreme value prior distributions. He concludes that the risk of the Bayes point estimator under squared-error loss is more sensitive to the central tendency of λ than it is to the form of the assigned prior distribution. That is, knowledge of only a broad neighborhood of the central tendency is sufficient for the Bayes estimator to maintain risk advantage over the corresponding MVU estimator. In summary, the Bayes point estimator of the failure rate λ in a Poisson sampling model using squared-error loss is fairly robust to the form of the prior distribution, being more sensitive to the location of the prior distribution than to its form.

Higgins and Tsokos (1977) investigate the sensitivity of Bayesian point estimators for λ to incorrectly identified prior distributions. They consider prior distributions whose parameters were empirically estimated through the marginal distribution of the observed number of failures in a sequence of experiments. Numerical examples are given for a $\mathcal{G}_1(\alpha, \beta)$, $\mathcal{U}(a, b)$, and inverted uniform prior distribution on λ. Each of the three prior distributions is fitted to five sets of actual failure data taken from Schafer, et al. (1971). The various estimates are then compared using the expected relative error as a criterion, which measures how closely two researchers agree in their estimates of λ when they use different estimators. It is found that the greatest discrepancies occur for small values of s, and expected relative errors as large as 45 percent are observed. However, even incorrect prior distributions are found to yield estimators that are superior to the classical ML estimators.

Tsokos and Rao (1980) also investigate the robustness of the Bayesian point estimator of λ using squared-error loss to a $\mathcal{G}_1(\phi, \alpha)$ prior distribution on λ. In particular, they investigate the robustness with respect to $\mathcal{U}(a, b)$, $\mathcal{E}(a_2^{-1})$, $\mathcal{B}(a_3, b_3 + a_3)$, and $\mathcal{W}(a_4^{1/2}, 2)$ prior distributions on λ. They find that the Bayes estimator is quite robust to the form of the prior distribution, provided that the first two prior moments are unchanged. It is also found that the Bayes estimator is moderately insensitive to the prior parameter values themselves, and that moderate changes in the values of the prior parameters (up to 20%) induce only moderate changes (up to 20%) in the Bayes estimates. Hence they conclude that the Bayes estimator of λ is largely insensitive to the exact specification of the prior parameter values for the classes of prior distributions given above.

Higgins and Tsokos (1980) study the sensitivity of the Bayesian estimates of λ and MTTF to five classes of loss functions, including squared-error, squared relative error, linear, exponential, and a loss function proposed by Harris (1977). They find that the commonly used squared-error loss function is *not* robust compared to the other four loss functions. They also find that the Bayes estimates of λ and MTTF are quite sensitive to the loss function, being affected by such things as the symmetry or asymmetry of the loss function and the relative weights placed on overestimation and underestimation.

7.4.4 Other Prior Distributions

Various other families of prior distributions on the range $(0, \infty)$, such as the $\mathcal{LN}(\xi, \sigma^2)$ and $\mathcal{W}(\alpha, \beta)$ families, may be used as prior distributions for λ.

Log Normal Prior Distribution. The Reactor Safety Study (1975) uses $\mathcal{LN}(\xi, \sigma^2)$ prior distributions for nuclear reactor component failure rates in a Bayesian analysis and justifies the selection on the following basis. If X represents an r.v. which can vary by factors in its error, for example, having a possible range between x_0/f and $x_0 \cdot f$, where x_0 is some midpoint reference value and f is some factor, then an $\mathcal{LN}(\xi, \sigma^2)$ distribution becomes a natural candidate. The factor range x_0/f to $x_0 \cdot f$ transforms to $\log x_0 \pm \log f$ upon taking logarithms, which agrees with the range for an $\mathcal{N}(\mu, \sigma^2)$ distribution. For example, a failure rate estimated to be 10^{-6} could vary from 10^{-7} $(10^{-6}/10)$ to 10^{-5} $(10^{-6} \cdot 10)$. When, as in the Reactor Safety Study, the data are expressed in exponents of 10, the use of a $\mathcal{LN}(\xi, \sigma^2)$ prior distribution implies that the exponent is normally distributed. That is, the exponent is viewed as the significant variable in the analysis and if this exponent is normally distributed then the corresponding failure rate follows an $\mathcal{LN}(\xi, \sigma^2)$ distribution. Further, the $\mathcal{LN}(\xi, \sigma^2)$ distribution can assume a near normal-type shape or a near exponential-type shape and is thus quite flexible in its description of prior knowledge about λ.

The $\mathcal{LN}(\xi, \sigma^2)$ distribution was presented in Section 4.2.4 and the prior pdf here becomes

$$g(\lambda; \xi, \sigma^2) = \frac{1}{\sigma\lambda\sqrt{2\pi}} \exp\left[-\frac{1}{2}\left(\frac{\ln\lambda - \xi}{\sigma}\right)^2\right], \qquad \begin{array}{c} \lambda, \sigma^2 > 0, \\ -\infty < \xi < \infty, \end{array} \tag{7.97}$$

where $E(\ln \Lambda) = \xi$ and $\text{Var}(\ln \Lambda) = \sigma^2$. The prior mean and variance of Λ are given in (4.33) and (4.34), respectively. The corresponding posterior pdf of Λ given s becomes

$$g(\lambda | s; \xi, \sigma^2) = \frac{\lambda^{s-1} e^{-\lambda t - [(\ln\lambda - \xi)/\sigma]^2/2}}{\int_0^\infty \lambda^{s-1} e^{-\lambda t - [(\ln\lambda - \xi)/\sigma]^2/2} \, d\lambda}, \qquad \lambda > 0, \tag{7.98}$$

which can be numerically evaluated using the Gauss-Laguerre quadrature formulas. The posterior mean of (7.98) is

$$E(\Lambda | s; \xi, \sigma^2) = \frac{\int_0^\infty \lambda^{s} e^{-\lambda t - [(\ln\lambda - \xi)/\sigma]^2/2} \, d\lambda}{\int_0^\infty \lambda^{s-1} e^{-\lambda t - [(\ln\lambda - \xi)/\sigma]^2/2} \, d\lambda}, \tag{7.99}$$

which also must be numerically computed.

Example 7.24. Again let us consider Example 7.3. For numerical accuracy and convenience let us redefine the units on λ to be failures per 10^6 hours. Suppose we consider the prior mean and variance to be 13.4 f/10^6 h and 73.0 $f^2/10^{12}$ h^2, respectively, as in Example 7.20, and wish to fit an $\mathcal{LN}(\xi, \sigma^2)$ prior distribution. This implies that $\xi = 2.42468$ and $\sigma^2 = 0.34114$, as it is easily shown by use of (4.33) and (4.34) that, in terms of the prior mean and variance,

$$\xi = \ln(\text{Prior Mean}) - \frac{1}{2}\ln\left[1 + \frac{\text{Prior Variance}}{(\text{Prior Mean})^2}\right]$$

and

$$\sigma^2 = \ln\left[1 + \frac{\text{Prior Variance}}{(\text{Prior Mean})^2}\right].$$

Since $t = 20.253120 \times 10^6$ h, the posterior pdf becomes

$$g(\lambda \mid s=32; 2.42468, 0.58407) = \frac{\lambda^{31} e^{-20.253120\lambda - [(\ln\lambda - 2.42468)/0.58407]^2/2}}{\int_0^\infty \lambda^{31} e^{-20.253120\lambda - [(\ln\lambda - 2.42468)/0.58407]^2/2}\, d\lambda},$$

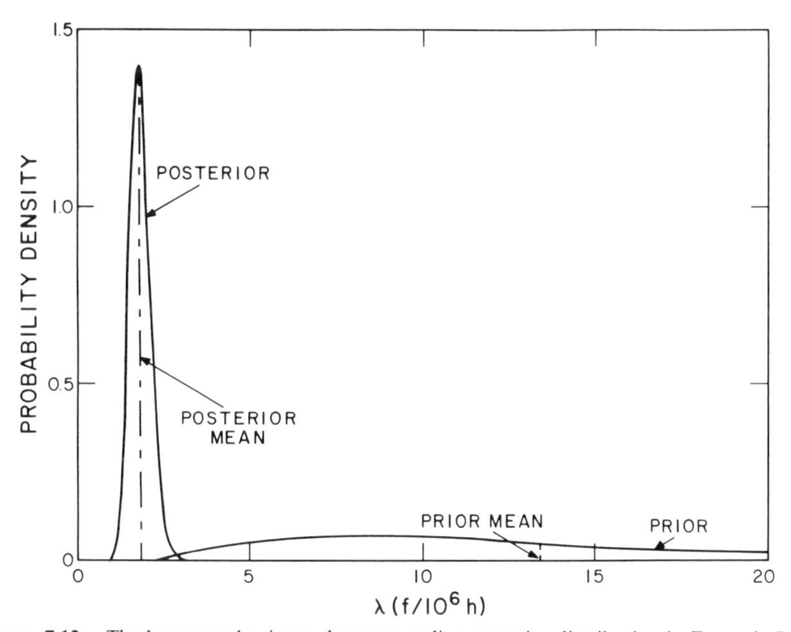

Figure 7.12. The log normal prior and corresponding posterior distribution in Example 7.24.

which is plotted in Figure 7.12 along with the prior distribution. The posterior mean is numerically computed to be 1.84 f/10^6 h, which is also shown in Figure 7.12. It is observed that the diffuse prior distribution is well dominated by the likelihood in this example, due to the large quantity of observed data. ■

Weibull Prior Distribution. The two-parameter $\mathcal{W}(\alpha, \beta)$ distribution was presented in Section 4.2.2, where its general properties were discussed. It has many of the same shape characteristics, although it does not yield a mathematically convenient posterior model, as does the $\mathcal{G}_1(\alpha, \beta)$ distribution. Recall that the $\mathcal{W}(\alpha, \beta)$ distribution has a pdf which is given by

$$g(\lambda; \alpha, \beta) = \frac{\beta}{\alpha}\left(\frac{\lambda}{\alpha}\right)^{\beta - 1} \exp\left[-\left(\frac{\lambda}{\alpha}\right)^{\beta}\right], \qquad \alpha, \beta, \lambda > 0, \quad (7.100)$$

where α and β are the scale and shape parameters, respectively. Recall that, for $\beta < 1$, the Weibull distribution is L-shaped; for $\beta = 1$, it reduces to an $\mathcal{E}(\lambda)$ distribution; and for $\beta > 1$, the Weibull distribution is unimodal and positively skewed. For $\beta = 3$, the Weibull has a near normal shape. The prior mean and variance of (7.100) are given in (4.20) and (4.21), respectively, in which $\theta = 0$.

The corresponding posterior pdf of Λ given s is given by

$$g(\lambda | s; \alpha, \beta) = \frac{\lambda^{s+\beta-1} e^{-\lambda t - (\lambda/\alpha)^{\beta}}}{\int_0^{\infty} \lambda^{s+\beta-1} e^{-\lambda t - (\lambda/\alpha)^{\beta}} \, d\lambda}, \qquad \lambda > 0, \quad (7.101)$$

which also must be numerically evaluated. The posterior mean of (7.101) is

$$E(\Lambda | s; \alpha, \beta) = \frac{\int_0^{\infty} \lambda^{s+\beta} e^{-\lambda t - (\lambda/\alpha)^{\beta}} \, d\lambda}{\int_0^{\infty} \lambda^{s+\beta-1} e^{-\lambda t - (\lambda/\alpha)^{\beta}} \, d\lambda}. \quad (7.102)$$

7.5. RELIABILITY ESTIMATION—POISSON SAMPLING

In the Poisson sampling model (7.4), the probability that no failures occur in a time period t is

$$f(0 | \lambda) = r(t) = e^{-\lambda t}, \qquad \lambda, t > 0, \quad (7.103)$$

where the reliability function is purposely denoted by $r(t)$. When λ is an r.v., the corresponding reliability function is also an r.v., which will consequently be denoted by $R(t)$ in conformance to our convention of using upper case letters to represent r.v.'s.

Since $r(t)$ is a monotonically decreasing function of λ for fixed t, this insures a one-to-one relationship between r and λ. Therefore, the unique inverse of (7.103) is given by

$$\lambda = \frac{-\ln r}{t}, \qquad 0<r>1, t>0, \tag{7.104}$$

and the pdf of $R=R(t)$, for fixed t, is given by [see e.g., Ang and Tang (1975)]

$$\begin{aligned} g_r(r) &= g_\lambda\left(\frac{-\ln r}{t}\right)\left|\frac{d\lambda}{dr}\right| \\ &= g_\lambda\left(\frac{-\ln r}{t}\right)\left(\frac{1}{rt}\right), \end{aligned} \tag{7.105}$$

where $g_r(r)$ denotes either the prior or posterior pdf of R and $g_\lambda(\lambda)$ denotes the corresponding prior or posterior pdf of Λ. Thus (7.105) can be used to determine either the prior or posterior distribution of R from the corresponding distribution of Λ.

On the other hand, suppose that the known or specified prior or posterior pdf of R is to be used to determine the corresponding implied pdf on Λ. In this case we have that

$$\begin{aligned} g_\lambda(\lambda) &= g_r(e^{-\lambda t})\left|\frac{dr}{d\lambda}\right| \\ &= g_r(e^{-\lambda t})(te^{-\lambda t}). \end{aligned} \tag{7.106}$$

We shall now consider uniform priors on both Λ and R, as well as the prior distribution on R implied by a $\mathcal{G}_1(\alpha, \beta)$ prior on Λ. In all cases, we are interested in both point and interval estimation. In the process, we shall observe some interesting facts regarding the implied relationship between the prior distributions on Λ and R.

7.5.1 Uniform Prior Distribution on the Failure Rate

Suppose that Λ has a $\mathcal{U}(0, \lambda_1)$ prior distribution. According to (7.105), the corresponding prior distribution implied on R is given by

$$g_r(r; \lambda_1) = \frac{1}{\lambda_1 tr}, \qquad e^{-\lambda_1 t}<r<1, \tag{7.107}$$

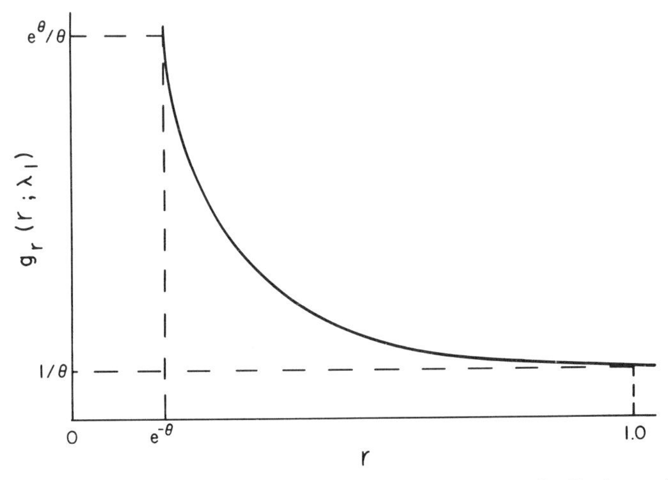

Figure 7.13. The prior distribution implied on R by a uniform prior distribution on $\lambda(\theta=\lambda_1 t)$.

which is observed to be different for different values of t. In general, (7.107) has the L-shape indicated in Figure 7.13. This illustrates that a $\mathcal{U}(0,\lambda_1)$ prior on Λ leads to a nonuniform prior on R and that sometimes seemingly innocuous assumptions, such as a uniform prior distribution on Λ, have important implications, such as a nonuniform prior distribution on R. Similarly, the posterior distribution of R, given the observed number of Poisson failures s in time t, is computed to be

$$g_r(r|s;\lambda_1)=g_\lambda\left(-\frac{\ln r}{t}\middle|s;\lambda_1\right)\left(\frac{1}{rt}\right)$$

$$=\frac{(\mathrm{t}/t)^{s+1}(-\ln r)^s r^{\mathrm{t}/t-1}}{\Gamma(s+1,\lambda_1\mathrm{t})},\qquad e^{-\lambda_1 t}<r<1. \quad (7.108)$$

Point Estimation. The posterior mean of (7.108) is the Bayesian point estimator for r under a squared-error loss function and becomes

$$\tilde{r}=E(R|s;\lambda_1)=\int_{e^{-\lambda_1 t}}^{1} r g_r(r|s;\lambda_1)\,dr. \quad (7.109)$$

It is observed that, as $t\to 0$, $E(R|s;\lambda_1)\to 1$ and also that, as $t\to\infty$, $E(R|s;\lambda_1)\to 0$.

The posterior variance is likewise given by

$$\mathrm{Var}(R|s;\lambda_1)=$$

$$\frac{1}{\Gamma(s+1,\lambda_1\mathrm{t})}\left\{\frac{\Gamma[s+1,\lambda_1(\mathrm{t}+2t)]}{(1+2t/\mathrm{t})^{s+1}}-\frac{\Gamma^2[s+1,\lambda_1(\mathrm{t}+t)]}{\Gamma(s+1,\lambda_1\mathrm{t})(1+t/\mathrm{t})^{2s+2}}\right\},$$

(7.110)

from which the posterior risk of (7.109) becomes

$$\phi(s)=a\mathrm{Var}(R|s;\lambda_1). \tag{7.111}$$

The minimum Bayes risk of (7.109) may be expressed in terms of the difference between two infinite series; namely,

$$r(g)=\frac{a}{\mathrm{t}}\left\{\sum_{s=0}^{\infty}\frac{\Gamma[s+1,\lambda_1(\mathrm{t}+2t)]}{s!(1+2t/\mathrm{t})^{s+1}}-\sum_{s=0}^{\infty}\frac{\Gamma^2[s+1,\lambda_1(\mathrm{t}+t)]}{s!\Gamma(s+1,\lambda_1\mathrm{t})(1+t/\mathrm{t})^{2s+2}}\right\},$$

(7.112)

which results upon directly taking the expectation of (7.110) with respect to the marginal distribution of S given in (7.66).

Finally, the mean and variance of (7.109) become

$$E_s(\tilde{r})=E_s\left[E_{r|s}(R|S;\lambda_1)\right]=E(R;\lambda_1)=\frac{1}{\lambda_1 t}(1-e^{-\lambda_1 t}), \tag{7.113}$$

and

$$\mathrm{Var}_s(\tilde{r})=\mathrm{Var}_s\left[E_{r|s}(R|S;\lambda_1)\right]=\frac{1}{2\lambda_1 t}(1-e^{-2\lambda_1 t})$$

$$-\frac{1}{\lambda_1^2t^2}(1-e^{-\lambda_1 t})^2-\frac{r(g)}{a}, \tag{7.114}$$

where $r(g)$ is given by (7.112).

Example 7.25. Again let us consider Example 7.3. Also suppose that the prior on Λ is a $\mathcal{U}(0,\ 10\ \mathrm{f}/10^6\ \mathrm{h})$ distribution. Further suppose that we are interested in the valve reliability for $t=10$ years $=0.0876\times10^6$ h. The prior

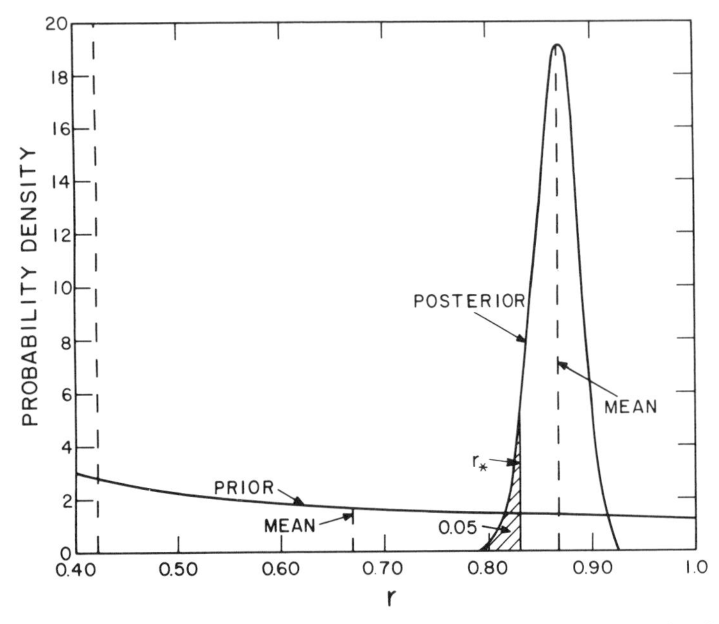

Figure 7.14. The prior and posterior distribution of reliability in Example 7.25.

and posterior distributions of R thus become

$$g_r(r;10) = \frac{1}{10(0.0876)r} = \frac{1}{0.876r}, \qquad e^{-0.876} < r < 1,$$

and

$$g_r(r \mid s=32;10) = \frac{(231.2)^{33}(-\ln r)^{32} r^{230.2}}{\Gamma(33,202.5312)}, \qquad e^{-0.876} < r < 1,$$

respectively. These distributions are shown in Figure 7.14. The posterior mean is

$$E(R \mid s=32;10) = \frac{\Gamma(33,203.4072)}{\Gamma(33,202.5312)(1.00433)^{33}} = 0.867,$$

and the posterior variance is likewise computed to be 0.0007. The prior and posterior means are also shown in Figure 7.14. ■

Interval Estimation. A Bayesian interval estimate for $r(t)$ is easily constructed from the corresponding interval estimate for λ as follows. Suppose that (λ_*, λ^*) have been computed according to (7.73) and (7.74) such that

$$\Pr(\lambda_* \leq \Lambda \leq \lambda^* | s; \lambda_1) = 1 - \gamma.$$

Since $r(t) = e^{-\lambda t}$ is a monotonically decreasing function of λ for a given value of t, we have that

$$\Pr(\lambda_* \leq \Lambda \leq \lambda^* | s; \lambda_1) = \Pr(e^{-\lambda^* t} \leq e^{-\Lambda t} \leq e^{-\lambda_* t} | s; \lambda_1)$$
$$= \Pr(e^{-\lambda^* t} \leq R(t) \leq e^{-\lambda_* t} | s; \lambda_1) = 1 - \gamma. \tag{7.115}$$

Thus, a symmetric $100(1-\gamma)\%$ TBPI estimate for $r(t)$ is given by

$$r_*(t) = e^{-\lambda^* t}, \qquad r^*(t) = e^{-\lambda_* t}, \tag{7.116}$$

where it may be claimed that $\Pr(r_* \leq R \leq r^* | s; \lambda_1) = 1 - \gamma$. One-sided Bayesian interval estimates may be calculated from the appropriate one-sided interval estimate for λ. The above discussion shows that the posterior distribution of Λ is sufficient for constructing Bayesian interval estimates for reliability due to the simple monotonic closed form relationship between λ and $r(t)$. This fact will again be exploited in Chapters 8 and 9.

Example 7.26. In the preceding example let us compute a 95% LBPI estimate of $r(t)$ for $t = 10$ years. In Example 7.18 we found that $\Pr(\Lambda \leq 2.12 \text{ f}/10^6 \text{ h} | s = 32; 10^{-5}) = 0.95$. Thus

$$r_*(0.0876) = e^{-2.12(0.0876)} = 0.83,$$

and hence $\Pr(0.83 \leq R | s = 32; 10^{-5}) = 0.95$. This interval is also shown in Figure 7.14. ■

7.5.2 Gamma Prior Distribution on the Failure Rate

If Λ has a $\mathcal{G}_1(\alpha, \beta)$ prior distribution given in (7.83), then, according to (7.105), this implies the prior distribution on R given by [see Waller, et al. (1977)]

$$g_r(r;\alpha,\beta)=\frac{\beta^\alpha}{\Gamma(\alpha)}\left(-\frac{\ln r}{t}\right)^{\alpha-1}e^{-\beta(-\ln r/t)}\left(\frac{1}{rt}\right)$$

$$=\frac{(\beta/t)^\alpha r^{\beta/t-1}(-\ln r)^{\alpha-1}}{\Gamma(\alpha)},\qquad 0<r<1,\qquad (7.117)$$

which is an $\mathcal{NLG}(\alpha,\beta/t)$ distribution.

Again it is observed that the actual prior depends upon the choice of mission time t. Figure 6.3 shows a plot of (7.117) for $\alpha=0.35$, $\beta=1/\beta_0=1/0.001=1000$, and $t=50$, 100, 500, and 4000 h. It is observed that $g_r(r;\alpha,\beta)$ can assume several different shapes. By differentiating (7.117) with respect to r, the nine possible cases for different combinations of α and $\delta=t/\beta$ values are summarized in Table 7.1.

Notice that when $0<\alpha<1$, the shape of $g_r(r;\alpha,\beta)$ changes from an increasing function of r (concentrated near 1) for $0<t/\beta\leq 1$ to a U-shaped distribution for $t/\beta>1$. Thus, a $\mathcal{G}_1(\alpha,\beta)$ prior on Λ implies a prior on R that is U-shaped for those mission times t such that $t>\beta$. This is an unexpected consequence and again illustrates that, although a $\mathcal{G}_1(\alpha,\beta)$ prior

Table 7.1[a]. Shape Properties of a Negative-Log Gamma Prior Distribution on R

δ	α		
	$0<\alpha<1$	$\alpha=1$	$\alpha>1$
$\delta>1$	U-shaped with antimode at $r=\exp\left[\frac{\delta(\alpha-1)}{\delta-1}\right]$	L-shaped or decreasing	L-shaped or decreasing
$\delta=1$	J-shaped or increasing	Uniform	L-shaped or decreasing
$0<\delta<1$	J-shaped or increasing	J-shaped or increasing	Unimodal with mode at $r=\exp\left[\frac{\delta(\alpha-1)}{\delta-1}\right]$

[a]Adapted, with permission, from *Nuclear Systems Reliability Engineering and Risk Assessment*, J. B. Fussell and G. R. Burdick, Editors.

on Λ may be desirable, the implied prior on R may be inconsistent with prior belief concerning r. This again reinforces our contention that priors must be carefully selected and the resulting implications thoroughly examined before being adopted.

Similarly, the implied posterior distribution of R is also a negative-log gamma distribution given by

$$g_r(r|s;\alpha,\beta)=\frac{[(\mathrm{t}+\beta)/t]^{s+\alpha}r^{(\mathrm{t}+\beta)/t-1}(-\ln r)^{s+\alpha-1}}{\Gamma(s+\alpha)},\qquad 0<r<1, \tag{7.118}$$

with mean and variance easily computed by successive transformations of the integrand to be

$$\tilde{r}=E(R|s;\alpha,\beta)=\left(\frac{\mathrm{t}+\beta}{\mathrm{t}+\beta+t}\right)^{s+\alpha} \tag{7.119}$$

and

$$\mathrm{Var}(R|s;\alpha,\beta)=\left(\frac{\mathrm{t}+\beta}{\mathrm{t}+\beta+2t}\right)^{s+\alpha}-\left(\frac{\mathrm{t}+\beta}{\mathrm{t}+\beta+t}\right)^{2s+2\alpha}, \tag{7.120}$$

respectively.

Point Estimation. Under a squared-error loss function, the Bayesian point estimator of r is the posterior mean given in (7.119) with associated posterior risk proportional to (7.120). The minimum Bayes risk of (7.119) is proportional to the marginal expectation of (7.120) with respect to (7.86) and may be computed as follows:

Now (5.32) and (5.33) give

$$r(g)=aE_s\left[\mathrm{Var}_{r|s}(R|S;\alpha,\beta)\right]=aE_\lambda\left\{E_{s|\lambda}\left[\mathrm{Var}_{r|s}(R|S;\alpha,\beta)\right]\right\},$$

using the laws of conditional expectation. Thus

$$r(g)=aE_\lambda\left\{E_{s|\lambda}\left(\frac{\mathrm{t}+\beta}{\mathrm{t}+\beta+2t}\right)^{s+\alpha}-E_{s|\lambda}\left(\frac{\mathrm{t}+\beta}{\mathrm{t}+\beta+t}\right)^{2s+2\alpha}\right\}$$

$$=aE_\lambda\left\{\left(\frac{\mathrm{t}+\beta}{\mathrm{t}+\beta+2t}\right)^{\alpha}E_{s|\lambda}\left(\frac{\mathrm{t}+\beta}{\mathrm{t}+\beta+2t}\right)^{s}-\left(\frac{\mathrm{t}+\beta}{\mathrm{t}+\beta+t}\right)^{2\alpha}E_{s|\lambda}\left(\frac{\mathrm{t}+\beta}{\mathrm{t}+\beta+t}\right)^{2s}\right\}.$$

Upon taking the inner expectations with respect to the Poisson sampling distribution we obtain the following

$$r(g) = a\left(\frac{\mathrm{t}+\beta}{\mathrm{t}+\beta+2t}\right)^{\alpha} E_{\Lambda}\left\{\exp\left(\frac{-2\lambda \mathrm{t}t}{\mathrm{t}+\beta+2t}\right)\right\}$$

$$-\,a\left(\frac{\mathrm{t}+\beta}{\mathrm{t}+\beta+t}\right)^{2\alpha} E_{\Lambda}\left\{\exp\left(\frac{-\lambda \mathrm{t}t[2\mathrm{t}+2\beta+t]}{[\mathrm{t}+\beta+t]^2}\right)\right\}.$$

Finally, upon taking the expectations with respect to the gamma prior distribution on Λ and simplifying, we find that the minimum Bayes risk of (7.119) becomes

$$r(g) = a\left[\frac{\beta}{\beta+2t}\right]^{\alpha} - a\left[\frac{\beta(\mathrm{t}+\beta)^2}{\mathrm{t}t(2\mathrm{t}+2\beta+t)+\beta(\mathrm{t}+\beta+t)^2}\right]^{\alpha}. \quad (7.121)$$

An alternate expression for $r(g)$ can be obtained by taking the expectation of the posterior variance directly with respect to the marginal distribution of S. Upon doing this we find that

$$r(g) = \frac{a\beta^{\alpha}}{\Gamma(\alpha)}\left[\sum_{s=0}^{\infty}\frac{\mathrm{t}^s\Gamma(s+\alpha)}{(\mathrm{t}+\beta+2t)^{s+\alpha}s!} - \sum_{s=0}^{\infty}\frac{(\mathrm{t}+\beta)^{s+\alpha}\mathrm{t}^s\Gamma(s+\alpha)}{(\mathrm{t}+\beta+t)^{2s+2\alpha}s!}\right]. \quad (7.122)$$

Finally, the mean and variance of (7.119) are

$$E_s\left[E_{r|s}(R|S;\alpha,\beta)\right] = \left(\frac{\beta}{\beta+t}\right)^{\alpha} \quad (7.123)$$

and

$$\mathrm{Var}_s\left[E_{r|s}(R|S;\alpha,\beta)\right] = \left(\frac{\beta}{\beta+2t}\right)^{\alpha} - \left(\frac{\beta}{\beta+t}\right)^{2\alpha} - \frac{r(g)}{a}, \quad (7.124)$$

where $r(g)$ is given in (7.121).

Example 7.27. Again consider Example 7.20 where we are now interested in estimating the valve reliability for $t = 10$ years. The prior and posterior distributions of R become

$$g_r(r;2.45,0.1832\times10^6)=\frac{(2.0913)^{2.45}r^{1.0913}(-\ln r)^{1.45}}{\Gamma(2.45)},\qquad 0<r<1,$$

and

$$g_r(r|s=32;2.45,0.1832\times10^6)=\frac{(233.29132)^{34.45}r^{231.29132}(-\ln r)^{33.45}}{\Gamma(34.45)},$$

$$0<r<1,$$

respectively, which are both plotted in Figure 7.15. The posterior mean and variance are

$$E(R|s=32;2.45,0.1832\times10^6)=\left(\frac{20.43632}{20.52392}\right)^{34.45}=0.86$$

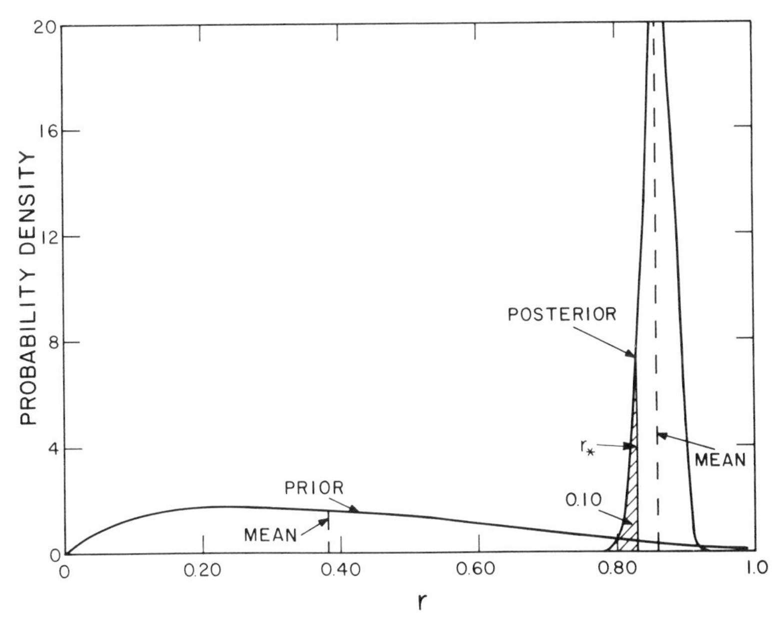

Figure 7.15. The prior and posterior negative-log gamma distribution of reliability in Example 7.27.

and

$$\mathrm{Var}\left(R \mid s=32;2.45,0.1832\times 10^{6}\right)=\left(\frac{20.43632}{20.61152}\right)^{34.45}-\left(\frac{20.43632}{20.52392}\right)^{68.9}$$

$$=0.00047.$$

The minimum Bayes risk is computed from (7.121) to be 0.00049. ■

Interval Estimation. Suppose that (λ_*, λ^*) is a symmetric $100(1-\gamma)\%$ TBPI estimate for λ given by (7.95) and (7.96). As in the preceding section, a corresponding Bayesian interval estimate for $r(t)$ is given by

$$r_*(t)=e^{-\lambda^* t}, \qquad r^*(t)=e^{-\lambda_* t}, \tag{7.125}$$

where it may be claimed that $\Pr(r_* \leqslant R \leqslant r^* \mid s;\, \alpha, \beta)=1-\gamma$. One-sided interval estimators are similarly constructed.

Example 7.28. In the preceding example let us construct a 90% LBPI estimate of $r(t)$ for $t=10$ years. In Example 7.22, λ^* was found to be 2.06 $\mathrm{f}/10^{6}$ h where $\Pr(\Lambda \leqslant 2.06 \mid s=32;2.45,0.1832)=0.90$. Thus

$$r_*(0.0876)=e^{-2.06(0.0876)}=0.83$$

and hence $\Pr(0.83 \leqslant R \mid s=32;2.45,0.1832)=0.90$. This interval is also shown in Figure 7.15. ■

Uniform Prior Distribution on *R*. Suppose now that we assume a $\mathcal{U}(0,1)$ prior distribution on R; that is

$$g_r(r)=1, \qquad 0<r<1. \tag{7.126}$$

This is a special case of the $\mathcal{NLG}(\alpha, \beta/t)$ distribution given by (7.117) in which $\alpha=1$ and $\beta=t$. Schafer (1970) points out that if $g_r(r)$ is uniform for some value of t_1 where $r_1=r(t_1)$, then this implies that there will be another distribution of $r_2=r(t_2)$, where $t_1 \neq t_2$. Thus the $\mathcal{U}(0,1)$ prior distribution must be judiciously used.

According to (7.106) it follows that the corresponding prior distribution implied on Λ will be a $\mathcal{G}_1(1,t)$ distribution. Thus, the posterior distribution of Λ will be a $\mathcal{G}_1(s+1, \mathrm{t}+t)$ distribution. It is observed that the prior distribution of Λ depends on time t. This is a consequence of the functional relationship between r and λ given by (7.103), which involves time, and may be undesirable. Similarly, if one were to consider some other prior distribution on R defined on the range (0, 1), such as a $\mathcal{B}(x_0, n_0)$ distribution, the implied prior distribution on Λ would also depend on time, which may be equally undesirable. On the other hand, a prior distribution on Λ that does not depend on time induces a prior distribution on R that properly depends on time, but that may also possess somewhat rigid, and perhaps undesirable, shape characteristics. Such was observed to be the case for the $\mathcal{NLG}(\alpha, \beta/t)$ prior distribution induced on R by a $\mathcal{G}_1(\alpha, \beta)$ prior on Λ.

Sensitivity to a Gamma Prior Distribution on Λ. Canavos (1977) investigates the robustness of (7.119) with respect to the assumed $\mathcal{G}_1(\alpha, \beta)$ prior distribution on Λ when the true prior distribution on Λ is $\mathcal{U}(0, \lambda_1)$. He considers the increase in Bayes risk as a result of the incorrect assumption. Canavos finds that, for the range of parameters considered, the Bayes risk increases approximately 20–200% as a result of the incorrect assumption. Thus, the Bayesian point estimator for r is considerably less robust to the form of the prior distribution than the Bayesian point estimator for λ [see Section 7.4.3]. However, in all cases the actual risk under the incorrect assumption is still observed to be smaller than the corresponding risk of the MVU estimator of reliability given by

$$\hat{r}(t) = \begin{cases} \left(1 - \dfrac{t}{\mathrm{t}}\right)^{S}, & \mathrm{t} > t \\ 0, & \mathrm{t} \leq t, \end{cases} \tag{7.127}$$

[see Zacks and Even (1966)].

Tsokos and Rao (1980) investigate the sensitivity of (7.119) when the true prior distribution on Λ is $\mathcal{U}(0, \lambda_1)$, $\mathcal{E}(\lambda)$, $\mathcal{B}(x_0, n_0)$, or Rayleigh. They find that, unlike the case where λ is being estimated as discussed in Section 7.4.3, the Bayes estimator of $r(t)$ is quite sensitive to the prior distribution selected. In fact, even when the first two prior moments match those of the conjugate $\mathcal{G}_1(\alpha, \beta)$ prior on Λ, the average mean squared-error is found to deviate by as much as 40% in the case of a $\mathcal{U}(0, \lambda_1)$, and by nearly 15 times in the case of an $\mathcal{E}(\lambda)$, prior distribution. They therefore conclude that, in Bayesian estimation of the reliability function corresponding to a Poisson sampling distribution, one must be very careful in selecting a prior distribution.

Higgins and Tsokos (1980) study the effect of the loss function on Bayesian estimates of reliability. As in the case of estimation of λ and

MTTF, the Bayesian estimator of $r(t)$ given by (7.119) is found to be quite sensitive to the loss function employed.

7.6. SELECTING BETA PRIOR DISTRIBUTIONS BASED ON OBSERVED BINOMIAL SAMPLING DATA

In Section 6.5 several methods were considered for determining $\mathcal{B}(x_0, n_0)$ prior distributions that were consistent with a variety of statements regarding prior belief about p. In all of these methods no attempt was made to make explicit use of any available previously observed test data from similar life test experiments as the prior was assumed to be based on degree of belief. Thus engineering judgment and experience was used to determine the prior parameter values x_0 and n_0 based on desired moments or percentiles of the prior distribution.

Suppose now that the prior distribution has a frequency interpretation. We now wish to calculate the values for x_0 and n_0 from the results of previous binomial life tests that have been conducted on the same or similar items. Consider the following situation: a sequence of N binomial life test experiments has previously been conducted in which, for the jth experiment, x_j survivors were observed out of a sample of size n_j. Further, let p_j denote the unknown reliability in the jth experiment. We assume that (1) the life test experiments are statistically independent of each other conditional on p_j and (2) the underlying sequence of reliability values $p_1, p_2, \ldots, p_N$ are statistically independent realizations of a $\mathcal{B}(x_0, n_0)$ r.v. with constant x_0 and n_0 values for the sequence of tests. Such a situation might arise when conducting routine life tests for purposes of acceptance qualification on successive production lots of the same item. As another example, the sequence of life tests may consist of the failure data results for a population of nuclear power reactors as in Example 7.1. Similarly, the sequence may contain the results of failure data on a certain device reported by different failure data reporting systems that report on various system applications of the device. Thus, we have the sequence of observed data given by $(x_1, n_1), (x_2, n_2), \ldots, (x_N, n_N)$ available for use in estimating x_0 and n_0.

Two basic methods are used to estimate x_0 and n_0: the method of matching moments and ML based on the marginal distribution. For the matching moments procedure, both weighted and unweighted sample moments are considered.

7.6.1 Method of Moments

Recall that the classical MVU estimator for the survival probability p_j is $\hat{p}_j = X_j/n_j$. Weiler (1965) presents simple moment estimators for x_0 and n_0

under the assumption that the sample size n_j is the same in each experiment. Under the assumption that the sample sizes do not vary too greatly from experiment to experiment, Copas (1972) presents simple (unweighted) moment estimators for the prior mean and variance. Let us first consider these unweighted sample moment estimators.

Unweighted Sample Moment Estimators. Now

$$E(\hat{p}_j | p_j) = p_j \tag{7.128}$$

and

$$\mathrm{Var}(\hat{p}_j | p_j) = \frac{p_j(1-p_j)}{n_j}, \tag{7.129}$$

are the conditional mean and variance of $\hat{p}_j$, conditioned upon the unknown value of p_j in the jth experiment. What we desire is the unconditional or marginal expectation and variance of $\hat{p}_j$, since p_j is unknown. Now

$$\begin{aligned} E(\hat{p}_j; x_0, n_0) &= E_{P_j}\left[E_{\hat{p}_j | p_j}(\hat{p}_j | P_j)\right] \\ &= E_{P_j}\left[P_j; x_0, n_0\right] = \frac{x_0}{n_0}, \end{aligned} \tag{7.130}$$

using the laws of conditional expectation. Also the relation between unconditional and conditional variance is given by

$$\mathrm{Var}(\hat{p}_j) = E_{P_j}\left[\mathrm{Var}_{\hat{p}_j | p_j}(\hat{p}_j | P_j)\right] + \mathrm{Var}_{P_j}\left[E_{\hat{p}_j | p_j}(\hat{p}_j | P_j)\right].$$

Substituting (7.128) and (7.129) we have

$$\begin{aligned} \mathrm{Var}(\hat{p}_j) &= E\left[\frac{P_j(1-P_j)}{n_j}\right] + \mathrm{Var}(P_j) \\ &= \frac{1}{n_j}E(P_j) - \frac{1}{n_j}E(P_j^2) + \mathrm{Var}(P_j) \\ &= \frac{1}{n_j}E(P_j) - \frac{1}{n_j}\left[\mathrm{Var}(P_j) + E^2(P_j)\right] + \mathrm{Var}(P_j) \\ &= \frac{1}{n_j}E(P_j)\left[1 - E(P_j)\right] + \left(\frac{n_j - 1}{n_j}\right)\mathrm{Var}(P_j). \end{aligned}$$

Using the fact that $E(P_j; x_0, n_0) = x_0/n_0$ and $\text{Var}(P_j; x_0, n_0) = [x_0(n_0 - x_0]/[n_0^2(n_0+1)]$, this expression reduces to

$$\text{Var}(\hat{p}_j; x_0, n_0) = \frac{x_0(n_0 - x_0)(n_0 + n_j)}{n_j^2 n_0^2 (n_0 + 1)}. \tag{7.131}$$

Now the unweighted sample mean and second sample moment about the origin of the sequence $\hat{p}_1, \ldots, \hat{p}_N$ are defined as

$$\bar{p}_u = \sum_{j=1}^{N} \frac{\hat{p}_j}{N}, \qquad m_u^2 = \sum_{j=1}^{N} \frac{\hat{p}_j^2}{N}, \tag{7.132}$$

respectively. It now follows that

$$E(\bar{P}_u; x_0, n_0) = \sum_{j=1}^{N} E(\hat{p}_j; x_0, n_0)/N = \frac{x_0}{n_0} \tag{7.133}$$

and

$$\begin{aligned} E(M_u^2; x_0, n_0) &= \sum_{j=1}^{N} E(\hat{p}_j^2; x_0, n_0)/N \\ &= \sum_{j=1}^{N} \left[\text{Var}(\hat{p}_j; x_0, n_0) + E^2(\hat{p}_j; x_0, n_0)\right]/N \\ &= \sum_{j=1}^{N} \left[\frac{x_0(n_0 - x_0)(n_0 + n_j)}{n_j^2 n_0^2 (n_0 + 1)} + \frac{x_0^2}{n_0^2}\right] \Big/ N. \end{aligned} \tag{7.134}$$

The method of moments procedure [see e.g., Mann, Schafer, and Singpurwalla (1974)] is to equate the sample moments to their expected values and to solve for the parameters x_0 and n_0. That is, a second-order match between the sample and theoretical moments is used to determine the values of the parameters that insure this match. By equating $\bar{p}_u$ to (7.133) and m_u^2 to (7.134), and solving the resulting pair of equations for x_0 and n_0, we obtain the unweighted moment estimators for x_0 and n_0 given by

$$\hat{n}_0 = \frac{N(\bar{P}_u - M_u^2)}{NM_u^2 - K\bar{P}_u - (N-K)\bar{P}_u^2}, \qquad K = \sum_{j=1}^{N} n_j^{-1}, \tag{7.135}$$

and

$$\hat{x}_0 = \hat{n}_0 \overline{P}_u. \tag{7.136}$$

It is noted that K is inversely proportional to the harmonic mean of the sample sizes n_j. These estimators reduce to those given by Weiler (1965) for the case of a constant sample size, and they provide good estimates of x_0 and n_0 when N is large.

Example 7.29. Suppose that we consider each of the eight PWRs in Example 7.1 to be an independent life test experiment. Now $\bar{p}_u = 0.985$, $m_u^2 = 0.9706$, $\bar{p}_u^2 = 0.9702$, and $K = 0.16$ from which

$$\hat{n}_0 = \frac{8(0.985 - 0.9706)}{8(0.9706) - 0.16(0.985) - (8 - 0.16)(0.9702)} = 138.46$$

and

$$\hat{x}_0 = (138.46)(0.985) = 136.38.$$

■

Weighted Sample Moment Estimators. If the difference between the values of n_j are severe, the weighted sample moment about the origin of the sequence $\hat{p}_1, \ldots, \hat{p}_N$ defined by

$$\bar{p}_w = \sum_{j=1}^{N} n_j \hat{p}_j \Big/ \sum_{j=1}^{N} n_j = \sum_{j=1}^{N} x_j / N^*, \qquad N^* = \sum_{j=1}^{N} n_j, \tag{7.137}$$

and

$$m_w^2 = \sum_{j=1}^{N} n_j \hat{p}_j^2 / N^*, \tag{7.138}$$

respectively, should be used in the method of moments procedure. Now

$$E(\overline{P}_w; x_0, n_0) = \sum_{j=1}^{N} E(X_j; x_0, n_0)/N^* = \sum_{j=1}^{N} \left(\frac{n_j x_0}{n_0} \right) \Big/ N^* = \frac{x_0}{n_0} \tag{7.139}$$

and

$$E(M_w^2; x_0, n_0) = \sum_{j=1}^{N} n_j E(\hat{p}_j^2; x_0, n_0)/N^*$$

$$= \sum_{j=1}^{N} n_j\left[\mathrm{Var}(\hat{p}_j; x_0, n_0) + E^2(\hat{p}_j; x_0, n_0)\right]/N^*$$

$$= \sum_{j=1}^{N}\left[\frac{x_0(n_0 - x_0)(n_0 + n_j)}{n_j n_0^2(n_0+1)} + \frac{n_j x_0^2}{n_0^2}\right]\bigg/ N^*. \quad (7.140)$$

Equating (7.139) and (7.140) to $\bar{p}_w$ and m_w^2, respectively, and solving for x_0 and n_0, yields the weighted moment estimators for x_0 and n_0 given by

$$\hat{n}_0 = \frac{N\bar{P}_w(1-\bar{P}_w) - N^*(M_w^2 - \bar{P}_w^2)}{N^* M_w^2 - K\bar{P}_w - (N^* - K)\bar{P}_w^2} \quad (7.141)$$

and

$$\hat{x}_0 = \hat{n}_0 \bar{P}_w. \quad (7.142)$$

If $n_j = n$, $j = 1,2,\ldots,N$, then it is easily shown that (7.141) reduces to the unweighted form in (7.135).

7.6.2 Marginal Maximum Likelihood

The method of marginal ML yields those estimators of x_0 and n_0 that maximize the marginal likelihood function. The marginal distribution of X_j, $j = 1,2,\ldots,N$ is given in (7.33) as

$$f(x_j, x_0, n_0) = \frac{n_j!\Gamma(n_0)\Gamma(x_j + x_0)\Gamma(n_j + n_0 - x_j - x_0)}{x_j!(n_j - x_j)!\Gamma(n_j + n_0)\Gamma(x_0)\Gamma(n_0 - x_0)},$$

and the marginal likelihood function $L(x_0, n_0)$ is the product of these. Thus

$$L = L(x_0, n_0) = \prod_{j=1}^{N} \frac{n_j!\Gamma(n_0)\Gamma(x_j + x_0)\Gamma(n_j + n_0 - x_j - x_0)}{x_j!(n_j - x_j)!\Gamma(n_j + n_0)\Gamma(x_0)\Gamma(n_0 - x_0)}. \quad (7.143)$$

This quantity is to be maximized by suitable choice of x_0 and n_0 or equivalently, $\ln L$ can be maximized. Hence we consider the more convenient form

$$\ln L = \sum_{j=1}^{N} \ln n_j! + N \ln \Gamma(n_0) + \sum_{j=1}^{N} \ln \Gamma(x_j + x_0)$$
$$+ \sum_{j=1}^{N} \ln \Gamma(n_j + n_0 - x_j - x_0) - \sum_{j=1}^{N} \ln x_j! - \sum_{j=1}^{N} \ln(n_j - x_j)!$$
$$- \sum_{j=1}^{N} \ln \Gamma(n_j + n_0) - N \ln \Gamma(x_0) - N \ln \Gamma(n_0 - x_0). \qquad (7.144)$$

Differentiation of (7.144) with respect to x_0 yields

$$\frac{\partial \ln L}{\partial x_0} = \sum_{j=1}^{N} \psi(x_j + x_0) - \sum_{j=1}^{N} \psi(n_j + n_0 - x_j - x_0)$$
$$- N\psi(x_0) + N\psi(n_0 - x_0), \qquad (7.145)$$

where $\psi(z)$ is the psi (digamma) function defined as

$$\psi(z) = \frac{d[\ln \Gamma(z)]}{dz} = \frac{\Gamma'(z)}{\Gamma(z)}. \qquad (7.146)$$

Using the well-known recurrence formulas for the psi function [Abramowitz and Stegun (1965), p. 258], (7.145) can be written as

$$\frac{\partial \ln L}{\partial x_0} = \sum_{j=1}^{N} \sum_{i=0}^{x_j - 1} \left(\frac{1}{x_0 + i} \right) - \sum_{j=1}^{N} \sum_{i=0}^{n_j - x_j - 1} \left(\frac{1}{n_0 - x_0 + i} \right), \qquad n_j > x_j \geqslant 1.$$
$$(7.147)$$

If either $x_j = 0$ or $x_j = n_j$, the corresponding summation on i in (7.147) is defined to be zero. Differentiating (7.144) with respect to n_0 yields

$$\frac{\partial \ln L}{\partial n_0} = N\psi(n_0) + \sum_{j=1}^{N} \psi(n_j + n_0 - x_j - x_0)$$
$$- \sum_{j=1}^{N} \psi(n_j + n_0) - N\psi(n_0 - x_0). \qquad (7.148)$$

Again, using the psi function recurrence formulas, this expression can be written as

$$\frac{\partial \ln L}{\partial n_0} = \sum_{j=1}^{N} \sum_{i=0}^{n_j - x_j - 1} \left(\frac{1}{n_0 - x_0 + i} \right) - \sum_{j=1}^{N} \sum_{i=0}^{n_j - 1} \left(\frac{1}{n_0 + i} \right), \qquad n_j > x_j. \tag{7.149}$$

Equating (7.147) and (7.149) to zero yields the pair of likelihood equations given by

$$\sum_{j=1}^{N} \sum_{i=0}^{x_j - 1} \left(\frac{1}{x_0 + i} \right) - \sum_{j=1}^{N} \sum_{i=0}^{n_j - x_j - 1} \left(\frac{1}{n_0 - x_0 + i} \right) = 0 \tag{7.150}$$

and

$$\sum_{j=1}^{N} \sum_{i=0}^{n_j - x_j - 1} \left(\frac{1}{n_0 - x_0 + i} \right) - \sum_{j=1}^{N} \sum_{i=0}^{n_j - 1} \left(\frac{1}{n_0 + i} \right) = 0, \tag{7.151}$$

which, when solved simultaneously, yield the maximum likelihood estimates $\hat{x}_0$ and $\hat{n}_0$. Starting values for the numerical solution may be taken to be either the weighted or unweighted method of moments estimates.

Example 7.30. Again let us consider each of the eight PWRs in Example 7.1 to be an independent life test experiment. Using $\hat{x}_0 = 136.38$ and $\hat{n}_0 = 138.46$ as starting values, (7.150) and (7.151) are iteratively solved by means of a nonlinear root solving code to yield the final marginal ML estimates $\hat{x}_0 = 136.36$ and $\hat{n}_0 = 138.46$, which are in excellent agreement with the method of moments estimates. ■

7.7. SELECTING GAMMA PRIOR DISTRIBUTIONS BASED ON OBSERVED POISSON SAMPLING DATA

Several methods were presented in Section 6.6 for fitting $\mathcal{G}(\alpha, \beta)$ and $\mathcal{G}_1(\alpha, \beta)$ prior distributions based on information derived from engineering judgment and experience. The prior parameter values α and β were determined from information regarding either the moments or percentiles of the prior distribution.

Let us now suppose that the gamma prior distribution has a frequency interpretation and consider the estimation of α and β from objective test results from a sequence of tests on the same or similar item. Consider the following situation: a sequence of N Poisson sampling experiments has previously been conducted in which, for the jth experiment, s_j failures were observed in total test time t_j. Further let λ_j denote the unknown failure rate in the jth experiment. We assume that (1) the experiments, conditional on λ_j, are statistically independent of each other, and (2) the underlying sequence of failure rates is a statistically independent realization of a $\mathcal{G}_1(\alpha, \beta)$ r.v. with constant α and β values for the sequence of tests. Thus, the sequence of observed data given by $(s_1, t_1), (s_2, t_2), \ldots, (s_N, t_N)$ is available for use in estimating α and β.

As in the preceding section, two basic methods are used to estimate α and β: the method of matching moments and ML based on the marginal distribution. Both weighted and unweighted sample moments are considered in the matching moments procedure. Tillman and Grosh (1979) have developed a computer program, called GAMMA, that uses five methods to estimate α and β: weighted and unweighted matching moments to the prior, weighted and unweighted matching moments to the marginal, and marginal ML. The shape parameter α can also be restricted to be greater than one for the ML method. In addition, the program computes and plots the fitted gamma distribution using the parameter estimates of α and β for each method.

7.7.1 Method of Moments

The classical MVU estimator for the failure rate λ_j is $\hat{\lambda}_j = S_j / t_j$. We begin by considering the unweighted procedure.

Unweighted Sample Moment Estimators. Now

$$E(\hat{\lambda}_j | \lambda_j) = t_j^{-1} E(S_j | \lambda_j) = \lambda_j \tag{7.152}$$

and

$$\mathrm{Var}(\hat{\lambda}_j | \lambda_j) = t_j^{-2} \mathrm{Var}(S_j | \lambda_j) = \lambda_j / t_j, \tag{7.153}$$

are the conditional mean and variance of $\hat{\lambda}_j$ given λ_j. As the failure rate λ_j is unknown, the observed $\hat{\lambda}_j$ values are observations from the marginal (unconditional) distribution and we desire the unconditional mean and variance of $\hat{\lambda}_j$. From (7.87) and (7.88) it follows that

$$E(\hat{\lambda}_j; \alpha, \beta) = t_j^{-1} E(S_j; \alpha, \beta) = \frac{\alpha}{\beta}, \tag{7.154}$$

and

$$\mathrm{Var}(\hat{\lambda}_j; \alpha, \beta) = t_j^{-2} \mathrm{Var}(S_j; \alpha, \beta) = \frac{\alpha(t_j + \beta)}{\beta^2 t_j}. \tag{7.155}$$

Now the unweighted sample mean and second sample moment about the origin of the sequence $\hat{\lambda}_1, \ldots, \hat{\lambda}_N$ are defined as

$$\bar{\lambda}_u = \sum_{j=1}^{N} \frac{\hat{\lambda}_j}{N}, \qquad m_u^2 = \sum_{j=1}^{N} \frac{\hat{\lambda}_j^2}{N}, \tag{7.156}$$

respectively. It follows that

$$E(\bar{\Lambda}_u; \alpha, \beta) = \sum_{j=1}^{N} \frac{E(\hat{\lambda}_j; \alpha, \beta)}{N} = \frac{\alpha}{\beta} \tag{7.157}$$

and

$$E(M_u^2; \alpha, \beta) = \sum_{j=1}^{N} E(\hat{\lambda}_j^2; \alpha, \beta)/N = \sum_{j=1}^{N} \left[\mathrm{Var}(\hat{\lambda}_j; \alpha, \beta) + E^2(\hat{\lambda}_j; \alpha, \beta)\right]/N$$

$$= \sum_{j=1}^{N} \left[\frac{\alpha(t_j + \beta)}{\beta^2 t_j} + \frac{\alpha^2}{\beta^2}\right] \Big/ N$$

$$= \frac{1}{N} \sum_{j=1}^{N} \left\{\frac{\alpha[t_j(1+\alpha) + \beta]}{\beta^2 t_j}\right\}. \tag{7.158}$$

By equating $\bar{\lambda}_u$ to (7.157) and m_u^2 to (7.158), and solving the resulting pair of equations for α and β, we obtain the unweighted moment estimators for α and β given by

$$\hat{\alpha} = \frac{N\bar{\Lambda}_u^2}{N(M_u^2 - \bar{\Lambda}_u^2) - H\bar{\Lambda}_u}, \qquad H = \sum_{j=1}^{N} t_j^{-1}, \tag{7.159}$$

and

$$\hat{\beta} = \frac{\hat{\alpha}}{\overline{\Lambda}_u} = \frac{N\overline{\Lambda}_u}{N\left(M_u^2 - \overline{\Lambda}_u^2\right) - H\overline{\Lambda}_u}. \tag{7.160}$$

It is noted that H is inversely proportional to the harmonic mean of the total test times t_j. These estimators reduce to those of Tillman and Grosh (1979) if the sample variance is used in place of the second sample moment about the origin.

Example 7.31. Again let us consider each of the eight PWRs in Example 7.1 to be an independent life test experiment. The column headed "$\mathrm{t}(h)$" gives the total test time (in h) during which the s_j failures occurred. Now $\overline{\lambda}_u = 1.7 \times 10^{-6}$ f/h, $\overline{\lambda}_u^2 = 2.93 \times 10^{-12}$, $m_u^2 = 7.82 \times 10^{-12}$, and $H = 1.83 \times 10^{-5}$ from which

$$\hat{\alpha} = \frac{8(2.93 \times 10^{-12})}{8(7.82 \times 10^{-12} - 2.93 \times 10^{-12}) - (1.83 \times 10^{-5})(1.7 \times 10^{-6})} = 2.93$$

and

$$\hat{\beta} = \frac{2.93}{1.7 \times 10^{-6}} = 1{,}723{,}529 \text{ h}.$$

■

Weighted Sample Moment Estimators. If the t_j values are severely different, the weighted sample mean and second sample moment defined by

$$\overline{\lambda}_w = \frac{\sum_{j=1}^{N} \mathrm{t}_j \hat{\lambda}_j}{\sum_{j=1}^{N} \mathrm{t}_j} = \frac{\sum_{j=1}^{N} s_j}{\mathrm{t}^*}, \qquad \mathrm{t}^* = \sum_{j=1}^{N} \mathrm{t}_j, \tag{7.161}$$

and

$$m_w^2 = \frac{\sum_{j=1}^{N} \mathrm{t}_j \hat{\lambda}_j^2}{\mathrm{t}^*}, \tag{7.162}$$

respectively, should be used in the method of moments procedure. Now

$$E(\bar{\Lambda}_w; \alpha, \beta) = \frac{\sum_{j=1}^{N} E(S_j; \alpha, \beta)}{t^*} = \frac{\sum_{j=1}^{N} (\alpha t_j/\beta)}{t^*} = \frac{\alpha}{\beta} \tag{7.163}$$

and

$$E(M_w^2; \alpha, \beta) = \sum_{j=1}^{N} t_j \left[\operatorname{Var}(\hat{\lambda}_j; \alpha, \beta) + E^2(\hat{\lambda}_j; \alpha, \beta)\right] / t^*$$

$$= \sum_{j=1}^{N} t_j \left[\frac{\alpha(t_j + \beta)}{\beta^2 t_j} + \frac{\alpha^2}{\beta^2}\right] \Big/ t^* = \sum_{j=1}^{N} \left\{\frac{\alpha[t_j(1+\alpha)+\beta]}{\beta^2}\right\} \Big/ t^*. \tag{7.164}$$

From (7.161)–(7.164), and using the same procedure as before, the weighted moment estimators for α and β become

$$\hat{\alpha} = \frac{\bar{\Lambda}_w^2 t^*}{t^*(M_w^2 - \bar{\Lambda}_w^2) - N\bar{\Lambda}_w} \tag{7.165}$$

and

$$\hat{\beta} = \frac{\hat{\alpha}}{\bar{\Lambda}_w}, \tag{7.166}$$

which are the same as the estimators given by Tillman and Grosh (1979).

7.7.2 Marginal Maximum Likelihood

The marginal distribution of S_j, $j = 1, 2, \ldots, N$ is given in (7.86) as

$$f(s_j; \alpha, \beta) = \frac{\beta^\alpha t_j^{s_j} \Gamma(s_j + \alpha)}{\Gamma(\alpha) s_j! (t_j + \beta)^{s_j + \alpha}},$$

and the marginal likelihood function $L(\alpha, \beta)$ is the product of these; namely,

$$L = L(\alpha, \beta) = \prod_{j=1}^{N} \frac{\beta^\alpha t_j^{s_j} \Gamma(s_j + \alpha)}{\Gamma(\alpha) s_j! (t_j + \beta)^{s_j + \alpha}}. \tag{7.167}$$

As in Section 7.6.2 we consider the values of α and β that maximize $\ln L$, which are the ML estimators of α and β, respectively. Thus

$$\ln L = N\alpha \ln \beta + \sum_{j=1}^{N} s_j \ln \mathrm{t}_j + \sum_{j=1}^{N} \ln \Gamma(s_j + \alpha)$$

$$- N \ln \Gamma(\alpha) - \sum_{j=1}^{N} \ln s_j! - \sum_{j=1}^{N} (s_j + \alpha) \ln(\mathrm{t}_j + \beta). \quad (7.168)$$

Differentiating (7.168) with respect to α yields

$$\frac{\partial \ln L}{\partial \alpha} = N \ln \beta + \sum_{j=1}^{N} \left[\psi(s_j + \alpha) - \psi(\alpha)\right] - \sum_{j=1}^{N} \ln(\mathrm{t}_j + \beta)$$

$$= N \ln \beta + \sum_{j=1}^{N} \sum_{i=0}^{s_j - 1} \left(\frac{1}{\alpha + i}\right) - \sum_{j=1}^{N} \ln(\mathrm{t}_j + \beta), \qquad s_j \geqslant 1,$$

$$(7.169)$$

by means of the well-known recurrence formulas for the psi function. If $s_j = 0$, the corresponding summation on i in (7.169) is defined to be zero.

Similarly,

$$\frac{\partial \ln L}{\partial \beta} = \frac{N\alpha}{\beta} - \sum_{j=1}^{N} \left(\frac{s_j + \alpha}{\mathrm{t}_j + \beta}\right). \quad (7.170)$$

Equating (7.169) and (.170) to zero yields the pair of likelihood equations

$$\alpha = \frac{\beta \sum_{j=1}^{N} \left[s_j / (\mathrm{t}_j + \beta)\right]}{N - \beta \sum_{j=1}^{N} \left[1 / (\mathrm{t}_j + \beta)\right]}, \quad (7.171)$$

and

$$N \ln \beta + \sum_{j=1}^{N} \sum_{i=0}^{s_j - 1} \left(\frac{1}{\alpha + i}\right) - \sum_{j=1}^{N} \ln(\mathrm{t}_j + \beta) = 0, \quad (7.172)$$

which must be iteratively solved for the maximum likelihood estimates $\hat{\alpha}$

and $\hat{\beta}$. Either the weighted or unweighted method of moments estimates may be used as starting values.

Example 7.32. Again let us consider Example 7.1. Using $\hat{\beta} = 1{,}723{,}529$ as a starting value in (7.171) we obtain $\hat{\alpha} = 2.952$ and $\hat{\beta} = 1{,}725{,}175$ as the final ML estimates of α and β. A nonlinear root solving code was used to solve (7.172) for $\hat{\beta}$. It is observed that the ML solution does not differ significantly from the method of moments solution. ■

EXERCISES

7.1. Suppose ten 60 W light bulbs are placed on a 1000 h life test. Of the ten there were five survivors (successes); that is, $x = 5$. Assume that the prior model selected for the survival probability P is a $\mathcal{B}(3, 5)$ distribution.

(a) Find the posterior pdf of P.

(b) Graph the prior and posterior pdf's and compare.

(c) Compute the usual Bayes point estimate of p; that is, the posterior mean of R. Compute a 95% TBPI estimate of p.

(d) Compute the classical estimate of p. Compute a 95% TCI estimate of p.

(e) Compare the limits in (c) and (d).

(f) Compute the minimum Bayes risk of the Bayes estimate in (c).

(g) Compute the mean and variance of the Bayes estimator for the given data.

7.2. Work Exercise 7.1 using a $\mathcal{B}(60, 100)$ prior distribution on P.

7.3. Determine the posterior variance of the beta distribution in (7.32) and discuss it for large values of n.

7.4. Assuming Pascal sampling and a $\mathcal{B}(x_0, n_0)$ prior distribution, the posterior mean of reliability is given by (7.57). Show that as $n - s$ tends to infinity the posterior mean approaches the classical estimate, $(n - s)/n$, of the proportion of survivors (successes).

7.5. Show that the mean of (7.57) is x_0/n_0.

7.6. The Reactor Safety Study (1975) reported the following data on the number of pump failures observed in 1972 in nine BWRs. Out of 450 pumps in the nine BWRs there were a total of 18 failures. Assume a $\mathcal{U}(0,1)$ prior for the parameter P.

(a) Find the posterior distribution of the annual BWR pump reliability.

(b) Find the Bayes point estimate of the annual BWR pump reliability, assuming a squared-error loss function with $a=1$.

(c) Find the associated Bayes risk.

(d) Find a 95% LBPI estimate of p.

(e) Suppose it is known that the annual BWR pump reliability is no less than 0.94; that is, the prior is a truncated uniform with $p_0=0.94$ and $p_1=1$. Find the posterior distribution of P. Find the posterior mean of P. Find the 90% LBPI estimate of p. Find the 95% LBPI estimate of p.

7.7. Assuming that the prior for P in Exercise 7.6 is $\mathcal{B}(25,30)$.

(a) Find the posterior distribution of P.

(b) Find the Bayes point estimate of p.

(c) Find the associated Bayes risk.

(d) Find a 95% TBPI estimate of p.

7.8. Suppose a company that manufactures a very expensive device is interested in studying the reliability of the device. The device is tested, resulting in either a survival or a failure, in which case the device is destroyed. In order to control the number of devices destroyed, the company decides to test the devices sequentially until there are three failures. The test is conducted and the third failure occurs on the fifteenth trial. Previously collected data suggests the use of a $\mathcal{B}(45,60)$ prior distribution.

(a) Find the posterior distribution of P.

(b) Find the posterior mean.

(c) Find the posterior variance.

(d) Find the minimum Bayes risk of the Bayes estimate in (b).

(e) Find the generalized ML estimate of p.

(f) Find a symmetric 95% TBPI estimate of p.

7.9. Show that the posterior mean and variance of (7.65) are given by (7.67) and (7.68), respectively.

7.10. Show that the Bayes point estimator of reliability in the case of Poisson sampling, given in (7.109), may be obtained directly by taking the expectation of $R(t)=e^{-\Lambda t}$ with respect to the posterior distribution $g(\lambda|s;\lambda_1)$ given in (7.65).

7.11. Show that (7.87) and (7.88) are true. (Hint: Recognize that the marginal distribution of S is a negative binomial distribution).

7.12. Determine the noninformative prior for reliability in the case of a Poisson sampling model.

7.13. If Λ has a noninformative prior distribution, show that the implied prior distribution on $R = e^{-\Lambda t}$ is a $\mathcal{NLG}_1(0.5, 0)$ distribution. Is this a proper prior distribution?

7.14. Show the equivalence of (7.121) and (7.122).

7.15. Show that the posterior risk of (7.67) is given by (7.69).

7.16. Derive (7.79) and (7.80).

7.17. Determine the noninformative prior for the survival probability P based on Pascal sampling. Compare this prior to the noninformative prior for P based on binomial sampling.

REFERENCES

Abramowitz, M. and Stegun, I. A. (1965). *Handbook of Mathematical Functions*, National Bureau of Standards Applied Mathematics Series· 55, U.S. Government Printing Office, Washington, DC.

Amos, D. E. and Daniel, S. L. (1969). *Significant Digit Incomplete Beta Ratios*, SC-DR-69-591. Sandia Laboratories, Albuquerque, NM.

Amos, D. E. and Daniel, S. L. (1972). *Significant Digit Incomplete Gamma Ratios*, SC-DR-72-0303. Sandia Laboratories, Albuquerque, NM.

Ang, A. H. and Tang, W. H. (1975). *Probability Concepts in Engineering Planning and Design, Vol. 1 — Basic Principles*, Wiley, New York.

Apostolakis, G. and Mosleh, A. (1979). Expert Opinion and Statistical Evidence: An Application to Reactor Core Melt Frequency, *Nuclear Science and Engineering*, Vol. 70, pp. 135–149.

Bartlett, M. S. (1937). Properties of Sufficiency and Statistical Test, *Proceedings of the Royal Statistical Society, Series A*, Vol. 160, p. 268.

Box, G. E. P. and Tiao, G. C. (1973). *Bayesian Inference in Statistical Analysis*, Addison-Wesley, Reading, MA.

Canavos, G. C. (1975). Bayesian Estimation: A Sensitivity Analysis, *Naval Research Logistics Quarterly*, Vol. 22, pp. 543–552.

Canavos, G. C. (1977). Robustness and the Prior Distribution in the Bayesian Model, *The Theory and Applications of Reliability, Vol. II*, Academic, New York, pp. 173–180.

Chiu, W. K. (1974). A New Prior Distribution for Attributes Sampling, *Technometrics*, Vol. 16, pp. 93–102.

Copas, J. B. (1972). Empirical Bayes Methods and the Repeated Use of a Standard, *Biometrika*, Vol. 59, pp. 349–360.

Grohowski, G., Hausman, W. C., and Lamberson, L. R. (1976). A Bayesian Statistical Inference Approach to Automotive Reliability Estimation, *Journal of Quality Technology*, Vol. 8, pp. 197–208.

Harris, B. (1977). A Survey of Statistical Methods in Systems Reliability Using Bernoulli Sampling of Components, *The Theory and Applications of Reliability, Vol. II*, Academic, New York, pp. 275–297.

Higgins, J. J. and Tsokos, C. P. (1977). Comparison of Bayesian Estimates of Failure Intensity for Fitted Priors of Life Data, *The Theory and Applications of Reliability, Vol. II*, Academic, New York, pp. 75–92.

Higgins, J. J. and Tsokos, C. P. (1980). A Study of the Effect of the Loss Function on Bayes Estimates of Failure Intensity, MTBF, and Reliability, *Applied Mathematics and Computations*, Vol. 6, pp. 145–166.

Mann, N. R., Schafer, R. E., and Singpurwalla, N. D. (1974). *Methods for Statistical Analysis of Reliability and Life Data*, Wiley, New York.

Papadopoulos, A. and Rao, A. N. V. (1977). Bayesian Confidence Bounds for the Poisson Failure Model, *The Theory and Applications of Reliability, Vol. II*, Academic, New York, pp. 107–121.

Pearson, K. (1934). *Tables of the Incomplete Beta-Function*, Cambridge University Press, London.

Raiffa, H. and Schlaifer, R. (1961). *Applied Statistical Decision Theory*, Harvard University Press, Cambridge, MA.

Reactor Safety Study, Appendix III — Failure Data (1975). WASH-1400 (NUREG-75/014), U.S. Nuclear Regulatory Commission.

Schafer, R. E. (1970). A Note on the Uniform Prior Distribution for Reliability, *IEEE Transactions on Reliability*, Vol. R-19, pp. 76–77.

Schafer, R. E. et al. (1971). *Bayesian Reliability Demonstration, Phase II — Development of A Priori Distribution*, RADC-TR-71-209, Rome Air Development Center, Rome, NY.

Tillman, F. A. and Grosh, D. L. (1979). *A Discussion and Users Guide to the Program GAMMA* (Draft Report), Department of Industrial Engineering, Kansas State University, Manhattan, KS.

Tsokos, C. P. and Rao, A. N. V. (1980). Sensitivity Analysis of the Bayesian Estimators of the Intensity Parameter and Reliability Function of the Poisson Failure Model, to appear in *Statistica*.

Waller, R. A., Johnson, M. M., Waterman, M. S. and Martz, H. F. (1977). Gamma Prior Distribution Selection for Bayesian Analysis of Failure Rate and Reliability, *Nuclear Systems Reliability Engineering and Risk Assessment*, The Society for Industrial and Applied Mathematics, Philadelphia, PA, pp. 584–606.

Weiler, H. (1965). The Use of Incomplete Beta Functions for Prior Distributions in Binomial Sampling, *Technometrics*, Vol. 7 , pp. 335–347.

Zacks, S. and Even, M. (1966). The Efficiencies in Small Samples of the Maximum Likelihood and Best Unbiased Estimators of Reliability Functions, *Journal of the American Statistical Association*, Vol. 61, pp. 1033–1051.

CHAPTER 8

The Exponential Distribution

The purpose of this chapter is to show how certain Bayesian estimation concepts discussed in Chapter 5 can be applied to life test data from the $\mathcal{E}(\lambda, \mu)$ distribution that was discussed in Section 4.2.1. Bayesian point and interval estimation procedures will be given for various parameters related to both the $\mathcal{E}(\lambda)$ and the $\mathcal{E}(\lambda, \mu)$ distribution.

Section 8.1 reviews the basic types of exponential sampling (with corresponding sampling distributions) typically used in Bayesian estimation. Section 8.2 is concerned with Bayesian estimation of the constant failure rate λ, for the $\mathcal{E}(\lambda)$ distribution, under a variety of prior distributions and sampling plans. Similarly, Sections 8.3, 8.4, and 8.5 deal respectively with estimation of the MTTF $\theta = 1/\lambda$, the reliability function $R(t) = e^{-\lambda t}$, and the reliable life $t_R = -(\ln R)/\lambda$. Note that θ, $R(t)$ and t_R are all one-to-one functions of the failure rate λ. The choice of reliability measure to be estimated depends on how available prior information can be most accurately expressed and on the conclusions desired from the life test experiment.

Example 8.1. In this example we present situations where each of the four reliability measures considered in this chapter would be the one of most interest.

1. A large plant uses a number of identical, repairable mechanisms that function independently. In order to forecast production rates, the management needs to know the frequency with which the units are likely to break down, causing delays in production. In this case the *failure rate* would likely be the quantity of most interest.

2. Refer to 1 above. Suppose that each mechanism contains an expensive subassembly that must be completely replaced upon failure, and that management is attempting to forecast maintenance costs over the next 5 years. In this situation knowledge of the *MTTF* would be useful.

3. In an attempt to remain competitive, a major United States auto manufacturer considers increasing the warranty period on major drivetrain components to 3 years or 36,000 miles, whichever comes first. A major factor in the decision is the predicted cost of warranty repairs to the manufacturer, which is directly proportional to the percentage of cars requiring warranty work. The parameter of most interest here would likely be the *reliability*, $R(t)$.

4. Refer to 3 above. Suppose that the manufacturer would like to determine the warranty period in order to achieve a certain desired repair cost, which can be equated to a desired proportion of autos requiring warranty work. For example, the manufacturer may wish to find the warranty period that leads to the expectation that only 5% of all autos sold will require warranty work. In this case, an estimate of *reliable life* t_R would be most appropriate. ■

Section 8.6 gives a brief treatment of Bayesian point and interval estimation when sampling from the $\mathcal{E}(\lambda, \mu)$ distribution.

In cases where the $\mathcal{E}(\lambda)$ or $\mathcal{E}(\lambda, \mu)$ distribution does not seem appropriate, the reader is referred to Chapter 4 for guidance regarding the choice of alternative failure distributions. Some of these alternative distributions will be considered from a Bayesian point of view in Chapter 9.

8.1. LIKELIHOOD AND SUFFICIENCY

Suppose that during a life test s failures are observed for units that have been on test for ordered times $t_1, \ldots, t_s$, and $(n-s)$ units are tested for times $t_{s+1}^*, \ldots, t_n^*$ without failing. Let $\underset{\sim}{z}$ denote the "information" (values of s, n, $t_1, \ldots, t_s$, and $t_{s+1}^*, \ldots, t_n^*$) obtained from the life test. Soland (1969) gives the likelihood function in terms of the $\mathcal{W}_1(\lambda, \beta)$ distribution, although his technique is equally valid for any failure time distribution with pdf $f(t)$. Assuming that failure times are statistically independent, the likelihood of $\underset{\sim}{z}$ is

$$L(\underset{\sim}{z}) \propto \left[\prod_{i=1}^{s} f(t_i)\right]\left[\prod_{i=s+1}^{n} [1-F(t_i^*)]\right], \tag{8.1}$$

where $F(t)$ is the cdf of the failure time T. This likelihood function is extremely flexible in that its form does not depend on the life test procedure used to obtain $\underset{\sim}{z}$. Any of the four procedures presented in Section 4.4 may be used, and the withdrawal of units that have not failed, prior to termination of the test, is permitted. The flexibility of the likelihood function enables us to develop posterior distributions and Bayes point and interval estimators that are applicable to a wide variety of life test situations. The only time that the particular type of life test used is relevant is when performing calculations that require the marginal distribution of the observed data, (e.g., when computing Bayes risk).

Four basic kinds of $\mathcal{E}(\lambda)$ life test procedures were presented in Section 4.4. In all cases $T_1, T_2, \ldots, T_s$ are random failure times. Recall that for Type II/item-censored testing, s is fixed, whereas T_s is an r.v. For Type I/time-truncated testing, S is random, whereas the test duration t_0 is fixed.

Although (8.1) is valid for any failure time distribution, this chapter is concerned with sampling from the $\mathcal{E}(\lambda)$ distribution having cdf

$$F(t|\lambda) = \begin{cases} 1 - e^{-\lambda t}, & t > 0 \\ 0, & t \leqslant 0 \end{cases} \tag{8.2}$$

and pdf

$$f(t|\lambda) = \lambda e^{-\lambda t}, \qquad t > 0. \tag{8.3}$$

Substituting (8.2) and (8.3) into (8.1) gives

$$L(\lambda|\underset{\sim}{z}) \propto \lambda^s e^{-\lambda\left[\sum_{i=1}^{s} t_i + \sum_{i=s+1}^{n} t_i^*\right]} = \lambda^s e^{-\lambda \mathrm{t}}, \tag{8.4}$$

where

$$\mathcal{T} = \sum_{i=1}^{s} T_i + \sum_{i=s+1}^{n} T_i^*$$

represents the total time on test and t is its observed value. Since the likelihood depends on the life test information only through the values s and t, estimation procedures given in Sections 8.2–8.5 depend only on s and/or t and not on individual failure times. Estimators of parameters in the $\mathcal{E}(\lambda, \mu)$ distribution [Section 8.6] will be seen to depend on the observed time of the first failure, as well as s and/or t. Note that when evaluating the likelihood function for the four sampling plans of Section 4.4, it is convenient to compute the total time on test according to (4.72). However, if withdrawals are present, then the above expression for $\mathcal{T}$ should be used.

Point and interval estimation procedures presented in this chapter are applicable to a variety of life test procedures. Computation of the moments of Bayes estimators and of minimum Bayes risk requires the marginal distribution of the life test information or of the sufficient statistics that summarize the information. These moments and risks are given, when mathematically tractable, for the case of Type II/item-censored testing. In this case, it is known that $\mathfrak{T}$ is sufficient for estimating λ, as discussed in Section 4.4, and that $\mathfrak{T}$ has a $\mathcal{G}_1(s, \lambda)$ distribution. Consequently, we will adopt the terminology used by Mann, Schafer, and Singpurwalla (1974) and refer to Type II/item-censored testing as *gamma sampling*.

For the case of Type I/time-truncated testing with replacement, the random number of failures S is sufficient for estimating λ and Mann, Schafer, and Singpurwalla (1974) show that S has a Poisson distribution with parameter $(\mathrm{t}\lambda)$. Consequently, estimation procedures given in Section 7.4 for Poisson sampling apply here as well.

Example 8.2. Suppose that $n=100$ identical valves are tested with replacement for $t_0=4000$ h, and that $s=2$ failures occur. This situation corresponds to Poisson sampling with observed total time on test $\mathrm{t}=nt_0=100(4000\text{ h})=4\times10^5$ h and $s=2$. Consequently, the results of Sections 7.4 and 7.5 are applicable here. For example if a $\mathcal{G}_1(2.45, 181{,}818\text{ h})$ prior distribution is assigned to Λ, then the usual Bayes point estimate of λ, using (7.89), is

$$E(\Lambda|2;2.45,181{,}818)=\frac{s+\alpha}{\mathrm{t}+\beta}=\frac{4.45}{5.8182\times10^5}=7.65\times10^{-6}\text{ f/h}.$$

Similarly, interval estimates can be found using the results of Section 7.4.3. ■

Moments of the Bayes estimator and the Bayes risk calculations for Type I/time-truncated testing without replacement are quite tedious and will not be discussed further.

8.2. FAILURE RATE ESTIMATION

This section considers Bayesian point and interval estimators and their properties for the constant failure rate λ. Of primary concern will be

$\mathcal{U}(\alpha_0, \beta_0)$, noninformative, and natural conjugate $\mathcal{G}(\alpha_0, \beta_0)$ prior distributions. Gamma sampling (Type II/item-censored testing either with or without replacement) will be assumed throughout.

By combining the likelihood function (8.4) with $g(\lambda)$, the prior distribution of Λ, we obtain the posterior distribution of Λ given t:

$$g(\lambda|\mathrm{t}; s) = \frac{\lambda^s e^{-\lambda \mathrm{t}} g(\lambda)}{\int_0^\infty \lambda^s e^{-\lambda \mathrm{t}} g(\lambda)\, d\lambda}. \tag{8.5}$$

8.2.1 Uniform Prior Distribution

Harris and Singpurwalla (1968) considered the case of a $\mathcal{U}(\alpha_0, \beta_0)$ prior distribution on Λ given by

$$g(\lambda; \alpha_0, \beta_0) = \begin{cases} \dfrac{1}{\beta_0 - \alpha_0}, & \alpha_0 < \lambda < \beta_0 \\ 0, & \text{elsewhere.} \end{cases} \tag{8.6}$$

Substituting this prior into (8.5) we obtain the posterior distribution of Λ given t

$$g(\lambda|\mathrm{t}; s, \alpha_0, \beta_0) = \frac{\lambda^s e^{-\lambda \mathrm{t}}}{\int_{\alpha_0}^{\beta_0} \lambda^s e^{-\lambda \mathrm{t}}\, d\lambda}, \qquad \alpha_0 < \lambda < \beta_0. \tag{8.7}$$

Letting $y = \lambda \mathrm{t}$ the denominator of (8.7) becomes

$$\int_{\alpha_0}^{\beta_0} \lambda^s e^{-\lambda \mathrm{t}}\, d\lambda = \int_{\alpha_0 \mathrm{t}}^{\beta_0 \mathrm{t}} \frac{y^s e^{-y}\, dy}{\mathrm{t}^{s+1}} = \frac{1}{\mathrm{t}^{s+1}} \left[\Gamma(s+1, \beta_0 \mathrm{t}) - \Gamma(s+1, \alpha_0 \mathrm{t}) \right], \tag{8.8}$$

where $\Gamma(a, z)$ represents the standard incomplete gamma function. Substituting (8.8) into (8.7) the posterior distribution of Λ given t is seen to be

$$g(\lambda|\mathrm{t}; s, \alpha_0, \beta_0) = \frac{\mathrm{t}^{s+1} \lambda^s e^{-\lambda \mathrm{t}}}{\Gamma(s+1, \beta_0 \mathrm{t}) - \Gamma(s+1, \alpha_0 \mathrm{t})}, \qquad \alpha_0 < \lambda < \beta_0. \tag{8.9}$$

The marginal distribution of $\mathcal{T}$ is given by

$$f(\mathrm{t}; s, \alpha_0, \beta_0) = \frac{\Gamma(s+1, \beta_0 \mathrm{t}) - \Gamma(s+1, \alpha_0 \mathrm{t})}{\Gamma(s)(\beta_0 - \alpha_0)\mathrm{t}^2}, \qquad 0 < \mathrm{t} < \infty. \tag{8.10}$$

Example 8.3. Bott and Haas (1978) discuss data collection efforts and give historical data for sodium valve failures in nuclear reactor applications. As an example, consider 0.5–3 in. manually operated freeze seal gate valves. Historical data shows a mean failure rate of 3.8×10^{-6} f/h, with 5% and 95% values of 1.9×10^{-6} f/h and 6.1×10^{-6} f/h, respectively.

Suppose that a new type of freeze seal gate valve is to be subjected to an accelerated life test in order to estimate the failure rate. Twenty valves are placed on test, without replacement, until the fifth failure occurs. The stress level of the accelerated test is expected to increase the failure rate by a factor of approximately eight. The test results in an observed total time on test to the fifth failure of t = 160,000 h.

A $\mathcal{U}(1.4\times10^{-5}, 5.0\times10^{-5})$ prior distribution is assigned to Λ, taking into account the historical data and the increased stress of the test. The resulting posterior distribution is

$$g\left(\lambda \mid t=160{,}000; 5, 1.4\times10^{-5}, 5.0\times10^{-5}\right)$$

$$= \frac{(160{,}000)^6 \lambda^5 e^{-160{,}000\lambda}}{\Gamma(6, 8.00) - \Gamma(6, 2.24)}, \qquad 1.4\times10^{-5} < \lambda < 5.0\times10^{-5}.$$

The prior and posterior distributions are plotted in Figure 8.1.

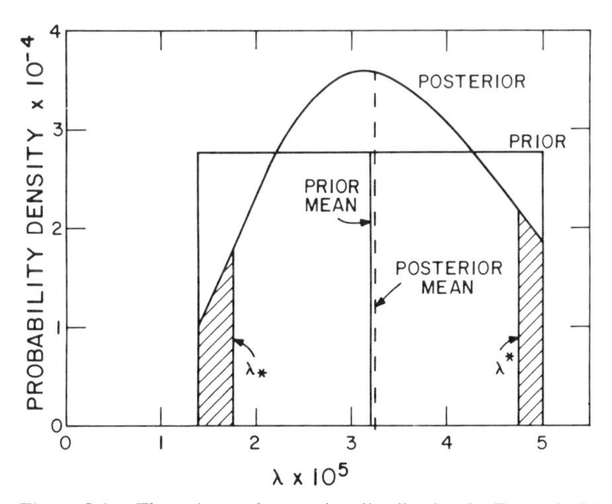

Figure 8.1. The prior and posterior distribution in Example 8.3. ■

Point Estimation. The posterior mean of (8.9) is computed to be

$$E(\Lambda|\mathrm{t}; s, \alpha_0, \beta_0) = \frac{\Gamma(s+2, \beta_0 \mathrm{t}) - \Gamma(s+2, \alpha_0 \mathrm{t})}{\mathrm{t}[\Gamma(s+1, \beta_0 \mathrm{t}) - \Gamma(s+1, \alpha_0 \mathrm{t})]} \tag{8.11}$$

and the second moment is given by

$$E(\Lambda^2|\mathrm{t}; s, \alpha_0, \beta_0) = \frac{\Gamma(s+3, \beta_0 \mathrm{t}) - \Gamma(s+3, \alpha_0 \mathrm{t})}{\mathrm{t}^2[\Gamma(s+1, \beta_0 \mathrm{t}) - \Gamma(s+1, \alpha_0 \mathrm{t})]}. \tag{8.12}$$

The posterior variance can be calculated using the relationship

$$\mathrm{Var}(\Lambda|\mathrm{t}; s, \alpha_0, \beta_0) = E(\Lambda^2|\mathrm{t}; s, \alpha_0, \beta_0) - E^2(\Lambda|\mathrm{t}; s, \alpha_0, \beta_0). \tag{8.13}$$

If a squared-error loss function is assumed, then the Bayes estimator is given by (8.11) and the Bayes risk is proportional to the expectation of (8.13) with respect to (8.10), the marginal distribution of $\mathfrak{T}$. However, this expectation is not available in closed form.

The generalized ML estimator for λ is the mode of (8.9), which is s/t. Finally, the mean of the squared-error loss Bayes estimator (8.11) is given by

$$E_{\mathrm{t}}\left[E_{\lambda|\mathrm{t}}(\Lambda|\mathfrak{T}; s, \alpha_0, \beta_0\right] = E(\Lambda; \alpha_0, \beta_0) = \left(\frac{\alpha_0 + \beta_0}{2}\right). \tag{8.14}$$

Example 8.4. In reference to Example 8.3, the Bayes point estimate of λ under squared-error loss is

$$E(\Lambda|\mathrm{t} = 160{,}000; 5, 1.4 \times 10^{-5}, 5.0 \times 10^{-5})$$
$$= \frac{\Gamma(7, 8) - \Gamma(7, 2.24)}{160{,}000[\Gamma(6, 8) - \Gamma(6, 2.24)]} = 3.25 \times 10^{-5},$$

which is shown in Figure 8.1. The posterior variance is computed to be 8.45×10^{-11}. The generalized maximum likelihood estimate is $5/160{,}000 = 3.125 \times 10^{-5}$. ■

Interval Estimation. A symmetric $100(1-\gamma)\%$ TBPI estimate for λ is found according to (5.45) and (5.46), using (8.9), for λ_* and λ^* respectively.

The equations are given by

$$\Pr(\Lambda \leq \lambda_* | t; s, \alpha_0, \beta_0) = \int_{\alpha_0}^{\lambda_*} \frac{t^{s+1}\lambda^s e^{-\lambda t}\, d\lambda}{\Gamma(s+1, \beta_0 t) - \Gamma(s+1, \alpha_0 t)}$$

$$= \frac{\Gamma(s+1, \lambda_* t) - \Gamma(s+1, \alpha_0 t)}{\Gamma(s+1, \beta_0 t) - \Gamma(s+1, \alpha_0 t)} = \frac{\gamma}{2} \tag{8.15}$$

and

$$\Pr(\Lambda \geq \lambda^* | t; s, \alpha_0, \beta_0) = \frac{\Gamma(s+1, \beta_0 t) - \Gamma(s+1, \lambda^* t)}{\Gamma(s+1, \beta_0 t) - \Gamma(s+1, \alpha_0 t)} = \frac{\gamma}{2}. \tag{8.16}$$

Equations (8.15) and (8.16) can be solved using an incomplete gamma subroutine and a simple search program. $100(1-\gamma)\%$ LBPI and UBPI estimates are found by solving (8.15) and (8.16), respectively, with $\gamma/2$ replaced by γ.

Example 8.5. Again let us consider Example 8.3. A direct computer search using the incomplete gamma subroutine INCGAM [Amos and Daniel (1972)], gives the 90% TBPI estimate of λ as $\lambda_* = 1.765 \times 10^{-5}$ f/h, $\lambda^* = 4.752 \times 10^{-5}$ f/h. These values are illustrated in Figure 8.1. The 90% UBPI estimate has upper limit $\lambda^* = 4.534 \times 10^{-5}$. ■

8.2.2 Noninformative Prior Distribution

The technique presented in Section 6.1 will be applied here to find a noninformative prior distribution for the failure rate λ. Taking the logarithm of (8.4), we obtain for the log likelihood function,

$$\ln L(\lambda | \underset{\sim}{z}) = s \ln \lambda - \lambda t + c. \tag{8.17}$$

Consequently,

$$\frac{\partial \ln L}{\partial \lambda} = \frac{s}{\lambda} - t, \qquad \frac{\partial^2 \ln L}{\partial \lambda^2} = \frac{-s}{\lambda^2},$$

and the Fisher information discussed in Section 6.1 is

$$J(\hat{\lambda}) = \frac{s}{\hat{\lambda}^2}. \tag{8.18}$$

Note that the value $n=1$ is used since the Type II/item-censored life test results in a *single* observation t. According to (6.2), a noninformative prior distribution for Λ is given by

$$g(\lambda) \propto J^{1/2}(\lambda) \propto 1/\lambda, \qquad \lambda > 0, \tag{8.19}$$

and $\phi(\lambda) = \ln \lambda$ is the transformation for which the approximately noninformative prior distribution is locally uniform. It is also interesting to compare this prior with the noninformative prior for λ based on Poisson sampling. In Section 7.4.2 the noninformative prior based on Poisson sampling was found to be proportional to $\lambda^{-1/2}$, *not* λ^{-1}. The difference is due to the fact that, in Poisson sampling, S is a sufficient statistic for estimating λ and t is fixed, whereas the opposite is true in the case of gamma sampling. The Fisher information thus depends on the sampling scheme employed and, consequently, so does the noninformative prior. A similar situation occurs in the case of binomial versus Pascal sampling [see Exercise 7.17].

Note that the prior distribution given above is improper. The posterior distribution is proper, however, and is calculated to be

$$g(\lambda | \mathrm{t}; s) = \frac{\mathrm{t}^s}{\Gamma(s)} \lambda^{s-1} e^{-\lambda \mathrm{t}}, \qquad 0 < \lambda < \infty, \tag{8.20}$$

which is observed to be a $\mathcal{G}_1(s, \mathrm{t})$ distribution, whereas the marginal distribution of $\mathcal{T}$ is improper and is given by

$$f(\mathrm{t}; s) \propto \frac{1}{\mathrm{t}}. \tag{8.21}$$

The posterior mean and variance are

$$E(\Lambda | \mathrm{t}; s) = \frac{s}{\mathrm{t}} \tag{8.22}$$

and

$$\mathrm{Var}(\Lambda | \mathrm{t}; s) = \frac{s}{\mathrm{t}^2}, \tag{8.23}$$

respectively. The generalized Bayes estimator with respect to squared-error loss is given by (8.22) and is seen to be equal to the classical ML estimator. Moments of the estimator and Bayes risk do not exist.

Since the conditional distribution of Λ given t is $\mathcal{G}_1(s, \mathrm{t})$, the transformed variable $2\mathrm{t}\Lambda$ follows a $\chi^2(2s)$ distribution. This fact can be used to obtain symmetric $100(1-\gamma)\%$ TBPI estimates for λ by solving the equations

$$\Pr(\Lambda \leqslant \lambda_* | t; s) = \Pr(2t\Lambda \leqslant 2t\lambda_* | t; s) = \Pr(\chi^2 \leqslant 2t\lambda_*) = \frac{\gamma}{2} \quad (8.24)$$

and

$$\Pr(\chi^2 \geqslant 2t\lambda^*) = \frac{\gamma}{2}, \quad (8.25)$$

where χ^2 has a $\chi^2(2s)$ distribution. The solutions to (8.24) and (8.25) are given by

$$\lambda_* = \frac{\chi^2_{\gamma/2}(2s)}{2t} \quad (8.26)$$

and

$$\lambda^* = \frac{\chi^2_{1-\gamma/2}(2s)}{2t}. \quad (8.27)$$

It should be noted that this interval is identical to the $100(1-\gamma)\%$ TCI estimate of λ presented in Section 4.5.1. Also, $100(1-\gamma)\%$ LBPI and UBPI estimators can be found by substituting γ for $\gamma/2$ in (8.26) and (8.27), respectively.

This interval can also be directly compared to the corresponding interval for λ based on Poisson sampling given in (7.81) and (7.82). The Poisson interval is observed to involve χ^2 percentiles with one more degree of freedom and thus the Poisson interval is slightly wider. It was mentioned in the introduction to Chapter 7 that attribute data generally produce inferior estimates compared to those based on variables data as clearly illustrated here.

Example 8.6. Again let us consider Example 8.3. The posterior distribution is $\mathcal{G}_1(5, 160{,}000\text{ h})$, with mean and variance

$$E(\Lambda | 160{,}000; 5) = 3.125 \times 10^{-5}\text{ f/h},$$

$$\text{Var}(\Lambda | 160{,}000; 5) = 1.953 \times 10^{-10}.$$

The 90% TBPI estimate of λ is given by

$$\lambda_* = \frac{\chi^2_{0.05}(10)}{2(160{,}000)} = \frac{3.940}{320{,}000} = 1.23 \times 10^{-5}\text{ f/h}$$

and

$$\lambda^* = \frac{\chi^2_{0.95}(10)}{320{,}000} = \frac{18.307}{320{,}000} = 5.72 \times 10^{-5} \text{ f/h},$$

where $\Pr(1.23 \times 10^{-5} \text{ f/h} \leq \Lambda \leq 5.72 \times 10^{-5} \text{ f/h} \mid 160{,}000; 5) = 0.90$. ■

8.2.3 Gamma Prior Distribution

The $\mathcal{G}(\alpha, \beta)$ distribution was discussed in Section 4.2.6, and its usefulness as a prior distribution for Λ in the case of Poisson sampling was discussed in Section 7.4.3. The reasoning that justifies the widespread use of a $\mathcal{G}(\alpha, \beta)$ prior in Poisson sampling (primarily the mathematical tractability of a conjugate prior combined with the versatility available through various choices of the prior parameters α and β) is equally valid in the case of gamma sampling. This is not surprising, as both gamma and Poisson sampling assume an underlying $\mathcal{E}(\lambda)$ failure time distribution, and in both cases our intent is to make inferences regarding the failure rate λ.

Substituting the $\mathcal{G}(\alpha_0, \beta_0)$ prior distribution with pdf

$$g(\lambda; \alpha_0, \beta_0) = \frac{1}{\Gamma(\alpha_0)\beta_0^{\alpha_0}} \lambda^{\alpha_0 - 1} e^{-\lambda/\beta_0} \tag{8.28}$$

into (8.5), the posterior distribution of Λ given t is found to be

$$g(\lambda \mid \mathrm{t}; s, \alpha_0, \beta_0) = \frac{1}{\Gamma(\alpha_0 + s)(\beta_0/(\beta_0 \mathrm{t} + 1))^{\alpha_0 + s}} \lambda^{\alpha_0 + s - 1} e^{-\lambda(\mathrm{t} + 1/\beta_0)},$$

$$0 < \lambda < \infty, \tag{8.29}$$

which is a $\mathcal{G}(\alpha_0 + s, \beta_0/(\beta_0 \mathrm{t} + 1))$ distribution. This is the same posterior as (7.85) found in Section 7.4.3 for the case of Poisson sampling. However, note that in the Poisson case the number of failures S is an r.v. and total time on test t is fixed, whereas in the case of gamma sampling, $\mathcal{T}$ is random and s is fixed. The marginal distribution of $\mathcal{T}$ is given by

$$f(\mathrm{t}; s, \alpha_0, \beta_0) = \frac{\Gamma(\alpha_0 + s)}{\Gamma(s)\Gamma(\alpha_0)\beta_0^{\alpha_0}} \mathrm{t}^{s-1} \left(\frac{\beta_0}{\beta_0 \mathrm{t} + 1} \right)^{\alpha_0 + s}, \qquad 0 < \mathrm{t} < \infty.$$

$$\tag{8.30}$$

Note that the interpretation of prior and posterior parameters given in Section 7.4.3 also holds here: α_0 and $1/\beta_0$ can be thought of as pseudo failures and pseudo time on test, respectively, whereas $(\alpha_0 + s)$ and $t + 1/\beta_0$ can be thought of as combined failures and combined time on test.

Example 8.7. Consider once again the situation presented in Example 8.3. A combination of the historical values and the increased stress level leads to desired prior 5th and 95th percentiles of 1.5×10^{-5} and 4.9×10^{-5}, respectively. Using the techniques of Chapter 6, we obtain $\mathcal{G}(\alpha_0, \beta_0)$ prior parameters $\alpha_0 = 8.5$, $\beta_0 = 3.5\times10^{-6}$. The corresponding posterior distribution is $\mathcal{G}(13.5, 2.24\times10^{-6})$. The prior and posterior distributions are plotted in Figure 8.2.

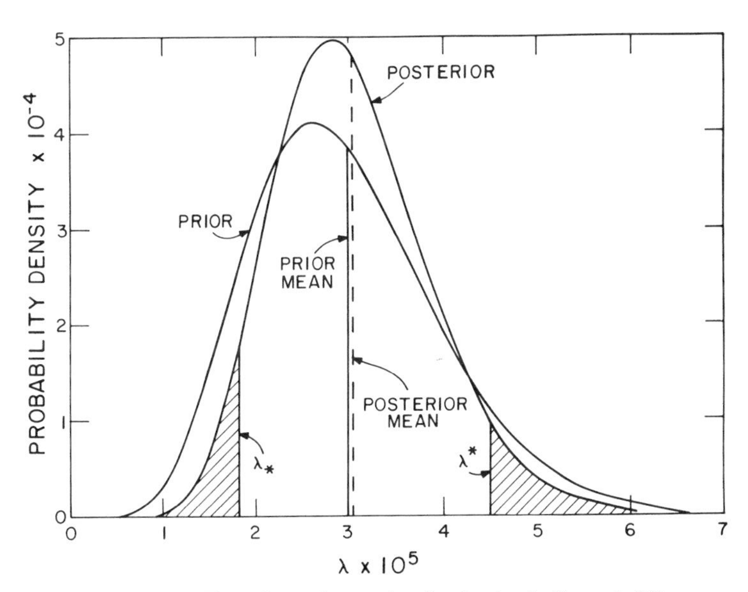

Figure 8.2. The prior and posterior distribution in Example 8.7. ■

Point Estimation. The posterior mean and variance of Λ given t are

$$E(\Lambda \mid t; s, \alpha_0, \beta_0) = \frac{\beta_0(\alpha_0 + s)}{\beta_0 t + 1} \tag{8.31}$$

and

$$\operatorname{Var}(\Lambda|\mathrm{t}; s, \alpha_0, \beta_0) = \frac{\beta_0^2(\alpha_0 + s)}{(\beta_0 \mathrm{t} + 1)^2}, \tag{8.32}$$

respectively. Under squared-error loss, (8.31) is the Bayes point estimator of λ. The generalized ML estimator is the mode of (8.29), given by $\beta_0(\alpha_0 + s - 1)/(\beta_0 \mathfrak{T} + 1)$. These estimators both converge almost surely to the classical ML estimator $s/\mathfrak{T}$ as s approaches infinity. Thus, as was seen in Section 7.4.3, the likelihood tends to dominate the prior as the number of failures specified (and corresponding total time on test) increases.

Under squared-error loss, the posterior risk is proportional to (8.32), and, according to (5.32), the Bayes risk of (8.31) is found by taking the expectation of (8.32) with respect to (8.30), the marginal distribution of $\mathfrak{T}$:

$$r(g) = aE\left\{\frac{\beta_0^2(\alpha_0 + s)}{(\beta_0 \mathfrak{T} + 1)^2}\right\} = \frac{a\beta_0^{\alpha_0 + s + 2}(\alpha_0 + s)\Gamma(\alpha_0 + s)}{\Gamma(s)\Gamma(\alpha_0)\beta_0^{\alpha_0}}$$

$$\times \int_0^\infty \frac{\mathrm{t}^{s-1}}{(\beta_0 \mathrm{t} + 1)^{\alpha_0 + s + 2}} d\mathrm{t}$$

$$= \frac{a\alpha_0(\alpha_0 + 1)\beta_0^2}{\alpha_0 + s + 1}. \tag{8.33}$$

The mean and variance of (8.31) are given by

$$E_{\mathrm{t}}\left[E_{\lambda|\mathrm{t}}(\Lambda|\mathfrak{T}; s, \alpha_0, \beta_0)\right] = \alpha_0\beta_0 \tag{8.34}$$

and

$$\operatorname{Var}_{\mathrm{t}}\left[E_{\lambda|\mathrm{t}}(\Lambda|\mathfrak{T}; s, \alpha_0, \beta_0)\right] = \frac{s\alpha_0\beta_0^2}{\alpha_0 + s + 1}. \tag{8.35}$$

As the $\mathcal{G}(\alpha_0, \beta_0)$ prior is so widely used in Bayesian estimation of λ, it is of interest to compare the resulting risk with that obtained using classical estimation techniques. The MVU estimator of λ is given by $\hat{\lambda} = (s-1)/\mathfrak{T}$, with mean squared-error $\lambda^2/(s-2)$. The Bayes risk of $\hat{\lambda}$ is found by taking the expectation of $\lambda^2/(s-2)$ with respect to the $\mathcal{G}(\alpha_0, \beta_0)$ prior on Λ which gives

$$r(g, \hat{\lambda}) = \frac{\alpha_0(\alpha_0 + 1)\beta_0^2}{s - 2}. \tag{8.36}$$

The ratio of the minimum Bayes risk of (8.31) to the Bayes risk of $\hat{\lambda}$ is

$$\frac{r(g)}{r(g,\hat{\lambda})}=\frac{s-2}{\alpha_0+s+1}. \tag{8.37}$$

This ratio never exceeds one, as expected, as (8.31) is the minimum risk estimator. Note that the risk advantage of the Bayes estimator over the MVU estimator tends to increase as the pseudo number of failures (α_0) increases relative to the observed number of failures (s).

Example 8.8. In regard to Example 8.7, the Bayes point estimator under squared-error loss is

$$E(\Lambda|160{,}000; 5, 8.5, 3.5\times10^{-6})=\frac{3.5\times10^{-6}(13.5)}{0.56+1}$$

$$=3.03\times10^{-5} \text{ f/h}.$$

The posterior and prior means are illustrated in Figure 8.2. The posterior variance and minimum Bayes risk ($a=1$) are computed to be 6.80×10^{-11} and 6.82×10^{-11}, respectively. The generalized maximum likelihood estimate is 2.80×10^{-5} f/h. The MVU estimate is $\hat{\lambda}=4/160{,}000=2.50\times10^{-5}$ f/h, with Bayes risk $r(g,\hat{\lambda})=3.30\times10^{-10}$. The ratio $r(g)/r(g,\hat{\lambda})$ is calculated to be 0.21, so the Bayes risk of the Bayes point estimate is about 1/5 as large as that of the MVU estimate. ■

Interval Estimation. As the posterior distribution (8.29) of Λ given t in the case of gamma sampling is the same as the posterior (7.85) for Poisson sampling, the 100(1 − γ)% TBPI estimation formulas (7.95) and (7.96) developed in Section 7.4.3 also apply here. A symmetric 100(1 − γ)% TBPI estimate of λ is given by

$$\lambda_*=\frac{\beta_0}{2(\beta_0 t+1)}\chi^2_{\gamma/2}(2s+2\alpha_0) \tag{8.38}$$

and

$$\lambda^*=\frac{\beta_0}{2(\beta_0 t+1)}\chi^2_{1-\gamma/2}(2s+2\alpha_0). \tag{8.39}$$

100(1 − γ)% LBPI and UBPI are computed using (8.38) and (8.39), respectively, with $\gamma/2$ replaced by γ.

Example 8.9. Again let us refer to Example 8.7. The 90% TBPI estimate of λ becomes

$$\lambda_* = \frac{3.5\times 10^{-6}}{3.12}\chi^2_{0.05}(27) = \frac{3.5\times 10^{-6}}{3.12}(16.151) = 1.812\times 10^{-5} \text{ f/h}$$

and

$$\lambda^* = \frac{3.5\times 10^{-6}}{3.12}\chi^2_{0.95}(27) = \frac{3.5\times 10^{-6}}{3.12}(40.113) = 4.500\times 10^{-5} \text{ f/h}.$$

Thus, it may be claimed that

$$\Pr\left(1.812\times 10^{-5} \text{ f/h} \leqslant \Lambda \leqslant 4.500\times 10^{-5} \text{ f/h} \mid 160{,}000; 5, 8.5, 3.5\times 10^{-6}\right) = 0.90.$$

The 90% TBPI is illustrated in Figure 8.2. ■

8.2.4 Other Prior Distributions

Section 7.4.4 considers the use of $\mathcal{LN}(\xi, \sigma^2)$ and $\mathcal{W}(\alpha, \beta)$ prior distributions for Λ in the case of Poisson sampling. For both priors, the mathematics involved is sufficiently complex that the only results given are equations involving integrals that must be numerically evaluated to find the posterior distribution and its mean.

Note that (8.5) is the same as (7.62). Consequently, the results given in Section 7.4.4 are equally applicable here, where $\mathcal{T}$ represents total time on test. In using those results for the case of gamma sampling, one need only keep in mind that $\mathcal{T}$ is now the r.v., whereas the number of failures s is fixed.

8.3. MEAN TIME TO FAILURE ESTIMATION

This section presents Bayesian point and interval estimation techniques for estimating the $\mathcal{E}(\lambda)$ MTTF parameter $\theta = 1/\lambda$. The main prior distributions considered are the $\mathcal{U}(\alpha_0, \beta_0)$, noninformative, and natural conjugate $\mathcal{IG}(\alpha_0, \beta_0)$ distributions. In addition, the case of MTTF estimation when a prior distribution has been assigned to the failure rate Λ is also discussed.

When expressed as a function of the MTTF parameter $\theta = 1/\lambda$, the likelihood function (8.4) for the life test information becomes

$$L(\theta|\underset{\sim}{z}) \propto \frac{1}{\theta^s} e^{-t/\theta}. \tag{8.40}$$

If $g(\theta)$ represents the prior distribution of Θ, the posterior distribution of Θ given t in the case of gamma sampling is

$$g(\theta|t; s) = \frac{e^{-t/\theta} g(\theta)}{\theta^s \int_0^\infty \frac{1}{y^s} e^{-t/y} g(y)\, dy}. \tag{8.41}$$

8.3.1 Uniform Prior Distribution

Consider a $\mathcal{U}(\alpha_0, \beta_0)$ prior distribution for Θ given by

$$g(\theta; \alpha_0, \beta_0) = \begin{cases} \dfrac{1}{\beta_0 - \alpha_0}, & \alpha_0 \leqslant \theta \leqslant \beta_0 \\ 0, & \text{elsewhere.} \end{cases} \tag{8.42}$$

Substituting (8.42) into (8.41) yields the posterior distribution for Θ given t:

$$g(\theta|t; s, \alpha_0, \beta_0) = \frac{t^{s-1} e^{-t/\theta}}{\theta^s [\Gamma(s-1, t/\alpha_0) - \Gamma(s-1, t/\beta_0)]}, \qquad \alpha_0 < \theta < \beta_0. \tag{8.43}$$

The marginal distribution of $\mathcal{T}$ in the case of gamma sampling is

$$f(t; s, \alpha_0, \beta_0) = \frac{\Gamma(s-1, t/\alpha_0) - \Gamma(s-1, t/\beta_0)}{\Gamma(s)(\beta_0 - \alpha_0)}, \qquad t > 0. \tag{8.44}$$

Example 8.10. In regard to Example 8.3, consider a $\mathcal{U}(2\times 10^4, 7\times 10^4)$ prior distribution for Θ. The resulting posterior distribution is

$$g(\theta|1.6\times 10^5; 5, 2\times 10^4, 7\times 10^4) = \frac{(1.6\times 10^5)^4 e^{-1.6\times 10^5/\theta}}{\theta^5[\Gamma(4, 8.00) - \Gamma(4, 2.29)]},$$

$$2\times 10^4 < \theta < 7\times 10^4.$$

The prior and posterior distributions are plotted in Figure 8.3.

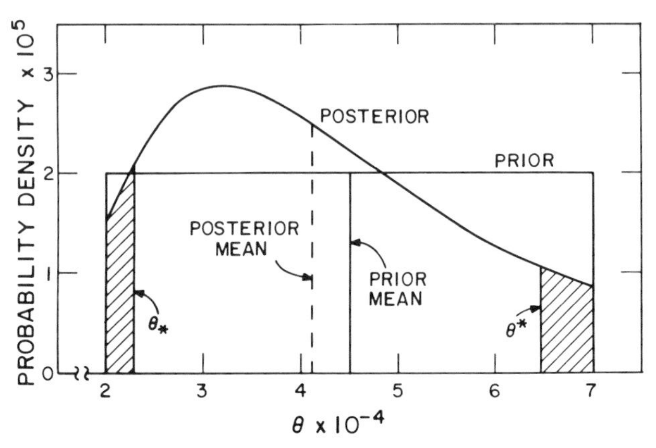

Figure 8.3. The prior and posterior distribution in Example 8.10. ■

Point Estimation. The posterior mean and the second moment of Θ are found, using (8.43), to be

$$E(\Theta|\text{t}; s, \alpha_0, \beta_0) = \text{t}\left\{\frac{\Gamma(s-2, \text{t}/\alpha_0) - \Gamma(s-2, \text{t}/\beta_0)}{\Gamma(s-1, \text{t}/\alpha_0) - \Gamma(s-1, \text{t}/\beta_0)}\right\} \tag{8.45}$$

and

$$E(\Theta^2|\text{t}; s, \alpha_0, \beta_0) = \text{t}^2\left\{\frac{\Gamma(s-3, \text{t}/\alpha_0) - \Gamma(s-3, \text{t}/\beta_0)}{\Gamma(s-1, \text{t}/\alpha_0) - \Gamma(s-1, \text{t}/\beta_0)}\right\}. \tag{8.46}$$

If a squared-error loss function is used, the Bayes point estimator is given by (8.45). The posterior variance can be found using (8.45), (8.46), and the usual definition

$$\text{Var}(\Theta|\text{t}; s, \alpha_0, \beta_0) = E(\Theta^2|\text{t}; s, \alpha_0, \beta_0) - E^2(\Theta|\text{t}; s, \alpha_0, \beta_0). \tag{8.47}$$

The Bayes risk under squared-error loss is proportional to the expectation of (8.47) with respect to (8.44), the marginal distribution of $\mathfrak{T}$. This expectation is not generally available in closed form.

The generalized ML estimator of Θ is the mode of (8.43), which is

$$\tilde{\theta} = \begin{cases} \alpha_0, & \dfrac{\mathfrak{T}}{s} < \alpha_0 \\ \dfrac{\mathfrak{T}}{s}, & \alpha_0 \leqslant \dfrac{\mathfrak{T}}{s} \leqslant \beta_0 \\ \beta_0, & \dfrac{\mathfrak{T}}{s} > \beta_0. \end{cases} \tag{8.48}$$

Finally, the mean of the Bayes estimator (8.45) is

$$E_{\mathrm{t}}\left[E_{\theta|\mathrm{t}}(\Theta|\mathfrak{T}; s, \alpha_0, \beta_0)\right] = \frac{\alpha_0 + \beta_0}{2}. \tag{8.49}$$

Example 8.11. Refer to Example 8.10. The Bayes point estimate of θ under squared-error loss, is

$$E(\Theta|1.6\times10^5; 5, 2\times10^5, 7\times10^4) = (1.6\times10^5)\cdot\left[\frac{\Gamma(3,8.00)-\Gamma(3,2.29)}{\Gamma(4,8.00)-\Gamma(4,2.29)}\right]$$

$$= 4.11\times10^4 \text{ h},$$

which is shown in Figure 8.3. The posterior variance is calculated to be 1.67×10^8. The generalized ML estimate of θ is $1.6\times10^5/5 = 3.2\times10^4$ h. ■

Interval Estimation. A symmetric $100(1-\gamma)\%$ TBPI estimate of θ is found by solving (5.45) and (5.46), using posterior distribution (8.43), for θ_* and θ^* respectively. The equations are given by

$$\Pr(\Theta \leqslant \theta_*|\mathrm{t}; s, \alpha_0, \beta_0) = \int_{\alpha_0}^{\theta_*} \frac{\mathrm{t}^{s-1}e^{-\mathrm{t}/\theta}}{\theta^s[\Gamma(s-1), \mathrm{t}/\alpha_0) - \Gamma(s-1, \mathrm{t}/\beta_0)]}\, d\theta$$

$$= \frac{\Gamma(s-1, \mathrm{t}/\alpha_0) - \Gamma(s-1, \mathrm{t}/\theta_*)}{\Gamma(s-1, \mathrm{t}/\alpha_0) - \Gamma(s-1, \mathrm{t}/\beta_0)} = \frac{\gamma}{2} \tag{8.50}$$

and

$$\Pr(\Theta \geqslant \theta^*|\mathrm{t}; s, \alpha_0, \beta_0) = \frac{\Gamma(s-1, \mathrm{t}/\theta^*) - \Gamma(s-1, \mathrm{t}/\beta_0)}{\Gamma(s-1, \mathrm{t}/\alpha_0) - \Gamma(s-1, \mathrm{t}/\beta_0)} = \frac{\gamma}{2}. \tag{8.51}$$

Equations (8.50) and (8.51) can be solved using INCGAM and a simple search program. Correspondingly, $100(1-\gamma)\%$ LBPI or UBPI estimates are found by solving (8.50) or (8.51), respectively, with $\gamma/2$ replaced by γ.

Example 8.12. Reconsidering Example 8.10, a direct search routine using INCGAM is used to find the 90% TBPI estimate of θ: $\theta_* = 2.28\times10^4$ h, $\theta^* = 6.48\times10^4$ h. These values are illustrated in Figure 8.3. Also, the 90% LBPI has lower bound $\theta_* = 2.50\times10^4$ h. ■

Bhattacharya (1967) derives the posterior mean and variance for a more general class of priors having finite support. He considers the prior distribution

$$g(\theta; a, \alpha_0, \beta_0) = \frac{(a-1)(\alpha_0\beta_0)^{a-1}}{\beta_0^{a-1} - \alpha_0^{a-1}} \frac{1}{\theta^a}, \qquad 0 < \alpha_0 \leqslant \theta \leqslant \beta_0. \quad (8.52)$$

The $\mathcal{U}(\alpha_0, \beta_0)$ prior is seen to be a special case of (8.52), obtained by setting $a=0$. Bhattacharya's expression for the posterior mean reduces to (8.45) when $a=0$.

8.3.2 Noninformative Prior Distribution

Sinha and Guttman (1976) state that a noninformative prior for Θ is

$$g(\theta) \propto 1/\theta, \qquad \theta > 0. \quad (8.53)$$

This choice of a noninformative prior can be verified through use of the techniques of Section 6.1. or by transforming the invariant prior (8.19) given for $\Lambda = 1/\Theta$. Substituting the improper prior (8.53) into (8.41) gives the proper posterior distribution

$$g(\theta|\mathrm{t}; s) = \frac{1}{\Gamma(s)\theta^{s+1}} \mathrm{t}^s e^{-\mathrm{t}/\theta}, \qquad 0 < \theta < \infty, \quad (8.54)$$

which is an $\mathcal{IG}(s, \mathrm{t})$ distribution. Since a noninformative prior for Θ is just a transformation of a noninformative prior for the failure rate Λ, the marginal distribution of $\mathcal{T}$ is the same as that given by (8.21):

$$f(\mathrm{t}; s) \propto \frac{1}{\mathrm{t}}. \quad (8.55)$$

The posterior mean and variance are

$$E(\Theta|\mathrm{t}; s) = \mathrm{t}/(s-1), \qquad s > 1, \quad (8.56)$$

and

$$\mathrm{Var}(\Theta|\mathrm{t}; s) = \frac{\mathrm{t}^2}{(s-1)^2(s-2)}, \qquad s > 2. \quad (8.57)$$

The generalized Bayes estimator with respect to squared-error loss is given by (8.56), and is seen to be asymptotically equivalent to the classical ML estimator $\hat{\theta} = \mathcal{T}/s$. Moments of the Bayes estimator and Bayes risk do not exist.

As the posterior distribution of Θ given t is $\mathcal{IG}(s, \mathrm{t})$, the transformed variable $2\mathrm{t}/\Theta$ has a $\chi^2(2s)$ distribution. Consequently, a symmetric $100(1-\gamma)\%$ TBPI estimate of θ is found by solving

$$\Pr(\Theta \leqslant \theta_* | \mathrm{t}; s) = \Pr\left(\frac{2\mathrm{t}}{\Theta} \geqslant \frac{2\mathrm{t}}{\theta_*}\middle| \mathrm{t}; s\right) = \Pr\left(\chi^2 \geqslant \frac{2\mathrm{t}}{\theta_*}\right) = \frac{\gamma}{2}, \quad (8.58)$$

where χ^2 has a $\chi^2(2s)$ distribution, and

$$\Pr(\Theta \geqslant \theta^* | \mathrm{t}; s) = \Pr\left(\chi^2 \leqslant \frac{2\mathrm{t}}{\theta^*}\right) = \frac{\gamma}{2}, \quad (8.59)$$

for θ_* and θ^*, respectively. The solutions are

$$\theta_* = 2\mathrm{t}/\chi^2_{1-\gamma/2}(2s), \quad (8.60)$$

$$\theta^* = 2\mathrm{t}/\chi^2_{\gamma/2}(2s). \quad (8.61)$$

$100(1-\gamma)\%$ LBPI or UBPI estimates are obtained in the usual manner.

Example 8.13. Regarding Example 8.3, the posterior distribution of Θ given t is

$$g(\theta | 1.6\times 10^5; 5) = \frac{(1.6\times 10^5)^5}{\Gamma(5)} \frac{e^{-1.6\times 10^5/\theta}}{\theta^6}, \qquad 0<\theta<\infty,$$

an $\mathcal{IG}(5, 1.6\times 10^5)$ distribution. The posterior mean and variance are

$$E(\Theta | 1.6\times 10^5; 5) = \frac{1.6\times 10^5}{4} = 4\times 10^4 \text{ h}$$

and

$$\mathrm{Var}(\Theta | 1.6\times 10^5; 5) = 5.33\times 10^8.$$

The 90% TBPI estimate of θ is calculated to be

$$\theta_* = \frac{2(1.6\times10^5)}{\chi^2_{0.95}(10)} = \frac{2(1.6\times10^5)}{18.307} = 1.75\times10^4\text{ h},$$

$$\theta^* = \frac{2(1.6\times10^5)}{\chi^2_{0.05}(10)} = \frac{2(1.6\times10^5)}{3.940} = 8.12\times10^4\text{ h},$$

and thus $\Pr(1.75\times10^4\text{ h} \leqslant \Theta \leqslant 8.12\times10^4\text{ h}\,|\,1.6\times10^5; 5) = 0.90$. ■

Moore and Bilikam (1978) consider a more general class of improper prior distributions given by $g(\theta)=1/\theta^{\alpha+1}$, which includes the noninformative prior as a special case ($\alpha=0$). Generalized Bayes estimators are given for θ and the reliability function.

8.3.3 Inverted Gamma Prior Distribution

The $\mathcal{IG}(\alpha,\beta)$ distribution, which is discussed in Section 4.2.7, is the natural conjugate prior for Θ in the case of gamma sampling. As a conjugate prior, it has an advantage in terms of mathematical tractability when compared with other priors we consider for Θ.

Note that the choice of an $\mathcal{IG}(\alpha,\beta)$ prior for Θ is equivalent to choosing a $\mathcal{G}_1(\alpha,\beta)$ prior for the failure rate Λ. It is easily verified that if Θ has the $\mathcal{IG}(\alpha_0,\beta_0)$ distribution with pdf

$$g(\theta;\alpha_0,\beta_0) = \frac{\beta_0^{\alpha}}{\Gamma(\alpha_0)}\left(\frac{1}{\theta}\right)^{\alpha_0+1} e^{-\beta_0/\theta}, \qquad \theta>0, \tag{8.62}$$

then $\Lambda=1/\Theta$ has $\mathcal{G}_1(\alpha_0,\beta_0)$ distribution.

Bhattacharya (1967) derives the posterior distribution of Θ given t and gives the posterior mean and variance. The posterior distribution, which can be obtained by substituting (8.62) into (8.41), is

$$g(\theta\,|\,\mathrm{t}; s,\alpha_0,\beta_0) = \frac{(\mathrm{t}+\beta_0)^{s+\alpha_0}}{\Gamma(s+\alpha_0)}\left(\frac{1}{\theta}\right)^{s+\alpha_0+1} e^{-(\mathrm{t}+\beta_0)/\theta}, \qquad 0<\theta<\infty, \tag{8.63}$$

which is an $\mathcal{IG}(s+\alpha_0, \mathrm{t}+\beta_0)$ distribution. The marginal distribution of $\mathcal{T}$ is found to be

$$f(t; s, \alpha_0, \beta_0) = \frac{\Gamma(s+\alpha_0)}{\Gamma(s)\Gamma(\alpha_0)} \beta_0^{\alpha_0} \frac{t^{s-1}}{(t+\beta_0)^{s+\alpha_0}}, \qquad t > 0. \qquad (8.64)$$

Example 8.14. Again let us consider Example 8.7. The $\mathcal{IG}(8.5, 2.86\times 10^5)$ prior distribution expresses the same prior information regarding Θ as the $\mathcal{G}_1(8.5, 2.86\times 10^5)$ prior does for Λ. Thus, using $\alpha_0 = 8.5$, $\beta_0 = 2.86\times 10^5$, we obtain an $\mathcal{IG}(13.5, 4.46\times 10^5)$ posterior distribution. The prior and posterior distributions are plotted in Figure 8.4.

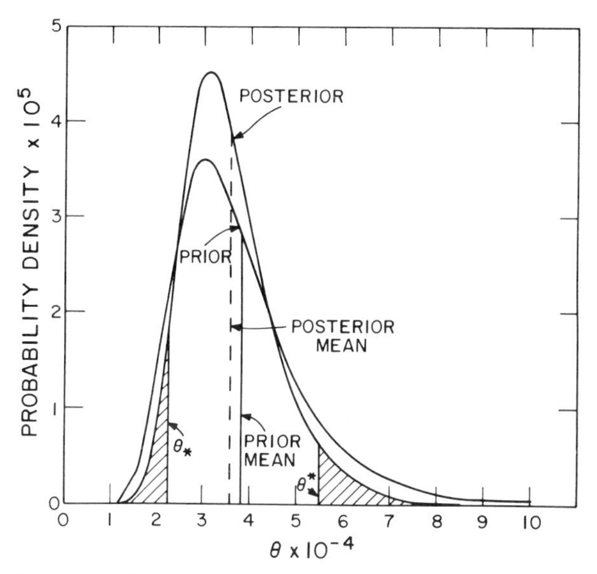

Figure 8.4. The prior and posterior distribution in Example 8.14. ■

Point Estimation. The posterior mean and variance of Θ given t are

$$E(\Theta | t; s, \alpha_0, \beta_0) = (t+\beta_0)/(s+\alpha_0-1), \qquad s+\alpha_0 > 1, \qquad (8.65)$$

and

$$\mathrm{Var}(\Theta | t; s, \alpha_0, \beta_0) = \frac{(t+\beta_0)^2}{(s+\alpha_0-1)^2(s+\alpha_0-2)}, \qquad s+\alpha_0 > 2. \qquad (8.66)$$

Under squared-error loss, (8.65) is the Bayes point estimator of θ. The generalized maximum likelihood estimator is the mode of (8.63), which is $(\mathcal{T}+\beta_0)/(s+\alpha_0+1)$. As in the case of failure rate estimation, the generalized ML estimator approaches the classical ML estimator $\mathcal{T}/s$ as $s\to\infty$ and, consequently, the likelihood tends to increasingly dominate the prior.

The posterior risk of (8.65) is proportional to (8.66) and according to (5.32) the Bayes risk is found by taking the expectation of (8.66) with respect to (8.64), the marginal distribution of $\mathcal{T}$:

$$r(g)=aE\left\{\frac{(\mathcal{T}+\beta_0)^2}{(s+\alpha_0-1)^2(s+\alpha_0-2)}\right\}$$

$$=\frac{a\beta_0^2}{(s+\alpha_0-1)(\alpha_0-1)(\alpha_0-2)},\qquad \alpha_0>2. \tag{8.67}$$

The mean of (8.65) is

$$E_t\left[E_{\theta|t}(\Theta|\mathcal{T};s,\alpha_0,\beta_0)\right]=\frac{\beta_0}{\alpha_0-1} \tag{8.68}$$

and the variance is

$$\mathrm{Var}_t\left[E_{\theta|t}(\Theta|\mathcal{T};s,\alpha_0,\beta_0)\right]=\frac{s\beta_0^2}{(\alpha_0-1)^2(\alpha_0-2)(s+\alpha_0-1)}. \tag{8.69}$$

As in Section 8.2.3, it may be of interest to compare the risks, using squared-error loss, obtained through the use of Bayesian and classical point estimators. The classical MVU estimator is just the ML estimator $\hat{\theta}=\mathcal{T}/s$, with mean squared-error θ^2/s. Taking the expectation of Θ^2/s with respect to the $\mathcal{IG}(\alpha_0,\beta_0)$ prior distribution, we obtain the Bayes risk of $\hat{\theta}$:

$$r(g,\hat{\theta})=\frac{a\beta_0^2}{s(\alpha_0-1)(\alpha_0-2)},\qquad \alpha_0>2. \tag{8.70}$$

The ratio of the minimum Bayes risk (8.67) to (8.70) is

$$\frac{r(g)}{r(g,\hat{\theta})}=\frac{s}{s+\alpha_0-1},\qquad \alpha_0>2. \tag{8.71}$$

Notice that the ratio never exceeds one, as $\alpha_0>2$, but that it does approach one as $s \to \infty$. In other words, the risk advantage of the Bayes estimator decreases as more and more test data become available.

Example 8.15. Reconsidering Example 8.14, the Bayes point estimate under squared-error loss is

$$E(\Theta|1.6\times10^5;5,8.5,2.86\times10^5)=\frac{1.6\times10^5+2.86\times10^5}{5+8.5-1}=3.57\times10^4\text{ h}.$$

This estimate is illustrated in Figure 8.4. The posterior variance and minimum Bayes risk ($a=1$) are computed to be 1.11×10^8 and 1.34×10^8, respectively. The MVU estimate is $\hat{\theta}=1.6\times10^5/5=3.2\times10^4$ h, having Bayes risk $r(g,\hat{\theta})=3.36\times10^8$. The ratio $r(g)/r(g,\hat{\theta})$ is found to be .40, so the Bayes risk of the Bayes estimate is 40% as large as the risk of the classical estimate. ■

Interval Estimation. Interval estimates of θ are found by transforming Θ given t to a χ^2 r.v. Since the posterior distribution of Θ given t is $\mathcal{IG}(s+\alpha_0,\mathrm{t}+\beta_0)$, we see that $2(\mathrm{t}+\beta_0)/\Theta$ has a $\chi^2(2s+2\alpha_0)$ distribution. A $100(1-\gamma)\%$ symmetric TBPI estimate of θ is found by solving

$$\Pr(\Theta\leqslant\theta_*|\mathrm{t};s,\alpha_0,\beta_0)=\Pr\left[\frac{2(\mathrm{t}+\beta_0)}{\Theta}\geqslant\frac{2(\mathrm{t}+\beta_0)}{\theta_*}\middle|\mathrm{t};s,\alpha_0,\beta_0\right]$$

$$=\Pr\left[\chi^2\geqslant\frac{2(\mathrm{t}+\beta_0)}{\theta_*}\right]=\frac{\gamma}{2} \tag{8.72}$$

and

$$\Pr(\Theta\geqslant\theta^*|\mathrm{t};s,\alpha_0,\beta_0)=\Pr\left[\chi^2\leqslant\frac{2(\mathrm{t}+\beta_0)}{\theta^*}\right]=\frac{\gamma}{2}, \tag{8.73}$$

for θ_* and θ^* where χ^2 has a $\chi^2(2s+2\alpha_0)$ distribution. The solution is

$$\theta_*=\frac{2(\mathrm{t}+\beta_0)}{\chi^2_{1-\gamma/2}(2s+2\alpha_0)}, \tag{8.74}$$

$$\theta^*=\frac{2(\mathrm{t}+\beta_0)}{\chi^2_{\gamma/2}(2s+2\alpha_0)}. \tag{8.75}$$

$100(1-\gamma)\%$ LBPI or UBPI estimates of θ are obtained by substituting γ for $\gamma/2$ in (8.74) or (8.75), respectively.

Example 8.16. Refer to Example 8.14. The 90% TBPI estimate of θ, illustrated in Figure 8.4, is computed to be

$$\theta_{*}=\frac{2(4.46\times 10^{5})}{\chi^{2}_{0.95}(27)}=\frac{8.92\times 10^{5}}{40.113}=2.22\times 10^{4}\text{ h},$$

$$\theta^{*}=\frac{2(4.46\times 10^{5})}{\chi^{2}_{0.05}(27)}=\frac{8.92\times 10^{5}}{16.151}=5.52\times 10^{4}\text{ h},$$

so, $\Pr(2.22\times 10^{4}\text{ h}\leqslant\Theta\leqslant 5.52\times 10^{4}\text{ h}|160{,}000;5,8.5,2.86\times 10^{5})=0.90$. ■

8.3.4 Implied Prior Distributions

Consider a situation where we wish to make inferences regarding θ, but the prior information available, whether in terms of subjective probability assessments or historical data, is expressed in terms of the failure rate λ. For example, the manufacturer of a valve assembly used in pressurized water reactors may have information from customers regarding the failure rate λ for the product. Suppose that the manufacturer wishes to use these data as prior information when estimating the MTTF θ of a new valve design.

The $\mathcal{U}(\alpha_0,\beta_0)$, noninformative, and conjugate $\mathcal{G}(\alpha_0,\beta_0)$ priors for Λ are all discussed in Section 8.2. Consideration of the situation described above leads to an analysis of the effects of these prior distributions on MTTF estimation.

The noninformative prior $g(\lambda)\propto 1/\lambda$ for Λ is equivalent to the noninformative prior $g(\theta)\propto 1/\theta$ for Θ, so this case is covered in Section 8.3.2. Also, the $\mathcal{G}_1(\alpha_0,\beta_0)$ prior for Λ is equivalent to an $\mathcal{IG}(\alpha_0,\beta_0)$ prior for Θ, which is covered in Section 8.3.3. The $\mathcal{U}(\alpha_0,\beta_0)$ prior distribution (8.6) for Λ is *not*, however, equivalent to any of the priors for Θ that have been considered. Making the transformation $\theta=1/\lambda$, the $\mathcal{U}(\alpha_0,\beta_0)$ prior for Λ corresponds to the prior

$$g(\theta;\alpha_0,\beta_0)=\frac{1}{(\beta_0-\alpha_0)}\frac{1}{\theta^2},\qquad \frac{1}{\beta_0}\leqslant\theta\leqslant\frac{1}{\alpha_0},\tag{8.76}$$

for Θ. The prior mean and variance are

$$E(\Theta;\alpha_0,\beta_0)=\frac{\ln\beta_0-\ln\alpha_0}{\beta_0-\alpha_0} \tag{8.77}$$

and

$$\operatorname{Var}(\Theta;\alpha_0,\beta_0)=\frac{1}{\alpha_0\beta_0}-\left[\frac{\ln\beta_0-\ln\alpha_0}{\beta_0-\alpha_0}\right]^2, \tag{8.78}$$

respectively. Note that, as in Section 7.5, a $\mathcal{U}(\alpha_0,\beta_0)$ prior for one parameter (failure rate) does not correspond to a uniform prior for a related parameter (MTTF).

Substituting (8.76) into (8.41), or transforming (8.9), gives the posterior distribution of Θ given t as

$$g(\theta|\mathrm{t};s,\alpha_0,\beta_0)=\frac{\mathrm{t}^{s+1}e^{-\mathrm{t}/\theta}}{\theta^{s+2}[\Gamma(s+1,\beta_0\mathrm{t})-\Gamma(s+1,\alpha_0\mathrm{t})]},\qquad \frac{1}{\beta_0}\leqslant\theta\leqslant\frac{1}{\alpha_0}. \tag{8.79}$$

The marginal distribution of $\mathcal{T}$ is given by (8.10). The squared-error loss Bayes estimator is the posterior mean

$$E(\Theta|\mathrm{t};s,\alpha_0,\beta_0)=\frac{\mathrm{t}[\Gamma(s,\beta_0\mathrm{t})-\Gamma(s,\alpha_0\mathrm{t})]}{\Gamma(s+1,\beta_0\mathrm{t})-\Gamma(s+1,\alpha_0\mathrm{t})} \tag{8.80}$$

and the second moment is

$$E(\Theta^2|\mathrm{t};s,\alpha_0,\beta_0)=\frac{\mathrm{t}^2[\Gamma(s-1,\beta_0\mathrm{t})-\Gamma(s-1,\alpha_0\mathrm{t})]}{\Gamma(s+1,\beta_0\mathrm{t})-\Gamma(s+1,\alpha_0\mathrm{t})}. \tag{8.81}$$

Equations (8.80) and (8.81) can be used to find the posterior variance. The Bayes risk, which is proportional to the expectation of the posterior variance with respect to the marginal distribution of $\mathcal{T}$, is not generally available in closed form. Interval estimates of θ are found by (1) substituting the posterior distribution (8.79) into (5.45) and (5.46) and solving the resulting equations for θ_* and θ^*, respectively, or (2) by transforming interval estimates of λ found according to (8.15) and (8.16), using the relationships $\theta_*=1/\lambda^*$, $\theta^*=1/\lambda_*$.

Example 8.17. Again let us refer to Example 8.3. The $\mathcal{U}(1.4\times10^{-5}, 5\times10^{-5})$ prior for Λ corresponds to the prior

$$g(\theta; 1.4\times10^{-5}, 5\times10^{-5}) = \frac{1}{(3.6\times10^{-5})\theta^2} = \frac{2.78\times10^4}{\theta^2},$$

$$2\times10^4 < \theta < 7.14\times10^4,$$

for Θ. The prior mean is 3.54×10^4 h and prior variance is 1.78×10^8. The corresponding posterior distribution is

$$g(\theta|1.6\times10^5; 5, 1.4\times10^{-5}, 5\times10^{-5}) =$$

$$\frac{(1.6\times10^5)^6 e^{-1.6\times10^5/\theta}}{\theta^7[\Gamma(6,8)-\Gamma(6,2.24)]}, \qquad 2\times10^4 < \theta < 7.14\times10^4.$$

The prior and posterior distributions are illustrated in Figure 8.5. The Bayes point estimate under squared-error loss, also illustrated in Figure 8.5, is given by

$$E(\Theta|1.6\times10^5; 5, 1.4\times10^{-5}, 5\times10^{-5})$$

$$= (1.6\times10^5)\left[\frac{\Gamma(5,8)-\Gamma(5,2.24)}{\Gamma(6,8)-\Gamma(6,2.24)}\right] = 3.37\times10^4 \text{ h}.$$

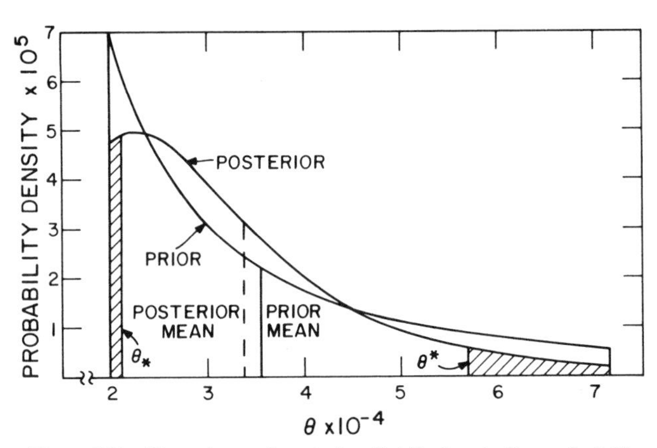

Figure 8.5. The prior and posterior distribution in Example 8.17.

The posterior variance is also calculated to be 1.23×10^8. The 90% TBPI estimate of θ is found, using the results of Example 8.5, to be

$$\theta_* = \frac{1}{4.752 \times 10^{-5}} = 2.10 \times 10^4 \text{ h},$$

$$\theta^* = \frac{1}{1.756 \times 10^{-5}} = 5.69 \times 10^4 \text{ h}.$$

This estimate is also shown in Figure 8.5. ■

8.3.5 Linear Bayes Estimators

We restrict our attention to the case where $s = n$, that is, the complete sample case. From (8.65) we observe that the Bayes estimator of θ under squared-error loss may be written as

$$E(\Theta | \mathrm{t}; n, \alpha_0, \beta_0) = (1 - v)\theta_0 + \frac{v\mathrm{t}}{n},$$

where

$$v = \frac{n}{\alpha_0 + n - 1}, \qquad \theta_0 = \frac{\beta_0}{\alpha_0 - 1}.$$

It is noted that θ_0 is the mean of the $\mathcal{IG}(\alpha_0, \beta_0)$ prior distribution and thus the posterior mean is a linear combination of the prior mean θ_0 and sample mean t/n.

Suppose now that T has an *arbitrary* failure time distribution with pdf $f(t | \theta)$, where θ is the MTTF. Further suppose that we let

$$v = \psi^2 \Big/ \left\{ \frac{1}{n} E_\theta\left[\mathrm{Var}_{t|\theta}(T | \Theta)\right] + \psi^2 \right\},$$

where $\psi^2 = \mathrm{Var}(\Theta) =$ prior variance, and define

$$\hat{\theta}_D = (1 - v)\theta_0 + \frac{v\mathrm{t}}{n},$$

where $\theta_0 = E(\Theta) =$ prior mean. Bühlmann (1970) shows that $\hat{\theta}_D$ is the *least squares linear, in observations*, approximation to the posterior mean for *any* failure time distribution and *any* prior distribution $g(\theta)$. For the $\mathcal{E}(\lambda)$

failure time model and $\mathcal{IG}(\alpha_0, \beta_0)$ prior distribution on $\Theta = 1/\Lambda$, it is easily shown that $\hat{\theta}_D$ is the *exact* posterior mean, as $E_\theta[\text{Var}_{t|\theta}(T|\Theta)] = E(\Theta^2) = \beta_0^2/[(\alpha_0 - 1)(\alpha_0 - 2)]$.

Since $E(T|\theta) = \theta$ for any failure time distribution, and since $E(\Theta) = \theta_0$, we find that $E(\hat{\theta}_D) = \theta_0$. Thus $\hat{\theta}_D$ is Bayes unbiased for *any* sampling model $f(t|\theta)$ and *any* prior $g(\theta)$.

Wu (1980) shows that if $f(t|\theta)$ is IFR or IFRA, then the Bayes risk of $\hat{\theta}_D$ is *less* than that under an $\mathcal{E}(\lambda)$ sampling distribution for *any* prior distribution $g(\theta)$. Consequently, the $\mathcal{E}(\lambda)$ model provides an upper bound on the Bayes risk when using linear estimators of the form $\hat{\theta}_D$. Wu also shows that the Bayes risk of $\hat{\theta}_D$ under squared-error loss tends to zero as n tends to ∞ for *any* sampling model and *any* prior distribution.

Barlow and Wu (1980) show that, based on an $\mathcal{E}(\lambda)$ model and arbitrary prior on $\Theta = 1/\Lambda$, the posterior mean is an increasing function of sample size n when the true sampling model is IFR and certain fixed stopping rules are used.

8.4. RELIABILITY ESTIMATION

In this section point and interval estimation techniques are developed for estimating the $\mathcal{E}(\lambda)$ reliability function $r(t_0)$. Recall that $r(t_0)$ represents the probability that a unit will survive a specified mission time t_0. Recall that, in the case of the $\mathcal{E}(\lambda)$ distribution, $r(t_0) = 1 - F(t_0) = e^{-\lambda t_0}$. As is our custom, $R(t_0)$ will be used to explicitly denote the fact that reliability is an r.v. for a given time t_0, while $r(t_0)$ will be used to denote a value of $R(t_0)$.

$\mathcal{B}_1(\alpha_0, \beta_0)$ and noninformative priors for $R(t_0)$ will be considered, with a special treatment of the $\mathcal{U}(\alpha_0, \beta_0)$ prior. We also consider prior distributions induced on $R(t_0)$ by $\mathcal{U}(\alpha_0, \beta_0)$ priors for the failure rate Λ and the MTTF Θ, as well as a $\mathcal{G}(\alpha_0, \beta_0)$ prior for Λ. Finally, implications of a $\mathcal{B}_1(\alpha_0, \beta_0)$ prior at $R(t_0)$ on the reliability at some other time t_1, $R(t_1)$, are considered.

The likelihood function (8.4), written in terms of $r = r(t_0)$, is given by

$$L(r|\underset{\sim}{z}) \propto \left(-\frac{1}{t_0}\ln r\right)^s r^{\mathrm{t}/t_0}, \qquad 0 \leqslant r \leqslant 1. \tag{8.82}$$

Letting $g(r)$ represent the prior distribution of $R = R(t_0)$, the posterior distribution of R given t is

$$g(r|\mathrm{t}; s) = \frac{(-\ln r)^s r^{\mathrm{t}/t_0} g(r)}{\int_0^1 (-\ln y)^s y^{\mathrm{t}/t_0} g(y)\, dy}, \qquad 0 \leqslant r \leqslant 1. \tag{8.83}$$

8.4.1 Beta Prior Distribution

Consider the $\mathcal{B}_1(\alpha_0, \beta_0)$ prior distribution for R given by

$$g(r; \alpha_0, \beta_0) = \frac{1}{B(\alpha_0, \beta_0)} r^{\alpha_0 - 1}(1-r)^{\beta_0 - 1}, \qquad 0 \leqslant r \leqslant 1, \qquad \alpha_0, \beta_0 > 0. \tag{8.84}$$

Canfield (1970) finds the Bayes estimator of reliability using this prior and an asymmetric loss function. Substituting (8.84) into (8.83) yields the following posterior distribution of R given t:

$$g(r|\mathrm{t}; s, t_0, \alpha_0, \beta_0) = \frac{(-\ln r)^s r^{\alpha_0 + \mathrm{t}/t_0 - 1}(1-r)^{\beta_0 - 1}}{\int_0^1 (-\ln y)^s y^{\alpha_0 + \mathrm{t}/t_0 - 1}(1-y)^{\beta_0 - 1}\, dy}, \qquad 0 \leqslant r \leqslant 1. \tag{8.85}$$

If β_0 is assumed to be an integer, the $(1-y)^{\beta_0 - 1}$ term in the denominator can be expanded using the binomial theorem. Using this expansion and making the transformation $w = -\ln y$, the denominator of (8.85) becomes

$$\sum_{k=0}^{\beta_0 - 1} \binom{\beta_0 - 1}{k} (-1)^k \int_0^\infty w^s e^{-w(\alpha_0 + \mathrm{t}/t_0 + k)}\, dw$$

$$= \Gamma(s+1) \sum_{k=0}^{\beta_0 - 1} \frac{\binom{\beta_0 - 1}{k}(-1)^k}{(\alpha_0 + \mathrm{t}/t_0 + k)^{s+1}}. \tag{8.86}$$

Substituting (8.86) into (8.85) gives the posterior distribution

$$g(r|\mathrm{t}; s, t_0, \alpha_0, \beta_0) = \frac{(-\ln r)^s r^{\alpha_0 + \mathrm{t}/t_0 - 1}(1-r)^{\beta_0 - 1}}{\Gamma(s+1) \sum_{k=0}^{\beta_0 - 1} \frac{\binom{\beta_0 - 1}{k}(-1)^k}{(\alpha_0 + \mathrm{t}/t_0 + k)^{s+1}}}, \qquad 0 \leqslant r \leqslant 1, \tag{8.87}$$

for integer values of β_0. The marginal distribution of $\mathcal{T}$ is

$$f(\mathrm{t}; s, t_0, \alpha_0, \beta_0) = \frac{s}{B(\alpha_0, \beta_0)} \frac{\mathrm{t}^{s-1}}{t_0^s} \sum_{k=0}^{\beta_0 - 1} \frac{\binom{\beta_0 - 1}{k}(-1)^k}{(\alpha_0 + \mathrm{t}/t_0 + k)^{s+1}},$$

$$0 < \mathrm{t} < \infty. \qquad (8.88)$$

Example 8.18. Refer to Example 8.3. Suppose that we wish to estimate reliability at time $t_0 = 4000$ h. Consider a $\mathcal{B}_1(27, 3)$ prior distribution for $R(t_0)$. These values were chosen to give a prior mean of 0.9 with a 5th percentile of approximately 0.8 [see Section 6.5]. The corresponding posterior distribution is given by

$$g(r | 1.6 \times 10^5; 5, 4 \times 10^3, 27, 3) = (9.0589 \times 10^{10})(-\ln r)^5 r^{66}(1-r)^2,$$

$$0 < r < 1.$$

The prior and posterior distributions are plotted in Figure 8.6.

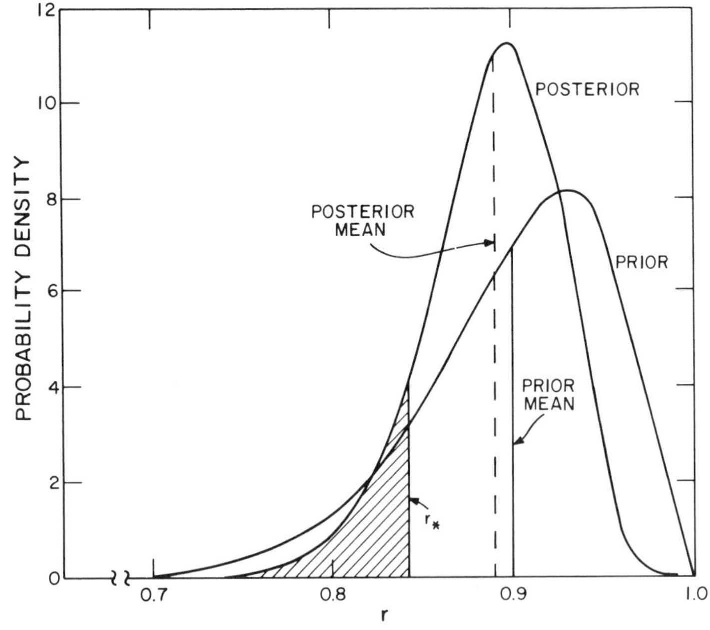

Figure 8.6. The prior and posterior distribution in Example 8.18. ■

Point Estimation. The posterior mean and second moment of R given t are found, using (8.87), to be

$$E(R|\mathrm{t}; s, t_0, \alpha_0, \beta_0) = \frac{\sum_{k=0}^{\beta_0 - 1} \left[\binom{\beta_0 - 1}{k} (-1)^k \right] \Big/ (\alpha_0 + \mathrm{t}/t_0 + k + 1)^{s+1}}{\sum_{k=0}^{\beta_0 - 1} \left[\binom{\beta_0 - 1}{k} (-1)^k \right] \Big/ (\alpha_0 + \mathrm{t}/t_0 + k)^{s+1}} \tag{8.89}$$

and

$$E(R^2|\mathrm{t}; s, t_0, \alpha_0, \beta_0) = \frac{\sum_{k=0}^{\beta_0 - 1} \left[\binom{\beta_0 - 1}{k} (-1)^k \right] \Big/ (\alpha_0 + \mathrm{t}/t_0 + k + 2)^{s+1}}{\sum_{k=0}^{\beta_0 - 1} \left[\binom{\beta_0 - 1}{k} (-1)^k \right] \Big/ (\alpha_0 + \mathrm{t}/t_0 + k)^{s+1}}. \tag{8.90}$$

The Bayes point estimator under squared-error loss is given by (8.89). The posterior variance is found by means of the usual expression involving (8.89) and (8.90). The Bayes risk of (8.89) is found by taking the expectation of the posterior variance with respect to the marginal distribution (8.88); however, this is not readily available in closed form. The mean of (8.89) is

$$E_{\mathrm{t}}\left[E_{r|\mathrm{t}}(R|\mathfrak{T}; s, t_0, \alpha_0, \beta_0) \right] = \frac{\alpha_0}{\alpha_0 + \beta_0}. \tag{8.91}$$

Example 8.19. Let us reconsider Example 8.18. The Bayes point estimate of r under squared-error loss, is

$$E(R|1.6 \times 10^5; 5, 4 \times 10^3, 27, 3) = \frac{1/(68)^6 - 2/(69)^6 + 1/(70)^6}{1/(67)^6 - 2/(68)^6 + 1/(69)^6} = .8897,$$

which is also shown in Figure 8.6. The posterior variance is calculated to be 1.33×10^{-3}. ■

Interval Estimation. A symmetric $100(1-\gamma)\%$ TBPI estimate of r is found by solving (5.45) and (5.46), using the posterior distribution (8.87), for r_*

and r^*, respectively. The resulting equations are given by

$$\Pr(R \geqslant r_* | t; s, t_0, \alpha_0, \beta_0) = \int_{r_*}^{1} g(r|t; s, t_0, \alpha_0, \beta_0)\, dr =$$

$$\frac{\sum_{k=0}^{\beta_0-1} \left[\binom{\beta_0-1}{k}(-1)^k \Big/ (\alpha_0 + t/t_0 + k)^{s+1}\right] \Gamma\left[s+1, -(\alpha_0 + t/t_0 + k)\ln r_*\right]}{\Gamma(s+1) \sum_{k=0}^{\beta_0-1} \binom{\beta_0-1}{k}(-1)^k \Big/ (\alpha_0 + t/t_0 + k)^{s+1}}$$

$$= 1 - \frac{\gamma}{2} \tag{8.92}$$

and

$$\Pr(R \geqslant r^* | t; s, t_0, \alpha_0, \beta_0) = \int_{r^*}^{1} g(r|t; s, t_0, \alpha_0, \beta_0)\, dr =$$

$$\frac{\sum_{k=0}^{\beta_0-1} \left[\binom{\beta_0-1}{k}(-1)^k \Big/ (\alpha_0 + t/t_0 + k)^{s+1}\right] \Gamma\left[s+1, -(\alpha_0 + t/t_0 + k)\ln r^*\right]}{\Gamma(s+1) \sum_{k=0}^{\beta_0-1} \binom{\beta_0-1}{k}(-1)^k \Big/ (\alpha_0 + t/t_0 + k)^{s+1}}$$

$$= \frac{\gamma}{2}, \tag{8.93}$$

where $\Gamma(a, x)$ is the standard incomplete gamma function. Correspondingly, $100(1-\gamma)\%$ LBPI or UBPI estimators are found by solving (8.92) or (8.93), respectively, with $\gamma/2$ replaced by γ.

Example 8.20. Again let us refer to Example 8.18. A direct computer search using the Amos and Daniel (1972) subroutine INCGAM is used to find the 90% LBPI estimate of r given by [0.841, 1.00]. The interval is shown in Figure 8.6. Also, the 90% TBPI estimate of r is found to be [.8241, .9431]. ■

8.4.2 Uniform Prior Distribution

The $\mathcal{U}(0,1)$ prior distribution on R is a special case of the $\mathcal{B}_1(\alpha_0, \beta_0)$ prior distribution discussed in Section 8.4.1 obtained by setting $\alpha_0 = \beta_0 = 1$. We treat this case separately due to the frequent use of the $\mathcal{U}(0,1)$ prior, and the fact that the results of the previous section are considerably simplified.

Setting $\alpha_0 = \beta_0 = 1$, the posterior distribution (8.87) of R given t becomes

$$g(r|\mathrm{t}; s, t_0) = \frac{(1+\mathrm{t}/t_0)^{s+1}(-\ln r)^s r^{\mathrm{t}/t_0}}{\Gamma(s+1)}, \qquad 0 \leqslant r \leqslant 1, \qquad (8.94)$$

which is recognized as an $\mathcal{NLG}_1(s, \mathrm{t}/t_0)$ distribution. The marginal distribution of $\mathcal{T}$ in the case of gamma sampling is

$$f(\mathrm{t}; s, t_0) = \frac{s\mathrm{t}^{s-1}}{t_0^s(1+\mathrm{t}/t_0)^{s+1}}, \qquad 0 < \mathrm{t} < \infty, \qquad (8.95)$$

and the posterior mean and variance of R are

$$E(R|\mathrm{t}; s, t_0) = \left(\frac{1+\mathrm{t}/t_0}{2+\mathrm{t}/t_0}\right)^{s+1} \qquad (8.96)$$

and

$$\mathrm{Var}(R|\mathrm{t}; s, t_0) = \left(\frac{1+\mathrm{t}/t_0}{3+\mathrm{t}/t_0}\right)^{s+1} - \left(\frac{1+\mathrm{t}/t_0}{2+\mathrm{t}/t_0}\right)^{2(s+1)}. \qquad (8.97)$$

Equation (8.96) gives the Bayes point estimator under squared-error loss. Bayes risk is not readily obtainable in closed form. The generalized ML estimator is the mode of (8.94), which is $\exp(-t_0 s/\mathcal{T})$. Note that this estimator is the same as the classical ML estimator.

A symmetric $100(1-\gamma)\%$ TBPI estimate of r is found by solving the equations

$$\Pr(R \geqslant r_*|\mathrm{t}; s, t_0) = \frac{\Gamma[s+1, -(1+\mathrm{t}/t_0)\ln r_*]}{\Gamma(s+1)} = 1 - \frac{\gamma}{2} \qquad (8.98)$$

and

$$\Pr(R \geqslant r^*|\mathrm{t}; s, t_0) = \frac{\Gamma[s+1, -(1+\mathrm{t}/t_0)\ln r^*]}{\Gamma(s+1)} = \frac{\gamma}{2} \qquad (8.99)$$

for r_* and r^*, respectively. $100(1-\gamma)\%$ LBPI or UBPI estimates are similarly obtained.

Example 8.21. Refer to Example 8.18. A $\mathcal{U}(0,1)$ prior for R yields the posterior distribution

$$g(r|1.6\times10^5;5,4\times10^3)=\frac{41^6}{120}(-\ln r)^5 r^{40}$$

$$=(3.9584\times10^7)(-\ln r)^5 r^{40}, \qquad 0<r<1.$$

The prior and posterior distributions are plotted in Figure 8.7. The Bayes estimate of r under squared-error loss is $E(R|1.6\times10^5;5,4\times10^3)=(41/42)^6=0.8654$. The posterior variance is calculated to be 2.6×10^{-3}. Interval estimates, found using subroutine INCGAM and a direct search program, are:

90% LBPI: [0.798, 1.0],
90% TBPI: [0.774, 0.938].

The Bayes point estimate and 90% LBPI are also illustrated in Figure 8.7.

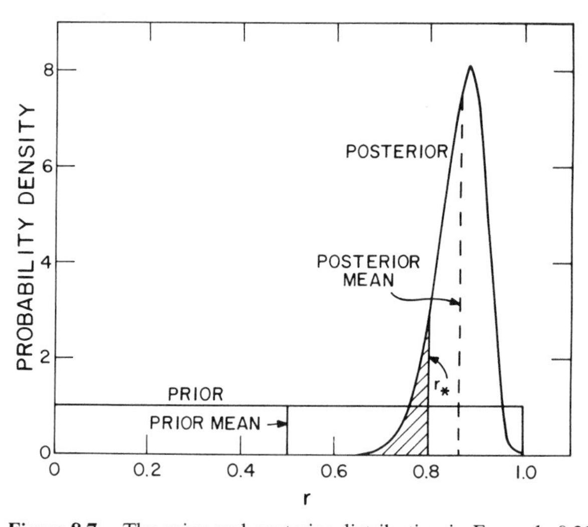

Figure 8.7. The prior and posterior distribution in Example 8.21. ■

8.4.3 Noninformative Prior Distribution

The noninformative prior distribution for $R(t_0)$, obtained either through the technique of Section 6.1 or by transforming the noninformative prior (8.19) for $\Lambda = -(1/t_0)\ln R$ is given by

$$g(r) \propto -\frac{1}{r \ln r}, \qquad 0 \leq r \leq 1. \tag{8.100}$$

Substituting this improper prior into (8.83) (or transforming (8.20)) gives the proper $\mathcal{NLG}(s, \mathrm{t}/t_0)$ posterior distribution

$$g(r|\mathrm{t}; s, t_0) = \frac{(-\ln r)^{s-1} \mathrm{t}^s r^{\mathrm{t}/t_0 - 1}}{t_0^s \Gamma(s)}, \qquad 0 \leq r \leq 1, \tag{8.101}$$

whereas the improper marginal distribution of $\mathcal{T}$, as in Sections 8.2.2 and 8.3.2, is $f(\mathrm{t}; s, t_0) \propto 1/\mathrm{t}$. The posterior mean and variance are

$$E(R|\mathrm{t}; s, t_0) = \frac{1}{(1 + t_0/\mathrm{t})^s} \tag{8.102}$$

and

$$\mathrm{Var}(R|\mathrm{t}; s, t_0) = \frac{1}{(1 + 2t_0/\mathrm{t})^s} - \frac{1}{(1 + t_0/\mathrm{t})^{2s}}, \tag{8.103}$$

respectively. The generalized Bayes point estimator with respect to squared-error loss is given by (8.102), and is seen to be asymptotically equal to the classical ML estimator $\hat{r} = \exp(-t_0 s/\mathcal{T})$. The moments and Bayes risk of (8.102) do not exist.

Equations for the $100(1-\gamma)\%$ TBPI estimate of the failure rate λ under a noninformative prior are given by (8.26) and (8.27). Inasmuch as r is a monotone decreasing function of λ, interval estimates for r are found by transforming λ_* and λ^*. The transformation results in the estimate

$$r_* = \exp(-\lambda^* t_0) = \exp\left[-t_0 \chi^2_{1-\gamma/2}(2s)/(2\mathrm{t})\right], \tag{8.104}$$

$$r^* = \exp(-\lambda_* t_0) = \exp\left[-t_0 \chi^2_{\gamma/2}(2s)/(2\mathrm{t})\right]. \tag{8.105}$$

$100(1-\gamma)\%$ LBPI or UBPI estimators are found by substituting γ for $\gamma/2$ in (8.104) or (8.105), respectively.

Example 8.22. Again refer to Example 8.18. The posterior distribution of R is

$$g(r|1.6 \times 10^5; 5, 4 \times 10^3) = (4.2667 \times 10^6)(-\ln r)^4 r^{39}, \qquad 0 < r < 1.$$

The Bayes point estimate of r under squared-error loss is

$$E(R|1.6\times10^5;5,4\times10^3)=1\Big/\left(1+\frac{4\times10^3}{1.6\times10^5}\right)^5=0.8839.$$

The posterior variance is calculated to be 2.33×10^{-3}. The classical ML estimate of r is $e^{-5/40}=.8825$. A 90% LBPI estimate of r has lower bound

$$r_*=\exp\left\{-\frac{4\times10^3}{2(1.6\times10^5)}\chi^2_{0.9}(10)\right\}=\exp\{-(0.0125)(15.987)\}=0.8189.$$

■

8.4.4 Implied Prior Distributions

Consider a situation where an estimate of reliability at some specified mission time t_1 is required, while the prior information available is most easily quantified in terms of some other reliability parameter. We may have prior information, for example, regarding λ, θ, or $r(t_0)$, where t_0 may be different from t_1. In these cases we develop estimation procedures for $r(t_1)$ based on the prior distribution implied by the original priors assigned to Λ, Θ, or $R(t_0)$. In particular, we consider procedures implied from $\mathcal{U}(\alpha_0,\beta_0)$ priors for Λ and Θ, a $\mathcal{G}_1(\alpha_0,\beta_0)$ prior for Λ (which is equivalent to an $\mathcal{IG}(\alpha_0,\beta_0)$ prior for Θ), and a $\mathcal{B}_1(\alpha_0,\beta_0)$ prior for $R(t_0)$. Notice that we do *not* consider the "reverse" process of implied priors for Λ or Θ based on original priors for $R(t_1)$. This process would lead to assigning priors for Λ or Θ that depend on the specified mission time t_1. It is natural for a prior for $R(t_1)$ to depend on the mission time t_1, as reliability is time-dependent. However, it is quite unnatural for a prior assigned to a parameter such as Λ to depend on a specified mission time.

The transformation of distributions for other parameters to a distribution for $R(t_1)$ is fairly straightforward. If $g_\lambda(\lambda)$ represents the prior or posterior distribution for Λ, the corresponding or implied distribution for $R(t_1)=\exp(-\Lambda t_1)$, using (7.105), is given by

$$g_r[r(t_1)]=g_\lambda[-\ln r(t_1)/t_1]\frac{1}{t_1 r(t_1)}. \tag{8.106}$$

Letting $g_\theta(\theta)$ represent the prior or posterior distribution for Θ, the corresponding distribution for $R(t_1)=\exp(-t_1/\Theta)$ is

$$g_r[r(t_1)] = g_\theta[-t_1/\ln r(t_1)]\frac{t_1}{r(t_1)[\ln r(t_1)]^2}. \tag{8.107}$$

Finally, if $g_{r(t_0)}[r(t_0)]$ represents the prior or posterior distribution for the reliability at time t_0, then the corresponding distribution for the reliability at time t_1, $R(t_1)=R(t_0)^{t_1/t_0}$, is

$$g_{r(t_1)}[r(t_1)] = g_{r(t_0)}\left[r(t_1)^{t_0/t_1}\right]\left(\frac{t_0}{t_1}\right)r(t_1)^{t_0/t_1-1}. \tag{8.108}$$

Uniform Prior for Λ. Harris and Singpurwalla (1968) consider the problem of reliability estimation when a $\mathcal{U}(\alpha_0,\beta_0)$ prior distribution is assigned to Λ. Substituting (8.6) into (8.106) gives the implied prior distribution

$$g(r;t_1,\alpha_0,\beta_0)=\frac{1}{\beta_0-\alpha_0}\frac{1}{t_1 r}, \qquad e^{-\beta_0 t_1}\leqslant r\leqslant e^{-\alpha_0 t_1}, \tag{8.109}$$

for $R=R(t_1)$.

The resulting prior mean and variance of R are

$$E(R;t_1,\alpha_0,\beta_0)=\frac{e^{-\alpha_0 t_1}-e^{-\beta_0 t_1}}{t_1(\beta_0-\alpha_0)} \tag{8.110}$$

and

$$\operatorname{Var}(R;t_1,\alpha_0,\beta_0)=\frac{e^{-2\alpha_0 t_1}-e^{-2\beta_0 t_1}}{2t_1(\beta_0-\alpha_0)}-\frac{(e^{-\alpha_0 t_1}-e^{-\beta_0 t_1})^2}{t_1^2(\beta_0-\alpha_0)^2}. \tag{8.111}$$

The posterior distribution, obtained by transforming (8.9), is

$$g(r|\mathrm{t};s,t_1,\alpha_0,\beta_0)=\frac{\mathrm{t}^{s+1}\left(-\dfrac{1}{t_1}\ln r\right)^s r^{\mathrm{t}/t_1-1}}{t_1[\Gamma(s+1,\beta_0\mathrm{t})-\Gamma(s+1,\alpha_0\mathrm{t})]},$$

$$e^{-\beta_0 t_1}\leqslant r\leqslant e^{-\alpha_0 t_1}. \tag{8.112}$$

The implied prior cdf for $R(t_1)$ is

$$G(r; t_1, \alpha_0, \beta_0) = \begin{cases} 0, & r < e^{-\beta_0 t_1} \\ \dfrac{\beta_0 t_1 + \ln r}{t_1(\beta_0 - \alpha_0)}, & e^{-\beta_0 t_1} \leqslant r \leqslant e^{-\alpha_0 t_1} \\ 1, & e^{-\alpha_0 t_1} < r, \end{cases} \tag{8.113}$$

and the corresponding posterior cdf is

$$G(r|\mathrm{t}; s, t_1, \alpha_0, \beta_0) = \begin{cases} 0, & r < e^{-\beta_0 t_1} \\ \dfrac{\Gamma(s+1, \beta_0 \mathrm{t}) - \Gamma(s+1, -\mathrm{t}/t_1 \ln r)}{\Gamma(s+1, \beta_0 \mathrm{t}) - \Gamma(s+1, \alpha_0 \mathrm{t})}, & e^{-\beta_0 t_1} \leqslant r \leqslant e^{-\alpha_0 t_1} \\ 1, & e^{-\alpha_0 t_1} < r. \end{cases} \tag{8.114}$$

The posterior mean of R given t is

$$E(R|\mathrm{t}; s, t_1, \alpha_0, \beta_0) = \left(\frac{\mathrm{t}}{\mathrm{t} + t_1}\right)^{s+1} \times \frac{\Gamma[s+1, \beta_0(\mathrm{t}+t_1)] - \Gamma[s+1, \alpha_0(\mathrm{t}+t_1)]}{\Gamma(s+1, \beta_0 \mathrm{t}) - \Gamma(s+1, \alpha_0 \mathrm{t})}. \tag{8.115}$$

The second moment of R given t is

$$E(R^2|\mathrm{t}; s, t_1, \alpha_0, \beta_0) = \left(\frac{\mathrm{t}}{\mathrm{t} + 2t_1}\right)^{s+1} \times \frac{\Gamma[s+1, \beta_0(\mathrm{t}+2t_1)] - \Gamma[s+1, \alpha_0(\mathrm{t}+2\mathrm{t}_1)]}{\Gamma(s+1, \beta_0 \mathrm{t}) - \Gamma(s+1, \alpha_0 \mathrm{t})}, \tag{8.116}$$

and the posterior variance can be found by means of the usual expression. Equation (8.115) is the Bayes point estimator for $r(t_1)$ under squared-error loss. Interval estimators are found by transforming values of λ_* and λ^* found according to (8.15) and (8.16). A symmetric $100(1-\gamma)\%$ TBPI estimate for $r(t_1)$ is given by

$$r_*(t_1) = e^{-\lambda^* t_1} \tag{8.117}$$

and

$$r^*(t_1) = e^{-\lambda_* t_1}. \tag{8.118}$$

Example 8.23. Suppose that a $\mathcal{U}(1.4\times 10^{-5},\ 5\times 10^{-5})$ prior has been assigned to Λ as in Example 8.3, and that we wish to estimate the reliability at time $t_1 = 5\times 10^3$ h. The implied prior distribution of $R(t_1)$ is

$$g(r; 5\times 10^3, 1.4\times 10^{-5}, 5\times 10^{-5}) = \frac{5.5556}{r}, \qquad 0.78 < r < 0.93.$$

The prior mean and variance are

$$E(R; 5\times 10^3, 1.4\times 10^{-5}, 5\times 10^{-5}) = \frac{e^{-0.07} - e^{-0.25}}{5000(3.6\times 10^{-5})} = 0.8533$$

and

$$\begin{aligned}\text{Var}(R; 5\times 10^3, 1.4\times 10^{-5}, 5\times 10^{-5}) &= \frac{e^{-0.14} - e^{-0.50}}{2(5000)(3.6\times 10^{-5})} - (0.8533)^2 \\ &= 2.0\times 10^{-3}.\end{aligned}$$

The corresponding posterior distribution is

$$\begin{aligned}g(r \mid 1.6\times 10^5; 5, 5\times 10^3, 1.4\times 10^{-5}, 5\times 10^{-5}) &= \frac{1.073\times 10^9}{\Gamma(6,8) - \Gamma(6,2.24)} \\ &\quad \times (-\ln r)^5 r^{31}, \qquad 0.78 < r < 0.93.\end{aligned}$$

Figure 8.8 illustrates the prior and posterior distributions, along with their respective means. The posterior mean is calculated to be

$$\left(\frac{1.6\times 10^5}{1.65\times 10^5}\right)^6 \frac{\Gamma(6,8.25) - \Gamma(6,2.31)}{\Gamma(6,8.00) - \Gamma(6,2.24)} = 0.8508,$$

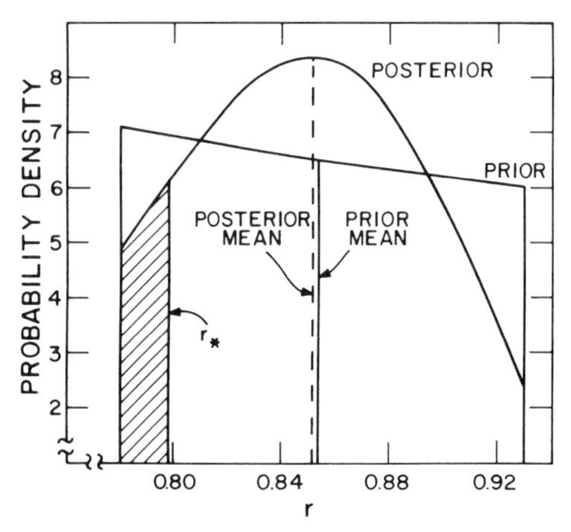

Figure 8.8. The prior and posterior distribution in Example 8.23.

and the posterior variance is calculated to be 1.53×10^{-3}. Interval estimates are found by transforming the results of Example 8.5. The 90% LBPI estimate of $r(t_1)$ has lower bound $r_*(5\times10^3)=e^{-(4.534\times10^{-5})(5\times10^3)}=0.7972$, and is also shown in Figure 8.8. ■

Uniform Prior for Θ. Bhattacharya (1967) considers the case of a $\mathcal{U}(\alpha_0, \beta_0)$ prior distribution for Θ, and obtains the corresponding prior and posterior distributions for $R=R(t_1)$. The prior distribution is

$$g(r; t_1, \alpha_0, \beta_0)=\frac{t_1}{(\beta_0-\alpha_0)}\frac{1}{r(\ln r)^2}, \qquad e^{-t_1/\alpha_0}\leqslant r\leqslant e^{-t_1/\beta_0}, \tag{8.119}$$

whereas the posterior distribution is found to be

$$g(r|\mathrm{t}; s, t_1, \alpha_0, \beta_0)=\frac{\mathrm{t}^{s-1}r^{\mathrm{t}/t_1-1}(-\ln r)^{s-2}}{t_1^{s-1}\Gamma^*(s-1,\mathrm{t})}, \qquad e^{-t_1/\alpha_0}\leqslant r\leqslant e^{-t_1/\beta_0}, \tag{8.120}$$

where $\Gamma^*(s-1,\mathrm{t})\equiv\Gamma(s-1,\mathrm{t}/\alpha_0)-\Gamma(s-1,\mathrm{t}/\beta_0)$. Prior moment calculations require the evaluation of an infinite series whose terms depend on the choice of (α_0, β_0), and will not be presented here. Bhattacharya also gives the posterior mean and variance:

$$E(R|\mathrm{t};s,t_1,\alpha_0,\beta_0)=\left(\frac{\mathrm{t}}{\mathrm{t}+t_1}\right)^{s-1}\frac{\Gamma^*(s-1,\mathrm{t}+t_1)}{\Gamma^*(s-1,\mathrm{t})}, \tag{8.121}$$

$$\mathrm{Var}(R|\mathrm{t};s,t_1,\alpha_0,\beta_0)=$$

$$\frac{1}{[\Gamma^*(s-1,\mathrm{t})]^2}\left\{\frac{\Gamma^*(s-1,\mathrm{t}+2t_1)\Gamma^*(s-1,\mathrm{t})}{(1+2t_1/\mathrm{t})^{s-1}}-\frac{[\Gamma^*(s-1,\mathrm{t}+t_1)]^2}{(1+t_1/\mathrm{t})^{2s-2}}\right\}. \tag{8.122}$$

Calculation of the variance of (8.121), in the case of gamma sampling, requires the calculation of expectations with respect to the marginal distribution (8.44).

Interval estimators for $R(t_1)$ are found by transforming the estimators developed in Section 8.3.1 for Θ. A symmetric $100(1-\gamma)\%$ TBPI estimate of $r(t_1)$ is given by

$$r_*(t_1)=e^{-t_1/\theta_*} \tag{8.123}$$

$$r^*(t_1)=e^{-t_1/\theta^*}, \tag{8.124}$$

where θ_* and θ^* are the solutions to (8.50) and (8.51), respectively.

Example 8.24. Suppose that a $\mathcal{U}(2\times10^4,7\times10^4)$ prior has been assigned to Θ, as in Example 8.10, and that once again we wish to estimate reliability at time $t_1=5000$ h. The implied prior distribution of $R(t_1)$ is

$$g(r;5\times10^3,2\times10^4,7\times10^4)=\frac{0.1}{r(\ln r)^2},\qquad 0.78<r<0.93,$$

and the corresponding posterior is

$$g(r|1.6\times10^5;5,5\times10^3,2\times10^4,7\times10^4)=\frac{(32)^4}{\Gamma^*(4,1.6\times10^5)}r^{31}(-\ln r)^3,$$

$$0.78<r<0.93.$$

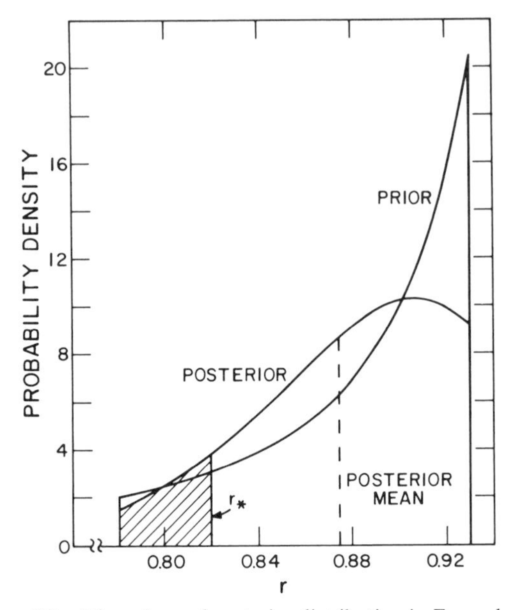

Figure 8.9. The prior and posterior distribution in Example 8.24.

The posterior mean is

$$E\left(R \mid 1.6\times 10^5; 5, 5\times 10^3, 2\times 10^4, 7\times 10^4\right) =$$

$$(0.9697)^4 \frac{\Gamma^*\left(4, 1.65\times 10^5\right)}{\Gamma^*\left(4, 1.6\times 10^5\right)} = 0.8749,$$

whereas the posterior variance is calculated to be 1.41×10^{-3}. Interval estimators are found by transforming the results of Example 8.12. The 90% LBPI estimate of $r(t_1)$ has lower bound $r_*(t_1) = e^{-5\times 10^3/2.5\times 10^4} = 0.8187$. Figure 8.9 illustrates the prior distribution, posterior distribution, posterior mean, and 90% LBPI estimate. ■

Gamma Prior for Λ. Suppose that a $\mathcal{G}(\alpha_0, \beta_0)$ prior distribution is assigned to Λ. This case has been considered by Bhattacharya (1967), Harris and Singpurwalla (1968), and Moore and Bilikam (1978). Harris and Singpurwalla derive the implied prior distribution on $R = R(t_1)$ as

$$g(r; t_1, \alpha_0, \beta_0) = \frac{1}{\Gamma(\alpha_0)(\beta_0 t_1)^{\alpha_0}} (-\ln r)^{\alpha_0 - 1} r^{1/\beta_0 t_1 - 1}, \qquad 0 \leqslant r \leqslant 1, \tag{8.125}$$

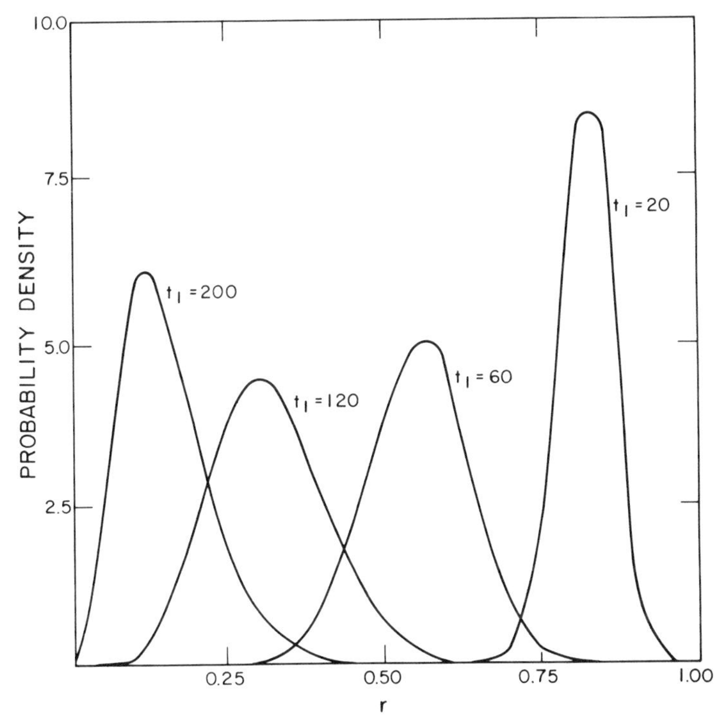

Figure 8.10. The implied prior distribution for $R(t_1)$ based on a $\mathcal{G}(16, 6.25\times10^{-4})$ prior for Λ.

which is an $\mathcal{NLG}[\alpha_0, 1/(\beta_0 t_1)]$ distribution. Figure 8.10 illustrates shapes of implied priors for various values of t_1 based on a $\mathcal{G}(16, 6.25\times10^{-4})$ prior for Λ. We compute the corresponding prior mean and variance to be

$$E(R; t_1, \alpha_0, \beta_0,) = \frac{1}{(1+\beta_0 t_1)^{\alpha_0}} \tag{8.126}$$

and

$$\mathrm{Var}(R; t_1, \alpha_0, \beta_0) = \frac{1}{(1+2\beta_0 t_1)^{\alpha_0}} - \frac{1}{(1+\beta_0 t_1)^{2\alpha_0}}. \tag{8.127}$$

The posterior distribution, obtained by transforming (8.29), is

$$g(R|\mathrm{t}; s, t_1, \alpha_0, \beta_0) = \frac{1}{\Gamma(\alpha_0+s)t_1^{\alpha_0+s}}(\mathrm{t}+1/\beta_0)^{\alpha_0+s}(-\ln r)^{\alpha_0+s-1} r^{(\mathrm{t}+1/\beta_0)/t_1-1},$$

$$0 \leqslant r \leqslant 1, \tag{8.128}$$

which is also recognized as being a $\mathcal{NLG}[\alpha_0 + s, (t + 1/\beta_0)/t_1]$ distribution. Bhattacharya derives the posterior mean and variance given by

$$E(R|\mathrm{t}; s, t_1, \alpha_0, \beta_0) = \frac{1}{[1 + t_1/(\mathrm{t} + 1/\beta_0)]^{\alpha_0 + s}} \tag{8.129}$$

and

$$\mathrm{Var}(R|\mathrm{t}; s, t_1, \alpha_0, \beta_0) = \frac{1}{[1 + 2t_1/(\mathrm{t} + 1/\beta_0)]^{\alpha_0 + s}} - \frac{1}{[1 + t_1/(\mathrm{t} + 1/\beta_0)]^{2\alpha_0 + 2s}}. \tag{8.130}$$

Calculation of the actual variance of (8.129) requires the calculation of moments of (8.129) with respect to the marginal distribution (8.30). The Bayes risk of (8.129), in the case where α_0 is an integer, is

$$r(g) = \left\{ \frac{1}{(1 + 2t_1\beta_0)^{\alpha_0}} - \sum_{k=0}^{\alpha_0 + s} \frac{\binom{\alpha_0 + s}{k} \Gamma(\alpha_0 + s)\Gamma(s + k)\Gamma(2\alpha_0 + s - k)}{\Gamma(s)\Gamma(\alpha_0)\Gamma(2\alpha_0 + 2s)(1 + \beta_0 t_1)^{2\alpha_0 + s - k}} \right\}. \tag{8.131}$$

Interval estimators are found by transforming the estimators λ_* and λ^* given in Section 8.2.3. A symmetric $100(1-\gamma)\%$ TBPI estimate of $r(t_1)$ is given by

$$r_*(t_1) = \exp\left\{ \frac{-t_1\beta_0}{2(1 + \beta_0 \mathrm{t})} \chi^2_{1-\gamma/2}(2\alpha_0 + 2s) \right\}, \tag{8.132}$$

$$r^*(t_1) = \exp\left\{ \frac{-t_1\beta_0}{2(1 + \beta_0 \mathrm{t})} \chi^2_{\gamma/2}(2\alpha_0 + 2s) \right\}. \tag{8.133}$$

Example 8.25. Consider the $\mathcal{G}(8.5, 3.5 \times 10^{-6})$ prior distribution for Λ used in Example 8.7 and suppose that we wish to estimate reliability at time $t_1 = 5000$ h. The implied prior distribution for $R = R(t_1)$ is

$$g(r; 5 \times 10^3, 8.5, 3.5 \times 10^{-6}) = (6.1232 \times 10^{10})(-\ln r)^{7.5} r^{56.14}, \qquad 0 < r < 1,$$

with prior mean

$$E(R;5\times10^{3},8.5,3.5\times10^{-6})=\frac{1}{[1+(3.5\times10^{-6})(5\times10^{3})]^{8.5}}=0.8629$$

and prior variance calculated to be 1.875×10^{-3}. The posterior distribution is

$$g(r|1.6\times10^{5};5,5\times10^{3},8.5,3.5\times10^{-6})=(1.2389\times10^{17})(-\ln r)^{12.5}r^{88.14},\qquad 0<r<1,$$

with posterior mean

$$E(R|1.6\times10^{5};5,5\times10^{3},8.5,3.5\times10^{-6})=\left[1+\frac{5\times10^{3}}{4.4571\times10^{5}}\right]^{-13.5}$$
$$=0.8602$$

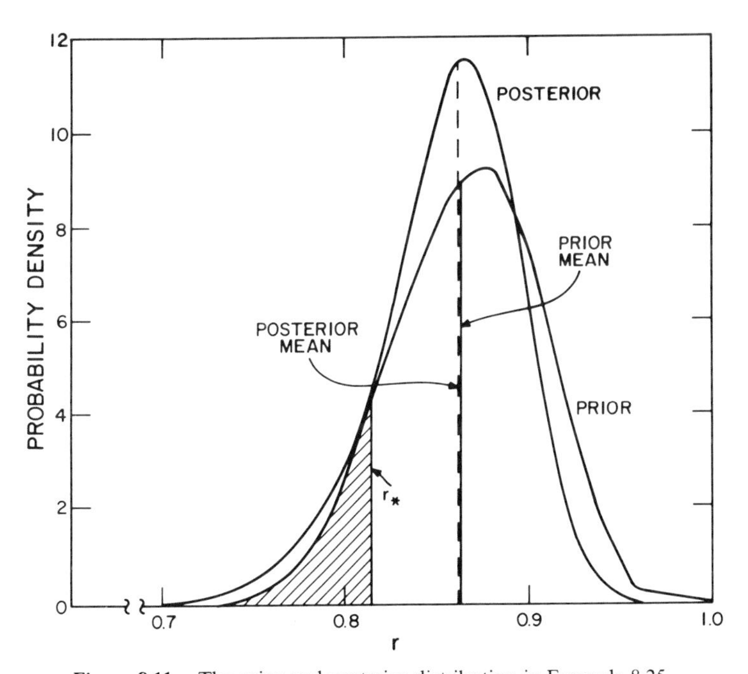

Figure 8.11. The prior and posterior distribution in Example 8.25.

and posterior variance calculated to be 1.230×10^{-3}. The 90% LBPI estimate of $r(t_1)$ has lower bound

$$r_*(t_1) = \exp\left\{ -\frac{(5\times 10^3)(3.5\times 10^{-6})}{2(1.56)} \chi^2_{0.9}(27) \right\}$$

$$= \exp\{-(5.609\times 10^{-3})(36.741)\} = 0.8138.$$

Figure 8.11 illustrates the prior and posterior distributions, along with their respective means, and the 90% LBPI estimate. ■

Beta Prior for $R(t_0)$. Consider the situation where a $\mathcal{B}_1(\alpha_0, \beta_0)$ prior has been assigned to the reliability at time t_0, $R(t_0)$, and we wish to estimate the reliability at time t_1, $R(t_1)$. Substituting (8.84) into the transformation formula (8.108) gives the implied prior distribution

$$g(r; t_0, t_1, \alpha_0, \beta_0) = \frac{1}{B(\alpha_0, \beta_0)} r^{t_0\alpha_0/t_1 - 1}(1 - r^{t_0/t_1})^{\beta_0 - 1}\left(\frac{t_0}{t_1}\right), \qquad 0 \leqslant r \leqslant 1, \tag{8.134}$$

for $R = R(t_1)$. The prior mean and variance are calculated to be

$$E(R; t_0, t_1, \alpha_0, \beta_0) = \frac{B(\alpha_0 + t_1/t_0, \beta_0)}{B(\alpha_0, \beta_0)} \tag{8.135}$$

and

$$\mathrm{Var}(R; t_0, t_1, \alpha_0, \beta_0) = \frac{B(\alpha_0 + 2t_1/t_0, \beta_0)}{B(\alpha_0, \beta_0)} - \frac{B^2(\alpha_0 + t_1/t_0, \beta_0)}{B^2(\alpha_0, \beta_0)}. \tag{8.136}$$

Figure 8.12 illustrates several implied prior distributions on $R(t_1)$ for various values of t_1 based on a $\mathcal{U}(0,1)$ prior for $R(100)$. The posterior distribution, in the case where β_0 is an integer, is

$$g(r|\mathrm{t}; s, t_0, t_1, \alpha_0, \beta_0) = \frac{(t_0/t_1)^{s+1}(-\ln r)^s r^{\mathrm{t}/t_1 + t_0\alpha_0/t_1 - 1}(1 - r^{t_0/t_1})^{\beta_0 - 1}}{\Gamma(s+1)\sum_{k=0}^{\beta_0 - 1}\binom{\beta_0 - 1}{k}(-1)^k \Big/ (\mathrm{t}/t_0 + \alpha_0 + k)^{s+1}},$$

$$0 \leqslant r \leqslant 1. \tag{8.137}$$

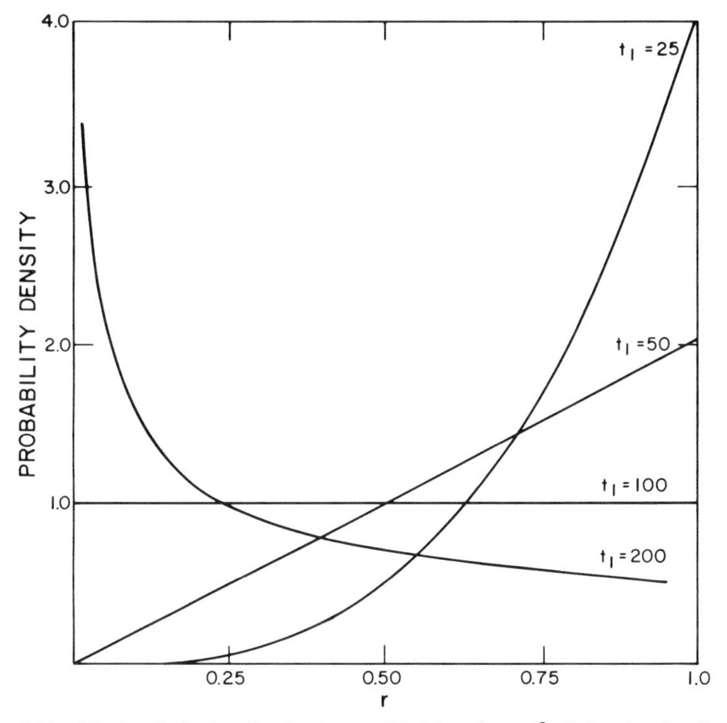

Figure 8.12. The implied prior distribution on $R(t_1)$ based on a $\mathcal{U}(0,1)$ prior distribution for $R(100)$.

The posterior mean and second moment of $R(t_1)$ are

$$E(R|\mathrm{t}; s, t_0, t_1, \alpha_0, \beta_0) = \frac{\sum_{k=0}^{\beta_0-1} \binom{\beta_0-1}{k} (-1)^k \Big/ (\mathrm{t}/t_0 + t_1/t_0 + \alpha_0 + k)^{s+1}}{\sum_{k=0}^{\beta_0-1} \binom{\beta_0-1}{k} (-1)^k \Big/ (\mathrm{t}/t_0 + \alpha_0 + k)^{s+1}} \tag{8.138}$$

and

$$E(R^2|\mathrm{t}; s, t_0, t_1, \alpha_0, \beta_0) = \frac{\sum_{k=0}^{\beta_0-1} \binom{\beta_0-1}{k} (-1)^k \Big/ (\mathrm{t}/t_0 + 2t_1/t_0 + \alpha_0 + k)^{s+1}}{\sum_{k=0}^{\beta_0-1} \binom{\beta_0-1}{k} (-1)^k \Big/ (\mathrm{t}/t_0 + \alpha_0 + k)^{s+1}}. \tag{8.139}$$

Note that in the case of the $\mathcal{U}(0,1)$ prior distribution for $R(t_0)$ the results simplify considerably, and, in particular, the Bayes estimator (8.138) reduces to

$$E(R|\mathrm{t};s,t_0,t_1)=\left(\frac{\mathrm{t}/t_0+1}{\mathrm{t}/t_0+t_1/t_0+1}\right)^{s+1} \tag{8.140}$$

Interval estimators are found by transforming the solutions $r_*(t_0)$ and $r^*(t_0)$ of (8.92) and (8.93). A symmetric $100(1-\gamma)\%$ TBPI estimate of $r(t_1)$ is given by

$$r_*(t_1)=r_*(t_0)^{t_1/t_0}, \tag{8.141}$$

$$r^*(t_1)=r^*(t_0)^{t_1/t_0}. \tag{8.142}$$

$100(1-\gamma)\%$ LBPI or UBPI estimators are found by solving (8.92) or (8.93), respectively, with $\gamma/2$ replaced by γ, and then making the appropriate transformation.

Example 8.26. Consider the $\mathcal{B}_1(27,3)$ prior assigned to $R(t_0)$ at $t_0=4000$ h in Example 8.18, and suppose we wish to estimate reliability at $t_1=5000$ h. The implied prior distribution on $R=R(t_1)$ is

$$g(r;4\times10^3,5\times10^3,27,3)=\frac{0.8}{B(27,3)}r^{20.6}(1-r^{0.8})^2,\qquad 0<r<1,$$

with prior mean

$$E(R;4\times10^3,5\times10^3,27,3)=\frac{B(28.25,3)}{B(27,3)}=0.8771$$

and prior variance calculated to be 4.232×10^{-3}. The corresponding posterior distribution is

$$g(r|1.6\times10^5;5,4\times10^3,5\times10^3,27,3)$$

$$=(2.3747\times10^{10})(-\ln r)^5 r^{52.6}(1-r^{0.8})^2,\qquad 0<r<1,$$

with posterior mean

$$E(R|1.6\times10^5;5,4\times10^3,5\times10^3,27,3)=\frac{7.9512\times10^{-14}}{9.1999\times10^{-14}}=0.8644,$$

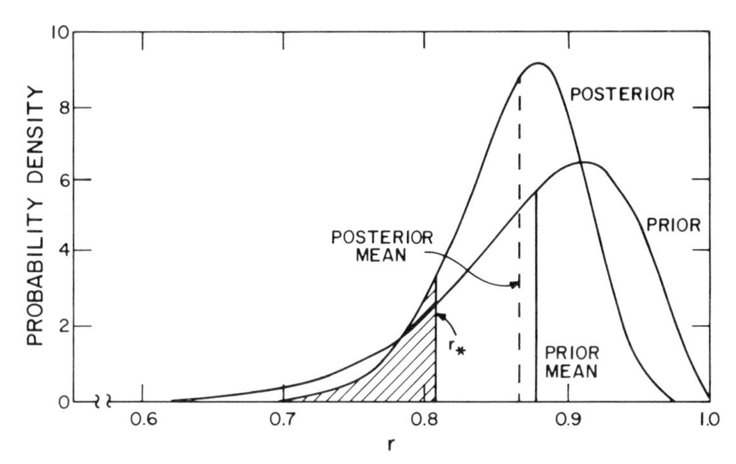

Figure 8.13. The prior and posterior distribution in Example 8.26.

and posterior variance calculated to be 1.951×10^{-3}. The lower bound of the 90% LBPI estimate of $r(t_1)$ is found by transforming the corresponding result of Example 8.20:

$$r_*(t_1) = (0.8241)^{1.25} = 0.8054.$$

Figure 8.13 illustrates the prior and posterior distributions along with their corresponding means, and the 90% LBPI estimate. ■

8.5. RELIABLE LIFE ESTIMATION

The reliable life t_R was defined in Chapter 4 as the time t_R for which $100R\%$ of the population will survive. For the $\mathcal{E}(\lambda)$ distribution, t_R is given by (4.14). Reliable life estimation in the case of an $\mathcal{E}(\lambda)$ lifetime distribution is easily done by relating reliable life to the MTTF θ. Recall that, in the case of an $\mathcal{E}(\lambda)$ distribution, the reliable life in terms of θ is given by

$$t_R = (-\ln R)\theta. \tag{8.143}$$

Thus, for a specified value of R, t_R is seen to be the product of the positive constant $(-\ln R)$ and the MTTF θ. Estimation procedures for t_R will be based on estimation procedures for θ developed in Section 8.3.

Letting $g_\theta(\theta)$ represent the prior or posterior distribution of Θ, the corresponding prior or posterior distribution for T_R is

$$g_{t_R}(t_R) = g_\theta\left(-\frac{t_R}{\ln R}\right)\left(-\frac{1}{\ln R}\right). \tag{8.144}$$

Prior and posterior moments of T_R are related to the corresponding moments of Θ according to

$$E(T_R) = (-\ln R)E(\Theta) \tag{8.145}$$

and

$$\mathrm{Var}(T_R) = (-\ln R)^2 \mathrm{Var}(\Theta). \tag{8.146}$$

Equation (8.145) can be used to obtain the squared-error loss Bayes point estimator of t_R based on the estimator of Θ. Furthermore, the squared-error loss minimum Bayes risks can be related by

$$r_{t_R}(g) = (-\ln R)^2 r_\theta(g). \tag{8.147}$$

Interval estimators are also obtained by transforming results previously derived for θ. A symmetric $100(1-\gamma)\%$ TBPI estimate of t_R is found by first finding the symmetric $100(1-\gamma)\%$ TBPI estimate of θ and then letting

$$t_{R*} = (-\ln R)\theta_*,$$

$$t_R{}^* = (-\ln R)\theta^*. \tag{8.148}$$

Example 8.27. Refer to Examples 8.14, 8.15, and 8.16. Suppose that we wish to estimate reliable life for $R=0.95$ based on an $\mathcal{IG}(8.5, 2.86\times10^5)$ prior distribution on Θ. The Bayes point estimate of $t_{0.95}$ under squared-error loss is

$$E(T_R|1.6\times10^5; 5, 8.5, 2.86\times10^5) = (-\ln 0.95)(3.57\times10^4) = 1.83\times10^3 \text{ h}$$

and the posterior variance is $(-\ln 0.95)^2(1.11\times10^8) = 2.92\times10^5$. The minimum Bayes risk ($a=1$) is computed to be 3.53×10^5, and the 90% TBPI estimate of t_R is

$$t_{R*} = (-\ln 0.95)(2.22\times10^4) = 1.14\times10^3,$$

$$t_R{}^* = (-\ln 0.95)(5.52\times10^4) = 2.83\times10^3. \qquad ■$$

8.6. ESTIMATION FOR THE TWO-PARAMETER EXPONENTIAL DISTRIBUTION

In this section we consider MTTF, threshold, and reliability estimation in the $\mathcal{E}_1(\theta, \mu)$ distribution with pdf given by

$$f(t; \theta, \mu) = \frac{1}{\theta} e^{-(t-\mu)/\theta}, \qquad t \geqslant \mu. \tag{8.149}$$

Sinha and Guttman (1976) and Varde (1969) consider MTTF, threshold, and reliability estimation for the case where both μ and θ are r.v.'s. Note that if μ is known, the simple transformation $t' = t - \mu$ reduces (8.149) to the $\mathcal{E}_1(\theta)$ distribution for which the results of Section 8.3 are applicable.

Varde assumes the four-parameter natural conjugate prior distribution for (Θ, μ) given by

$$g(\theta, \mu; \delta_0, \lambda_0, \eta_0, \nu_0) \propto \frac{1}{\theta^{\nu_0+1}} e^{-(\delta_0 - \lambda_0 \mu)/\theta}, \qquad 0 < \mu \leqslant \eta_0, \theta > 0, \tag{8.150}$$

where $\delta_0 > 0$, $\eta_0 > 0$, and ν_0 and λ_0 are nonnegative integers such that $\nu_0 \leqslant \lambda_0 \leqslant \delta_0 / \eta_0$. The results given in this section differ from those of Varde, even though the assumed prior distribution is the same, due to an algebraic error in Varde's derivation. Sinha and Guttman consider the noninformative prior distribution $g(\theta, \mu; \nu_0) \propto 1/\theta^{\nu_0+1}$. Pierce (1973) considers reliability estimation using the prior $g(\theta) = 1/\theta$, which is a special case of the prior used by Sinha and Guttman.

Varde and Sinha and Guttman assume that Type II/item-censored testing without replacement is the structure of the life test experiment. Due to the complexity of the likelihood function and particularly the marginal distributions, the results of this section will not apply to as large a class of life testing procedures as will the results of Sections 8.2 through 8.5. Throughout the section we will assume that the life testing procedure used is either Type I/time-truncated or Type II/item-censored testing without replacement.

The likelihood function for the "information" $\underset{\sim}{z}$ obtained from the life test is

$$L(\theta, \mu | \underset{\sim}{z}) \propto \frac{1}{\theta^s} e^{-(\text{t} - n\mu)/\theta}, \qquad \mu \leqslant t_1, \tag{8.151}$$

where (t, t_1) represents the observed total time on test and the first ordered failure time, respectively, and s is the observed number of failures in the sample of size n. Letting $g(\theta, \mu)$ represent the prior distribution of (Θ, μ), the posterior distribution is given by

$$g(\theta, \mu | \underset{\sim}{z}) = \frac{e^{-(t-n\mu)/\theta} g(\theta, \mu)}{\theta^s \int_0^\infty \int_0^{t_1} \frac{1}{x^s} e^{-(t-ny)/x} g(x, y)\, dy\, dx}, \qquad \mu \leqslant t_1. \quad (8.152)$$

8.6.1 Conjugate Prior Distribution

Consider the four-parameter natural conjugate prior distribution (8.150) assumed by Varde. Substituting (8.150) into (8.152) gives the corresponding posterior distribution

$$g(\theta, \mu | \underset{\sim}{z}; \delta_0, \lambda_0, \eta_0, \nu_0) = \frac{(n+\lambda)}{\theta^{s+\nu_0+1}\Gamma(s+\nu_0-1)C_0} e^{-[t+\delta_0-\mu(n+\lambda_0)]/\theta},$$

$$0 < \mu \leqslant M, \theta > 0, \quad (8.153)$$

where we define

$$M = \min(t_1, \eta_0)$$

and

$$C_k = \frac{1}{[t+\delta_0 - M(n+\lambda_0)]^{s+\nu_0-k-1}} - \frac{1}{(t+\delta_0)^{s+\nu_0-k-1}}. \quad (8.154)$$

Integrating (8.153) with respect to μ and θ gives the marginal posterior distributions of Θ and μ, respectively. Thus

$$g(\theta | \underset{\sim}{z}; \delta_0, \lambda_0, \eta_0, \nu_0) = \frac{e^{-(t+\delta_0)/\theta}}{C_0 \theta^{s+\nu_0}\Gamma(s+\nu_0-1)} [e^{M(n+\lambda_0)/\theta} - 1], \qquad \theta > 0,$$

$$(8.155)$$

and

$$g(\mu | \underset{\sim}{z}; \delta_0, \lambda_0, \eta_0, \nu_0) = \frac{(n+\lambda_0)(s+\nu_0-1)}{C_0[t+\delta_0-\mu(n+\lambda_0)]^{s+\nu_0}}, \qquad 0 < \mu \leqslant M.$$

$$(8.156)$$

Point Estimation. The posterior mean and variance of Θ are

$$E(\Theta|\underset{\sim}{z};\delta_0,\lambda_0,\eta_0,\nu_0)=\frac{C_1}{C_0(s+\nu_0-2)},\qquad s+\nu_0>2, \tag{8.157}$$

$$\operatorname{Var}(\Theta|\underset{\sim}{z};\delta_0,\lambda_0,\eta_0,\nu_0)=\frac{1}{C_0(s+\nu_0-2)}\left[\frac{C_2}{s+\nu_0-3}-\frac{C_1^2}{C_0(s+\nu_0-2)}\right],\qquad s+\nu_0>3, \tag{8.158}$$

and the posterior mean and second moment of μ are

$$E(\mu|\underset{\sim}{z};\delta_0,\lambda_0,\eta_0,\nu_0)=$$

$$\frac{1}{C_0}\left\{\frac{M}{[\mathrm{t}+\delta_0-M(n+\lambda_0)]^{s+\nu_0-1}}-\frac{C_1}{(n+\lambda_0)(s+\nu_0-2)}\right\} \tag{8.159}$$

$$E(\mu^2|\underset{\sim}{z};\delta_0,\lambda_0,\eta_0,\nu_0)=$$

$$\frac{1}{C_0}\left\{\frac{M^2}{[\mathrm{t}+\delta_0-M(n+\lambda_0)]^{s+\nu_0-1}}-\frac{2}{(n+\lambda_0)(s+\nu_0-2)}\times\left[\frac{M}{[\mathrm{t}+\delta_0-M(n+\lambda_0)]^{s+\nu_0-2}}-\frac{C_2}{(n+\lambda_0)(s+\nu_0-3)}\right]\right\}. \tag{8.160}$$

Equations (8.157) and (8.159) are the Bayes point estimators of Θ and μ, respectively, with respect to squared-error loss.

The Bayes point estimator of reliability is found by taking the expectation of the reliability function

$$R=R(t_0)=\begin{cases}1, & t_0<\mu\\ e^{-(t_0-\mu)/\theta}, & t_0\geqslant\mu,\end{cases} \tag{8.161}$$

with respect to the posterior distribution (8.153). The resulting estimator, in the case where $t_0 \geqslant M$ is

$$E(R|\underset{\sim}{z}; \delta_0, \lambda_0, \eta_0, \nu_0) = \frac{n+\lambda_0}{C_0(n+\lambda_0+1)} \times \left\{ \frac{1}{[t_0+t+\delta_0 - M(n+\lambda_0+1)]^{s+\nu_0-1}} - \frac{1}{[t_0+t+\delta_0]^{s+\nu_0-1}} \right\}. \tag{8.162}$$

If $t_0 < M$, the Bayes point estimator of reliability is

$$\frac{1}{C_0}\left\{ \frac{(n+\lambda_0)}{(n+\lambda_0+1)} \left[\frac{1}{[t+\delta_0 - t_0(n+\lambda_0)]^{s+\nu_0-1}} - \frac{1}{[t+\delta_0+t_0]^{s+\nu_0-1}} \right] + \left[\frac{1}{[t+\delta_0 - M(n+\lambda_0)]^{s+\nu_0-1}} - \frac{1}{[t+\delta_0 - t_0(n+\lambda_0)]^{s+\nu_0-1}} \right] \right\}. \tag{8.163}$$

Interval Estimation. Interval estimators for Θ and μ are found by solving (5.45) and (5.46), using the appropriate posterior distribution. A symmetric $100(1-\gamma)\%$ TBPI estimate of θ is found by solving

$$\Pr\{\Theta > \theta_* | \underset{\sim}{z}; \delta_0, \lambda_0, \eta_0, \nu_0\} = \frac{1}{C_0 \Gamma(s+\nu_0-1)} \times \left\{ \frac{\Gamma[s+\nu_0-1, (t+\delta_0 - M(n+\lambda_0))/\theta_*]}{[t+\delta_0 - M(n+\lambda_0)]^{s+\nu_0-1}} - \frac{\Gamma[s+\nu_0-1, (t+\delta_0)/\theta_*]}{[t+\delta_0]^{s+\nu_0-1}} \right\} = 1 - \frac{\gamma}{2} \tag{8.164}$$

and

$$\Pr\{\Theta > \theta^* | \underset{\sim}{z}; \delta_0, \lambda_0, \eta_0, \nu_0\} = \frac{1}{C_0 \Gamma(s+\nu_0-1)} \times \left\{ \frac{\Gamma[s+\nu_0-1, (t+\delta_0 - M(n+\lambda_0))/\theta^*]}{[t+\delta_0 - M(n+\lambda_0)]^{s+\nu_0-1}} - \frac{\Gamma[s+\nu_0-1, (t+\delta_0)/\theta^*]}{[t+\delta_0]^{s+\nu_0-1}} \right\} = \frac{\gamma}{2} \tag{8.165}$$

for θ_* and θ^*, where $\Gamma(a, x)$ is the incomplete gamma function. A symmetric $100(1-\gamma)\%$ TBPI estimator for μ is found by solving

$$\Pr\{\mu < \mu_* | \underset{\sim}{z}; \delta_0, \lambda_0, \eta_0, \nu_0\}$$

$$= \int_0^{\mu_*} \frac{(n+\lambda_0)(s+\nu_0-1)}{C_0} \frac{d\mu}{[t+\delta_0-\mu(n+\lambda_0)]^{s+\nu_0}} = \frac{\gamma}{2} \tag{8.166}$$

and

$$\Pr\{\mu < \mu^* | \underset{\sim}{z}; \delta_0, \lambda_0, \eta_0, \nu_0\}$$

$$= \int_0^{\mu^*} \frac{(n+\lambda_0)(s+\nu_0-1)}{C_0} \frac{d\mu}{[t+\delta_0-\mu(n+\lambda_0)]^{s+\nu_0}} = 1-\frac{\gamma}{2}. \tag{8.167}$$

Integrating and solving algebraically, the solutions to (8.166) and (8.167) are found to be

$$\mu_* = \left(\frac{t+\delta_0}{n+\lambda_0}\right)\left\{1-\left[C_0\left(\frac{\gamma}{2}\right)(t+\delta_0)^{s+\nu_0-1}+1\right]^{-1/(s+\nu_0-1)}\right\} \tag{8.168}$$

and

$$\mu^* = \left(\frac{t+\delta_0}{n+\lambda_0}\right)\left\{1-\left[C_0\left(1-\frac{\gamma}{2}\right)(t+\delta_0)^{s+\nu_0-1}+1\right]^{-1/(s+\nu_0-1)}\right\}, \tag{8.169}$$

respectively.

8.6.2 Noninformative Prior Distribution

Sinha and Guttman (1976) consider the improper noninformative prior distribution

$$g(\theta, \mu; \nu_0) \propto \frac{1}{\theta^{\nu_0+1}}, \qquad \theta > 0, 0 < \mu \leq t_1, \tag{8.170}$$

which is a special case of the prior (8.150) used in the previous section, obtained by letting $\lambda_0 = \delta_0 = 0$ and $\eta_0 \to \infty$. Substituting (8.170) into (8.152) gives the posterior distribution

$$g(\theta, \mu \mid \underset{\sim}{z}; \nu_0) = \frac{ne^{-(\mathrm{t}-n\mu)/\theta}}{\theta^{s+\nu_0+1}\Gamma(s+\nu_0-1)C_0}, \qquad \theta > 0, \mu > 0, \quad (8.171)$$

where the $\{C_k\}$ sequence defined in Section 8.6.1 can now be expressed as

$$C_k = \frac{1}{(\mathrm{t}-nt_1)^{s+\nu_0-k-1}} - \frac{1}{\mathrm{t}^{s+\nu_0-k-1}}. \qquad (8.172)$$

The marginal posterior distributions of Θ and μ are subsequently found to be

$$g(\theta \mid \underset{\sim}{z}; \nu_0) = \frac{e^{-\mathrm{t}/\theta}(e^{nt_1/\theta}-1)}{C_0\theta^{s+\nu_0}\Gamma(s+\nu_0-1)}, \qquad \theta > 0, \qquad (8.173)$$

and

$$g(\mu \mid \underset{\sim}{z}; \nu_0) = \frac{n(s+\nu_0-1)}{C_0(\mathrm{t}-\mu n)^{s+\nu_0}}, \qquad 0 < \mu \leqslant t_1. \qquad (8.174)$$

Point Estimation. The posterior mean and variance of Θ are given by (8.157) and (8.158), where the $\{C_k\}$ sequence is computed according to (8.172). The posterior mean and second moment of μ are

$$E(\mu \mid \underset{\sim}{z}; \nu_0) = \frac{1}{C_0}\left\{\frac{t_1}{(\mathrm{t}-nt_1)^{s+\nu_0-1}} - \frac{C_1}{n(s+\nu_0-2)}\right\}, \qquad s+\nu_0 > 2, \qquad (8.175)$$

and

$$E(\mu^2 \mid \underset{\sim}{z}; \nu_0) = \frac{1}{C_0}\left\{\frac{t_1^2}{(\mathrm{t}-nt_1)^{s+\nu_0-1}} - \frac{2}{n(s+\nu_0-2)} \times \left[\frac{t_1}{(\mathrm{t}-nt_1)^{s+\nu_0-2}} - \frac{C_2}{n(s+\nu_0-3)}\right]\right\}, \qquad s+\nu_0 > 3. \qquad (8.176)$$

Equations (8.157) and (8.175) give the Bayes point estimators of Θ and μ, respectively, with respect to squared-error loss.

Sinha and Guttman also derive the Bayes point estimator of reliability which is the expectation of (8.161) with respect to (8.171). The estimator is

$$E\{R|\underset{\sim}{z};\nu_0\} = \frac{n}{C_0(n+1)}\left\{\frac{1}{[t_0+\mathrm{t}-(n-1)t_1]^{s+\nu_0-1}} - \frac{1}{[t_0+\mathrm{t}]^{s+\nu_0-1}}\right\}, \qquad t_0 \geqslant t_1, \tag{8.177}$$

$$E\{R|\underset{\sim}{z};\nu_0\} = \frac{1}{C_0}\left\{\left(\frac{n}{n+1}\right)\left[\frac{1}{(\mathrm{t}-nt_0)^{s+\nu_0-1}} - \frac{1}{(\mathrm{t}+t_0)^{s+\nu_0-1}}\right] + \left[\frac{1}{(\mathrm{t}-nt_1)^{s+\nu_0-1}} - \frac{1}{(\mathrm{t}-nt_0)^{s+\nu_0-1}}\right]\right\}, \qquad t_0 < t_1. \tag{8.178}$$

Pierce (1973) also considers Bayesian estimation of R for the case of a joint noninformative prior distribution on (Θ, μ).

Interval Estimation. Formulas for interval estimators of Θ and μ are found by substituting $\lambda_0 = \delta_0 = 0$ and letting $\eta_0 \to \infty$ in (8.164), (8.165), (8.166), and (8.167). The symmetric $100(1-\gamma)\%$ TBPI for Θ is found by solving

$$\Pr\{\Theta > \theta_* | \underset{\sim}{z}; \nu_0\} = \frac{1}{C_0\Gamma(s+\nu_0-1)} \times \left\{\frac{\Gamma\left[s+\nu_0-1,(\mathrm{t}-nt_1)/\theta_*\right]}{[\mathrm{t}-nt_1]^{s+\nu_0-1}} - \frac{\Gamma\left[s+\nu_0-1,\mathrm{t}/\theta_*\right]}{\mathrm{t}^{s+\nu_0-1}}\right\} = 1-\frac{\gamma}{2} \tag{8.179}$$

and

$$\Pr\{\Theta > \theta^* | \underset{\sim}{z}; \nu_0\} = \frac{1}{C_0\Gamma(s+\nu_0-1)} \times \left\{\frac{\Gamma[s+\nu_0-1,(\mathrm{t}-nt_1)/\theta^*]}{[\mathrm{t}-nt_1]^{s+\nu_0-1}} - \frac{\Gamma[s+\nu_0-1,\mathrm{t}/\theta^*]}{\mathrm{t}^{s+\nu_0-1}}\right\} = \frac{\gamma}{2} \tag{8.180}$$

for θ_* and θ^*.

The symmetric $100(1-\gamma)\%$ TBPI estimator for μ is found according to

$$\mu_* = \frac{\mathrm{t}}{n}\left\{1 - \left[C_0\left(\frac{\gamma}{2}\right)\mathrm{t}^{s+\nu_0-1} + 1\right]^{-1/(s+\nu_0-1)}\right\} \tag{8.181}$$

and

$$\mu^* = \frac{\mathrm{t}}{n}\left\{1 - \left[C_0\left(1-\frac{\gamma}{2}\right)\mathrm{t}^{s+\nu_0-1} + 1\right]^{-1/(s+\nu_0-1)}\right\}. \tag{8.182}$$

Example 8.28. In Example 4.10, data were reported on the cycles to failure of 15 reciprocating pumps ($n=15$) which were tested until the eighth failure occurred ($s=8$). Recall that $\mathrm{t}=15,962,989$ and $t_1=237,217$. Suppose that we consider a joint noninformative prior distribution on (Θ, μ) in which $\nu_0=0$. We desire Bayesian point estimates of both θ and μ. We also desire a Bayes point estimate of r for $t_0=10^6$ cycles. From (8.172) we find that $C_0=1.8339\times 10^{-50}$ and $C_1=2.1402\times 10^{-43}$. Using (8.157) we find that

$$E(\Theta|\underset{\sim}{z};0) = \frac{2.1402\times 10^{-43}}{1.8339\times 10^{-50}(6)} = 1,945,035.$$

According to (8.175),

$$E(\mu|\underset{\sim}{z};0) = \frac{1}{1.8339\times 10^{-50}} \times$$

$$\left\{\frac{237,217}{[15,962,989-15(237,217)]^7} - \frac{2.1402\times 10^{-43}}{15(6)}\right\} = 156,523,$$

which is somewhat larger than the estimate of μ computed in Example 4.10. From (8.177) we have

$$E(R|\underset{\sim}{z};0) = \frac{15}{1.8339\times 10^{-50}(16)}\left\{\frac{1}{(13,167,517)^7} - \frac{1}{(16,962,989)^7}\right\} = 0.62.$$

Thus it is estimated that 62% of such pumps should survive at least 10^6 operating cycles. ■

EXERCISES

8.1. A Type II/item-censored life test with replacement was conducted on a certain type of instrument in which six failures occurred in 22,425,600 unit-hours of testing. From past experience it is believed that the prior probability is 5% that the failure rate λ is less than 10^{-7}, whereas there is also a 5% probability that λ exceeds 10^{-5}. Assume an $\mathcal{E}(\lambda)$ sampling distribution and a $\mathcal{G}(\alpha_0, \beta_0)$ prior distribution on Λ.

(a) Determine the values of the prior parameters α_0 and β_0 consistent with the stated prior beliefs regarding λ.

(b) Find the posterior distribution of Λ.

(c) Find the Bayes point estimate of λ assuming a squared-error loss function with $a = 1$.

(d) Compute the generalized ML estimate of λ.

(e) Find the associated Bayes risk of the estimate in (c).

(f) Compare the risk in (e) with the Bayes risk of the usual ML estimator [see Exercise (8.10)].

(g) Find a 95% UBPI estimate of λ.

8.2. With regard to Exercise 8.1, suppose that it is also desired to estimate the reliability of such instruments for $t_0 = 20$ years.

(a) Find the Bayes point estimate of the desired reliability assuming a squared-error loss function with $a = 1$.

(b) Find the 90% LBPI estimate of the desired reliability.

8.3. With regard to Exercise 8.1, suppose that we wish to estimate the reliable life t_R for $R = 0.95$.

(a) Assuming squared-error loss with $a = 1$, find the Bayes point estimate of t_R.

(b) Determine the Bayes risk of the estimate in (a).

(c) Compute a 90% TBPI estimate of t_R.

8.4. Show that (8.138) reduces to (8.140) in the case of a $\mathcal{U}(0,1)$ prior distribution on $R(t_0)$.

8.5. Find a simplified integral representation of the Bayes risk associated with the Bayes estimator given in (8.11).

8.6. Failure data have been collected on 88 water pumps in the two gas cooled reactors of Saint-Laurent-des-Eaux since the beginning of operation of Saint-Laurent 1 in 1968 through June 1975 [Dorey and Gachot (1978)]. Suppose we consider only condenser extraction pumps. Based on past experience with similar pumps in coal and fuel-oil fired units, it is believed that the lower and upper bounds of the 90 percent prior probability interval on the failure rate λ are 6×10^{-5} f/h and 8×10^{-5} f/h, respectively. At Saint-Laurent-des-Eaux 24 failures have been observed on the 32 condenser extraction pumps which have been working for 948,400 h. Assume an $\mathcal{E}(\lambda)$ sampling distribution and Type II testing with replacement.

(a) For a $\mathcal{G}(\alpha_0, \beta_0)$ prior distribution on Λ, determine the Bayesian point estimate of λ for a squared-error loss function with $a=1$.

(b) For a $\mathcal{G}(\alpha_0, \beta_0)$ prior distribution on Λ, find a 90% TBPI estimate of λ and compare this interval to the prior interval.

8.7. With regard to Exercise 8.6, a specific investigation was carried out in 1975 in 19 light water reactors regarding the failure rate of condenser extraction pumps [Dorey and Gachot (1978)]. The specific investigation revealed eight failures among 42 such pumps for a total operating time of 146,000 h.

(a) How would you combine the prior data from the specific investigation with the prior data given in Exercise 8.6, assuming a $\mathcal{G}(\alpha_0, \beta_0)$ prior distribution?

(b) Using your combined $\mathcal{G}(\alpha_0, \beta_0)$ prior distribution found in (a), find the corresponding point and 90% TBPI estimates of λ based on the use of the observed Saint-Laurent-des-Eaux failure data. Compare these estimates to the corresponding estimates found in Exercise 8.6.

8.8. Barlow, Toland, and Freeman (1979) present failure data on Kevlar/epoxy spherical pressure vessels that have been subjected to constant sustained pressure. Twenty-four vessels were tested at 3700 psi of which the first 18 ordered failures were observed to occur after 225.2, 503.6, 1087.7, 1134.3, 1824.3, 1920.1, 2383.0, 2442.5, 3708.9, 3708.9, 4908.9, 5556.0, 6271.1, 7332.0, 7918.7, 7996.0, 9240.3, and 9973.0 h.

(a) Using the methods of Section 4.3, give assurance that the $\mathcal{E}_1(\theta)$ model is appropriate for describing this data.

(b) Assuming a noninformative prior distribution on Θ, compute a Bayesian point estimate of θ using a squared-error loss function with $a=1$.

(c) Compute a 95% TBPI estimate of θ assuming a noninformative prior distribution on Θ.

8.9. Data were obtained on the failure times (in hours) of the shaft seals of 10 recirculation pumps from a Type II/item-censored life test of

20 pumps which was terminated at the time of the tenth seal failure. The ordered data are:

10,930	14,033
11,698	15,139
13,668	15,751
13,675	16,193
13,701	17,694

Assume that the $\mathcal{E}_1(\theta, \mu)$ distribution adequately describes this data.

(a) Test the hypothesis H_0: $\mu = 0$ at the 5% level of significance.

(b) Assuming a joint noninformative prior distribution on (Θ, μ) in which $\nu_0 = 1$, compute the usual Bayes point estimates of θ and μ.

(c) For the same prior as in (b), compute the usual Bayes point estimate of the reliability of such pumps for $t_0 = 20{,}000$ h.

8.10. Compute the Bayes risk associated with the ML estimator $\hat{\lambda} = S/\mathcal{T}$ of the failure rate λ in an $\mathcal{E}(\lambda)$ distribution for a $\mathcal{G}(\alpha_0, \beta_0)$ prior distribution on Λ. Compare this to the risk of the Bayes estimator for a squared-error loss function.

8.11. [Canfield (1970)] Suppose that we are interested in Bayesian estimation of the reliability function $r(t_0) = \exp(-\lambda t_0)$ in the case of Type II/item-censored testing. Consider the class of loss functions defined by

$$L(r, r_\gamma) = \frac{1}{\lambda}\left(\frac{r_\gamma}{r} - 1\right)^2,$$

where $r_\gamma = \exp[-t_0 \chi^2_{1-\gamma}(2s)/(2t)]$ for a given mission time t_0. It is noted that r_γ is the usual (classical) $100(1-\gamma)\%$ LCI estimate of $r(t_0)$ given by (4.80). Show that for a $\mathcal{U}(0,1)$ prior distribution on $R(t_0)$, the Bayes estimator for $r(t_0)$ is exactly the MVU estimator of $r(t_0)$ given by

$$\tilde{r}_\gamma = \left(1 - \frac{t_0}{t}\right)^s.$$

8.12. Show that the minimum Bayes risk of the Bayes estimator given in (8.65) is given by (8.67).

8.13. Determine the least squares linear estimator $\hat{\theta}_D$ for θ given in Section 8.3.5 for an $\mathcal{E}_1(\theta)$ sampling model and a $\mathcal{U}(\alpha_0, \beta_0)$ prior distribution on Θ. Compare this estimator to the usual ML estimator $\hat{\theta} = \mathcal{T}/n$.

8.14. [Moore and Bilikam (1978)] Show that the Bayes point estimator for θ in the $\mathcal{E}_1(\theta)$ sampling distribution using squared-error loss and improper prior distribution on Θ given by

$$g(\theta;\alpha_0)=\theta^{-\alpha_0-1}, \qquad \theta>0,$$

is

$$\tilde{\theta}=\frac{\mathcal{T}}{\alpha_0+s-1},$$

for the case of Type II/item-censored testing. Note that if $\alpha_0=0$ the prior is noninformative and $\tilde{\theta}$ reduces to (8.56), whereas if $\alpha_0=1$, $\tilde{\theta}$ reduces to the usual ML estimator.

8.15. [Moore and Bilikam (1978)] With regard to Exercise 8.14, show that the Bayes point estimator for the reliability function using squared-error loss is given by

$$\tilde{r}(t_0)=\left(\frac{\mathcal{T}}{\mathcal{T}+t_0}\right)^{\alpha_0+s}.$$

8.16. Using the method of Section 6.1, show that (8.100) is the noninformative prior distribution for reliability.

REFERENCES

Amos, D. E. and Daniel, S. L. (1972). *Significant Digit Incomplete Gamma Ratios*, SC-DR-72 0303, Sandia Laboratories, Albuquerque, NM.

Barlow, R. E., Toland, R. H., and Freeman, T. (1979). *Stress–Rupture Life of Kevlar/Epoxy Spherical Pressure Vessels*, UCID-17755 Part 3, Lawrence Livermore Laboratory, Livermore, CA.

Barlow, R. E. and Wu, A. S. (1980). *Preposterior Analysis of Bayes Estimators of Mean Life I*, ORC 80-4, University of California, Berkeley, CA.

Bhattacharya, S. K. (1967). Bayesian Approach to Life Testing and Reliability Estimation, *Journal of the American Statistical Association*, Vol. 62, pp. 49–62.

Bott, T. F. and Haas, P. M. (1978). *Initial Data Collection Efforts of CREDO: Sodium Value Failures*, NCSR R20, National Center of Systems Reliability.

Bühlmann, H. (1970). *Mathematical Methods in Risk Theory*, Springer-Verlag, New York.

Canfield, R. V. (1970). A Bayesian Approach to Reliability Estimation Using a Loss Function, *IEEE Transactions on Reliability*, Vol. R-19, pp. 13–16.

Dorey, J. and Gachot, B. (1978). Pump Reliability Data Derived from Electricité de France

Operating Experience, *Inservice Data Reporting and Analysis* (Edited by J. T. Fong), PVP-PB-032, The American Society of Mechanical Engineers, pp. 67–76.

Harris, C. M. and Singpurwalla, N. D. (1968). Life Distributions Derived from Stochastic Hazard Functions, *IEEE Transactions on Reliability*, Vol. R-17, pp. 70–79.

Mann, N. R., Schafer, R. E. and Singpurwalla, N. D. (1974). *Methods for Statistical Analysis of Reliability and Life Data*, Wiley, New York.

Moore, A. H. and Bilikam, J. E. (1978). Bayesian Estimation of Parameters of Life Distributions and Reliability from Type-II Censored Samples, *IEEE Transactions on Reliability*, Vol. R-27, pp. 64–67.

Pierce, D. A. (1973). Fiducial Frequency and Bayesian Inference on Reliability for the Two-Parameter Negative Exponential Distribution, *Technometrics*, Vol. 15, pp. 249–253.

Sinha, S. K. and Guttman, I. (1976). Bayesian Inference About the Reliability Function for the Exponential Distributions, *Communications in Statistics—Theory and Methods*, Vol. A5, pp. 471–479.

Soland, R. M. (1969). Bayesian Analysis of the Weibull Process with Unknown Scale and Shape Parameters, *IEEE Transactions on Reliability*, Vol. R-18, pp. 181–184.

Varde, S. D. (1969). Life Testing and Reliability Estimation for the Two-Parameter Exponential Distribution, *Journal of the American Statistical Association*, Vol. 64, pp. 621–631.

Wu, A. S. (1980). *Bayesian Evaluation of Life Test Sampling Plans*, Ph.D. Dissertation, University of California, Berkeley, CA.

CHAPTER 9

Estimation in the Weibull, Normal, Log Normal, Inverse Gaussian, and Gamma Distributions

Bayesian reliability estimation methods for the $\mathcal{E}(\lambda)$ model were presented in Chapter 8. In this chapter we consider Bayesian estimators for certain parameters and associated reliability measures of some other important, but less commonly used, failure time models. In particular, the $\mathcal{W}(\alpha, \beta)$ distribution is considered in Section 9.1. Sections 9.2–9.4 consider the $\mathcal{N}(\mu, \sigma^2)$, $\mathcal{LN}(\xi, \sigma^2)$, and $\mathcal{IN}(\mu, \lambda)$ distributions, respectively. Finally, the $\mathcal{G}(\alpha, \beta)$ failure time distribution is discussed in Section 9.5. Each of these distributions and their basic properties were introduced in Section 4.2.

9.1. THE WEIBULL DISTRIBUTION

The $\mathcal{W}(\alpha, \beta, \theta)$ distribution was introduced in Section 4.2.2. A $\mathcal{W}(\alpha, \beta, \theta)$ sampling process is defined as one that generates independent r.v.'s $T_1, T_2, \ldots, T_j, \ldots$ having identical $\mathcal{W}(\alpha, \beta, \theta)$ distributions with pdf given by (4.18). In this section we restrict consideration to the case in which the guaranteed life $\theta = 0$.

We will also consider a reparameterized version of (4.18) given by

$$f(t; \lambda, \beta) = \lambda\beta t^{\beta-1}\exp(-\lambda t^{\beta}), \qquad t \geq 0, \qquad \lambda, \beta > 0, \tag{9.1}$$

which is obtained from (4.18) by letting $\alpha = \lambda^{-1/\beta}$. This version of the Weibull distribution "separates" the two parameters and often simplifies the algebra in the subsequent Bayesian manipulations. In order to distinguish between the two forms of the two-parameter Weibull distribution, in which $\theta = 0$ in both forms, the form given in (9.1) will be referred to as the $\mathcal{W}_1(\lambda, \beta)$ distribution. The parameter λ will also be referred to as the *scale parameter*. Recall that β controls the shape of either the $\mathcal{W}(\alpha, \beta)$ or the $\mathcal{W}_1(\lambda, \beta)$ distribution and is thus referred to as the *shape parameter*. A slight reparameterization of (9.1) is also considered in which $\lambda = \theta^{-1}$. This form of the Weibull distribution will be denoted as $\mathcal{W}_2(\theta, \beta)$.

In this section we consider Bayesian estimation of the Weibull process. The available literature has sharply distinguished two fundamental cases of interest. The first is the case in which the shape parameter value is known and only the scale parameter is the unknown value of an r.v. that must be estimated. This case will be presented in Section 9.1.1. The second considers the case in which both parameters are unknown values of r.v.'s that must be estimated. Section 9.1.2 considers this case. In both cases various prior distributions and loss functions will be considered, and both point and interval estimators will be obtained for the appropriate parameter(s) of interest. Bayesian estimation of the $\mathcal{W}(\alpha, \beta)$ reliability function given in (4.19), or its equivalent reparameterized form, is discussed in Section 9.1.3. In Section 9.1.4 the $\mathcal{W}_1(\lambda, \beta)$ MTTF is of interest and Bayes point and interval estimators are presented. Finally, the robustness of some Bayes estimators in a Weibull process to the form of the prior distribution is discussed in Section 9.1.5. In addition, Section 9.1.5 considers the sensitivity of some Bayes estimators to the values of the prior parameters in certain classes of prior distributions.

9.1.1 Scale Parameter Is a Random Variable

In Weibull reliability analysis it is frequently the case that the value of the shape parameter is known. For example, the $\mathcal{E}(\lambda)$ and Raleigh distributions are obtained when $\beta = 1$ and 2, respectively. Soland (1968) gives a justification for this situation. The earliest references to Bayesian estimation of the unknown random scale parameter are Harris and Singpurwalla (1968, 1969) and Soland (1966, 1968). Since that time this case has been considered by numerous authors, such as Canavos (1974), Canavos and Tsokos (1971a, 1973), Jayaram (1974), Moore and Bilikam (1978), Papadopoulos and Tsokos (1975), Tsokos (1972), Tsokos and Canavos (1972), and Tummala and Sathe (1978).

It is easily shown that, if T has a $\mathcal{W}_1(\lambda, \beta)$ distribution, then T^β follows an $\mathcal{E}(\lambda)$ distribution. Hence, for the Weibull case, the Bayesian estimators follow directly from this transformation and the results of Chapter 8.

Consider a life test of n items in which s items have failed at ordered times $t_1, \ldots, t_s$ and $n-s$ items have operated for times $t_{s+1}^*, \ldots, t_n^*$ without failing; thus $T_{s+1} > t_{s+1}^*, \ldots, T_n > t_n^*$. The times $t_{s+1}^*, \ldots, t_n^*$ are the withdrawal times of the nonfailed items. As in Chapter 8, let $\underset{\sim}{z}$ denote the "information" (values of s, n, $t_1, \ldots, t_s$ and $t_{s+1}^*, \ldots, t_n^*$) obtained from the life test. Soland (1968) also considers this situation. The statistic

$$\mathcal{W} = \sum_{i=1}^{s} T_i^{\beta} + \sum_{i=s+1}^{n} T_i^{*\beta} \tag{9.2}$$

is sufficient for estimating λ (or θ). This may be seen by examination of the likelihood corresponding to the above sampling scheme which is given by

$$L(\lambda | \underset{\sim}{z}) \propto \lambda^s \beta^s \left(\prod_{i=1}^{s} t_i \right)^{\beta - 1} \exp\left[-\lambda \left(\sum_{i=1}^{s} t_i^{\beta} + \sum_{i=s+1}^{n} t_i^{*\beta} \right) \right] \tag{9.3}$$

and applying (5.42). If there are no withdrawals prior to test termination, we find that

$$\begin{aligned} \mathcal{W} &= nT_s^{\beta} && \text{(testing with replacement)} \\ &= \sum_{i=1}^{s} T_i^{\beta} + (n-s) T_s^{\beta} && \text{(testing without replacement),} \end{aligned} \tag{9.4}$$

which represents the usual Type II/item-censored situation in which n items are simultaneously tested, either with or without replacement, until s failures occur. It is important to recognize that $\mathcal{W}$ is the *rescaled total time on test* in which the lifetime or survival time of each item tested is "rescaled" by raising it to a β power and the rescaled test times are then added. This implies that records on the failure or withdrawal time of each item tested must be maintained in contrast to the $\mathcal{E}(\lambda)$ case in which only the number of failures and accumulated total time on test had to be recorded. In more complicated testing schemes, such as when an item is removed and replaced several times during the test, it may be somewhat cumbersome to express $\mathcal{W}$ in a convenient mathematical form. In such cases, by recognizing that $\mathcal{W}$ is the rescaled total time on test, this fact can be used to calculate the value of $\mathcal{W}$ *without* the explicit use of an expression such as (9.4).

For the testing scheme described above, the sampling distribution of $\mathcal{W}$ given λ is $\mathcal{G}_1(s, \lambda)$ [see Soland (1968)], whereas the sampling distribution of $\mathcal{W}$ given θ is $\mathcal{G}(s, \theta)$. Thus the sampling distribution of $\mathcal{W}$ given λ is the same as the sampling distribution of the total time on test statistic $\mathcal{T}$, in the case of gamma sampling, discussed in Section 8.1.

This may be clearly seen as follows: Letting $g(\lambda)$ represent the prior distribution on Λ, the posterior distribution of Λ given $\mathcal{W}=w$ is given by

$$g(\lambda|w)=\frac{\lambda^s e^{-\lambda w}g(\lambda)}{\int_0^\infty \lambda^s e^{-\lambda w}g(\lambda)\,d\lambda}, \tag{9.5}$$

which is precisely the same form as the posterior distribution of Λ given $\mathcal{T}=t$, in the case of gamma sampling, given by (8.5). Thus the Bayesian results of Section 8.2 may be directly applied here by replacing the observed total time on test t by w. The situation here is thus gamma sampling as defined in Section 8.1.

If $\theta=\lambda^{-1}$ is to be estimated, then the results of Section 8.3 apply here as well, in which case t is again replaced by w. In this case the posterior distribution of Θ given $\mathcal{W}=w$ becomes

$$g(\theta|w)=\frac{\theta^{-s}e^{-w/\theta}g(\theta)}{\int_0^\infty \theta^{-s}e^{-w/\theta}g(\theta)\,d\theta}, \tag{9.6}$$

where $g(\theta)$ represents the prior distribution on Θ.

Example 9.1. Nine failure times were observed for a heat exchanger used in the alkylation unit of a gasoline refinery. The observed failure times were 0.41, 0.58, 0.75, 0.83, 1.00, 1.08, 1.17, 1.25, and 1.35 years. Based on a preliminary analysis of these data [see Example 4.2], a $\mathcal{W}_1(\lambda,\beta)$ distribution, in which $\beta=3.5$, is used to describe this data. Manufacturer's reliability specifications for this particular unit were translated into a $\mathcal{U}(0.5,1.5)$ prior distribution on Λ. We wish to compute both a point estimate and a 95% UBPI estimate of Λ using a squared-error loss function.

Now $w=\Sigma_{i=1}^{9}t_i^{3.5}=10.16$, $s=n=9$, $\alpha_0=0.5$, and $\beta_0=1.5$. From (9.5) and (8.9), the posterior distribution of Λ given $w=10.16$ becomes

$$g(\lambda|10.16;0.5,1.5)=\frac{(10.16)^{10}\lambda^9 e^{-10.16\lambda}}{\Gamma(10,15.24)-\Gamma(10,5.08)},\qquad 0.5<\lambda<1.5,$$

which is shown in Figure 9.1 along with the prior distribution.

From (8.11), the Bayesian point estimate of λ here becomes

$$E(\Lambda|10.16;0.5,1.5)=\frac{\Gamma(11,15.24)-\Gamma(11,5.08)}{10.16[\Gamma(10,15.24)-\Gamma(10,5.08)]}=0.96$$

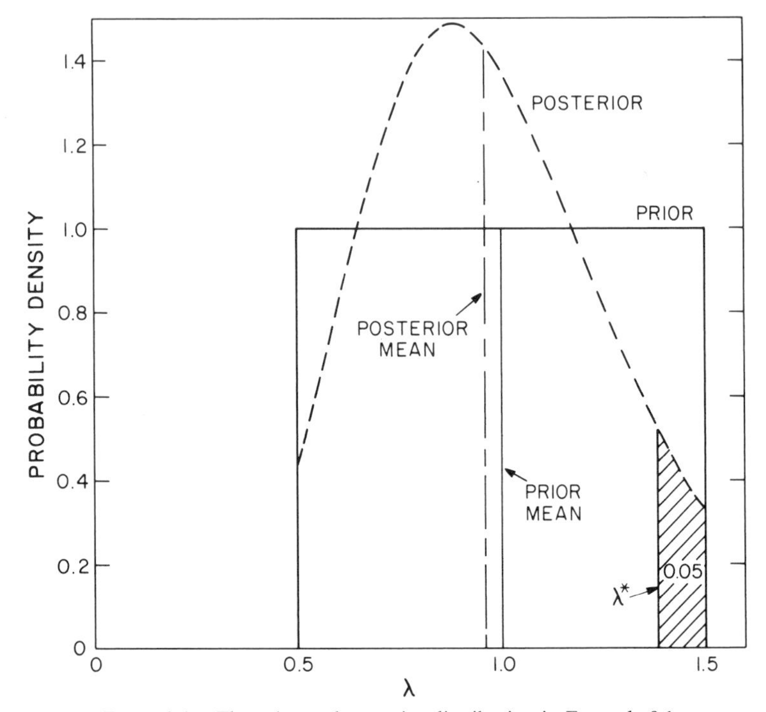

Figure 9.1. The prior and posterior distribution in Example 9.1.

and the 95% UBPI estimate of λ is obtained from (8.16) by solving the equation

$$\frac{\Gamma(10, 15.24) - \Gamma(10, 10.16\lambda^*)}{\Gamma(10, 15.24) - \Gamma(10, 5.08)} = 0.05$$

for λ^*. The solution is obtained for $\lambda^* = 1.381$ and thus $\Pr(\Lambda \leqslant 1.381 | 10.16; 0.5, 1.5) = 0.95$. The point and interval estimates are shown in Figure 9.1. ■

Bayesian interval estimates for θ are also explicitly presented by Papadopoulos and Tsokos (1975) in the case of either a $\mathcal{U}(a, b)$ or $\mathcal{IG}(\nu, \mu)$ prior distribution on Θ.

Tsokos (1972b) and Canavos and Tsokos (1973) use Monte Carlo simulation to study the mean squared-error performance of the usual Bayesian point estimators for θ in the case of $\mathcal{U}(\alpha, \beta)$, $\mathcal{E}(\lambda^{-1})$, and $\mathcal{IG}(\nu, \mu)$ prior distributions on Θ. The mean squared-error of the Bayesian estimators is found to be uniformly smaller than that of the classical MVU estimator. It was also discovered that the particular prior distribution, as well as the value of the shape parameter, had little effect on the general nature of the results.

Weighted Squared-Error Loss Function. Tummala and Sathe (1978) generalized the foregoing Bayesian estimators of λ by considering the weighted squared-error loss function given by (5.24) in which $\gamma(\lambda-\hat{\lambda})=(\lambda-\hat{\lambda})^2$. Minimizing the posterior expected loss yields the Bayesian point estimator for λ given by

$$\tilde{\lambda}=\frac{E[\Lambda h(\Lambda)|\mathcal{W}]}{E[h(\Lambda)|\mathcal{W}]}, \tag{9.7}$$

which Tummala and Sathe designate as the *minimum expected loss* (MELO) estimator. In particular, they consider the case where $h(\lambda)=\lambda^{-2}$ and in this special case $\tilde{\lambda}$ becomes

$$\tilde{\lambda}=\frac{s+\alpha_0-2}{\beta_0+\mathcal{W}}, \tag{9.8}$$

for a $\mathcal{G}_1(\alpha_0,\beta_0)$ prior distribution on Λ. Recall that in this case the usual Bayes estimator of λ, obtained when $h(\lambda)=$ constant in (9.7), is given by $(s+\alpha_0)/(\beta_0+\mathcal{W})$, where the ML estimator is simply $s/\mathcal{W}$.

Other Prior Distributions. Canavos and Tsokos (1973) give the Bayesian point estimator for θ for squared-error loss and the corresponding posterior risk for the case of an $\mathcal{E}(\lambda)$ prior distribution on Θ. The estimator is not available in closed form and involves the use of modified Bessel functions of the third kind.

Moore and Bilikam (1978) derive the Bayesian point estimator of θ for the improper prior distribution $g(\theta;\alpha_0)=\theta^{-\alpha_0-1}$ on Θ. In this case the Bayesian point estimator for squared-error loss becomes

$$\tilde{\theta}=\frac{\mathcal{W}}{\alpha_0+s-1}. \tag{9.9}$$

It is noted that, if $\alpha_0=0$, the prior distribution is noninformative and (9.9) reduces to (8.56). If $\alpha_0=1$, (9.9) reduces to the usual ML estimator for θ, whereas if $\alpha_0=2$, one obtains the *admissible optimum biased* estimator discussed by Moore (1974).

9.1.2 Scale and Shape Parameters Are Random Variables

In many cases it is desirable to consider uncertainty about both the Weibull scale and shape parameters in a reliability analysis. For convenience both the Weibull shape parameter β, when considered as an r.v., and its value will be denoted by the same symbol β. In the Bayesian framework both parameters are treated as unknown values of jointly distributed r.v.'s. For the $\mathcal{W}_1(\lambda, \beta)$ distribution, Soland (1966) searched for a conjugate family of continuous joint prior distributions on Λ and β and came to the conclusion that such a family does not exist. Consequently, Soland (1969) considered a family of joint prior distributions that places continuous distributions on the scale parameter and discrete distributions on the shape parameter. In addition, this family is also conjugate in the sense that the posterior distribution is a member of this family.

Specifically, suppose that β can assume any of m values in $(0, \infty)$. Soland forms the joint prior distribution on (Λ, β) as follows. Let

$$\Pr(\beta = \beta_i) = p_i', \qquad i = 1, 2, \ldots, m, \tag{9.10}$$

where $\sum_{i=1}^m p_i' = 1$. The conditional prior distribution of Λ given β_i, call it $g(\lambda | \beta_i)$, is taken to be a $\mathcal{G}_1(s_{i0}, w_{i0})$ distribution, where it is noted that $g(\lambda | \beta_i)$ depends on β_i only through the dependence of the prior parameters s_{i0} and w_{i0} on the index i.

For the sampling scheme considered in Section 9.1.1., the likelihood given by (9.3) can be considered to be a joint function of λ and β_i conditional on the sample evidence. In the most general case, the sample evidence, which we refer to as $\underset{\sim}{z}$, is jointly contained in the sufficient statistics S,

$$V = \prod_{j=1}^{s} T_j \qquad \text{and} \qquad \mathcal{W}_i = \sum_{j=1}^{s} T_j^{\beta_i} + \sum_{j=s+1}^{n} T_j^{*\beta_i}, \qquad i = 1, 2, \ldots, m.$$

Applying Bayes' theorem, the marginal posterior distribution of β given $\underset{\sim}{z}$ is given by

$$\Pr(\beta = \beta_i | \underset{\sim}{z}) = p_i'' = \frac{p_i' \beta_i^s v^{\beta_i} w_{i0}^{s_{i0}} \Gamma(s_{i0} + s) / \left[(w_{i0} + \mathcal{W}_i)^{s_{i0} + s} \Gamma(s_{i0}) \right]}{\sum_{i=1}^{m} p_i' \beta_i^s v^{\beta_i} w_{i0}^{s_{i0}} \Gamma(s_{i0} + s) / \left[(w_{i0} + \mathcal{W}_i)^{s_{i0} + s} \Gamma(s_{i0}) \right]},$$

$$i = 1, 2, \ldots, m. \tag{9.11}$$

Also, the posterior distribution of Λ conditional on β_i is given by

$$g(\lambda | \beta_i, \underset{\sim}{z}) = \mathcal{G}_1(s_{i0} + s, w_{i0} + \mathcal{W}_i). \tag{9.12}$$

The joint posterior distribution of (Λ, β) is thus the product of (9.11) and (9.12). For a squared-error loss function the Bayesian point estimator of λ is given by

$$\tilde{\lambda} = E(\Lambda | \underset{\sim}{z}) = \int_0^\infty \lambda \left[\sum_{i=1}^{m} p_i'' g(\lambda | \beta_i, \underset{\sim}{z}) \right] d\lambda$$

$$= \sum_{i=1}^{m} p_i'' \left(\frac{s_{i0} + s}{w_{i0} + \mathcal{W}_i} \right). \tag{9.13}$$

Unfortunately, the Bayes risk of this estimator is difficult to obtain as the marginal distribution of $\underset{\sim}{Z}$ is unavailable in any simple form.

Example 9.2. [Soland (1969)] A certain long-life component was developed for use in a communications satellite. Based on experience with similar components, a $\mathcal{W}_1(\lambda, \beta)$ failure model in which $\beta < 1$ was selected. Three possible values of β were chosen: $\beta_1 = 0.7$, $\beta_2 = 0.8$, and $\beta_3 = 0.9$ with prior probabilities $p_1' = 0.3$, $p_2' = 0.4$, and $p_3' = 0.3$, respectively. Based on additional prior information regarding the reliability of the component, it was determined that $s_{10} = 26.12$, $w_{10} = 2.665 \times 10^4$ h, $s_{20} = 3.44$, $w_{20} = 8.218 \times 10^3$ h, $s_{30} = 1.91$, and $w_{30} = 1.074 \times 10^4$ h.

A life test was conducted in which 25 items were initially placed on test. The first failure occurred after 600 h and the second occurred after 1500 h. After each failure a new item was substituted for the failed one. The life test was terminated at 2000 h. The values of the sufficient statistics are $s = 2$, $v = (600)(1500) = 9 \times 10^5$, $\mathcal{W}_1 = (23)(2000)^{0.7} + (600)^{0.7} + (1400)^{0.7} + (1500)^{0.7} + (500)^{0.7} = 5.192 \times 10^3$, $\mathcal{W}_2 = 11.046 \times 10^3$, and $\mathcal{W}_3 = 23.496 \times 10^3$. From (9.11) we find that $p_1'' = 0.191$, $p_2'' = 0.430$, and $p_3'' = 0.379$. Thus, the parameter λ is estimated to be

$$\tilde{\lambda} = 0.191 \left(\frac{26.12 + 2}{2.665 \times 10^4 + 5.192 \times 10^3} \right) + 0.430 \left(\frac{3.44 + 2}{8.218 \times 10^3 + 11.046 \times 10^3} \right)$$

$$+ 0.379 \left(\frac{1.91 + 2}{1.074 \times 10^4 + 23.496 \times 10^3} \right) = 3.33 \times 10^{-4} \text{ f/h}.$$

■

Bayesian estimation in the case of both random scale and shape parameters has also been considered by Bury (1972b), Canavos and Tsokos (1973), Papadopoulos and Tsokos (1976), Tsokos (1972a, 1972b), and Tsokos and Rao (1976). Canavos and Tsokos (1973) assume independent prior distributions for Θ and β in the $\mathcal{W}_2(\theta, \beta)$ distribution. Specifically, they consider both $\mathcal{IG}(\nu_0, \mu_0)$ and $\mathcal{U}(0, \delta_0)$ prior distributions on Θ, while an independent $\mathcal{U}(\alpha_0, \beta_0)$ prior distribution is placed on β.

Inverted Gamma Distribution on Θ; Uniform Distribution on β. First consider the case of an $\mathcal{IG}(\nu_0, \mu_0)$ distribution on Θ. For the same sampling scheme considered in Section 9.1.1., the joint posterior distribution of (Θ, β) given $\underset{\sim}{z}$ is given by

$$g(\theta, \beta | \underset{\sim}{z}) = \frac{\left(\beta^s / \theta^{s+\nu_0+1}\right) v^\beta e^{-(1/\theta)\mathcal{W}_1}}{\int_{\alpha_0}^{\beta_0} \beta^s v^\beta \left[\int_0^\infty \left(1/\theta^{s+\nu_0+1}\right) e^{-(1/\theta)\mathcal{W}_1} d\theta\right] d\beta}, \tag{9.14}$$

where

$$v = \prod_{i=1}^{s} t_i \qquad \text{and} \qquad \mathcal{W}_1 = \sum_{i=1}^{s} t_i^\beta + \sum_{i=s+1}^{n} t_i^{*\beta} + \mu_0.$$

The inner integral in the denominator of (9.14) is evaluated as

$$\int_0^\infty \left(\frac{1}{\theta^{s+\nu_0+1}}\right) e^{-(1/\theta)\mathcal{W}_1} d\theta = \frac{\Gamma(s+\nu_0)}{\mathcal{W}_1^{s+\nu_0}}. \tag{9.15}$$

Thus, (9.14) reduces to

$$g(\theta, \beta | \underset{\sim}{z}) = \frac{\left(\beta^s / \theta^{s+\nu_0+1}\right) v^\beta e^{-(1/\theta)\mathcal{W}_1}}{\Gamma(s+\nu_0) J_1}, \tag{9.16}$$

where

$$J_1 = \int_{\alpha_0}^{\beta_0} \left[\frac{\beta^s v^\beta}{\mathcal{W}_1^{s+\nu_0}}\right] d\beta. \tag{9.17}$$

The integral J_1 cannot be integrated analytically and thus the posterior distribution does not have a closed form solution. This is the usual case when both Weibull parameters are considered to be r.v.'s.

With respect to a squared-error loss function, the Bayesian point estimator for θ is given by

$$E(\Theta|\underset{\sim}{z}) = \int_{\alpha_0}^{\beta_0}\int_0^{\infty} \theta g(\theta, \beta|\underset{\sim}{z})\, d\theta\, d\beta$$

$$= J_2/[(s+\nu_0-1)J_1], \tag{9.18}$$

where

$$J_2 = \int_{\alpha_0}^{\beta_0}\left[\frac{\beta^s v^\beta}{\mathcal{W}_1^{s+\nu_0-1}}\right] d\beta. \tag{9.19}$$

Similarly, the Bayesian point estimator for β becomes

$$E(\beta|\underset{\sim}{z}) = \int_{\alpha_0}^{\beta_0}\int_0^{\infty} \beta g(\theta, \beta|\underset{\sim}{z})\, d\theta\, d\beta = \frac{J_3}{J_1}, \tag{9.20}$$

where

$$J_3 = \int_{\alpha_0}^{\beta_0}\left[\frac{\beta^{s+1} v^\beta}{\mathcal{W}_1^{s+\nu_0}}\right] d\beta. \tag{9.21}$$

The integrals J_2 and J_3 have no closed form solution and thus the Bayesian estimates must be computed by numerical integration techniques. Canavos and Tsokos (1973) use the 10-point Gauss–Legendre quadrature formula to evaluate these integrals based on a technique proposed by Scarborough (1957).

Uniform Distributions on Θ and β. Let us now examine the case in which Θ has a $\mathcal{U}(0, \delta_0)$ distribution, while β has an independent $\mathcal{U}(\alpha_0, \beta_0)$ distribution. As in the above case, the joint posterior distribution of Θ and β may be expressed by

$$g(\theta, \beta|\underset{\sim}{z}) = \frac{(\beta/\theta)^s v^\beta e^{-(1/\theta)\mathcal{W}}}{\int_{\alpha_0}^{\beta_0} \beta^s v^\beta \left[\int_0^{\delta_0} \frac{1}{\theta^s} e^{-(1/\theta)\mathcal{W}}\, d\theta\right] d\beta}, \tag{9.22}$$

where

$$\mathcal{W} = \sum_{i=1}^{s} t_i^\beta + \sum_{i=s+1}^{n} t_i^{*\beta}.$$

The inner integral in the denominator of (9.22) is evaluated to be

$$\int_0^{\delta_0}\left(\frac{1}{\theta^s}\right)e^{-(1/\theta)\mathscr{W}}\,d\theta=\frac{\Gamma^c(s-1,\mathscr{W}/\delta_0)}{\mathscr{W}^{s-1}},\tag{9.23}$$

where $\Gamma^c(a,z)$ is the complement of the incomplete gamma function defined by

$$\begin{aligned}\Gamma^c(a,z)&=\int_z^{\infty}y^{a-1}e^{-y}\,dy\\&=\Gamma(a)-\Gamma(a,z).\end{aligned}\tag{9.24}$$

Therefore, the joint posterior distribution of Θ and β reduces to

$$g(\theta,\beta\,|\,\underset{\sim}{z})=\frac{(\beta/\theta)^s\upsilon^{\beta}e^{-(1/\theta)\mathscr{W}}}{J_4},\tag{9.25}$$

where

$$J_4=\int_{\alpha_0}^{\beta_0}\left(\frac{\beta^s\upsilon^{\beta}}{\mathscr{W}^{s-1}}\right)\Gamma^c\left(s-1,\frac{\mathscr{W}}{\delta_0}\right)d\beta.\tag{9.26}$$

As in the previous case, (9.25) does not have a closed-form representation.

For squared-error loss functions, the Bayesian point estimators for θ and β are given by

$$E(\Theta\,|\,\underset{\sim}{z})=\int_{\alpha_0}^{\beta_0}\int_0^{\delta_0}\theta g(\theta,\beta\,|\,\underset{\sim}{z})\,d\theta\,d\beta=\frac{J_5}{J_4}\tag{9.27}$$

and

$$E(\beta\,|\,\underset{\sim}{z})=\int_{\alpha_0}^{\beta_0}\int_0^{\delta_0}\beta g(\theta,\beta\,|\,\underset{\sim}{z})\,d\theta\,d\beta=\frac{J_6}{J_4},\tag{9.28}$$

where

$$J_5=\int_{\alpha_0}^{\beta_0}\left(\frac{\beta^s\upsilon^{\beta}}{\mathscr{W}^{s-2}}\right)\Gamma^c\left(s-2,\frac{\mathscr{W}}{\delta_0}\right)d\beta\tag{9.29}$$

and

$$J_6=\int_{\alpha_0}^{\beta_0}\left(\frac{\beta^{s+1}\upsilon^{\beta}}{\mathscr{W}^{s-1}}\right)\Gamma^c\left(s-1,\frac{\mathscr{W}}{\delta_0}\right)d\beta.\tag{9.30}$$

Numerical integration must also be used to evaluate (9.27) and (9.28).

Canavos and Tsokos (1973) use Monte Carlo simulation to show that the mean squared-errors of the Bayesian point estimators given in both cases above are significantly smaller than those of the ML estimators of θ and β. In fact, differences of an order of magnitude or more were found to be common.

Bury (1972a, 1972b) considers a transformation of the $\mathcal{W}(\alpha, \beta)$ distribution into a location-scale model from which he then deduces the conjugate joint prior distribution on the parameters. However, he is not able to express the prior distribution in closed form and, consequently, the corresponding joint posterior distribution is also unavailable in closed form. The method is unique in several ways. First, it considers a joint conjugate prior distribution. Second, a prior censored life test is envisioned as producing the "pseudo" data used to fit the prior distribution and at least two pseudo prior observations are required to fit the prior distribution.

9.1.3 Reliability Estimation

We now consider Bayesian estimation of the Weibull reliability function given by either

$$r(t_0; \lambda, \beta) = \exp(-\lambda t_0^{\beta}) \tag{9.31}$$

or

$$r(t_0; \theta, \beta) = \exp\left(-\frac{t_0^{\beta}}{\theta}\right), \tag{9.32}$$

in the case of either a $\mathcal{W}_1(\lambda, \beta)$ or a $\mathcal{W}_2(\theta, \beta)$ distribution, respectively. We will consider the Bayesian estimation of r for the case of a random scale and known fixed shape parameter and the case where both parameters are r.v.'s.

Scale Parameter Is a Random Variable. Numerous papers have considered the case where Λ (or Θ) is an r.v. and r is to be estimated using Bayesian techniques. Harris and Singpurwalla (1968) present a point estimator of r in the case of a two-point, $\mathcal{U}(a, b)$, and $\mathcal{G}(\alpha, \beta)$ prior distribution on Λ. Canavos and Tsokos (1971a, 1973), Tsokos (1972a, 1972b), and Tsokos and Canavos (1972) give point estimators for r in the case of a $\mathcal{U}(\alpha, \beta)$, $\mathcal{E}(\lambda^{-1})$, or $\mathcal{IG}(\nu, \mu)$ prior distribution on Θ. Jayaram (1974) considers the point estimation of r using a specialized asymmetric loss function and a $\mathcal{B}(x_0, n_0)$ prior distribution on R. A piecewise linear prior distribution on R is used by Lian (1975) and Martz and Lian (1977) to derive Bayesian point and interval estimators for r. Papadopoulos and Tsokos (1975) obtain LBPI estimates for r in the case of both $\mathcal{IG}(\nu, \mu)$ and $\mathcal{U}(a, b)$ prior distributions on Θ. Moore and Bilikam (1978) derive Bayesian point estimators for r in the case of an $\mathcal{IG}(\alpha, \gamma)$, as well as a class of improper prior distributions on Θ. They restrict their discussion to Type II/item-censored testing. For a weighted squared-error loss function on λ, Tummala and Sathe (1978) consider Bayesian point estimation of r. Padgett and Tsokos (1978) consider both point and interval estimation of r in the case of mixtures of $\mathcal{W}_2(\theta, \beta)$ distributions. Bury (1972a) and Yatsu (1977) also consider Bayesian estimation of r.

The prior (posterior) distribution of R may be obtained directly from the prior (posterior) distributions of Λ or Θ by means of a transformation. Since r is a monotonic function of λ (or θ), the unique inverse exists which we may represent as

$$\lambda = -\frac{\ln r}{t_0^\beta}, \qquad 0 < r < 1, \quad \beta, t_0 > 0, \tag{9.33}$$

or

$$\theta = -\frac{t_0^\beta}{\ln r}, \qquad 0 < r < 1, \quad \beta, t_0 > 0. \tag{9.34}$$

Letting $g_\lambda(\cdot)$ and $g_\theta(\cdot)$ represent either the prior or posterior distribution of Λ and Θ, respectively, the corresponding prior or posterior distribution of R, denoted by $g_r(\cdot)$, may be obtained as

$$g_r(r) = g_\lambda\left(-\frac{\ln r}{t_0^\beta}\right)\left(\frac{1}{t_0^\beta}\right) \tag{9.35}$$

or

$$g_r(r) = g_\theta\left(-\frac{t_0^\beta}{\ln r}\right)\left(\frac{t_0^\beta}{r \ln^2 r}\right). \tag{9.36}$$

The Bayesian point estimator of r is the mean of the posterior distribution of R under squared-error loss. There are two techniques that can be used to determine this mean. As the rescaled total time on test $\mathcal{W}$ given by (9.2) is sufficient for estimating λ (or θ), we can find the posterior mean of R given w either by directly calculating the mean of the posterior distribution of R or by taking the expectation of R with respect to the posterior distribution of Λ (or Θ) given w. The latter method is most often used, except when the explicit posterior distribution of R is required. Both methods will be considered below.

As in Section 9.1.1, the results of Section 8.4 can be directly applied here by replacing the observed total time on test t by $\mathscr{W}$ and t_0 by t_0^β in all of the equations in Section 8.4. Thus, the Bayesian reliability estimators for a noninformative, $\mathscr{U}(\lambda_1, \lambda_2)$, or $\mathscr{G}(\alpha_0, \beta_0)$ prior distribution on Λ follow directly from the results in Section 8.4.4.

Example 9.3. Again let us consider Example 9.1. Suppose we desire a point and 95% LBPI estimate on the reliability of the heat exchanger for $t_0 = 1.5$ years. The induced prior and posterior distributions on R for $t_0 = 1.5$, as a consequence of the $\mathscr{U}(0.5, 1.5)$ prior distribution on Λ, are obtained according to (9.35) as

$$g_r(r; \alpha_0, \beta_0) = \frac{1}{(\beta_0 - \alpha_0) r (1.5)^\beta}, \qquad e^{-\beta_0 (1.5)^\beta} < r < e^{-\alpha_0 (1.5)^\beta},$$

and

$$g_r(r \mid \mathscr{W}; \alpha_0, \beta_0) = \frac{\mathscr{W}^{s+1}\left[-\ln r/(1.5)^\beta\right]^s e^{\mathscr{W} \ln r/(1.5)^\beta}}{\left[\Gamma(s+1, \beta_0 \mathscr{W}) - \Gamma(s+1, \alpha_0 \mathscr{W})\right] r (1.5)^\beta},$$

$$e^{-\beta_0 (1.5)^\beta} < r < e^{-\alpha_0 (1.5)^\beta}.$$

These distributions are plotted in Figure 9.2, where $\beta = 3.5$, $\mathscr{W} = 10.16$, $s = 9$, $\alpha_0 = 0.5$, and $\beta_0 = 1.5$. Using (8.115), the Bayesian point estimate of $r(1.5)$ is

$$E[R(1.5) \mid 10.16; 0.5, 1.5] = \left[\frac{10.16}{10.16 + (1.5)^{3.5}}\right]^{10} \left[\frac{\Gamma(10, 21.44) - \Gamma(10, 7.15)}{\Gamma(10, 15.24) - \Gamma(10, 5.08)}\right]$$

$$= 0.030.$$

Since $\Pr(\Lambda \leqslant 1.381 \mid 10.16; 0.5, 1.5) = 0.95$ [see Example 9.1], it follows from (9.31) that

$$\Pr\left\{R(1.5) \geqslant \exp\left[-1.38(1.5)^{3.5}\right] \mid 10.16; 0.5, 1.5\right\}$$

$$= \Pr\{R(1.5) \geqslant 0.0033 \mid 10.16; 0.5, 1.5\} = 0.95.$$

Thus the 95% LBPI estimate of $r(1.5)$ is 0.0033. The point and interval estimates are also shown in Figure 9.2.

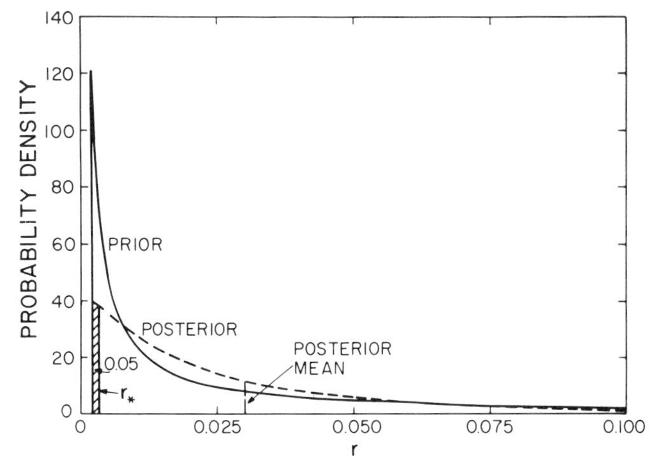

Figure 9.2. The prior and posterior distribution of Weibull reliability for $t_0 = 1.5$ years in Example 9.3.

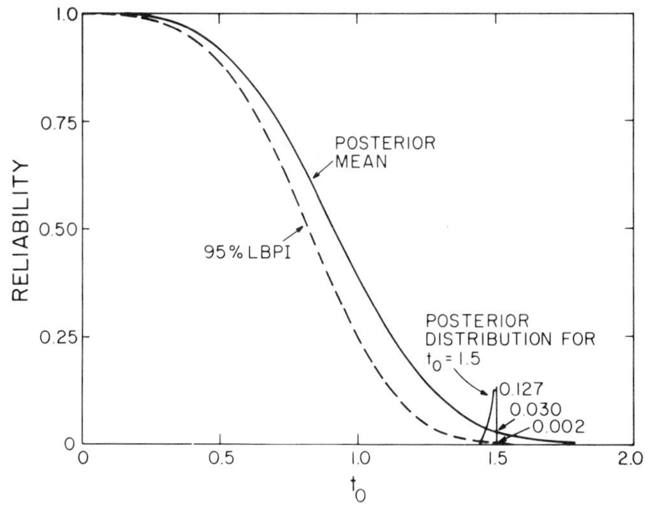

Figure 9.3. The posterior mean and 95% LBPI for $r(t_0)$ versus t_0 in Example 9.3.

In Figure 9.3 we have plotted the posterior mean and 95% LBPI estimate of $r(t_0)$ as a function of t_0. This plot of the posterior mean is the Bayesian point estimate of $r(t_0)$ for a squared-error loss function. We have also indicated the posterior distribution of R for $t_0 = 1.5$, which is shown in Figure 9.2. ■

Canavos and Tsokos (1973) and Tsokos (1972b) find by use of Monte Carlo simulation that the Bayesian point estimator of r, in the case of a $\mathcal{U}(\alpha, \beta)$, $\mathcal{E}(\lambda^{-1})$, or $\mathcal{IG}(\nu, \mu)$ prior distribution on Θ, has uniformly smaller mean squared-error than the MVU estimator. The form of the prior distribution was also observed to have little effect on the results.

Papadopoulos and Tsokos (1975) use Monte Carlo simulation to study the performance of the 90% and 95% LBPI estimates for r in the case of either an $\mathcal{IG}(\nu, \mu)$ or $\mathcal{U}(a, b)$ prior distribution on Θ. They find that the average of ratios of the Bayes to ML average squared-error for the reliability bounds varies from around 0.5 to 0.75, for a $\mathcal{U}(a, b)$ prior distribution, and from around 0.5 to 0.8, for an $\mathcal{IG}(\nu, \mu)$ prior on Θ. This represents a fairly significant decrease in the average squared-error of the Bayes estimator compared to the usual reliability bounds based on the ML estimator.

Martz and Lian (1977) consider a piecewise linear class of prior distributions on R for a fixed time t_0. The main motivation for use of this family is that, while permitting a variety of prior "shapes" depending upon the number and positioning of the line segments, the resulting Bayes point and interval estimators for $r(t_0)$ can be expressed in terms of standard incomplete gamma functions.

The piecewise linear family of prior distributions consists of $k-1$ adjoining straight line segments over the range 0 to 1. The segments have coordinates (a_i, b_i), $i=1,2,\ldots,k$, where $0 \leqslant a_1 \leqslant \cdots \leqslant a_k \leqslant 1$ and $b_i \geqslant 0$. For a given set of coordinates let

$$A = \sum_{i=1}^{k-1} \frac{(b_{i+1}+b_i)(a_{i+1}-a_i)}{2}. \tag{9.37}$$

Further define $v_i = b_i/A$, $i=1,\ldots,k$. The general form of the prior distribution on R is

$$g(r) = \lambda_i r + \tau_i, \qquad a_i < r \leqslant a_{i+1}, \qquad i=1,2,\ldots,k-1, \tag{9.38}$$

where

$$\lambda_i = \frac{v_{i+1}-v_i}{a_{i+1}-a_i}, \qquad \tau_i = v_i - \lambda_i a_i, \qquad \text{and} \qquad r = r(t_0).$$

The prior mean and variance of (9.38) are easily found to be

$$E(R) = \sum_{i=1}^{k-1} (a_{i+1}-a_i)\left[a_{i+1}v_{i+1} + a_i v_i + (v_{i+1}+v_i)(a_{i+1}+a_i)\right]/6 \tag{9.39}$$

and

$$\operatorname{Var}(R)=\sum_{i=1}^{k-1}(a_{i+1}-a_i)\times$$

$$\left[2\left(a_{i+1}^2 v_{i+1}+a_i^2 v_i\right)+(v_{i+1}+v_i)(a_{i+1}+a_i)^2\right]/12-E^2(R). \tag{9.40}$$

It is easily shown that $g(r)$ given by (9.38) is a proper prior distribution.

Martz and Lian (1977) consider a Type II/item-censored life test without replacement as the sampling experiment. In this case the statistic

$$\mathcal{W}=\sum_{i=1}^{s} T_i^{\beta}+(n-s)T_s^{\beta}$$

has a $\mathcal{G}_1(s,\lambda)$ sampling distribution. By a simple reparameterization, the sampling distribution of $\mathcal{W}$, given $R(t_0)=r$ for a fixed t_0, becomes

$$f(\mathcal{W}\,|\,r)=\frac{\mathcal{W}^{s-1}t_0^{-s\beta}}{\Gamma(s)}(-\ln r)^s r^{\mathcal{W}t_0^{-\beta}}, \qquad \mathcal{W}>0. \tag{9.41}$$

In this equation β is assumed to be known and t_0 is specified.

By use of (9.38) and (9.41) and Bayes' theorem, the posterior distribution of R is given by

$$g(r\,|\,\mathcal{W})=\frac{\lambda_i(-\ln r)^s r^{\mathcal{W}t_0^{-\beta}+1}+\tau_i(-\ln r)^s r^{\mathcal{W}t_0^{-\beta}}}{\sum_{i=1}^{k-1}\left[\lambda_i\Gamma_1^*(a_i,a_{i+1})+\tau_i\Gamma_0^*(a_i,a_{i+1})\right]},$$

$$a_i<r\leqslant a_{i+1}, \qquad i=1,2,\ldots,k-1, \tag{9.42}$$

where

$$\Gamma_q^*(x,y)=\left(\mathcal{W}t_0^{-\beta}+q+1\right)^{-s-1}\left\{\Gamma^c\left[s+1,-\left(\mathcal{W}t_0^{-\beta}+q+1\right)\ln y\right]\right.$$

$$\left.-\Gamma^c\left[s+1,-\left(\mathcal{W}t_0^{-\beta}+q+1\right)\ln x\right]\right\},$$

and $\Gamma^c(a,z)$ is the complement of the incomplete gamma function defined in (9.24).

The posterior mean of (9.42) is easily computed to be

$$E(R|\mathscr{W})=\frac{\sum_{i=1}^{k-1}\left[\lambda_i\Gamma_2^*(a_i,a_{i+1})+\tau_i\Gamma_1^*(a_i,a_{i+1})\right]}{\sum_{i=1}^{k-1}\left[\lambda_i\Gamma_1^*(a_i,a_{i+1})+\tau_i\Gamma_0^*(a_i,a_{i+1})\right]}. \tag{9.43}$$

Let us now consider a symmetric $100(1-\gamma)\%$ TBPI estimate for r. We desire two functions r_* and r^* of t_0 and $\mathscr{W}$ such that $\Pr(r_*\leqslant R\leqslant r^*|\mathscr{W})=1-\gamma$, subject to the restriction that $\Pr(R\leqslant r_*|\mathscr{W})=\Pr(R\geqslant r^*|\mathscr{W})=\gamma/2$, for fixed t_0. The desired interval estimate is obtained from (9.42) by numerically solving the two equations given by

$$\frac{\gamma}{2}=\frac{\sum_{i=1}^{j-1}\left[\lambda_i\Gamma_1^*(a_i,a_{i+1})+\tau_i\Gamma_0^*(a_i,a_{i+1})\right]+\lambda_j\Gamma_1^*(a_j,r_*)+\tau_j\Gamma_0^*(a_j,r_*)}{\sum_{i=1}^{k-1}\left[\lambda_i\Gamma_1^*(a_i,a_{i+1})+\tau_i\Gamma_0^*(a_i,a_{i+1})\right]}, \tag{9.44}$$

where $a_j<r_*\leqslant a_{j+1}$, $j=1,2,\ldots,k-2$ and

$$\frac{\gamma}{2}=\frac{\sum_{i=\nu+1}^{k-1}\left[\lambda_i\Gamma_1^*(a_i,a_{i+1})+\tau_i\Gamma_0^*(a_i,a_{i+1})\right]+\lambda_\nu\Gamma_1^*(r^*,a_{\nu+1})+\tau_\nu\Gamma_0^*(r^*,a_{\nu+1})}{\sum_{i=1}^{k-1}\left[\lambda_i\Gamma_1^*(a_i,a_{i+1})+\tau_i\Gamma_0^*(a_i,a_{i+1})\right]}, \tag{9.45}$$

where $a_\nu<r^*\leqslant a_{\nu+1}$, $\nu=1,2,\ldots,k-2$. Martz and Lian (1977) provide two algorithms that may be used to numerically solve (9.44) and (9.45) for the desired interval estimate (r_*,r^*). The algorithms are easily programmed on a computer for which an incomplete gamma function subroutine is available. If a $100(1-\gamma)\%$ LBPI estimate of r is desired, (9.44) can be solved with $\gamma/2$ replaced by γ.

Example 9.4. [Martz and Lian (1977)] Suppose that a certain device of interest has an assigned prior distribution of reliability for $t_0=100$ h given

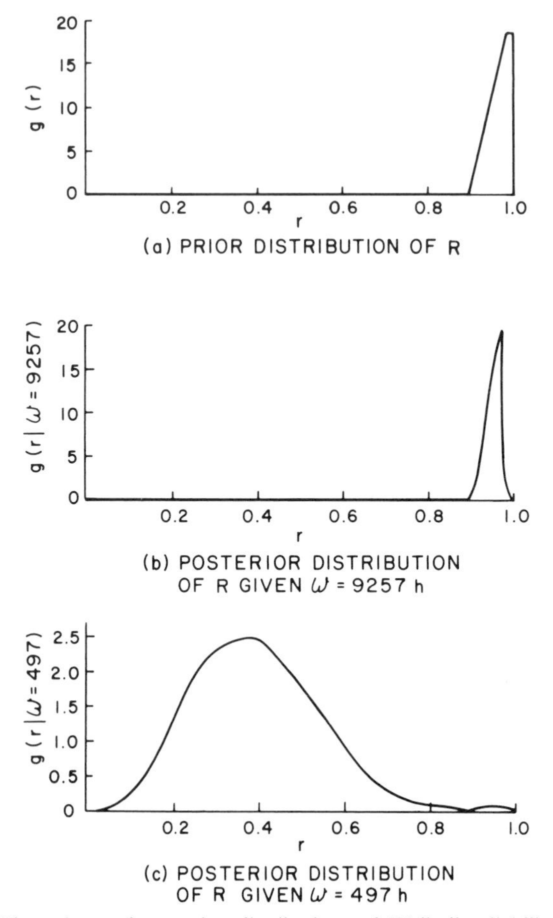

Figure 9.4. The prior and posterior distributions of Weibull reliability in Example 9.4 [Adapted, with permission, from Martz and Lian (1977)].

by

$$g(r)=\begin{cases}0.04, & 0\leqslant r\leqslant 0.90\\ 194.9r-175.4, & 0.90<r\leqslant 0.99\\ 17.6, & 0.99<r\leqslant 1.\end{cases}$$

The prior distribution consists of three segments and thus $k=4$. A life test is performed in which 20 devices are tested, and the test is terminated at the time of the fifth failure. Further suppose that an $\mathcal{E}(\lambda)$ distribution adequately describes the failure data. Thus $\beta=1$. The rescaled total time on test is computed to be $\mathcal{W}=9257$ h. Figure 9.4a and 9.4b give a plot of the prior and posterior distributions of $R = R(100)$. The posterior distribution is sharply peaked near $r = 0.95$, as the test results confirm the prior distribution. The posterior mean is computed according to (9.43) to be 0.95 and a 90% TBPI estimate on $r(100)$ is found to be (0.92, 0.98). The corresponding ML estimates are also computed to be 0.95 and (0.91, 0.98), which compare well with the Bayesian estimates. Figure 9.4c shows a plot of the resulting posterior distribution if $\mathcal{W} = 497$ h had been observed instead. In this case, the test results contradict the prior distribution, and the mode of the posterior distribution is now located near 0.4. The small "bump", located near 0.95, is due to the remaining influence of the prior distribution on the posterior. In this case, the Bayesian point and 90% interval estimates are 0.40 and (0.17, 0.65), respectively, while the corresponding ML estimates are 0.37 and (0.16, 0.67). ■

Scale and Shape Parameters Are Random Variables. Soland (1969) considers the point estimation of r in the case of a discrete prior distribution on β and a conditional $\mathcal{G}_1(s_{i0}, w_{i0})$ prior distribution on Λ. Canavos and Tsokos (1971a, 1973), Tsokos (1972a, 1972b), and Tsokos and Canavos (1972) give point estimators for r in the case of either an $\mathcal{IG}(\nu_0, \mu_0)$ or a $\mathcal{U}(0, \delta_0)$ prior distribution on Θ and an independent $\mathcal{U}(\alpha_0, \beta_0)$ prior distribution on β. Tsokos and Rao (1976) give point estimators for r in the $\mathcal{W}_2(\theta, \beta)$ distribution in the case of either an $\mathcal{IG}(\nu, \mu)$ or a $\mathcal{U}(a, b)$ prior distribution on Θ and an independent truncated $\mathcal{N}(\alpha_0, \sigma^2)$ prior distribution on β. The resultant estimators are derived based on the use of the first few terms in appropriate Taylor series expansions. Consequently, the estimators are approximations that involve the use of numerical integration techniques for their evaluation and thus are somewhat difficult to apply in practice.

First let us consider the work by Soland. For the joint posterior distribution given by the product of (9.11) and (9.12), the Bayesian point estimator of $r(t_0)$ is the posterior mean of (9.31) given by

$$E\left[\exp(-\Lambda t_0^{\beta}) \mid \underset{\sim}{z}\right] = \sum_{i=1}^{m} p_i'' \left[\frac{w_{i0} + \mathcal{W}_i}{w_{i0} + \mathcal{W}_i + t_0^{\beta_i}}\right]^{s_{i0}+s}, \tag{9.46}$$

where p_i'' is given by (9.11). Similarly, the prior mean of (9.31) is given by

$$E\left[\exp(-\Lambda t_0^{\beta})\right] = \sum_{i=1}^{m} p_i' \left[\frac{w_{i0}}{w_{i0} + t_0^{\beta_i}}\right]^{s_{i0}}, \tag{9.47}$$

where p_i' is defined in (9.10).

Example 9.5. [Soland (1969)] Again let us consider Example 9.2. The prior expected reliability that the component will survive 12 months is

$$E\{\exp[-\Lambda(8760)^{\beta}]\} = 0.3\left[\frac{2.665\times 10^{4}}{2.665\times 10^{4}+(8760)^{0.7}}\right]^{26.12}$$

$$+0.4\left[\frac{8.218\times 10^{3}}{8.218\times 10^{3}+(8760)^{0.8}}\right]^{3.44} +0.3\left[\frac{1.074\times 10^{4}}{1.074\times 10^{4}+(8760)^{0.9}}\right]^{1.91}$$

$$=0.577,$$

and the corresponding posterior expected reliability is

$$E\{\exp[-\Lambda(8760)^{\beta}]\,|\,\underset{\sim}{z}\} = 0.191\left[\frac{2.665\times 10^{4}+5.192\times 10^{3}}{2.665\times 10^{4}+5.192\times 10^{3}+(8760)^{0.7}}\right]^{28.12}$$

$$+0.430\left[\frac{8.218\times 10^{3}+11.046\times 10^{3}}{8.218\times 10^{3}+11.046\times 10^{3}+(8760)^{0.8}}\right]^{5.44}$$

$$+0.379\left[\frac{1.074\times 10^{4}+23.496\times 10^{3}}{1.074\times 10^{4}+23.496\times 10^{3}+(8760)^{0.9}}\right]^{3.91}$$

$$=0.665,$$

an increase of 15% over the prior estimate. ■

Now let us consider the prior distributions proposed by Canavos and Tsokos (1973). With respect to the joint posterior distribution given in (9.16), resulting from an $\mathcal{IG}(\nu_0, \mu_0)$ prior on Θ and a $\mathcal{U}(\alpha_0, \beta_0)$ prior on β, the posterior expectation of R becomes

$$E[\exp(-t_0^{\beta}/\Theta)\,|\,\underset{\sim}{z}] = \int_{\alpha_0}^{\beta_0}\int_0^{\infty}\exp\left(-\frac{t_0^{\beta}}{\theta}\right)g(\theta,\beta\,|\,\underset{\sim}{z})\,d\theta\,d\beta$$

$$=\frac{J_7}{J_1}, \tag{9.48}$$

where

$$J_7 = \int_{\alpha_0}^{\beta_0} \frac{(\beta^s v^\beta)}{\left(\mathcal{W}_1 + t_0^\beta\right)^{s+\nu_0}} d\beta. \tag{9.49}$$

As before, (9.48) must be evaluated using numerical integration techniques.

When both Θ and β have independent $\mathcal{U}(\cdot)$ distributions as in Section 9.1.2, the posterior expectation of R is taken with respect to the joint distribution given in (9.25). Thus

$$E\left[\exp(-t_0^\beta/\Theta)\,|\,\underset{\sim}{z}\right] = \int_{\alpha_0}^{\beta_0}\int_0^{\delta_0} \exp\left(-\frac{t_0^\beta}{\theta}\right) g(\theta, \beta\,|\,\underset{\sim}{z})\, d\theta\, d\beta$$

$$= \frac{J_8}{J_4}, \tag{9.50}$$

where

$$J_8 = \int_{\alpha_0}^{\beta_0} \left(\frac{\beta^s v^\beta}{\mathcal{W}^{s-1}}\right) \Gamma^c\left[s-1, \frac{\left(\mathcal{W} + t_0^\beta\right)}{\delta_0}\right] d\beta. \tag{9.51}$$

The Monte Carlo simulation results of Tsokos (1972b) and Canavos and Tsokos (1973) confirm the previous results that indeed the Bayesian estimators of r given above have uniformly smaller mean squared-errors than the ML estimator.

9.1.4 MTTF Estimation

Less attention has been directed in the literature at Bayesian estimation of the $\mathcal{W}_1(\lambda, \beta)$ mttf which is given by

$$\text{mttf} = E(T; \lambda, \beta) = \lambda^{-1/\beta}\Gamma\left(1 + \frac{1}{\beta}\right). \tag{9.52}$$

Again it is noted that we have used lower case letters, in this case mttf, to represent the value of an r.v., while MTTF will be used to denote a random mean time to failure. We restrict our consideration here to the case where β is known, and present both point and interval estimators for mttf for $\mathcal{U}(\alpha_0, \beta_0)$, noninformative, and $\mathcal{G}(\alpha_0, \beta_0)$ prior distributions on Λ.

We consider the life testing situation described in Section 9.1.1, with likelihood given by (9.3).

Uniform Prior Distribution on Λ. Suppose that Λ has a $\mathcal{U}(\alpha_0, \beta_0)$ prior distribution. Combining (9.3) and (8.6) by use of Bayes' theorem, the posterior distribution of Λ given $\mathcal{W}$ is given by

$$g(\lambda | \mathcal{W}; \alpha_0, \beta_0) = \frac{\mathcal{W}^{s+1}\lambda^s e^{-\lambda \mathcal{W}}}{\Gamma(s+1, \beta_0 \mathcal{W}) - \Gamma(s+1, \alpha_0 \mathcal{W})}, \qquad \alpha_0 < \lambda < \beta_0, \quad (9.53)$$

where

$$\mathcal{W} = \sum_{i=1}^{s} t_i^{\beta} + \sum_{i=s+1}^{n} t_i^{*\beta},$$

and s is the observed number of failures. The posterior expectation of MTTF, which is the Bayesian point estimator of mttf using squared-error loss, becomes

$$E[\text{MTTF} | \mathcal{W}; \alpha_0, \beta_0] = \int_{\alpha_0}^{\beta_0} \lambda^{-1/\beta} \Gamma(1 + 1/\beta) g(\lambda | \mathcal{W}; \alpha_0, \beta_0)\, d\lambda$$

$$= \frac{\mathcal{W}^{1/\beta}\Gamma(1+1/\beta)[\Gamma(s+1-1/\beta, \beta_0 \mathcal{W}) - \Gamma(s+1-1/\beta, \alpha_0 \mathcal{W})]}{\Gamma(s+1, \beta_0 \mathcal{W}) - \Gamma(s+1, \alpha_0 \mathcal{W})}. \quad (9.54)$$

The posterior risk of (9.54) is proportional to the posterior variance of MTTF given by

$$\text{Var}[\text{MTTF} | \mathcal{W}; \alpha_0, \beta_0] = \int_{\alpha_0}^{\beta_0} \lambda^{-2/\beta} \Gamma^2(1 + 1/\beta) g(\lambda | \mathcal{W}; \alpha_0, \beta_0)\, d\lambda$$

$$- E^2[\text{MTTF} | \mathcal{W}; \alpha_0, \beta_0]$$

$$= \frac{\mathcal{W}^{2/\beta}\Gamma^2(1+1/\beta)\Gamma(s+1-2/\beta, \beta_0 \mathcal{W}) - \Gamma(s+1-2/\beta, \alpha_0 \mathcal{W})}{\Gamma(s+1, \beta_0 \mathcal{W}) - \Gamma(s+1, \alpha_0 \mathcal{W})}$$

$$- E^2[\text{MTTF} | \mathcal{W}; \alpha_0, \beta_0]. \quad (9.55)$$

A symmetric $100(1-\gamma)\%$ TBPI estimate for mttf is easily obtained from the corresponding $100(1-\gamma)\%$ TBPI estimate for λ given by (8.15) and (8.16). Since

$$\Pr(\Lambda \leq \lambda_* | \mathcal{W}; \alpha_0, \beta_0) = \Pr\left[\Gamma(1+1/\beta)\Lambda^{-1/\beta} \geq \Gamma(1+1/\beta)\lambda_*^{-1/\beta} | \mathcal{W}; \alpha_0, \beta_0\right]$$

$$= \Pr\left[\text{MTTF} \geq \Gamma(1+1/\beta)\lambda_*^{-1/\beta} | \mathcal{W}; \alpha_0, \beta_0\right] = \frac{\gamma}{2},$$

the upper limit of the $100(1-\gamma)\%$ TBPI estimate for mttf is

$$\text{mttf}^* = \Gamma(1+1/\beta)(\lambda_*)^{-1/\beta}, \tag{9.56}$$

where λ_* is the corresponding lower limit on λ obtained by solving (8.15). Similarly, the lower limit of the desired TBPI estimate is

$$\text{mttf}_* = \Gamma(1+1/\beta)(\lambda^*)^{-1/\beta}, \tag{9.57}$$

and $\Pr(\text{mttf}_* \leqslant \text{MTTF} \leqslant \text{mttf}^* \mid \mathscr{W}; \alpha_0, \beta_0) = 1-\gamma$. In a similar way, $100(1-\gamma)\%$ LBPI and UBPI estimates are found from the corresponding UBPI and LBPI estimates, respectively, on λ.

Example 9.6. Again let us consider Example 9.1. Suppose we desire a Bayesian point and 95% TBPI estimate of the mttf of the heat exchanger. From (9.54), the point estimate is computed to be

$$\text{mttf} = \frac{(10.16)^{1/3.5}(0.8997)(181457.70 - 8326.32)}{(340221.64 - 12636.57)} = 0.92.$$

A "naive" Bayesian point estimate of mttf can be obtained by replacing λ in (9.52) by the Bayesian point estimate of λ. In this example, the naive point estimate becomes $(0.96)^{-1/3.5}(0.8997) = 0.91$, which agrees well with the above Bayesian estimate. Such an approach is not uncommon in Bayesian estimation and has been used by Tummala and Sathe (1978) to estimate the $\mathscr{W}(\lambda, \beta)$ reliability function.

The 95% TBPI estimate on λ is computed according to (8.15) and (8.16) to be $(0.548, 1.434)$. Thus,

$$\text{mttf}^* = (0.8997)(0.548)^{-1/3.5} = 1.068$$

and

$$\text{mttf}_* = (0.8997)(1.434)^{-1/3.5} = 0.812,$$

and $\Pr(0.812 \leqslant \text{MTTF} \leqslant 1.068 \mid 10.16; 0.5, 1.5) = 0.95$. ■

Noninformative Prior Distribution on Λ. In the case where $g(\lambda)$ is noninformative, as in (8.19), the posterior distribution of Λ given $\mathscr{W}$ is given by

$$g(\lambda \mid \mathscr{W}) = \frac{\mathscr{W}^s}{\Gamma(s)} \lambda^{s-1} e^{-\lambda \mathscr{W}}, \tag{9.58}$$

which is a $\mathcal{G}_1(s, \mathcal{W})$ distribution. The posterior expected MTTF thus becomes

$$E[\text{MTTF}|\mathcal{W}] = \int_0^\infty \lambda^{-1/\beta}\Gamma(1+1/\beta)\frac{\mathcal{W}^s}{\Gamma(s)}\lambda^{s-1}e^{-\lambda\mathcal{W}}\,d\lambda$$

$$= \frac{\mathcal{W}^{1/\beta}\Gamma(1+1/\beta)\Gamma(s-1/\beta)}{\Gamma(s)}. \tag{9.59}$$

The posterior variance is likewise given by

$$\text{Var}[\text{MTTF}|\mathcal{W}] = \frac{\mathcal{W}^{2/\beta}\Gamma^2(1+1/\beta)\Gamma(s-2/\beta)}{\Gamma(s)} - E^2[\text{MTTF}|\mathcal{W}]. \tag{9.60}$$

Using (8.26) and (8.27), a symmetric $100(1-\gamma)\%$ TBPI estimate for mttf is given by

$$\text{mttf}^* = \Gamma\left(1+\frac{1}{\beta}\right)\left[\frac{\chi^2_{\gamma/2}(2s)}{2\mathcal{W}}\right]^{-1/\beta} \tag{9.61}$$

and

$$\text{mttf}_* = \Gamma\left(1+\frac{1}{\beta}\right)\left[\frac{\chi^2_{1-\gamma/2}(2s)}{2\mathcal{W}}\right]^{-1/\beta}. \tag{9.62}$$

Corresponding LBPI or UBPI estimators are obtained from (9.62) or (9.61), respectively, by replacing $\gamma/2$ by γ.

Example 9.7. Suppose that in Example 9.1 a noninformative prior distribution is placed on Λ. The Bayesian point estimate of λ is

$$\text{mttf} = \frac{(10.16)^{1/3.5}(0.8997)(21978.98)}{(40320.00)} = 0.95.$$

A symmetric 95% TBPI estimate of mttf is given by

$$\text{mttf}^* = (0.8997)[(8.23)/(20.32)]^{-1/3.5} = 1.16,$$

$$\text{mttf}_* = (0.8997)[(31.5)/(20.32)]^{-1/3.5} = 0.79,$$

and thus, for a noninformative prior distribution on Λ, $\Pr(0.79 \leqslant \text{MTTF} \leqslant 1.16|10.16) = 0.95$. This interval is somewhat wider than the interval estimate in Example 9.6, since less prior knowledge is assumed about Λ. ■

Gamma Prior Distribution on Λ. If Λ has a $\mathcal{G}(\alpha_0, \beta_0)$ prior distribution, then the posterior distribution of Λ given $\mathcal{W}$ is $\mathcal{G}[\alpha_0 + s, \beta_0/(\beta_0\mathcal{W} + 1)]$. Hence the posterior MTTF is given by

$$E[\text{MTTF}|\mathcal{W}; \alpha_0, \beta_0] = \int_0^\infty \lambda^{-1/\beta}\Gamma(1+1/\beta)\frac{1}{\Gamma(\alpha_0+s)[\beta_0/(\beta_0\mathcal{W}+1)]^{\alpha_0+s}}$$

$$\times \lambda^{\alpha_0+s-1}\exp\left[-\lambda\left(\frac{\beta_0\mathcal{W}+1}{\beta_0}\right)\right] d\lambda$$

$$= \frac{\Gamma(\alpha_0+s-1/\beta)\Gamma(1+1/\beta)}{\Gamma(\alpha_0+s)[\beta_0/(\beta_0\mathcal{W}+1)]^{1/\beta}}, \tag{9.63}$$

and the posterior variance of MTTF is

$$\text{Var}[\text{MTTF}|\mathcal{W}; \alpha_0, \beta_0]$$

$$= \frac{\Gamma(\alpha_0+s-2/\beta)\Gamma(1+1/\beta)}{\Gamma(\alpha_0+s)[\beta_0/(\beta_0\mathcal{W}+1)]^{2/\beta}} - E^2[\text{MTTF}|\mathcal{W}; \alpha_0, \beta_0]. \tag{9.64}$$

From the Bayesian interval estimates presented in Section 8.2.3, a symmetric $100(1-\gamma)\%$ TBPI estimate for mttf is given by

$$\text{mttf}^* = \Gamma(1+1/\beta)\left[\frac{\beta_0\chi^2_{\gamma/2}(2s+2\alpha_0)}{2\beta_0\mathcal{W}+2}\right]^{-1/\beta} \tag{9.65}$$

and

$$\text{mttf}_* = \Gamma(1+1/\beta)\left[\frac{\beta_0\chi^2_{1-\gamma/2}(2s+2\alpha_0)}{2\beta_0\mathcal{W}+2}\right]^{-1/\beta}. \tag{9.66}$$

One-sided Bayesian interval estimates are similarly constructed by replacing $\gamma/2$ with γ.

Example 9.8. Again let us consider Example 9.1 in which, instead of a $\mathcal{U}(0.5, 1.5)$ prior on Λ, a $\mathcal{G}(2, 0.5)$ prior distribution is placed on Λ. Thus $\alpha_0 = 2$ and $\beta_0 = 0.5$ which implies that the prior mean is 1.0. The Bayesian point estimate of mttf in this case becomes

$$\text{mttf} = \frac{\Gamma(11 - 1/3.5)\Gamma(1 + 1/3.5)}{\Gamma(11)[0.5/6.08]^{1/3.5}} = \frac{(1860588.26)(0.8997)}{(3628800.00)(0.4898)} = 0.94.$$

A 95% TBPI estimate of mttf has upper and lower endpoints given by

$$\text{mttf}^* = (0.8997)[(0.5)(11.0)/(12.16)]^{-1/3.5} = 1.13,$$

$$\text{mttf}_* = (0.8997)[(0.5)(36.8)/(12.16)]^{-1/3.5} = 0.80.$$

Thus $\Pr(0.80 \leqslant \text{MTTF} \leqslant 1.13 | 10.16; 2.0, 0.5) = 0.95$. ■

9.1.5 Reliable Life Estimation

The reliable life was defined in Chapter 4 as the time t_R for which $100R\%$ of the population will survive. For the $\mathcal{W}_1(\lambda, \beta)$ distribution the reliable life t_R is defined as

$$t_R = \lambda^{-1/\beta}(-\ln R)^{1/\beta}, \tag{9.67}$$

where R is a specified proportion. We restrict our consideration here to the case where β is known, and consider point and interval estimators for t_R for $\mathcal{U}(\alpha_0, \beta_0)$, noninformative, and $\mathcal{G}(\alpha_0, \beta_0)$ prior distributions on Λ. Again we consider the life testing situation described in Section 9.1.1.

It is observed that (9.67) is of the same functional form in λ as the $\mathcal{W}_1(\lambda, \beta)$ mttf given in (9.52). Thus, all of the estimators for mttf given in Section 9.1.4 apply here as well, the only change being that the term $\Gamma(1 + 1/\beta)$ must now be replaced by the term $(-\ln R)^{1/\beta}$ in all of the point and interval estimation equations. In addition, the posterior variance of T_R may be obtained by making the same replacement.

For example, for a $\mathcal{G}(\alpha_0, \beta_0)$ prior distribution on Λ, it follows directly from (9.63) that the Bayesian point estimator for t_R is

$$\tilde{t}_R = E[T_R | \mathcal{W}; \alpha_0, \beta_0] = \frac{\Gamma(\alpha_0 + s - 1/\beta)(-\ln R)^{1/\beta}}{\Gamma(\alpha_0 + s)[\beta_0/(\beta_0 \mathcal{W} + 1)]^{1/\beta}}, \tag{9.68}$$

where $\tilde{}$ denotes a Bayesian point estimate. The symmetric $100(1-\gamma)\%$ TBPI estimate of t_R becomes

$$t_R^* = (-\ln R)^{1/\beta}\left[\beta_0\chi^2_{\gamma/2}(2s+2\alpha_0)/(2\beta_0\mathcal{W}+2)\right]^{-1/\beta} \tag{9.69}$$

and

$$t_{R*} = (-\ln R)^{1/\beta}\left[\beta_0\chi^2_{1-\gamma/2}(2s+2\alpha_0)/(2\beta_0\mathcal{W}+2)\right]^{-1/\beta}. \tag{9.70}$$

Thus $\Pr(t_{R*} \leqslant T_R \leqslant t_R{}^* | \mathcal{W}; \alpha_0, \beta_0) = 1-\gamma$.

Example 9.9. In Example 9.1, suppose that Λ has a $\mathcal{G}(2.0, 0.5)$ prior distribution and that we desire a Bayesian point and 95% LBPI estimate of t_R for $R=0.90$. According to (9.68) we have

$$\tilde{t}_{0.90} = \frac{\Gamma(11-1/3.5)(-\ln 0.90)^{1/3.5}}{\Gamma(11)(0.5/6.08)^{1/3.5}} = \frac{(1860588.26)(0.5257)}{(3628800.00)(0.4898)} = 0.55 \text{ years.}$$

Calculating (9.70) with $1-\gamma/2=0.95$, we find

$$t_{0.90*} = (0.5257)\left[\frac{(0.5)(33.9)}{(12.16)}\right]^{-1/3.5} = 0.478 \text{ years,}$$

and thus $\Pr(0.478 \leqslant T_{0.90} | 10.16; 2.0, 0.5) = 0.95$. ■

9.1.6 Robustness of the Bayes Estimators

Canavos (1974) examines the robustness of the $\mathcal{IG}(\nu_0, \mu_0)$ natural conjugate prior distribution for Θ in the $\mathcal{W}_2(\theta, \beta)$ distribution with known shape parameter β. Using Monte Carlo simulation, Canavos investigates the mean squared-error (MSE) sensitivity of the posterior mean $(\mathcal{W}+\mu_0)/(n+\nu_0-1)$, in which $s=n$, when the true prior distribution is *not* $\mathcal{IG}(\nu_0, \mu_0)$. His simulations are conducted using $\mathcal{U}(a,b)$, $\mathcal{G}(k,c)$, $\mathcal{N}(d,\sigma^2)$, and $\mathcal{W}_2(2\gamma^2, 2)$ prior distributions for Θ. The MSE ratio of the Bayes to the classical MVU estimators was formed and plotted as a function of the ratio of the variance of the true prior distribution to the variance of the assumed $\mathcal{IG}(\nu_0, \mu_0)$ prior.

Canavos also investigates the sensitivity of the above posterior mean to departures from the assumed $\mathcal{W}_2(\theta, \beta)$ sampling distribution, when the actual sampling distribution is $\mathcal{G}(\alpha, \theta)$.

Canavos finds that, when compared to the MVU estimator, the Bayes estimator is quite robust to the form of the prior distribution as long as the variance of the true prior distribution is approximately equal to that of the assumed $\mathcal{IG}(\nu_0, \mu_0)$ prior. However, for variance ratios of 1/5 and 5, the MVU estimator sometimes has a smaller MSE. Canavos' results also support the fact that the Bayes estimator is superior to the MVU estimator for samples of size $n = 3$ and 5, more so than for larger samples such as $n = 15$. Canavos also finds that the MVU estimator is more sensitive to the assigned $\mathcal{W}_2(\theta, \beta)$ sampling distribution than the Bayes estimator, and MSE ratios less than 1 were also observed.

Papadopoulos and Tsokos (1975) investigate the robustness of the $100(1-\gamma)\%$ LBPI for θ and $R(t_0)$ when the true prior distribution of θ is $\mathcal{U}(a, b)$, but an $\mathcal{IG}(\nu, \mu)$ was used. They find that the Bayesian interval estimates are still superior to the classical interval estimates.

Based on the above studies, it appears that the Bayesian estimation of the Weibull scale parameter with known shape parameter is indeed robust to the form of the prior distribution.

Tsokos and Rao (1979) investigate the robustness of the Bayesian point estimator for the $\mathcal{W}_1(\lambda, \beta)$ reliability function (9.31), using squared-error loss and a natural conjugate $\mathcal{G}_1(\nu, \mu)$ prior distribution on Λ, when the "true" prior distribution is either $\mathcal{B}(x_0, n_0)$, $\mathcal{IG}(\alpha_0, \beta_0)$, truncated $\mathcal{N}(\mu_0, \sigma_0^2)$, $\mathcal{LN}(\xi_0, \sigma_0^2)$, $\mathcal{U}(\alpha_0, \beta_0)$, $\mathcal{E}(\lambda_0)$, or $\mathcal{W}(\alpha_0, \beta_0)$. A few other distributions are also considered. It is found that, in many cases, there is a significant variation (20% or more) in average MSE even when the priors are chosen so that their first two moments agree with that of the conjugate prior distribution. The single exception to this conclusion is for the case of a $\mathcal{B}(x_0, n_0)$ prior. In this case the Bayesian reliability estimator is found to be quite robust. They conclude that, in Bayesian estimation of the $\mathcal{W}_1(\lambda, \beta)$ reliability function with known shape parameter, one must be very careful in choosing a prior distribution for the random scale parameter.

Tsokos and Rao (1976) examine the robustness of the Bayes reliability estimator with respect to the assumption that β has a fixed and known value, versus the assumption that it is an r.v. with a truncated $\mathcal{N}(\alpha_0, \sigma^2)$ distribution. They find that the Bayes reliability estimate, obtained under the assumption that β is a known constant, does not differ significantly from the estimate obtained by assuming that β is an r.v. Consequently, they conclude that in a $\mathcal{W}_2(\theta, \beta)$ failure model it is reasonable to assume that β is a known constant.

9.2. THE NORMAL DISTRIBUTION

As mentioned in Chapter 4, the $\mathcal{N}(\mu, \sigma^2)$ distribution is sometimes considered as a failure time model in which the mean μ is a sufficiently large positive value relative to σ. Bayesian estimators for the parameters μ and σ are well-known for noninformative prior distributions [Box and Tiao (1973), Chapter 2] and the natural conjugate distributions [Raiffa and Schlaifer (1961), Chapter 11]. We shall primarily focus our attention on Bayesian estimation of reliability and reliable life given by (4.27) and (4.30), respectively, for both noninformative and natural conjugate prior distributions. We consider two cases: μ an r.v. and σ known (Section 9.2.1) and both μ and σ r.v.'s (Section 9.2.2). Sections 9.2.3 and 9.2.4 consider reliability and reliable life estimation, respectively. Throughout this section we use the symbols μ and σ to denote both r.v.'s and their values.

The *precision* τ of a $\mathfrak{N}(\mu,\sigma^2)$ distribution is defined to be the reciprocal of the variance; that is,

$$\tau=\frac{1}{\sigma^2}. \tag{9.71}$$

In much of the discussion that follows, it will be more convenient to specify an $\mathfrak{N}(\mu,\sigma^2)$ distribution by its mean μ and its precision τ, rather than by its mean and variance. Since the variance σ^2 of an $\mathfrak{N}(\mu,\sigma^2)$ distribution is a measure of its dispersion, the precision τ is a measure of its concentration about the mean.

9.2.1 μ Is a Random Variable

First let us consider the case in which the mean μ in an $\mathfrak{N}(\mu,\sigma^2)$ distribution is an r.v. with prior distribution $g(\mu)$ and the standard deviation σ is known. We shall further restrict our consideration to complete (nontruncated/noncensored) life test data in which a sample of n complete failure times $t_1, t_2,\ldots,t_n$ is observed. Define the usual sufficient statistic $\bar{T}=\sum_{i=1}^n T_i/n$; thus, $\bar{T}$ has an $\mathfrak{N}(\mu,\sigma^2/n)$ distribution.

Noninformative Prior Distribution on μ. Box and Tiao (1973) derive the noninformative prior distribution on μ as

$$g(\mu)\propto c, \qquad c=\text{constant}, \tag{9.72}$$

and consequently, the corresponding posterior distribution on μ given $\bar{t}$ is

$$g(\mu|\bar{t};\sigma^2)\cong\left(\frac{2\pi\sigma^2}{n}\right)^{-1/2}\exp\left[-\frac{n(\mu-\bar{t})^2}{2\sigma^2}\right], \qquad -\infty<\mu<\infty,\quad \sigma>0, \tag{9.73}$$

which is an $\mathfrak{N}(\bar{t},\sigma^2/n)$ distribution.

Normal Prior Distribution on μ. Raiffa and Schlaifer (1961) show that the natural conjugate prior distribution of μ is an $\mathcal{N}(\lambda_0, \psi_0^2)$ distribution, where ψ_0 is the prior standard deviation, and that the posterior distribution of μ given $\bar{t}$ is thus an $\mathcal{N}(\lambda, \psi^2)$ distribution, where the posterior mean λ is given by

$$\lambda = \frac{\sigma^2\lambda_0 + n\psi_0^2\bar{t}}{\sigma^2 + n\psi_0^2} = \frac{\tau_0\lambda_0 + n\tau\bar{t}}{\tau_0 + n\tau}, \tag{9.74}$$

and the posterior variance ψ^2 is

$$\psi^2 = \frac{\sigma^2\psi_0^2}{\sigma^2 + n\psi_0^2} = (\tau_0 + n\tau)^{-1}, \tag{9.75}$$

where $\tau_0 = 1/\psi_0^2$ is the prior precision and $\tau = 1/\sigma^2$ is the sampling precision. Thus the posterior precision is $\tau_0 + n\tau$. It is observed that the posterior mean λ is a weighted average of the prior mean λ_0 and the sample mean $\bar{t}$. The larger the prior or sampling precision, the greater will be the weight that is given to λ_0 or $\bar{t}$, respectively. Also the larger the sample size n, the greater will be the weight that is given to $\bar{t}$.

The Bayes point estimate of μ, using squared-error loss, is λ and a symmetric $100(1-\gamma)\%$ TBPI on μ is given by

$$\mu^* = \lambda + z_{1-\gamma/2}\psi \tag{9.76}$$

and

$$\mu_* = \lambda - z_{1-\gamma/2}\psi, \tag{9.77}$$

where, according to (4.26), $\Pr(Z \leq z_{1-\gamma/2}) = 1-\gamma/2$. In other words, we may write $z_{1-\gamma/2} = \Phi^{-1}(1-\gamma/2)$. Thus, $z_{1-\gamma/2}$ is the $100(1-\gamma/2)$th percentile of a standard normal r.v. and may be found in Table B1 of Appendix B. Therefore, $\Pr(\mu_* \leq \mu \leq \mu^* | \bar{t}; \sigma^2, \lambda_0, \psi_0^2) = 1-\gamma$.

Example 9.10. In Example 4.7, the failure times (in years) for 22 bushings of a 115 kV power generator were found to be appropriately described by an $\mathcal{N}(\mu, \sigma^2)$ distribution. Further suppose that, based on manufacturer's specifications, $\sigma = 3$ years. If an $\mathcal{N}(15, 4)$ prior distribution is assigned to μ, the Bayes point estimate of μ, for $\bar{t} = 14.01$ years, becomes

$$\lambda = \frac{9(15) + 22(4)(14.01)}{9 + 22(4)} = 14.10 \text{ years.}$$

A 95% TBPI estimate of μ is computed to be

$$\mu^* = 14.10 + 1.96\left[\frac{9(4)}{9+22(4)}\right]^{1/2} = 15.29 \text{ years},$$

$$\mu_* = 14.10 - 1.96\left[\frac{9(4)}{9+22(4)}\right]^{1/2} = 12.91 \text{ years},$$

since, from Table B1 in Appendix B, $z_{0.975} = 1.96$. Thus $\Pr(12.91 \leqslant \mu \leqslant 15.29 \,|\, 14.01; 9, 15, 4) = 0.95$. ■

9.2.2 μ and σ Are Random Variables

Now let us consider the case where both the mean μ and standard deviation σ (or precision τ) are r.v.'s whose values are unknown. Define the statistic $S^2 = \nu^{-1}\Sigma_{i=1}^{n}(T_i - \bar{T})^2$, where $\nu = n-1$. It is well known [see Box and Tiao (1973), Section 1.4] that the sample mean $\bar{T}$ and sample variance S^2 are jointly sufficient for estimating (μ, σ^2), and are distributed independently as $\mathfrak{N}(\mu, \sigma^2/n)$ and $\sigma^2\chi^2(\nu)/\nu$, respectively.

Noninformative Prior Distribution for (μ, σ). If we assume that, a priori, μ and σ are approximately independent so that $g(\mu, \sigma) \approx g(\mu)g(\sigma)$, then Box and Tiao (1973) give the joint noninformative prior distribution on (μ, σ) as

$$g(\mu, \sigma) \propto \sigma^{-1}. \tag{9.78}$$

Using Bayes' theorem, the corresponding joint posterior distribution on (μ, σ) given $\underset{\sim}{t}' = (t_1, t_2, \ldots, t_n)$ is

$$g(\mu, \sigma | \underset{\sim}{t}) = \sqrt{\frac{n}{2\pi}}\left[\frac{1}{2}\Gamma\left(\frac{\nu}{2}\right)\right]^{-1}\left(\frac{\nu s^2}{2}\right)^{\nu/2}\sigma^{-(n+1)}\exp\left\{-\frac{1}{2\sigma^2}\left[\nu s^2 + n(\mu - \bar{t})^2\right]\right\},$$

$$-\infty < \mu < \infty, \quad \sigma > 0. \tag{9.79}$$

Box and Tiao (1973), Section 2.4 develop $100(1-\gamma)\%$ Bayesian highest posterior density regions for μ and σ. In addition, they also obtain the marginal posterior distributions of σ and μ given $\underset{\sim}{t}$. The marginal posterior distribution of σ is given by

$$g(\sigma | \underset{\sim}{t}) = \left[\frac{1}{2}\Gamma\left(\frac{\nu}{2}\right)\right]^{-1}\left(\frac{\nu s^2}{2}\right)^{\nu/2}\sigma^{-(\nu+1)}\exp\left(-\frac{\nu s^2}{2\sigma^2}\right), \qquad \sigma > 0, \tag{9.80}$$

which is known as an *inverted-gamma-2* distribution [see Raiffa & Schlaifer (1961), p. 228], and the marginal posterior distribution of μ becomes

$$g(\mu|\underset{\sim}{t}) = \frac{\Gamma(\nu/2+1/2)\left(s/\sqrt{n}\right)^{-1}}{\Gamma(\nu/2)(\pi\nu)^{1/2}}\left[1+\frac{n(\mu-\bar{t})^2}{\nu s^2}\right]^{-(\nu+1)/2},$$

$$-\infty<\mu<\infty. \quad (9.81)$$

Thus $\sqrt{n}(\mu-\bar{t})/s$ given $\underset{\sim}{t}$ has a $t(\nu)$ distribution, which is defined in (3.12).

Normal-Gamma Prior Distribution on (μ, τ). Raiffa and Schlaifer (1961) derive the joint conjugate prior distribution on (μ,τ) as the product of a conditional $\mathfrak{N}[\lambda_0,(\tau_0\tau)^{-1}]$ prior distribution, given τ, on μ and a marginal $\mathcal{G}_1(\alpha_0,\beta_0)$ prior distribution on τ. Thus, the prior distribution of (μ,τ) is given by

$$g(\mu,\tau;\lambda_0,\tau_0,\alpha_0,\beta_0) \propto \tau^{1/2}\exp\left[-\left(\frac{\tau_0\tau}{2}\right)(\mu-\lambda_0)^2\right]\tau^{\alpha_0-1}\exp(-\beta_0\tau),$$

$$-\infty<\mu<\infty, \qquad \tau>0, \quad (9.82)$$

which is known as a *normal-gamma* distribution. They then show that the joint posterior distribution of (μ,τ) given $\underset{\sim}{t}$ is also a normal-gamma distribution in which the conditional posterior distribution of μ given τ and $\underset{\sim}{t}$ is an $\mathfrak{N}\{\lambda',[(\tau_0+n)\tau]^{-1}\}$ distribution where

$$\lambda' = \frac{\tau_0\lambda_0+n\bar{t}}{\tau_0+n}, \quad (9.83)$$

and the marginal posterior distribution of τ given $\underset{\sim}{t}$ is a $\mathcal{G}_1(\alpha_0+n/2,\beta)$ distribution, where

$$\beta = \beta_0+\frac{\nu s^2}{2}+\frac{\tau_0 n(\bar{t}-\lambda_0)^2}{2(\tau_0+n)}. \quad (9.84)$$

Thus, the joint posterior distribution of (μ,τ) given $\underset{\sim}{t}$ is the normal-gamma distribution given by

$$g(\mu,\tau|\underset{\sim}{t};\lambda_0,\tau_0,\alpha_0,\beta_0) \propto \tau^{1/2}\exp\left[-\frac{(\tau_0+n)\tau}{2}(\mu-\lambda')^2\right]\tau^{\alpha_0+n/2-1}\exp(-\beta\tau),$$

$$-\infty<\mu<\infty, \qquad \tau>0. \quad (9.85)$$

De Groot (1970) shows that the marginal posterior distribution of μ given $\underset{\sim}{t}$ becomes

$$g(\mu|\underset{\sim}{t};\lambda_0,\tau_0,\alpha_0,\beta_0)\propto\left[\beta+\frac{(\tau_0+n)}{2}(\mu-\lambda')^2\right]^{-\alpha_0-n/2-1/2},$$

$$-\infty<\mu<\infty. \quad (9.86)$$

It can be shown that $[(\alpha_0+n/2)(\tau_0+n)/\beta]^{1/2}(\mu-\lambda')$ given $\underset{\sim}{t}$ has a $t(2\alpha_0+n)$ distribution. It is also noted that μ and τ are dependent r.v.'s. In fact, there is no joint conjugate prior distribution such that μ has an $\mathcal{N}(\mu_0,\sigma_0^2)$ distribution, τ has a $\mathcal{G}_1(\alpha_0,\beta_0)$ distribution, and μ and τ are independently distributed. In any case, the posterior distribution induces dependency between μ and τ.

The Bayes point estimator for μ, under squared-error loss, is the posterior mean of (9.86) which is λ'. A symmetric $100(1-\gamma)\%$ TBPI estimate of μ is easily obtained from (9.86) as

$$\mu^*=\lambda'+\left[\left(\alpha_0+\frac{n}{2}\right)\left(\frac{\tau_0+n}{\beta}\right)\right]^{-1/2}t_{1-\gamma/2}(2\alpha_0+n) \quad (9.87)$$

and

$$\mu_*=\lambda'-\left[\left(\alpha_0+\frac{n}{2}\right)\left(\frac{\tau_0+n}{\beta}\right)\right]^{-1/2}t_{1-\gamma/2}(2\alpha_0+n), \quad (9.88)$$

where $t_{1-\gamma/2}(k)$ is the $100(1-\gamma/2)$th percentile of a $t(k)$ distribution. Table B2 in Appendix B gives values of $t_\gamma(k)$ for selected values of γ and k.

Example 9.11. Again consider Example 4.7, where we now suppose that μ and σ are both unknown. From the failure data, $\bar{t}=14.01$ and $s^2=11.31$. Further suppose that a normal-gamma prior distribution is desired in which $E(\mu)=15$, $\text{Var}(\mu)=4$, $E(\tau)=2$, and $\text{Var}(\tau)=1$. Since $E(\tau)=\alpha_0\beta_0$, $\text{Var}(\tau)=\alpha_0\beta_0^2$, $E(\mu)=\lambda_0$, and $\text{Var}(\mu)=\beta_0/[\tau_0(\alpha_0-1)]$, this implies that $\lambda_0=15$, $\tau_0=0.04$, $\alpha_0=4$, and $\beta_0=0.5$. Thus

$$\lambda'=\frac{0.04(15)+22(14.01)}{0.04+22}=14.012$$

and

$$\beta=\frac{1}{2}+21(11.31)/2+\frac{0.04(22)(14.01-15)^2}{2(0.04+22)}=119.275.$$

A symmetric 95% TBPI estimate of μ is given by

$$\mu^* = 14.012 + [(4+11)(0.04+22)/119.275]^{-1/2}(2.042) = 15.239$$

and

$$\mu_* = 14.012 - [(4+11)(0.04+22)/119.275]^{-1/2}(2.042) = 12.785,$$

since, from Table B2 in Appendix B, $t_{0.975}(30) = 2.042$. Hence

$$\Pr(12.785 \leqslant \mu \leqslant 15.239 | \underset{\sim}{t}; 15, 0.04, 4, 0.5) = 0.95. \quad \blacksquare$$

9.2.3 Reliability Estimation

We will now consider Bayesian estimation of the $\mathfrak{N}(\mu, \sigma^2)$ reliability function given by

$$r(t_0; \mu, \sigma^2) = 1 - \Phi\left(\frac{t_0 - \mu}{\sigma}\right), \tag{9.89}$$

where $\Phi(\cdot)$ is the $\mathfrak{N}(0,1)$ cdf defined in (4.26).

μ Is a Random Variable. Let us consider the case where σ is known and μ has either a noninformative prior distribution given in (9.72) or an $\mathfrak{N}(\lambda_0, \psi_0^2)$ prior distribution.

In the case of a noninformative prior distribution, the Bayesian point estimator of $r = r(t_0; \mu, \sigma^2)$, under squared-error loss, is the posterior expectation of R with respect to (9.73). Thus

$$E[R | \bar{t}; \sigma^2] = 1 - \int_{-\infty}^{\infty} \Phi\left(\frac{t_0 - \mu}{\sigma}\right) g(\mu | \bar{t}; \sigma^2)\, d\mu$$

$$= 1 - \int_{-\infty}^{t_0} \int_{-\infty}^{\infty} \frac{1}{\sigma\sqrt{2\pi}} \exp\left[-\frac{1}{2\sigma^2}(x-\mu)^2\right] \times$$

$$\frac{1}{\sqrt{2\pi\sigma^2/n}} \exp\left[-\frac{n}{2\sigma^2}(\mu - \bar{t})^2\right] d\mu\, dx,$$

upon interchanging the order of integration according to Fubini's theorem. Upon completing the square in μ and integrating with respect to μ, we obtain

$$E[R|\bar{t};\sigma^2]=1-\int_{-\infty}^{t_0}\frac{1}{\sigma\sqrt{2\pi}}\sqrt{\frac{n}{n+1}}\times$$

$$\exp\left[\frac{1}{2\sigma^2}\frac{(x+n\bar{t})^2}{n+1}\right]\exp\left[-\frac{1}{2\sigma^2}(x^2+n\bar{t}^2)\right]dx,$$

which, upon simplification, becomes

$$E(R|\bar{t};\sigma^2]=1-\int_{-\infty}^{t_0}\frac{1}{\sigma\sqrt{2\pi}}\sqrt{\frac{n}{n+1}}\exp\left[-\frac{n}{2(n+1)\sigma^2}(x-\bar{t})^2\right]dx.$$

The integrand is recognized as the pdf of an $\mathfrak{N}[\bar{t},(n+1)\sigma^2/n]$ distribution and thus

$$\tilde{r}=E[R|\bar{t};\sigma^2]=1-\Phi\left[\frac{t_0-\bar{t}}{\sigma\sqrt{(n+1)/n}}\right]. \tag{9.90}$$

It is interesting to compare the Bayes estimator, $\tilde{r}$, with the ML estimator which is $\hat{r}=1-\Phi[(t_0-\bar{T})/\sigma]$. It is observed that, if $t_0=\bar{t}$, then $\tilde{r}=\hat{r}$; otherwise, $|\tilde{r}-0.5|<|\hat{r}-0.5|$. This slight bias is the result of the noninformative prior on μ which tends to "shrink" the Bayes estimate towards $\Phi(0)=0.5$.

It is noted that the $\mathfrak{N}[\bar{t},(n+1)\sigma^2/n]$ distribution is often referred to as the *predictive distribution* of X [Guttman and Tiao (1964)], which is the posterior distribution of a "future" observation X given the current sample statistic $\bar{t}$.

Since, for a given t_0, $r(t_0;\mu,\sigma^2)$ is a monotonically increasing function of μ, we can easily construct a $100(1-\gamma)\%$ TBPI estimate of r. The upper and lower limits on r become

$$r^*=1-\Phi\left[\frac{t_0-\bar{t}-z_{1-\gamma/2}\sigma/\sqrt{n}}{\sigma}\right] \tag{9.91}$$

and

$$r_*=1-\Phi\left[\frac{t_0-\bar{t}+z_{1-\gamma/2}\sigma/\sqrt{n}}{\sigma}\right], \tag{9.92}$$

since, according to (9.73), the corresponding upper and lower limits on μ are $\mu^* = \bar{t} + z_{1-\gamma/2}\sigma/\sqrt{n}$ and $\mu_* = \bar{t} - z_{1-\gamma/2}\sigma/\sqrt{n}$, respectively. In this case it is observed that the Bayesian limits on r are identical to the classical $(1-\gamma)$ confidence limits. If a $100(1-\gamma)\%$ LBPI estimate of r is desired, (9.92) yields the desired estimator in which $\gamma/2$ is replaced by γ.

Now suppose that an $\mathfrak{N}(\lambda_0, \psi_0^2)$ prior distribution is placed on μ. The usual Bayesian point estimator of r, under squared-error loss, is the expectation of R in the posterior $\mathfrak{N}(\lambda, \psi^2)$ distribution, where λ and ψ^2 are given by (9.74) and (9.75), respectively. Thus

$$E\left[R|\bar{t}; \sigma^2, \lambda_0, \psi_0^2\right] = 1 - \int_{-\infty}^{\infty} \Phi\left(\frac{t_0 - \mu}{\sigma}\right) g\left(\mu|\bar{t}; \sigma^2, \lambda_0, \psi_0^2\right) d\mu$$

$$= 1 - \int_{-\infty}^{t_0} \int_{-\infty}^{\infty} \frac{1}{\sigma\sqrt{2\pi}} \exp\left[-\frac{1}{2\sigma^2}(x-\mu)^2\right] \times$$

$$\frac{1}{\psi\sqrt{2\pi}} \exp\left[-\frac{1}{2\psi^2}(\mu-\lambda)^2\right] d\mu\, dx.$$

Upon completing the square in μ, integrating with respect to μ, and simplifying, we obtain

$$E\left[R|\bar{t}; \sigma^2, \lambda_0, \psi_0^2\right] = 1 - \int_{-\infty}^{t_0} \frac{1}{(\sigma^2+\psi^2)^{1/2}\sqrt{2\pi}} \times$$

$$\exp\left[-\frac{1}{2(\sigma^2+\psi^2)}(x-\lambda)^2\right] dx.$$

The integrand is the pdf of an $\mathfrak{N}[\lambda, (\sigma^2+\psi^2)]$ distribution (the predictive distribution) and thus

$$E\left[R|\bar{t}; \sigma^2, \lambda_0, \psi_0^2\right] = 1 - \Phi\left[\frac{t_0 - \lambda}{\sqrt{\sigma^2+\psi^2}}\right]. \tag{9.93}$$

A prior estimator of r would also be given by (9.93) in which λ and ψ^2 are replaced by λ_0 and ψ_0^2, respectively.

A $100(1-\gamma)\%$ TBPI estimate of r has upper and lower limits given by

$$r^* = 1 - \Phi\left[\frac{t_0 - \mu^*}{\sigma}\right] \tag{9.94}$$

and

$$r_* = 1 - \Phi\left[\frac{t_0 - \mu_*}{\sigma}\right], \tag{9.95}$$

respectively, where (μ_*, μ^*) is the corresponding $100(1-\gamma)\%$ TBPI estimate of μ given by (9.76) and (9.77). A $100(1-\gamma)\%$ LBPI estimate of r may be obtained from (9.95) by replacing $\gamma/2$ with γ when computing μ_* given in (9.77).

Example 9.12. Let us again consider Example 9.10. Suppose that we desire a Bayesian point and 95% LBPI estimate of the bushing reliability for $t_0 = 5$ years. Recall that $\sigma^2 = 9$, $\lambda = 14.10$, and $\psi^2 = 0.37$. Thus,

$$\tilde{r} = 1 - \Phi\left[\frac{5 - 14.01}{\sqrt{9 + 0.37}}\right] = 1 - \Phi(-2.97) = 0.9985$$

is the desired point estimate. A 95% LBPI estimate of μ is $\mu_* = 14.01 - 1.645(0.37)^{1/2} = 13.009$, and, hence, a 95% LBPI estimate of $r(5)$ is

$$r_* = 1 - \Phi\left[\frac{5 - 13.009}{3}\right] = 1 - \Phi(-2.67) = 0.9962.$$

Thus $\Pr[R(5) \geq 0.9962 | 14.01; 3, 15, 2] = 0.95$. Figure 9.5 gives a plot of the

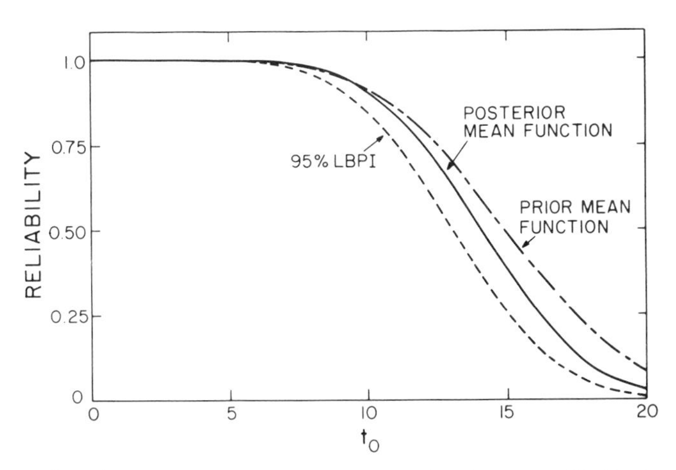

Figure 9.5. The prior mean function, posterior mean function, and 95% LBPI estimate of $r(t_0)$ in Example 9.12.

posterior point estimate and 95% LBPI estimate of the normal reliability function $r(t_0)$ as a function of t_0, assuming a squared-error loss function. Also shown is the corresponding prior point estimate of the reliability function. ■

μ and σ Are Random Variables. In the case of a noninformative prior distribution on (μ, σ), the Bayesian point estimator of r, under squared-error loss, is the expectation of R in the joint posterior distribution given in (9.79). Thus

$$E[R|\underset{\sim}{t}]=1-\int_0^{\infty}\int_{-\infty}^{\infty}\Phi\left(\frac{t_0-\mu}{\sigma}\right)g(\mu,\sigma|\underset{\sim}{t})\,d\mu\,d\sigma$$

$$=1-\int_{-\infty}^{t_0}\int_0^{\infty}\int_{-\infty}^{\infty}\frac{1}{\sigma\sqrt{2\pi}}\exp\left[-\frac{1}{2\sigma^2}(x-\mu)^2\right]\sqrt{\frac{n}{2\pi}}\left[\frac{1}{2}\Gamma\left(\frac{\nu}{2}\right)\right]^{-1}\times$$

$$\left(\frac{\nu s^2}{2}\right)^{\nu/2}\sigma^{-(n+1)}\exp\left\{-\frac{1}{2\sigma^2}\left[\nu s^2+n(\mu-\bar{t})^2\right]\right\}d\mu\,d\sigma\,dx.$$

Completing the square on μ, integrating with respect to μ, and simplifying, yields

$$E[R|\underset{\sim}{t}]=1-\int_{-\infty}^{t_0}\int_0^{\infty}\left[\frac{1}{2}\Gamma\left(\frac{\nu}{2}\right)\right]^{-1}\left(\frac{\nu s^2}{2}\right)^{\nu/2}\sqrt{\frac{n}{n+1}}\,\frac{1}{\sqrt{2\pi}}\sigma^{-(n+1)}\times$$

$$\exp\left\{-\frac{1}{\sigma^2}\left[\frac{\nu s^2}{2}+\frac{n(x-\bar{t})^2}{2(n+1)}\right]\right\}d\sigma\,dx.$$

Using the well-known identity,

$$\int_0^{\infty}z^{-(p+1)}\exp\left(-\frac{a}{z^2}\right)dz=\frac{1}{2}a^{-p/2}\Gamma\left(\frac{p}{2}\right),$$

we have upon integration and simplification that

$$E[R|\underset{\sim}{t}]=1-\int_{-\infty}^{t_0}\frac{B^{1/2}\Gamma[(\nu+1)/2]}{\Gamma(\nu/2)(\pi\nu)^{1/2}}\left[1+\frac{B(x-\bar{t})^2}{\nu}\right]^{-(\nu+1)/2}dx,$$

where

$$B=\frac{n}{(n+1)s^2}. \tag{9.96}$$

Now letting $y=B^{1/2}(x-\bar{t})$ we have

$$E[R|\underset{\sim}{t}]=1-\int_{-\infty}^{B^{1/2}(t_0-\bar{t})}\frac{\Gamma[(\nu+1)/2]}{\Gamma(\nu/2)(\pi\nu)^{1/2}}\left[1+\frac{y^2}{\nu}\right]^{-(\nu+1)/2}dy.$$

The integrand, which is the predictive distribution, is observed to be the pdf of a $t(\nu)$ distribution and thus

$$\tilde{r}=E[R|\underset{\sim}{t}]=1-\Pr\left[t\leqslant\sqrt{\frac{n}{n+1}}\,\frac{(t_0-\bar{t})}{s}\right]=t\left[B^{1/2}(t_0-\bar{t});\nu\right], \tag{9.97}$$

where $t(x;\nu)$ denotes the cdf of t r.v. with ν degrees of freedom evaluated at x. This result can also be obtained using a theorem of Ellison (1964). Padgett and Wei (1977) also obtain a similar result for the $\mathcal{LN}(\xi,\sigma^2)$ distribution. Numerous software routines exist for evaluating $t(x;\nu)$, some of which are available on hand held calculators.

Interval estimation of r when both μ and σ are unknown is somewhat complicated, since r is a bivariate function of both of these quantities. Thus it is not considered here.

Example 9.13. In Example 4.7, suppose we desire a point estimate of the bushing reliability for $t_0=5$ years, in which a noninformative prior distribution on (μ,σ) is specified. From Example 4.7, $n=22$, $\bar{t}=14.01$, and $s^2=11.31$. Thus

$$\tilde{r}=1-\Pr\left[t\leqslant\sqrt{\frac{22}{23}}\,\frac{(5-14.01)}{(11.31)^{1/2}}\right]=1-\Pr(t\leqslant-2.62)$$

$$=1-t(-2.62;21)=1-0.0080=0.9920.$$

■

Now let us consider the case of a normal-gamma prior distribution on (μ,τ). In this case, we must find the expectation of R in the joint normal-gamma posterior distribution given by (9.85). We have that

$$E[R|\underset{\sim}{t}; \lambda_0, \tau_0, \alpha_0, \beta_0] = 1 - \int_0^\infty \int_{-\infty}^\infty \Phi[\tau^{1/2}(t_0 - \mu)] \times$$

$$g(\mu, \tau|\underset{\sim}{t}; \lambda_0, \tau_0, \alpha_0, \beta_0)\, d\mu\, d\tau$$

$$= 1 - \int_{-\infty}^{t_0} \int_0^\infty \int_{-\infty}^\infty \frac{\tau^{1/2}}{\sqrt{2\pi}} \exp\left[-\frac{\tau}{2}(x-\mu)^2\right] \times$$

$$\frac{(\tau_0 + n)^{1/2}\tau^{1/2}}{\sqrt{2\pi}} \exp\left[-\frac{(\tau_0 + n)\tau}{2}(\mu - \lambda')^2\right] \times$$

$$\frac{\beta^{\alpha_0 + n/2}}{\Gamma(\alpha_0 + n/2)} \tau^{\alpha_0 + n/2 - 1} \exp[-\beta\tau]\, d\mu\, d\tau\, dx.$$

Completing the square on μ, integrating with respect to μ, and simplifying, yields

$$E[R|\underset{\sim}{t}; \lambda_0, \tau_0, \alpha_0, \beta_0] = 1 - \int_{-\infty}^{t_0} \int_0^\infty \frac{1}{\sqrt{2\pi}} \left(\frac{\tau_0 + n}{\tau_0 + n + 1}\right)^{1/2} \times$$

$$\frac{\beta^{\alpha_0 + n/2}}{\Gamma(\alpha_0 + n/2)} \tau^{\alpha_0 + n/2 - 1/2} \times$$

$$\exp\left\{-\tau\left[\frac{(\tau_0 + n)}{2(\tau_0 + n + 1)}(x - \lambda')^2 + \beta\right]\right\} d\tau\, dx.$$

Using the identity

$$\int_0^\infty z^{p-1} \exp(-az)\, dz = a^{-p}\Gamma(p),$$

we have upon integration with respect to τ and simplification that

$$E[R|\underset{\sim}{t}; \lambda_0, \tau_0, \alpha_0, \beta_0] = 1 - \int_{-\infty}^{t_0} \frac{C^{1/2}\Gamma[(w+1)/2]}{\Gamma(w/2)(\pi w)^{1/2}} \times$$

$$\left[1 + \frac{C(x - \lambda')^2}{w}\right]^{-(w+1)/2} dx,$$

where

$$C = \frac{(\tau_0 + n)(2\alpha_0 + n)}{2\beta(\tau_0 + n + 1)} \tag{9.98}$$

and

$$w = 2\alpha_0 + n. \tag{9.99}$$

Letting $y = C^{1/2}(x - \lambda')$ we have

$$E[R|\underset{\sim}{t}; \lambda_0, \tau_0, \alpha_0, \beta_0] = 1 - \int_{-\infty}^{C^{1/2}(t_0 - \lambda')} \frac{\Gamma[(w+1)/2]}{\Gamma(w/2)(\pi w)^{1/2}} \times$$

$$\left[1 + \frac{y^2}{w}\right]^{-(w+1)/2} dy.$$

The predictive distribution is again observed to be the pdf of a $t(w)$ distribution, and thus the Bayesian point estimator of r, for squared-error loss, becomes

$$E[R|\underset{\sim}{t}; \lambda_0, \tau_0, \alpha_0, \beta_0] = 1 - t\left\{\left[\frac{(\tau_0 + n)(2\alpha_0 + n)}{2\beta(\tau_0 + n + 1)}\right]^{1/2}(t_0 - \lambda'); w\right\}. \tag{9.100}$$

This estimator is analogous to that of Padgett and Wei (1977) for the $\mathcal{LN}(\xi, \sigma^2)$ distribution.

Example 9.14. Reconsidering Example 9.11, suppose we desire a point estimate of the reliability for $t_0 = 5$ years. Recall that $\lambda' = 14.012$ and $\beta = 119.275$. Now $w = 2(4) + 22 = 30$ and

$$C = \frac{(0.04 + 22)(30)}{2(119.275)(0.04 + 22 + 1)} = 0.120.$$

Thus

$$\tilde{r} = 1 - t\left[(0.120)^{1/2}(5 - 14.012); 30\right] = 1 - t(-3.12; 30) = 0.998.$$

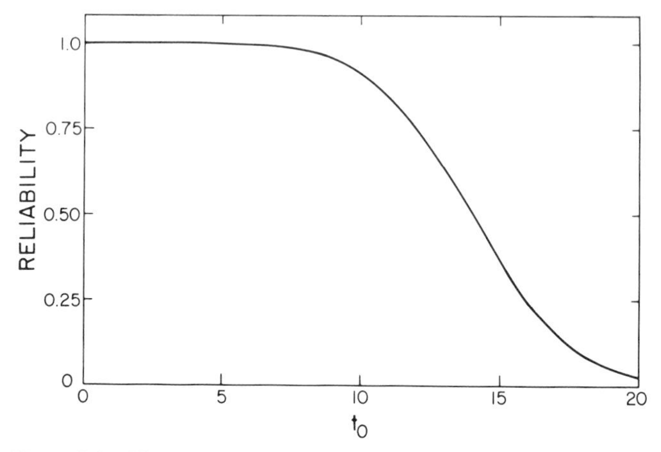

Figure 9.6. The estimated normal reliability function in Example 9.14.

Figure 9.6 shows a plot of the estimated $\mathcal{N}(\mu, \sigma^2)$ reliability function based on (9.100). ■

9.2.4 Reliable Life Estimation

The reliable life corresponding to an $\mathcal{N}(\mu, \sigma^2)$ distribution is given by

$$t_R = \sigma\Phi^{-1}(1 - R) + \mu, \tag{9.101}$$

where R is specified. Since t_R is a linear combination of σ and μ the estimation of t_R is straightforward.

μ Is a Random Variable. Let us first restrict our attention to the case where σ is known and consider both a noninformative and an $\mathcal{N}(\lambda_0, \psi_0^2)$ prior distribution on μ.

For a noninformative prior distribution on μ, the Bayesian point estimator of t_R, under squared-error loss, is the expectation of T_R in the posterior distribution given in (9.73). That is,

$$\begin{aligned} E\left[T_R \mid \bar{t}; \sigma^2\right] &= \sigma\Phi^{-1}(1 - R) + \int_{-\infty}^{\infty} \mu \frac{\sqrt{n}}{\sigma\sqrt{2\pi}} \exp\left[-\frac{n}{2\sigma^2}(\mu - \bar{t})^2\right] d\mu \\ &= \sigma\Phi^{-1}(1 - R) + E\left[\mu \mid \bar{t}; \sigma^2\right] \\ &= \sigma\Phi^{-1}(1 - R) + \bar{t}. \end{aligned} \tag{9.102}$$

A symmetric $100(1-\gamma)\%$ TBPI estimate of t_R is given as

$$t_R^* = \sigma\Phi^{-1}(1-R) + \bar{t} + z_{1-\gamma/2}\sigma/\sqrt{n} \tag{9.103}$$

and

$$t_{R*} = \sigma\Phi^{-1}(1-R) + \bar{t} - z_{1-\gamma/2}\sigma/\sqrt{n}, \tag{9.104}$$

where $\Pr(t_{R*} \leqslant T_R \leqslant t_R^* | \bar{t};\, \sigma^2) = 1-\gamma$. Frequently, a $100(1-\gamma)\%$ LBPI estimate of t_R is desired and this may be calculated using (9.104) in which $\gamma/2$ is replaced by γ.

Now consider an $\mathcal{N}(\lambda_0, \psi_0^2)$ prior distribution on μ. In this case the expectation of T_R is taken with respect to the posterior $\mathcal{N}(\lambda, \psi^2)$ distribution, where λ and ψ^2 are given by (9.74) and (9.75), respectively, and thus

$$\begin{aligned}\tilde{t}_R = E\left[T_R | \bar{t};\, \sigma^2, \lambda_0, \psi_0^2\right] &= \sigma\Phi^{-1}(1-R) + E\left[\mu | \bar{t};\, \sigma^2, \lambda_0, \psi_0^2\right] \\ &= \sigma\Phi^{-1}(1-R) + \lambda. \end{aligned} \tag{9.105}$$

A symmetric $100(1-\gamma)\%$ TBPI estimate of t_R in this case becomes

$$t_R^* = \sigma\Phi^{-1}(1-R) + \lambda + z_{1-\gamma/2}\psi \tag{9.106}$$

and

$$t_{R*} = \sigma\Phi^{-1}(1-R) + \lambda - z_{1-\gamma/2}\psi, \tag{9.107}$$

where $\Pr(t_{R*} \leqslant T_R \leqslant t_R^* | \bar{t};\, \sigma^2, \lambda_0, \psi_0^2) = 1-\gamma$.

Example 9.15. In Example 4.7, suppose we consider an $\mathcal{N}(15,4)$ prior distribution on μ. Further suppose, as before, that $\sigma = 3$. Let us compute both a Bayesian point and 95% LBPI estimate of the reliable life for which 80% of the population will survive. Thus $R = 0.80$. As in Example 9.10, $\lambda = 14.10$ and $\psi^2 = 0.37$. Thus

$$\tilde{t}_{0.80} = 3(-0.841) + 14.10 = 11.58 \text{ years},$$

since $\Phi^{-1}(0.20) = -0.841$. Thus, on the average, it is estimated that 80% of the bushings will survive at least 11.58 years. Also,

$$t_{0.80*} = 3(-0.841) + 14.10 - 1.645(0.37)^{1/2} = 10.58 \text{ years}$$

and thus it is estimated that 80% of the bushings will survive at least 10.58 years with probability 0.95. ■

μ and σ Are Random Variables. In the case of a noninformative prior on (μ, σ), the usual Bayesian point estimator of t_R is the posterior mean of T_R with respect to (9.79). Thus

$$E[T_R|\underset{\sim}{t}] = \Phi^{-1}(1-R)E[\sigma|\underset{\sim}{t}] + E[\mu|\underset{\sim}{t}], \tag{9.108}$$

where the expectations on the right-hand side of the expression are taken with respect to the marginal distributions of σ and μ given by (9.80) and (9.81). Since

$$E[\sigma|\underset{\sim}{t}] = s\sqrt{\nu/2}\,\frac{\Gamma[(\nu-1)/2]}{\Gamma(\nu/2)}, \qquad \nu > 1,$$

and

$$E[\mu|\underset{\sim}{t}] = \bar{t},$$

upon substituting into (9.108) we find that

$$E[T_R|\underset{\sim}{t}] = \Phi^{-1}(1-R)s\sqrt{\nu/2}\,\Gamma\left(\frac{\nu-1}{2}\right)\Big/\Gamma\left(\frac{\nu}{2}\right) + \bar{t}, \tag{9.109}$$

where $\nu = n-1$ and $s^2 = \nu^{-1}\sum_{i=1}^{n}(t_i - \bar{t})^2$.

As in the case of reliability estimation, interval estimation of t_R when both μ and σ are unknown is complicated and is not considered here.

Now consider a normal-gamma prior distribution on (μ, σ). With regard to the posterior distribution given by (9.85), we find that

$$\begin{aligned} E[T_R|\underset{\sim}{t}; \lambda_0, \tau_0, \alpha_0, \beta_0] &= E\left[\tau^{-1/2}\Phi^{-1}(1-R) + \mu|\underset{\sim}{t}; \lambda_0, \tau_0, \alpha_0, \beta_0\right] \\ &= \Phi^{-1}(1-R)E\left[\tau^{-1/2}|\underset{\sim}{t}; \lambda_0, \tau_0, \alpha_0, \beta_0\right] \\ &\quad + E[\mu|\underset{\sim}{t}; \lambda_0, \tau_0, \alpha_0, \beta_0], \end{aligned} \tag{9.110}$$

where the expectations on the right-hand side of (9.110) are taken with respect to the marginal posterior distributions of τ and μ, respectively. Since

τ given $\underset{\sim}{t}$ has a $\mathcal{G}_1(\alpha_0 + n/2, \beta)$ distribution we have that

$$E\left[\tau^{-1/2}|\underset{\sim}{t}; \lambda_0, \tau_0, \alpha_0, \beta_0\right] = \frac{\beta^{1/2}\Gamma(\alpha_0 + \nu/2)}{\Gamma(\alpha_0 + n/2)},$$

where β is given by (9.84). Also

$$E[\mu|\underset{\sim}{t}; \lambda_0, \tau_0, \alpha_0, \beta_0] = \lambda',$$

where λ' is given in (9.83). Upon substituting into (9.110) we find that

$$\tilde{t}_R = E[T_R|\underset{\sim}{t}; \lambda_0, \tau_0, \alpha_0, \beta_0]$$

$$= \Phi^{-1}(1-R)\beta^{1/2}\Gamma(\alpha_0 + \nu/2)/\Gamma(\alpha_0 + n/2) + \lambda'. \qquad (9.111)$$

Example 9.16. Consider again Example 9.11 and suppose we now desire a point estimate of the reliable life for $R = 0.80$. Recall $\alpha_0 = 4$, $\lambda' = 14.012$, and $\beta = 119.275$. Thus

$$\tilde{t}_{0.80} = \Phi^{-1}(0.20)(119.275)^{1/2}\Gamma(4+10.5)/\Gamma(4+11) + 14.012$$

$$= (-0.841)(10.92)(2.3092 \times 10^{10})/(8.7178 \times 10^{10}) + 14.012$$

$$= 11.58 \text{ years.}$$

■

9.3. THE LOG NORMAL DISTRIBUTION

The $\mathcal{LN}(\xi, \sigma^2)$ distribution was introduced in Section 4.2.4. Recall that if $Z = \ln T$ has an $\mathcal{N}(\xi, \sigma^2)$ distribution, then T has, by definition, an $\mathcal{LN}(\xi, \sigma^2)$ distribution with pdf given by (4.31). Recall that the mean of T is $\theta = \exp(\xi + \sigma^2/2)$. Since Z has an $\mathcal{N}(\xi, \sigma^2)$ distribution, many of the results in this section follow directly from corresponding results for the $\mathcal{N}(\mu, \sigma^2)$ distribution in the preceding section.

We consider Bayesian estimation of θ and the reliability function given in (4.35) for both noninformative and conjugate prior distributions. As in the case of an $\mathcal{N}(\mu, \sigma^2)$ distribution, we distinguish two cases: (1) ξ an r.v. (σ known), and (2) both ξ and σ are r.v.'s, where ξ and σ are used to denote both r.v.'s and their values.

We consider only complete samples in which n items are tested and failures (or repairs in the case of repair time data) are observed at ordered times $t_1 \leqslant t_2 \leqslant \cdots \leqslant t_n$. Let $z_i = \ln t_i$ and thus $\underset{\sim}{z}' = (z_1, z_2, \ldots, z_n)$. However, many of these results can be extended to the case of incomplete testing [Padgett and Tsokos (1977)]. By replacing $\bar{t}$ and s^2 with

$$\bar{z} = \sum_{i=1}^{n} \ln t_i / n \qquad \text{and} \qquad s_z^2 = \nu^{-1} \sum_{i=1}^{n} (\ln t_i - \bar{z})^2,$$

Sections 9.2.1 and 9.2.2 provide the required posterior distributions for ξ and σ given $\underset{\sim}{z}$ where now $\mu \equiv \xi$.

9.3.1 Estimating the Mean

In estimating θ we consider the case where ξ is an r.v., with σ known, and also the case where both ξ and σ are r.v.'s.

ξ Is a Random Variable. First let us consider a noninformative prior distribution on ξ. When σ is known, it follows from Section 9.2.1 that $\ln \Theta = \xi + \sigma^2/2$ given $\bar{z}$ has an $\mathfrak{N}(\bar{z} + \sigma^2/2, \sigma^2/n)$ posterior distribution. Thus Θ has an $\mathcal{L}\mathfrak{N}(\bar{z} + \sigma^2/2, \sigma^2/n)$ posterior distribution.

Under the squared-error loss, the Bayesian point estimator of θ is the mean of the posterior distribution of Θ and thus

$$E[\Theta | \bar{z}; \sigma^2] = \exp\left(\bar{z} + \frac{\sigma^2}{2} + \frac{\sigma^2}{2n} \right) = \exp\left[\bar{z} + \frac{\sigma^2}{2}\left(1 + \frac{1}{n}\right)\right]. \tag{9.112}$$

As pointed out by Zellner (1971), this estimate minimizes the posterior risk defined by (5.28); however, with the improper (noninformative) prior distribution on ξ, the Bayes risk is not finite and thus (9.112) cannot minimize the Bayes risk. Nevertheless, we continue to refer to (9.112) as the Bayes estimator, keeping in mind that in reality it is only the minimum posterior risk estimator of θ.

For the so-called "quasi" prior distribution on ξ given by

$$g(\xi) = \exp(d\xi), \qquad -\infty < \xi, d < \infty, \tag{9.113}$$

which includes the noninformative prior as a special case when $d = 0$, the Bayes estimator of θ under squared-error loss is

$$E[\Theta | \bar{z}; \sigma^2] = \exp\left[\bar{z} + \frac{\sigma^2}{2}\left(1 + \frac{1 + 2d}{n}\right)\right] \tag{9.114}$$

[Padgett and Tsokos (1977)]. If $d=-1$, this estimator reduces to the well-known MVU estimator for θ.

For a noninformative prior distribution on ξ, Zellner (1971) shows that the Bayes estimator for θ, in the case of a relative squared-error loss function $L=[(\hat{\theta}-\theta)/\theta]^2$, is the minimal mean squared-error estimator given by

$$\tilde{\theta}_m = \exp\left[\bar{Z} + \frac{\sigma^2}{2}\left(1-\frac{3}{n}\right)\right]. \tag{9.115}$$

Zellner (1971) then shows that (9.115) has a much smaller mean squared-error for a broad range of conditions than either the ML estimator given by $\exp(\bar{Z}+\sigma^2/2)$ or the MVU estimator. It is noted that the median of the posterior distribution of Θ, which is the Bayes estimator for an absolute-error loss function, is $\exp(\bar{Z}+\sigma^2/2)$, which is the ML estimator. As $n\to\infty$, (9.112), (9.114), and (9.115) all converge almost surely to the ML estimator, so that the ML estimator is a limiting case of a Bayes estimator, which often is the case.

A $100(1-\gamma)\%$ TBPI estimate of θ for a noninformative prior distribution on ξ is easily computed to be

$$\theta^* = \exp\left[\bar{z} + z_{1-\gamma/2}\frac{\sigma}{\sqrt{n}} + \frac{\sigma^2}{2}\left(1+\frac{1}{n}\right)\right] \tag{9.116}$$

and

$$\theta_* = \exp\left[\bar{z} - z_{1-\gamma/2}\frac{\sigma}{\sqrt{n}} + \frac{\sigma^2}{2}\left(1+\frac{1}{n}\right)\right], \tag{9.117}$$

where $\Pr(\theta_* \leqslant \Theta \leqslant \theta^* | \bar{z};\, \sigma^2)=1-\gamma$. One-sided probability intervals can be similarly constructed.

Example 9.17. In an examination of large piping in a power plant, numerous drill-induced blemishes were discovered. A total of 75 such blemishes were found to be adequately described by an $\mathcal{LN}(\xi,\sigma^2)$ distribution in which $\bar{z}=-5.72$ and in which it was assumed that $\sigma^2=0.40$. For a noninformative prior distribution on ξ, compute a Bayesian point estimate, using squared-error loss, and a 90% TBPI estimate of ξ.

According to (9.112),

$$\tilde{\theta}=\exp\left[-5.72+\frac{0.40}{2}\left(1+\frac{1}{75}\right)\right]=0.0040''$$

is the required point estimate. Also

$$\theta^*=\exp\left[-5.72+1.645\left(\frac{0.40}{75}\right)^{1/2}+\frac{0.40}{2}\left(1+\frac{1}{75}\right)\right]=0.0045''$$

and

$$\theta_*=\exp\left[-5.72-1.645\left(\frac{0.40}{75}\right)^{1/2}+\frac{0.40}{2}\left(1+\frac{1}{75}\right)\right]=0.0036'',$$

and thus $\Pr(0.0036''\leqslant\Theta\leqslant 0.0045''|-5.72;\,0.40)=0.90$. ∎

Now suppose that ξ has a natural conjugate $\mathcal{N}(\lambda_0,\psi_0^2)$ distribution. It follows from Section 9.2.1 that the posterior distribution of ξ given $\bar{z}$ is an $\mathcal{N}(\lambda,\psi^2)$ distribution, where λ is given by (9.74), in which $\bar{z}$ replaces $\bar{t}$, and ψ^2 is defined in (9.75). It follows that the posterior distribution of Θ given $\bar{z}$ is an $\mathcal{LN}(\lambda+\sigma^2/2,\psi^2)$ distribution.

The posterior mean and median of Θ given $\bar{z}$ are thus given by

$$E\left[\Theta|\bar{z};\,\sigma^2,\lambda_0,\psi_0^2\right]=\exp\left[\frac{\sigma^2\lambda_0+n\psi_0^2\bar{z}}{\sigma^2+n\psi_0^2}+\frac{\sigma^2}{2}\left(1+\frac{\psi_0^2}{\sigma^2+n\psi_0^2}\right)\right] \tag{9.118}$$

and

$$\exp\left[\frac{\sigma^2\lambda_0+n\psi_0^2\bar{z}}{\sigma^2+n\psi_0^2}\right], \tag{9.119}$$

respectively. Again as $n\to\infty$, (9.118) converges almost surely to the ML estimator.

Padgett and Tsokos (1977) also consider a $\mathcal{U}(\alpha_0,\beta_0)$ prior distribution on ξ and derive the corresponding posterior mean of Θ given $\bar{z}$. Unfortunately, this posterior mean is unavailable in closed form and involves rather complicated integrals that must be numerically evaluated. Padgett and Tsokos also compare the estimated mean squared-error of (9.118) and the ML estimator by means of Monte Carlo simulation. They find that the

Bayes estimator is generally superior to the ML estimator. For small sample sizes, the Bayes estimator performs much better. They also find that the Bayes estimators based on informative priors, such as $\mathcal{U}(\alpha, \beta)$ and $\mathcal{N}(\mu, \gamma^2)$, have significantly smaller mean squared-errors than for noninformative priors.

A $100(1-\gamma)\%$ TBPI estimate of θ is again readily computed from the results in Section 9.2.1 to be

$$\theta^* = \exp\left[\frac{\sigma^2\lambda_0 + n\psi_0^2\bar{z}}{\sigma^2 + n\psi_0^2} + z_{1-\gamma/2}\left(\frac{\sigma^2\psi_0^2}{\sigma^2 + n\psi_0^2}\right)^{1/2} + \frac{\sigma^2}{2}\left(1 + \frac{\psi_0^2}{\sigma^2 + n\psi_0^2}\right)\right] \tag{9.120}$$

and

$$\theta_* = \exp\left[\frac{\sigma^2\lambda_0 + n\psi_0^2\bar{z}}{\sigma^2 + n\psi_0^2} - z_{1-\gamma/2}\left(\frac{\sigma^2\psi_0^2}{\sigma^2 + n\psi_0^2}\right)^{1/2} + \frac{\sigma^2}{2}\left(1 + \frac{\psi_0^2}{\sigma^2 + n\psi_0^2}\right)\right], \tag{9.121}$$

where $\Pr(\theta_* \leqslant \Theta \leqslant \theta^* | \bar{z}; \sigma^2, \lambda_0, \psi_0^2) = 1 - \gamma$.

ξ and σ Are Random Variables. For the joint noninformative prior distribution on ξ and σ given in (9.78), the corresponding joint posterior distribution of Θ and σ given $\underset{\sim}{z}$ is a product of the $\mathcal{LN}(\bar{z} + \sigma^2/2, \sigma^2/n)$ posterior conditional distribution of Θ given σ and the inverted-gamma-2 marginal posterior distribution of σ given in (9.80). Thus, the marginal posterior distribution of Θ given $\underset{\sim}{z}$ becomes

$$g(\theta | \underset{\sim}{z}) \propto \theta^{(n-2)/2} \int_0^\infty \sigma^{-(\nu+2)} \exp\left\{-\left[\frac{\nu s_z^2 + n(\ln\theta - \bar{z})^2}{2\sigma^2} + \frac{n\sigma^2}{8}\right]\right\} d\sigma, \tag{9.122}$$

where $\nu = n-1$. Zellner (1971) expresses this distribution in terms of modified Bessel functions. However, as pointed out by Zellner, the posterior mean of Θ given $\underset{\sim}{z}$ does not exist, although the posterior distribution is unimodal and positively skewed. Since (9.122) is only numerically expressible, Bayesian estimation based on (9.122) is rather cumbersome and will not be pursued further.

A similar result holds for the case of a natural conjugate normal-gamma prior distribution on (ξ, τ); consequently, Bayesian estimation of θ in this case is not considered further.

9.3.2 Reliability Estimation

We now consider estimation of the $\mathcal{LN}(\xi, \sigma^2)$ reliability function given by $r(t_0;\ \xi, \sigma) = 1 - \Phi[(\ln t_0 - \xi)/\sigma]$ for the case where σ is both known and unknown.

ξ Is a Random Variable. Again let us first suppose that ξ has a noninformative prior distribution and that σ is a known constant. It follows from (9.90) that the posterior mean of $r = r(t_0;\ \xi, \sigma)$ is given by

$$E\left[R \mid \bar{z};\ \sigma^2\right] = 1 - \int_{-\infty}^{\infty} \Phi\left(\frac{\ln t_0 - \xi}{\sigma}\right) g\left(\xi \mid \bar{z};\ \sigma^2\right) d\xi$$

$$= 1 - \Phi\left[\frac{\ln t_0 - \bar{z}}{\sigma\sqrt{(n+1)/n}}\right]. \tag{9.123}$$

A $100(1-\gamma)\%$ TBPI estimate of r may be computed according to (9.91) and (9.92) in which t_0 and $\bar{t}$ are replaced by $\ln t_0$ and $\bar{z}$, respectively.

If ξ has a natural conjugate $\mathcal{N}(\lambda_0, \psi_0^2)$ prior distribution, it again follows directly from (9.93) that the posterior mean of R is

$$\tilde{r} = E\left[R \mid \bar{z};\ \sigma^2, \lambda_0, \psi_0^2\right] = 1 - \Phi\left[\frac{\ln t_0 - \lambda}{\sqrt{\sigma^2 + \psi^2}}\right], \tag{9.124}$$

where λ is defined in (9.74) in which $\bar{z}$ now replaces $\bar{t}$. The quantity ψ^2 is given by (9.75).

A $100(1-\gamma)\%$ TBPI estimate of r is likewise given by (9.94) and (9.95) in which t_0 is now replaced by $\ln t_0$, μ_* and μ^* are replaced by ξ_* and ξ^*, and $\bar{t}$ is replaced by $\bar{z}$ in (9.76) and (9.77).

Example 9.18. Suppose that in Example 9.17, an $\mathcal{N}(-5, 1)$ prior distribution is considered on ξ. For $\bar{z} = -5.72$ and $\sigma^2 = 0.40$, let us compute a Bayesian point and 95% LBPI estimate of $r(0.005)$. Now

$$\lambda = \frac{0.40(-5) + 75(1)(-5.72)}{0.40 + 75(1)} = -5.717$$

and

$$\psi^2 = \frac{0.40(1)}{0.40+75(1)} = 0.0053.$$

Thus

$$\tilde{r} = 1 - \Phi\left[\frac{\ln 0.005 + 5.717}{\sqrt{0.40 + 0.0053}}\right] = 1 - \Phi(0.658) = 0.255$$

is the required point estimate of r and thus it is estimated that the probability that a drill-induced blemish exceeds 0.005 in. is approximately 0.26. Also

$$\xi_* = -5.717 - 1.645\,(0.0053)^{1/2} = -5.8368,$$

from which

$$r_* = 1 - \Phi\left[\frac{\ln 0.005 + 5.8368}{(0.40)^{1/2}}\right] = 1 - \Phi(0.851) = 0.197.$$

Thus $\Pr(R \geqslant 0.197 \mid -5.72;\, 0.40, -5, 1) = 0.95$. ■

Padgett and Tsokos (1977) study the mean squared-error performance of (9.124) compared to the ML estimator and find that (9.124) is generally superior to the ML estimator. The Bayes estimator is much better than ML for samples of size 5 than for samples of size 10. Also, the average Bayes estimates of reliability are generally closer to the true average value of reliability (averaged over 100 values of ξ) than the average ML estimates. The average mean squared-error performance is likewise observed to be superior to the ML estimator, particularly for large values of σ^2 or for large values of t_0.

ξ and σ Are Random Variables. For the joint noninformative prior distribution on ξ and σ given by (9.78), the marginal posterior mean of R given $\underset{\sim}{z}$, according to (9.97), becomes

$$E[R \mid \underset{\sim}{z}] = 1 - t\left[\sqrt{\frac{n}{n+1}}\,\frac{(\ln t_0 - \bar{z})}{s_z};\, \nu\right], \tag{9.125}$$

where $t(x;\nu)$ again denotes the cdf of a t r.v. with ν degrees of freedom evaluated at x. Padgett and Wei (1977) show that (9.125) converges almost surely to the $\mathcal{LN}(\xi,\sigma^2)$ reliability function.

In the case of a normal-gamma prior distribution on (ξ,τ) given by (9.82), in which μ is replaced by ξ, the marginal posterior mean of R given $\underset{\sim}{z}$ according to (9.100) is given by

$$E[R|\underset{\sim}{z};\lambda_0,\tau_0,\alpha_0,\beta_0]=1-t\left\{\left[\frac{(\tau_0+n)(2\alpha_0+n)}{2\beta(\tau_0+n+1)}\right]^{1/2}(\ln t_0-\lambda');\,2\alpha_0+n\right\}, \tag{9.126}$$

where λ' and β are now given by (9.83) and (9.84) in which $\bar{t}$ and s^2 are replaced by $\bar{z}$ and s_z^2, respectively.

As pointed out by Padgett and Wei (1977), as $n\to\infty$ $\lambda'\to\xi_0$ and $\beta/(\alpha_0+n/2)\to\sigma_0^2$ almost surely, where ξ_0 and σ_0^2 are the true values of ξ and σ^2, respectively, and $t(2\alpha_0+n)$ has an $\mathcal{N}(0,1)$ limiting distribution. Thus (9.126) converges almost surely to the $\mathcal{LN}(\xi,\sigma^2)$ reliability function.

Padgett and Wei also compare some of the small sample performance properties of (9.125) and (9.126) with those of the corresponding ML estimator given by $\hat{R}=1-\Phi[\sqrt{n/(n-1)}\,(\ln t_0-\bar{Z})/S_z]$, and the MVU estimator given by Ellison (1964). Using Monte Carlo simulation, they find that the Bayes estimators have smaller variance than either the ML or the MVU estimators when the assumed prior is actually the true prior used in the simulation. However, the Bayes estimators are observed to be slightly biased. They also observe that the Bayes estimator given in (9.126) has almost uniformly smaller mean squared-error than either the ML or MVU estimator for a large range of values of t_0. However, for $n\geqslant 35$, all of the estimators perform equally well.

Finally, Bayesian point and interval estimation of reliable life t_r, given by (4.37), parallels the development in Section 9.3.1, since t_r has a form similar to the mean θ. These results are deferred to the exercises at the end of this chapter.

9.4. THE INVERSE GAUSSIAN DISTRIBUTION

Banerjee and Bhattacharyya (1979) and Padgett (1981) consider Bayesian estimation in the $\mathcal{IN}(\mu,\lambda)$ distribution, which was introduced in Chapter 4. Banerjee and Bhattacharyya (1979) consider noninformative and conjugate prior distributions for both parameters in a reparameterized form of (4.38), and obtain both Bayesian point and highest posterior density interval estimators for the parameters. However, as they point out, Bayesian inference about the reliability in this model is extremely complex and requires the use of numerical integration in order to plot the posterior pdf of reliability or to determine highest posterior density intervals for reliability.

On the other hand, Padgett (1981) considers Bayesian point estimation of reliability given by (4.41) for the case where the mean life μ is known and only λ is considered to be an r.v. Both noninformative and conjugate prior distributions are considered. In the case where both μ and λ are r.v.'s, a simple estimator of (4.41) is proposed which is shown by Padgett to possess good average mean squared-error performance properties compared to the MVU or ML estimators.

9.4.1 MTTF Estimation

Banerjee and Bhattacharyya (1979) consider a reparameterized form of (4.38) in which $\mu = 1/\psi$. Thus they consider the pdf of the inverse Gaussian distribution given by

$$f(t;\psi,\lambda)=\left(\frac{\lambda}{2\pi t^3}\right)^{1/2}\exp\left[-\frac{\lambda t}{2}(\psi - t^{-1})^2\right],\qquad t,\psi,\lambda>0, \tag{9.127}$$

where λ is a shape parameter and ψ is the reciprocal of the mean. This distribution will be referred to as the $\mathcal{IN}_1(\psi, \lambda)$ distribution. They consider both noninformative and conjugate joint prior distributions on (Ψ, Λ) and derive corresponding Bayesian highest posterior density interval estimates for ψ and λ.

Noninformative Prior Distribution. Following Box and Tiao (1973), consider the noninformative locally uniform joint prior distribution on (Ψ, Λ) given by

$$g(\psi,\lambda)\propto\lambda^{-1}. \tag{9.128}$$

Based on a complete random sample $\underset{\sim}{t}'=(t_1, t_2,\ldots,t_n)$ from (9.127), the likelihood function can be expressed as

$$L(\psi,\lambda|\underset{\sim}{t})\propto\exp\left\{-\frac{nu}{2}\left[1+\frac{\bar{t}}{u}\left(\psi-\frac{1}{\bar{t}}\right)^2\right]\lambda\right\}\lambda^{n/2}, \tag{9.129}$$

where

$$\bar{t} = \frac{\sum_{i=1}^{n} t_i}{n}, \qquad u = \bar{t}_r - \frac{1}{\bar{t}}, \qquad \bar{t}_r = \frac{\sum_{i=1}^{n} t_i^{-1}}{n}. \tag{9.130}$$

Using (9.128) and (9.129) in conjunction with Bayes' theorem, the joint posterior distribution of (Ψ, Λ) given $\underset{\sim}{t}$ becomes

$$g(\psi, \lambda \mid \underset{\sim}{t}) = \frac{(\bar{t}/u)^{1/2}(nu/2)^{n/2}\lambda^{n/2-1}}{t(\xi;\nu)B(\nu/2,1/2)\Gamma(n/2)} \exp\left\{-\frac{nu}{2}\left[1+\frac{\bar{t}}{u}\left(\psi-\frac{1}{\bar{t}}\right)^2\right]\lambda\right\}, \tag{9.131}$$

where $\xi = [u\bar{t}/(n-1)]^{-1/2}$ and $B(x, y)$ denotes the complete beta function defined as $B(x, y) = \Gamma(x)\Gamma(y)/\Gamma(x+y)$. If we let $\nu = n-1$ and $q = (\xi\bar{t})^{-1}$, integrating (9.131) with respect to λ yields the marginal posterior distribution of Ψ given $\underset{\sim}{t}$ as

$$g(\psi \mid \underset{\sim}{t}) = \frac{(\bar{t}/u)^{1/2}}{t(\xi;\nu)B(\nu/2,1/2)}\left[1+\frac{1}{\nu q^2}\left(\psi-\frac{1}{\bar{t}}\right)^2\right]^{-(\nu+1)/2}, \qquad \psi > 0. \tag{9.132}$$

Thus $(\Psi - 1/\bar{t})/q$ given $\underset{\sim}{t}$ has a left-truncated $t(\nu)$ distribution, which is truncated to the left at $-\xi$.

Similarly, by integrating over ψ in (9.131) we obtain the marginal posterior distribution of Λ given $\underset{\sim}{t}$ as

$$g(\lambda \mid \underset{\sim}{t}) = \frac{(nu/2)^{\nu/2}\Phi\left[\left(\frac{n\lambda}{\bar{t}}\right)^{1/2}\right]}{\Gamma(\nu/2)t(\xi;\nu)}\lambda^{\nu/2-1}\exp\frac{(-nu\lambda)}{2}, \qquad \lambda > 0, \tag{9.133}$$

where $\Phi(\cdot)$ is the standard normal cdf defined in (4.26). Equation (9.133) has a structural similarity to a $\mathcal{G}[\nu/2, 2/(nu)]$ distribution, but with added $\Phi(\cdot)$ and $t(\cdot;\nu)$ terms. It is referred to as a "modified" gamma distribution.

If $\xi \leq t_{1/(1+\gamma)}(\nu)$, the $100(1-\gamma)\%$ highest posterior density interval estimate of ψ is given by

$$\psi_* = 0 \tag{9.134}$$

and

$$\psi^* = \frac{1}{\bar{t}} + qt_{1-\gamma t(\xi;\,\nu)}(\nu). \tag{9.135}$$

On the other hand, if $\xi > t_{1/(1+\gamma)}(\nu)$, the interval is

$$\psi_* = \frac{1}{\bar{t}} - qt_{[1+(1-\gamma)t(\xi;\,\nu)]/2}(\nu) \tag{9.136}$$

and

$$\psi^* = \frac{1}{\bar{t}} + qt_{[1+(1-\gamma)t(\xi;\,\nu)]/2}(\nu). \tag{9.137}$$

In the case of a conjugate prior distribution, the marginal posterior distribution of Ψ given $\underset{\sim}{t}$ is again a truncated t distribution, whereas the marginal posterior distribution of Λ given $\underset{\sim}{t}$ is also a modified gamma distribution. The derivation is quite similar to the above and will not be presented here.

Example 9.19. [Nadas (1973)] Ten electronic devices were tested to failure under high stress conditions from which $\bar{t} = 1.352$ and $\bar{t}_r = 0.948$. Assuming an $\mathcal{IN}_1(\psi, \lambda)$ failure time model with a noninformative prior distribution for these data, compute a 90% TBPI estimate of the mttf. Now $u = 0.208$, $\nu = 9$, $\xi = 5.657$, and $q = 0.131$. Since $\xi > t_{0.909}(9) = 1.446$, we use (9.136) and (9.137); thus, since $t(5.657;\, 9) = 0.9998$, we have

$$\psi_* = \frac{1}{1.352} - (0.131)t_{0.94991}(9)$$

$$= 0.740 - (0.131)(1.832) = 0.500$$

and

$$\psi^* = 0.740 + (0.131)(1.832) = 0.980.$$

Hence $\Pr(0.500 \leqslant \Psi \leqslant 0.980 \,|\, \underset{\sim}{t}) = 0.90$, and inasmuch as $\text{mttf} = \psi^{-1}$, this statement implies that $\Pr(1.020 \leqslant \text{MTTF} \leqslant 2.000 \,|\, \underset{\sim}{t}) = 0.90$. ∎

9.4.2 Reliability Estimation

Let us now consider Bayesian estimators for the $\mathcal{IN}(\mu, \lambda)$ reliability function given by (4.41). As pointed out by Padgett (1981), the use of a noninformative joint prior distribution for (μ, Λ) produces either an intractable or an improper joint posterior distribution on R. Thus, following Padgett (1981), we shall assume here that μ, the mean life, is known and consider both a noninformative and a conjugate prior distribution on Λ.

For a complete random sample from (4.38), the likelihood function may be expressed as

$$L(\lambda, \mu | \underset{\sim}{t}) \propto \left[\prod_{i=1}^{n} t_i^{-3/2}\right] \exp\left\{-\frac{\lambda}{2}\left[\frac{n(\bar{t}-2\mu)}{\mu^2} + n\bar{t}_r\right]\right\}. \quad (9.138)$$

Noninformative Prior on Λ. For the improper noninformative prior distribution $g(\lambda) \propto \lambda^{-1}$, the posterior distribution of Λ given $\underset{\sim}{t}$ is

$$g(\lambda | \underset{\sim}{t}; \mu) = \frac{\left[v/(2\mu^2)\right]^{n/2}}{\Gamma(n/2)} \lambda^{n/2-1} \exp\left(-\frac{\lambda v}{2\mu^2}\right), \qquad \lambda > 0, \quad (9.139)$$

where

$$v = n(\bar{t} - 2\mu) + n\mu^2 \bar{t}_r. \quad (9.140)$$

Hence $g(\lambda | \underset{\sim}{t}; \mu)$ is a $\mathcal{G}(n/2, 2\mu^2/v)$ distribution. With respect to a squared-error loss function, the Bayesian point estimator of λ is $\tilde{\lambda} = n\mu^2/V$, which is also the ML estimator of λ when μ is known.

The Bayesian point estimator for $r = r(t_0;\, \mu, \lambda)$ given in (4.41) is

$$\tilde{R} = E\left\{\Phi\left[\left(\frac{\Lambda}{t_0}\right)^{1/2}\left(1 - \frac{t_0}{\mu}\right)\right] \Big| \underset{\sim}{t};\, \mu\right\}$$
$$- E\left\{\exp\left(\frac{2\Lambda}{\mu}\right)\Phi\left[-\left(\frac{\Lambda}{t_0}\right)^{1/2}\left(1 + \frac{t_0}{\mu}\right)\right] \Big| \underset{\sim}{t};\, \mu\right\}. \quad (9.141)$$

Using Lemma 1 of Padgett and Wei (1977) to evaluate the expected values in (9.141) yields

$$\tilde{R} = t\left[\left(1 - \frac{t_0}{\mu}\right)\left(\frac{n\mu^2}{vt_0}\right)^{1/2};\, n\right]$$
$$- \left(1 - \frac{4\mu}{v}\right)^{-n/2} t\left[-\left(1 + \frac{t_0}{\mu}\right)\left(\frac{n\mu^2}{t_0 v - 4t_0\mu}\right)^{1/2};\, n\right], \quad (9.142)$$

where it must be required that $\upsilon \geqslant 4\mu$, since if $\upsilon < 4\mu$ the posterior mean of R given $\underset{\sim}{t}$ does not exist.

Example 9.20. In Example 9.19, suppose we desire a point estimate of the reliability for $t_0 = 0.5$ where will now assume that $\mu = 2.0$. Thus $\upsilon = 10(1.352 - 4.0) + 10(4.0)(0.948) = 11.440$. As $\upsilon > 4\mu$, we have that

$$\tilde{r} = t\left\{\left(1 - \frac{0.5}{2.0}\right)\left[\frac{10(4.0)}{11.440(0.50)}\right]^{1/2}; 10\right\}$$

$$-\left[1 - \frac{4(2.0)}{11.440}\right]^{-5} t\left\{-\left(1 + \frac{0.5}{2.0}\right)\left[\frac{10(4.0)}{0.5(11.440) - 4(0.5)(2.0)}\right]^{1/2}; 10\right\}$$

$$= t(1.983;\ 10) - (406.760)t(-6.028;\ 10)$$

$$= 0.962 - (406.760)(6.364 \times 10^{-5}) = 0.936.$$

■

Conjugate Prior on Λ. If a natural conjugate $\mathcal{G}(\gamma_0, \delta_0)$ prior distribution is placed on Λ, then Lemma 1 of Padgett and Wei (1977) may again be used to obtain the Bayesian point estimator of R. In this case, we have that

$$\tilde{R} = t\left[c_1\left(\frac{\gamma}{\delta}\right)^{1/2}; 2\gamma\right] - \left(\frac{\delta}{\delta - 2/\mu}\right)^{\gamma} t\left[c_2\left(\frac{\gamma}{\delta - 2/\mu}\right)^{1/2}; 2\gamma\right], \quad (9.143)$$

where it is required that $\delta \geqslant 2/\mu$, and where

$$\gamma = \gamma_0 + \frac{n}{2}, \qquad \delta = \delta_0^{-1} + \frac{\upsilon}{2\mu^2}, \qquad c_1 = t_0^{-1/2}\left(1 - \frac{t_0}{\mu}\right),$$

and

$$c_2 = -t_0^{-1/2}\left(1 + \frac{t_0}{\mu}\right).$$

9.5. THE GAMMA DISTRIBUTION

A justification of the $\mathcal{G}(\kappa, \theta)$ distribution as a failure time model was given in Section 4.2.6. Canavos (1971), Canavos and Tsokos (1971b), Lwin and

Singh (1974), and Tummala and Sathe (1978) consider Bayesian estimation in the $\mathcal{G}(\kappa, \theta)$ failure time distribution. Again, for convenience, both the r.v. κ and its value will be denoted by the same symbol.

Canavos and Tsokos (1971b) consider Bayesian estimation of both the random scale parameter Θ and reliability function given in (4.47) for known (fixed) shape parameter κ. Their estimators are based on the use of $\mathcal{U}(\alpha, \beta)$, $\mathcal{E}_1(\lambda)$, and $\mathcal{IG}(\nu, \mu)$ prior distributions on Θ. In addition, Canavos (1971) considers the case where both Θ and κ are r.v.'s having independent marginal $\mathcal{IG}(\nu, \mu)$ and $\mathcal{U}(\alpha, \beta)$ prior distributions, respectively. He also considers the case where both parameters have independent uniform prior distributions.

Lwin and Singh (1974) derive Bayesian point estimators of both θ and the reliability function under both squared-error and absolute-error loss functions. For κ known, they consider a certain three-parameter prior distribution for Θ, as well as an $\mathcal{IG}(\nu, \mu)$ and an $\mathcal{E}_1(\phi)$ prior distribution. In the case where both κ and Θ are r.v.'s, they consider a natural conjugate mixture of a discrete prior distribution on κ and a conditional $\mathcal{IG}(\nu_i, \mu_i)$ prior distribution on Θ given $\kappa = \kappa_i$, analogous to that used by Soland (1969) for the $\mathcal{W}_1(\lambda, \beta)$ distribution. Tummala and Sathe (1978) consider Bayesian point estimation of λ in the $\mathcal{G}_1(\beta_0, \lambda)$ distribution. They consider a $\mathcal{G}_1(k, \alpha)$ prior distribution on Λ and a proportional squared-error loss function.

In Section 9.5.1 we consider the case where Θ is random and κ is known, whereas Section 9.5.2 considers the case where both parameters are r.v.'s. Bayesian estimation of the reliability function given in (4.47) is undertaken in Section 9.5.3.

9.5.1 Scale Parameter Is a Random Variable

We restrict our attention to the case of a simple random sample from a complete life test as otherwise, analytical tractability is nearly impossible as a result of the absence of a closed form expression for the reliability function. Thus let $\underset{\sim}{T}' = (T_1, T_2, \ldots, T_n)$ denote the failure times from a complete random sample of n items according to a $\mathcal{G}(\kappa, \theta)$ distribution in which κ is known. Recall that $\kappa = 1$ corresponds to an $\mathcal{E}(\theta)$ failure time distribution.

It may be easily shown by use of the Neyman factorization criterion [see Section 3.6.1] that $\mathcal{T} = \Sigma_{i=1}^{n} T_i$ is a sufficient statistic for estimating θ and that $\mathcal{T}$ also has a $\mathcal{G}(n\kappa, \theta)$ distribution. Consequently, we refer to this situation as *gamma sampling*. In fact, if each sample r.v. has a $\mathcal{G}(\kappa_i, \theta)$ distribution, where the known shape parameter values are not necessarily the same for each observation, then $\mathcal{T}$ follows a $\mathcal{G}(\kappa, \theta)$ distribution where

$\kappa=\sum_{i=1}^{n}\kappa_i$. This is known as the *reproductive property* of the $\mathcal{G}(\kappa,\theta)$ distribution [Johnson and Kotz (1970)]. When κ is known, the results based on Type II/item-censored testing in the $\mathcal{E}(\lambda)$ distribution considered in Chapter 8 are a special case of the more general gamma sampling model considered here. Thus the results of Section 8.3 may be applied here by everywhere replacing s in the equations and text of Section 8.3 with $n\kappa$. That this is true may be seen as follows: if we let $g(\theta)$ represent the prior distribution of Θ, then according to Bayes' theorem the posterior distribution of Θ given t is given by

$$g(\theta|\mathrm{t};\kappa)=\frac{(1/\theta)^{n\kappa}e^{-\mathrm{t}/\theta}g(\theta)}{\int_0^\infty(1/\theta)^{n\kappa}e^{-\mathrm{t}/\theta}g(\theta)\,d\theta},\qquad 0<\theta<\infty,\tag{9.144}$$

which is the same form as (8.41) where $s=n\kappa$. The Bayesian estimators of θ and accompanying properties derived in Section 8.3 are thus appropriate for estimating the $\mathcal{G}(\kappa,\theta)$ scale parameter θ, when κ is known, once the indicated replacement has been made.

Example 9.21. Suppose that the failure time to leaking of containment quality welds in nuclear reactors follows a $\mathcal{G}(\kappa,\theta)$ distribution with $\kappa=2$. Based on the Reactor Safety Study (1975) results, suppose that an $\mathcal{IG}$(4.0, 1.17×10^8 h) prior distribution is placed on Θ [Waller and Martz (1977)]. That is, based on (4.55) and (4.56), the prior mean and variance of Θ are 3.9×10^7 h and 7.605×10^{14} h^2, respectively.

Further suppose that an accelerated life test was conducted on 10 such welded test specimens whose extrapolated accumulated total time on test until all 10 welds failed was observed to be $\mathrm{t}=1.0\times10^9$ h.

Compute both a Bayes point and a 90% TBPI estimate of θ (assuming a squared-error loss). Upon substituting $n\kappa=10(2)=20$ for s in (8.65), the Bayes point estimate, $\tilde{\theta}$, of θ becomes

$$\tilde{\theta}=\frac{(1.0\times10^9+1.17\times10^8)}{10(2)+4-1}=4.86\times10^7\ \mathrm{h}.$$

According to (8.74) and (8.75), the required 90% TBPI estimate of θ is given by

$$\theta_*=\frac{2(1.0\times10^9+1.17\times10^8)}{65.171}=3.43\times10^7\ \mathrm{h}$$

and

$$\theta^* = \frac{2(1.0\times10^9+1.17\times10^8)}{33.098} = 6.75\times10^7 \text{ h},$$

since $\chi^2_{0.95}[2(n\kappa+\alpha_0)] = \chi^2_{0.95}(48) = 65.171$ and $\chi^2_{0.05}(48) = 33.098$. Thus $\Pr(3.43\times10^7 \leqslant \Theta \leqslant 6.75\times10^7 | 1.0\times10^9; 2.0, 4.0, 1.17\times10^8) = 0.95$. ■

9.5.2 Scale and Shape Parameters Are Random Variables

Let us now consider the case where both κ and Θ are r.v.'s in the $\mathcal{G}(\kappa,\theta)$ sampling distribution. Again we restrict our attention to a complete life test in which the failure times of n items are observed. As before, let $\underset{\sim}{T}' = (T_1, T_2, \ldots, T_n)$ denote the random failure times of the n observations.

The likelihood function is given by

$$L(\kappa,\theta|\underset{\sim}{t}) \propto \left[\prod_{i=1}^{n} t_i^{\kappa-1}\right] e^{-\mathrm{t}/\theta}, \tag{9.145}$$

where $\mathrm{t} = \sum_{i=1}^{n} t_i$ is the observed total time on test. Based on (9.145) it is clear from the Neyman factorization criterion given in Section 3.6.1 that

$$\mathcal{T} = \sum_{i=1}^{n} T_i \qquad \text{and} \qquad V = \prod_{i=1}^{n} T_i$$

are jointly sufficient statistics for estimating κ and θ.

Lwin and Singh (1974) consider the natural conjugate prior distribution for κ and Θ defined as follows. Let

$$\Pr(\kappa=\kappa_i) = p_i, \qquad i=1,2,\ldots,m, \tag{9.146}$$

where $\sum_{i=1}^{m} p_i = 1$ and where κ_i, $i=1,2,\ldots,m$, are m specified values of κ in $(0,\infty)$. The conditional prior distribution of Θ given κ_i, call it $g(\theta|\kappa_i)$, is taken to be an $\mathcal{IG}(\nu_i,\mu_i)$ distribution, where it is noted that $g(\theta|\kappa_i)$ depends on κ_i only through the dependence of ν_i and μ_i on the index i.

Based on the above joint prior distributions and the likelihood given by (9.145), the posterior distribution of κ given $\underset{\sim}{t}$ is given by

$$\Pr(\kappa=\kappa_i|\underset{\sim}{t}) = p_i^* = \frac{p_i \mu_i^{\nu_i} v^{\kappa_i-1}\Gamma(n\kappa_i+\nu_i)/\left[(\mu_i+\mathrm{t})^{n\kappa_i+\nu_i}\Gamma(\nu_i)\right]}{\sum_{i=1}^{m} p_i \mu_i^{\nu_i} v^{\kappa_i-1}\Gamma(n\kappa_i+\nu_i)/\left[(\mu_i+\mathrm{t})^{n\kappa_i+\nu_i}\Gamma(\nu_i)\right]},$$

$$i=1,2,\ldots,m. \tag{9.147}$$

Also, the conditional posterior distribution of Θ given κ_i becomes

$$g(\theta|\kappa_i,\underset{\sim}{t})=\mathcal{IG}(\nu_i+n\kappa_i,\mu_i+\mathrm{t}). \tag{9.148}$$

The joint posterior distribution of (κ,Θ) given $\underset{\sim}{t}$ is thus the product of (9.147) and (9.148).

The (unconditional) posterior mean of Θ given $\underset{\sim}{t}$ is the Bayesian point estimator of θ under a squared-error loss function and thus

$$E(\Theta|\underset{\sim}{t};\,\underset{\sim}{\kappa},\underset{\sim}{\nu},\underset{\sim}{\mu})=\int_0^\infty\theta\left[\sum_{i=1}^{m}p_i^*g(\theta|\kappa_i,\underset{\sim}{t})\right]d\theta$$

$$=\sum_{i=1}^{m}p_i^*\left(\frac{\mu_i+\mathrm{t}}{\nu_i+n\kappa_i-1}\right),\qquad \nu_i+n\kappa_i>1, \tag{9.149}$$

where $\underset{\sim}{\kappa}'=(\kappa_1,\kappa_2,\ldots,\kappa_m)$, $\underset{\sim}{\nu}'=(\nu_1,\nu_2,\ldots,\nu_m)$, and $\underset{\sim}{\mu}'=(\mu_1,\mu_2,\ldots,\mu_m)$.

Canavos (1971) also considers the case in which both κ and Θ are r.v.'s. Suppose that κ has a $\mathcal{U}(\alpha,\beta)$ prior distribution and that Θ has an independent $\mathcal{IG}(\nu,\mu)$ prior distribution. It follows that the joint posterior pdf of (κ,Θ) given $\underset{\sim}{t}$ may be expressed as

$$g(\kappa,\theta|\underset{\sim}{t};\,\alpha,\beta,\nu,\mu)=\frac{\upsilon^\kappa e^{-(\mathrm{t}+\mu)/\theta}}{I_1^*\Gamma^n(\kappa)\theta^{n\kappa+\nu+1}},\qquad \alpha\leqslant\kappa\leqslant\beta,\quad 0<\theta<\infty, \tag{9.150}$$

where

$$I_1^*=\int_\alpha^\beta\frac{\upsilon^\kappa\Gamma(n\kappa+\nu)}{\Gamma^n(\kappa)(\mathrm{t}+\mu)^{n\kappa+\nu}}\,d\kappa. \tag{9.151}$$

As in the case of the $\mathcal{W}_2(\theta,\beta)$ distribution in Section 9.1.2, closed form solutions are not possible and numerical integration techniques must be used to evaluate (9.151).

Since κ and Θ are independently distributed, the posterior expectations of Θ and κ become

$$E(\Theta|\underset{\sim}{t};\,\alpha,\beta,\nu,\mu)=\int_\alpha^\beta\int_0^\infty\theta g(\kappa,\theta|\underset{\sim}{t};\,\alpha,\beta,\nu,\mu)\,d\theta\,d\kappa$$

$$=\frac{1}{I_1^*}\int_\alpha^\beta\frac{\upsilon^\kappa\Gamma(n\kappa+\nu-1)}{\Gamma^n(\kappa)(\mathrm{t}+\mu)^{n\kappa+\nu-1}}\,d\kappa \tag{9.152}$$

and

$$E(\kappa|\underset{\sim}{t};\alpha,\beta,\nu,\mu)=\int_{\alpha}^{\beta}\int_{0}^{\infty}\kappa g(\kappa,\theta|\underset{\sim}{t};\alpha,\beta,\nu,\mu)\,d\theta\,d\kappa$$

$$=\frac{1}{I_1^*}\int_{\alpha}^{\beta}\frac{\kappa v^{\kappa}\Gamma(n\kappa+\nu)}{\Gamma^n(\kappa)(\mathrm{t}+\mu)^{n\kappa+\nu}}\,d\kappa, \tag{9.153}$$

respectively.

Now let us suppose that κ and Θ have independent $\mathcal{U}(\alpha,\beta)$ and $\mathcal{U}(0,\sigma)$ prior distributions, respectively. Such prior distributions would be appropriate if an investigator had vague prior information concerning only the range of each parameter. In this case, upon simplification, the joint posterior pdf of (κ,Θ) given $\underset{\sim}{t}$ becomes

$$g(\kappa,\theta|\underset{\sim}{t};\alpha,\beta,\sigma)=\frac{v^{\kappa}e^{-\mathrm{t}/\theta}}{I_2^*\Gamma^n(\kappa)\mathrm{t}^{n\kappa}},\qquad \begin{matrix}\alpha\leqslant\kappa\leqslant\beta,\\ 0<\theta\leqslant\sigma,\end{matrix} \tag{9.154}$$

where

$$I_2^*=\int_{\alpha}^{\beta}\frac{v^{\kappa}\Gamma^c(n\kappa-1,\mathrm{t}/\sigma)}{\Gamma^n(\kappa)\mathrm{t}^{n\kappa-1}}\,d\kappa, \tag{9.155}$$

and where $\Gamma^c(a,z)$ is defined in (9.24).

The posterior expectations of Θ and κ are given by

$$E(\Theta|\underset{\sim}{t};\alpha,\beta,\sigma)=\int_{\alpha}^{\beta}\int_{0}^{\sigma}\theta g(\kappa,\theta|\underset{\sim}{t};\alpha,\beta,\sigma)\,d\theta\,d\kappa$$

$$=\frac{1}{I_2^*}\int_{\alpha}^{\beta}\frac{v^{\kappa}\Gamma^c(n\kappa-2,\mathrm{t}/\sigma)}{\Gamma^n(\kappa)\mathrm{t}^{n\kappa-2}}\,d\kappa,\quad n\kappa>2, \tag{9.156}$$

and

$$E(\kappa|\underset{\sim}{t};\alpha,\beta,\sigma)=\int_{\alpha}^{\beta}\int_{0}^{\sigma}\kappa g(\kappa,\theta|\underset{\sim}{t};\alpha,\beta,\sigma)\,d\theta\,d\kappa$$

$$=\frac{1}{I_2^*}\int_{\alpha}^{\beta}\frac{\kappa v^{\kappa}\Gamma^c(n\kappa-1,\mathrm{t}/\theta)}{\Gamma^n(\kappa)\mathrm{t}^{n\kappa-1}}\,d\kappa,\quad n\kappa>1, \tag{9.157}$$

respectively.

9.5.3 Reliability Estimation

According to (4.47) the reliability function for the $\mathcal{G}(\kappa, \theta)$ failure time distribution is given by

$$r(t_0;\, \kappa, \theta) = \frac{\Gamma(\kappa) - \Gamma(\kappa, t_0/\theta)}{\Gamma(\kappa)}, \tag{9.158}$$

which, if κ is an integer, becomes

$$r(t_0;\, \kappa, \theta) = \sum_{j=0}^{\kappa-1} \frac{(t_0/\theta)^j \exp(-t_0/\theta)}{j!}. \tag{9.159}$$

We consider the Bayesian estimation of $r = r(t_0;\ \kappa, \theta)$ for the case of a random scale parameter and the case where both parameters are r.v.'s. Both point and interval estimators will be considered in the former case, whereas only point estimation will be considered in the latter.

Scale Parameter Is a Random Variable. Canavos and Tsokos (1971b) derive Bayesian point estimators for r, using squared-error loss functions, under the assumption that Θ has either a $\mathcal{U}(\alpha, \beta)$, $\mathcal{E}_1(\lambda)$, or $\mathcal{IG}(\nu, \mu)$ prior distribution and where κ is restricted to have a known integer value.

For a $\mathcal{U}(\alpha, \beta)$ prior distribution on Θ, the posterior mean of R in (9.159) given t can be obtained by taking the expectation of R with respect to (8.43), once the appropriate substitutions discussed in Section 9.5.1 have been made. The final expression becomes

$$E(R|\mathrm{t};\, \kappa, \alpha, \beta) = \int_\alpha^\beta r g(\theta|\mathrm{t};\, \kappa, \alpha, \beta)\, d\theta$$

$$= \frac{1}{[\Gamma(n\kappa-1, \mathrm{t}/\alpha) - \Gamma(n\kappa-1, \mathrm{t}/\beta)](1+t_0/\mathrm{t})^{n\kappa-1}} \times$$

$$\sum_{j=0}^{\kappa-1} \frac{[\Gamma(n\kappa+j-1, \mathrm{t}/\alpha + t_0/\alpha) - \Gamma(n\kappa+j-1, \mathrm{t}/\beta + t_0/\beta)]}{j!} (1+\mathrm{t}/t_0)^{-j}, \tag{9.160}$$

after integrating term by term and rearranging terms.

For an $\mathcal{IG}(\nu, \mu)$ prior distribution on Θ, the posterior mean of R in (9.159) given t can also be found by taking the expectation of R with respect to (8.63), once the appropriate substitutions have been made and where $\nu = \alpha_0$ and $\mu = \beta_0$. The resulting expression is given by

$$E(R|\mathrm{t}; \kappa, \nu, \mu) = \int_0^\infty r g(\theta|\mathrm{t}; \kappa, \nu, \mu)\, d\theta$$

$$= \frac{1}{\Gamma(n\kappa+\nu)[1+t_0/(\mathrm{t}+\mu)]^{n\kappa+\nu}} \sum_{j=0}^{\kappa-1} \frac{\Gamma(n\kappa+\nu+j)}{j!}\left(1+\frac{\mathrm{t}+\mu}{t_0}\right)^{-j}. \tag{9.161}$$

The posterior mean of R for the $\mathcal{E}_1(\lambda)$ prior distribution on Θ is left as an exercise at the end of the chapter.

Example 9.22. Reconsidering Example 9.21, let us compute a Bayesian point estimate of the weld reliability for $t_0 = 10^6$ h. Using (9.161) and the data in Example 9.21, we have

$$\tilde{r} = \frac{1}{\Gamma(24)[1+10^6/(10^9+1.17\times10^8)]^{24}} \times$$

$$\left[\frac{\Gamma(24)}{0!} + \frac{\Gamma(25)}{1!}\left(1+\frac{10^9+1.17\times10^8}{10^6}\right)^{-1}\right] = 0.99976.$$

Thus, on the average, such welds should survive 10^6 h with estimated probability 0.99976. ■

Lwin and Singh (1974) relax the restriction that κ must be an integer and consider Bayes estimation of r given by (9.158) for an $\mathcal{E}_1(\phi)$, $\mathcal{IG}(\nu, \mu)$, and a certain other three-parameter prior distribution on Θ. Although they consider both squared-error and absolute-error loss functions, we restrict our attention here to a squared-error loss function and the $\mathcal{IG}(\nu, \mu)$ prior distribution on Θ. Other cases are deferred to the exercises at the end of the chapter. Both point and interval estimation of r are now considered.

In the case of an $\mathcal{IG}(\nu, \mu)$ prior distribution on Θ, the posterior mean of R in (9.158) given t can be obtained from the predictive distribution of T as follows:

$$E(R|\mathrm{t}; \kappa, \nu, \mu) = \int_0^\infty r g(\theta|\mathrm{t}; \kappa, \nu, \mu)\, d\theta$$

$$= \int_0^\infty \left[\int_{t_0}^\infty f(t|\theta; \kappa)\, dt\right] g(\theta|\mathrm{t}; \kappa, \nu, \mu)\, d\theta,$$

where $f(t|\theta;\ \kappa)$ is the $\mathcal{G}(\kappa, \theta)$ distribution. As the conditions of Fubini's theorem hold here, we can interchange the order of the integration and get

$$E(R|\mathrm{t};\ \kappa, \nu, \mu) = \int_{t_0}^{\infty} f(t|\mathrm{t};\ \kappa, \nu, \mu)\, dt, \tag{9.162}$$

where $f(t|\mathrm{t};\ \kappa, \nu, \mu)$ is the predictive pdf of T defined by

$$f(t|\mathrm{t};\ \kappa, \nu, \mu) = \int_{0}^{\infty} f(t|\theta;\ \kappa) g(\theta|\mathrm{t};\ \kappa, \nu, \mu)\, d\theta. \tag{9.163}$$

This same technique was used in Section 9.2.3 when estimating the reliability function for an $\mathcal{N}(\mu, \sigma^2)$ sampling distribution.

For the $\mathcal{IG}(\nu, \mu)$ prior distribution considered here, the predictive pdf of the failure time T becomes

$$f(t|\mathrm{t};\ \kappa, \nu, \mu) = \frac{1}{B(n\kappa+\nu, \kappa)} \frac{t^{\kappa-1}(\mathrm{t}+\mu)^{n\kappa+\nu}}{(t+\mathrm{t}+\mu)^{n\kappa+\kappa+\nu}}, \qquad 0<t<\infty, \tag{9.164}$$

where $B(a, b)$ is the complete beta function. Integrating (9.164) according to (9.162), and by using the transformation $y = t/(t+\mathrm{t}+\mu)$, yields

$$\tilde{r} = E(R|\mathrm{t};\ \kappa, \nu, \mu) = 1 - I\left(\frac{t_0}{t_0+\mathrm{t}+\mu};\ \kappa, n\kappa+\nu\right), \tag{9.165}$$

where $I(z;\ a, b)$ is the standard incomplete beta function given in (7.26).

Example 9.23. Reconsidering Example 9.22 and using (9.165), we find that

$$\tilde{r} = 1 - I\left(\frac{10^6}{10^6+10^9+1.17\times 10^8};2,24\right)$$

$$= 1 - I(8.945\times 10^{-4};2,24) = 1 - 0.000237 = 0.999763,$$

which agrees with the answer in Example 9.22. ■

Now let us consider Bayesian interval estimation of r following the procedure given by Lwin and Singh (1974). From (9.158), the point $t_{0,p}$ for which $r = p$ is

$$t_{0,p} = 0.5\chi^2_{1-p}(2\kappa)\theta,$$

where $\chi^2_{1-p}(2\kappa)$ is the $100(1-p)$th percentile of a $\chi^2(2\kappa)$ distribution. Since r is an increasing function of θ, it follows that

$$\Pr(R \geqslant p) = \Pr\left[\Theta \geqslant \frac{2t_{0,p}}{\chi^2_{1-p}(2\kappa)}\right]. \tag{9.166}$$

Thus to find the $100(1-\gamma)$th percentile of the posterior distribution of R given t we must find p such that

$$\Pr(R \geqslant p|\mathrm{t}) = \gamma,$$

which, using (9.166), is equivalent to solving

$$\int_{2t_{0,p}/\chi^2_{1-p}(2\kappa)}^{\infty} g(\theta|\mathrm{t})\,d\theta = \gamma$$

for p. Since $t_{0,p} = t_0$ is specified, p is the solution to the equation given by

$$\int_{2t_0/\chi^2_{1-p}(2\kappa)}^{\infty} g(\theta|\mathrm{t})\,d\theta = \gamma. \tag{9.167}$$

In the case of an $\mathcal{IG}(\nu, \mu)$ prior distribution on Θ, (9.167) becomes

$$\Pr\left[\chi^2 \leqslant \frac{(\mathrm{t}+\mu)\chi^2_{1-p}(2\kappa)}{t_0}\right] = \chi^2\left[\frac{(\mathrm{t}+\mu)\chi^2_{1-p}(2\kappa)}{t_0};\ 2n\kappa + 2\nu\right] = \gamma, \tag{9.168}$$

where $\chi^2(x;\,n)$ denotes the cdf of a $\chi^2(n)$ distribution evaluated at x. Thus, by definition, (9.168) states that

$$\frac{(\mathrm{t}+\mu)\chi^2_{1-p}(2\kappa)}{t_0} = \chi^2_{\gamma}(2n\kappa + 2\nu),$$

from which

$$\chi^2_{1-p}(2\kappa) = \frac{t_0\chi^2_{\gamma}(2n\kappa+2\nu)}{\mathrm{t}+\mu}. \tag{9.169}$$

Hence, the solution to (9.169) for p is given by

$$p = 1 - \Pr\left[\chi^2 \leqslant \frac{t_0\chi^2_{\gamma}(2n\kappa+2\nu)}{\mathrm{t}+\mu}\right] = 1 - \chi^2\left[\frac{t_0\chi^2_{\gamma}(2n\kappa+2\nu)}{\mathrm{t}+\mu};\ 2\kappa\right]. \tag{9.170}$$

A symmetric $100(1-\gamma)\%$ TBPI estimate of r in the case of an $\mathcal{IG}(\nu, \mu)$ prior distribution on Θ becomes

$$r_{*}=1-\chi^{2}\left[\frac{t_{0}\chi_{1-\gamma/2}^{2}(2n\kappa+2\nu)}{\mathrm{t}+\mu};\ 2\kappa\right] \tag{9.171}$$

and

$$r^{*}=1-\chi^{2}\left[\frac{t_{0}\chi_{\gamma/2}^{2}(2n\kappa+2\nu)}{\mathrm{t}+\mu};\ 2\kappa\right], \tag{9.172}$$

where $\Pr(r_{*}\leqslant R\leqslant r^{*}|\mathrm{t};\ \kappa,\nu,\mu)=1-\gamma$. One-sided Bayesian interval estimates may be constructed in the usual way.

Example 9.24. In Example 9.21, suppose that we desire a 95% LBPI estimate of the weld reliability for $t_0=10^6$ h. Based on the data in Example 9.21, according to (9.171) we find that

$$r_{*}=1-\chi^{2}\left[\frac{(10^{6})\chi_{0.95}^{2}(48)}{10^{9}+1.17\times10^{8}};\ 4\right]=1-\chi^{2}\left[\frac{(10^{6})(65.1710)}{10^{9}+1.17\times10^{8}};\ 4\right]$$

$$=1-\chi^{2}(0.0583;\ 4)=1-0.000355=0.999645,$$

and thus it may be claimed that $\Pr(R\geqslant0.999645|10^{9};2.0,4.0,1.17\times10^{8})=0.95$. ■

Scale and Shape Parameters Are Random Variables. Suppose now that both κ and Θ are r.v.'s. We consider only the case where κ is a discrete r.v. with prior distribution given by (9.146) and where the conditional prior distribution of Θ given κ_i is an $\mathcal{IG}(\nu_i, \mu_i)$ distribution. This case was first introduced in Section 9.5.2. Lwin and Singh (1974) give the corresponding posterior mean of R in (9.158) given $\underset{\sim}{t}$ as

$$E(R|\underset{\sim}{t};\ \underset{\sim}{\kappa},\underset{\sim}{\nu},\underset{\sim}{\mu})=1-\sum_{i=1}^{m}p_{i}^{*}I\left(\frac{t_{0}}{t_{0}+\mathrm{t}+\mu_{i}};\ \kappa_{i},n\kappa_{i}+\nu_{i}\right). \tag{9.173}$$

In computing the posterior mean of R in (9.159) with respect to either (9.150) or (9.154), double numerical integration must be used. Consequently, the estimation of r with respect to the prior distributions proposed by Canavos (1971) is somewhat computationally tedious and is not considered here.

EXERCISES

9.1. Barlow, Toland, and Freeman (1979) present several sets of failure data on Kevlar/epoxy spherical pressure vessels that were tested at constant sustained pressure. Twenty-one vessels were tested at 3400 psi of which the first 5 ordered failures occurred after 4000, 5376, 7320, 8616, and 9120 h. Assume that a $\mathscr{W}_1(\lambda, \beta)$ failure model adequately describes these data in which $\beta = 1.7$.

(a) For a noninformative prior distribution on Λ, compute a Bayesian point and 95% TBPI estimate of the MTTF at this pressure assuming a squared-error loss function.

(b) For the same situation as in (a), compute a Bayesian point and 90% LBPI estimate on the reliability of these vessels for a 10000 h period.

9.2. With regard to Exercise 9.1, suppose that it is believed a priori that $\Pr(\Lambda > 10^{-7}) = \Pr(\Lambda < 10^{-8}) = 0.05$.

(a) Find a $\mathscr{G}(\alpha_0, \beta_0)$ prior distribution for Λ that satisfies the prior considerations.

(b) Using the prior distribution found in (a), compute a Bayesian point and 95% TBPI estimate of the MTTF at this pressure assuming a squared-error loss function. Compare these estimates to those found in Exercise 9.1(a).

(c) Based on (a), compute a Bayesian point and 90% LBPI estimate on the reliability of these vessels for a 10000 h period. Compare these estimates to those found in Exercise 9.1(b).

9.3. Again from Barlow, Toland, and Freeman (1979), failure times were also reported on pressure vessels that were tested at 4300 psi. A complete sample of size $n = s = 39$ vessels was tested until each vessel failed. The ordered failure times were reported to be 2.2, 4.0, 4.0, 4.6, 6.1, 6.7, 7.9, 8.3, 8.5, 9.1, 10.2, 12.5, 13.3, 14.0, 14.6, 15.0, 18.7, 22.1, 45.9, 55.4, 61.2, 87.5, 98.2, 101.0, 111.4, 144.0, 158.7, 243.9, 254.1, 444.4, 590.4, 638.2, 755.2, 952.2, 1108.2, 1148.5, 1569.3, 1750.6, and 1802.1 h.

(a) Based on the use of a total time on test plot and/or Weibull probability plot [Section 4.3], discuss the claim that the underlying failure model is strongly DFR.

(b) For a $\mathscr{W}_1(\lambda, \beta)$ failure model in which $\beta = 0.5$ and a noninformative prior on Λ, compute a Bayesian point and 95% TBPI estimate of the MTTF assuming a squared-error loss function.

9.4. Plot the data in Exercise 9.3 on log normal probability paper. On the basis of the plot would you conclude that an $\mathcal{LN}(\xi, \sigma^2)$ failure model adequately describes these data?

9.5. Use the goodness-of-fit procedures discussed in Section 3.8 to determine whether a $\mathcal{G}(\alpha, \beta)$ failure model adequately represents the data in Exercise 9.3.

9.6. Suppose that the failure time of a certain type of rotary gear-type positive displacement pump follows a $\mathcal{G}(\kappa, \theta)$ distribution with $\kappa = 3$. Further suppose that an $\mathcal{IG}(3.0,\ 2.5 \times 10^4\ \text{h})$ prior distribution is placed on Θ and that 15 such pumps have accumulated 200000 h of total test time prior to failure.

(a) Compute a Bayes point estimate of θ assuming a squared-error loss function.

(b) Compute a 90% TBPI estimate of θ.

9.7. In Exercise 9.6, compute a Bayesian point estimate of the pump reliability for $t_0 = 10^5$ h.

9.8. A certain device has an $\mathcal{N}(\mu, \sigma^2)$ failure time distribution in which $\sigma^2 = 12$. For a noninformative prior distribution on μ and an observed sample of 10 failure times in which $\bar{t} = 20.5$ h, compute a Bayesian point and 90% LBPI estimate of the reliability of this device for $t_0 = 25$ h. Based on this information, what is the probability that the next failure time will exceed 25 h?

9.9. Repeat Exercise 9.8 for the case in which an $\mathcal{N}(25, 4)$ prior distribution is placed on μ.

9.10. Suppose in Exercise 9.8 that an $\mathcal{N}(25, 4)$ prior distribution is placed on μ. Compute a Bayesian point and 90% TBPI estimate of the reliable life t_R for $R = 0.95$.

9.11. [Canavos and Tsokos (1971b)] Show that the posterior mean of R in (9.159), in the case of an $\mathcal{E}_1(\lambda)$ prior distribution on Θ, is given by

$$E(R|\mathrm{t}; \kappa, \lambda) = \frac{\left(\sqrt{\lambda \mathrm{t}}\right)^{n\kappa - 1}}{K_{n\kappa - 1}\left(2\sqrt{\mathrm{t}/\lambda}\right)\left(1 + t_0/\mathrm{t}\right)^{(n\kappa - 1)/2}} \times \sum_{j=0}^{\kappa - 1} \frac{t_0^j K_{n\kappa + j - 1}\left(2\sqrt{(\mathrm{t} + t_0)/\lambda}\right)}{j!\left[\sqrt{\lambda(\mathrm{t} + t_0)}\right]^j},$$

where κ is a known integer and $K_m(az)$ is the modified Bessel function of the third kind of order m.

9.12. Following the procedure leading to (9.170), obtain a similar expression involving $\chi^2(x; n)$ probabilities for the $100(1-\gamma)\%$ LBPI estimate of r in (9.158) when the prior distribution on Θ is

$$g(\theta; a, \alpha, \beta) = \frac{(a-1)(\alpha\beta)^{a-1}}{\beta^{a-1} - \alpha^{a-1}}\left(\frac{1}{\theta^a}\right),$$

$$0 < \alpha \leqslant \theta \leqslant \beta,\ a \geqslant 0 (a \neq 1),$$

and where κ is known. Hint: the pdf $g(\theta|\mathrm{t})$ in (9.166) is given by

$$g(\theta|\mathrm{t}) = \frac{\mathrm{t}^{n\kappa+a-1}e^{-\mathrm{t}/\theta}}{\theta^{n\kappa+a}[\Gamma(n\kappa+a-1, \mathrm{t}/\alpha) - \Gamma(n\kappa+a-1, \mathrm{t}/\beta)]},$$

$$\alpha \leqslant \theta \leqslant \beta.$$

9.13. Show that $\hat{\lambda} = n\mu^2/V$ is the ML estimator of the parameter λ in the $\mathcal{IN}(\mu, \lambda)$ distribution when μ is known and where $V = n(\overline{T} - 2\mu) + n\mu^2\overline{T}_r$ with $(\overline{T}, \overline{T}_r)$ defined in (9.130).

9.14. [Moore and Bilikam (1978)] Show that $\tilde{\theta}$ given by (9.9) is the usual Bayesian point estimator of θ in the $\mathcal{W}_2(\theta, \beta)$ distribution when β is known and when the improper prior distribution $g(\theta; \alpha_0) = \theta^{-\alpha_0 - 1}$, $\theta > 0$, is placed on Θ. Consider Type II/item-censored testing without replacement.

9.15. [Moore and Bilikam (1978)] Show that the Bayes point estimator for the $\mathcal{W}_2(\theta, \beta)$ reliability function given in (9.32) using squared-error loss and improper prior distribution on Θ given by $g(\theta; \alpha_0) = \theta^{-\alpha_0 - 1}, \theta > 0$, is

$$\tilde{r}(t_0) = \left[\frac{\mathcal{W}}{\mathcal{W} + t_0^\beta}\right]^{\alpha_0 + s}$$

for the case of Type II/item-censored testing without replacement and where $\mathcal{W}$ is given in (9.4). Assume that β is known.

9.16. Show that the usual Bayesian point estimator of reliable life t_R in the $\mathcal{LN}(\xi, \sigma^2)$ distribution for a noninformative prior distribution on ξ (σ known) is

$$E[T_R|\bar{z}; \sigma^2] = \exp\left[\bar{z} + \sigma\Phi^{-1}(1-R) + \frac{\sigma^2}{2n}\right].$$

9.17. Show that the posterior expectation of reliable life T_R, in the case of an $\mathcal{LN}(\xi, \sigma^2)$ sampling distribution and an $\mathcal{N}(\lambda_0, \psi_0^2)$ prior distribution on ξ (σ known), is given by

$$E\left[T_R \mid \bar{z}; \sigma^2, \lambda_0, \psi_0^2\right] = \exp\left[\lambda + \sigma\Phi^{-1}(1-R) + \frac{\psi^2}{2}\right],$$

where λ is given by (9.74) in which $\bar{z}$ replaces $\bar{t}$, and ψ^2 is defined in (9.75).

9.18. Show that a $100(1-\gamma)\%$ TBPI on reliable life t_R in the $\mathcal{LN}(\xi, \sigma^2)$ sampling distribution for a noninformative prior distribution on ξ (σ known) is given by

$$t_R^* = \exp\left[\bar{z} + z_{1-\gamma/2}\frac{\sigma}{\sqrt{n}} + \sigma\Phi^{-1}(1-R) + \frac{\sigma^2}{2n}\right]$$

and

$$t_{R*} = \exp\left[\bar{z} - z_{1-\gamma/2}\frac{\sigma}{\sqrt{n}} + \sigma\Phi^{-1}(1-R) + \frac{\sigma^2}{2n}\right].$$

9.19. For the $\mathcal{N}(\mu, \sigma^2)$ sampling distribution, show that the joint posterior distribution of (μ, σ), in the case of a joint noninformative prior distribution $g(\mu, \sigma) \propto \sigma^{-1}$, is given by (9.79).

REFERENCES

Banerjee, A. K. and Bhattacharyya, G. K. (1979). Bayesian Results for the Inverse Gaussian Distribution with an Application, *Technometrics*, Vol. 21, pp. 247–251.

Barlow, R. E., Toland, R. H., and Freeman, T. (1979). *Stress-Rupture Life of Kevlar/Epoxy Spherical Pressure Vessels*, UCID-17755 Part 3, Lawrence Livermore Laboratory, Livermore, CA.

Box, G. E. P. and Tiao, G. C. (1973). *Bayesian Inference in Statistical Analysis*, Addison-Wesley, Reading, MA.

Bury, K. V. (1972a). On the Reliability Analysis of a Two-Parameter Weibull Process, *Canadian Journal of OR and Information Processing*, Vol. 10, pp. 129–139.

Bury, K. V. (1972b). Bayesian Decision Analysis of the Hazard Rate for a Two-Parameter Weibull Process, *IEEE Transactions on Reliability*, Vol. R-21, pp. 159–169.

Canavos, G. C. (1971). *A Bayesian Approach to Parameter and Reliability Estimation in Failure Distributions*, Ph.D. Dissertation, Virginia Polytechnic Institute and State University, Blacksburg, VA.

Canavos, G. C. (1974). On the Robustness of a Bayes Estimate, *Proceedings 1974 Annual Reliability and Maintainability Symposium*, pp. 432–435.

Canavos, G. C. and Tsokos, C. P. (1971a). Ordinary and Empirical Bayes Approach to Estimation of Reliability in the Weibull Life Testing Model, *Proceedings of the 16th Conference on the Design of Experiments in Army Research Development and Testing*, pp. 379–392.

Canavos, G. C. and Tsokos, C. P. (1971b). A Study of an Ordinary and Empirical Bayes Approach to Reliability Estimation in the Gamma Life Testing Model, *Proceedings 1971 Annual Reliability and Maintainability Symposium*, pp. 343–349.

Canavos, G. C. and Tsokos, C. P. (1973). Bayesian Estimation of Life Parameters in the Weibull Distribution, *Operations Research*, Vol. 21, pp. 755–763.

De Groot, M. H. (1970). *Optimal Statistical Decisions*, McGraw–Hill, New York.

Ellison, B. E. (1964). Two Theorems for Inference about the Normal Distribution with Applications in Acceptance Sampling, *Journal of the American Statistical Association*, Vol. 59, pp. 89–95.

Guttman, I. and Tiao, G. C. (1964). A Bayesian Approach to Some Best Population Problems, *Annals of Mathematical Statistics*, Vol. 35, pp. 825–835.

Harris, C. M. and Singpurwalla, N. D. (1968). Life Distributions Derived from Stochastic Hazard Functions, *IEEE Transactions on Reliability*, Vol. R-17, pp. 70–79.

Harris, C. M. and Singpurwalla, N. D. (1969). On Estimation in Weibull Distribution with Random Scale Parameters, *Naval Research Logistics Quarterly*, Vol. 16, pp. 405–410.

Jayaram, Y. G. (1974). A Bayesian Estimate of Reliability in the Weibull Distribution, *Microelectronics and Reliability*, Vol. 13, pp. 29–32.

Johnson, N. L. and Kotz, S. (1970). *Continuous Univariate Distributions — 1*, Houghton Mifflin, Boston.

Lian, M. G. (1975). *Bayes and Empirical Bayes Estimation of Reliability for the Weibull Model*, Ph.D. Dissertation, Texas Tech University, Lubbock, TX.

Lwin, T. and Singh, N. (1974). Bayesian Analysis of the Gamma Distribution Model in Reliability Engineering, *IEEE Transactions on Reliability*, Vol. R-23, pp. 314–319.

Martz, H. F. and Lian, M. G. (1977). Bayes and Empirical Bayes Point and Interval Estimation of Reliability for the Weibull Model, *The Theory and Applications of Reliability, Vol. II*, Academic, New York, pp. 203–233.

Moore, A. H. (1974). Admissible, Minimax, and Equivariant Estimators in Certain Life Testing Models, *Bulletin of the Institute of Mathematical Statistics*, Vol. 3, p. 191.

Moore, A. H. and Bilikam, J. E. (1978). Bayesian Estimation of Parameters of Life Distributions and Reliability from Type II Censored Samples. *IEEE Transactions on Reliability*, Vol. R-27, pp. 64–67.

Nadas, A. (1973). Best Tests for Zero Drift Based on First Passage Times in Brownian Motion, *Technometrics*, Vol. 15, pp. 125–132.

Padgett, W. J. (1981). Bayes Estimation of Reliability for the Inverse Gaussian Model, *IEEE Transactions on Reliability*, Vol. R-30, pp. 384–385.

Padgett, W. J. and Tsokos, C. P. (1977). Bayes Estimation of Reliability for the Lognormal Failure Model, *The Theory and Applications of Reliability, Vol. II*, Academic, New York, pp. 133–161.

Padgett, W. J. and Tsokos, C. P. (1978). On Bayes Estimation of Reliability for Mixtures of Life Distributions, *SIAM Journal of Applied Mathematics*, Vol. 34, pp. 692–703.

Padgett, W. J. and Wei, L. J. (1977). Bayes Estimation of Reliability for the Two-Parameter Lognormal Distribution, *Communications in Statistics—Theory and Methods*, Series A, Vol. A6, pp. 443–457.

Papadopoulos, A. and Tsokos, C. P. (1975). Bayesian Confidence Bounds for the Weibull Failure Model, *IEEE Transactions on Reliability*, Vol. R-24, pp. 21–25.

Papadopoulos, A. and Tsokos, C. P. (1976). Bayesian Analysis of the Weibull Failure Model with Unknown Scale and Shape Parameters, *Statistica*.

Raiffa, H. and Schlaifer, R. (1961). *Applied Statistical Decision Theory*, Harvard University Press, Cambridge, MA.

Reactor Safety Study, Appendix III—Failure Data (1975), WASH-1400 (NUREG-75/014), U.S. Nuclear Regulatory Commission.

Scarborough, J. B. (1957). *Numerical Mathematical Analysis*, Johns Hopkins Press, Baltimore, MD.

Soland, R. M. (1966). *Use of the Weibull Distribution in Bayesian Decision Theory*, Report No. RAC-TP-225, Research Analysis Corporation, McLean, VA.

Soland, R. M. (1968). Bayesian Analysis of the Weibull Process with Unknown Scale Parameter and Its Application to Acceptance Sampling, *IEEE Transactions on Reliability*, Vol. R-17, pp. 84–90.

Soland, R. M. (1969). Bayesian Analysis of the Weibull Process with Unknown Scale and Shape Parameters, *IEEE Transactions on Reliability*, Vol. R-18, pp. 181–184.

Tsokos, C. P. (1972a). Bayesian Approach to Reliability Using the Weibull Distribution with Unknown Parameters and Its Computer Simulation, *Report of Applied Statistical Research, Japanese Union of Scientists and Engineers*, Vol. 19, pp. 123–134.

Tsokos, C. P. (1972b). A Bayesian Approach to Reliability Theory and Simulation, *Proceedings 1972 Annual Reliability and Maintainability Symposium*, pp. 78–87.

Tsokos, C. P. and Canavos, G. C. (1972). Bayesian Concepts for the Estimation of Reliability in the Weibull Life Testing Model, *International Statistical Review*, Vol. 40, pp. 153–160.

Tsokos, C. P. and Rao, A. N. V. (1976). Bayesian Analysis of the Weibull Failure Model under Stochastic Variation of the Shape and Scale Parameters, *Metron*, Vol. XXXIV, pp. 201–217.

Tsokos, C. P. and Rao, A. N. V. (1979). Robustness Studies for Bayesian Developments in Reliability, *Decision Information*, Academic, New York, pp. 239–258.

Tummala, V. M. R. and Sathe, P. T. (1978). Minimum Expected Loss Estimators of Reliability and Parameters of Certain Lifetime Distributions, *IEEE Transactions on Reliability*, Vol. R-27, pp. 283–285.

Waller, R. A. and Martz, H. F. (1977). *A Bayesian Zero-Failure (BAZE) Reliability Demonstration Procedure for Components of Nuclear Reactor Safety Systems*, LA-6813-MS, Los Alamos Scientific Laboratory, Los Alamos, NM.

Yatsu, S. (1977). *Bayesian Approach to Reliability Estimation for the Weibull Distribution*, Memoirs No. 12, Tamagawa University Faculty of Engineering, pp. 161–177.

Zellner, A. (1971). Bayesian and Non-Bayesian Analysis of the Log-Normal Distribution and Log-Normal Regression, *Journal of the American Statistical Association*, Vol. 66, pp. 327–330.

CHAPTER 10

Reliability Demonstration Testing

In previous chapters we have dealt with the problem of estimating the reliability of a component or system under a variety of assumed lifetime distributions. A related problem is that of demonstrating whether a specified reliability has been achieved in a newly designed component or system, rather than trying to estimate the actual reliability. A reliability demonstration test is frequently used for this purpose as discussed in Section 4.6.

Recall that the reliability demonstration test (RDT) problem consists of the determination of the amount and type of testing that will be required in order to demonstrate a certain level of reliability with specified assurance.

The concept of demonstration testing is related to that of acceptance sampling, wherein a testing or sampling procedure is developed in order to determine whether to accept a particular lot or shipment. In acceptance sampling, it is assumed that a basically satisfactory production design exists, and we are only checking to see if an occasional bad lot has been encountered. However, demonstration testing is generally done on initial production samples or prototypes, and the decision of whether to accept a design, or to go into production, may rest on the results of the demonstration test. Consequently, the test will usually be more rigid than that for acceptance sampling, as expressed by the degree and nature of risks that the producer and consumer are willing to undertake in the testing process. These risks will be discussed in depth in the following section.

A common problem of reliability demonstration testing is the magnitude of total time on test required to demonstrate reliability to the consumer's satisfaction, particularly in the case of high reliability components. One solution is the use of accelerated life testing techniques. Another is to incorporate prior beliefs, engineering experience, or previous data into the testing framework. This may have the effect of reducing the amount of testing required in the RDT in order to reach a decision regarding conformance to the reliability specification. It is in this spirit that the use of a Bayesian approach can, in many cases, significantly reduce the amount of testing required.

This chapter is intended to provide an introduction to various *Bayesian reliability demonstration test* (BRDT) philosophies and procedures. One of the earliest references to Bayesian reliability demonstration plans is that of Bonis (1966). Since that time various techniques have been developed. Easterling (1970) presents a somewhat modified Bayesian demonstration procedure. Schafer and Singpurwalla (1970) develop sequential Bayes procedures. Schafer (1969, 1973) and Schafer and Sheffield (1971) do extensive work on three types of Bayesian plans: Bayesian fixed time tests, mixed Classical/Bayesian plans, and sequential plans. Goel and Joglekar (1976) compile a comprehensive study of Bayesian acceptance sampling procedures. Martz and Waller (1979) develop a zero-failure demonstration plan, and Martz and Waterman (1978) give a Bayesian approach to the problem of determining optimal test stress in a demonstration test when only a single test unit is available. Other developments include those of Balaban (1969, 1975), Blumenthal (1973), Goel and Coppola (1979), Gonzáles (1970), Guild (1968, 1973), Higgins and Tsokos (1976), Joglekar (1975), and Schafer (1975).

Several aspects of demonstration testing are discussed in the remaining sections. Section 10.1 deals with the criteria for determining the nature of the test, based on protecting both the consumer's and producer's interests. Section 10.2 gives methods for finding BRDT procedures in the case of attribute sampling. Section 10.3 describes Schafer's and Goel and Joglekar's techniques for determining BRDTs for testing MTTF when the failure time follows an $\mathcal{E}(\lambda)$ distribution and Section 10.4 outlines the zero-failure demonstration plan of Martz and Waller, again for the $\mathcal{E}(\lambda)$ distribution. Finally, sequential test procedures, which constantly update and check test results in order to decrease expected sample size, are considered in Section 10.5.

10.1. DEMONSTRATION TESTING CRITERIA

As in the case of any decision based on limited testing or sampling information, the BRDT admits the possibility of error. An unreliable design may by chance pass the test, whereas a reliable design may fail it. The only way to eliminate such errors is to test or sample indefinitely. Consequently, the approach taken is for producer and consumer to jointly arrive at a

BRDT procedure for which each party feels that the risks incurred are acceptably small. The type and degree of acceptable risks will be specified by consumer and producer, and the BRDT will be chosen so as to minimize the amount of testing required while not exceeding the specified risks.

Suppose, for example, that a product's reliability specification is in terms of MTTF (the cases of specified failure rates or failure probabilities are completely analogous). Most BRDTs discussed in this chapter would require the specification of two values, θ_0 and θ_1, where θ_0 represents the contract specified or desired MTTF, while θ_1 represents the minimum MTTF that the consumer is willing to accept. The region $\theta_1 \leqslant \theta \leqslant \theta_0$ is sometimes referred to as the *indifference region*. Several possible risk criteria have been proposed for BRDTs and are defined below, along with a brief discussion of each.

10.1.1 Risks

Let $\mathcal{A}$ represent the action of accepting the design (passing the BRDT) and let $\mathcal{R}$ represent the action of rejecting the design (failing the BRDT).

Classical Risks.

$$\gamma = \Pr(\mathcal{R} \mid \theta_0), \qquad \delta = \Pr(\mathcal{A} \mid \theta_1). \tag{10.1}$$

Here γ is the producer's risk that a design having acceptable MTTF will fail the BRDT, whereas δ represents the consumer's risk that a design having minimum acceptable MTTF will pass the BRDT. Procedures based on these risks are not "Bayesian" in that they do not consider θ to be an r.v. with a prior distribution. Their use in mixed classical/Bayesian RDTs will be discussed at the end of this section.

Average Risks.

$$\bar{\gamma} = \Pr(\mathcal{R} \mid \Theta \geqslant \theta_0), \qquad \bar{\delta} = \Pr(\mathcal{A} \mid \Theta \leqslant \theta_1). \tag{10.2}$$

These risks are conceptually similar to classical risks; however, prior information is taken into consideration. Averaging with respect to the prior distribution is done over satisfactory ($\bar{\gamma}$) or unsatisfactory ($\bar{\delta}$) values of Θ.

These risks seem more appropriate for the case of acceptance sampling rather than reliability demonstration. Specification of small values for $\bar{\gamma}$ and $\bar{\delta}$ would guarantee acceptable performance on the average or in the long run, where the average is over a sequence of RDTs. However, in reliability demonstration testing, more emphasis is placed on the results of a *single* test so knowing that the test performs well on the average may not be reassuring. Also, due to the averaging property, no risk protection is provided for any particular values of Θ. Consequently, there may be a range of unsatisfactory values that have a relatively high probability of passing the BRDT, or a range of satisfactory values that have a low probability of passing the BRDT.

Posterior Risks.

$$\gamma^* = \Pr(\Theta \geqslant \theta_0 | \mathcal{R}), \qquad \delta^* = \Pr(\Theta \leqslant \theta_1 | \mathcal{A}). \tag{10.3}$$

The posterior risks γ^* and δ^* convey a different outlook for both consumer and producer. Specification of classical or average risks provides assurance that satisfactory units will pass the test and that unsatisfactory units will fail it. However, the producer may prefer assurance that the designs that *do* fail the test are indeed unsatisfactory, and the consumer may prefer assurance that the designs that *pass* the test are indeed satisfactory. This can be achieved through specification of small values for γ^* and δ^*, respectively. Posterior risks, like average risks, are Bayesian in the sense that they are influenced by the prior distribution.

Rejection Probability.

$$\Pr(\mathcal{R}) = \int_{\Omega_\theta} \Pr(\mathcal{R} | \theta) g(\theta)\, d\theta, \tag{10.4}$$

where Ω_θ denotes the possible values of θ [see Section 5.4.1]. Specification of $\Pr(\mathcal{R})$ allows the producer control over the unconditional probability of the design passing the BRDT. From the producer's point of view, this could be of primary importance. As the rejection probability is averaged with respect to the prior distribution over all values of θ, there is no guarantee that there is a high probability of passing the BRDT at any specific value.

Alternate Posterior Consumer's Risk.

$$\delta^{**} = \Pr(\Theta \leqslant \theta_0 | \mathcal{A}). \tag{10.5}$$

This consumer's risk is conceptually similar to δ^*, representing the probability that a design that passed the BRDT has actual MTTF below the contract specified value θ_0. δ^{**} would be more appropriate than δ^* if the consumer has more to lose when the design has true MTTF in the indifference region.

10.1.2 Recommendations

The choice of criteria used to develop the BRDT is of course up to the producer and consumer, and they need not specify the same risk types (classical, average, posterior, etc.). Balaban (1975) favors the mixed classical/Bayesian pair (γ, δ^*). Easterling (1970) favors average risks $(\bar{\gamma}, \bar{\delta})$, but his primary emphasis is on acceptance sampling rather than reliability demonstration. Schick and Drnas (1972) use the posterior pair (γ^*, δ^*). Schafer and Sheffield (1971) compile extensive tables giving BRDT plans for $[\Pr(\mathcal{R}), \delta^*]$. Goel and Joglekar (1976) provide FORTRAN subroutines for determining BRDTs for various combinations of Bayesian risks, and also give examples where more than two risks can be specified [e.g., $\{\Pr(\mathcal{R}), \gamma^*, \bar{\delta}\}$]. Use of the plans given by Schafer and by Goel and Joglekar will be discussed in Section 10.3.

In some cases, the consumer and producer will not agree on a prior distribution for Θ. As the producer presumably has more experience with his or her own design success than the consumer, the most common case would be one in which the producer specifies a prior that the consumer will not accept (e.g., the producer may have a higher regard for the a priori probability that the design will be successful). A mixed Bayesian/classical procedure could be used here, where the producer specifies a Bayesian risk [$\bar{\gamma}$, γ^*, or $\Pr(\mathcal{R})$] based on the prior, while the consumer chooses to ignore the prior and specifies a classical risk δ. Schafer and Sheffield (1971) also compile tables of BRDTs for the case [$(\Pr(\mathcal{R}), \delta$], to be discussed in Section 10.3. Another possibility would be for producer and consumer to each propose a separate prior. In this case a Bayesian risk [$\bar{\gamma}$, γ^*, or $\Pr(\mathcal{R})$] could be specified based on the producer's prior, whereas a Bayesian risk [$\bar{\delta}$, δ^*, or δ^{**}] could be specified based on the consumer's prior.

Note that any combination of producer and consumer risks is feasible and potentially meaningful, but that some plans do have advantages in terms of mathematical tractability, shorter required test times, or better availability of predetermined test plans. Plans that are readily available are discussed in the following sections.

10.2. BRDT PLANS WITH ATTRIBUTE TESTING

The first case considered in the determination of BRDT plans will be that of single fixed sample attribute testing. Recall from Section 7.1 that an attribute test is one in which the only observed outcome is survival or nonsurvival of the unit under test. For example, a battery may be tested to determine whether it will start an engine, or an electronic component may be tested for a fixed time t_0 where the only observation made is whether the component survives the test (as opposed to observing the actual failure time). The purpose of the BRDT is to demonstrate that units of a particular design have an acceptable probability of surviving the test.

Let the r.v. R represent the reliability or probability that the component survives a specified attribute test, where R is assumed to have a $\mathcal{B}_1(\alpha_0, \beta_0)$ prior distribution. Guidelines for determining the applicability of this assumption and for choosing prior parameter values can be found in Chapters 6 and 7.

The BRDT plan can be defined by specifying values (n, s^*) where n represents the number of units to be tested and s^* represents the maximum allowable number of failures (nonsurvivors). Values of n and s^* are chosen so as to minimize the amount of testing required while not exceeding the specified consumer's and producer's risks. Note that the sampling distribution of the number of failures S in the test, $f(s|R)$, is binomial with parameters n and $1-R$. Also, it is easily shown that the posterior distribution $g(r|s)$ is $\mathcal{B}_1(\alpha_0+n-s, \beta_0+s)$. Using this information we obtain the following expressions for the risks defined in Section 10.1:

Classical.

$$1-\gamma=\sum_{s=0}^{s^*}\frac{\Gamma(n+1)}{\Gamma(s+1)\Gamma(n-s+1)}R_0^{n-s}(1-R_0)^s, \tag{10.6}$$

$$\delta=\sum_{s=0}^{s^*}\frac{\Gamma(n+1)}{\Gamma(s+1)\Gamma(n-s+1)}R_1^{n-s}(1-R_1)^s, \tag{10.7}$$

where R_0 is the specified reliability and R_1 is the minimum acceptable reliability.

Pr($\mathcal{R}$).

$$\Pr(\mathcal{R})=1-\sum_{s=0}^{s^*}\frac{\Gamma(n+1)\Gamma(\alpha_0+\beta_0)}{\Gamma(s+1)\Gamma(n-s+1)\Gamma(\alpha_0)\Gamma(\beta_0)}\times$$

$$\frac{\Gamma(n+\alpha_0-s)\Gamma(\beta_0+s)}{\Gamma(n+\alpha_0+\beta_0)}. \tag{10.8}$$

Posterior.

$$\gamma^* = \left\{ P_0 - \sum_{s=0}^{s^*} \frac{\Gamma(n+1)\Gamma(\alpha_0+\beta_0)}{\Gamma(s+1)\Gamma(n-s+1)\Gamma(\alpha_0)\Gamma(\beta_0)} \frac{\Gamma(n+\alpha_0-s)\Gamma(\beta_0+s)}{\Gamma(n+\alpha_0+\beta_0)} \right.$$

$$\left. \times [1 - I(R_0; n+\alpha_0-s, \beta_0+s)] \right\} \Big/ \Pr(\mathfrak{R}), \quad (10.9)$$

$$\delta^* = \left\{ \sum_{s=0}^{s^*} \frac{\Gamma(n+1)\Gamma(\alpha_0+\beta_0)}{\Gamma(s+1)\Gamma(n-s+1)\Gamma(\alpha_0)\Gamma(\beta_0)} \frac{\Gamma(n+\alpha_0-s)\Gamma(\beta_0+s)}{\Gamma(n+\alpha_0+\beta_0)} \right.$$

$$\left. \times I(R_1; n+\alpha_0-s, \beta_0+s) \right\} \Big/ [1 - \Pr(\mathfrak{R})], \quad (10.10)$$

where $I(x; a, b)$ is the incomplete beta function defined in (7.26) and $P_0 = \Pr(R \geqslant R_0) = 1 - I(R_0; \alpha_0, \beta_0)$.

Average.

$$\bar{\gamma} = \frac{\gamma^* \Pr(\mathfrak{R})}{P_0}, \qquad \bar{\delta} = \frac{\delta^*[1 - \Pr(\mathfrak{R})]}{P_1}, \quad (10.11)$$

where $P_1 = \Pr(R \leqslant R_1) = I(R_1; \alpha_0, \beta_0)$.

Alternate Posterior. δ^{**} is given by (10.10), replacing R_1 by R_0.

The pair (n, s^*) that defines the desired BRDT plan can be found by setting the appropriate risk formulas equal to the specified risks then solving for n and s^*. Since only integer values for n and s^* are meaningful, a simple computer search technique is recommended.

Example 10.1. An attribute test is to be used for determining the effectiveness of a newly designed starting mechanism for a diesel engine associated with an auxiliary power plant. The contract calls for a reliability of $R_0 = 0.95$, with a minimum acceptable value of $R_1 = 0.90$. The $\mathfrak{B}_1(\alpha_0, \beta_0)$ prior parameters, using techniques from Section 6.5, have been determined to be $\alpha_0 = 47$, $\beta_0 = 3$. The consumer has chosen the posterior risk criterion with a maximum allowable value of $\delta^* = 0.10$, whereas the producer has chosen the rejection probability criterion with a maximum value of $\Pr(\mathfrak{R}) = 0.05$. By means of a simple direct search the required plan is found to be

$n=19$, $s^*=3$. Thus, 19 independent starts should be conducted and the contract reliability 0.95 will have been demonstrated at the 90 percent level if three or fewer failures to start occur.

For the plan (19,3) the risks given in Section 10.1.1 are all computed to be

$$\gamma^* = 0.0407, \qquad \bar{\gamma} = 0.0043, \qquad \Pr(\mathcal{R}) = 0.0471,$$

$$\delta^* = 0.0992, \qquad \bar{\delta} = 0.7876, \qquad \delta^{**} = 0.5337,$$

$$P_0 = \Pr(R \geqslant R_0) = 0.4463, \qquad P_1 = \Pr(R \leqslant R_1) = 0.1200. \qquad \blacksquare$$

Note that the value of the consumer's risk $\bar{\delta}$ found in the above example is quite high (0.7876). This serves to illustrate the inappropriateness of the $\bar{\delta}$ criterion in demonstration testing, as discussed in Section 10.1.1. Consider the following facts:

1. The probability that $R \leqslant R_1$ is only 0.1200.
2. If in fact $R \leqslant R_1$, the expected value $E(R|R \leqslant R_1)$ is computed to be 0.8749, which is very near the minimum acceptable value 0.9.

So, although the high value of $\bar{\delta}$ shows that unsatisfactory designs *are* likely to be accepted under the (19,3) plan, the likelihood of unsatisfactory designs P_1 is low and the designs that are unsatisfactory tend to have reliabilities near the minimum acceptable value R_1. Consequently, a large value of $\bar{\delta}$ is no cause for concern in this case. The consumer insures, through specification of δ^*, that a design that the consumer accepts has a high probability (in this case $1-0.0992=0.90$) of being satisfactory.

10.3. TESTING THE MEAN OF AN EXPONENTIAL DISTRIBUTION

The determination of test plans for BRDTs has primarily focused on the case of an assumed $\mathcal{E}(\lambda)$ failure distribution, having MTTF $\theta = 1/\lambda$. Recall from Chapter 8 that the natural conjugate prior distribution for Θ is the $\mathcal{IG}(\alpha_0, \beta_0)$ distribution.

Goel and Joglekar (1976) develop formulas and appropriate subroutines for computing the various risks discussed in Section 10.1.1 under the $\mathcal{E}_1(\theta)/\mathcal{IG}(\alpha_0, \beta_0)$ assumption. The formulas are based on a truncated single sample test plan defined by the pair (t_0, s^*). One or more units are tested with replacement for total time on test t_0, and s^* is the maximum allowable number of failures. Letting s represent the observed number of failures, we say that the specified reliability has been satisfactorily demonstrated if and only if $s \leqslant s^*$ once the total time on test t_0 has been accumulated. Relationships that can be used to determine other test plans equivalent to (t_0, s^*) will be given later in this section.

To determine the appropriate BRDT, we must find the values (t_0, s^*) that minimize the amount of testing required while not exceeding the specified consumer's and producer's risks. For example, suppose that the $[\Pr(\mathfrak{R}), \delta^*]$ criteria is chosen and the specified risks are $\Pr(\mathfrak{R}) = 0.10$, $\delta^* = 0.05$. (t_0, s^*) could be found by setting the appropriate expressions given by Goel and Joglekar (1976) equal to their respective values and then solving the resulting 2×2 system of nonlinear equations for t_0 and s^*. Since s^* is restricted to be an integer, an exact solution will probably not exist. (t_0, s^*) are then chosen by either modifying the specified risks or explicitly examining the risks of the candidate plans.

Example 10.2. Suppose that we require a demonstration test to determine whether a unit has a contract specified MTTF of $\theta_0 = 100$ h. The $[\Pr(\mathfrak{R}), \delta^*]$ criteria has been chosen, with specified risks $\Pr(\mathfrak{R}) = 0.1$, $\delta^* = 0.05$. Minimum acceptable MTTF is $\theta_1 = 50$ h. Prior shape and scale parameters have been determined to be $\alpha_0 = 3.0$ and $\beta_0 = 300$. Using the appropriate subroutine given in Goel and Joglekar (1976) the solution is found to be

$$t_0 = 39.847, \qquad s^* = 1.$$

Thus, the BRDT would require one or more items to be tested with replacement for approximately 40 h, and the test would be passed if no more than one failure occurs.

For the plan (39.847, 1) the risks given in Section 10.1.1 are computed to be

$$\gamma^* = 0.2288, \qquad \bar{\gamma} = 0.0278, \qquad \Pr(\mathfrak{A}) = 0.9298 \ [\text{or } \Pr(\mathfrak{R}) = 0.0702],$$

$$\delta^* = 0.0497, \qquad \bar{\delta} = 0.7462, \qquad \delta^{**} = 0.3969. \qquad \blacksquare$$

In addition to providing the appropriate subroutines, Goel and Joglekar also present graphical techniques for determining BRDTs. These techniques allow better explanation and understanding of the test design region, the use of multiple risk criteria, and designs with engineering constraints or with partial information (only one risk specified).

Schafer and Sheffield (1971) assume the $[\Pr(\mathcal{R}), \delta^*]$ risk criteria and provide actual BRDT plans for 2400 combinations of the parameters $E(\Theta)$ (the prior mean, expressed in units of θ_0), α_0, δ^*, $\Pr(\mathcal{A}) = 1 - \Pr(\mathcal{R})$, and the discrimination ratio θ_0/θ_1. The following range of values are provided in the tables:

1. $E(\Theta) = \theta_0, 1.1\theta_0, \ldots, 2\theta_0$.
2. $\alpha_0 = 0.5, 1.0, 1.5, 2.0, 2.5, 3.0, 3.5, 4.0, 4.5, 5.0$.
3. $\delta^* = 0.05, 0.10, 0.15, 0.20$.
4. $\Pr(\mathcal{A}) = 0.80, 0.90$.
5. $d = \theta_0/\theta_1 = 1.5, 2.0, 2.5$.

Schafer and Sheffield's tables are easy to use, but are too extensive to be reproduced in this text.

Example 10.3. Consider the problem described in the previous example. The prior mean $E(\Theta)$ is given by $300/(3-1) = 150$ h $= 1.5\theta_0$. Using the appropriate table $[E(\Theta) = 1.5\theta_0$, $\alpha_0 = 3$, $\delta^* = 0.05$, $\Pr(\mathcal{A}) = 0.90$, and $d = 2.0]$ we obtain the BRDT plan $t_0/\theta_0 = 0.499$ or $t_0 = 49.9$ h, $s^* = 1$. ■

Note that the value of t_0 obtained in Example 10.3 is significantly different from the value produced in Example 10.2, in which 25% less total test time was required. We have already discussed the fact that solutions that satisfy the two specified risk criteria exactly do not generally exist, and consequently most solutions are compromises of some sort. Goel and Joglekar's solution satisfies the δ^* criterion almost exactly, while giving a $\Pr(\mathcal{R})$ value that is significantly lower than that specified. Schafer's solution does the opposite: the $\Pr(\mathcal{A})$ criterion is exact while δ^* is smaller than specified. Both solutions are valid, but Goel and Joglekar's requires significantly less test time in this example.

10.3.1 Bayesian/Classical Test Situation

Schafer and Sheffield (1971) also provide tables for the mixed Bayesian/classical test situation described in Section 10.1.2. This situation occurs when the producer has specified a prior distribution that the consumer is not willing to accept. Schafer and Sheffield's tables assume that the producer specifies a Bayesian risk $\Pr(\mathcal{A})$ based on the prior distribution, whereas the consumer disregards the prior and specifies a classical risk δ. As in the case of producer/consumer agreement on the prior, Schafer and Sheffield provide RDT plans for a wide selection of parameter values.

10.3.2 Equivalent Test Procedures

The test plan (t_0, s^*) given by Schafer and Sheffield's tables or by Goel and Joglekar's procedure is a truncated life test with replacement. Although this is the only plan for which values are given, equivalent plans are easily obtained once t_0 and s^* are known. A number of plans are described below, along with examples showing how each could be made equivalent to the (39.847, 1) plan found in Example 10.2. Note that the equivalences shown only hold in the case of $\mathcal{E}(\lambda)$ distributed failure times.

Plans with Replacement.

1. Truncated plan for a lot $[(t_0, s')]$:
   n units are tested for fixed time t_0, and the test is passed if $s \leq s'$. This is equivalent to the (t_0, s^*) test plan when $t_0 = \mathrm{t}_0/n$ and $s' = s^*$. Example: Four units are tested with replacement for time $t_0 = 39.847/4 = 9.96$ h. The plan is (9.96, 1).
2. Censored plan for a single sample $[(\mathrm{t}'_0, s')]$:
   A single sample is tested until s' failures occur. The test is passed if the total test time is greater than t'_0. Equivalence to (t_0, s^*) holds when $\mathrm{t}'_0 = \mathrm{t}_0$ and $s' = s^* + 1$. Example: Test one unit with replacement until two failures occur. The plan is (39.847, 2).
3. Censored plan for a lot $[(\mathrm{t}'_0, s')]$:
   n units are tested with replacement until $s' \leq n$ failures occur. Again, the test is passed if the total test time is greater than t'_0. Equivalence holds when $\mathrm{t}'_0 = \mathrm{t}_0$ and $s' = s^* + 1$. The plan, once again, would be (39.847, 2).

Plans without Replacement.

1. Truncated plan for a lot $[(t_0, s')]$:
   n units are life tested without replacement for time t_0, and the test is passed if $s \leq s'$. Equivalence to (t_0, s^*) holds (approximately) when

$$t_0 = \frac{\mathrm{t}_0}{s^*} \ln\left(\frac{n}{n - s^*}\right) \qquad \text{and} \qquad s' = s^*.$$

   Example: Test four units without replacement for time

$$t_0 = \frac{39.847}{1} \ln\left(\frac{4}{4-1}\right) = 11.46 \text{ h}.$$

   The plan is (11.46, 1).

2. Censored plan for a lot $[(\mathrm{t}'_0, s')]$:
   n units are life tested until $s' \leqslant n$ failures occur. The test is passed if the total test time is greater than t'_0. Equivalence holds when $\mathrm{t}'_0 = \mathrm{t}_0$ and $s' = s^* + 1$. Example: Test four units without replacement until two failures occur. The plan is (39.847, 2).

10.4. TESTING THE FAILURE RATE OF AN EXPONENTIAL DISTRIBUTION

The techniques for finding BRDTs presented in the previous section, all of which require a two-risk criteria, tend to be somewhat difficult to apply. Martz and Waller (1979) present an alternative approach called the Bayesian Zero-Failure (BAZE) reliability demonstration testing procedure. BRDT plans using the BAZE procedure can be enumerated using a relatively small set of tables and the required calculations are easily done on a pocket calculator. The BAZE procedure is used to test whether a specified failure rate (rather than MTTF) has been satisfactorily demonstrated. Note that the question of whether the contract reliability specification is given in terms of MTTF (θ) or failure rate (λ) should *not* be used to determine whether to use the BAZE plan as opposed to the two-risk plans given in the preceding section. As failure rate is simply the inverse of MTTF [assuming $\mathcal{E}(\lambda)$ failure times] the two specifications are completely interchangeable.

Under the BAZE plan, n units are each tested for time t_0 and the test is passed if and only if no failures are observed ($s = 0$). This is similar to the (t_0, s^*) criteria described in Section 10.3, but the maximum allowable number of failures ($s^* = 0$) is predetermined and does not depend on specified risks. The test is determined by the choice of t_0 alone, (or total time on test $\mathrm{t}_0 = nt_0$) so only one risk need be specified and one equation solved. The BAZE plan bases the test on a modification of the posterior consumer's risk δ^*.

As the test plan is determined solely by specification of a consumer's risk, the producer may feel a loss of control over the BRDT chosen. However, the rejection probability $\Pr(\mathcal{R})$ is easily computed for any BAZE plan, and if the producer finds this value unacceptable an adjustment may be requested in the specified δ^* in order to make the corresponding $\Pr(\mathcal{R})$ acceptable.

The BAZE approach assumes, as in Section 10.3, an $\mathcal{E}(\lambda)$ failure time distribution. The procedure is based on values of the failure rate $\Lambda = 1/\Theta$ rather than MTTF Θ, where Λ is assumed to have a $\mathcal{G}_1(\alpha_0, \beta_0)$ prior distribution.

Let λ_0 and λ_1 respectively represent the contract specified and test criterion failure rates, where λ_1 may be equal to λ_0. Then the posterior distribution of Λ, given that the test has been passed (zero failures), is $\mathcal{G}_1(\alpha_0, \beta_0 + nt_0)$. Consequently, we have

$$\Pr(\Lambda \leqslant \lambda_1 | \mathcal{A}) = \int_0^{\lambda_1} \frac{(\beta_0 + nt_0)^{\alpha_0}}{\Gamma(\alpha_0)} \lambda^{\alpha_0 - 1} e^{-(\beta_0 + nt_0)\lambda} \, d\lambda$$

$$= \frac{\Gamma[\alpha_0, (\beta_0 + nt_0)\lambda_1]}{\Gamma(\alpha_0)}, \tag{10.12}$$

where $\Gamma(a, x)$ is the usual incomplete gamma function. To satisfy the specified δ^* risk criterion, we require that

$$1 - \delta^* = \frac{\Gamma[\alpha_0, (\beta_0 + nt_0)\lambda_1]}{\Gamma(\alpha_0)}. \tag{10.13}$$

We wish to solve (10.13) for (nt_0), the required total time on test. Table B7 in Appendix B gives the solution to this equation for the quantity $\nu = (\beta_0 + nt_0)\lambda_1$ for selected choices of α_0 and $(1 - \delta^*)$. Since δ^* is the posterior risk, $1 - \delta^*$ may be thought of as the posterior assurance that $\Lambda \leqslant \lambda_1$ in a test that is passed (zero failures occur). Then using simple algebra we obtain

$$nt_0 = \frac{\nu - \beta_0 \lambda_1}{\lambda_1}. \tag{10.14}$$

Note that the solution is given in terms of the product (nt_0) rather than the value for t_0 alone. Every combination of n and t_0 that yields the same product gives equivalent tests; for example, testing four units for 50 h each $(nt_0 = 200 \text{ h})$ is equivalent to testing two units for 100 h each, and so on. In general, this equivalence only holds in the $\mathcal{E}(\lambda)$ sampling distribution.

Once the product (nt_0) has been determined, the rejection probability $\Pr(\mathcal{R})$ can be found using the formula

$$\Pr(\mathcal{R}) = 1 - \left(\frac{\beta_0}{\beta_0 + nt_0} \right)^{\alpha_0}. \tag{10.15}$$

The test can be suitably modified if the producer finds the value of $\Pr(\mathcal{R})$ to be unsatisfactory.

Example 10.4. Consider a certain component whose failure rate must be demonstrated. Assume that the techniques of Chapter 6 have been used to determine prior distribution parameters $\alpha_0 = 1.0$, $\beta_0 = 0.6 \times 10^6$ h. Contract specified failure rate is $\lambda_0 = 2.0 \times 10^{-6}$ f/h, which must be demonstrated with 70 percent consumer's posterior assurance ($1-\delta^* = 0.70$). We further assume that $\lambda_1 = \lambda_0$.

From Appendix Table B7 with $\alpha_0 = 1.0$, $1-\delta^* = 0.70$ we obtain $\nu = 1.203973$. Thus (10.14) gives

$$nt_0 = \frac{1.203973 - (0.6 \times 10^6 \text{ h})(2.0 \times 10^{-6} \text{ f/h})}{2.0 \times 10^{-6} \text{ f/h}}$$

$$= 1987 \text{ unit-h.}$$

So, for example, one unit could be tested for 1987 h, or 10 units could be tested for 198.7 h each. The failure rate is considered satisfactory if and only if no failures occur during the test.

The producer's risk is computed from (10.15) to be

$$\Pr(\mathscr{R}) = 1 - \left[\frac{0.6 \times 10^6 \text{ h}}{0.6 \times 10^6 \text{ h} + 1987 \text{ h}}\right]^1 = 1 - 0.9967 = 0.0033.$$

In a case where $\Pr(\mathscr{R})$ turns out to be unacceptably large, either δ^* could be increased or λ_1 increased until both consumer and producer are satisfied. ■

10.5. SEQUENTIAL TEST PLANS

The RDT plans presented in Sections 10.3 and 10.4 all involve fixed sample sizes that are chosen in advance of the actual demonstration test. In many cases it is possible to reduce the amount of testing required, without increasing the risks, through the use of a sequential test plan. With this type of plan, one unit at a time is tested and the testing process continues until enough information is accumulated in order to determine whether the desired reliability has been satisfactorily demonstrated. After each single unit test, the following outcomes can occur: (1) conclude that reliability is satisfactory and terminate the test; (2) conclude that reliability is unsatisfactory and terminate the test; and (3) continue the testing process. The third outcome occurs when the cumulative test results do not contain enough information to make a decision at that stage.

In a sequential test plan the number of units tested, N, is an r.v. rather than a predetermined constant. Although one cannot predict in advance just how many units will have to be tested before a decision is reached, the following statements generally hold for sequential test plans:

1. The test will always terminate. That is, there is no possibility that the test will require an infinite number of test units. (Obviously, this result is of more theoretical than practical value, as time and money constraints would prevent an unreasonably large number of test units anyway.)
2. The expected number of test units required, $E(N)$, will be smaller than the value of n needed in a fixed sample size test having equivalent risks.
3. The realized value of N will be smaller than the n mentioned in 2 with high probability.

Consequently, although it is possible for a sequential test to require more test units than the equivalent fixed sample size test, one can generally expect a savings. This is the primary motivation for sequential testing procedures. Note that truncated sequential test plans are also a possibility, where the testing process continues sequentially until either a decision or an upper bound on the sample size is reached. If the upper bound is reached, an accept/reject decision is made at that point according to established criteria.

The use of Bayesian rather than classical RDTs has the same advantage as in the fixed sample size case: incorporation of prior beliefs, engineering experience, and previous data into a prior distribution reduces the amount of information that the actual RDT must provide. Once again, these considerations may have the effect of reducing the amount of testing required.

Sequential Bayesian RDTs have been proposed by Barnett (1972), MacFarland (1971), and Schafer and Singpurwalla (1970). MacFarland's technique is a Bayesian analog of the classical sequential probability ratio test [Wald (1947)]. Schafer and Singpurwalla's approach is now discussed.

Suppose, for example, that Θ is an r.v. representing MTTF, with prior distribution $g(\theta)$. Note that the technique presented here is general and is equally applicable for testing either reliability or failure rate parameters.

Let θ_1 represent the minimum acceptable MTTF, and let $x_1,\dots,x_n$ represent the results from the first n test units. Using techniques from Chapter 5, the posterior distribution $g(\theta|x_1,\dots,x_n)$ can be calculated at each stage of the sequential test process. This distribution quantifies our knowledge regarding the probability of Θ, based on the original prior $g(\theta)$ and the observed data.

If $g(\theta|x_1,\dots,x_n)$ assigns a large mass to values of Θ greater than θ_1, then we have high assurance that the minimum MTTF has been achieved, and the RDT is considered successful. If the posterior distribution assigns a small mass to values of Θ greater than θ_1, then the RDT is considered unsuccessful. Consequently, a sequential RDT may be determined by specifying values of K_0 and K_1, where we consider reliability to have been demonstrated satisfactorily if

$$P_n = \Pr(\Theta \geqslant \theta_1 | x_1, \ldots, x_n) \geqslant K_0, \qquad (\text{accept } \Theta \geqslant \theta_1), \qquad (10.16)$$

and we consider reliability to have been demonstrated unsatisfactorily if

$$P_n \leqslant K_1, \qquad (\text{reject } \Theta \geqslant \theta_1). \qquad (10.17)$$

If neither criterion holds, the information available is not sufficient to draw a conclusion and further testing is required. The criterion test probability P_n can be found by integrating the posterior pdf over the appropriate region.

Note that the acceptance criteria is equivalent to determining whether the value of θ_1 is less than the $100\times K_0\%$ LBPI for MTTF, and the rejection criteria is equivalent to determining whether or not θ_1 is greater than the $100(1-K_1)\%$ UBPI for MTTF. Typically, K_0 is near one (values like 0.90, 0.95, 0.99), and K_1 is near zero (0.10,0.05,0.01).

The procedure described above involves the computation of the posterior probability P_n at each stage of the testing process. In practice, it is often more convenient to establish accept, reject, and continue-to-sample regions for each stage prior to the performance of the test. The probability P_n is generally a monotone function of either the number of failures, s_n, occurring in the n samples (attribute testing) or the total time on test for n samples, t_n (variables testing). For example, in the case of attribute testing values $\underline{s}_n$ and $\bar{s}_n$ could be found where at stage n we accept if $s_n \leqslant \underline{s}_n$, reject if $s_n \geqslant \bar{s}_n$, and continue to sample otherwise. An analogous situation holds for the case of variables testing.

10.5.1 Attribute Testing

Suppose that units are to be tested for survival/nonsurvival only, and that a minimum acceptable survival probability R_1 is to be demonstrated. Assume that the survival probability R is an r.v. having a $\mathcal{B}_1(\alpha_0, \beta_0)$ prior distribution. Let s_n represent the number of failures occurring in the first n trials. Then we have

$$\begin{aligned} P_n &= \int_{R_1}^{\infty} g(r | s_n)\, dr \\ &= \int_{R_1}^{\infty} \frac{\Gamma(\alpha_0+\beta_0+n)}{\Gamma(\alpha_0+n-s_n)\Gamma(\beta_0+s_n)} r^{\alpha_0+n-s_n-1}(1-r)^{\beta_0+s_n-1}\, dr. \qquad (10.18) \end{aligned}$$

P_n could be computed at each stage of the process and compared to K_0 and K_1. However, P_n is strictly decreasing in s_n, so

1. $P_n \geq K_0$ if and only if s_n is less than or equal to some value $\underline{s}_n$;
2. $P_n \leq K_1$ if and only if s_n is greater than or equal to some value $\bar{s}_n$.

By means of an incomplete beta function subroutine, it is easy to compute the sequences $\{\underline{s}_n\}$ and $\{\bar{s}_n\}$ for specified values of α_0, β_0, K_0, K_1, and R_1. At any stage, the formulas used are

$$\underline{s}_n = \max\{s_n;\ P_n \geq K_0\},$$

$$\bar{s}_n = \min\{s_n;\ P_n \leq K_1\}, \tag{10.19}$$

where we note that P_n is a function of s_n.

Example 10.5. Suppose that $\alpha_0 = 47$, $\beta_0 = 3$ (giving a prior mean of 0.94 and prior variance of 0.0011), $K_0 = 0.95$, $K_1 = 0.10$, and $R_1 = 0.90$. The first 25 values of $\{\underline{s}_n\}$ and $\{\bar{s}_n\}$ are found to be

n	1	2	3	4	5	6	7	8	9	10	11	12	13
$\underline{s}_n$	*	*	*	*	*	*	*	*	*	*	*	0	0
$\bar{s}_n$	*	*	*	*	*	6	7	7	7	7	7	7	7

n	14	15	16	17	18	19	20	21	22	23	24	25
$\underline{s}_n$	0	0	0	0	0	0	0	0	0	0	0	0
$\bar{s}_n$	7	8	8	8	8	8	8	8	8	9	9	9

The prior probability of exceedence represents

$$\Pr(R \geq R_1) = \int_{R_1}^{1} g(r)\, dr.$$

If this value is greater than K_0 or smaller than K_1 then no testing would be necessary as the conclusion is evident based on the assumed prior distribution. The * values in the above table represent those cases where the number of test units is insufficient to draw a conclusion. Note that acceptance cannot possibly occur until at least 12 units have been tested, whereas rejection cannot occur until at least six units have been tested. ■

10.5.2 Risk Calculations

Computations related to risks and expected sample sizes for the sequential RDT described above are in general quite complex (K_0 and K_1 do *not* represent risks directly—they are comparable to critical values in hypothesis testing). Mann, Schafer, and Singpurwalla (1974) give the following approximations and bounds, which they claim are fairly accurate for values of K_1 less than 0.1, K_0 greater than 0.9, and P_0 "not too near" K_0 or K_1, where $P_0 = \Pr\{\Theta \geqslant \theta_1\} = \int_{\theta_1}^{\infty} g(\theta)\,d\theta$ (in the MTTF case).

1. Probability of a satisfactory demonstration (accept $\Theta \geqslant \theta_1$):

$$\Pr(\mathscr{A}) = 1, \qquad \text{if } P_0 \geqslant K_0,$$

$$\Pr(\mathscr{A}) = \left(\frac{P_0 - K_1}{K_0 - K_1}\right), \qquad \text{if } K_1 < P_0 < K_0,$$

$$\Pr(\mathscr{A}) = 0, \qquad \text{if } P_0 \leqslant K_1. \tag{10.20}$$

2. Probability of accepting $(\Theta \geqslant \theta_1)$ when $(\Theta \geqslant \theta_1)$ is indeed true:

$$\Pr(\mathscr{A} \mid \Theta \geqslant \theta_1) = 1, \qquad \text{if } P_0 \geqslant K_0,$$

$$\Pr(\mathscr{A} \mid \Theta \geqslant \theta_1) \cong \left(\frac{K_0}{P_0}\right)\left(\frac{P_0 - K_1}{K_0 - K_1}\right), \qquad \text{if } K_1 < P_0 < K_0,$$

$$\Pr(\mathscr{A} \mid \Theta \geqslant \theta_1) = 0, \qquad \text{if } P_0 \leqslant K_1. \tag{10.21}$$

3. Probability that N, the required sample size, is less than any fixed n:

$$\Pr\{N \leqslant n\} \geqslant \Pr\{P_n \geqslant K_0\} + \Pr\{P_n \leqslant K_1\}. \tag{10.22}$$

4. Probability of accepting $(\Theta \geqslant \theta_1)$ when $(\Theta \geqslant \theta_1)$ is false:

$$\Pr(\mathscr{A} \mid \Theta < \theta_1) = 1, \qquad P_0 \geqslant K_0,$$

$$\Pr(\mathscr{A} \mid \Theta < \theta_1) \cong \left(\frac{1 - K_0}{1 - P_0}\right)\left(\frac{P_0 - K_1}{K_0 - K_1}\right), \qquad K_1 < P_0 < K_0,$$

$$\Pr(\mathscr{A} \mid \Theta < \theta_1) = 0, \qquad P_0 \leqslant K_1. \tag{10.23}$$

Note that the cases where a test is not required ($P_0 \geqslant K_0$ or $P_0 \leqslant K_1$) are unlikely to occur in practice, as it would be unusual for both consumer and producer to agree on a prior that presupposes that the unit to be produced is already either satisfactory or already unsatisfactory.

Example 10.6. Continuing with Example 10.5, the given test plan has values $K_0 = 0.95$, $K_1 = 0.10$, and $P_0 = 0.88$. Using (10.20), (10.21), and (10.23) we have

$$\Pr(\mathcal{A}) \cong \frac{0.88 - 0.10}{0.95 - 0.10} = 0.9176 \qquad \text{or} \quad \Pr(\mathcal{R}) \cong 0.0824,$$

$$\Pr(\mathcal{A} \mid \Theta \geqslant \theta_1) \cong \left(\frac{0.95}{0.88}\right)(0.9176) = 0.9906 \quad \text{or} \quad \Pr(\mathcal{R} \mid \Theta \geqslant \theta_1) \cong 0.0094,$$

$$\Pr(\mathcal{A} \mid \Theta < \theta_1) \cong \left(\frac{1 - 0.95}{1 - 0.88}\right)(0.9176) = 0.3823.$$

■

It is noted that $\Pr(\mathcal{A} \mid \Theta < \theta_1)$ is a consumer's risk, whereas $\Pr(\mathcal{R})$ and $\Pr(\mathcal{R} \mid \Theta \geqslant \theta_1)$ are producer's risks. The parameters K_0 and K_1 can be adjusted accordingly if the computed approximate risks are found to be unsatisfactory. Note that acceptance probabilities decrease as either K_0 or K_1 increases.

As in the case of fixed sample size tests [Section 10.3], Schafer and Sheffield (1971) have compiled tables of sequential RDTs for testing MTTF with $\mathcal{E}_1(\theta)$ failure times and an $\mathcal{IG}(\alpha_0, \beta_0)$ prior distribution on Θ. The tables are restricted to the case where $K_0 + K_1 = 1$.

EXERCISES

10.1. Can you suggest a situation where θ_0 (the contract MTTF) and θ_1 (the minimum consumer acceptable MTTF) are such that $\theta_1 < \theta < \theta_0$? What is the interpretation of the indifference region for your example? Can you suggest a situation where $\theta_0 = \theta_1$?

10.2. Assume that the failure time T for an automobile head lamp follows an $\mathcal{E}_1(\theta)$ distribution. A random sample of lamps are placed on life test with replacement for a total test time of 1000 h. The test is passed if no failures occur. Find the classical risks γ and δ if $\theta_0 = 1000$ h and $\theta_1 = 750$ h. Is this a reasonable test plan? Why?

10.3. Suppose that a random sample of $\mathcal{E}_1(\theta)$ distributed items is life tested with replacement such that the test is passed if no more than one failure is observed in 6000 h of total test time. Further suppose that Θ has an $\mathcal{IG}_1(\alpha_0, \beta_0)$ prior distribution. Write expressions for $\bar{\gamma}$, $\bar{\delta}$, $\Pr(\mathcal{R})$, γ^*, and δ^{**}.

10.4. Refer to Example 10.1 and verify the values for $\Pr(\mathcal{R})$, δ^*, P_1, and $\bar{\delta}$.

10.5. Suppose the test plan $n = 20$ and $s^* = 2$ is used in Example 10.1 instead of the plan $n = 19$ and $s^* = 3$. Find $\Pr(\mathcal{R})$, δ^*, P_1, and $\bar{\delta}$. Discuss your answers relative to the corresponding quantities in Example 10.1.

10.6. Suppose that a given brand of fuses has an $\mathcal{E}(\lambda)$ life distribution and that Λ has a $\mathcal{G}_1(\alpha_0, \beta_0)$ prior distribution where we have determined $\alpha_0 = 1.5$ and $\beta_0 = 1.0 \times 10^4$ h. The contract failure rate is $\lambda_0 = 2.0 \times 10^{-4}$ f/h which must be demonstrated with 80% consumer's posterior assurance. Assume $\lambda_1 = \lambda_0$ and find a BAZE test plan to satisfy the given conditions. Find the producer's risk, the probability of rejecting a good lot of fuses, for this situation.

10.7. Solve Example 10.4 for each of three posterior assurances: $1 - \delta^* =$ 0.50, 0.80, 0.95. Discuss the effects of changing the posterior assurance. Use (10.15) to find $\Pr(\mathcal{R})$ for each of the three levels of posterior assurance.

10.8. Electrical switching units are tested to demonstrate a reliability $R_1 = 0.95$. Assume that R has a $\mathcal{B}_1(49, 1)$ prior distribution. Use (10.18) and (10.19) to calculate $\underline{s}_n$ and $\bar{s}_n$ for $K_0 = 0.90$, $K_1 = 0.10$, and $n = 1, 2, \ldots, 25$.

10.9. In Exercise 10.8, compute $\Pr(\mathcal{A})$, $\Pr(\mathcal{R} \mid R \geq R_1)$, and $\Pr(\mathcal{A} \mid R < R_1)$.

REFERENCES

Balaban, H. (1969). A Bayesian Approach to Reliability Demonstration, *Annals of Assurance Sciences*, Vol. 8, pp. 497–506.

Balaban, H. (1975). Reliability Demonstration: Purposes, Practices, and Value, *Proceedings 1975 Annual Reliability and Maintainability Symposium*, pp. 246–248.

Barnett, V. D. (1972). A Bayesian Sequential Life Test, *Technometrics*, Vol. 14, pp. 453–467.

Blumenthal, S. (1973). *Reliability Demonstration*, Technical Report No. 183, Department of Operations Research, Cornell University, Ithaca, NY.

Bonis, A. J. (1966). Bayesian Reliability Demonstration Plans, *Annals of Reliability and Maintainability*, Vol. 5, pp. 861–873.

Easterling, R. G. (1970). On the Use of Prior Distribution in Acceptance Sampling, *Annals of Reliability and Maintainability*, Vol. 9, pp. 31–35.

Goel, A. L. and Coppola, A. (1979). Design of Reliability Acceptance Plans Based upon Prior Distributions, *Proceedings 1979 Annual Reliability and Maintainability Symposium*, pp. 34–38.

Goel, A. L. and Joglekar, A. M. (1976). *Reliability Acceptance Sampling Plans Based upon Prior Distribution*, Technical Reports 76-1 to 76-5, Department of Industrial Engineering and Operations Research, Syracuse University, Syracuse, NY.

Gonzáles, J. R. (1970). *Development of Bayesian Life Test Sampling Plans Assuming a Failure Rate with a Gamma Prior Distribution*, M. S. Thesis, Pennsylvania State University, University Park, PA.

Guild, R. D. (1968). *Reliability Testing and Equipment Design Using Bayesian Models*, Ph.D. Dissertation, Northwestern University, Evanston, IL.

Guild, R. D. (1973). Bayesian MFR Life Test Sampling Plans, *Journal of Quality Technology*, Vol. 5, pp. 11–15.

Higgins, J. J. and Tsokos, C. P. (1976). On the Behavior of Some Quantities Used in Bayesian Reliability Demonstration Tests, *IEEE Transactions on Reliability*, Vol. R-25, pp. 261–264.

Joglekar, A. M. (1975). Reliability Demonstration Based on Prior Distributions—Sensitivity Analysis and Multi Sample Plans, *Proceedings 1975 Annual Reliability and Maintainability Symposium*, pp. 251–252.

MacFarland, W. J. (1971). Sequential Analysis and Bayesian Demonstration, *Proceedings 1971 Annual Reliability and Maintainability Symposium*, pp. 24–38.

Mann, N. R., Schafer, R. E., and Singpurwalla, N. D. (1974). *Methods for Statistical Analysis of Reliability and Life Data*, Wiley, New York.

Martz, H. F. and Waller, R. A. (1979). A Bayesian Zero-Failure (BAZE) Reliability Demonstration Testing Procedure, *Journal of Quality Technology*, Vol. 11, pp. 128–138.

Martz, H. F. and Waterman, M. S. (1978). A Bayesian Model for Determining the Optimal Test Stress for a Single Test Unit, *Technometrics*, Vol. 20, pp. 179–185.

Schafer, R. E. (1975). Some Approaches to Bayesian Reliability Demonstration, *Proceedings 1975 Annual Reliability and Maintainability Symposium*, pp. 253–254.

Schafer, R. E. (1969). *Bayesian Reliability Demonstration: Phase I—Data for the A Priori Distribution*, RADC-TR-69-389, Rome Air Development Center, Rome, NY.

Schafer, R. E. (1973). *Bayesian Reliability Demonstration: Phase III—Development of Test Plans*, RADC-TR-139, Rome Air Development Center, Rome, NY.

Schafer, R. E. and Sheffield, T. S. (1971). *Bayesian Reliability Demonstration, Phase II—Development of a Prior Distribution*, RADC-TR-71-209, Rome Air Development Center, Rome, NY.

Schafer, R. E. and Singpurwalla, N. D. (1970). A Sequential Bayes Procedure for Reliability Demonstration, *Naval Research Logistics Quarterly*, Vol. 17, pp. 55–67.

Schick, G. J. and Drnas, T. M. (1972). Bayesian Reliability Demonstration, *AIIE Transactions*, Vol. 4, pp. 92–102.

Wald, A. (1947). *Sequential Analysis*, Wiley, New York.

CHAPTER 11

System Reliability Assessment

The notion of a "system" was introduced in Chapter 4. There it was stated that the assessment of the reliability of a system is frequently required in industrial, military, and everyday life situations. For such an assessment, it is necessary to specify the configuration of components, the failure mode of each component, and the states in which the system is classified as failed. In this chapter, we are primarily interested in the time to the first failure of a system. A failed component is neither repaired nor replaced during the time in which the system is required to function. System maintenance will be discussed in Chapter 12.

11.1. INTRODUCTION

In order to study the relationship between the reliability of a system and the reliabilities of its components, we need to examine how the performance of the various components affects the performance of the system.

11.1.1 Coherent Systems

Consider a system composed of k components. We associate with the ith component a state variable x_i such that, at any specified time,

$$x_i = \begin{cases} 1, & \text{if the } i\text{th component is in a functioning state} \\ 0, & \text{if the } i\text{th component is in a failed state.} \end{cases}$$

The *state of all components* in the system will be described by the vector $\underset{\sim}{x} = (x_1, x_2, \ldots, x_k)$. Some of the 2^k different states correspond to the system in a functioning state, and others correspond to the system in a failed state.

Thus, the state of the system may be written as a function of $\underset{\sim}{x}$, that is,

$$\phi(\underset{\sim}{x})=\begin{cases}1, & \text{if the system is in a functioning state}\\ 0, & \text{if the system is in a failed state.}\end{cases}$$

The function ϕ will be called the *structure function* of the system. Any vector $\underset{\sim}{x}$ for which $\phi(\underset{\sim}{x})=1$ will be called a *path* for the structure ϕ. A *minimal path* is a path $\underset{\sim}{x}$ such that for every $\underset{\sim}{y}<\underset{\sim}{x}$ we have $\phi(\underset{\sim}{y})=0$, where $\underset{\sim}{y}<\underset{\sim}{x}$ means $y_i \leqslant x_i$, $i=1,2,\ldots,k$, with $y_i<x_i$ for at least one i. Corresponding to a minimal path is a *minimal path set*, which was introduced in Section 4.7.5, and which is taken to be the subset of components each of which is in a functioning state as so indicated by the minimal path. Recall that a minimal path set is a minimal subset of components which, if they are all in a functioning state, insures the system is in a functioning state, but if any of them is in a failed state (and all other components outside the subset have failed), the system is in a failed state. Any vector $\underset{\sim}{x}$ for which $\phi(\underset{\sim}{x})=0$ will be called a *cut* for the structure ϕ. A *minimal cut* is a cut $\underset{\sim}{x}$ such that for every $\underset{\sim}{y}>\underset{\sim}{x}$ we have $\phi(\underset{\sim}{y})=1$. Recall that a *minimal cut set* is a minimal subset of components which, if they are all in a failed state, cause the system to be in a failed state, but if any of them is in a functioning state (and all other components outside the subset function), the system is in a functioning state.

A system is said to be *coherent* if its structure function is such that

1. $\phi(0,0,\ldots,0)=0$,
2. $\phi(1,1,\ldots,1)=1$,
3. $\phi(x_1,x_2,\ldots,x_k)$ is nondecreasing in each argument.

A coherent system is, therefore, a system which is in a failed state when all of its components are in failed states; is in a functioning state when all of its components are in functioning states; and, if initially in a functioning state for a state of its components, remains in a functioning state whenever some components that were initially in failed states are restored to functioning states. The structure function of a coherent system may be determined from either the minimal path sets or minimal cut sets of the system. The following two methods are given by Kaufmann, Grouchko, and Cruon (1977, pp. 92–93).

Method of Minimal Path Sets. Suppose $\{a_1,\ldots,a_m\}$ is the collection of all minimal path sets of a coherent system. Let x_i be the state variable of the

ith component. The structure function of the system is

$$\phi(\underset{\sim}{x}) = 1 - \prod_{j=1}^{m} \left(1 - \prod_{i \in a_j} x_i\right). \tag{11.1}$$

Method of Minimal Cut Sets. Suppose $\{b_1, \ldots, b_n\}$ is the collection of all minimal cut sets of a coherent system. Let x_i be the state variable of the ith component. The structure function of the system is

$$\phi(\underset{\sim}{x}) = \prod_{j=1}^{n} \left[1 - \prod_{i \in b_j} (1 - x_i)\right]. \tag{11.2}$$

The literature on coherent systems is extensive. The basic ideas were first formulated by Birnbaum, Esary, and Saunders (1961). Further detailed expositions are given by Barlow and Proschan (1975, Chapters 1 and 2), and Kaufmann, Grouchko, and Cruon (1977, Chapters 3 and 4).

Example 11.1. Consider the coherent system of components as diagrammed in Figure 11.1. Note that the subset of components given by {1, 2} forms a coherent subsystem that can be treated as if it were a component. The subset {1, 2} is often called a module. Diagrammatically speaking, a module is a collection with one wire leading into it and one wire leading out of it.

The state of all components is described by $\underset{\sim}{x} = (x_1, x_2, x_3, x_4)$. There are three minimal paths: (1,0,1,0), (0,1,1,0), and (0,0,0,1). The minimal path sets are:

$$a_1 = \{1,3\}, \qquad a_2 = \{2,3\}, \qquad a_3 = \{4\}.$$

There are two minimal cuts: (1,1,0,0) and (0,0,1,0). The minimal cut sets

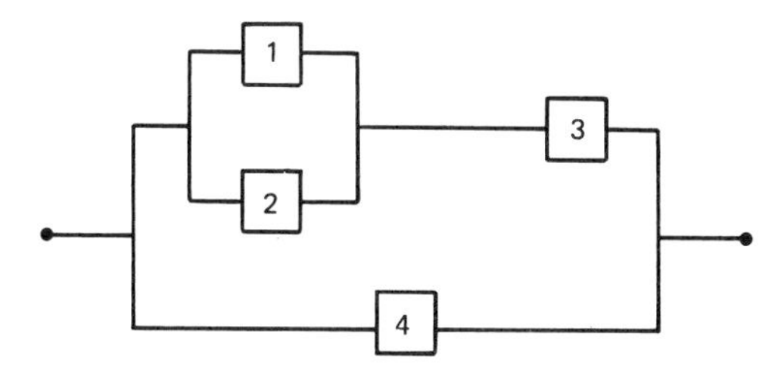

Figure 11.1. The system configuration for Example 11.1.

are:

$$b_1 = \{3,4\} \quad \text{and} \quad b_2 = \{1,2,4\}.$$

The structure function of the system is (noting that $x_i = x_i^2$ for all i)

1. By the method of minimal path sets:

$$\begin{aligned}\phi(x_1, x_2, x_3, x_4) &= 1 - \prod_{j=1}^{3}\left(1 - \prod_{i \in a_j} x_i\right)\\ &= 1-(1-x_1x_3)(1-x_2x_3)(1-x_4)\\ &= x_1x_3 + x_2x_3 - x_1x_2x_3 + x_4 - x_1x_3x_4\\ &\quad - x_2x_3x_4 + x_1x_2x_3x_4,\end{aligned}$$

2. By the method of minimal cut sets:

$$\begin{aligned}\phi(x_1, x_2, x_3, x_4) &= \prod_{j=1}^{2}\left[1 - \prod_{i \in b_j}(1-x_i)\right]\\ &= [1-(1-x_3)(1-x_4)][1-(1-x_1)(1-x_2)(1-x_4)]\\ &= x_1x_3 + x_2x_3 - x_1x_2x_3 + x_4 - x_1x_3x_4\\ &\quad - x_2x_3x_4 + x_1x_2x_3x_4.\end{aligned}$$

■

We next introduce the reliability function of a coherent system. It is customary to assume that the components of a given system may function or fail in a random manner and do so independently of one another. Let the state $\underset{\sim}{x} = (x_1, x_2, \ldots, x_k)$ be a possible value of the random state $\underset{\sim}{X} = (X_1, X_2, \ldots, X_k)$. The *reliability of the* ith *component* is the probability that the ith component is in a functioning state and is denoted by $R_i = \Pr\{X_i = 1\}$, $i = 1, 2, \ldots, k$. The *reliability function of the system* is the probability that the system is in a functioning state and is given by

$$R = h(R_1, R_2, \ldots, R_k) = \Pr[\phi(\underset{\sim}{X}) = 1].$$

As the components are assumed to perform *independently*, the reliability function of a coherent system is obtained simply by replacing the state variables $x_1, x_2, \ldots, x_k$ by the component reliabilities $R_1, R_2, \ldots, R_k$ in the structure function, [i.e., $R = \phi(R_1, R_2, \ldots, R_k)$]. Thus, as an illustration, the reliability function of the system in Example 11.1 is $R = R_1R_3 + R_2R_3 - R_1R_2R_3 + R_4 - R_1R_3R_4 - R_2R_3R_4 + R_1R_2R_3R_4$. In the event that the components are identical with reliability $\tilde{R}$, then $R = \tilde{R}^4 - 3\tilde{R}^3 + 2\tilde{R}^2 + \tilde{R}$.

11.1.2 Basic System Configurations

There are numerous coherent system configurations. The following basic types of systems are frequently encountered.

Series System. The notion of a *series system* was introduced in Section 4.7.1. Recall that a series system is a configuration of components such that the system functions if and only if all of its components function. Figure 4.18 is a graphical representation of a series system.

Each component is a minimal cut set. The set of all components is the only minimal path set. By either (11.1) or (11.2), the structure function of the system is

$$\phi(\underset{\sim}{x}) = \prod_{i=1}^{k} x_i. \tag{11.3}$$

The reliability function of the system is

$$R = \Pr[\phi(\underset{\sim}{X}) = 1] = \Pr\left(\prod_{i=1}^{k} X_i = 1\right) = \prod_{i=1}^{k} \Pr(X_i = 1) = \prod_{i=1}^{k} R_i. \tag{11.4}$$

Parallel System. Recall from Section 4.7.2 that a parallel system is a configuration of components such that the system functions if and only if at least one component functions. Such a system is sometimes said to be *active-redundant* as all components are initially put into operation yet only the functioning of any one of them is required. Figure 4.19 depicts a parallel system.

Each component is a minimal path set. The set of all components is the only minimal cut set. The structure function of the system is

$$\phi(\underset{\sim}{x}) = 1 - \prod_{i=1}^{k} (1 - x_i). \tag{11.5}$$

The reliability function of the system is

$$R = \Pr[\phi(\underset{\sim}{X}) = 1] = \Pr\left[\prod_{i=1}^{k} (1 - X_i) = 0\right]$$

$$= 1 - \Pr\left[\prod_{i=1}^{k} (1 - X_i) = 1\right] = 1 - \prod_{i=1}^{k} \Pr(X_i = 0)$$

$$= 1 - \prod_{i=1}^{k} (1 - R_i). \tag{11.6}$$

***r*-Out-Of-*k* System.** Recall that an r-out-of-k system is a configuration of k components such that the system functions if and only if at least r of its k components function. For $r = 1$, the r-out-of-k system reduces to a parallel system; for $r = k$, a series system is obtained. Thus, series and parallel systems are special cases of an r-out-of-k system. The structure function of the system is

$$\phi(\underset{\sim}{x}) = \sum_{j} \left(\prod_{i \in A_j} x_i \right) \left[\prod_{i \in A_j^c} (1 - x_i) \right], \tag{11.7}$$

where A_j is any subset of $\{1, 2, \ldots, k\}$ with at least r elements and the sum is over all such subsets. Any collection of r or more functioning components is a path. The reliability function of the system is

$$R = \sum_{j} \left(\prod_{i \in A_j} R_i \right) \left[\prod_{i \in A_j^c} (1 - R_i) \right]. \tag{11.8}$$

If all the components are identical with reliability $\tilde{R}$, (11.8) reduces to

$$R = \sum_{j=r}^{k} \binom{k}{j} \tilde{R}^j (1 - \tilde{R})^{k-j}, \tag{11.9}$$

which is equivalent to (4.110).

The notational convention adhered to mainly in preceding chapters, in which upper case letters are used to denote r.v.'s while corresponding lower case letters denote values of r.v.'s, is not followed in either this or the succeeding chapter. For example, the letter R will be used here to denote both reliability when considered as an r.v., as well as its value. This departure from previous convention reduces the general level of notational complexity of many already cumbersome expressions. In this same spirit, the lower case Greek alphabet will also be used to denote random parameters and their corresponding values if the parameter is an r.v..

11.2. ASSIGNMENT OF PRIOR DISTRIBUTIONS

Of fundamental importance in the Bayesian approach to system reliability is the assignment of the prior distribution(s). Such an assignment may take place at either or both the component and system level. In any case, the goal is to determine the posterior distribution (or possibly it moments) of the reliability of the system, from which Bayesian point and interval estimates may be found. Generally speaking, test data are required at the level to which the prior is assigned. Consequently, prior assignment is frequently at the component level as this usually allows for the derivation of the posterior distribution of system reliability based only on component test data and without assembling and testing the system per se. It is then possible to evaluate the merits of a system before its assembly, in the event that component reliabilities need to be reassessed in order to attain a required level of system reliability.

11.2.1 Component Level Priors

Prior assignment at the component level is often made for the reliability of each component in the system. Based on test data obtained from the components, the posterior distributions of component reliabilities can then be found. More importantly, for many systems it is possible to derive the posterior distribution of system reliability from the posterior distributions of component reliabilities. A key requirement in such a derivation is that system reliability be expressed as essentially a product of independent r.v.'s each of which corresponds to either component reliability or component unreliability. The basic types of systems considered in Section 11.1.2 have this feature. The method of derivation utilizes Mellin integral transforms which, for the convenience of the reader, is now briefly described.

Let X be a nonnegative r.v. with pdf $f(x)$. The *Mellin transform* of $f(x)$ with respect to the complex parameter u is defined as

$$M\{f(x); u\} = \int_0^\infty x^{u-1} f(x)\, dx. \tag{11.10}$$

It is convenient to think of a Mellin transform in terms of the moments of X; specifically,

$$M\{f(x); r+1\} = E(X^r). \tag{11.11}$$

There is an inversion integral [Springer (1979), p. 96] which enables one to recover the pdf $f(x)$ from the transform $M\{f(x); u\}$. The inversion integral can often be evaluated using the method of residues [Churchill (1960), pp. 206–213]. In addition, there are available extensive tables of inverse Mellin transforms [Erdélyi (1954); Oberhettinger (1974)] from which $f(x)$ may be obtained directly.

Let $X_1, X_2, \ldots, X_k$ be independent nonnegative r.v.'s with pdf's $f_1(x_1), f_2(x_2), \ldots, f_k(x_k)$, respectively. Springer and Thompson (1966a) show that

$$M\{h_k(y); u\} = \prod_{i=1}^{k} M\{f_i(x_i); u\}, \tag{11.12}$$

where $h_k(y)$ is the pdf of the product $Y = X_1 X_2 \cdots X_k$. Thus, by evaluating the inversion integral of the product of the Mellin transforms of the pdf's of independent nonnegative r.v.'s, the pdf of the product of these r.v.'s is obtained.

Throughout this chapter, repeated use of the Mellin transform will be made when determining Bayesian point and interval estimates of system reliability. As we shall see, implementation of the technique requires an initial specification of the failure type of each component in the system. Such considerations will be limited to the following two types of situations.

Pass/Fail Component. No assumptions are made with regard to the distribution of the lifetime of the component. All that is important is whether the component functions when required to do so. Such a component is said to have a pass/fail failure type and will be called a pass/fail component.

Suppose the ith component of a system is of the pass/fail type. The reliability of the component is $R_i = \Pr(X_i = 1)$. Suppose n_i units are tested and that s_i survivors are observed on the ith component. We thus have attribute (pass/fail) test data. The sampling distribution is the binomial distribution with parameter R_i. By choosing the natural conjugate beta prior pdf of R_i given by $\mathfrak{B}_1(s_{i0}+1, n_{i0}-s_{i0}+1)$, the posterior pdf of R_i becomes

$$g_i(R_i|s_i) = \frac{\Gamma(n_i'+2)}{\Gamma(s_i'+1)\Gamma(n_i'-s_i'+1)} R_i^{s_i'}(1-R_i)^{n_i'-s_i'}, \qquad 0 < R_i < 1, \tag{11.13}$$

where $n_i' = n_{i0} + n_i$ and $s_i' = s_{i0} + s_i$. Note that we have also notationally suppressed the explicit dependency of $g_i(R_i|s_i)$ on the prior parameters, a procedure we will follow for the remainder of this chapter. In the prior pdf of R_i, s_{i0} may be considered to be the total number of successful units in a prior (pseudo) test of n_{i0} units. In this case, the posterior pdf may be viewed as an updated prior pdf and it retains the same form as the prior pdf. This situation can only be achieved with the use of the natural conjugate prior pdf. By taking $n_{i0} = s_{i0} = 0$ (no pseudo test data are used), we obtain the $\mathcal{U}(0,1)$ prior pdf used by Breipohl, Prairie, and Zimmer (1965).

The Mellin transform of the posterior pdf of R_i is easily shown to be

$$\begin{aligned} M\{g_i(R_i|s_i);u\} &= \frac{\Gamma(n_i'+2)}{\Gamma(s_i'+1)}\,\frac{\Gamma(s_i'+u)}{\Gamma(n_i'+u+1)} \\ &= \frac{(n_i'+1)!}{s_i'!}\,\frac{1}{(u+s_i')(u+s_i'+1)\cdots(u+n_i')}, \end{aligned} \tag{11.14}$$

where $\mathrm{Re}(u) > -s_i'$.

The kth moment of the posterior pdf of R_i is, by (11.11),

$$E\left(R_i^k|s_i\right) = M\{g_i(R_i|s_i);u=k+1\} = \frac{\Gamma(n_i'+2)}{\Gamma(s_i'+1)}\,\frac{\Gamma(s_i'+k+1)}{\Gamma(n_i'+k+2)}. \tag{11.15}$$

Exponential Component. The failure time of the component, say X, is assumed to have an $\mathcal{E}(\lambda)$ distribution. Such a component is characterized by its constant rate of failure and will be called an $\mathcal{E}(\lambda)$ component.

Suppose the ith component of a system is such that its failure time pdf is $\mathcal{E}(\lambda_i)$; that is, its reliability at time t_i is $R_i = \exp(-\lambda_i t_i)$, where λ_i is a constant failure rate. Let $X_{i1} \leqslant X_{i2} \leqslant \cdots \leqslant X_{ir_i}$ represent the ordered first r_i (predetermined) failure times from a Type II/item-censored life test of size n_i. We thus have variables (time-to-failure) test data. The total time on test is

$$V_i = \sum_{j=1}^{r_i} X_{ij} + (n_i - r_i)X_{ir_i},$$

which has a $\mathcal{G}_1(r_i, \lambda_i)$ sampling distribution with pdf

$$f_i(v_i|\lambda_i) = \frac{\lambda_i^{r_i}}{\Gamma(r_i)} v_i^{r_i-1}\exp(-\lambda_i v_i), \qquad v_i > 0. \tag{11.16}$$

As $\lambda_i = (\ln R_i^{-1})/t_i$, reparameterizing (11.16) gives

$$f_i(v_i|R_i) = \frac{(v_i/t_i)^{r_i-1}}{\Gamma(r_i)t_i} R_i^{v_i/t_i}(\ln R_1^{-1})^{r_i}, \quad v_i > 0. \tag{11.17}$$

Treating R_i as an r.v., the posterior pdf of R_i is proportional to $R_i^{v_i/t_i}(\ln R_i^{-1})^{r_i} g_i(R_i)$, where $g_i(R_i)$ is the prior pdf of R_i. By choosing the natural conjugate prior pdf of R_i [Springer and Thompson (1967)] given by

$$g_i(R_i) = \frac{(v_{i0}/t_i + 1)^{r_{i0}+1}}{\Gamma(r_{i0}+1)} R_i^{v_{i0}/t_i}(\ln R_i^{-1})^{r_{i0}}, \qquad 0 < R_i < 1, \tag{11.18}$$

the posterior pdf of R_i becomes

$$g_i(R_i|v_i) = \frac{(\beta_i + 1)^{\alpha_i+1}}{\Gamma(\alpha_i+1)} R_i^{\beta_i}(\ln R_i^{-1})^{\alpha_i}, \qquad 0 < R_i < 1, \tag{11.19}$$

where $\beta_i = (v_i + v_{i0})/t_i$ and $\alpha_i = r_i + r_{i0}$.

Since $(-\ln R_i)$ has a $\mathcal{G}_1(r_{i0}+1, v_{i0}/t_i + 1)$ distribution, (11.18) is observed to be an $\mathcal{NLG}(r_{i0}+1, v_{i0}/t_i+1)$ prior distribution. In addition, v_{i0} may be considered as the total accumulated test time to observe r_{i0} failures in a prior (pseudo) life test. The uniform prior pdf is obtained by taking $v_{i0} = r_{i0} = 0$ (no prior test data are available). Alternately, by taking $v_{i0}/t_i = r_{i0} = -1$, (11.18) reduces to the noninformative prior pdf, $g_i(R_i) = R_i^{-1}(\ln R_i^{-1})^{-1}$, which has been used by Burnett and Wales (1961).

Frequently a $\mathcal{G}(a_{i0}, b_{i0})$ pdf is assigned to the constant failure rate λ_i. Suppose the prior pdf of λ_i is of the form

$$g_{\lambda_i}(\lambda_i) = \frac{1}{b_{i0}^{a_{i0}}\Gamma(a_{i0})} \lambda_i^{a_{i0}-1} \exp\left(\frac{-\lambda_i}{b_{i0}}\right), \qquad \lambda_i > 0.$$

The derived prior pdf of $R_i = \exp(-\lambda_i t_i)$ is

$$\begin{aligned} g_{R_i}(R_i) &= g_{\lambda_i}\left(\frac{\ln R_i^{-1}}{t_i}\right)(R_i t_i)^{-1} \\ &= \frac{1}{\Gamma(a_{i0})(t_i b_{i0})^{a_{i0}}} R_i^{(1/t_i b_{i0})-1}(\ln R_i^{-1})^{a_{i0}-1}, \qquad 0 < R_i < 1, \end{aligned}$$

which is the same prior as given by (11.18) with $v_{i0} = (1/b_{i0}) - t_i$ and $r_{i0} = a_{i0} - 1$. Thus, assigning a $\mathcal{G}(a_{i0}, b_{i0})$ prior pdf to the constant failure rate λ_i is equivalent to assigning an $\mathcal{NLG}[a_{i0}, 1/(t_i b_{i0})]$ prior pdf to the reliability R_i. It should also be noted that the $\mathcal{U}(0,1)$ prior pdf on $R_i(v_{i0}=r_{i0}=0)$ leads to a nonuniform prior pdf on λ_i, namely, $g_{\lambda_i}(\lambda_i)=t_i\exp(-\lambda_i t_i)$. This prior pdf appears somewhat paradoxical as for different mission times one would obtain different prior pdfs for λ_i (on the same component). This situation also occurred in Chapter 8. For further discussion of the matter, see Schafer (1970). The point to be made is that it is judicious to examine the repercussion of the choice of a prior pdf for R_i.

The Mellin transform of the posterior pdf of R_i is, by straightforward integration,

$$M\{g_i(R_i|v_i);u\}=\left(\frac{\beta_i+1}{\beta_i+u}\right)^{\alpha_i+1}, \qquad \mathrm{Re}(u)>-\beta_i. \tag{11.20}$$

The kth moment of the posterior pdf of R_i is, by (11.11),

$$E(R_i^k|v_i)=M\{g_i(R_i|v_i);u=k+1\}=\left(\frac{\beta_i+1}{\beta_i+1+k}\right)^{\alpha_i+1}. \tag{11.21}$$

11.2.2 System Level Prior

We now consider assignment of the prior pdf to system reliability. By using the Mellin transform technique, specification of component prior pdf's can often lead to the determination of the corresponding system prior pdf. On the other hand, for a system with identical components, specification of the system prior pdf can lead to the determination of the (common) component prior pdf. It is, therefore, often possible to examine the consequences of the choice of the component (system) prior pdf by assessing the merits of the corresponding system (component) prior pdf. Parker (1972) shows that for a series system with k identical pass/fail components: (1) a $\mathcal{U}(0,1)$ prior pdf on component reliability leads to a nonuniform prior pdf on system reliability of the form $g(R)=[\Gamma(k)]^{-1}(\ln R^{-1})^{k-1}$ which has a mean of 2^{-k}, and (2) a $\mathcal{U}(0,1)$ prior pdf on system reliability produces a nonuniform prior pdf on component reliability of the form $g^*(R)=[\Gamma(1/k)]^{-1}(\ln R^{-1})^{-(k-1)/k}$ which has a mean of $2^{-1/k}$. The low average system reliability in (1) may initially be disturbing unless viewed in its proper perspective; namely, that under a uniform prior pdf on component reliability (no prior test data), a component is, on the average, as likely to fail as not.

In the event that only system level test data are available, by using Bayes' theorem we can determine the posterior pdf of system reliability. In this case, the system is viewed essentially as a single component. On occasion, test data may have been gathered both at the component level of a multicomponent system and at the system level. Depending on the system configuration, a system test resulting in a success does not necessarily imply that all components in the system were successful, nor does a system test failure necessarily imply that all components have failed. In other words, system test data usually provide no information on component performances. Mastran (1976) and Mastran and Singpurwalla (1978) develop a Bayesian procedure that allows for the reliability assessment of a coherent system using both component and system test data. The steps of this procedure are outlined below.

1. First assign a prior pdf to system reliability R, say $g(R)$. By using Bayes' theorem, incorporate the system test data into $g(R)$ and obtain what is called the *preliminary* posterior (or *updated* prior) pdf of R, say $g_p(R|\cdot)$.
2. By using $g_p(R|\cdot)$, the system configuration, and any prior knowledge that is available regarding the component reliabilities R_i, derive either what is called the *consistent* prior pdf of each R_i, say $g_i(R_i)$, or its moments $E(R_i^m)$, whichever is feasible. The consistent prior pdf's are those pdf's that lead to $g_p(R|\cdot)$, given the system configuration.
3. By using the test data on the components of the system, obtain either the posterior pdf of each R_i, say $g_i(R_i|\cdot)$, or its moments $E(R_i^m|\cdot)$, whichever is feasible.
4. By using $g_i(R_i|\cdot)$ or $E(R_i^m|\cdot)$ with the system configuration, derive either the (final) posterior pdf of R, say $g_f(R|\cdot)$, or its moments $E_f(R^m|\cdot)$, whichever is feasible.

The above steps are quite general and can be used with either attribute or variables test data. In Section 11.3 we give an illustrative example of the use of this method for a series system. Dostal and Iannuzzelli (1977, pp. 531–552) propose a technique similar to the above but do not assume that system test data are available. Cole (1975) devises a Bayesian system reliability assessment procedure that combines system and component attribute test data with partial system attributes (simulated tests on systems under environments that would in no way hinder the system's future operation). $\mathcal{B}(x_0, n_0)$ prior pdf's are used on all reliabilities both before and after sampling. Thompson and Haynes (1980) present an excellent survey article on Bayesian interval estimation methods for the reliability, as well as the availability, of various systems.

11.3. SERIES SYSTEMS

For a series system consisting of k independent components, system reliability R is the product of component reliabilities R_i; that is, $\prod_{i=1}^{k} R_i$. We shall discuss the cases in which the components are: (1) each pass/fail, (2) each $\mathcal{E}(\lambda_i)$ or (3) a mixture of pass/fail and $\mathcal{E}(\lambda_i)$. Unless otherwise indicated, the only available test data are on the components; thus, assignment of prior pdf's will be to component reliabilities. By (11.12), the posterior pdf of system reliability has a Mellin transform which is the product of the Mellin transforms of the posterior pdf's of component reliabilities. Once the posterior pdf of system reliability is found, it can be used to determine Bayesian point and interval estimates of system reliability. Numerical examples are given to illustrate the use of the theory.

11.3.1 Pass/Fail Components

Consider a series system consisting of k independent pass/fail components. Suppose n_i units are tested and that s_i survivors are observed on the ith component. Assuming $\mathcal{B}_1(s_{i0}+1, n_{i0}-s_{i0}+1)$ prior pdf's on component reliabilities, the posterior pdf of system reliability, $g(R|\underset{\sim}{s})$, where $\underset{\sim}{s}'=(s_1,\ldots,s_k)$, has Mellin transform [see (11.14)] given by

$$M\{g(R|\underset{\sim}{s});u\} = \prod_{i=1}^{k}\left[\frac{(n_i'+1)!}{s_i'!}\,\frac{1}{(u+s_i')(u+s_i'+1)\cdots(u+n_i')}\right]$$

$$=\left[\prod_{i=1}^{k}\frac{(n_i'+1)!}{s_i'!}\right]\frac{1}{(u+s)^{c_0}(u+s+1)^{c_1}\cdots(u+n)^{c_{n-s}}}, \tag{11.22}$$

where $\mathrm{Re}(u) > -s$, $n=\max(n_i'=n_{i0}+n_i)$, $s=\min(s_i'=s_{i0}+s_i)$, and c_j, $j=0,1,\ldots,n-s$ is the collective exponent for the factor $(u+s+j)^{c_j}$ since $(u+s+j)$ may occur in several (or none) of the component Mellin transforms.

For a squared-error loss function, the Bayes point estimator of system reliability is the mean of the posterior pdf $g(R|\underset{\sim}{s})$:

$$E(R|\underset{\sim}{s}) = M\{g(R|\underset{\sim}{s}); u=2\} = \prod_{i=1}^{k}\left(\frac{s_{i0}+s_i+1}{n_{i0}+n_i+2}\right). \tag{11.23}$$

By expanding (11.22) in partial fractions and then using the Mellin inversion integral, a closed form expression for $g(R|\underset{\sim}{s})$ can be found [Springer and Thompson, (1966b)]. The posterior cdf, $G(R|\underset{\sim}{s})$, can be found by integrating $g(R|\underset{\sim}{s})$ directly. This gives

$$G(R|\underset{\sim}{s}) = R^{s+1}\sum_{j=0}^{n-s} R^j\left[A_{j1}(\ln R^{-1})^0 + \cdots + A_{jc_j}(\ln R^{-1})^{c_j-1}\right], \quad (11.24)$$

where

$$A_{j\ell} = \frac{1}{(\ell-1)!}\sum_{r=0}^{c_j-\ell}\frac{K_{j,\ell+r}}{(s+j+1)^{r+1}}, \qquad \ell = 1,2,\ldots,c_j, \quad (11.25)$$

and

$$K_{jm} = \begin{cases} \dfrac{1}{(c_j-m)!}\dfrac{d^{c_j-m}}{du^{c_j-m}}\left[(u+s+j)^{c_j}M\{g(R|\underset{\sim}{s});u\}\right]\Big|_{u=-(s+j)}, \\ \qquad\qquad m = 1,2,\ldots,c_j \\ 0, \quad c_j = 0. \end{cases} \quad (11.26)$$

A symmetric $100(1-\gamma)\%$ TBPI for system reliability, (R_L, R_U), is determined by solving the equations $G(R_L|\underset{\sim}{s}) = \gamma/2$ for R_L and $G(R_U|\underset{\sim}{s}) = 1-\gamma/2$ for R_U. A $100(1-\gamma)\%$ LBPI for R is given by $(R_\gamma, 1)$, where R_γ is the solution to the equation $G(R_\gamma|\underset{\sim}{s}) = \gamma$.

An alternate way to determine $g(R|\underset{\sim}{s})$ has been given by Wolf (1976). This method does not use Mellin transforms but instead expresses $g(R|\underset{\sim}{s})$ as a $(k-1)$-tuple integral which, at least for small values of k, can be evaluated in a rather straightforward manner [see Exercise 11.8].

Example 11.2. [Springer and Thompson (1966b)] A system consists of three independent pass/fail components in series. Testing on the components produced the following results:

	Successes	Failures	Units Tested
Component 1	8	2	10
Component 2	7	2	9
Component 3	3	1	4

Under the $\mathcal{U}(0,1)$ prior pdf ($n_{i0}=s_{i0}=0, i=1,2,3$) on each component reliability, we have

$$n_1'=n_{10}+n_1=10, \qquad s_1'=s_{10}+s_1=8,$$

$$n_2'=n_{20}+n_2=9, \qquad s_2'=s_{20}+s_2=7,$$

$$n_3'=n_{30}+n_3=4, \qquad s_3'=s_{30}+s_3=3;$$

thus, $n=\max(n_i')=10$ and $s=\min(s_i')=3$.

The posterior pdf's of the component reliabilities R_1, R_2, and R_3 [see (11.13)] are, upon simplification,

$$g_1(R_1|s_1)=495R_1^8(1-R_1)^2, \qquad 0<R_1<1,$$

$$g_2(R_2|s_2)=360R_2^7(1-R_2)^2, \qquad 0<R_2<1,$$

$$g_3(R_3|s_3)=\ 20R_3^3(1-R_3), \qquad 0<R_3<1.$$

The corresponding Mellin transforms, (11.14), are given by

$$M\{g_1(R_1|s_1);u\}=990/(u+8)(u+9)(u+10), \qquad \text{Re}(u)>-8,$$

$$M\{g_2(R_2|s_2);u\}=720/(u+7)(u+8)(u+9), \qquad \text{Re}(u)>-7,$$

$$M\{g_3(R_3|s_3);u\}=\ 20/(u+3)(u+4), \qquad \text{Re}(u)>-3.$$

The Mellin transform of the posterior pdf of system reliability is, by (11.22),

$$M\{g(R|\underset{\sim}{s});u\}=\frac{990\times720\times20}{(u+3)(u+4)(u+7)(u+8)^2(u+9)^2(u+10)},$$

$$\text{Re}(u)>-3,$$

so that $c_0=1$, $c_1=1$, $c_2=0$, $c_3=0$, $c_4=1$, $c_5=2$ $c_6=2$, $c_7=1$. Upon evaluating the K_{jm}'s from (11.26), we can obtain the $A_{j\ell}$'s by (11.25). Consequently, the posterior cdf of R is

$$\begin{aligned} G(R|\underset{\sim}{s})=&(990/7)R^4-396R^5+12375R^8\\ &+(37180+39600\ \ln R)R^9\\ &-(46728-23760\ \ln R)R^{10}-(18000/7)R^{11}. \end{aligned}$$

From (11.23), a Bayes point estimate of system reliability is $E(R|\underset{\sim}{s}) = (9/12)(8/11)(4/6) = 0.364$. By solving the nonlinear equations $G(R_L|\underset{\sim}{s}) = 0.05$ and $G(R_U|\underset{\sim}{s}) = 0.95$, we find $R_L = 0.157$ and $R_U = 0.594$. Thus, a symmetric 90% TBPI for system reliability is (0.157, 0.594).

Example 11.3. [Mastran (1976)] A system consists of three independent pass/fail components in series. Life tests were made on the components as well as the system with the following results:

	Successes	Failures	Units Tested
Component 1	5	1	6
Component 2	6	0	6
Component 3	9	1	10
System	10	2	12

We will determine a Bayes point estimate of system reliability that is based on both component and system test data. It is also of interest to compare this result with the corresponding results using only the system test data or the component test data. A squared-error loss function is used to measure the loss incurred by using an estimator $\hat{R}$ to estimate system reliability R. For the squared-error loss function, the mean and variance of the posterior pdf of R are the Bayes point estimator of R and the corresponding posterior risk $(a=1)$.

Under the $\mathcal{U}(0,1)$ prior pdf $(n_0 = s_0 = 0)$ on system reliability in conjunction with the system test data $(n=12,\ s=10)$, we have

$$n' = n_0 + n = 12, \qquad s' = s_0 + s = 10.$$

By (11.13), the preliminary posterior pdf of system reliability R is

$$g_p(R|s) = 858R^{10}(1-R)^2, \qquad 0 < R < 1.$$

The mth moment of $g_p(R|s)$ is, by (11.15),

$$E(R^m|s) = 1716/(m+13)(m+12)(m+11).$$

In particular, $E(R|s) = 0.7857$ and $E(R^2|s) = 0.6286$; thus, $\mathrm{Var}(R|s) = 0.0112$. If only the system test data are used, the Bayes point estimate of system reliability is 0.7857 with a posterior risk of 0.0112.

We forego the difficult problem of determining the forms of the prior pdf's of component reliabilities, say $g_i(R_i)$, $i = 1,2,3$ which are consistent with $g_p(R|s)$. Instead, we find the mth moments of $g_i(R_i)$, say $E(R_i^m)$, which are related to the mth moment of $g_p(R|s)$ by the equation $E(R^m|s) = \prod_{i=1}^{3} E(R_i^m)$. To simplify the example we assume that $E(R_1^m) = E(R_2^m) = E(R_3^m)$. This assumption, as noted by Mastran, is not restrictive and does not imply that the components themselves are identical. Thus,

$$E(R_i^m) = [E(R^m|s)]^{1/3} = [1716/(m+13)(m+12)(m+11)]^{1/3}, \qquad i=1,2,3. \quad (11.27)$$

The next step is to incorporate the component test data into the procedure in order to obtain the mth moments of the posterior pdfs of component reliabilities, $E(R_i^m|s, s_i)$, $i=1,2,3$. The following general relationship will be used:

$$E(R_i^m|s, s_i) = \frac{\sum_{j=0}^{f_i} \binom{f_i}{j} (-1)^j E(R_i^{s_i+m+j})}{\sum_{j=0}^{f_i} \binom{f_i}{j} (-1)^j E(R_i^{s_i+j})}, \quad (11.28)$$

where s_i and f_i are the number of observed successes and failures among the n_i units tested for the ith component. Using the component test data in (11.28), we find

$$E(R_1|s, s_1) = \frac{E(R_1^6) - E(R_1^7)}{E(R_1^5) - E(R_1^6)} = 0.8944,$$

$$E(R_1^2|s, s_1) = \frac{E(R_1^7) - E(R_1^8)}{E(R_1^5) - E(R_1^6)} = 0.8047,$$

$$E(R_2|s, s_2) = \frac{E(R_2^7)}{E(R_2^6)} = 0.9473,$$

$$E(R_2^2|s, s_2) = \frac{E(R_2^8)}{E(R_2^6)} = 0.8998,$$

$$E(R_3|s, s_3) = \frac{E(R_3^{10}) - E(R_3^{11})}{E(R_3^9) - E(R_3^{10})} = 0.9129,$$

$$E(R_3^2|s, s_3) = \frac{E(R_3^{11}) - E(R_3^{12})}{E(R_3^9) - E(R_3^{10})} = 0.8367.$$

It follows that the mean of the (final) posterior pdf of R is

$$E(R|s, s_1, s_2, s_3) = \prod_{i=1}^{3} E(R_i|s, s_i) = (0.8944)(0.9473)(0.9129) = 0.7734;$$

the second moment is

$$E(R^2|s, s_1, s_2, s_3) = \prod_{i=1}^{3} E(R_i^2|s, s_i) = (0.8047)(0.8998)(0.8367) = 0.6058,$$

and $\mathrm{Var}(R|s, s_1, s_2, s_3) = 0.0077$. Using both component and system test data, the Bayes point estimate of system reliability is 0.7734 with a posterior risk of 0.0077.

If the system test data are disregarded, the preliminary posterior pdf of system reliability is $g_p(R) = 1, 0 < R < 1$. The mth moment of $g_p(R)$ is $E(R^m) = 1/(m+1)$. Following the same steps as before, we find that if only component test data are used, the Bayes point estimate of system reliability is 0.6993 with a posterior risk of 0.0198. A summary of the overall results is given below. The benefits of using both sets of test data are realized by the lower posterior risk achieved.

Data Used	$\hat{R}$	Posterior Risk
System test data only	0.7857	0.0112
Component test data only	0.6993	0.0198
Both system and component test data	0.7734	0.0077

■

11.3.2 Exponential Components

Consider a series system consisting of k independent components, each of which has an $\mathcal{E}(\lambda_i)$ failure time distribution. Suppose that n_i units are placed on a life test for the ith component, and the test is terminated at the time of the r_ith failure. The observed total time on test is

$$v_i = \sum_{j=1}^{r_i} x_{ij} + (n_i - r_i)x_{ir_i},$$

where x_{ij} is the jth ordered failure time. Assuming $\mathfrak{NLG}(r_{i0}+1, v_{i0}/t+1)$ prior pdf's on component reliabilities at time t, the posterior pdf of system reliability at time t, $g(R|\underset{\sim}{v})$, where $\underset{\sim}{v}' = (v_1, \ldots, v_k)$, has Mellin transform [see (11.20)] given by

$$M\{g(R|\underset{\sim}{v}); u\} = \prod_{i=1}^{k} \left(\frac{\beta_i + 1}{\beta_i + u} \right)^{\alpha_i + 1}, \qquad \text{Re}(u) > -\min(\beta_i), \quad (11.29)$$

where $\beta_i = (v_i + v_{i0})/t$ and $\alpha_i = r_i + r_{i0}$.

For a squared-error loss function, the Bayes point estimator of system reliability at time t is the mean of the posterior pdf $g(R|\underset{\sim}{v})$:

$$E(R|\underset{\sim}{v}) = M\{g(R|\underset{\sim}{v}); u = 2\} = \prod_{i=1}^{k} \left(\frac{\beta_i + 1}{\beta_i + 2} \right)^{\alpha_i + 1}. \quad (11.30)$$

From an earlier discussion, (11.30) may also be obtained by assigning the $\mathcal{G}[r_{i0} + 1, 1/(v_{i0} + t)]$ prior pdf to the constant failure rate λ_i [Zacks (1977), pp. 55–74].

By expanding (11.29) in partial fractions and then using the Mellin inversion integral, a closed form expression for $g(R|\underset{\sim}{v})$ can be obtained [(Springer and Thompson (1967)]. The posterior cdf $G(R|\underset{\sim}{v})$ can be determined by integrating $g(R|\underset{\sim}{v})$ directly, in which case the result is

$$G(R|\underset{\sim}{v}) = \sum_{i=1}^{k} R^{\beta_i + 1} \left[A_{i\alpha_i} (\ln R^{-1})^0 + \cdots + A_{i0} (\ln R^{-1})^{\alpha_i} \right], \quad (11.31)$$

where

$$A_{ij} = \frac{1}{(\alpha_i - j)!} \sum_{r=0}^{j} \frac{K_{i, j-r}}{(\beta_i + 1)^{r+1}}, \qquad j = 0, 1, \ldots, \alpha_i, \quad (11.32)$$

and

$$K_{i\ell} = \frac{1}{\ell!} \frac{d^\ell}{du^\ell} \left[(u + \beta_i)^{\alpha_i + 1} M\{g(R|\underset{\sim}{v}); u\} \right] \Big|_{u = -\beta_i},$$

$$\ell = 0, 1, \ldots, \alpha_i. \quad (11.33)$$

A symmetric $100(1-\gamma)\%$ TBPI for system reliability, (R_L, R_U), is determined by solving the equations $G(R_L|\underset{\sim}{v}) = \gamma/2$ for R_L and $G(R_U|\underset{\sim}{v}) = 1 - \gamma/2$ for R_U. A $100(1-\gamma)\%$ LBPI for R is $(R_\gamma, 1)$, where R_γ is the solution to the equation $G(R_\gamma|\underset{\sim}{v}) = \gamma$. Using Monte Carlo techniques, Berkbigler and Byers (1975) note that for a variety of situations, the value of

R_γ based on noninformative prior pdf's on component reliabilities is consistently larger than that based on $\mathcal{U}(0,1)$ prior pdf's. This occurrence is also reported by Mann (1970) who, in addition, finds the value of R_γ based on noninformative prior pdf's on the R_i's to be conservative in the classical sense. The question as to whether there exist prior pdf's on the R_i's that yield the classical lower confidence bound on R is considered by Fertig (1972). An optimum property of these prior pdf's would be that in the absence of previous information, they yield uniformly most accurate unbiased classical confidence intervals for R. It is shown that no such set of prior pdf's exist that are independent of the current test data.

Example 11.4. [Springer and Thompson (1967)] A system consisting of three independent $\mathcal{E}(\lambda_i)$ components in series has a mission time of $t = 20$ h. Life tests on the components produced the following results:

	Units Tested	Test Terminated at rth Failure	Failure Times (in h)
Component 1	10	$r = 3$	20, 32, 41
Component 2	8	$r = 2$	60, 80
Component 3	5	$r = 1$	44

Based on the above data, we will determine a Bayes point estimate and a 90% LBPI for system reliability at $t = 20$ h using a noniformative prior pdf $(v_{i0}/t = r_{i0} = -1, i = 1,2,3)$ on each component reliability. As

$$\alpha_1 = r_1 + r_{10} = 2, \quad \beta_1 = (v_1 + v_{10})/t = 18,$$
$$\alpha_2 = r_2 + r_{20} = 1, \quad \beta_2 = (v_2 + v_{20})/t = 30,$$
$$\alpha_3 = r_3 + r_{30} = 0, \quad \beta_3 = (v_3 + v_{30})/t = 10,$$

the posterior pdf's of component reliabilities R_1, R_2, and R_3 [see (11.19)] are

$$g_1(R_1|v_1) = (19^3/2)R_1^{18}(\ln R_1^{-1})^2, \qquad 0 < R_1 < 1,$$
$$g_2(R_2|v_2) = 31^2 R_2^{30}(\ln R_2^{-1}), \qquad 0 < R_2 < 1,$$
$$g_3(R_3|v_3) = 11 R_3^{10}, \qquad 0 < R_3 < 1.$$

The corresponding Mellin transforms, (11.20), are given by

$$M\{g_1(R_1|v_1);u\} = 19^3/(u+18)^3, \qquad \text{Re}(u) > -18,$$
$$M\{g_2(R_2|v_2);u\} = 31^2/(u+30)^2, \qquad \text{Re}(u) > -30,$$
$$M\{g_3(R_3|v_3);u\} = 11/(u+10), \qquad \text{Re}(u) > -10.$$

The Mellin transform of the posterior pdf of system reliability at $t = 20$ h is, by (11.29),

$$M\{g(R|\underset{\sim}{v}); u\} \frac{19^3 \times 31^2 \times 11}{(u+18)^3(u+30)^2(u+10)}, \qquad \text{Re}(u) > -10.$$

Upon evaluating the $K_{i\ell}$'s given by (11.33), one can obtain the A_{ij}'s from (11.32). The posterior cdf of R is then found to be

$$\begin{aligned} G(R|\underset{\sim}{v}) = \big[&-53.6713 + 36.3225 \ln R \\ &-1656.3069(\ln R)^2\big] R^{19} \\ &+(22.4862 - 67.6771 \ln R) R^{31} \\ &+32.1851\, R^{11}. \end{aligned}$$

From (11.30), a Bayes point estimate of system reliability at $t = 20$ h is $E(R|\underset{\sim}{v}) = (19/20)^3(31/32)^2(11/12) = 0.738$. Solving the nonlinear equation $G(R_{0.10}|\underset{\sim}{v}) = 0.10$ yields $R_{0.10} = 0.610$; thus a 90% LBPI for R is (0.610, 1). ■

Mastran (1976) has given an example that is similar to Example 11.3 but considers independent $\mathcal{E}(\lambda_i)$ components in series. System reliability assessment is made based on both component and system life test data. A slightly different approach to determining an LBPI for system reliability R is considered by Zacks (1977, pp. 55–74) using natural conjugate $\mathcal{G}(\alpha_0, \beta_0)$ prior pdf's on component failure rates λ_i. Since system reliability, $R = \exp[-(\sum_{i=1}^{k} \lambda_i)t]$, is a decreasing function in $\sum_{i=1}^{k} \lambda_i$ for fixed t, a simple transformation of the $100(1-\gamma)$th percentile of the posterior pdf of $\sum_{i=1}^{k} \lambda_i$ gives a $100(1-\gamma)\%$ LBPI for R. The distribution of the sum of independent $\mathcal{G}(\alpha_0, \beta_0)$ r.v.'s having different scale parameters can be represented by a mixture of $\mathcal{G}(\cdot,\cdot)$ distributions with negative binomial mixing probabilities [Neuts and Zacks (1967)]. Unfortunately, even for the case of $k = 2$ components, computational difficulties arise. Zacks (1977, pp. 67–69) proposes a conservative solution; in addition, an approximation is given when the number of units tested (n_i) for each component is large.

11.3.3 Mixture of Pass/Fail and Exponential Components

For a series system consisting of k independent components of which k_1 are $\mathcal{E}(\lambda_i)$ and $k_2 = k - k_1$ are pass/fail, the methods of the two previous subsections may be jointly applied to determine Bayes point and interval estimates of system reliability. By combining (11.22) and (11.29), the Mellin transform of the posterior pdf of system reliability at time t, $g(R|\underset{\sim}{s}, \underset{\sim}{v})$, is given by

$$M\{g(R|\underset{\sim}{s}, \underset{\sim}{v}); u\} = \left[\prod_{i=1}^{k_1}\left(\frac{\beta_i+1}{\beta_i+u}\right)^{\alpha_i+1}\right]\left[\prod_{j=1}^{k_2}\frac{(n'_j+1)!}{s'_j!}\right]$$

$$\times\frac{1}{(u+s)^{c_0}(u+s+1)^{c_1}\cdots(u+n)^{c_{n-s}}}, \quad (11.34)$$

where $\text{Re}(u) > -\min[s, \min(\beta_i)]$. Each symbol has been defined previously in the appropriate subsection.

For a squared-error loss function, the Bayes point estimator of system reliability at time t is

$$E(R|\underset{\sim}{s}, \underset{\sim}{v}) = M\{g(R|\underset{\sim}{s}, \underset{\sim}{v}); u=2\} = \prod_{i=1}^{k_1}\left(\frac{\beta_i+1}{\beta_i+2}\right)^{\alpha_i+1} \times \prod_{j=1}^{k_2}\left(\frac{s'_j+1}{n'_j+2}\right). \quad (11.35)$$

The posterior pdf $g(R|\underset{\sim}{s}, \underset{\sim}{v})$ can be determined by evaluating the corresponding Mellin inversion integral using the method of residues. The result is [Springer and Byers (1971)]

$$g(R|\underset{\sim}{s}, \underset{\sim}{v}) = \sum_{i=1}^{k_1} h(\beta_i, \alpha_i+1) + \sum_{m=0}^{n-s} h(m+s, c_m), \quad (11.36)$$

where

$$h(p, q+1) = \frac{1}{q!}\frac{d^q}{du^q}\left[R^{-u}(u+p)^{q+1}M\{g(R|\underset{\sim}{s}, \underset{\sim}{v}); u\}\right]\Big|_{u=-p}. \quad (11.37)$$

The function h denotes the residues at the relevant poles: the pole at $u = -\beta_i$ for the $\mathcal{E}(\lambda_i)$ components is of order $\alpha_i + 1$, whereas the pole at $u = -(m+s)$ for the pass/fail components is of order c_m, $m = 0, 1, \ldots, n-s$. Certain modifications are required on the order of the poles when the values of β_i are not distinct or when there is duplication of the values of β_i with those of $s, s+1, \ldots, n$. If $\beta_i = \beta_r$, $i \neq r$, $i \in \{1, 2, \ldots, k_1\}$, the pole at $u = -\beta_i$ is of order $\alpha_i + \alpha_r + 2$. If $\beta_i = m+s$, $i \in \{1, 2, \ldots, k_1\}$, the order of the pole at $u = -\beta_i$ is $\alpha_i + c_m + 1$. The evaluation of (11.37) can be quite cumbersome;

consequently, Springer and Byers (1971) give expressions from which h may be determined recursively in terms of lower order derivatives.

The posterior cdf $G(R|\underset{\sim}{s}, \underset{\sim}{v})$ can be obtained by integrating $g(R|\underset{\sim}{s}, \underset{\sim}{v})$. A symmetric $100(1-\gamma)\%$ TBPI for system reliability, (R_L, R_U), is determined by solving equations $G(R_L|\underset{\sim}{s}, \underset{\sim}{v})=\gamma/2$ for R_L and $G(R_U|\underset{\sim}{s}, \underset{\sim}{v})=1-\gamma/2$ for R_U. A $100(1-\gamma)\%$ LBPI for R is $(R_\gamma, 1)$, where R_γ is the solution to the equation $G(R_\gamma|\underset{\sim}{s}, \underset{\sim}{v})=\gamma$.

Example 11.5. [Springer and Byers (1971)] A series system consists of six independent components, three of which are $\mathcal{E}(\lambda_i)$ and three of which are pass/fail. Assume that test data and prior information yield the following parameter values:

$\mathcal{E}(\lambda_i)$ *components:*

$$\alpha_1=0, \qquad \beta_1=90,$$

$$\alpha_2=1, \qquad \beta_2=90,$$

$$\alpha_3=1, \qquad \beta_3=87.$$

Pass/fail components:

$$n_1'=60, \qquad s_1'=58,$$

$$n_2'=61, \qquad s_2'=60,$$

$$n_3'=58, \qquad s_3'=57.$$

From (11.35), a Bayes point estimate of system reliability is $E(R|\underset{\sim}{s}, \underset{\sim}{v})=(91/92)^3(88/89)^2(59/62)(61/63)(58/60)=0.843$. From (11.36), the posterior pdf of system reliability is

$$g(R|\underset{\sim}{s}, \underset{\sim}{v})=h(90,3)+h(87,2)+h(57,1)+h(58,2)$$

$$+h(59,1)+h(60,2)+h(61,1)$$

$$=\sum_{i=1}^{12} a_i R^{b_i}(\ln R^{-1})^{d_i},$$

where the constant a_i, b_i, and d_i are tabled in Springer and Byers (1971). By

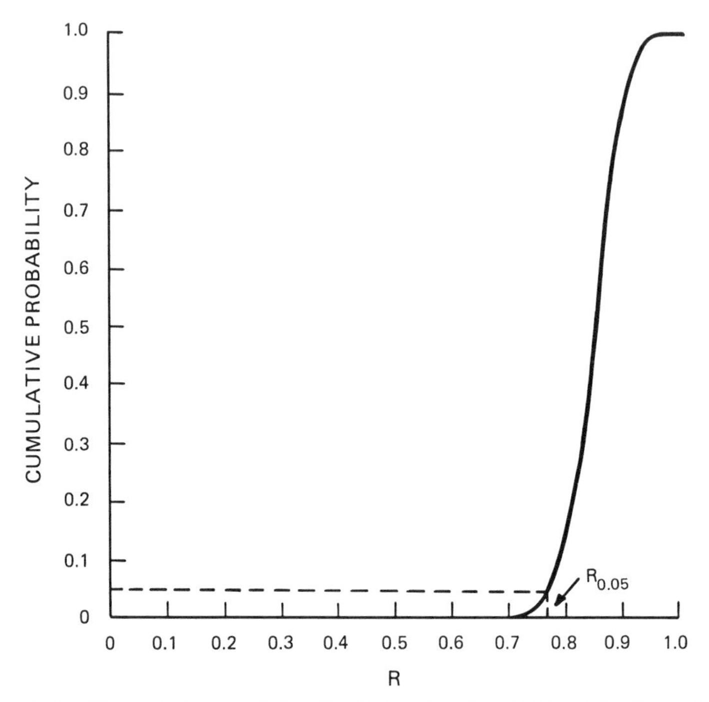

Figure 11.2. The posterior cumulative distribution function $G(R|\underset{\sim}{s}, \underset{\sim}{v})$ for Example 11.5.

using integration by parts, the posterior cdf is

$$G(R|\underset{\sim}{s}, \underset{\sim}{v}) = \int_0^R g(t|\underset{\sim}{s}, \underset{\sim}{v})\, dt$$

$$= \sum_{i=1}^{12} \frac{a_i d_i!}{(b_i+1)^{d_i+1}} \left(\sum_{j=0}^{d_i} \frac{\phi^j e^{-\phi}}{j!} \right),$$

where $\phi = (b_i + 1)(\ln R^{-1})$. The expression for $G(R|\underset{\sim}{s}, \underset{\sim}{v})$ has been evaluated for $R = 0.50(0.50)0.95$. Based on these tabulations, the graph of $G(R|\underset{\sim}{s}, \underset{\sim}{v})$ is shown in Figure 11.2. From the graph, a 95% LBPI for system reliability is $(.76, 1)$. ■

11.4. PARALLEL SYSTEMS

For a parallel system consisting of k independent components, system reliability R is given by $R = 1 - \prod_{i=1}^{k}(1 - R_i)$, where R_i is the reliability of the ith component. Equivalently, system unreliability $Q=1-R$ is the product of component unreliabilities $Q_i=1-R_i$, (that is, $Q=\prod_{i=1}^{k}Q_i$). We shall consider the cases in which the components are: (1) each pass/fail or (2) each $\mathcal{E}(\lambda_i)$. In cases where a parallel system is used as an active redundant system, it seems unlikely that a mixture of pass/fail and $\mathcal{E}(\lambda_i)$ components in parallel would occur in practice. Based on test data available only on the components, Bayesian point and interval estimates of system reliability are given. An example of a Bayesian assessment of a 2-dependent-component parallel system using component and system test data is given by Mastran and Singpurwalla (1978).

11.4.1 Pass/Fail Components

Consider a parallel system consisting of k independent pass/fail components. In this case, the use of the Mellin transform technique in deriving the posterior pdf of parallel system unreliability is entirely analogous to its use, as discussed in Section 11.3.1, in deriving the posterior pdf of series system reliability. Assuming $\mathcal{B}_1(n_{i0}-s_{i0}+1, s_{i0}+1)$ prior pdf's on component unreliabilities, it follows that the Mellin transform of the posterior pdf of system unreliability, say $\bar{g}(Q|\underset{\sim}{s})$, is [see (11.22)]

$$M\{\bar{g}(Q|\underset{\sim}{s});u\}=\left[\prod_{i=1}^{k}\frac{(n_i'+1)!}{f_i'!}\right]\frac{1}{(u+n)^{d_n}(u+n-1)^{d_{n-1}}\cdots(u+f)^{d_f}},\tag{11.38}$$

where $\operatorname{Re}(u)>-f$, $f=\min(f_i'=n_i'-s_i')$, and the exponents $d_n, d_{n-1},\ldots,d_f$ are the multiplicities of the factors that arise due to taking products of component Mellin transforms. Some of the d_i's might be zero.

Under a squared-error loss function, the Bayes point estimator of system reliability is

$$\begin{aligned}E(R|\underset{\sim}{s})&=1-E(Q|\underset{\sim}{s})=1-M\{\bar{g}(Q|\underset{\sim}{s});u=2\}\\&=1-\prod_{i=1}^{k}\left(\frac{f_i'+1}{n_i'+2}\right)=1-\prod_{i=1}^{k}\left(1-\frac{s_i'+1}{n_i'+2}\right).\end{aligned}\tag{11.39}$$

The posterior pdf of system unreliability, obtained by evaluating the Mellin inversion integral by the method of residues, is [Smith and Springer (1976)]

$$\bar{g}(Q|\underset{\sim}{s})=\left[\prod_{i=1}^{k}\frac{(n_i'+1)!}{f_i'!}\right]Q^f\sum_{\ell=f}^{n}\sum_{j=0}^{d_\ell-1}\frac{c_{\ell j}Q^{n-\ell}(\ln Q^{-1})^{d_\ell-1-j}}{(d_\ell-1-j)!j!},\tag{11.40}$$

where

$$c_{\ell 0} = \sum_{\substack{q=f \\ q \neq \ell}}^{n} (\ell - q)^{-d_q}, \qquad j=0, \tag{11.41}$$

$$c_{\ell j} = \sum_{r=0}^{j-1} \sum_{\substack{q=f \\ q \neq \ell}}^{n} \binom{j-1}{r} \frac{r!d_q}{(\ell - q)^{r+1}} c_{\ell, j-1-r}, \qquad j>0. \tag{11.42}$$

It follows that the posterior pdf of system reliability is $g(R|\underset{\sim}{s}) = \bar{g}(1-R|\underset{\sim}{s})$. The posterior cdf of system reliability is determined by

$$G(R|\underset{\sim}{s}) = \int_0^R g(t|\underset{\sim}{s})\,dt = \int_0^R \bar{g}(1-t|\underset{\sim}{s})\,dt = 1 - \int_0^{1-R} \bar{g}(Q|\underset{\sim}{s})\,dQ. \tag{11.43}$$

A symmetric $100(1-\gamma)\%$ TBPI for system reliability, (R_L, R_U), is determined by solving the equations $G(R_L|\underset{\sim}{s}) = \gamma/2$ for R_L and $G(R_U|\underset{\sim}{s}) = 1-\gamma/2$ for R_U. A $100(1-\gamma)\%$ LBPI for R is given by $(R_\gamma, 1)$, where R_γ is the solution to the equation $G(R_\gamma|\underset{\sim}{s}) = \gamma$.

For a parallel system consisting of a large number of pass/fail components and/or large values of the attribute test data n_i and s_i, computational difficulties may arise in the evaluation of $G(R|\underset{\sim}{s})$. Smith (1977, pp. 93–106) develops an approximate method that circumvents inverting the Mellin transform of system unreliability. From the Mellin transform of $\bar{g}(Q|\underset{\sim}{s})$ given by (11.38), compute $M\{\bar{g}(Q|\underset{\sim}{s}); u=2\} = E(Q|\underset{\sim}{s})$ and $M\{\bar{g}(Q|\underset{\sim}{s}); u=3\} = E(Q^2|\underset{\sim}{s})$, (i.e., the first two moments of the posterior pdf of system unreliability Q). The corresponding moments of system reliability R are then determined from the equations $E(R|\underset{\sim}{s}) = 1 - E(Q|\underset{\sim}{s})$ and $E(R^2|\underset{\sim}{s}) = 1 - 2E(Q|\underset{\sim}{s}) + E(Q^2|\underset{\sim}{s})$. The posterior pdf of system reliability, $g(R|\underset{\sim}{s})$, is then approximated by a $\mathcal{B}_1(a, b)$ distribution having the same first two moments as that of $g(R|\underset{\sim}{s})$. The parameters a and b are thus given by

$$b = [1 - E(R|\underset{\sim}{s})]\{E(R|\underset{\sim}{s})[1 - E(R|\underset{\sim}{s})]/\sigma^2_{R|\underset{\sim}{s}} - 1\} \tag{11.44}$$

and

$$a = bE(R|\underset{\sim}{s})/[1 - E(R|\underset{\sim}{s})], \tag{11.45}$$

where $\sigma^2_{R|\underset{\sim}{s}} = E(R^2|\underset{\sim}{s}) - [E(R|\underset{\sim}{s})]^2$. The approximate posterior cdf of system reliability is, therefore,

$$G^*(R|\underset{\sim}{s}) = \int_0^R \frac{\Gamma(a+b)}{\Gamma(a)\Gamma(b)} t^{a-1}(1-t)^{b-1}\,dt, \tag{11.46}$$

which may be determined by numerical integration.

Example 11.6. [Smith (1977), pp. 93–106] A system consists of two independent pass/fail components in parallel. Four units of the first component were tested of which two units were successful. Three units of the second component were tested of which two units were successful. Suppose the parameters of the $\mathscr{B}_1(n_{i0} - s_{i0} + 1, s_{i0} + 1)$ prior pdf's are $n_{10} = 2$, $s_{10} = 0$, $n_{20} = 1$, $s_{20} = 0$. Then

$$\begin{aligned} n_1' &= n_{10} + n_1 = 6, & n_2' &= n_{20} + n_2 = 4, \\ s_1' &= s_{10} + s_1 = 2, & s_2' &= s_{20} + s_2 = 2, \\ f_1' &= n_1' - s_1' = 4, & f_2' &= n_2' - s_2' = 2, \\ n &= \max(n_1', n_2') = 6, & f &= \min(f_1', f_2') = 2. \end{aligned}$$

The Mellin transform of the posterior pdf of system unreliability is, by (11.38),

$$M\{\bar{g}(Q|\underset{\sim}{s}); u\} = \frac{12600}{(u+6)(u+5)(u+4)^2(u+3)(u+2)}, \qquad \operatorname{Re}(u) > -2.$$

The exponents are $d_6 = 1$, $d_5 = 1$, $d_4 = 2$, $d_3 = 1$, $d_2 = 1$. By (11.40), the posterior pdf of system unreliability is

$$\bar{g}(Q|\underset{\sim}{s}) = 12600Q^2\left[c_{20}Q^4 + c_{30}Q^3 + c_{40}Q^2(\ln Q^{-1}) + c_{41}Q^2 + c_{50}Q + c_{60}\right],$$

where

$$c_{20} = \prod_{q=3}^{6} (2-q)^{-d_q} = -\tfrac{1}{48}, \qquad c_{30} = \prod_{\substack{q=2\\ q\neq 3}}^{6} (3-q)^{-d_q} = \tfrac{1}{6},$$

$$c_{40} = \prod_{\substack{q=2\\ q\neq 4}}^{6} (4-q)^{-d_q} = \tfrac{1}{4}, \qquad c_{50} = \prod_{\substack{q=2\\ q\neq 5}}^{6} (5-q)^{-d_q} = -\tfrac{1}{6},$$

$$c_{60} = \prod_{q=2}^{5} (6-q)^{-d_q} = \tfrac{1}{48}, \qquad c_{41} = \sum_{\substack{q=2\\ q\neq 4}}^{6} \frac{d_q}{(4-q)} c_{40} = 0;$$

thus,

$$\bar{g}(Q|\underset{\sim}{s}) = -262.5Q^6 + 2100Q^5 - 3150(\ln Q^{-1})Q^4$$

$$-2100Q^3 + 262.5Q^2.$$

The posterior cdf of system reliability is determined by $G(R|\underset{\sim}{s}) = 1 - \int_0^{1-R} \bar{g}(Q|\underset{\sim}{s})\,dQ$ and is tabulated below for various values of R.

For the $\mathcal{B}_1(a, b)$ approximation, the first two moments of system unreliability are found to be

$$E(Q|\underset{\sim}{s}) = M\{\bar{g}(Q|\underset{\sim}{s}); u=2\} = 0.3125,$$

$$E(Q^2|\underset{\sim}{s}) = M\{\bar{g}(Q|\underset{\sim}{s}); u=3\} = 0.1190,$$

from which $E(R|\underset{\sim}{s}) = 0.6875$, $E(R^2|\underset{\sim}{s}) = 0.4940$, and $\sigma^2_{R|\underset{\sim}{s}} = 0.0213$. By (11.44) and (11.45), $a = 6.21739$ and $b = 2.82608$. The tabulation of the approximate posterior cdf of system reliability, $G^*(R|\underset{\sim}{s})$, is shown below adjacent to the exact result. The $\mathcal{B}_1(a, b)$ approximation appears to be quite satisfactory.

From (11.39), a Bayes point estimate of system reliability is $E(R|\underset{\sim}{s}) = 1 - (5/8)(3/6) = 0.688$. From computer tabulations, a 95% LBPI for system reliability is (0.425, 1), for the exact form, and (0.424, 1) for the $\mathcal{B}_1(a, b)$ approximation.

R	$G(R\|\underset{\sim}{s})$ (exact form)	$G^*(R\|\underset{\sim}{s})$ (approximation)
0.10	0.000015	0.000012
0.20	0.000843	0.000760
0.30	0.008111	0.007744
0.40	0.037665	0.037056
0.50	0.111538	0.115214
0.60	0.266425	0.267462
0.70	0.493709	0.495341
0.80	0.753297	0.753376
0.90	0.958887	0.947396
0.95	0.991709	0.990941
0.99	0.999917	0.999889

This table adapted, with permission, from Smith (1977).

■

11.4.2 Exponential Components

Consider a parallel system consisting of k independent components, each of which has an $\mathcal{E}(\lambda_i)$ failure time distribution. Suppose that n_i units are placed on a life test for the ith component, and the test is terminated at the time of the r_ith failure. The corresponding observed total time on test is v_i. We assume the $\mathcal{NLG}(r_{i0}+1, v_{i0}/t+1)$ prior pdf on R_i, the ith component reliability at time t. The posterior pdf of unreliability, $Q_i = 1 - R_i$, is after a simple change of variable in (11.19), given by

$$\bar{g}(Q_i|v_i) = \frac{(\beta_i+1)^{\alpha_i+1}}{\Gamma(\alpha_i+1)}(1-Q_i)^{\beta_i}\left[\ln(1-Q_i)^{-1}\right]^{\alpha_i}, \qquad 0<Q_i<1, \quad (11.47)$$

where $\beta_i = (v_i + v_{i0})/t$ and $\alpha_i = r_i + r_{i0}$.

The mth moment of $\bar{g}(Q_i|v_i)$ is

$$E(Q_i^m|v_i) = \sum_{j=0}^{m}(-1)^j\binom{m}{j}\left(\frac{\beta_i+1}{j+\beta_i+1}\right)^{\alpha_i+1}. \qquad (11.48)$$

Since $E(Q_i|v_i) = 1 - [(\beta_i+1)/(\beta_i+2)]^{\alpha_i+1}$, the Bayes point estimator of system reliability under a squared-error loss function is

$$E(R|\underset{\sim}{v}) = 1 - \prod_{i=1}^{k}\left[1 - \left(\frac{\beta_i+1}{\beta_i+2}\right)^{\alpha_i+1}\right]. \qquad (11.49)$$

The Mellin transform of the posterior pdf of system unreliability is the product of the Mellin transforms of $\bar{g}(Q_i|v_i)$, $i=1,2,\ldots,k$, which do not appear to be simple functions. Consequently, evaluation of the Mellin inversion integral has been solved only in the special case when $r_i = 1$ and $r_{i0}=0$, $i=1,2,\ldots,k$; that is, each life test is terminated at the first failure [Springer and Thompson (1968)]. An alternate approach is given by Thompson and Chang (1975) which avoids the use of Mellin transforms. Instead, the posterior cdf of system reliability, $G(R|\underset{\sim}{v})$, is written in terms of shifted Chebyshev polynomials of the second kind. The results are

$$G(R|\underset{\sim}{v}) = R + \left(\frac{8}{\pi}\right)\sqrt{R(1-R)}\times$$

$$\left[a_0 + a_1U_1^*(1-R) + \cdots + a_mU_m^*(1-R) + \cdots\right], \qquad 0<R<1, \quad (11.50)$$

where $U_m^*(x)$ satisfies the recursion relation

$$U_{m+1}^*(x) = (4x-2)U_m^*(x) - U_{m-1}^*(x), \tag{11.51}$$

with $U_0^*(x)=1$ and $U_1^*(x)=4x-2$. Note that $U_m^*(x)$ may be written as an mth order polynomial in x, $U_m^*(x) = c_{0m} + c_{1m}x + \cdots + c_{mm}x^m$. The coefficients a_m are computed by the formula

$$a_m = \sum_{i=0}^{m} c_{im}\left[\frac{M_{i+1}}{(i+1)} - \frac{1}{(i+1)(i+2)}\right], \tag{11.52}$$

where

$$M_m = \prod_{i=1}^{k}\left\{\sum_{j=0}^{m}(-1)^j\binom{m}{j}\left(\frac{\beta_i+1}{j+\beta_i+1}\right)^{\alpha_i+1}\right\}.$$

The expansion for $G(R|\underset{\sim}{v})$ can be computed recursively. However, the numerical evaluation requires some caution because of the rapid increase in magnitudes of the c_{im}'s. If computations are carried out using a fixed precision machine, it is impractical to carry an excessive number of terms because the result will be dominated by numerical round-off errors. Numerical examples are given by Thompson and Chang (1975) to illustrate the above procedures.

11.4.3 Identical Components and Reliability Design

One of the basic ways to improve system reliability is through redundancy, (i.e., the inclusion of identical backup components into the system). As previously noted, a parallel system features active-redundant components; all components are initially put into operation yet only the functioning of any one of them is required for the system to function. Obviously, the reliability of a single component system can be improved by placing two such components in parallel, if feasible [Bazovsky (1961), p. 101]. It, therefore, seems appropriate to assess the reliability of parallel systems with identical pass/fail or $\mathcal{E}(\lambda)$ components. And whereas these systems are simply special cases of the systems discussed in the previous two subsections, the reliability assessment results in the general cases do not readily yield results for the special cases.

In addition, we shall consider another type of problem not heretofore discussed. Inasmuch as active redundancy increases cost, size, weight, and maintenance of a system, important consideration as to the design of the system should be given. To achieve the required performance with the minimum number of components is a sound approach to system reliability design. Thus, the art of reliability design is not a matter of designing highly sophisticated and complex systems, but rather of designing the simplest system that will yield satisfactory performance. Specifically, given the component test data, suppose we wish to determine the minimum number of active-redundant components required so that, with probability at least ξ ($\xi > 0$ fixed), the parallel system reliability R is at least η ($\eta > 0$ fixed). Thus, we shall seek the smallest integer k such that $\Pr(R \geqslant \eta \mid \cdot) \geqslant \xi$.

Identical Pass/Fail Components. Consider a parallel system consisting of k independent and identical pass/fail components. The test data consist of s successful units out of n (common) components tested. Assuming the $\mathcal{B}_1(s_0+1, n_0-s_0+1)$ prior pdf on component reliability $\tilde{R}$, the posterior pdf of $\tilde{R}$ is [see (11.13)]

$$g(\tilde{R}\mid s)=\frac{\Gamma(n'+2)}{\Gamma(s'+1)\Gamma(n'-s'+1)}\tilde{R}^{s'}(1-\tilde{R})^{n'-s'}, \qquad 0<\tilde{R}<1, \quad (11.53)$$

where $n' = n_0 + n$ and $s' = s_0 + s$.

Under a squared-error loss function, the Bayes point estimator of system reliability R, where $R = 1-(1-\tilde{R})^k$, is

$$\begin{aligned} E(R\mid s) &= 1 - E\left[(1-\tilde{R})^k \mid s\right] \\ &= 1 - \frac{\Gamma(n'+2)\Gamma(n'-s'+k+1)}{\Gamma(n'+k+2)\Gamma(n'-s'+1)}. \end{aligned} \quad (11.54)$$

A $100(1-\gamma)\%$ LBPI for system reliability is $(R_\gamma, 1)$, where R_γ is the solution to the equation $\Pr(R < R_\gamma \mid s) = \gamma$. However,

$$\begin{aligned} \Pr(R<R_\gamma\mid s) &= \Pr\left[1-(1-\tilde{R})^k<R_\gamma\mid s\right] \\ &= \Pr\left[\tilde{R}<1-(1-R_\gamma)^{1/k}\mid s\right] = \mathcal{B}_1\left[1-(1-R_\gamma)^{1/k}; s'+1, n'-s'+1\right], \end{aligned}$$

where $\mathcal{B}_1(x; p, q)$ is the cdf of the $\mathcal{B}_1(p, q)$ distribution evaluated at x. By using the relationship between the percentiles of the $\mathcal{B}_1(p, q)$ and $\mathcal{F}(m, n)$ distributions discussed in Chapter 7, it follows that

$$R_\gamma = 1 - [H(1-\gamma; n', s')]^k, \quad (11.55)$$

where

$$H(\theta; n', s') = \frac{(n'-s'+1)\mathcal{F}_\theta[2(n'-s'+1), 2(s'+1)]}{s'+1+(n'-s'+1)\mathcal{F}_\theta[2(n'-s'+1), 2(s'+1)]}. \quad (11.56)$$

Finally, we consider the reliability design problem. By an analysis similar to that used in deriving (11.55), the inequality $\Pr(R \geqslant \eta | s) \geqslant \xi$ is equivalent to $(1-\eta)^{1/k} \geqslant H(\xi; n', s')$. Thus the smallest integer k such that $\Pr(R \geqslant \eta | s) \geqslant \xi$ holds is given by

$$k = [\![\ln(1-\eta)/\ln H(\xi; n', s')]\!] + 1, \quad (11.57)$$

where $[\![x]\!]$ is the greatest integer function of x.

Example 11.7. A parallel system consists of two independent and identical pass/fail components. Fifteen units of the (common) component were tested with one failure occurring. No prior (pseudo) testing was performed; therefore, we assume a $\mathcal{U}(0,1)$ prior pdf ($n_0 = s_0 = 0$) on component reliability.

Since $n' = n_0 + n = 15$ and $s' = s_0 + s = 14$, a Bayes point estimate of system reliability is [see (11.54)]

$$E(R|s) = 1 - \frac{\Gamma(17)\Gamma(4)}{\Gamma(19)\Gamma(2)} = 0.980.$$

In addition, $\mathcal{F}_{0.95}(4, 30) = 2.69$; thus by (11.56),

$$H(0.95; 15, 14) = \frac{2(2.69)}{15 + 2(2.69)} = 0.264.$$

A 95% LBPI for system reliability is $(0.930, 1)$.

Suppose we require a parallel system with a minimum of k independent and identical pass/fail components to be such that, with probability at least 0.95, system reliability is at least 0.99. Based on the previously given test data, by (11.57),

$$k = [\![\ln(0.01)/\ln H(0.95; 15, 14)]\!] + 1 = 4.$$

Thus, a minimum of four components is required. It would, of course, be possible to reduce this number, yet attain the required level of system reliability, simply by developing a more reliable (common) component to be used in the system. Along these lines, suppose it is required that a parallel system consist of two independent and identical pass/fail components and that, with probability at least 0.95, system reliability is at least 0.99. In order that this be achieved, what is the minimum number, say n, of components that would need to be tested without a failure (disregard the original test data)?

By (11.57), $[\![\ln(0.01)/\ln H(0.95;\, n,\, n)]\!] + 1 = 2$; thus, $1 \leqslant \ln(0.01/\ln H(0.95; n, n) < 2$ or, equivalently,

$$0.01 \leqslant \frac{\mathcal{F}_{0.95}[2, 2(n+1)]}{n+1+\mathcal{F}_{0.95}[2, 2(n+1)]} < 0.1.$$

However, by straightforward integration of the pdf of the $\mathcal{F}[2, 2(n+1)]$ distribution, it can be shown that $\mathcal{F}_{0.95}[2, 2(n+1)] = (n+1)[(1/0.05)^{1/(n+1)} - 1]$. It then follows that $n = 28$. ■

Identical Exponential Components. Consider a parallel system consisting of k independent and identical $\mathcal{E}(\lambda)$ components. A life test on the (common) component consists of testing n units with the test terminated at the time of the rth failure. Let $v = \sum_{i=1}^{r} x_i + (n-r)x_r$ be the observed total time on test, where x_i is the ith ordered failure time. Assuming the $\mathcal{NLG}(r_0+1, v_0/t+1)$ prior pdf on component reliability $\tilde{R}$ at time t, the posterior pdf of $\tilde{R}$ is [see (11.19)]

$$g(\tilde{R} \mid v) = \frac{(\beta+1)^{\alpha+1}}{\Gamma(\alpha+1)} \tilde{R}^{\beta} (\ln \tilde{R}^{-1})^{\alpha}, \qquad 0 < \tilde{R} < 1, \tag{11.58}$$

where $\beta = (v + v_0)/t$ and $\alpha = r + r_0$.

Under a squared-error loss function, the Bayes point estimator of system reliability R at time t is, by (11.48),

$$E(R \mid v) = 1 - E(\tilde{Q}^k \mid v)$$

$$= 1 - \sum_{j=0}^{k} (-1)^j \binom{k}{j} \left(\frac{\beta+1}{j+\beta+1} \right)^{\alpha+1}. \tag{11.59}$$

A $100(1-\gamma)\%$ LBPI for system reliability is $(R_\gamma, 1)$, where R_γ satisfies the equation $\Pr(R < R_\gamma | v) = \gamma$. As

$$\Pr(R < R_\gamma | v) = \Pr\left[\tilde{R} < 1 - (1 - R_\gamma)^{1/k} | v\right]$$
$$= \frac{(\beta+1)^{\alpha+1}}{\Gamma(\alpha+1)} \int_0^{1-(1-R_\gamma)^{1/k}} t^\beta (\ln t^{-1})^\alpha \, dt$$
$$= \int_{-2(\beta+1)\ln[1-(1-R_\gamma)^{1/k}]}^{\infty} \frac{1}{\Gamma(\alpha+1)2^{\alpha+1}} y^\alpha \exp(-y/2)\, dy,$$

by the change of variable $t = \exp[-y/2(\beta+1)]$, then $-2(\beta+1)\ln[1-(1-R_\gamma)^{1/k}] = \chi^2_{1-\gamma}[2(\alpha+1)]$, where $\chi^2_\gamma(n)$ is the 100γth percentile of the $\chi^2(n)$ distribution. Consequently,

$$R_\gamma = 1 - \left(1 - \exp\left\{-\chi^2_{1-\gamma}[2(\alpha+1)]/2(\beta+1)\right\}\right)^k. \qquad (11.60)$$

In the reliability design problem, the inequality $\Pr(R \geqslant \eta | v) \geqslant \xi$ is equivalent to $-2(\beta+1)\ln[1-(1-\eta)^{1/k}] \geqslant \chi^2_\xi[2(\alpha+1)]$. The smallest integer k such that $\Pr(R \geqslant \eta | v) \geqslant \xi$ holds is given by

$$k = [\![\ln(1-\eta)/\ln\left(1 - \exp\left\{-\chi^2_\xi[2(\alpha+1)]/2(\beta+1)\right\}\right)]\!] + 1. \quad (11.61)$$

Example 11.8. A system consisting of three independent and identical $\mathcal{E}(\lambda)$ components in parallel has a mission time of $t = 100$ h. The life test on the (common) component consisted of testing 10 units until the time of occurrence of the sixth failure. The failure times (in h) are 43, 65, 90, 102, 130, 150. The total time on test is 1180 h. Using a noninformative prior pdf $(v_0/t = r_0 = -1)$ on component reliability, we will obtain a Bayes point estimate and a 95% LBPI for system reliability at $t = 100$ h.

As $\alpha = r + r_0 = 5$ and $\beta = (v + v_0)/t = 10.8$, by (11.59), a Bayes point estimate of system reliability is

$$E(R|v) = 1 - \sum_{j=0}^{3} (-1)^j \binom{3}{j} \left[\frac{11.8}{j+11.8}\right]^6 = 0.926.$$

Since $\chi^2_{0.95}(12) = 21.026$, then by (11.60)

$$R_{0.05} = 1 - \left\{1 - \exp\left[-\chi^2_{0.95}(12)/2(11.8)\right]\right\}^3 = 0.795;$$

thus, a 95% LBPI for system reliability at $t = 100$ h is $(0.795, 1)$.

Suppose we wish to design a parallel system by using the fewest number of independent and identical exponential components such that $\Pr(R \geqslant 0.95 \mid v) \geqslant 0.90$, (i.e., with probability at least 0.90, system reliability R is at least 0.95). Using (11.61) with $\eta = 0.95$ and $\xi = 0.90$,

$$k = [\![\ln(0.05)/\ln\{1 - \exp[-\chi^2_{0.90}(12)/2(11.8)]\}]\!] + 1 = 5.$$

Thus, a minimum of five components is required. ■

11.5 r-OUT-OF-k SYSTEMS

For an r-out-of-k system consisting of k independent and identical components, by (11.9), system reliability R is given by

$$R = \sum_{j=r}^{k} \binom{k}{j} \tilde{R}^j (1 - \tilde{R})^{k-j},$$

where $\tilde{R}$ is the reliability of the (common) component. For fixed r and k, R is an increasing function of $\tilde{R}$. Consequently, Bayesian assessment of system reliability may be made through the posterior pdf of $\tilde{R}$. In particular, the 100γth percentile, say R_γ, of the posterior distribution of system reliability R is given by

$$R_\gamma = \sum_{j=r}^{k} \binom{k}{j} \tilde{R}_\gamma^j (1 - \tilde{R}_\gamma)^{k-j}, \qquad (11.62)$$

where $\tilde{R}_\gamma$ is the solution to the equation $G(\tilde{R}_\gamma \mid \cdot) = \gamma$ and $G(\tilde{R} \mid \cdot)$ is the posterior cdf of component reliability. We consider the two cases where the (common) component is either pass/fail or $\mathcal{E}(\lambda)$.

11.5.1 Pass/Fail (Identical) Components

Consider an r-out-of-k system consisting of k independent and identical pass/fail components. The test data consist of s successful units out of n (common) components tested. Assuming the $\mathcal{B}_1(s_0 + 1, n_0 - s_0 + 1)$ prior pdf on component reliability $\tilde{R}$, the posterior pdf of $\tilde{R}$ is [see (11.13)]

$$g(\tilde{R} \mid s) = \frac{\Gamma(n'+2)}{\Gamma(s'+1)\Gamma(n'-s'+1)} \tilde{R}^{s'} (1 - \tilde{R})^{n'-s'}, \qquad 0 < \tilde{R} < 1, \quad (11.63)$$

where $n' = n_0 + n$ and $s' = s_0 + s$.

Under a squared-error loss function, the Bayes point estimator of system reliability is

$$E(R|s) = \sum_{j=r}^{k} \binom{k}{j} \int_0^1 \tilde{R}^j (1-\tilde{R})^{k-j} g(\tilde{R}|s)\, d\tilde{R}$$

$$= \frac{\Gamma(n'+2)}{\Gamma(s'+1)\Gamma(n'-s'+1)} \sum_{j=r}^{k} \binom{k}{j} \times$$

$$\frac{\Gamma(j+s'+1)\Gamma(k-j+n'-s'+1)}{\Gamma(k+n'+2)}$$

$$= \frac{n'+1}{n'+k} \sum_{j=r}^{k} \frac{\binom{j+s'}{s'}\binom{k+n'-j-s'}{n'-s'}}{\binom{k+n'}{n'}}. \qquad (11.64)$$

Since component reliability $\tilde{R}$ has a $\mathcal{B}_1(s'+1, n'-s'+1)$ distribution, it follows from (7.18) and (11.62) that the 100γth percentile of the posterior distribution of system reliability R is

$$R_\gamma = \sum_{j=r}^{k} \binom{k}{j} [H(\gamma; n', s')]^j [1 - H(\gamma; n', s')]^{k-j}, \qquad (11.65)$$

where

$$H(\gamma; n', s') = \frac{s'+1}{s'+1+(n'-s'+1)\mathcal{F}_{1-\gamma}[2(n'-s'+1), 2(s'+1)]}. \qquad (11.66)$$

A symmetric $100(1-\gamma)\%$ TBPI for system reliability is given by $(R_{\gamma/2}, R_{1-\gamma/2})$. A $100(1-\gamma)\%$ LBPI for R is $(R_\gamma, 1)$.

11.5.2 Exponential (Identical) Components

Consider an r-out-of-k system consisting of k independent and identical $\mathcal{E}(\lambda)$ components. A life test on the (common) component consists of testing n units with the test terminated at the time of the sth failure. Let $v = \sum_{i=1}^{s} x_i + (n-s)x_s$ be the observed total time on test, where x_i is the ith ordered failure time. Assuming the $\mathcal{NLG}(s_0+1, v_0/t+1)$ prior pdf on component reliability $\tilde{R}$ at time t, the posterior pdf of $\tilde{R}$ is [see (11.19)]

$$g(\tilde{R}|v) = \frac{(\beta+1)^{\alpha+1}}{\Gamma(\alpha+1)} \tilde{R}^{\beta}(\ln \tilde{R}^{-1})^{\alpha}, \qquad 0 < \tilde{R} < 1, \tag{11.67}$$

where $\beta = (v + v_0)/t$ and $\alpha = s + s_0$.

Under a squared-error loss function, the Bayes point estimator of system reliability at time t is, by (11.48),

$$\begin{aligned} E(R|v) &= \sum_{j=r}^{k} \binom{k}{j} \int_0^1 \tilde{R}^j (1-\tilde{R})^{k-j} g(\tilde{R}|v)\, d\tilde{R} \\ &= \sum_{j=r}^{k} \sum_{\ell=0}^{k-j} (-1)^{\ell} \binom{k}{j} \binom{k-j}{\ell} \left(\frac{\beta+1}{\ell+j+\beta+1} \right)^{\alpha+1}. \end{aligned} \tag{11.68}$$

As

$$\begin{aligned} \Pr(\tilde{R} < \tilde{R}_{\gamma}|v) &= \frac{(\beta+1)^{\alpha+1}}{\Gamma(\alpha+1)} \int_0^{\tilde{R}_{\gamma}} w^{\beta} (\ln w^{-1})^{\alpha}\, dw \\ &= \int_{-2(\beta+1)\ln \tilde{R}_{\gamma}}^{\infty} \frac{1}{\Gamma(\alpha+1) 2^{\alpha+1}} y^{\alpha} \exp(-y/2)\, dy, \end{aligned}$$

by the change of variable $w = \exp[-y/2(\beta+1)]$, then the equation $\Pr(\tilde{R} < \tilde{R}_{\gamma}|v) = \gamma$ is equivalent to $-2(\beta+1)\ln \tilde{R}_{\gamma} = \chi^2_{1-\gamma}[2(\alpha+1)]$. It follows from (11.62) that the 100γth percentile of the posterior distribution of system reliability at time t is

$$\begin{aligned} R_{\gamma} = \sum_{j=r}^{k} \binom{k}{j} &\left(\exp\{-\chi^2_{1-\gamma}[2(\alpha+1)]/2(\beta+1)\} \right)^{j} \times \\ &\left(1 - \exp\{-\chi^2_{1-\gamma}[2(\alpha+1)]/2(\beta+1)\}\right)^{k-j}. \end{aligned} \tag{11.69}$$

A symmetric $100(1-\gamma)\%$ TBPI for system reliability at time t is $(R_{\gamma/2}, R_{1-\gamma/2})$, whereas a $100(1-\gamma)\%$ LBPI for R is $(R_{\gamma}, 1)$.

Example 11.9. A 2-out-of-3 system consisting of independent and identical $\mathcal{E}(\lambda)$ components has a mission time of $t = 20$ h. A life test on the (common) component consisted of testing five units until all had failed. The observed failure times (in h) were 37, 62, 79, 93, and 104. The total time on test is 375 h. Suppose a previous test on the component consisted of testing three units until failure. The failure times (in h) in this prior (pseudo) test were 44, 50, and 72. Under the assumption of an $\mathcal{NLG}(s_0+1, v_0/t+1)$ prior pdf on component reliability, the parameters of the posterior pdf are

$$\beta = \frac{375+166}{20} = 27.05, \qquad \alpha = 5+3 = 8.$$

A Bayes point estimate of system reliability at $t=20$ h is, by (11.68),

$$E(R|v) = \sum_{j=2}^{3} \sum_{\ell=0}^{3-j} (-1)^{\ell} \binom{3}{j} \binom{3-j}{\ell} \left(\frac{28.05}{\ell + j + 28.05} \right)^{9} = 0.813.$$

As $\chi^2_{0.90}(18) = 25.989$, then $\tilde{R}_{0.10} = \exp[-25.989/2(28.05)] = 0.629$. Thus, by (11.69),

$$R_{0.10} = \sum_{j=2}^{3} \binom{3}{j} (0.629)^{j} (0.371)^{3-j} = 0.689,$$

and a 90% LBPI for system reliability at $t = 20$ h is $(0.689, 1)$. ■

11.5.3 Stress-Strength Model

A stress-strength r-out-of-k system is considered by Draper and Guttman (1978). The component strengths, $Y_{10}, \ldots, Y_{k0}$, are assumed to be i.i.d r.v.'s with (common) cdf $F_y(y)$. The system is subject to a stress, say X_0, which is an r.v. with cdf $F_X(x)$. The system reliability R is the probability that at least r of $Y_{10}, \ldots, Y_{k0}$ exceed X_0, so that

$$R = \sum_{j=r}^{k} \binom{k}{j} \int_{-\infty}^{\infty} [1 - F_Y(x)]^{j} [F_Y(x)]^{k-j} \, dF_X(x). \qquad (11.70)$$

Under the assumption of an $\mathcal{E}(\lambda_i)$ distribution for X and Y; that is, $F_X(x) = 1 - \exp(-\lambda_1 x)$, $F_Y(y) = 1 - \exp(-\lambda_2 y)$, (11.70) reduces to

$$R = 1 - \prod_{j=r}^{k} \left(\frac{j}{\phi + j} \right), \qquad (11.71)$$

where $\phi=\lambda_1/\lambda_2=E(Y)/E(X)$, is the strength-stress ratio of the system. Inference on R is based on a sample of $n=n_1+n_2$ independent observations $(x_1,\ldots,x_{n_1};\ y_1,\ldots,y_{n_2})$ where, typically, the strength data $y_1,\ldots,y_{n_2}$ can be generated from laboratory load tests on the (common) component and the stress data $x_1,\ldots,x_{n_1}$ can be obtained from a simulation of conditions for operating the system.

Under the $\mathcal{G}(a_{i0},1/b_{i0})$ prior pdf on λ_i $(i=1,2)$, the posterior pdf of λ_i is $\mathcal{G}[n_i+a_{i0},1/(v_i+b_{i0})]$, where $v_1=\sum_{i=1}^{n_1}x_i$ and $v_2=\sum_{i=1}^{n_2}y_i$. Since the posterior pdf of $2\lambda_i(v_i+b_{i0})$ is a $\chi^2[2(n_i+a_{i0})]$ distribution, and λ_1 and λ_2 are independent, then the posterior pdf of the ratio

$$\phi\frac{(n_2+a_{20})(v_1+b_{10})}{(n_1+a_{10})(v_2+b_{20})}$$

is an $\mathcal{F}[2(n_1+a_{10}),2(n_2+a_{20})]$ distribution. This result leads easily to inferences about system reliability R without explicit knowledge of the posterior pdf of R due to the fact that, for fixed r and k, R is an increasing function of ϕ [see (11.71)]. Thus a $100(1-\gamma)\%$ LBPI for system reliability is $(R_\gamma,1)$, where

$$R_\gamma=1-\prod_{j=r}^{k}\left(\frac{j}{\phi_\gamma+j}\right), \tag{11.72}$$

where ϕ_γ is such that $\Pr(\phi<\phi_\gamma|v_1,v_2)=\gamma$. Clearly, $\phi_\gamma(n_2+a_{20})(v_1+b_{10})/(n_1+a_{10})(v_2+b_{20})=\mathcal{F}_\gamma[2(n_1+a_{10}),2(n_2+a_{20})]$; therefore,

$$\phi_\gamma=\frac{(n_1+a_{10})(v_2+b_{20})}{(n_2+a_{20})(v_1+b_{10})\mathcal{F}_{1-\gamma}[2(n_2+a_{20}),2(n_1+a_{10})]}. \tag{11.73}$$

For a noninformative prior pdf $(a_{i0}=0,\ b_{i0}=\infty)$ on λ_i $(i=1,2)$, Draper and Guttman (1978) give a numerical example discussing various results for an r-out-of-5 system, where $r=3,4,5$.

11.6. COMPLEX SYSTEMS

Bayesian reliability assessment of a system whose configuration of components is more complex than that of a series or parallel system is now considered.

11.6.1 Mellin Transform Techniques

Certain types of complex systems that are basically mixtures of series and parallel systems may be dealt with by jointly considering the Mellin transform techniques discussed in Sections 11.3 and 11.4. Examples of such systems are series-parallel (modules in series, where each module is a parallel configuration of components), parallel-series (modules in parallel, where each module is a series configuration of components), or certain variations thereof. To illustrate the procedure, we give a numerical example.

Example 11.10. [Byers, Skeith, and Springer (1974)] Consider a system consisting of six independent $\mathcal{E}(\lambda_i)$ components as shown in Figure 11.3.

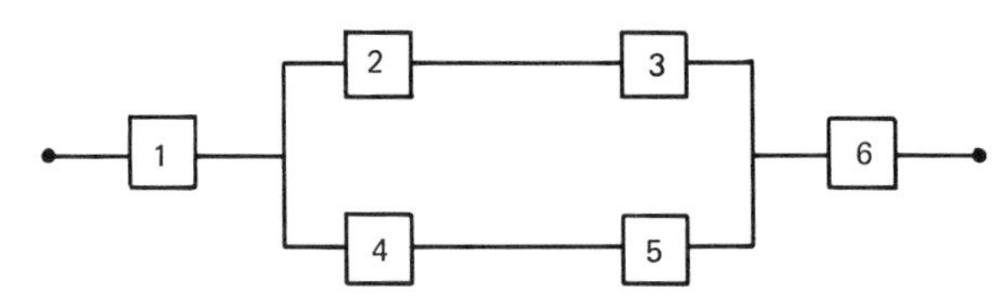

Figure 11.3. The system configuration for Example 11.10.

The test data and prior information (under the assumption of $\mathcal{NLG}(\cdot,\cdot)$ prior pdf's on component reliabilities) yield the following values:

$$\alpha_1=0,\quad \alpha_2=1,\quad \alpha_3=0,\quad \alpha_4=0,\quad \alpha_5=1,\quad \alpha_6=1,$$

$$\beta_1=20,\quad \beta_2=5,\quad \beta_3=4,\quad \beta_4=5,\quad \beta_5=6,\quad \beta_6=25.$$

We obtain a 95% LBPI for system reliability. Only a portion of the derivations will be given due to the increasing complexity of the expressions.

The analysis begins by finding the posterior pdf of the reliability of the parallel configuration. Components 2 and 3 are connected in series. For the top minimal path set, $p_1=\{2,3\}$, the posterior pdf of reliability, say $R_{p_1}=R_2R_3$, is, upon differentiating (11.31),

$$g_1(R_{p_1}|v_2,v_3)=180\left[R_{p_1}^4-R_{p_1}^5\left(1+\ln R_{p_1}^{-1}\right)\right],\qquad 0<R_{p_1}<1.$$

Similarly, components 4 and 5 are connected in series. For the bottom minimal path set, $p_2 = \{4,5\}$, the posterior pdf of reliability, say $R_{p_2} = R_4 R_5$, is

$$g_2(R_{p_2}|v_4, v_5) = 294\left[R_{p_2}^5 - R_{p_2}^6\left(1 + \ln R_{p_2}^{-1}\right)\right], \qquad 0 < R_{p_2} < 1.$$

Let Q_{p_i} be the unreliability corresponding to the minimal path set p_i, $i = 1,2$. Then the posterior pdf's of unreliabilities are

$$\bar{g}_1(Q_{p_1}|v_2, v_3) = 180\left\{(1 - Q_{p_1})^4 - (1 - Q_{p_1})^5\left[1 + \ln(1 - Q_{p_1})^{-1}\right]\right\},$$

$$0 < Q_{p_1} < 1,$$

and

$$\bar{g}_2(Q_{p_2}|v_4, v_5) = 294\left\{(1 - Q_{p_2})^5 - (1 - Q_{p_2})^6\left[1 + \ln(1 - Q_{p_2})^{-1}\right]\right\},$$

$$0 < Q_{p_2} < 1.$$

As the parallel configuration unreliability is $Q_p = Q_{p_1} Q_{p_2}$, the posterior pdf of reliability $R_p = 1 - Q_p$ is determined by the integral expression

$$g_p(R_p|v_2, v_3, v_4, v_5) = \int_{1-R_p}^{1} (1/y)\bar{g}_2\left[(1 - R_p)/y|v_4, v_5\right]\bar{g}_1(y|v_2, v_3)\, dy.$$

The closed form expression for $g_p(R_p|v_2, v_3, v_4, v_5)$, obtained by evaluating the integral, is extremely cumbersome and will not be given here.

Continuing the analysis, system reliability $R = R_1 R_p R_6$ has posterior pdf $g(R|\underset{\sim}{v})$, where $\underset{\sim}{v}' = (v_1, v_2, v_3, v_4, v_5, v_6)$, whose Mellin transform is

$$M\{g(R|\underset{\sim}{v}); u\} = M\{g_1(R_1|v_1); u\} \times$$

$$M\{g_p(R_p|v_2, v_3, v_4, v_5); u\} M\{g_6(R_6|v_6); u\}.$$

The Mellin inversion integral then produces $g(R|\underset{\sim}{v})$. The results of the inversion and further details are given by Byers (1970).

The expression for the posterior cdf of system reliability, $G(R|\underset{\sim}{v})$, is obtained by integrating $g(R|\underset{\sim}{v})$. The graph of $G(R|\underset{\sim}{v})$, which is based on the evaluation of $G(R|\underset{\sim}{v})$ for $R = 0.10(0.10)0.80(0.05)0.95$ by Byers, Skeith,

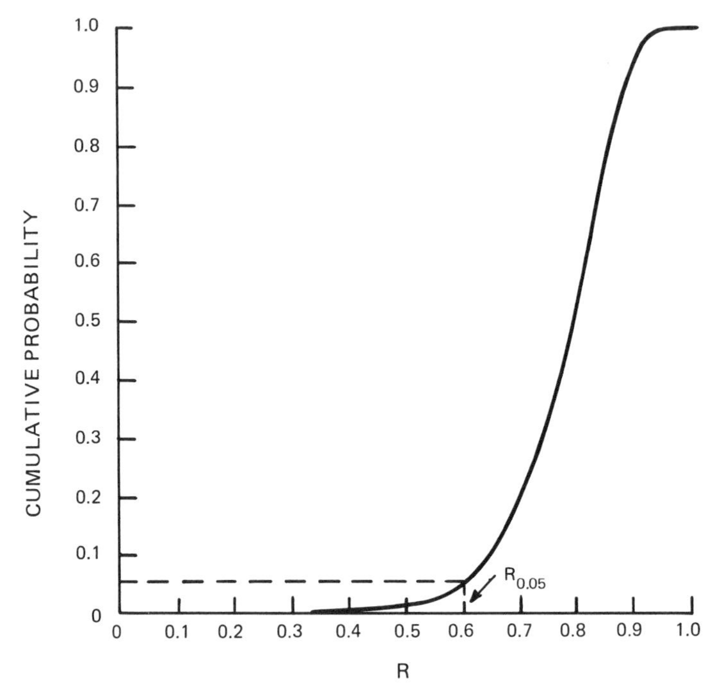

Figure 11.4. The posterior cumulative distribution function $G(R|\underset{\sim}{v})$ for Example 11.10.

and Springer (1974), is shown in Figure 11.4. From the graph, a 95% LBPI for system reliability is (0.60, 1). ■

It seems rather evident from the example that the use of the Mellin transform technique in this and other similar cases is greatly limited due to difficulties in making the various inversions. Two alternate, yet somewhat similar, approaches to determine the posterior pdf of system reliability have been given. Chang and Thompson (1976) consider a parallel-series system that features standby (passive-redundant) $\mathcal{E}(\lambda)$ components. The posterior pdf of system reliability is developed in terms of shifted Chebyshev polynomials of the second kind. Numerical examples are given along with suggestions to aid in the computer evaluations. Wolf (1976) gives a procedure that fits the posterior pdf of system reliability with shifted Legendre polynomials. Numerical examples that deal with a series-parallel system as well as a bridge system are given to illustrate the theory.

11.6.2 Monte Carlo Simulation Method

Based on a combination of Boolean logic, probability, and Bayesian and Monte Carlo methods, a PL/1 computer program (SPARCS-2) has been developed which assesses the reliability of highly complex systems [Locks (1978)]. The system can consist of any configuration of independently failing pass/fail and $\mathscr{E}(\lambda)$ components. [For a methodology similar to the one being discussed but for $\mathscr{W}(\alpha, \beta)$ components, see Kamat (1977), pp. 123–131]. Alternately, it can be a configuration of independently failing modules, where each module has either or both pass/fail and $\mathscr{E}(\lambda)$ components. An MTBF assessment option is also available.

The raw data requirements for reliability assessments are the failure history data on each component as well as the system configuration. The component failure history data are: (1) "number of successes and failures" for a pass/fail component, assuming a $\mathscr{B}_1(p, q)$ prior pdf on component reliability; and (2) "number of failures and total testing time" for an $\mathscr{E}(\lambda)$ component, assuming an $\mathscr{NLG}(\alpha, \beta)$ prior pdf on component reliability. To facilitate SPARCS-2, an $\mathscr{E}(\lambda)$ component is replaced by an "equivalent mission" pass/fail component. The configuration data consist of a list of minimal path sets (or else a list of minimal cut sets) for the system as a list of modules, and for each module as a list of components. From the list of minimal path sets, SPARCS-2 derives the system reliability equation as a polynomial in the component reliabilities. Dually, the system unreliability equation can be derived from the list of minimal cut sets as a polynomial in the component unreliabilities.

The Monte Carlo procedure is as follows. At each trial, for each component, a value of the reliability is generated from its natural conjugate prior pdf and substituted into the system reliability equation to obtain a value of the prior system reliability for that trial. The resulting "empirical" distribution of prior system reliabilities, obtained over a series of trials, provides the basis for the assessment. Percentage points for that distribution are interpreted as system reliability "confidence" limits.i

SPARCS-2 can process a system consisting of up to 128 modules or components in any configuration with up to 256 minimal path sets. Likewise, a module within the system can have 128 components and 256 minimal path sets. The system or module reliability equation can have up to 3500 terms.

Example 11.11. [Locks (1978)] The system in Figure 11.5 is subdivided into two modules, A with components $1, \ldots, 7$, and B with components

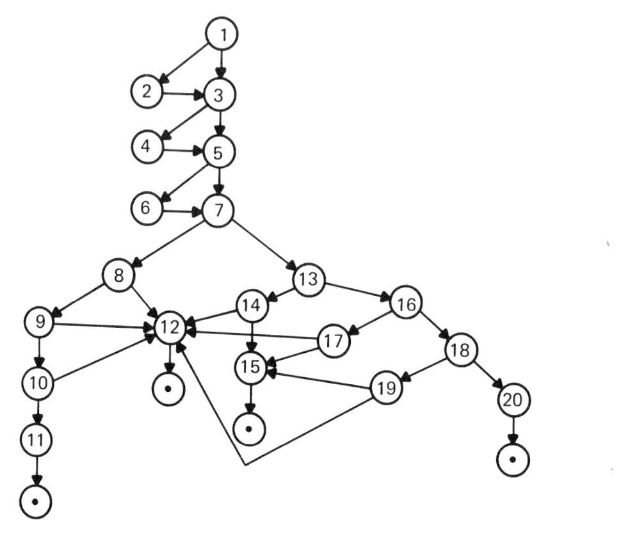

Figure 11.5. The network of node-to-node branching potential [Adapted, with permission, from Locks (1978)].

```
* S P A R C S *
SIMULATICN PROGRAM FOP THE ANALYSIS OF THE RELIABILITY OF COMPLEX SYSTEMS


SYSTEM IDENTIFICATION          SOFTWARE RELIABILITY COMPUTATION - J R BROWN AND LIPOW
NUMBER OF SIMULATIONS          20
NUMBER OF MODULES              2
NUMBER OF COMPONENTS           2
NUMBER OF MINIMAL PATHS        1
TYPE OF ANALYSIS               RELIABILITY


SPARCS :: EQUATION GENERATION ROUTINE

THE   1 MINIMAL PATH   FOR SYSTEM

<A,B>

 NUMBER OF TERMS IN EQUATION:   1


  R    = R R
   SYS    A B


THE   1 MINIMAL PATH   FOR MODULE A

<1,3,5,7>

 NUMBER OF TERMS IN EQUATION:   1


  R    = R R R R
   A      1 3 5 7
```

HISTORICAL INFORMATION FOR EACH COMPONENT IN MODULE A

COMPONENT	TYPE	SUCCESS (BETA) EQUIVALENT MISSIONS(GAMMA)	FAILURES	PRIOR DISTRIBUTION MEAN
1	TIME-TO-FAILURE(GAMMA)	96.50	2.00	0.96985
2	ATTRIBUTE(BETA)	99.00	1.00	0.98039
3	ATTRIBUTE(BETA)	99.00	1.00	0.98039
4	ATTRIBUTE(BETA)	99.00	1.00	0.98039
5	TIME-TO-FAILURE(GAMMA)	96.50	2.00	0.96985
6	ATTRIBUTE(BETA)	99.00	1.00	0.98039
7	ATTRIBUTE(BETA)	99.00	1.00	0.98039

THE 9 MINIMAL PATHS FOR MODULE B

<6,9,11,13>
<5,6,7>
<6,7,8>
<5,6,9,10>
<6,8,9,10>
<5,6,9,11,12>
<6,8,9,11,12>
<1,2,3,4>
<1,5>

NUMBER OF TERMS IN EQUATION: 91

$$R_B = R_6R_9R_{11}R_{13} + R_5R_6R_7 - R_5R_6R_7R_9R_{11}R_{13} + R_6R_7R_8 - R_6R_7R_8R_9R_{11}R_{13} - R_5R_6R_7R_8 + R_5R_6R_7R_8R_9R_{11}R_{13} + R_5R_6R_9R_{10} - R_5 *$$

$$R_6R_9R_{10}R_{11}R_{13} - R_5R_6R_7R_9R_{10} + R_5R_6R_7R_9R_{10}R_{11}R_{13} + R_6R_8R_9R_{10} - R_6R_8R_9R_{10}R_{11}R_{13} + R_5R_6R_7R_8R_9R_{10} - R_5R_6R_7R_8R_9R_{10}R_{11}R_{13}$$

$$- R_6R_7R_8R_9R_{10} + R_6R_7R_8R_9R_{10}R_{11}R_{13} - R_5R_6R_8R_9R_{10} + R_5R_6R_8R_9R_{10}R_{11}R_{13} + R_5R_6R_9R_{11}R_{12} - R_5R_6R_9R_{11}R_{12}R_{13} - R_5R_6R_7R_9R_{11} *$$

$$R_{12} + R_5R_6R_7R_9R_{11}R_{12}R_{13} - R_5R_6R_9R_{10}R_{11}R_{12} + R_5R_6R_9R_{10}R_{11}R_{12}R_{13} + R_5R_6R_7R_9R_{10}R_{11}R_{12} - R_5R_6R_7R_9R_{10}R_{11}R_{12}R_{13} + R_6R_8R_9 *$$

$$R_{11}R_{12} - R_6R_8R_9R_{11}R_{12}R_{13} + R_5R_6R_7R_8R_9R_{11}R_{12} - R_5R_6R_7R_8R_9R_{11}R_{12}R_{13} - R_6R_7R_8R_9R_{11}R_{12} + R_6R_7R_8R_9R_{11}R_{12}R_{13} + R_5R_6R_8R_9 *$$

$$R_{10}R_{11}R_{12} - R_5R_6R_8R_9R_{10}R_{11}R_{12}R_{13} - R_5R_6R_7R_8R_9R_{10}R_{11}R_{12} + R_5R_6R_7R_8R_9R_{10}R_{11}R_{12}R_{13} - R_6R_8R_9R_{10}R_{11}R_{12} + R_6R_8R_9R_{10}R_{11}R_{12} *$$

$$R_{13} + R_6R_7R_8R_9R_{10}R_{11}R_{12} - R_6R_7R_8R_9R_{10}R_{11}R_{12}R_{13} - R_5R_6R_8R_9R_{11}R_{12} + R_5R_6R_8R_9R_{11}R_{12}R_{13} + R_1R_2R_3R_4 - R_1R_2R_3R_4R_6R_9R_{11}R_{13}$$

$$- R_1R_2R_3R_4R_6R_7R_8 + R_1R_2R_3R_4R_6R_7R_8R_9R_{11}R_{13} + R_1R_2R_3R_4R_5R_6R_7R_8 - R_1R_2R_3R_4R_5R_6R_7R_8R_9R_{11}R_{13} - R_1R_2R_3R_4R_6R_8R_9R_{10} + R_1 *$$

$$R_2R_3R_4R_6R_8R_9R_{10}R_{11}R_{13} - R_1R_2R_3R_4R_5R_6R_7R_8R_9R_{10} + R_1R_2R_3R_4R_5R_6R_7R_8R_9R_{10}R_{11}R_{13} + R_1R_2R_3R_4R_6R_7R_8R_9R_{10} - R_1R_2R_3R_4R_6R_7R_8 *$$

$$R_9R_{10}R_{11}R_{13} + R_1R_2R_3R_4R_5R_6R_8R_9R_{10} - R_1R_2R_3R_4R_5R_6R_8R_9R_{10}R_{11}R_{13} - R_1R_2R_3R_4R_6R_8R_9R_{11}R_{12} + R_1R_2R_3R_4R_6R_8R_9R_{11}R_{12}R_{13} - R_1 *$$

$$R_2R_3R_4R_5R_6R_7R_8R_9R_{11}R_{12} + R_1R_2R_3R_4R_5R_6R_7R_8R_9R_{11}R_{12}R_{13} + R_1R_2R_3R_4R_6R_7R_8R_9R_{11}R_{12} - R_1R_2R_3R_4R_6R_7R_8R_9R_{11}R_{12}R_{13} - R_1R_2R_3 *$$

$$R_4R_5R_6R_8R_9R_{10}R_{11}R_{12} + R_1R_2R_3R_4R_5R_6R_8R_9R_{10}R_{11}R_{12}R_{13} + R_1R_2R_3R_4R_5R_6R_7R_8R_9R_{10}R_{11}R_{12} - R_1R_2R_3R_4R_5R_6R_7R_8R_9R_{10}R_{11}R_{12}R_{13}$$

$$+ R_1R_2R_3R_4R_6R_8R_9R_{10}R_{11}R_{12} - R_1R_2R_3R_4R_6R_8R_9R_{10}R_{11}R_{12}R_{13} - R_1R_2R_3R_4R_6R_7R_8R_9R_{10}R_{11}R_{12} + R_1R_2R_3R_4R_6R_7R_8R_9R_{10}R_{11}R_{12}R_{13}$$

$$+ R_1R_2R_3R_4R_5R_6R_8R_9R_{11}R_{12} - R_1R_2R_3R_4R_5R_6R_8R_9R_{11}R_{12}R_{13} + R_1R_5 - R_1R_5R_6R_9R_{11}R_{13} - R_1R_5R_6R_7 + R_1R_5R_6R_7R_9R_{11}R_{13} - R_1R_5 *$$

$$R_6R_9R_{10} + R_1R_5R_6R_9R_{10}R_{11}R_{13} + R_1R_5R_6R_7R_9R_{10} - R_1R_5R_6R_7R_9R_{10}R_{11}R_{13} - R_1R_5R_6R_9R_{11}R_{12} + R_1R_5R_6R_9R_{11}R_{12}R_{13} + R_1R_5R_6R_7 *$$

$$R_9R_{11}R_{12} - R_1R_5R_6R_7R_9R_{11}R_{12}R_{13} + R_1R_5R_6R_9R_{10}R_{11}R_{12} - R_1R_5R_6R_9R_{10}R_{11}R_{12}R_{13} - R_1R_5R_6R_7R_9R_{10}R_{11}R_{12} + R_1R_5R_6R_7R_9R_{10}R_{11} *$$

$$R_{12}R_{13} - R_1R_2R_3R_4R_5 + R_1R_2R_3R_4R_5R_6R_9R_{11}R_{13}$$

HISTORICAL INFORMATION FOR EACH COMPONENT IN MODULE 8

COMPONENT	TYPE	SUCCESS (BETA) EQUIVALENT MISSIONS(GAMMA)	FAILURES	PRIOR DISTRIBUTION MEAN
1	TIME-TO-FAILURE(GAMMA)	96.50	2.00	0.96985
2	ATTRIBUTE(BETA)	99.00	1.00	0.98039
3	TIME-TO-FAILURE(GAMMA)	96.50	2.00	0.96985
4	ATTRIBUTE(BETA)	99.00	1.00	0.98039
5	TIME-TO-FAILURE(GAMMA)	96.50	2.00	0.96985
6	ATTRIBUTE(BETA)	99.00	1.00	0.98039
7	TIME-TO-FAILURE(GAMMA)	96.50	2.00	0.96985
8	ATTRIBUTE(BETA)	99.00	1.00	0.98039
9	TIME-TO-FAILURE(GAMMA)	96.50	2.00	0.96985
10	ATTRIBUTE(BETA)	99.00	1.00	0.98039
11	TIME-TO-FAILURE(GAMMA)	96.50	2.00	0.96985
12	ATTRIBUTE(BETA)	99.00	1.00	0.98039
13	TIME-TO-FAILURE(GAMMA)	96.50	2.00	0.96985

SPARCS :: SYSTEM SIMULATION ROUTINE

SYSTEM RELIABILITY CALCULATED FROM MEAN COMPONENT RELIABILITIES IS 0.903486; SYSTEM UNRELIABILITY IS 0.096514

AVERAGE SYSTEM RELIABILITY FROM 20 MONTE CARLO TRIALS IS 0.901238 ; AVERAGE SYSTEM UNRELIABILTY IS 0.098762

VARIANCE 0.000671

STANDARD DEVIATION 0.025896

THE MISSION TIME IS 100.000 HOUR

THE ESTIMATED SYSTEM MTBF BASED UPON MEAN COMPONENT RELIABILITIES IS 9.85000000E+02

THE ESTIMATED SYSTEM MTBF BASED UPON MEAN SYSTEM RELIABILITY 20 MONTE CARLO TRIALS IS 9.61669678E+02

PERCENTILE	RELIABILITY PERCENTILE POINTS	MTBF PERCENTILE POINTS
5.0 PERCENT	0.848968	6.10748047E+02 HOUR
10.0 PERCENT	0.865494	6.92260010E+02 HOUR
20.0 PERCENT	0.878241	7.70209473E+02 HOUR
25.0 PERCENT	0.884024	8.11221191E+02 HOUR
50.0 PERCENT	0.904296	9.94052002E+02 HOUR
75.0 PERCENT	0.921494	1.22310010E+03 HOUR
80.0 PERCENT	0.924841	1.27986133E+03 HOUR
90.0 PERCENT	C.929575	1.36934155E+03 HOUR
95.0 PERCENT	0.933471	1.45253003E+03 HOUR
97.5 PERCENT	0.937314	1.54470239E+03 HOUR
99.0 PERCENT	C.939619	1.60563623E+03 HOUR

8,...,20. (In the computer output, module-B components were renumbered. For example, the component originally numbered 8 is 1 in the computer output, 9 is 2, etc.)

The component failure history data will be listed in the computer output. For the sake of brevity, we only consider the case where the configuration data input are the minimal path sets. The other optional alternatives are discussed by Locks (1978). Module A represents the top half of Figure 11.5 and module B the bottom half. The system may be viewed as two "successful" modules in series for reliability calculations. Thus, there is one minimal path set of modules

$$\{A, B\}.$$

Since all four components 1, 3, 5, and 7 are needed for module A to function, it has one minimal path set of components

$$\{1,3,5,7\}.$$

Note that components 2, 4, and 6 in module A are not essential, and do not affect the reliability calculations.

Module B has nine minimal path sets of components: $\{13,16,18,20\}$, $\{12,13,14\}$, $\{13,14,15\}$, $\{12,13,16,17\}$, $\{13,15,16,17\}$, $\{12,13,16,18,19\}$, $\{13,15,16,18,19\}$, $\{8,9,10,11\}$, $\{8,12\}$. As we shall see, the module-B reliability equation based on these minimal path sets contains 91 terms. It can be obtained from (11.1).

The computer output given above is reproduced, with permission, from Locks (1978). Following the system ID data, we have the minimal path sets; the system reliability equation; the module-A minimal path sets, reliability equation, and historical data; and the module-B minimal path sets, reliability equation, and historical data. Then follow various sets of statistical data relating to the output and finally the empirical distribution displaying the percentage points for use in the required assessment. ■

EXERCISES

11.1. For each of the coherent systems shown in Figure 11.6, determine
  (a) the minimal path sets,
  (b) the minimal cut sets,
  (c) the structure function of the system,
  (d) the reliability function of the system.

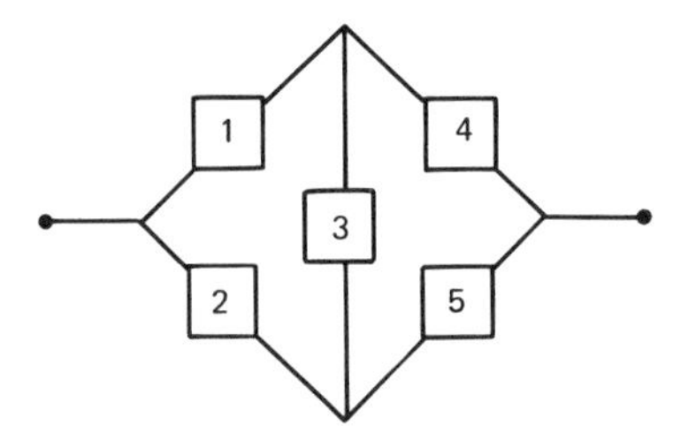

(a) BRIDGE SYSTEM

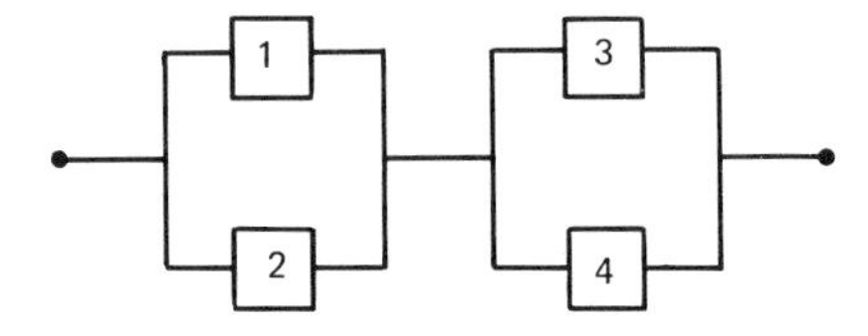

(b) SERIES-PARALLEL SYSTEM

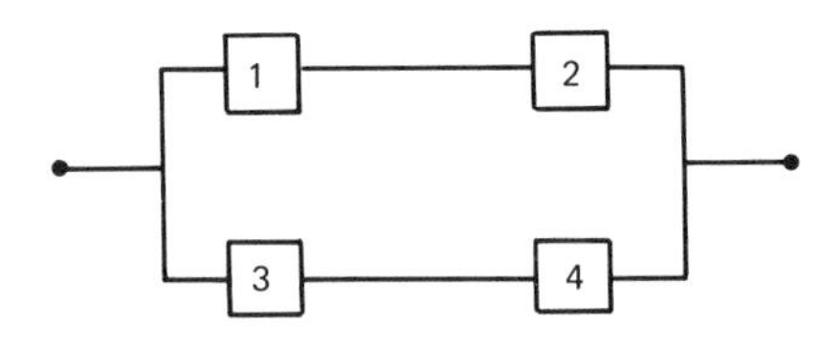

(c) PARALLEL-SERIES SYSTEM

Figure 11.6. A bridge, series-parallel, and parallel-series system configuration.

11.2. Suppose the structure function of a system is

$$\phi(x_1, x_2, x_3) = x_1 x_3 + x_2 x_3 - x_1 x_2 x_3.$$

(a) Show that the system is coherent.
(b) Determine the minimal paths.
(c) Diagram the system configuration.

11.3. Given a coherent system with structure function $\phi(x_1, x_2, \ldots, x_k)$, the *dual system* is such that its structure function $\phi^*(x_1, x_2, \ldots, x_k)$ is defined by the equation

$$\phi^*(x_1, x_2, \ldots, x_k) = 1 - \phi(1 - x_1, 1 - x_2, \ldots, 1 - x_k).$$

The concept of duality is often used in the study of the reliability of systems of electric relays where a relay is subject to two types of failures: open circuit (no current passes) or short circuit (current always passes).

Determine the dual system of

(a) a series system of k components,

(b) a parallel system of k components,

(c) an r-out-of-k system of k identical components.

11.4. The MTTF of a nonmaintained system is the expected time for the system to operate before reaching a failed state. Thus,

$$\text{MTTF}=\int_0^{\infty} t f_S(t)\,dt,$$

where $f_S(t)$ is the pdf of the lifetime of the system.

Determine the MTTF of the system configuration shown in Figure 11.1 assuming the ith component has a constant failure rate λ_i.

11.5. Consider a series system consisting of k independent components, where the ith component has a constant failure rate λ_i.

(a) Determine the MTTF of the system.

(b) Show that the posterior cdf of the series system MTTF, say $G_s(M\mid\cdot)$, is given by

$$G_s(M\mid\cdot)=1-G[\exp(-Mt)\mid\cdot],$$

where $G(\cdot\mid\cdot)$ is the posterior cdf of the series system reliability R at time t.

(c) (Continuation of Example 11.4) By using the result in (b), determine a symmetric 90% TBPI for the system MTTF. Hint: from the graph of $G_s(M\mid\cdot)$, find the 5th and 95th percentiles of the posterior distribution of MTTF.

11.6. Consider a parallel system consisting of k independent components, where the ith component has a constant failure rate λ_i.

(a) Show that the MTTF of the system is given by

$$\text{MTTF}=\sum_{\alpha_1=1}^{k}\lambda_{\alpha_1}^{-1}-\sum_{\substack{\alpha_1=1\\ \alpha_1<\alpha_2}}^{k}\sum_{\alpha_2=1}^{k}(\lambda_{\alpha_1}+\lambda_{\alpha_2})^{-1}+\cdots$$

$$+(-1)^{k-1}\sum_{\substack{\alpha_1=1\\ \alpha_1<\alpha_2<\cdots<\alpha_k}}^{k}\sum_{\alpha_2=1}^{k}\cdots\sum_{\alpha_k=1}^{k}(\lambda_{\alpha_1}+\lambda_{\alpha_2}+\cdots+\lambda_{\alpha_k})^{-1}.$$

(b) For the situation when the components are identical each with a common failure rate λ, show that the posterior cdf of the parallel system MTTF, say $G_p(M|\cdot)$, is given by

$$G_p(M|\cdot) = G\left[\exp\left(-\frac{ct}{M}\right)\middle|\cdot\right],$$

where $G(\tilde{R}|\cdot)$ is the posterior cdf of the (common) component reliability $\tilde{R}$ at time t and $c = 1 + (1/2) + (1/3) + \cdots + (1/k)$.

(c) (Continuation of Example 11.8) By using the result in (b), determine a symmetric 90% TBPI for the MTTF of the 3-component parallel system.

11.7. The Mellin convolution of two functions $f_1(x)$ and $f_2(x)$, $x \geqslant 0$, is defined as

$$h(x) = \int_0^\infty \left(\frac{1}{y}\right) f_2\left(\frac{x}{y}\right) f_1(y)\, dy.$$

Prove that the pdf of the product of two independent nonnegative r.v.'s with pdf's $f_1(x_1)$ and $f_2(x_2)$ is the Mellin convolution whose transform is the product of the Mellin transforms of $f_1(x_1)$ and $f_2(x_2)$.

11.8. (a) Show that the posterior cdf of the reliability R of a k-component series system may be determined by the integral expression

$$G(R|\cdot) = 1 - \int_R^1 \int_{R/R_1}^1 \cdots \int_{R/R_1 R_2 \cdots R_{k-1}}^1$$

$$\times g_1(R_1|\cdot) g_2(R_2|\cdot) \ldots g_k(R_k|\cdot)\, dR_k \ldots dR_2\, dR_1,$$

where $g_i(R_i|\cdot)$ is the posterior pdf of the ith component reliability.

(b) Consider the 3-component series system described in Example 11.2. Derive the posterior cdf of system reliability by using the result in (a).

11.9. One of the modules in a large system is a 2-component parallel subsystem where both components are identical with a failure rate λ. This particular module is required to operate without failing for (at least) 25 h and, per such a requirement, is considered as a pass/fail module. The results of testing on both the module and the (common) component are given below:

Module	5 units tested with 5 successes
Component	10 units tested until failure with failure times (in h): 13.2, 18.7, 22.6, 30.8, 32.0, 36.3, 38.5, 49.1, 65.3, 95.6

(a) Using only the module test data and a $\mathcal{U}(0,1)$ prior pdf on module reliability R, determine the preliminary posterior pdf of R. Give a 95% LBPI for R based on the module test data.

(b) Find the consistent prior pdf of component reliability $\tilde{R}$ [see Section 11.2.2].

(c) By using the component test data, derive the posterior pdf of $\tilde{R}$. Hint: Make use of (11.21).

(d) From the result in (c) determine a 95% LBPI for R based on both module and component test data.

11.10. Consider a parallel system consisting of two independent components, where the ith component has a constant failure rate λ_i. Life tests on the components produced the following results:

Component 1	7 units tested until failure with failure times (in h): 9.5, 21.0, 40.0, 74.6, 146.3, 148.3, 211.1
Component 2	10 units tested until failure with failure times (in h): 23.1, 29.6, 37.5, 44.3, 44.8, 71.6, 92.2, 101.7, 118.7, 189.9

Assume noninformative prior pdfs on component reliabilities.

(a) Determine the Bayes point estimate of system reliability at $t = 50$ h for a squared-error loss function.

(b) Based on shifted Chebyshev polynomials of the second kind [see Section 11.4.2], give an expression for the posterior cdf of system reliability at $t = 50$ h using only the first five terms in its series expansion.

(c) Graph the posterior cdf obtained in (b). From the graph, find a 90% LBPI for system reliability at $t = 50$ h.

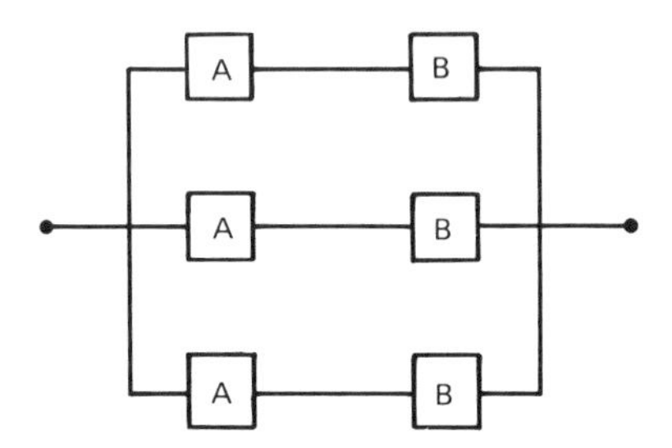

Figure 11.7. A parallel-series system for Exercise 11.11.

11.11. The parallel-series system shown in Figure 11.7 consists of three identical modules in parallel, where each module is a pass/fail component A in series with an $\mathcal{E}(\lambda)$ component B.

(a) Show that the system reliability R is given by

$$R = 1 - (1 - R_A R_B)^3,$$

where $R_A(R_B)$ is the reliability of a type A (type B) component.

(b) Testing on the (common) components produced the following results:

Component A	10 units tested with 9 successes
Component B	6 units tested until failure with failure times (in h): 13.4, 27.9, 60.0, 60.9, 231.3, 318.8

Determine the posterior pdf of module reliability at $t = 150$ h assuming $\mathcal{U}(0, 1)$ prior pdf's on component reliabilities. Hint: Make use of (11.36).

(c) By using the results in (a) and (b), find a 95% LBPI for system reliability at $t = 150$ h.

11.12. Standby redundancy is a method of increasing the reliability of a system by using additional backup or standby units. *A standby redundant system*, as shown in Figure 4.20, consists of one component in operation (on-line) at a time with standby (off-line) components each waiting to be sequentially activated when the on-line component fails. The exchange time of an off-line component for a failed on-line component is usually taken to be negligible.

We make the following assumptions:

1. The standby components are inactive; that is, they cannot fail before inclusion into the system.
2. All components are identical and nonrepairable.
3. The failure-sensing/switching device is reliable enough so that its influence on the overall system reliability is negligible.

Determine the MTTF of the system.

11.13. Consider the k-component standby redundant system described in Exercise 11.12. Suppose that n units of the (common) $\mathcal{E}(\lambda)$ component are tested until failure with the observed total time on test given by $v = \Sigma_{i=1}^{n} x_i$, where x_i is the ith failure time. Assume an $\mathcal{NLG}(n_0 + 1, v_0/t + 1)$ prior pdf on component reliability at time t [see Section 11.2.1].

(a) Derive the Bayes point estimator of system reliability at time t for a squared-error loss function.

(b) Show that a $100(1-\gamma)\%$ LBPI for system reliability at time t is given by $(R_\gamma, 1)$, where

$$R_\gamma = \sum_{i=0}^{k-1} \frac{\left(\ln \tilde{R}_\gamma^{-1}\right)^i \tilde{R}_\gamma}{i!},$$

and $\tilde{R}_\gamma = \exp\{-\chi^2_{1-\gamma}[2(\alpha+1)]/2(\beta+1)\}$ with $\alpha = n_0 + n$ and $\beta = (v_0 + v)/t$.

11.14. Suppose a 2-component standby redundant system is such that both the on-line and backup components are identical, nonrepairable, and have a constant failure rate λ. We assume the backup component cannot fail while on standby. In addition, the failure-sensing/switching device is a pass/fail component that has probability $q = 1 - p$ of failing to either sense the failure of the on-line component or switch in the backup component. Determine the MTTF of the system.

11.15. Consider the 2-component standby redundant system described in Exercise 11.14. Suppose that n switches are tested and s operate properly. In addition, m units of the (common) $\mathcal{E}(\lambda)$ component are tested until failure with the observed total time on test given by $v = \Sigma_{i=1}^{m} x_i$, where x_i is the ith failure time. Assume a $\mathcal{B}_1(s_0 + 1, n_0 - s_0 + 1)$ prior pdf on switch reliability and $\mathcal{NLG}(m_0 + 1, v_0/t + 1)$ prior pdf on component reliability at time t [see Section 11.2.1].

(a) Derive the Bayes point estimator of system reliability at time t for a squared-error loss function.

(b) Derive the Bayes point estimator of system MTTF for a squared-error loss function.

(c) Given the test data: $n = 10$, $s = 9$; $m = 8$, $v = 1200$ h. Under $\mathcal{U}(0,1)$ prior pdf's ($n_0 = s_0 = 0$, $m_0 = v_0 = 0$), evaluate (1) the Bayes point estimate of system reliability at $t = 200$ h, and (2) the system MTTF.

11.16. (Continuation of Example 11.10) Determine the Bayes point estimator of system reliability assuming the loss function to be (1) squared-error, (2) absolute-error.

REFERENCES

Barlow, R. E. and Proschan, F. (1975). *Statistical Theory of Reliability and Life-Testing*, Holt, Rinehart & Winston, New York.

Bazovsky, I. (1961). *Reliability Theory and Practice*, Prentice-Hall, Englewood Cliffs, NJ.

Berkbigler, K. P. and Byers, J. K. (1975). System Reliability: Exact Bayesian Intervals Compared with Fiducial Intervals, *IEEE Transactions on Reliability*, Vol. R-24, pp. 199–200.

Birnbaum, Z. W., Esary, J. D., and Saunders, S. C. (1961). Multi-Component Systems and Structures and Their Reliability, *Technometrics*, Vol. 3, pp. 55–77.

Breipohl, A. M., Prairie, R. R., and Zimmer, W. J. (1965). A Consideration of the Bayesian Approach in Reliability Evaluation, *IEEE Transactions on Reliability*, Vol. R-14, pp. 107–113.

Burnett, T. L. and Wales, B. A. (1961). System Reliability Confidence Limits, *Proceedings 7th National Symposium on Reliability and Quality Control*, pp. 118–128.

Byers, J. K. (1970). *Applications of GERT to Reliability Analysis*, Ph.D. Dissertation, University of Arkansas, Fayetteville, AR.

Byers, J. K., Skeith, R. W., and Springer, M. D. (1974). Bayesian Confidence Limits for the Reliability of Mixed Cascade and Parallel Independent Exponential Subsystems, *IEEE Transactions on Reliability*, Vol. R-23, pp. 104–107.

Chang, E. Y. and Thompson, W. E. (1976). Bayes Analyses of Reliability for Complex Systems, *Operations Research*, Vol. 24, pp. 156–168.

Churchill, R. V. (1960). *Complex Variables and Applications* (2nd ed.), McGraw-Hill, New York.

Cole, P. V. Z. (1975). A Bayesian Reliability Assessment of Complex Systems for Binomial Sampling, *IEEE Transactions on Reliability*, Vol. R-24, pp. 114–117.

Dostal, R. G. and Iannuzzelli, L. M. (1977). Confidence Limits for System Reliability When Testing Takes Place at the Component Level, *The Theory and Applications of Reliability, Vol. II*, Academic, New York.

Draper, N. R. and Guttman, I. (1978). Bayesian Analysis of Reliability in Multicomponent Stress-Strength Models, *Communications in Statistics*, Vol. A7, pp. 441–451.

Erdélyi, A. (1954). *Tables of Integral Transforms, Vol. I*, McGraw-Hill, New York.

Fertig, K. W. (1972). Bayesian Prior Distributions for Systems with Exponential Failure-Time Data, *Annals of Mathematical Statistics*, Vol. 43, pp. 1441–1448.

Kamat, S. J. (1977). Bayesian Estimation of System Reliability for Weibull Distribution Using Monte Carlo Simulation, *The Theory and Applications of Reliability, Vol. II*, Academic, New York.

Kaufmann, A., Grouchko, D., and Cruon, R. (1977). *Mathematical Models for the Study of the Reliability of Systems*, Academic, New York.

Locks, M. O. (1978). *Monte Carlo Bayesian System Reliability—and MTBF—Confidence Assessment, II, Vol. 1: Theory and Vol. 2: SPARCS-2 Users Manual*, AFFDL-TR-78-18, Air Force Flight Dynamics Laboratory, Wright–Patterson AFB, OH.

Mann, N. R. (1970). Computer-Aided Selection of Prior Distributions for Generating Monte Carlo Confidence Bounds on System Reliability, *Naval Research Logistics Quarterly*, Vol. 17, pp. 41–54.

Mastran, D. V. (1976). Incorporating Component and System Test Data into the Same Assessment: A Bayesian Approach, *Operations Research*, Vol. 24, pp. 491–499.

Mastran, D. V. and Singpurwalla, N. D. (1978). A Bayesian Estimation of the Reliability of Coherent Structures, *Operations Research*, Vol. 26, pp. 663–672.

Neuts, M. F. and Zacks, S. (1976). On Mixtures of χ^2 and F-Distributions Which Yield Distributions of the Same Family, *Annals of the Institute of Statistical Mathematics*, Vol. 19, pp. 527–536.

Oberhettinger, F. (1974). *Tables of Mellin Transforms*, Springer–Verlag, New York.

Parker, J. B. (1972). Bayesian Prior Distributions for Multi-Component Systems, *Naval Research Logistics Quarterly*, Vol. 19, pp. 509–515.

Schafer, R. E. (1970). A Note on the Uniform Prior Distribution for Reliability, *IEEE Transactions on Reliability*, Vol. R-19, pp. 76–77.

Smith, D. R. (1977). An Analysis Regarding the Determination of Bayesian Confidence Limits for the Reliability of Distribution-Free Parallel Subsystems, *The Theory and Applications of Reliability, Vol. II*, Academic, New York.

Smith, D. R. and Springer, M. D. (1976). Bayesian Limits for the Reliability of Pass/Fail Parallel Units, *IEEE Transactions on Reliability*, Vol. R-25, pp. 213–215.

Springer, M. D. (1979). *The Algebra of Random Variables*, Wiley, New York.

Springer, M. D. and Byers, J. K. (1971). Bayesian Confidence Limits for the Reliability of Mixed Exponential and Distribution-Free Cascade Subsystems, *IEEE Transactions on Reliability*, Vol. R-20, pp. 24–28.

Springer, M. D. and Thompson, W. E. (1966a). The Distribution of Products of Independent Random Variables, *SIAM Journal of Applied Mathematics*, Vol. 14, pp. 511–526.

Springer, M. D. and Thompson, W. E. (1966b). Bayesian Confidence Limits for the Product of *N* Binomial Parameters, *Biometrika*, Vol. 53, pp. 611–613.

Springer, M. D. and Thompson, W. E. (1967). Bayesian Confidence Limits for the Reliability of Cascade Exponential Subsystems, *IEEE Transactions on Reliability*, Vol. R-16, pp. 86–89.

Springer, M. D. and Thompson, W. E. (1968). Bayesian Confidence Limits for Reliability of Redundant Systems when Tests are Terminated at First Failure, *Technometrics*, Vol. 10, pp. 29–36.

Thompson, W. E. and Chang, E. Y. (1975). Bayesian Confidence Limits for Reliability of Redundant Systems, *Technometrics*, Vol. 17, pp. 89–93.

Thompson, W. E. and Haynes, R. D. (1980). On the Reliability, Availability, and Bayes Confidence Intervals for Multicomponent Systems, *Naval Research Logistics Quarterly*, Vol. 27, pp. 354–358.

Wolf, J. E. (1976). Bayesian Reliability Assessment from Test Data, *Proceedings 1976 Annual Reliability and Maintainability Symposium*, pp. 411–419.

Zacks, S. (1977). Bayes Estimation of the Reliability of Series and Parallel Systems of Independent Exponential Components, *The Theory and Applications of Reliability, Vol. II*, Academic, New York.

CHAPTER 12

Availability of Maintained Systems

Throughout Chapter 11 Bayesian estimation was considered only for nonmaintained systems, that is, it was assumed that no maintenance action was initiated during the interval $(0, t)$. Specifically, there was neither

1. *Preventive (periodic) maintenance*: Regularly scheduled maintenance performed on a functioning system (e.g., regular care, if required, such as lubricating, cleaning, and adjusting; replacement of failed redundant components if the system contains redundancy; identification of components nearing a failed condition the intent of which is to keep the system in an operating condition [Bazovsky (1961), Chapter 17; Smith (1972), Chapter 9], nor
2. *Corrective (repair) maintenance*: Unscheduled maintenance performed on a failed system in order to restore it to an operating condition by replacing or repairing the failed component(s) that caused the system failure [Von Alven (1964), Chapter 10].

In this chapter we shall consider system availability—a measure of the effectiveness of a maintained system that incorporates the concepts of both reliability and maintainability. The concept of system availability was introduced in Section 4.1.5. An excellent survey of the literature on availability of maintained systems is provided by Lie, Hwang, and Tillman (1977).

12.1. AVAILABILITY

Recall from Chapter 4 that the availability of a system at time $t > 0$ is the probability that the system is operating satisfactorily at time t when used under stated conditions, where the interval $(0, t)$ is essentially divided into two categories: uptime and downtime. Recall that *uptime* is the time during which a system is in an operating state and *downtime* is usually taken to be the time during which a system is not in acceptable operating condition. In this chapter we are largely concerned with the portion of downtime during which the system is undergoing active repair, that is, the *active repair time*. Section 4.1.5 discusses the relationship between this and the other downtime categories.

Corrective maintenance is frequently associated with a system's downtime. However, as in the case of redundant systems, it also may be performed during such a system's operating time [Shooman (1968)]. For example, in a 2-component parallel system for which there is one repair facility, corrective maintenance may take place during the system's uptime (in the event that exactly one component has failed) or during the system's downtime (in the event both components have failed). If a single component or system treated as a unit has no repair facility, then the component/system availability and reliability are equivalent. If repair is allowed, then the reliability is unchanged but the availability is improved. This is also true for a series system. By repairing failed components in a redundant system during the system's uptime, availability as well as reliability will be improved (often there is a substantial improvement).

Preventive maintenance is usually associated with a system's uptime but may be performed at any time and in a variety of ways [Lie, Hwang, and Tillman (1977)]. The effect of preventive maintenance on availability is often debatable. One point of contention is that a satisfactory operating system should be left alone. On the other hand, it is often felt that adherence to a preventive maintenance schedule will, generally speaking, reduce operational failures and thereby increase availability. Unfortunately, it is often difficult to decide *when* to perform such maintenance. Aside from this difficulty, Shooman (1968, pp. 389–392) has shown preventive maintenance to be worthwhile when the system failure rate increases with time or when preventive maintenance of the system costs less than the corrective maintenance.

Let $X(t)=1$ if the system is operating at time t, and 0 if it is not. Recall that *availability* $A(t)$, at time $t>0$ is defined to be $A(t)=\Pr[X(t)=1]$. Recall from Section 4.8.2 that *steady-state availability* is given by

$$A=\lim_{t\to\infty} A(t), \tag{12.1}$$

when the limit exists, and represents the expected proportion of time in the long run that the system operates satisfactorily [Barlow and Proschan

(1975), p. 191]. Because of the time parameter, expressions for availability in most situations are usually rather cumbersome. In most cases, interval estimates (Bayesian or otherwise) for this type of availability have not yet been determined. Bayesian point estimates of instantaneous availability in certain situations have been derived by Brender (1968a, 1968b). One of these is discussed in Exercise 12.5.

Throughout this chapter we are primarily concerned with steady-state availability. A general expression for the availability of a component is now given. Consider a component (or a system treated as a single unit) that cycles between alternate intervals of operation and repair. Such a component is said to be *repairable*. It is assumed that repair begins immediately upon failure and, once repaired, the component is returned to the operating state. Let the r.v. U represent the uptime for the component with $U_i, i = 1,2,\ldots,$ denoting the duration of the ith operating period. We assume the sequence $\{U_i, i=1,2,\ldots\}$ of successive uptimes to be i.i.d as U. Let the r.v. D represent the downtime (repair time) for the component with $D_i, i = 1,2,\ldots,$ denoting the time required to effect the ith repair. We assume the sequence $\{D_i, i=1,2,\ldots\}$ of successive downtimes to be i.i.d as D. We assume the sequence $\{U_i + D_i, i=1,2,\ldots\}$ of cycle-times to be independent; however, for fixed i, we do not assume that U_i and D_i are independent. Under these assumptions the component is characterized by a two-state renewal process, and its availability is given by [Karlin and Taylor (1975), pp. 201–202]

$$A = \frac{E(U)}{E(U) + E(D)}. \tag{12.2}$$

Note that the expression in (12.2) depends only on the mean failure time, $E(U)$, and the mean repair time, $E(D)$. When describing the availability of a component it is customary to specify two things: (1) the distribution of the failure time and (2) the distribution of the repair time. Expressions for system availability depend not only on component failure and repair distributions but also on such factors as the system configuration and number of repair facilities. Such expressions are presented throughout the chapter whenever the discussion pertains to systems.

12.2. GENERAL FAILURE TIMES/GENERAL REPAIR TIMES

Consider a repairable component for which both the failure time and repair time distributions are unknown. Based on periodic observations that reveal whether the component is in operation or under repair, we determine Bayesian point and interval estimates of component availability. The methodology is then extended to a series system.

12.2.1 Component Availability

Suppose n independent observations are taken on a repairable component at widely spaced points in time that reveal only the state (either in operation or under repair) that prevails at the instant sampling occurs. Such observations are called "snapshots" [Gaver and Mazumdar (1967)]. Let s be the total number of times the component is observed in the operating state. Based on these attribute data, the sampling distribution is the binomial distribution with parameter A. By choosing the prior distribution of A to be $\mathcal{B}_1(s_0 + 1, n_0 - s_0 + 1)$, the posterior pdf of A is given by $\mathcal{B}_1(s + s_0 + 1, n + n_0 - s - s_0 + 1)$. In the prior pdf of A, s_0 may be interpreted as the total number of times a similar repairable component was observed in the operating state out of a total of n_0 previous "snapshots".

For a squared-error loss function, the Bayes point estimate of component availability is the mean of the posterior pdf of A:

$$E(A|s) = \frac{s + s_0 + 1}{n + n_0 + 2}, \tag{12.3}$$

where, for convenience, we have suppressed the notation showing the explicit dependency of $E(A|s)$ on s_0, n_0 and n. Due to the cumbersome nature of most of the posterior distributions in this chapter this convention is adopted throughout the chapter.

A symmetric $100(1-\gamma)\%$ TBPI estimate of A, (A_L, A_U), is determined from the percentiles of the beta posterior pdf of A; thus, $\mathcal{B}_1(A_L;\ s + s_0 + 1, n + n_0 - s - s_0 + 1) = \gamma/2$ and $\mathcal{B}_1(A_U;\ s + s_0 + 1, n + n_0 - s - s_0 + 1) = 1 - \gamma/2$, where $\mathcal{B}_1(x; p, q)$ is the cdf of the $\mathcal{B}_1(p, q)$ distribution evaluated at x. By (7.42), we obtain

$$A_L = \frac{s + s_0 + 1}{s + s_0 + 1 + (n + n_0 - s - s_0 + 1)\mathcal{F}_{1-\gamma/2}[2(n + n_0 - s - s_0 + 1), 2(s + s_0 + 1)]},$$

$$A_U = \frac{(s + s_0 + 1)\mathcal{F}_{1-\gamma/2}[2(s + s_0 + 1), 2(n + n_0 - s - s_0 + 1)]}{n + n_0 - s - s_0 + 1 + (s + s_0 + 1)\mathcal{F}_{1-\gamma/2}[2(s + s_0 + 1), 2(n + n_0 - s - s_0 + 1)]}. \tag{12.4}$$

A $100(1-\gamma)\%$ LBPI estimate of A is given by $(A_\gamma, 1)$, where

$$A_\gamma = \frac{s+s_0+1}{s+s_0+1+(n+n_0-s-s_0+1)\mathcal{F}_{1-\gamma}[2(n+n_0-s-s_0+1),2(s+s_0+1)]}. \tag{12.5}$$

Example 12.1. Suppose 10 widely spaced "snapshots" were taken on a component that cycles between alternate intervals of operation and repair. The component was observed in the up state seven times. From previous tests, an identical component was observed to be up six times in eight "snapshots". In this case, $n_0=8$, $s_0=6$, $n=10$, $s=7$. A Bayes point estimate of component availability is $E(A|s)=\frac{14}{20}=0.70$. A symmetric 95% TBPI for A is

$$\begin{aligned}(A_L, A_U) &= \left(14/\left[14+6\,\mathcal{F}_{0.975}(12,28)\right],\ 14\mathcal{F}_{0.975}(28,12)/\left[6+14\mathcal{F}_{0.975}(28,12)\right]\right)\\ &= (0.488, 0.874).\end{aligned}$$

■

12.2.2 Series System Availability

By using methodology identical to that given for series system reliability [Section 11.3.1], Thompson and Springer (1972) carry out a Bayesian analysis of the availability of a series system. Consider a system consisting of k independently operating repairable components in series. The system is such that

1. while a failed component is undergoing repair, nonfailed components continue to operate (the system is down, of course, during this time), and
2. there is a repair facility for each component that effects a repair independently of that of any other repair facility. Figure 12.1 depicts this case.

System availability A is given by $A=\prod_{i=1}^{k} A_i$, where A_i is availability of the ith component [Barlow and Proschan (1975), pp. 192–193].

Suppose that n_i "snapshots" are taken on the ith component with s_i the number of times the component is observed in the operating state. Under the $\mathcal{B}_1(s_{i0}+1, n_{i0}-s_{i0}+1)$ prior distribution on component availabilities,

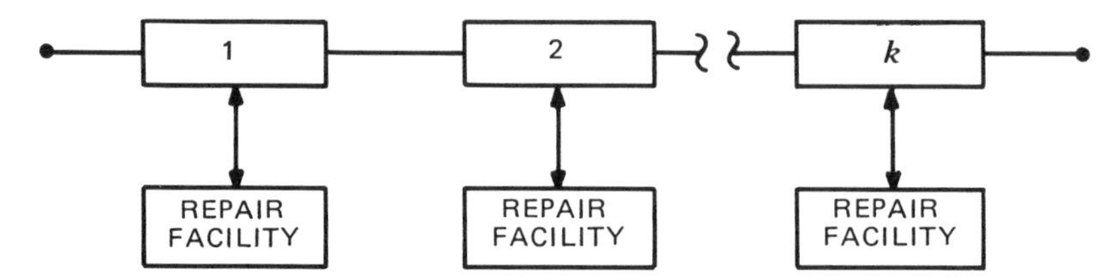

Figure 12.1. A k-component series system with k repair facilities.

the posterior pdf of system availability, $g(A|\underset{\sim}{s})$, where $\underset{\sim}{s}'=(s_1,\ldots,s_k)$, has Mellin transform [see (11.22)]

$$M\{g(A|\underset{\sim}{s});u\}=\left[\prod_{i=1}^{k}\frac{(n_i'+1)!}{s_i'!}\right]\frac{1}{(u+s)^{c_0}(u+s+1)^{c_1}\cdots(u+n)^{c_{n-s}}}, \tag{12.6}$$

where $\mathrm{Re}(u)>-s$, $n=\max(n_i'=n_{i0}+n_i)$, $s=\min(s_i'=s_{i0}+s_i)$, and c_j, $j=0,1,\ldots,n-s$ is the collective exponent of the factor $(u+s+j)^{c_j}$.

For a squared-error loss function, the Bayes point estimate of system availability is the mean of the posterior pdf $g(A|\underset{\sim}{s})$:

$$E(A|\underset{\sim}{s})=M\{g(A|\underset{\sim}{s});u=2\}=\prod_{i=1}^{k}\left(\frac{s_{i0}+s_i+1}{n_{i0}+n_i+2}\right). \tag{12.7}$$

By expanding (12.6) in partial fractions and then using the Mellin inversion integral, a closed form expression for $g(A|\underset{\sim}{s})$ can be determined [Thompson and Springer (1972)]. The posterior cdf, $G(A|\underset{\sim}{s})$, can be found by integrating $g(A|\underset{\sim}{s})$ directly and is given by

$$G(A|\underset{\sim}{s})=A^{s+1}\sum_{j=0}^{n-s}A^j\left[B_{j1}(\ln A^{-1})^0+\cdots+B_{jc_j}(\ln A^{-1})^{c_j-1}\right], \tag{12.8}$$

where

$$B_{j\ell}=\frac{1}{(\ell-1)!}\sum_{r=0}^{c_j-\ell}\frac{K_{j,\ell+r}}{(s+j+1)^{r+1}},\quad \ell=1,2,\ldots,c_j, \tag{12.9}$$

and

$$K_{jm} = \begin{cases} \dfrac{1}{(c_j - m)!} \dfrac{d^{c_j - m}}{du^{c_j - m}} \left[(u+s+j)^{c_j} M\{g(A|\underset{\sim}{s}); u\} \right] \Big|_{u=-(s+j)}, & m = 1, 2, \ldots, c_j \\ 0, & c_j = 0. \end{cases} \tag{12.10}$$

A symmetric $100(1-\gamma)\%$ TBPI for system availability, (A_L, A_U), is determined by solving the equations $G(A_L|\underset{\sim}{s}) = \gamma/2$ for A_L and $G(A_U|\underset{\sim}{s}) = 1-\gamma/2$ for A_U. A $100(1-\gamma)\%$ LBPI for system availability is $(A_\gamma, 1)$, where A_γ is the solution to the equation $G(A_\gamma|\underset{\sim}{s}) = \gamma$.

12.3. EXPONENTIAL FAILURE TIMES/EXPONENTIAL REPAIR TIMES

Consider a repairable component for which both the failure and repair times follow an $\mathcal{E}(\cdot)$ distribution. The choice of the $\mathcal{E}(\lambda)$ distribution to model a component's failure process has become rather commonplace. One rationale is that many electronic components do not fail from deterioration but from overstress, and these overstressed conditions are randomly distributed. On the other hand, the use of an $\mathcal{E}(\mu)$ distribution to model the repair process is often debatable. Under a constant repair rate, in effect the repairman learns nothing about what has caused the failure as his work proceeds. As noted by Shooman (1968, p. 348): "One situation which does seem to satisfy this assumption is the case of a poor repairman searching mainly by trial and error to replace a plug-in module until the correct one is found." Rohn (1959) contends that repair times of complex electronic equipment are often characterized by a high frequency of short repair times and a few long repair times; thus this sort of behavior suggests representation by $\mathcal{E}(\mu)$ distribution. In any event, $\mathcal{E}(\mu)$ distributed repair times are often assumed for reasons of mathematical tractability with the consolation that either the repair times are, in fact, $\mathcal{E}(\mu)$ distributed, or, if not, they have a positively skewed distribution that can be approximated by an $\mathcal{E}(\mu)$ distribution. Alternately, in Section 12.4 we consider the $\mathcal{G}_1(\kappa, \mu)$ distribution as a repair model.

12.3.1 Component Availability

The availability of a repairable component for which the failure time X has an $\mathcal{E}(\lambda)$ distribution, and the repair time Y has an $\mathcal{E}(\mu)$ distribution, is given by (12.2) as

$$A = \frac{E(X)}{E(X)+E(Y)} = \frac{\mu}{\lambda+\mu}. \tag{12.11}$$

It is assumed that X and Y are independent. Inferences on A depend on the manner in which the data are obtained. The following general sampling procedure is proposed by Gaver and Mazumdar (1967, 1969). Suppose a sequence of continuous observations are made on the up and downtimes of the component throughout intervals of random duration. These observations are called "patches". For example, suppose "snapshots" are made at *random* instants. It is the time remaining in a given state after the "snapshot" has been made (rather than the total time in the state, before and after the observation) that is $\mathcal{E}(\cdot)$ distributed with the parameter appropriate to the state observed. In this case, a particular "patch" observation contains two pieces of information: (1) the state (in operation or under repair) of the component when the "snapshot" was made and (2) the length of time the component remained in that state.

We define the following quantities:

- a Total number of operating intervals observed during "patches",
- x_+ Total operating time observed; $x_+ = \sum_{i=1}^{a} x_i$, where x_i is the observed failure time for the ith operating interval,
- b Total number of repair intervals observed during "patches",
- y_+ Total repair time observed; $y_+ = \sum_{i=1}^{b} y_i$, where y_i is the observed repair time for the ith repair interval,
- s Total number of "snapshots" showing the system to be in operation from n (fixed) "snapshots" taken.

The two most frequently considered sampling plans are as follows:

Case 1. A component's operating and repair history is continuously recorded through k cycles, each consisting of one operating period and one repair period (possibly during the developmental phase of the component). Thereafter, n "snapshots" are taken, s of which show the component to be in operation. In this case, $a = b = k$. In the event that no "snapshots" are taken, then $n = s = 0$.

Case 2. A component is observed n times at random intervals and, each time, the component's state and the remaining time in that state are recorded. If s is the number of "snapshots" showing the component to be in operation, then $a = s$ and $b = n - s$.

Under the general sampling procedure described above, the sampling distribution is given by

$$f(a, b, x_+, y_+, s|\mu, \lambda) =$$

$$\binom{n}{s}\left(\frac{\mu}{\lambda+\mu}\right)^s\left(\frac{\lambda}{\lambda+\mu}\right)^{n-s}\lambda^a \exp(-\lambda x_+)\mu^b \exp(-\mu y_+). \tag{12.12}$$

We assume that λ and μ (where λ and μ denote both r.v.'s and their values) each have independent gamma prior pdfs given by

$$g_1(\lambda) = [\Gamma(\alpha)]^{-1}\xi^\alpha\lambda^{\alpha-1}\exp(-\xi\lambda), \qquad \lambda > 0, \tag{12.13}$$

and

$$g_2(\mu) = [\Gamma(\beta)]^{-1}\eta^\beta\mu^{\beta-1}\exp(-\eta\mu), \quad \mu > 0. \tag{12.14}$$

The joint posterior pdf of λ and μ is given by

$$g_{\lambda,\mu}(\lambda, u|a, b, x_+, y_+, s) =$$

$$\frac{f(a, b, x_+, y_+, s|\mu, \lambda)g_1(\lambda)g_2(\mu)}{\int_0^\infty\int_0^\infty f(a, b, x_+, y_+, s|\mu, \lambda)g_1(\lambda)g_2(\mu)\,d\mu\,d\lambda}, \qquad \lambda > 0, \mu > 0. \tag{12.15}$$

The integration of the denominator of the right-hand side of (12.15) is readily carried out, although quite cumbersome, by making use of two integration formulas given by Gradshteyn and Ryzhik (1965, p. 319: 3.383.8 and p. 860: 7.621.3). Accordingly, it can be shown that (12.15) may be written as

$$g_{\lambda,\mu}(\lambda, \mu|a, b, x_+, y_+, s) = c_1\left(\frac{\mu}{\lambda+\mu}\right)^s\left(\frac{\lambda}{\lambda+\mu}\right)^{n-s}\lambda^{a+\alpha-1}\mu^{b+\beta-1}$$

$$\times\exp\{-[(x_+ + \xi)\lambda + (y_+ + \eta)\mu]\},$$

$$\lambda > 0, \mu > 0, \tag{12.16}$$

where

$$c_1 = \frac{(x_+ + \xi)^{n-s+a+\alpha}/(y_+ + \eta)^{n-s-b-\beta}}{B(s+b+\beta, n-s+a+\alpha)\Gamma(a+b+\alpha+\beta)_2F_1[n, n-s+a+\alpha; n+a+b+\alpha+\beta; 1-(y_+ + \eta)/(x_+ + \xi)]} \tag{12.17}$$

and

$${}_2F_1(\alpha, \beta; \gamma; z) = \sum_{j=0}^{\infty} \frac{(\alpha)_j(\beta)_j}{(\gamma)_j j!} z^j, \tag{12.18}$$

with

$$(\alpha)_0 = 1, (\alpha)_j = \alpha(\alpha+1)\cdots(\alpha+j-1), \quad j = 1, 2, 3, \ldots, \tag{12.19}$$

is the hypergeometric series of the variable z with parameters α, β, γ [Erdélyi (1953), Chapter II]. If $n=0$, then ${}_2F_1 \equiv 1$; thus $g_{\lambda,\mu}(\lambda, \mu|\cdot)$ is well defined. If $n>0$, then $g_{\lambda,\mu}(\lambda, \mu|\cdot)$ exists only provided $|1-(y_+ + \eta)/(x_+ + \xi)| < 1$ or, equivalently, $0 < (y_+ + \eta)/(x_+ + \xi) < 2$. This would not seem too restrictive as it is likely that the component's total operating time exceeds its total repair time.

The prior parameters, α, ξ, β, η are assumed to be positive constants that will likely be estimated from the prior data available about the distribution of λ and μ. In the event of no prior information, we take $\alpha = \beta = 1$, $\xi = \eta = 0$. For these values of the parameters, (12.16) reduces to the standardized likelihood function which is the joint posterior pdf of λ and μ corresponding to uniform prior pdf's on both λ and μ. By taking $\alpha = \beta = \xi = \eta = 0$, we get noninformative prior pdf's on both λ and μ.

The posterior pdf of $\rho = \lambda/\mu$ (which, in queueing theory, is sometimes referred to as the *service factor*) is given by

$$\begin{aligned} g_\rho(\rho|a, b, x_+, y_+, s) &= \int_0^\infty g_{\lambda,\mu}(\mu\rho, \mu|a, b, x_+, y_+, s)\mu\, d\mu \\ &= c_1(1+\rho)^{-n}\rho^{n-s+a+\alpha-1}\int_0^\infty \mu^{a+b+\alpha+\beta-1} \times \\ &\quad \exp\{-\mu[(y_+ + \eta) + \rho(x_+ + \xi)]\}\, d\mu \\ &= \frac{c_1\Gamma(a+b+\alpha+\beta)\rho^{n-s+a+\alpha-1}}{(1+\rho)^n[(y_+ + \eta) + \rho(x_+ + \xi)]^{a+b+\alpha+\beta}}, \quad \rho > 0. \end{aligned} \tag{12.20}$$

Consequently, the posterior pdf of component availability $A=1/(1+\rho)$ is given by $g(A|a, b, x_+, y_+, s)=g_\rho(A^{-1}-1|a, b, x_+, y_+, s)\cdot A^{-2}$. Thus

$$g(A|a,b,x_+,y_+,s)=c_2\frac{A^{s+b+\beta-1}(1-A)^{n-s+a+\alpha-1}}{\{1-A[1-(y_+ +\eta)/(x_+ +\xi)]\}^{a+b+\alpha+\beta}},$$

$$0<A<1, \quad (12.21)$$

where

$$c_2=\frac{[(x_+ +\xi)/(y_+ +\eta)]^{n-s-b-\beta}}{B(s+b+\beta,n-s+a+\alpha)\,{}_2F_1[n,n-s+a+\alpha;n+a+b+\alpha+\beta;1-(y_+ +\eta)/(x_+ +\xi)]}.$$

$$(12.22)$$

The pdf in (12.21) is called an *Euler distribution* [Brender (1968a)] due to Euler's integral representation of the hypergeometric function [Erdélyi (1953), p. 59].

Under a squared-error loss function, the Bayes point estimator of component availability A is the mean of the posterior pdf of A. Making use of Euler's integral representation of the hypergeometric function, we find that

$$E(A|a,b,x_+,y_+,s)=\frac{(s+b+\beta)}{(n+a+b+\alpha+\beta)}\times$$

$$\frac{{}_2F_1[n+1,n-s+a+\alpha;n+a+b+\alpha+\beta+1;1-(y_+ +\eta)/(x_+ +\xi)]}{{}_2F_1[n,n-s+a+\alpha;n+a+b+\alpha+\beta;1-(y_+ +\eta)/(x_+ +\xi)]},$$

$$(12.23)$$

which exists only provided $0<(y_+ +\eta)/(x_+ +\xi)<2$. Gaver and Mazumdar (1969) consider loss functions other than squared-error and give the corresponding expressions for Bayes point estimators of A. The existence of each, however, requires that $0<(y_+ +\eta)/(x_+ +\xi)<2$. For computational purposes, the hypergeometric functions in (12.23) may be approximated by a Padé rational approximation [see Luke (1977), Chapter 13].

The posterior cdf of component availability A is given by

$$G(A|a,b,x_+,y_+,s)=\int_0^A g(t|a,b,x_+,y_+,s)\,dt$$

$$=c_2\int_0^A\frac{t^{s+b+\beta-1}(1-t)^{n-s+a+\alpha-1}\,dt}{\{1-t[1-(y_+ +\eta)/(x_+ +\xi)]\}^{a+b+\alpha+\beta}}.$$

$$(12.24)$$

Let $z=1-(y_+ + \eta)/(x_+ + \xi)$. By the binomial series,

$$(1-zt)^{-(a+b+\alpha+\beta)} = \sum_{j=0}^{\infty} \frac{(a+b+\alpha+\beta)_j}{j!}(zt)^j.$$

Thus, we may write (12.24) as

$$G(A|a,b,x_+,y_+,s) = c_2 \sum_{j=0}^{\infty} \frac{(a+b+\alpha+\beta)_j}{j!} z^j$$

$$\times \int_0^A t^{j+s+b+\beta-1}(1-t)^{n-s+a+\alpha-1}\,dt$$

$$= \frac{[(x_+ + \xi)/(y_+ + \eta)]^{n-s-b-\beta}}{{}_2F_1[n, n-s+a+\alpha; n+a+b+\alpha+\beta; 1-(y_+ + \eta)/(x_+ + \xi)]}$$

$$\times \sum_{j=0}^{\infty} \frac{(a+b+\alpha+\beta)_j (s+b+\beta)_j}{(n+a+b+\alpha+\beta)_j j!} [1-(y_+ + \eta)/$$

$$(x_+ + \xi)]^j \mathcal{B}_1(A; j+s+b+\beta, n-s+a+\alpha). \tag{12.25}$$

A symmetric $100(1-\gamma)\%$ TBPI for component availability, (A_L, A_U), is determined by solving the equations $G(A_L|a,b,x_+,y_+,s)=\gamma/2$ for A_L and $G(A_U|a,b,x_+,y_+,s)=1-\gamma/2$ for A_U. A $100(1-\gamma)\%$ LBPI for component availability is $(A_\gamma, 1)$, where A_γ is the solution to the equation $G(A_\gamma|a,b,x_+,y_+,s)=\gamma$.

Example 12.2. A repairable component with $\mathcal{E}(\cdot)$ distributed failure and repair times is monitored continuously through five failure/repair cycles, after which six "snapshots" are taken at 12-hour intervals and four uptimes are observed. Based on simulation ($\lambda=0.111$, $\mu=0.667$; thus $A=0.857$) the data are

Cycle	Failure Times (h)	Repair Times (h)
1	2.96	1.55
2	23.43	1.91
3	2.90	1.09
4	8.02	1.84
5	11.62	0.28

Number of "Snapshots": 6 Number of Uptimes: 4

From the data: $a=b=5$, $x_+=48.93$, $y_+=6.67$, $n=6$, $s=4$. We wish to determine Bayes point and interval estimates of component availability under uniform prior pdfs on both the failure and repair rates ($\alpha=\beta=1$, $\xi=\eta=0$).

By (12.23), the Bayes point estimate of A under a squared-error loss function is

$$E(A|a,b,x_+,y_+,s)=\frac{10}{18}\times\frac{{}_2F_1(7,8;\ 19;\ 0.864)}{{}_2F_1(6,8;\ 18;\ 0.864)}=0.827.$$

Note that if only the "patch" observations are used ($n=s=0$), then

$$E(A|a,b,x_+,y_+,s)=0.5\times{}_2F_1(1,6;\ 13;\ 0.864)=0.866.$$

By (12.25), the posterior cdf of A is

$$G(A|a,b,x_+,y_+,s)=\frac{(0.136)^4}{{}_2F_1(6,8;\ 18;\ 0.864)}\times \sum_{j=0}^{\infty}\frac{(12)_j(10)_j}{(18)_j j!}(0.864)^j\mathcal{B}_1(A;j+10,8).$$

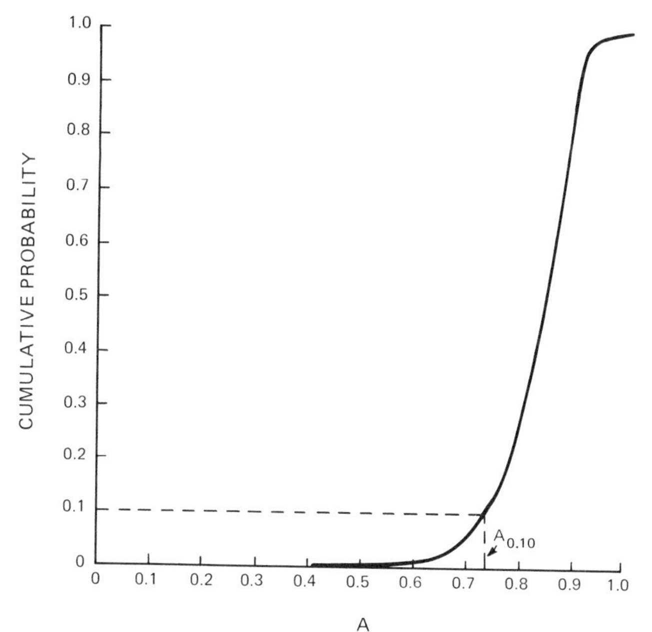

Figure 12.2. The posterior cumulative distribution function $G(A|a,b,x_+,y_+,s)$ for Example 12.2.

This expression has been evaluated for $A = 0.05(0.05)0.95$. Based on these tabulations, the graph of $G(A|a, b, x_+, y_+, s)$ is shown in Figure 12.2. From the graph, a 90% LBPI for component availability is (0.74, 1), which is also shown in Figure 12.2. ■

We now consider the important special case when a component's up and down history is continuously recorded through k cycles and no "snapshots" are taken thereafter (see Case 1); thus, $a = b = k$ and $n = s = 0$. In this situation there is a considerable simplification of the previously derived expressions for the Bayes point and interval estimates of A.

By (12.21), the posterior pdf of A reduces to

$$g(A|x_+, y_+) = \frac{[(y_+ + \eta)/(x_+ + \xi)]^{k+\beta}}{B(k+\beta, k+\alpha)} \times \frac{A^{k+\beta-1}(1-A)^{k+\alpha-1}}{\{1 - A[1-(y_+ + \eta)/(x_+ + \xi)]\}^{2k+\alpha+\beta}},$$

$$0 < A < 1. \quad (12.26)$$

Under a squared-error loss function, the Bayes point estimator of A is

$$E(A|x_+, y_+) = \frac{k+\beta}{2k+\alpha+\beta} \times {}_2F_1[1, k+\alpha; 2k+\alpha+\beta+1; 1-(y_+ + \eta)/(x_+ + \xi)], \quad (12.27)$$

provided $0 < (y_+ + \eta)/(x_+ + \xi) < 2$. For computational purposes, the hypergeometric function in (12.27) often may be very quickly and accurately approximated by a Padé rational approximation [Luke (1977), Chapter 14]. The general result is

$${}_2F_1(1, \beta; \gamma; -z) \cong \frac{A_n(z)}{B_n(z)}, \quad (12.28)$$

where

$$B_0(z) = 1, \ B_1(z) = 1 + (\beta+1)z/(\gamma+1),$$

$$A_0(z) = 1, \ A_1(z) = B_1(z) - \beta z/\gamma,$$

and both the polynomials $A_n(z)$ and $B_n(z)$ satisfy the same recursion formula

$$B_n(z) = (1 + F_1 z)B_{n-1}(z) + F_2 z^2 B_{n-2}(z), \qquad n \geqslant 2,$$

$$F_1 = \frac{2n^2 + 2n(\gamma - 2) + (\beta - 1)(\gamma - 1)}{(2n + \gamma - 1)(2n + \gamma - 3)},$$

$$F_2 = -\frac{(n-1)(n-2+\gamma-\beta)(n+\beta-1)(n+\gamma-2)}{(2n+\gamma-2)(2n+\gamma-3)^2(2n+\gamma-4)}.$$

By (12.24), the posterior cdf of A is

$$G(A|x_+, y_+) = \frac{[(y_+ + \eta)/(x_+ + \xi)]^{k+\beta}}{B(k+\beta, k+\alpha)} \times$$

$$\int_0^A \frac{t^{k+\beta-1}(1-t)^{k+\alpha-1}\,dt}{\{1 - t[1 - (y_+ + \eta)/(x_+ + \xi)]\}^{2k+\alpha+\beta}}$$

which, using the transformation

$$z = t(k+\alpha)(y_+ + \eta)/(1-t)(k+\beta)(x_+ + \xi),$$

may be written as

$$G(A|x_+, y_+) = \frac{\Gamma(2k+\alpha+\beta)}{\Gamma(k+\beta)\Gamma(k+\alpha)}\left(\frac{k+\beta}{k+\alpha}\right)^{k+\beta} \times$$

$$\int_0^{\frac{A(k+\alpha)(y_+ + \eta)}{(1-A)(k+\beta)(x_+ + \xi)}} \frac{z^{k+\beta-1}}{\{1 + [(k+\beta)/(k+\alpha)]z\}^{2k+\alpha+\beta}}\,dz$$

$$= \mathfrak{F}[A(k+\alpha)(y_+ + \eta)/(1-A)(k+\beta)(x_+ + \xi);\, 2(k+\beta), 2(k+\alpha)], \tag{12.29}$$

where $\mathfrak{F}(x; n_1, n_2)$ is the cdf of an $\mathfrak{F}(n_1, n_2)$ distribution evaluated at x.

A symmetric $100(1-\gamma)\%$ TBPI for component availability A, (A_L, A_U), is determined by solving the equations $G(A_L|x_+, y_+) = \gamma/2$ for A_L and $G(A_U|x_+, y_+) = 1 - \gamma/2$ for A_U. Consequently,

$$A_L = \frac{(k+\beta)(x_+ + \xi)}{(k+\beta)(x_+ + \xi) + (k+\alpha)(y_+ + \eta)\mathfrak{F}_{1-\gamma/2}[2(k+\alpha), 2(k+\beta)]} \tag{12.30}$$

and

$$A_U = \frac{(k+\beta)(x_+ + \xi)\mathcal{F}_{1-\gamma/2}[2(k+\beta), 2(k+\alpha)]}{(k+\alpha)(y_+ + \eta) + (k+\beta)(x_+ + \xi)\mathcal{F}_{1-\gamma/2}[2(k+\beta), 2(k+\alpha)]}. \tag{12.31}$$

A $100(1-\gamma)\%$ LBPI for component availability A is $(A_\gamma, 1)$, where A_γ is given by (12.30) with $\gamma/2$ replaced by γ.

The posterior cdf of A given by (12.29) may also be used to solve another type of problem which we now describe. Suppose a component (or system treated as a single unit) availability goal is specified as part of a contract specification. Specifically, it is required that component availability be at least γ_1, and the probability that this be attained be at least γ_2. Analytically we require that

$$\Pr(A \geqslant \gamma_1 | x_+, y_+) \geqslant \gamma_2, \tag{12.32}$$

or, equivalently,

$$G(\gamma_1 | x_+, y_+) \leqslant 1 - \gamma_2. \tag{12.33}$$

For example, it may be required to design a component such that $\gamma_1 = .99$ and where $\gamma_2 = .90$. In this case, the probability is at least .90 that the availability design value of .99 will be met in practice. In other words, at least 90% of the time in which the component is called upon for operation, its availability is at least 0.99.

By using (12.29) in (12.33), it follows that

$$\frac{\gamma_1(k+\alpha)(y_+ + \eta)}{(1-\gamma_1)(k+\beta)(x_+ + \xi)} \leqslant \mathcal{F}_{1-\gamma_2}[2(k+\beta), 2(k+\alpha)]$$

or, equivalently,

$$\frac{(y_+ + \eta)}{(x_+ + \xi)} \leqslant \frac{(1-\gamma_1)(k+\beta)\mathcal{F}_{1-\gamma_2}[2(k+\beta), 2(k+\alpha)]}{\gamma_1(k+\alpha)}. \tag{12.34}$$

The interpretation of (12.34) for uniform or noninformative prior pdfs on the failure and repair rates is as follows. The right-hand side is an upper bound on the ratio of the observed average repair time to the observed average failure time. If, based on the data, the ratio does not exceed this upper bound, then the availability design criteria is satisfied. Otherwise, consideration should be given to redesigning the component either through a reduction in the repair time or an increase in the operating time.

Example 12.3. [Thompson (1966)] During field testing of a communications system treated as a single unit, five failure/repair cycles were experienced. The average failure time was $\bar{x}_+ = (1/5)x_+ = 123.5$ h. The average repair time was $\bar{y}_+ = (1/5)y_+ = 2.7$ h. Previous experience with similar systems indicates that both failure and repair times are $\mathcal{E}(\cdot)$ distributed. It is assumed that the time required to effect a repair does not depend on how long the system operated prior to the failure. A Bayes point estimate and 90% LBPI for system availability are required.

Under uniform prior pdfs ($\alpha = \beta = 1, \xi = \eta = 0$), a Bayesian point estimate of A is given by (12.27) as

$$E(A|x_+, y_+) = 0.5 \times {}_2F_1(1,6;\ 13;0.979) = 0.976.$$

From (12.30),

$$A_L = \frac{6(123.5)}{6(123.5) + 6(2.7)(2.147)} = 0.955;$$

thus the required interval estimate of A is (0.955, 1).

Note that with probability 0.90, system availability is at least 0.955. This does not, however, satisfy a preset system availability goal of at least 0.99. Consequently, a redesign of the system should be considered. With regard to such a possible redesign, the following information should be useful. First note that for five failure/repair cycles, an availability goal of at least 0.99 with probability of at least 0.90 requires that [see (12.34)]

$$\bar{y}_+ \leqslant \frac{(0.01)(6)(1/2.147)}{0.99(6)} \bar{x}_+ = 0.005\bar{x}_+ .$$

For the original system, the estimated repair rate is about 46 times the estimated failure rate. For the availability goal to be attained, the estimated repair rate should have been at least 200 times the estimated failure rate. ■

12.3.2 Series System Availability

Consider a system consisting of N independently operating repairable components in series (see Figure 12.1) such that

1. While a failed component is undergoing repair, nonfailed components continue to operate.

2. There is a repair facility for each component that effects a repair independently of that of any other repair facility.
3. For the ith component, the failure time X_i and repair time Y_i are independent r.v.'s with $\mathcal{E}(\cdot)$ pdfs

$$f_{X_i}(x_i) = \lambda_i \exp(-\lambda_i x_i), \qquad x_i > 0,$$

$$f_{y_i}(y_i) = \mu_i \exp(-\mu_i y_i), \qquad y_i > 0.$$

The parameters λ_i and μ_i are the failure rate and repair rate, respectively, for the ith component.

System availability A is the product of component availabilities [Sandler (1963), pp. 125–126]; thus,

$$A = \prod_{i=1}^{N} \left(\frac{\mu_i}{\lambda_i + \mu_i} \right). \tag{12.35}$$

For the ith component, suppose k_i failure/repair cycles are observed: the total operating time observed is $x_i^+ = \sum_{j=1}^{k_i} x_{ij}$, where x_{ij} is the jth failure time; the total repair time observed is $y_i^+ = \sum_{j=1}^{k_i} y_{ij}$, where y_{ij} is the jth repair time. For $\mathcal{G}_1(\alpha_i, \xi_i)$ and $\mathcal{G}_1(\beta_i, \eta_i)$ prior distributions on λ_i and μ_i, respectively, the posterior pdf of the ith component availability, $A_i = \mu_i/(\lambda_i + \mu_i)$, is an Euler distribution [see (12.26)]

$$g_i(A_i | x_i^+, y_i^+) = \frac{[(y_i^+ + \eta_i)/(x_i^+ + \xi_i)]^{k_i + \beta_i}}{B(k_i + \beta_i, k_i + \alpha_i)} \times$$

$$\frac{A_i^{k_i + \beta_i - 1}(1 - A_i)^{k_i + \alpha_i - 1}}{\{1 - A_i[1 - (y_i^+ + \eta_i)/(x_i^+ + \xi_i)]\}^{2k_i + \alpha_i + \beta_i}},$$

$$0 < A_i < 1. \tag{12.36}$$

By making use of Euler's integral representation of the hypergeometric function [Erdélyi (1953), p. 59], it can be shown that the Mellin transform of $g_i(A_i | x_i^+, y_i^+)$ is

$$M\{g_i(A_i | x_i^+, y_i^+); u\} = \frac{B(k_i + \beta_i + u - 1, k_i + \alpha_i)}{B(k_i + \beta_i, k_i + \alpha_i)} \times$$

$${}_2F_1[u - 1, k_i + \alpha_i; 2k_i + \alpha_i + \beta_i + u - 1; 1 - (y_i^+ + \eta_i)/(x_i^+ + \xi_i)],$$

$$\tag{12.37}$$

provided $0<(y_i^+ + \eta_i)/(x_i^+ + \xi_i)<2$ and $\text{Re}(u)>1-k_i-\beta_i$. As the Mellin transform of the posterior pdf of system availability, $g(A|\underset{\sim}{x}^+, \underset{\sim}{y}^+)$, where $\underset{\sim}{x}^+=(x_1^+,\ldots,x_N^+)$ and $\underset{\sim}{y}^+=(y_1^+,\ldots,y_N^+)$, is the product of the Mellin transforms of component availabilities [see (11.12)]; then

$$M\{g(A|\underset{\sim}{x}^+, \underset{\sim}{y}^+); u\} = \prod_{i=1}^{N} M\{g_i(A_i|x_i^+, y_i^+); u\}, \qquad (12.38)$$

where $\text{Re}(u)>1-\min(k_i+\beta_i)$ and $\max[(y_i^+ + \eta_i)/(x_i^+ + \xi_i)]<2$.

Under a squared-error loss function, the Bayes point estimate of system availability A is

$$E(A|\underset{\sim}{x}^+, \underset{\sim}{y}^+) = M\{g(A|\underset{\sim}{x}^+, \underset{\sim}{y}^+); u=2\} =$$

$$\prod_{i=1}^{N}\left(\frac{k_i+\beta_i}{2k_i+\alpha_i+\beta_i}\right){}_2F_1[1, k_i+\alpha_i; 2k_i+\alpha_i+\beta_i+1; 1-(y_i^+ + \eta_i)/(x_i^+ + \xi_i)], \qquad (12.39)$$

provided $\max[(y_i^+ + \eta_i)/(x_i^+ + \xi_i)]<2$.

The Mellin inversion of (12.38) appears intractable [Thompson and Palicio (1975)]; thus, a closed form expression for $g(A|\underset{\sim}{x}^+, y^+)$ would not seem possible. Consequently, an alternate approach will be given to determine an approximation to the posterior cdf, $G(A\,|\,\underset{\sim}{x}^+, y^+)$, from which Bayes interval estimates of A can be found. An approximation to $G(A|\underset{\sim}{x}^+, y^+)$ is provided by a step function, say $V_m(A)$ where $0\leqslant A\leqslant 1$, having the same first $2m$ moments as $G(A|\underset{\sim}{x}^+, y^+)$. An alternate method for approximating $G(A|\underset{\sim}{x}^+, y^+)$ is given by Dyer (1982). Furthermore, the step function $V_m(A)$ is such that at m abscissa values $A_1^*, A_2^*,\ldots,A_m^*$ the value of $G(A|\underset{\sim}{x}^+, y^+)$ is contained in step intervals of length of $S_1, S_2,\ldots,S_m$, respectively. The technique is credited to Von Mises (1964), pp. 384–401, and the procedure is outlined below.

1. Determine the rth moment of A, $M_r = E(A^r\,|\,\underset{\sim}{x}^+, \underset{\sim}{y}^+) = M\{g(A|\underset{\sim}{x}^+, y^+); u=r+1\}$, $r=0,1,\ldots,2m-1$.
2. Solve the following system of linear equations simultaneously for c_j, $j=0,1,\ldots,m-1$:

$$\sum_{j=0}^{m-1} c_j M_{i+j} = -M_{m+i}, \quad i=0,1,\ldots,m-1. \qquad (12.40)$$

3. Having obtained the values of c_j, $j = 0, 1, \ldots, m-1$, the abscissas $A_1^*, A_2^*, \ldots, A_m^*$ are the roots of the polynomial equation

$$x^m + c_{m-1}x^{m-1} + c_{m-2}x^{m-2} + \cdots + c_1 x + c_0 = 0. \quad (12.41)$$

4. Having obtained the abscissas $A_1^*, A_2^*, \ldots, A_m^*$, the step sizes $S_1, S_2, \ldots, S_m$ are determined from the m equations

$$\sum_{j=1}^{m} (A_j^*)^i S_j = M_i, \quad i = 0, 1, \ldots, m-1. \quad (12.42)$$

The computation is easily carried out by a computer using library programs for the solution of the polynomial equation (12.41) and the systems of simultaneous linear equations, (12.40) and (12.42). The posterior cdf $G(A \mid \underset{\sim}{x}^+, \underset{\sim}{y}^+)$ is approximated with increasing accuracy as m increases.

Example 12.4. [Thompson and Palicio (1975)] Consider a series system consisting of two (nonidentical) repairable components each of which has $\mathcal{E}(\cdot)$ distributed failure and repair times. There are two independently operating repair facilities—one for each component. In addition, while a failed component is undergoing repair, a nonfailed component continues to operate; the system, however, is down during such a situation.

Suppose six failure/repair cycles are observed on component 1 and 12 on component 2 with the following results:

$$k_1 = 6, \; y_1^+/x_1^+ = 1.000,$$

$$k_2 = 12, \; y_2^+/x_2^+ = 0.250.$$

Under noninformative prior pdfs ($\alpha_i = \beta_i = \eta_i = \xi_i = 0, i = 1, 2$) on each component's failure and repair rates, the Bayes point estimate of system availability is, by (12.39),

$$E(A \mid \underset{\sim}{x}^+, \underset{\sim}{y}^+) = [0.5 \times {}_2F_1(1, 6; 13; 0)][0.5 \times {}_2F_1(1, 12; 25; 0.750)]$$

$$= (0.5)(1.0)(0.5)(1.58397) = 0.396.$$

To determine a Bayes interval estimate of system availability, we use an 18-step Von Mises step function to approximate the posterior cdf $G(A \mid \underset{\sim}{x}^+, \underset{\sim}{y}^+)$. First, the rth moments of A, $r = 0, 1, \ldots, 35$, are calculated from the

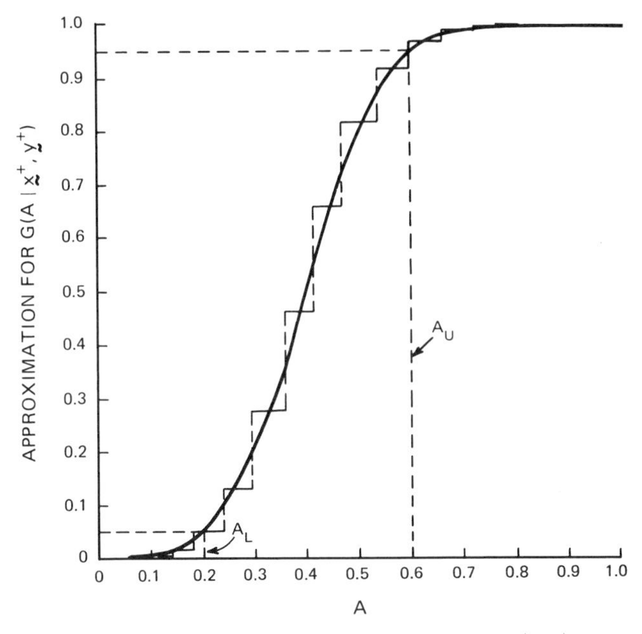

Figure 12.3. The posterior cumulative distribution function $G(A|\underset{\sim}{x}^+, \underset{\sim}{y}^+)$ for Example 12.4.

formula

$$M_r = E(A^r|\underset{\sim}{x}^+, \underset{\sim}{y}^+) = \frac{B(6+r,6)B(12+r,12)}{B(6,6)B(12,12)} \times {}_2F_1(r,12;\ 24+r;\ 0.750).$$

These values of M_r are then substituted into the system of equations given by (12.40) from which the coefficients c_j, $j=0,1,\ldots,17$, are found. These coefficients are then substituted into (12.41) which is then solved for its roots $A_1^*, A_2^*, \ldots, A_{18}^*$. These values of A_j^* are then substituted into the system of equations given by (12.42) from which the step sizes $S_1, S_2, \ldots, S_{18}$ are determined. The specific values of the M_r's, A_j^*'s are not given here but are tabulated in Springer (1979), pp. 272–273. The approximation for $G(A|\underset{\sim}{x}^+, \underset{\sim}{y}^+)$ may be obtained by connecting the midpoints of the steps of $\tilde{V}_{18}(A)$ and is shown in Figure 12.3. From the graph, a symmetric 90% TBPI for system availability is $0.20 \leqslant A \leqslant 0.60$, which is also shown in Figure 12.3.■

In the previous example, the 2-component series system was such that each component's availability was independent of that of the other. Specifically, while repair of a failed component was occurring, the other component continued to operate and, in the event it too failed during this repair period in which the system was down, its repair would be initiated immediately and independently of the other ongoing repair. This situation is usually referred to as a "2-component /2-repairmen" model [Sandler (1963), pp. 125–126].

Alternately, suppose there is only a single repairman for the system. Furthermore, while a failed component is undergoing repair, the other component remains in "suspended animation." When the repair of the failed component is completed, the remaining component resumes operation and the system is up once again. In this situation, system availability is given by [Barlow and Proschan (1975), pp. 194–197]

$$A=(1+\rho_1+\rho_2)^{-1}, \tag{12.43}$$

where $\rho_i=\lambda_i/\mu_i$ is the service factor for the ith component, $i=1,2$. For gamma prior pdf's, $\mathcal{G}_1(\alpha_i,\xi_i)$ and $\mathcal{G}_1(\beta_i,\eta_i)$ on λ_i and μ_i, respectively; the posterior pdf of ρ_i is a generalized $\mathcal{F}$-distribution [Springer (1979), p. 374] and, by (12.20), is given by

$$g_i(\rho_i|x_i^+,y_i^+)=\frac{c_i\rho_i^{k_i+\alpha_i-1}}{\{1+[(x_i^++\xi_i)/(y_i^++\eta_i)]\rho_i\}^{2k_i+\alpha_i+\beta_i}}, \qquad \rho_i>0, \tag{12.44}$$

where $c_i=[(x_i^++\xi_i)/(y_i^++\eta_i)]^{k_i+\alpha_i}/B(k_i+\beta_i,k_i+\alpha_i)$ and $0<(y_i^++\eta_i)/(x_i^++\xi_i)<2, i=1,2.$

Under a squared-error loss function, the Bayes point estimator of system availability may be determined by calculating

$$E(A|\underset{\sim}{x}^+,\underset{\sim}{y}^+)=\int_0^\infty\int_0^\infty(1+\rho_1+\rho_2)^{-1}g_2(\rho_2|x_2^+,y_2^+)\times$$

$$g_1(\rho_1|x_1^+,y_1^+)\,d\rho_2 d\rho_1.$$

However,

$$\int_0^\infty(1+\rho_1+\rho_2)^{-1}g_2(\rho_2|x_2^+,y_2^+)\,d\rho_2=\frac{(k_2+\beta_2)}{(2k_2+\alpha_2+\beta_2+1)}\frac{1}{(1+\rho_1)}$$

$$\times{}_2F_1[1,k_2+\alpha_2;2k_2+\alpha_2+\beta_2+1;1-(y_2^++\eta_2)/(x_2^++\xi_2)(1+\rho_1)]$$

[see Gradshteyn and Ryzhik (1965), p. 286, 3.197.1]. Upon using the series representation of the hypergeometric function (12.18), we find that

$$E(A|\underset{\sim}{x}^{+},\underset{\sim}{y}^{+})=\frac{c_1(k_2+\beta_2)}{(2k_2+\alpha_2+\beta_2+1)}\sum_{j=0}^{\infty}\frac{(k_2+\alpha_2)_j}{(2k_2+\alpha_2+\beta_2+1)_j}\times$$

$$\int_0^{\infty}\frac{[1-(y_2^{+}+\eta_2)/(x_2^{+}+\xi_2)+\rho_1]^j\rho_1^{k_1+\alpha_1-1}\,d\rho_1}{(1+\rho_1)^{j+1}[1+\rho_1(x_1^{+}+\xi_1)/(y_1^{+}+\eta_1)]^{2k_1+\alpha_1+\beta_1}}.$$

By making the transformation $t=1/(1+\rho_1)$, the last integral may be written as

$$\left(\frac{y_1^{+}+\eta_1}{x_1^{+}+\xi_1}\right)^{2k_1+\alpha_1+\beta_1}\int_0^1\frac{t^{k_1+\beta_1}(1-t)^{k_1+\alpha_1-1}[1-t(y_2^{+}+\eta_2)/(x_2^{+}+\xi_2)]^j\,dt}{\{1-[1-(y_1^{+}+\eta_1)/(x_1^{+}+\xi_1)]t\}^{2k_1+\alpha_1+\beta_1}}$$

$$=\left[\frac{y_1^{+}+\eta_1}{x_1^{+}+\xi_1}\right]^{2k_1+\alpha_1+\beta_1}B(k_1+\alpha_1,k_1+\beta_1+1)$$

$$\times F_1[k_1+\beta_1+1,-j,2k_1+\alpha_1+\beta_1;\,2k_1+\alpha_1+\beta_1+1;$$

$$(y_2^{+}+\eta_2)/(x_2^{+}+\xi_2),1-(y_1^{+}+\eta_1)/(x_1^{+}+\xi_1)],$$

with equality following from Picard's integral representation [Erdélyi (1953), p. 231] of the hypergeometric function of two variables

$$F_1(\alpha,\beta,\beta';\gamma;x,y)=\sum_{t=0}^{\infty}\sum_{s=0}^{\infty}\frac{(\alpha)_{s+t}(\beta)_t(\beta')_s}{(\gamma)_{s+t}s!t!}x^ty^s,\qquad |x|<1,\ |y|<1.\tag{12.45}$$

By using the transformation [Erdélyi (1953), p. 240]

$$F_1(\alpha,\beta,\beta';\gamma;x,y)=(1-x)^{-\beta}(1-y)^{\gamma-\alpha-\beta'}F_1\left(\gamma-\alpha,\beta,\gamma-\beta-\beta';\gamma;\frac{x-y}{x-1},y\right),\tag{12.46}$$

it now follows that the Bayes point estimator of A is given by

$$E(A|\underset{\sim}{x}^+, \underset{\sim}{y}^+) = \frac{(k_1+\beta_1)(k_2+\beta_2)}{(2k_1+\alpha_1+\beta_1)(2k_2+\alpha_2+\beta_2)}$$

$$\times \sum_{j=0}^{\infty} \frac{(k_2+\alpha_2)_j}{(2k_2+\alpha_2+\beta_2+1)_j}$$

$$\times \theta_2^j F_1\left[k_1+\alpha_1, -j, j+1; 2k_1+\alpha_1+\beta_1+1; 1-\frac{(1-\theta_1)}{\theta_2}, \theta_1\right], \tag{12.47}$$

where $\theta_i = 1-(y_i^+ + \eta_i)/(x_i^+ + \xi_i), i=1,2$.

The posterior cdf of system availability is determined by

$$G(A|\underset{\sim}{x}^+, \underset{\sim}{y}^+) = \Pr\left[(1+\rho_1+\rho_2)^{-1} \leqslant A|\underset{\sim}{x}^+, \underset{\sim}{y}^+\right]$$

$$= 1-\Pr\left(\rho_1+\rho_2 \leqslant A^{-1}-1|\underset{\sim}{x}^+, \underset{\sim}{y}^+\right)$$

$$= 1-\int_0^{A^{-1}-1}\int_0^{A^{-1}-1-\rho_2} g_1(\rho_1|x_1^+, y_1^+) g_2(\rho_2|x_2^+, y_2^+)\, d\rho_1 d\rho_2.$$

However, [Gradshteyn and Ryzhik (1965), p. 284, 3.194.1]

$$\int_0^{A^{-1}-1-\rho_2} g_1(\rho_1|x_1^+, y_1^+)\, d\rho_1 = c_1 \frac{\left(A^{-1}-1-\rho_2\right)^{k_1+\alpha_1}}{(k_1+\alpha_1)}$$

$$\times {}_2F_1\left[2k_1+\alpha_1+\beta_1, k_1+\alpha_1; k_1+\alpha_1+1; -\frac{\left(A^{-1}-1-\rho_2\right)(x_1^+ + \xi_1)}{y_1^+ + \eta_1}\right].$$

Upon making use of the transformation [Erdélyi (1953), p. 105]

$${}_2F_1(\alpha, \beta; \gamma; z) = (1-z)^{\gamma-\alpha-\beta}\, {}_2F_1(\gamma-\alpha, \gamma-\beta; \gamma; z), \tag{12.48}$$

and the fact that [see (12.18)]

$${}_2F_1(-m, \beta; \gamma; z) = \sum_{j=0}^{m} \frac{(-m)_j(\beta)_j}{(\gamma)_j j!} z^j,$$

it follows that

$$G(A|\underset{\sim}{x}^{+}, \underset{\sim}{y}^{+}) = 1 - \frac{c_1}{(k_1+\alpha_1)} \sum_{j=0}^{k_1+\beta_1-1} \frac{[-(k_1+\beta_1-1)]_j}{(k_1+\alpha_1+1)_j} \left[-\left(\frac{x_1^+ + \xi_1}{y_1^+ + \eta_1} \right) \right]^j$$

$$\times \int_0^{A^{-1}-1} \frac{(A^{-1}-1-\rho_2)^{k_1+\alpha_1+j} g_2(\rho_2 | x_2^+, y_2^+)\, d\rho_2}{\left[1 + \left(\frac{x_1^+ + \xi_1}{y_1^+ + \eta_1} \right) (A^{-1}-1-\rho_2) \right]^{2k_1+\alpha_1+\beta_1-1}}.$$

By making the transformation $\rho_2 = (A^{-1}-1)t$, then using Picard's integral representation of the hypergeometric function of two variables followed by the transformation given by (12.46); we obtain the posterior cdf of system availability

$$G(A|\underset{\sim}{x}^{+}, \underset{\sim}{y}^{+}) = 1 - \frac{\phi_1^{k_1+\alpha_1} \phi_2^{k_2+\alpha_2} (1-\phi_1)(1-\phi_2)^{k_2+\beta_2} B(k_1+\alpha_1, k_2+\alpha_2)}{(k_1+k_2+\alpha_1+\alpha_2) B(k_1+\beta_1, k_1+\alpha_1) B(k_2+\beta_2, k_2+\alpha_2)}$$

$$\times \sum_{j=0}^{k_1+\beta_1-1} \frac{[-(k_1+\beta_1-1)]_j}{(k_1+k_2+\alpha_1+\alpha_2+1)_j} (-\phi_1)^j \times F_1[k_1+\alpha_1+j+1, 2k_2+\alpha_2+\beta_2,$$

$$-k_1-k_2-\beta_1-\beta_2+j+2;\ k_1+k_2+\alpha_1+\alpha_2+j+1;\ 1-(1-\phi_1)(1-\phi_2), \phi_1], \tag{12.49}$$

where

$$\phi_i = 1 - \left[1 + \left(\frac{x_i^+ + \xi_i}{y_i^+ + \eta_i} \right) (A^{-1}-1) \right]^{-1}, \quad i = 1, 2.$$

A symmetric $100(1-\gamma)\%$ TBPI for system availability, (A_L, A_U), is determined by solving the equations $G(A_L|\underset{\sim}{x}^{+}, \underset{\sim}{y}^{+}) = \gamma/2$ for A_L and $G(A_U|\underset{\sim}{x}^{+}, \underset{\sim}{y}^{+}) = 1-\gamma/2$ for A_U.

Example 12.5. Again let us consider Example 12.4. Suppose that for the series system there is one repair facility instead of two. Moreover, while a failed component is undergoing repair, the other component remains in "suspended animation." In this case, the Bayes point estimate of system availability is, by (12.47),

$$E(A|\underset{\sim}{x}^{+}, \underset{\sim}{y}^{+}) = \frac{6(12)}{12(24)} \sum_{j=0}^{\infty} \frac{(12)_j}{(25)_j} (0.75)^j F_1\left(6, -j, j+1; 13; -\frac{1}{3}, 0\right).$$

However,

$$F_1\left(6, -j, j+1; 13; -\frac{1}{3}, 0\right) = {}_2F_1\left(6, -j; 13; -\frac{1}{3}\right) = {}_2F_1\left(-j, 6; 13; -\frac{1}{3}\right)$$

$$= \sum_{t=0}^{j} \frac{(-j)_t (6)_t}{(13)_t t!} \left(-\frac{1}{3}\right)^t;$$

thus

$$E(A|\underset{\sim}{x}^{+}, \underset{\sim}{y}^{+}) = 0.25 \sum_{j=0}^{\infty} \frac{(12)_j}{(25)_j} (0.75)^j \left[\sum_{t=0}^{j} \frac{(-j)_t (6)_t}{(13)_t t!} \left(-\frac{1}{3}\right)^t \right] = 0.429.$$

The posterior cdf of system availability is, by (12.49),

$$G(A|\underset{\sim}{x}^{+}, \underset{\sim}{y}^{+}) = 1 - 33649 \phi_1^6 \phi_2^{12} (1-\phi_1)(1-\phi_2)^{12}$$

$$\times \sum_{j=0}^{5} \frac{(-5)_j}{(19)_j} (-\phi_1)^j F_1[j+7, 24, j-16; j+19; 1-(1-\phi_1)(1-\phi_2), \phi_1],$$

where $\phi_1 = 1 - A$ and $\phi_2 = (4-4A)/(4-3A)$. For computational purposes, the hypergeometric function of two variables in the expression for $G(A|\underset{\sim}{x}^{+}, \underset{\sim}{y}^{+})$ may be written as

$$F_1[j+7, 24, j-16; j+19; 1-(1-\phi_1)(1-\phi_2), \phi_1]$$

$$= \sum_{t=0}^{\infty} \sum_{s=0}^{\infty} \frac{(j+7)_{s+t} (24)_t (j-16)_s}{(j+19)_{s+t} s! t!} [1-(1-\phi_1)(1-\phi_2)]^t \phi_1^s$$

$$= \sum_{s=0}^{16-j} \frac{[-(16-j)]_s (j+7)_s}{(j+19)_s s!} \phi_1^s \times {}_2F_1[j+7+s, 24; j+19+s; 1-(1-\phi_1)(1-\phi_2)]$$

$$= (1-\phi_1)^{-12} (1-\phi_2)^{-12} \sum_{s=0}^{16-j} \frac{[-(16-j)]_s (j+7)_s}{(j+19)_s s!} \phi_1^s \times {}_2F_1[12, j+s-5; j+s+19;$$

$$1-(1-\phi_1)(1-\phi_2)].$$

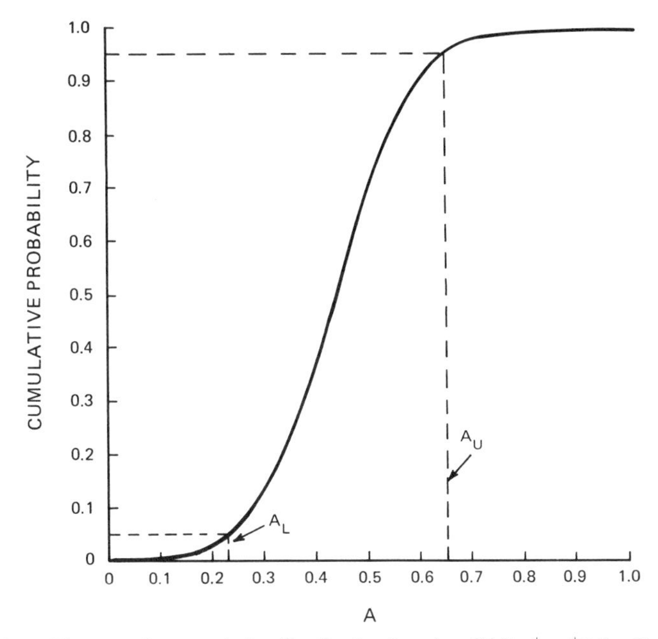

Figure 12.4. The posterior cumulative distribution function $G(A|\underset{\sim}{x}^+, \underset{\sim}{y}^+)$ for Example 12.5.

The expression for $G(A|\underset{\sim}{x}^+, \underset{\sim}{y}^+)$ has been evaluated for $A = 0.05(0.05)0.95$ and, based on these tabulations, is graphed in Figure 12.4. From the graph, a symmetric 90% TBPI for system availability is $0.23 \leq A \leq 0.64$, which is also shown in Figure 12.4. ■

12.3.3 Parallel System Availability

Consider a system consisting of N independently operating repairable components in parallel. We assume the failure and repair times for the ith component are $\mathcal{E}(\cdot)$ distributed with failure rate λ_i and repair rate μ_i, respectively. The components are simultaneously performing the same function, hence the system will be available if at least one of the components is available.

***N* Components/*N* Repair Facilities.** Suppose there is a repair facility for each component (see Figure 12.5) and that each component's availability is independent of the availability of all of the others. System unavailability, $\bar{A} = 1 - A$, is the product of component unavailabilities [Sandler (1963), p.

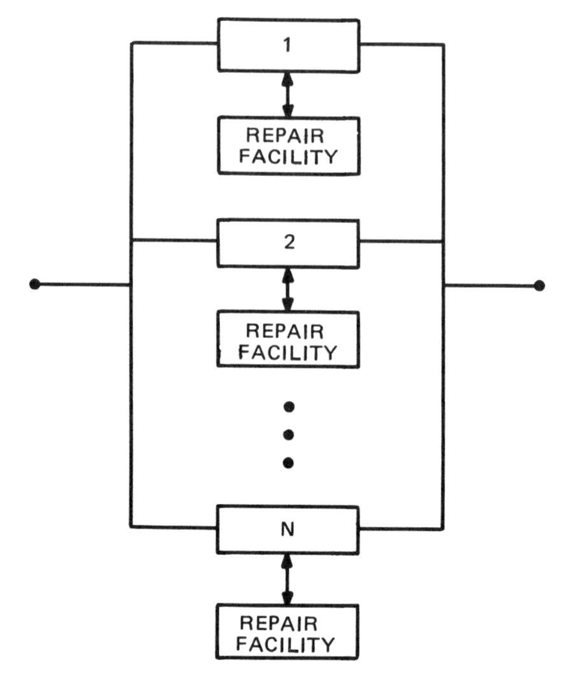

Figure 12.5. An N-component parallel system with N repair facilities.

133]; thus

$$\bar{A} = \prod_{i=1}^{N} \left(\frac{\lambda_i}{\lambda_i + \mu_i} \right). \tag{12.50}$$

For the ith component, suppose k_i failure/repair cycles are observed: the total operating time observed is $x_i^+ = \sum_{j=1}^{k_i} x_{ij}$, where x_{ij} is the jth failure time; the total repair time observed is $y_i^+ = \sum_{j=1}^{k_i} y_{ij}$, where y_{ij} is the jth repair time. Again using gamma prior pdf's, $\mathcal{G}_1(\alpha_i, \xi_i)$ and $\mathcal{G}_1(\beta_i, \eta_i)$ on λ_i and μ_i, respectively, the posterior pdf of the ith component unavailability, $\bar{A}_i = \lambda_i/(\lambda_i + \mu_i)$, is an Euler distribution and is obtained by a change of variable in (12.36). Thus,

$$\bar{g}_i(\bar{A}_i | x_i^+, y_i^+) = \frac{[(x_i^+ + \xi_i)/(y_i^+ + \eta_i)]^{k_i + \alpha_i}}{B(k_i + \alpha_i, k_i + \beta_i)} \times$$

$$\frac{\bar{A}_i^{k_i + \alpha_i - 1}(1 - \bar{A}_i)^{k_i + \beta_i - 1}}{\{1 - \bar{A}_i[1 - (x_i^+ + \xi_i)/(y_i^+ + \eta_i)]\}^{2k_i + \alpha_i + \beta_i}}, \quad 0 < \bar{A}_i < 1, \tag{12.51}$$

provided $0 < (y_i^+ + \eta_i)/(x_i^+ + \xi_i) < 2$.

By making use of the transformation

$$t = \frac{\bar{A}_i(x_i^+ + \xi_i)/(y_i^+ + \eta_i)}{1 - \bar{A}_i[1 - (x_i^+ + \xi_i)/(y_i^+ + \eta_i)]},$$

it can be shown that the rth moment of $\bar{A}_i$ is

$$E(\bar{A}_i^r | x_i^+, y_i^+) = \int_0^1 \bar{A}_i^r \bar{g}_i(\bar{A}_i | x_i^+, y_i^+)\, d\bar{A}_i$$

$$= \frac{[(y_i^+ + \eta_i)/(x_i^+ + \xi_i)]^r B(k_i + \alpha_i + r, k_i + \beta_i)}{B(k_i + \alpha_i, k_i + \beta_i)} \times$$

$$_2F_1[r, k_i + \alpha_i + r; 2k_i + \alpha_i + \beta_i + r; 1 - (y_i^+ + \eta_i)/(x_i^+ + \xi_i)], \tag{12.52}$$

provided $0 < (y_i^+ + \eta_i)/(x_i^+ + \xi_i) < 2$. Under a squared-error loss function, the Bayes point estimator of system availability is given by

$$E(A | \underset{\sim}{x}^+, \underset{\sim}{y}^+) = 1 - \prod_{i=1}^{N} E(\bar{A}_i | x_i^+, y_i^+)$$

$$= 1 - \prod_{i=1}^{N} \left(\frac{y_i^+ + \eta_i}{x_i^+ + \xi_i} \right) \left(\frac{k_i + \alpha_i}{2k_i + \alpha_i + \beta_i} \right)$$

$$\times {}_2F_1[1, k_i + \alpha_i + 1; 2k_i + \alpha_i + \beta_i + 1; 1 - (y_i^+ + \eta_i)/(x_i^+ + \xi_i)]. \tag{12.53}$$

In order to determine Bayes interval estimates of A, and approximation to the posterior cdf, $G(A | \underset{\sim}{x}^+, y^+)$, is found using the Von Mises (1964) step function procedure as described in Section 12.3.2. Specifically, $G(A | \underset{\sim}{x}^+, \underset{\sim}{y}^+) \cong 1 - \bar{V}_m(\bar{A})$, where $\bar{V}_m(\bar{A})$ is an m-step step function based on the rth moments of system unavailability, $\bar{M}_r = \prod_{i=1}^{N} E(\bar{A}_i^r | x_i^+, y_i^+)$, where $E(\bar{A}_i^r | x_i^+, y_i^+)$ is given by (12.52), which approximates the posterior cdf of system unavailability.

It is often the case in a parallel system that the components are identical—each component fails at a constant rate λ and can be repaired at a constant rate μ. In this case, system availability is given by

$$A = 1 - \tilde{A}^N, \tag{12.54}$$

where $\tilde{A} = \lambda / (\lambda + \mu)$ is the unavailability of each component.

For the (common) component, suppose k failure/repair cycles are observed: the total operating time observed is $x_+ = \Sigma_{i=1}^k x_i$, where x_i is the ith failure time; the total repair time observed is $y_+ = \Sigma_{i=1}^k y_i$, where y_i is the ith repair time. Let us assume independent gamma prior pdf's, $\mathcal{G}_1(\alpha, \xi)$ and $\mathcal{G}_1(\beta, \eta)$ on λ and μ, respectively. Under a squared-error loss function, the Bayes point estimator of system availability is, by (12.52),

$$E(A|x_+, y_+) = 1 - E(\tilde{A}^N | x_+, y_+)$$

$$= 1 - \frac{[(y_+ + \eta)/(x_+ + \xi)]^N B(k+\alpha+N, k+\beta)}{B(k+\alpha, k+\beta)} \times$$

$${}_2F_1\left[N, k+\alpha+N; 2k+\alpha+\beta+N; 1 - \frac{(y_+ + \eta)}{(x_+ + \xi)}\right]. \tag{12.55}$$

The posterior cdf of system availability is, by (12.29),

$$G(A|x_+, y_+) = \mathcal{F}\{[1-(1-A)^{1/N}](k+\alpha)(y_+ + \eta)/(1-A)^{1/N} \times$$

$$(k+\beta)(x_+ + \xi); 2(k+\beta), 2(k+\alpha)\}.$$

It follows that a symmetric $100(1-\gamma)\%$ TBPI for system availability is (A_L, A_U), where

$$A_L = 1 - \left[H\left(1 - \frac{\gamma}{2}\right)\right]^N, \qquad A_U = 1 - \left[H\left(\frac{\gamma}{2}\right)\right]^N, \tag{12.56}$$

and

$$H(\theta) = \frac{(k+\alpha)(y_+ + \eta)\mathcal{F}_\theta[2(k+\alpha), 2(k+\beta)]}{(k+\beta)(x_+ + \xi) + (k+\alpha)(y_+ + \eta)\mathcal{F}_\theta[2(k+\alpha), 2(k+\beta)]}. \tag{12.57}$$

A $100(1-\gamma)\%$ LBPI for system availability is $(A_\gamma, 1)$, where A_γ is determined by replacing $\gamma/2$ by γ in the expression for A_L.

Example 12.6. Consider a parallel system consisting of two identical components each of which has $\mathcal{E}(\cdot)$ distributed failure and repair times. There is a repair facility for each component, and it is assumed that the repair rates are identical. Each component's availability is independent of that of the other. The following failure/repair data were obtained by simulation using a failure rate of $\lambda = 1/60$ and a repair rate of $\mu = 1/4$. The true value of system availability is thus equal to 0.996. The system was observed through a total of eight component failure/repair cycles with the results given in the order obtained for each component. Note that the system was down for 5.8 h during the third repair period for component 1.

Component 1		Component 2	
Failure Times (h)	Repair Times (h)	Failure Times (h)	Repair Times (h)
74.3	0.5	128.3	11.8
19.0	10.1	17.8	4.8
26.7	5.8	47.8	3.6
88.5	1.2	5.2	5.0

From the pooled data: $k = 8$, $y_+/x_+ = 42.8/407.6 = 0.105$. Under uniform prior pdf's ($\alpha = \beta = 1$, $\xi = \eta = 0$) on both the failure and repair rates, a Bayes point estimate of system availability is, by (12.55),

$$E(A|x_+, y_+) = 1 - \frac{(0.105)^2 B(11,9)}{B(9,9)} {}_2F_1(2, 11; 20; 0.895)$$

$$= 1 - 0.0029013\ (4.38159) = 0.987.$$

By (12.56),

$$A_L = 1 - \left[\frac{9(42.8)(2.217)}{9(407.6) + 9(42.8)(2.217)} \right]^2 = 0.964;$$

thus a 95% LBPI for system availability is (0.964, 1). ■

Two Identical Components/One Repair Facility. In the previous example it was assumed that there was a repair facility for each of the two identical components in a parallel system. Suppose, however, there is a single repair facility to service both components (see Figure 12.6). In this situation system availability is given by [Sandler (1963), p. 133]

$$A = \frac{\mu^2 + 2\lambda\mu}{\mu^2 + 2\lambda\mu + 2\lambda^2}. \tag{12.58}$$

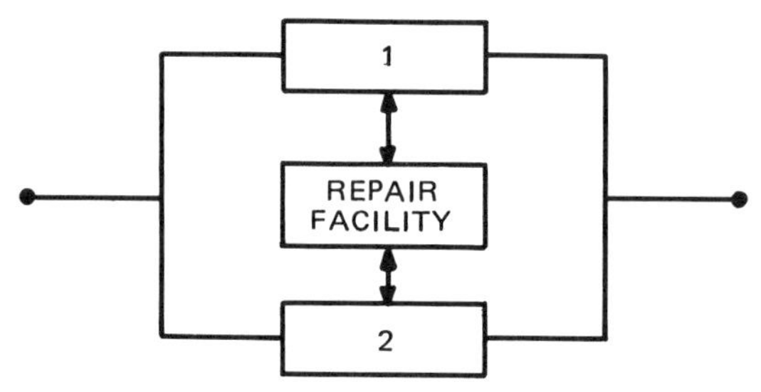

Figure 12.6. A two-component parallel system with one repair facility.

It is no longer true that each component's availability is independent of that of the other. Moreover, it is worthwhile to compare (12.58) with the corresponding expression for system availability when there are two repair facilities; that is, $A=1-[\lambda/(\lambda+\mu)]^2=(\mu^2+2\lambda\mu)/(\mu^2+2\lambda\mu+\lambda^2)$. The difference in the two availability expressions is due to a possible waiting time for repair in the one repair facility model brought about by a component failure occurring while repair is being made on the other component. There is no delayed repair in the two repair facility model and, consequently, system availability is higher in this case.

For the (common) component, suppose k failure/repair cycles are observed: the total operating time observed is $x_+ = \Sigma_{i=1}^k x_i$, where x_i is the ith failure time; the total repair time observed (exclusive of any waiting time for repair) is $y_+ = \Sigma_{i=1}^k y_i$, where y_i is the ith repair time. Gamma prior pdf's, $\mathcal{G}_1(\alpha, \xi)$ and $\mathcal{G}_1(\beta, \eta)$, are assigned to λ and μ, respectively. It will be convenient to rewrite (12.58) as

$$A=\frac{1-\tilde{A}^2}{1+\tilde{A}^2}=1-2\sum_{j=1}^{\infty}(-1)^{j+1}\tilde{A}^{2j}, \tag{12.59}$$

where $\tilde{A}=\lambda/(\lambda+\mu)$ is the unavailability of each component.

Under a squared-error loss function, the Bayes point estimator of system availability is, by (12.52),

$$\begin{aligned} E(A|x_+, y_+) &= 1-2\sum_{j=1}^{\infty}(-1)^{j+1}E(\tilde{A}^{2j}|x_+, y_+) \\ &= 1-2\sum_{j=1}^{\infty}\frac{(-1)^{j+1}[(y_+ + \eta)/(x_+ + \xi)]^{2j}B(k+\alpha+2j, k+\beta)}{B(k+\alpha, k+\beta)}\times \\ &\quad {}_2F_1\left[2j, k+\alpha+2j; 2k+\alpha+\beta+2j; 1-\frac{(y_+ + \eta)}{(x_+ + \xi)}\right]. \end{aligned} \tag{12.60}$$

The posterior cdf of system availability is, by (12.29),

$$G(A|x_+, y_+) = \mathcal{F}\Big\{\Big[1-\sqrt{(1-A)/(1+A)}\,\Big](k+\alpha)\times$$
$$(y_+ + \eta)/\sqrt{(1-A)/(1+A)}\,(k+\beta)(x_+ + \xi);\ 2(k+\beta), 2(k+\alpha)\Big\}.$$

Consequently, a symmetric $100(1-\gamma)\%$ TBPI for system availability is (A_L, A_U), where

$$A_L = \frac{1-[H(1-\gamma/2)]^2}{1+[H(1-\gamma/2)]^2}, \qquad A_U = \frac{1-[H(\gamma/2)]^2}{1+[H(\gamma/2)]^2}, \tag{12.61}$$

and $H(\theta)$ is given by (12.57).

Example 12.7. Again let us consider Example 12.6. Suppose that for the parallel system of Example 12.6 there is one repair facility instead of two. We determine Bayes point and interval estimates of system availability based on the failure/repair data in Example 12.6. Note that component 2 was undergoing repair when the third failure of component 1 occurred. There was then a delay of 9.5 h before the repair of component 1 was initiated, and the system was down during this time. Of course, the actual repair times are the same regardless of whether there are one or two repair facilities.

Under uniform prior pdf's ($\alpha=\beta=1, \xi=\eta=0$) on both the failure and repair rates, a Bayes point estimate of system availability is, by (12.60),

$$E(A|x_+, y_+) = 1-2\sum_{j=1}^{\infty}\frac{(-1)^{j+1}(0.105)^{2j}B(9+2j,9)}{B(9,9)}\times$$
$${}_2F_1(2j, 9+2j; 18+2j; 0.895) = 0.977.$$

The true value of system availability is equal to 0.992. As

$$H(0.95) = \frac{9(42.8)(2.217)}{9(407.6)+9(42.8)(2.217)} = 0.18884;$$

then, by (12.61),

$$A_L = \frac{1-(0.18884)^2}{1+(0.18884)^2} = 0.931.$$

Thus, a 95% LBPI for system availability is (0.931, 1). ■

12.3.4 Standby System Availability

Consider a two-component standby redundant system with one repair facility (see Figure 12.7). The on-line component operates until failure, then is replaced by the spare in negligible time (assume perfect switching); the failed component then immediately undergoes repair. When repair is completed, the repaired component then becomes the spare. A component in spare (off-line) status cannot age nor fail. The system fails when the on-line component fails and no spare is available to replace it (i.e., repair has not yet been completed on the previously failed component).

When both components are identical with $\mathcal{E}(\cdot)$ distributed failure and repair times (with failure rate λ and repair μ), system availability is given by [Barlow and Proschan (1975), pp. 202–206]

$$A = \frac{\mu^2 + \mu\lambda}{\mu^2 + \mu\lambda + \lambda^2} \tag{12.62}$$

or, more conveniently,

$$A = \frac{1 - \tilde{A}}{1 - \tilde{A}(1 - \tilde{A})} = \sum_{j=0}^{\infty} \tilde{A}^j (1 - \tilde{A})^{j+1}, \tag{12.63}$$

where $\tilde{A} = \lambda/(\lambda + \mu)$. The difference in (12.62) and the availability expression for the two component parallel system with one repair facility; that is, $A = (\mu^2 + 2\lambda\mu)/(\mu^2 + 2\lambda\mu + 2\lambda^2)$, is due to a shorter delayed repair time, when it exists, in the standby system. Consequently, system availability in the standby system is higher than in the parallel system with one repair facility but lower than in the case of two repair facilities.

For the (common) component, suppose k failure/repair cycles are observed: the total operating time observed is $x_+ = \Sigma_{i=1}^{k} x_i$, where x_i is the ith failure time; the total repair time observed (exclusive of any waiting time for

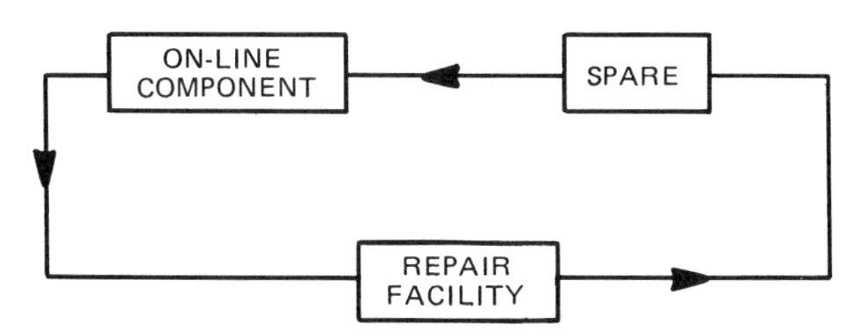

Figure 12.7. A two-component standby system with one repair facility.

repair) is $y_+ = \Sigma_{i=1}^{k} y_i$, where y_i is the i th repair time. Gamma prior pdf's, $\mathcal{G}_1(\alpha, \xi)$ and $\mathcal{G}_1(\beta, \eta)$), are assigned to λ and μ, respectively.

Under a squared-error loss function, the Bayes point estimator of system availability is determined by

$$E(A|x_+, y_+) = \int_0^1 \left[\sum_{j=0}^{\infty} \tilde{A}^j (1-\tilde{A})^{j+1} \right] \tilde{g}(\tilde{A}|x_+, y_+)\, d\tilde{A}$$

$$= \sum_{j=0}^{\infty} \int_0^1 \tilde{A}^j (1-\tilde{A})^{j+1} \tilde{g}(\tilde{A}|x_+, y_+)\, d\tilde{A},$$

where $\tilde{g}(\tilde{A}|x_+, y_+)$ is the posterior pdf of component unavailability given by (12.51). By making use of the transformation

$$t = \frac{\tilde{A}(x_+ + \xi)/(y_+ + \eta)}{1 - \tilde{A}[1 - (x_+ + \xi)/(y_+ + \eta)]}$$

when integrating $\int_0^1 \tilde{A}^j (1-\tilde{A})^{j+1} \tilde{g}(\tilde{A}|x_+, y_+)\, d\tilde{A}$, it can be shown that

$$E(A|x_+, y_+) = \sum_{j=0}^{\infty} \frac{[(y_+ + \eta)/(x_+ + \xi)]^j B(k+\alpha+j, k+\beta+j+1)}{B(k+\alpha, k+\beta)}$$

$$\times {}_2F_1\left[2j+1, k+\alpha+j; 2k+\alpha+\beta+2j+1; 1 - \frac{y_+ + \eta}{x_+ + \xi}\right], \qquad (12.64)$$

provided $0 < (y_+ + \eta)/(x_+ + \xi) < 2$.

The posterior cdf of system availability is determined by

$$G(A|x_+, y_+) = \Pr\left[\frac{(1-\tilde{A})}{1 - \tilde{A}(1-\tilde{A})} \leqslant A | x_+, y_+ \right]$$

$$= \Pr[\tilde{A}^2 + (A^{-1} - 1)\tilde{A} - (A^{-1} - 1) \geqslant 0 | x_+, y_+]$$

$$= \Pr[(\tilde{A} - a_1)(\tilde{A} - a_2) \geqslant 0 | x_+, y_+],$$

where

$$a_1 = \frac{-(1-A) + \sqrt{1 + 2A - 3A^2}}{2A} \quad \text{and} \quad a_2 = \frac{-(1-A) - \sqrt{1 + 2A - 3A^2}}{2A}.$$

Since $a_2 < 0 < a_1 < 1$, it follows from (12.29) that

$$G(A|x_+, y_+) = \mathcal{F}\left[\frac{(1-a_1)(k+\alpha)(y_+ + \eta)}{a_1(k+\beta)(x_+ + \xi)}; \; 2(k+\beta), 2(k+\alpha)\right].$$

A symmetric $100(1-\gamma)\%$ TBPI for system availability is (A_L, A_U), where

$$A_L = \frac{1 - H(1-\gamma/2)}{1 - H(1-\gamma/2)[1 - H(1-\gamma/2)]}, \quad A_U = \frac{1 - H(\gamma/2)}{1 - H(\gamma/2)[1 - H(\gamma/2)]}, \tag{12.65}$$

and $H(\theta)$ is given by (12.57).

Example 12.8. Let us further consider Example 12.7. For the system described in Example 12.7 suppose that the two components supplement rather than duplicate each other's efforts, that is, only one is on-line at a time. For such a situation, the results of this section may be used to obtain Bayes point and interval estimates of system availability based on the failure/repair data in Example 12.6. Component 1 is initially on-line and component 2 is the spare; accordingly, the data given below are presented in a different format than in Example 12.6. Note that the system was not down during the observation period.

Failure Number	Uptime (h) of On-Line Component	Repair Time (h) for Off-Line Component
1	74.3	0.5
2	128.3	11.8
3	19.0	10.1
4	17.8	4.8
5	26.7	5.8
6	47.8	3.6
7	88.5	1.2
8	5.2	5.0

Under uniform prior pdf's ($\alpha=\beta=1$, $\xi=\eta=0$) on both the failure and repair rates, a Bayes point estimate of system availability is, by (12.64),

$$E(A|x_+,y_+)=\sum_{j=0}^{\infty}\frac{(0.105)^j B(9+j,10+j)}{B(9,9)}{}_2F_1(2j+1,9+j;\,19+2j;\,0.895)$$

$$=0.986.$$

The true value of system availability is equal to 0.996. Since $H(.95)=0.18884$, then by (12.65),

$$A_L=\frac{1-0.18884}{1-0.18884(0.81116)}=0.958;$$

thus a 95% LBPI for system availability is (0.958, 1). ■

12.4. EXPONENTIAL FAILURE TIMES/GAMMA REPAIR TIMES

Throughout the previous section it was assumed that repair times followed an $\mathcal{E}(\mu)$ distribution. In some situations, however, a distribution having a mode away from the origin may provide a more realistic model for the repair time. For example, suppose the repair process goes through a certain number of $\mathcal{E}(\mu)$ phases each of the same average duration before repair is completed. In such cases the repair time Y may have a $\mathcal{G}_1(\kappa,\mu)$ distribution with pdf

$$f_Y(y)=\frac{1}{\Gamma(\kappa)}\mu^{\kappa}y^{\kappa-1}\exp(-\mu y), \qquad y>0. \tag{12.66}$$

The reciprocal of the repair rate, $w(t)$, is given by

$$[w(t)]^{-1}=\int_t^{\infty}\frac{f_Y(y)\,dy}{f_Y(t)}=\int_t^{\infty}\left(\frac{y}{t}\right)^{\kappa-1}\exp[-\mu(y-t)]\,dy$$

$$=\int_0^{\infty}\left(1+\frac{z}{t}\right)^{\kappa-1}\exp(-\mu z)\,dz$$

by the change of variable $z = y - t$. Since

$$\frac{d[w(t)]^{-1}}{dt} = -\frac{(\kappa-1)}{t^2}\int_0^\infty z\left(1+\frac{z}{t}\right)^{\kappa-2}\exp(-\mu z)\,dz,$$

then $[w(t)]^{-1}$ is increasing for $0<\kappa<1$ and decreasing for $\kappa>1$; thus the repair rate is decreasing for $0<\kappa<1$ and increasing for $\kappa>1$. The case $\kappa=1$ corresponds to a constant repair rate. A repair rate that increases with time would seem to be particularly appealing. As noted by Shooman (1968, pp. 348–349): "With an intelligent repairman who systematically tracks down errors, one would expect a rapid increase in the probability of repair as time increased." Another interesting and realistic feature is that for $\kappa>1, w(t)\to\mu$ as $t\to\infty$; in other words, after some time, the repair rate approaches a stable condition, and from then on repair is as likely to be completed in one time period as in another. It should also be noted that Butterworth and Nikolaisen (1973) suggest that the $\mathcal{G}_1(\kappa,\mu)$ distribution may be used in lieu of the $\mathcal{LN}(\xi,\sigma^2)$ distribution (often believed to model certain repair processes) for computing availability.

For a component with an $\mathcal{E}(\lambda)$ distributed failure time and a $\mathcal{G}_1(\kappa,\mu)$ distributed repair time the availability is given by

$$A = \frac{E(X)}{E(X)+E(Y)} = \frac{\mu}{\mu+\kappa\lambda}. \tag{12.67}$$

Suppose k failure/repair cycles are observed: The total operating time observed is $x_+ = \Sigma_{i=1}^k x_i$, where x_i is the ith failure time; the total repair time observed is $y_+ = \Sigma_{i=1}^k y_i$, where y_i is the ith repair time. We assume a gamma prior pdf, $\mathcal{G}_1(\alpha,\xi)$, for the failure rate λ. Similar to what was done in Chapter 9, the following mixture is taken as the joint prior pdf of κ and μ, where κ and μ denote both r.v.'s and their values [Brender (1968b); Lwin and Singh (1974)]:

$$\begin{aligned}\Pr\{\kappa=\kappa_i\} &= p_i, && i=1,2,\ldots,m,\\ g(\mu\,|\,\kappa=\kappa_i) &= [1/\Gamma(\beta_i)]\eta_i^{\beta_i}\mu^{\beta_i-1}\exp(-\eta_i\mu), && \mu>0,\\ &= \mathcal{G}_1(\beta_i,\eta_i).\end{aligned} \tag{12.68}$$

As we shall see, (12.68) is a natural conjugate prior pdf. The discrete nature of κ may arise as follows. Suppose there are m distinct repair facilities (or types of repairs), exactly one of which will be called upon to effect repairs of the component during the k failure/repair cycles. It is not known a priori which repair facility will be used, but from a previous history of repairs the proportion of the time that the ith repair facility was called upon was p_i. For the ith repair facility, the value of κ_i is the number of $\mathcal{E}(\mu)$ phases the repair process goes through in order to make the repair.

The sampling distribution for the observed repair times $\underset{\sim}{y}' = (y_1, y_2, \dots, y_k)$ is

$$f(\underset{\sim}{y}|\mu, \kappa_i) = \left[\frac{1}{\Gamma(\kappa_i)}\right]^k \mu^{k\kappa_i}\left(\prod_{i=1}^{k} y_i\right)^{\kappa_i - 1} \exp(-\mu y_+).$$

The conditional posterior pdf of μ, given $\kappa = \kappa_i$, is given by

$$g(\mu|\kappa_i, \underset{\sim}{y}) = \frac{f(\underset{\sim}{y}\,|\mu, \kappa_i)g(\mu|\kappa = \kappa_i)}{\int_0^\infty f(\underset{\sim}{y}|\mu, \kappa_i)g(\mu|\kappa = \kappa_i)\, d\mu}$$

$$= \frac{(y_+ + \eta_i)^{k\kappa_i + \beta_i}}{\Gamma(k\kappa_i + \beta_i)} \mu^{k\kappa_i + \beta_i - 1} \exp[-\mu(y_+ + \eta_i)], \qquad \mu > 0, \tag{12.69}$$

which is a $\mathcal{G}_1(k\kappa_i + \beta_i, y_+ + \eta_i)$ distribution.

Furthermore, the (marginal) posterior pdf of κ is

$$p_i^* = \Pr\{\kappa = \kappa_i | \underset{\sim}{y}\} = \frac{p_i'}{\sum_{i=1}^{m} p_i'}, \qquad i = 1, 2, \dots, m, \tag{12.70}$$

where

$$p_i' = p_i \int_0^\infty f(\underset{\sim}{y}|\mu, \kappa_i) g(\mu|\kappa = \kappa_i)\, d\mu$$

$$= p_i \frac{\Gamma(k\kappa_i + \beta_i)\eta_i^{\beta_i}\left(\prod_{i=1}^{k} y_i\right)^{\kappa_i - 1}}{[\Gamma(\kappa_i)]^k \Gamma(\beta_i)(y_+ + \eta_i)^{k\kappa_i + \beta_i}}. \tag{12.71}$$

Observe that the posterior pdf of κ is of the same discrete form as the prior pdf of κ. In addition, the conditional prior and posterior pdf's of μ, given $\kappa = \kappa_i$, are both gamma distributions.

The conditional posterior pdf of $\rho=\lambda/\mu$, given $\kappa=\kappa_i$, is given by

$$g_{\rho|\kappa}(\rho|\kappa=\kappa_i)=\int_0^\infty g_\lambda(\mu\rho)g_{\mu|\kappa}(\mu|\kappa=\kappa_i)\mu\,d\mu$$

$$=\frac{(x_+ +\xi)^{k+\alpha}}{\Gamma(k+\alpha)}\frac{(y_+ +\eta_i)^{k\kappa_i+\beta_i}}{\Gamma(k\kappa_i+\beta_i)}\rho^{k+\alpha-1}\times$$

$$\int_0^\infty \mu^{k+\alpha+k\kappa_i+\beta_i-1}\exp\{-\mu[(y_+ +\eta_i)+\rho(x_+ +\xi)]\}\,d\mu$$

$$=\frac{(x_+ +\xi)^{k+\alpha}(y_+ +\eta_i)^{k\kappa_i+\beta_i}\Gamma(k+\alpha+k\kappa_i+\beta_i)}{\Gamma(k+\alpha)\Gamma(k\kappa_i+\beta_i)}\times$$

$$\frac{\rho^{k+\alpha-1}}{[(y_+ +\eta_i)+\rho(x_+ +\xi)]^{k+\alpha+k\kappa_i+\beta_i}},\qquad \rho>0. \quad (12.72)$$

Consequently, the conditional posterior pdf of component availability $A=1/(1+\kappa_i\rho)$, given $\kappa=\kappa_i$, is determined by $g_{A|\kappa}(A|\kappa=\kappa_i)=g_{\rho|\kappa}[(A^{-1}-1)/\kappa_i|\kappa=\kappa_i](A^{-2}/\kappa_i)$. Thus,

$$g_{A|\kappa}(A|\kappa=\kappa_i)=\frac{[\kappa_i(y_+ +\eta_i)/(x_+ +\xi)]^{k\kappa_i+\beta_i}}{B(k\kappa_i+\beta_i,k+\alpha)}\times$$

$$\frac{A^{k\kappa_i+\beta_i-1}(1-A)^{k+\alpha-1}}{\{1-A[1-\kappa_i(y_+ +\eta_i)/(x_+ +\xi)]\}^{k+\alpha+k\kappa_i+\beta_i}},\qquad 0<A<1.$$

$$(12.73)$$

Under a squared-error loss function, the Bayes point estimator of component availability may be determined by the law of iterated expectations, that is,

$$E(A|x_+,\underset{\sim}{y})=E_\kappa\left[E_{A|\kappa}(A|\kappa)|\underset{\sim}{y}\right],$$

where the inner expectation is taken using the conditional posterior pdf of A, given $\kappa=\kappa_i$; the outer expectation is taken using the (marginal) posterior pdf of κ. By making use of Euler's integral representation of the hypergeometric function, we find that

$$E(A|\kappa=\kappa_i)=\left(\frac{k\kappa_i+\beta_i}{k\kappa_i+\beta_i+k+\alpha}\right)\times$$

$${}_2F_1[1,k+\alpha;k+\alpha+k\kappa_i+\beta_i+1;1-\kappa_i(y_++\eta_i)/(x_++\xi)],$$

provided $0<\kappa_i(y_++\eta_i)/(x_++\xi)<2$. Therefore, the Bayes point estimator of component availability is given by

$$E(A|x_+,\underset{\sim}{y})=\sum_{i=1}^{m}p_i^*\left(\frac{k\kappa_i+\beta_i}{k\kappa_i+\beta_i+k+\alpha}\right)\times$$

$${}_2F_1[1,k+\alpha;k+\alpha+k\kappa_i+\beta_i+1;1-\kappa_i(y_++\eta_i)/(x_++\xi)], \tag{12.74}$$

where p_i^* is given by (12.70). Note that when the posterior pdf of κ is degenerate at one, (12.74) reduces to the expression for the Bayes point estimate of component availability when the repair times are $\mathcal{E}(\mu)$ distributed [see (12.27)].

The conditional posterior cdf of A, given $\kappa=\kappa_i$, is

$$G_{A|\kappa}(A|\kappa=\kappa_i)=\int_0^A g_{A|\kappa}(t|\kappa=\kappa_i)\,dt$$

$$=\frac{[\kappa_i(y_++\eta_i)/(x_++\xi)]^{k\kappa_i+\beta_i}}{B(k\kappa_i+\beta_i,k+\alpha)}\times$$

$$\int_0^A\frac{t^{k\kappa_i+\beta_i-1}(1-t)^{k+\alpha-1}\,dt}{\{1-t[1-\kappa_i(y_++\eta_i)/(x_++\xi)]\}^{k+\alpha+k\kappa_i+\beta_i}}.$$

By making the change of variable

$$z=\frac{t\kappa_i(y_++\eta_i)(k+\alpha)}{(1-t)(x_++\xi)(k\kappa_i+\beta_i)},$$

we obtain

$$G_{A|\kappa}(A|\kappa=\kappa_i)=\frac{\Gamma(k+\alpha+k\kappa_i+\beta_i)}{\Gamma(k\kappa_i+\beta_i)\Gamma(k+\alpha)}\left(\frac{k\kappa_i+\beta_i}{k+\alpha}\right)^{k\kappa_i+\beta_i}$$

$$\times\int_0^{\frac{A\kappa_i(y_++\eta_i)(k+\alpha)}{(1-A)(x_++\xi)(k\kappa_i+\beta_i)}}\frac{z^{k\kappa_i+\beta_i-1}dz}{\{1+[(k\kappa_i+\beta_i)/(k+\alpha)]z\}^{k+\alpha+k\kappa_i+\beta_i}}$$

$$=\mathcal{F}\left[A\kappa_i(k+\alpha)(y_++\eta_i)/(1-A)(k\kappa_i+\beta_i)(x_++\xi);\right.$$

$$\left.2(k\kappa_i+\beta_i),2(k+\alpha)\right].$$

Thus the posterior cdf of component availability is given by

$$G(A|x_+,\underset{\sim}{y})=\sum_{i=1}^{m}G_{A|\kappa}(A|\kappa=\kappa_i)\Pr(\kappa=\kappa_i|\underset{\sim}{y})$$

$$=\sum_{i=1}^{m}p_i^*\mathcal{F}\left[A\kappa_i(k+\alpha)(y_++\eta_i)/(1-A)(k\kappa_i+\beta_i)(x_++\xi);\right.$$

$$\left.2(k\kappa_i+\beta_i),2(k+\alpha)\right]. \tag{12.75}$$

Note that by differentiating both sides of (12.75) with respect to A, the posterior pdf of A is a mixture of Euler pdf's which is also called a repetitive Euler pdf by Brender (1968b). A symmetric $100(1-\gamma)\%$ TBPI for component availability, (A_L, A_U), is determined by solving the equations $G(A_L|x_+,\underset{\sim}{y})=\gamma/2$ for A_L and $G(A_U|x_+,\underset{\sim}{y})=1-\gamma/2$ for A_U.

12.5. PERIODIC MAINTENANCE IN REDUNDANT SYSTEMS

As noted in the previous chapter, one of the basic ways to improve system reliability is through redundancy. A parallel system is an active-redundant system: all components are initially put into operation yet only the functioning of any one of them is required for the system to function. A standby system is a passive-redundant system: only one component is in operation (on-line) at a time with the off-line completely inactive components each waiting to be sequentially activated when the on-line component fails.

Improvements on system reliability through redundancy can be greatly enhanced by performing periodic maintenance on the system [Von Alven

(1964), pp. 262–264]. Suppose periodic maintenance is performed every T hours, starting at time zero, on a redundant system comprised of identical $\mathcal{E}(\lambda)$ components each with constant failure rate λ. The procedure is to check every component in the system, and any one that has failed is replaced by a new and identical component. Since the $\mathcal{E}(\lambda)$ distribution is assumed for all component failure times, no aging takes place and the system is restored to a new condition after each maintenance action.

Let $R(t;T)$ denote the reliability at time t of a redundant system for which maintenance is performed every T hours. In the event that no maintenance action is taken, the reliability at time t is denoted by simply $R(t)$. A time period of t hours can be written as $t=jT+\tau$, where $j=0,1,2,\ldots$ and $0\leqslant\tau<T$. To survive t hours, the system must survive j maintenance periods plus an additional τ hours. Thus

$$R(t;T)=[R(T)]^{j}R(\tau), \qquad j=0,1,2,\ldots \qquad \text{and} \qquad 0\leqslant\tau<T. \tag{12.76}$$

We give Bayes point and interval estimates of $R(t;T)$ for parallel and standby systems. The results are based on a life test on the (common) $\mathcal{E}(\lambda)$ component consisting of N units put on test with the test terminated at the rth failure; $V=\sum_{i=1}^{r}X_i+(N-r)X_r$ is the total time on test, where X_i is the ith ordered failure time. The sampling distribution of V is a gamma distribution, $\mathcal{G}_1(r,\lambda)$. For a gamma prior pdf, $\mathcal{G}_1(\alpha,\xi)$, on the failure rate λ, the posterior pdf of λ is

$$g(\lambda|v)=[\Gamma(\alpha+r)]^{-1}(v+\xi)^{\alpha+r}\lambda^{\alpha+r-1}\exp[-(v+\xi)\lambda], \qquad \lambda>0, \tag{12.77}$$

which is also a $\mathcal{G}_1[\alpha+r,v+\xi)$ distribution.

Parallel System. For a parallel system consisting of k identical $\mathcal{E}(\lambda)$ components each with constant failure rate λ, system reliability at time t under no maintenance is, by (11.6),

$$R(t)=1-[1-\exp(-\lambda t)]^{k}.$$

Consequently, by (12.76), the reliability at time t, $R_p(t;T)$, of the periodically maintained parallel system is

$$R_p(t;T)=\{1-[1-\exp(-\lambda T)]^{k}\}^{j}\{1-[1-\exp(-\lambda\tau)]^{k}\}. \tag{12.78}$$

As $[1-\exp(-\lambda T)]^k = \Sigma_{i=0}^{k}\binom{k}{i}[-\exp(-\lambda T)]^i$, then

$$1-[1-\exp(-\lambda T)]^k = \sum_{i=1}^{k}\binom{k}{i}(-1)^{i-1}\exp(-\lambda Ti)$$

$$= \exp(-\lambda T)\sum_{i=0}^{k-1}\binom{k}{i+1}(-1)^i\exp(-\lambda Ti).$$

Let

$$a_i = \begin{cases}\binom{k}{i+1}(-1)^i, & i=0,1,\ldots,k-1 \\ 0, & i \geqslant k.\end{cases} \tag{12.79}$$

By using the formula for a power series raised to a power [Gradshteyn and Ryzhik (1965), p. 14], we find that

$$\{1-[1-\exp(-\lambda T)]^k\}^j = \exp(-\lambda jT)\left[\sum_{i=0}^{k-1} a_i\exp(-\lambda Ti)\right]^j$$

$$= \exp(-\lambda jT)\sum_{m=0}^{j(k-1)} c_m\exp(-\lambda Tm),$$

where

$$c_0 = a_0^j, \qquad c_m = \left(\frac{1}{ma_0}\right)\sum_{\ell=1}^{\min(m,k-1)}(\ell j - m + \ell)a_\ell c_{m-\ell},$$

$$m = 1,2,\ldots,j(k-1). \tag{12.80}$$

It now follows that

$$R_p(t;T) = \left[\exp(-\lambda jT)\sum_{m=0}^{j(k-1)} c_m\exp(-\lambda Tm)\right]\times$$

$$\left[\exp(-\lambda\tau)\sum_{i=0}^{k-1} a_i\exp(-\lambda\tau i)\right]$$

$$= \exp[-\lambda(jT+\tau)]\left[\sum_{m=0}^{j(k-1)} c_m\exp(-\lambda Tm)\right]\left[\sum_{i=0}^{k-1} a_i\exp(-\lambda\tau i)\right]$$

$$= \exp(-\lambda t)\sum_{m=0}^{j(k-1)}\sum_{i=0}^{k-1} a_i c_m\exp[-\lambda(\tau i + Tm)].$$

Under a squared-error loss function, the Bayes point estimator of system reliability at time t is given by

$$E\left[R_p(t;T)|v\right] = [\Gamma(\alpha+r)]^{-1}(v+\xi)^{\alpha+r}\sum_{m=0}^{j(k-1)}\sum_{i=0}^{k-1} a_i c_m \times$$

$$\int_0^{\infty}\lambda^{\alpha+r-1}\exp[-\lambda(v+\xi+\tau i+Tm+t)]\,d\lambda$$

$$=\sum_{m=0}^{j(k-1)}\sum_{i=0}^{k-1} a_i c_m\left(\frac{v+\xi}{v+\xi+\tau i+Tm+t}\right)^{\alpha+r}, \qquad (12.81)$$

where a_i is given by (12.79) and c_m by (12.80).

By differentiating the expression for $R_p(t;T)$given by (12.78) with respect to λ, it can be shown that $R_p(t;T)$ is a decreasing function of λ. Consequently, the posterior cdf of $R_p(t;T)$ is given by

$$G_p(R|v) = \Pr\left[R_p(t;T) \leq R|v\right] = \Pr(\lambda > \lambda_R|v)$$

$$= 1 - \mathcal{G}_1[\lambda_R(v+\xi);\alpha+r,1], \qquad (12.82)$$

where $\mathcal{G}_1(x;\alpha,\beta)$ is the cdf of the $\mathcal{G}_1(\alpha,\beta)$ distribution evaluated at x and λ_R is the value of λ for which the equation $R_p(t;T) = R$ (fixed) holds. To determine a symmetric $100(1-\gamma)\%$ TBPI for system reliability at time t, (R_L, R_U), we solve the equations $G_p(R_L|v) = \gamma/2$ for R_L and $G_p(R_U|v) = 1-\gamma/2$ for R_u. A $100(1-\gamma)\%$ LBPI for $R_p(t;T)$ is $(R_\gamma, 1)$, where R_γ is the solution to the equation $G_p(R_\gamma|v) = \gamma$. For the special case when α is an integer, it follows that

$$R_\gamma = \left\{1-[1-\exp(-\lambda_R T)]^k\right\}^j\left\{1-[1-\exp(-\lambda_R \tau)]^k\right\}, \qquad (12.83)$$

where

$$\lambda_R = \frac{\chi^2_{1-\gamma}[2(\alpha+r)]}{2(v+\xi)}, \qquad (12.84)$$

and $\chi^2_\gamma(n)$ is the 100γth percentile of the $\chi^2(n)$ distribution.

Standby System. For a standby system (assuming perfect switching) consisting of k identical $\mathcal{E}(\lambda)$ components with one on-line unit and $k-1$ off-line units that are completely inactive, the reliability of the system at time t under no maintenance is [Sandler (1963), p. 78]

$$R(t) = \exp(-\lambda t) \sum_{i=0}^{k-1} \frac{(\lambda t)^i}{i!}.$$

By (12.76), the reliability at time t, $R_s(t; T)$, of the periodically maintained standby system is

$$R_s(t; T) = \exp(-\lambda t) \left[\sum_{i=0}^{k-1} \frac{(\lambda T)^i}{i!} \right]^j \sum_{i=0}^{k-1} \frac{(\lambda \tau)^i}{i!}. \tag{12.85}$$

Let

$$a_i = T^i / i! \qquad \text{and} \qquad b_i = \tau^i / i!, \qquad i = 0, 1, \ldots, k-1. \tag{12.86}$$

By using the formula for a power series raised to a power, we find that

$$\left(\sum_{i=0}^{k-1} a_i \lambda^i \right)^j = \sum_{i=0}^{j(k-1)} c_i \lambda^i,$$

where

$$c_0 = 1, \quad c_m = \left(\frac{1}{m} \right) \sum_{\ell=1}^{\min(m, k-1)} (\ell j - m + \ell) a_\ell c_{m-\ell}, \quad m = 1, 2, \ldots, j(k-1). \tag{12.87}$$

Thus

$$R_s(t; T) = \exp(-\lambda t) \left(\sum_{i=0}^{j(k-1)} c_i \lambda^i \right) \left(\sum_{i=0}^{k-1} b_i \lambda^i \right)$$

$$= \exp(-\lambda t) \sum_{i=0}^{(j+1)(k-1)} d_i \lambda^i, \tag{12.88}$$

where

$$d_i = \sum_{r=\max(0,\, i-k+1)}^{\min[i,\, j(k-1)]} c_r b_{i-r}, \qquad i = 0, 1, \ldots, (j+1)(k-1), \quad (12.89)$$

and c_m is given by (12.87) and b_i by (12.86).

Under a squared-error loss function, the Bayes point estimator of system reliability at time t is given by

$$E[R_s(t;T)|v] = [\Gamma(\alpha+r)]^{-1}(v+\xi)^{\alpha+r} \times \sum_{i=0}^{(j+1)(k-1)} d_i \int_0^{\infty} \lambda^{i+\alpha+r-1} \exp[-\lambda(t+v+\xi)]\, d\lambda$$

$$= \left(\frac{v+\xi}{v+\xi+t}\right)^{\alpha+r} \frac{1}{\Gamma(\alpha+r)} \sum_{i=0}^{(j+1)(k-1)} d_i \frac{\Gamma(\alpha+r+i)}{(v+\xi+t)^i}, \quad (12.90)$$

where d_i is given by (12.89).

By differentiating the expression for $R_s(t;T)$ given by (12.85) with respect to λ, it can be shown that $R_s(t;T)$ is a decreasing function of λ. The posterior cdf of $R_s(t;T)$ is given by

$$G_s(R|v) = 1 - \mathcal{G}_1[\lambda_R(v+\xi); \alpha+r, 1], \quad (12.91)$$

where λ_R is the value of λ for which the equation $R_s(t;T) = R$ (fixed) holds. To determine a symmetric $100(1-\gamma)\%$ TBPI for system reliability at time t, (R_L, R_U), we solve the equations $G_s(R_L|v) = \gamma/2$ for R_L and $G_s(R_U|v) = 1-\gamma/2$ for R_U. A $100(1-\gamma)\%$ LBPI for $R_s(t;T)$ is $(R_\gamma, 1)$, where R_γ is the solution to the equation $G_s(R_\gamma|v) = \gamma$. When α is an integer, it follows that

$$R_\gamma = \exp(-\lambda_R t)\left[\sum_{i=0}^{k-1} \frac{(\lambda_R T)^i}{i!}\right]^j \left[\sum_{i=0}^{k-1} \frac{(\lambda_R \tau)^i}{i!}\right], \quad (12.92)$$

where λ_R is given by (12.84).

Example 12.9. The following data are given by Shooman (1968, p. 162 and p. 460) and represent the results of a life test on 10 hypothetical (identical) electronic components. It was shown that the data are well fitted by an $\mathcal{E}(\lambda)$

distribution.

Failure Number	Operating Time (h)
1	8
2	20
3	34
4	46
5	63
6	86
7	111
8	141
9	186
10	266

In summary, $N=r=10$ and the observed total time on test is $v=961$ h.

Further suppose that to an electronic component of the type that was tested, an identical component is added in parallel. Based on the life test data, we wish to compare point estimates of system reliability at time t, where $0 \leqslant t < 200$, for the following situations: (1) periodic maintenance at $T=50$ h, (2) periodic maintenance at $T=100$ h; and (3) no periodic maintenance performed. We assume a uniform prior pdf ($\alpha=1$, $\xi=0$) on the failure rate λ.

T = 50 h

Following considerable simplification, the Bayes point estimate of $R_p(t;50)$ is, by (12.81),

$$E\left[R_p(t;50)\,|\,v\right] = 2\left(\frac{961}{961+t}\right)^{11} - \left(\frac{961}{961+2t}\right)^{11}, \qquad 0 \leqslant t < 50,$$

$$= 4\left(\frac{961}{961+t}\right)^{11} - 2\left(\frac{961}{1011+t}\right)^{11} - 2\left(\frac{961}{911+2t}\right)^{11} + \left(\frac{961}{961+2t}\right)^{11}, \qquad 50 \leqslant t < 100,$$

$$= 8\left(\frac{961}{961+t}\right)^{11} - 8\left(\frac{961}{1011+t}\right)^{11} + 2\left(\frac{961}{1061+t}\right)^{11} - 4\left(\frac{961}{861+2t}\right)^{11} + 4\left(\frac{961}{911+2t}\right)^{11} - \left(\frac{961}{961+2t}\right)^{11}, \qquad 100 \leqslant t < 150,$$

$$= 16\left(\frac{961}{961+t}\right)^{11} - 24\left(\frac{961}{1011+t}\right)^{11} + 12\left(\frac{961}{1061+t}\right)^{11} - 2\left(\frac{961}{1111+t}\right)^{11} - 8\left(\frac{961}{811+2t}\right)^{11} + 12\left(\frac{961}{861+2t}\right)^{11} - 6\left(\frac{961}{911+2t}\right)^{11} + \left(\frac{961}{961+2t}\right)^{11}, \qquad 150 \leqslant t < 200.$$

To illustrate the manner in which the above estimate was determined, consider the case where the time t is such that $100 \leqslant t < 150$. There are $j = 2$ maintenance periods plus an additional $\tau = t - 100$ h. By (12.79),

$$a_0 = \binom{2}{1}(-1)^0 = 2, \qquad a_1 = \binom{2}{2}(-1) = -1.$$

By (12.80),

$$c_0 = 2^2 = 4,$$

$$c_1 = (1/2)(2-1+1)a_1 c_0 = -4,$$

$$c_2 = (1/4)(2-2+1)a_1 c_1 = 1.$$

By (12.81),

$$E\left[R_p(t;50)|v\right] = \sum_{m=0}^{2} \sum_{i=0}^{1} a_i c_m \left(\frac{961}{961 + \tau i + 50m + t}\right)^{11}.$$

Upon writing out the double sum and replacing τ by $t - 100$, we obtain the previously given result.

T = 100 h

The Bayes point estimate of $R_p(t; 100)$ is given by

$$E\left[R_p(t;100)|v\right] = 2\left(\frac{961}{961+t}\right)^{11} - \left(\frac{961}{961+2t}\right)^{11}, \qquad 0 \leqslant t < 100,$$

$$= 4\left(\frac{961}{961+t}\right)^{11} - 2\left(\frac{961}{1061+t}\right)^{11} - 2\left(\frac{961}{861+2t}\right)^{11} + \left(\frac{961}{961+2t}\right)^{11},$$

$$100 \leqslant t < 200.$$

T = ∞ (no periodic maintenance)

The Bayes point estimate of $R_p(t)$ is given by

$$E\left[R_p(t)|v\right] = 2\left(\frac{961}{961+t}\right)^{11} - \left(\frac{961}{961+2t}\right)^{11}, \qquad 0 \leqslant t < 200.$$

The three point estimates are plotted in Figure 12.8. Note that from zero to 50 h, all three estimates are identical as no maintenance has been initiated. From the figure we can see the extent to which periodic maintenance

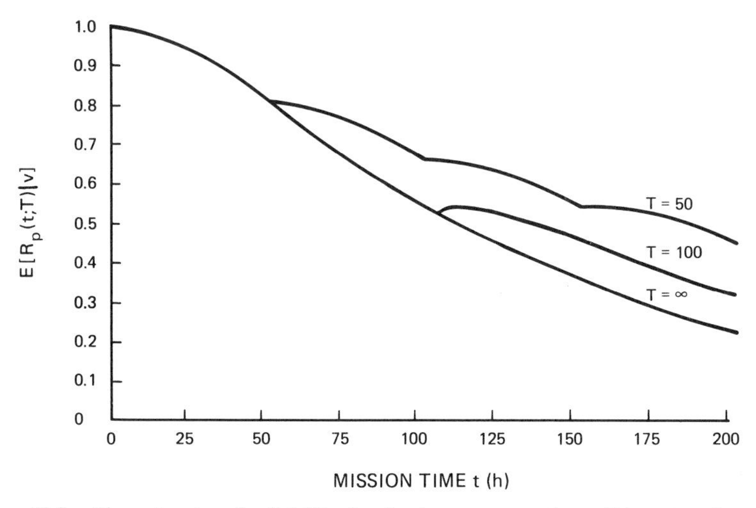

Figure 12.8. The estimates of reliability for the two-component parallel system for which periodic maintenance is performed every T hours in Example 12.9.

improves system reliability. For example, for a mission time of $t = 125$ h, periodic maintenance performed every 100 h, as opposed to no maintenance, improves system reliability from 0.442 to 0.517—a 17% increase. If, instead, periodic maintenance is performed every 50 h, system reliability improves from 0.442 to 0.622—a 41% increase.

Example 12.10. Again let us consider Example 12.9. For the parallel system described in Example 12.9 and a mission time of $t = 125$ h, suppose that a system reliability goal of at least 0.80 has been set, and the probability that it be attained is required to be at least 0.95. What periodic maintenance schedule should be carried out using the fewest number of maintenance periods to achieve this? Specifically, we require

$$\Pr\left[R_p(125; T) \geqslant 0.80 \,|\, v\right] \geqslant 0.95.$$

Based on the life test data and (12.84),

$$\lambda_{0.80} = \frac{\chi^2_{0.95}(22)}{2(961)} = 0.0177.$$

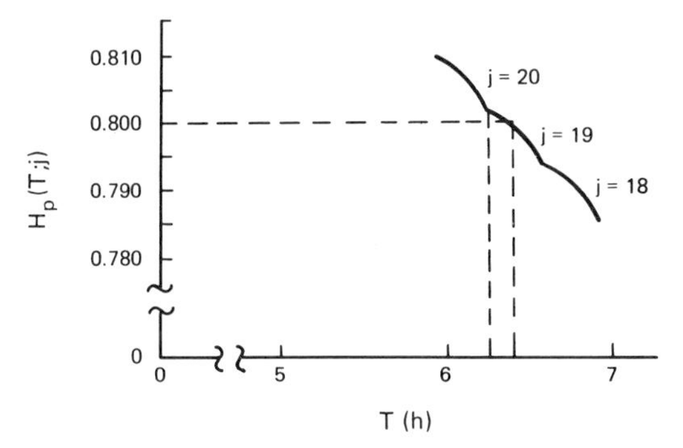

Figure 12.9. The effects on parallel system reliability due to various periodic maintenance schedules in Example 12.10.

By (12.83),

$$0.80 \leqslant \{1-[1-\exp(-0.0177T)]^2\}^j\{1-[1-\exp(-0.0177\tau)]^2\}$$

$$= \{1-[1-\exp(-0.0177T)]^2\}^j\left(1-\{1-\exp[-0.0177(125-jT)]\}^2\right)$$

$$= H_p(T;j), \text{ say.}$$

For specified values of j, Figure 12.9 is a graph of T versus $H_p(T;j)$. From the figure we see that the required number of maintenance periods is 19. The values of T may range from 6.26 to 6.41 h. Thus, if periodic maintenance is performed, for example, every 6 hours and 20 minutes, then with a probability of at least 0.95 system reliability at $t = 125$ h will be at least 0.80.

Example 12.11. Let us further consider Example 12.10. Suppose that to the original electronic component an identical component is added but in a standby rather than parallel configuration. If periodic maintenance is performed every 6 hours and 20 minutes on this system, let us calculate a 95% LBPI for system reliability at $t = 125$ h. Assume a uniform prior pdf ($\alpha = 1$, $\xi = 0$) on the failure rate λ. As before, $\lambda_R = 0.0177$. By (12.92),

$$R_{0.95} = \exp[-0.0177(125)] \times$$

$$\left\{\sum_{i=0}^{1} \frac{[0.0177(6.33)]^i}{i!}\right\}^{19}\left\{\sum_{i=0}^{1} \frac{[0.0177(4.73)]^i}{i!}\right\}$$

$$= 0.892;$$

thus the required interval estimate of $R_s(125; 6.33)$ is $(0.892, 1)$.

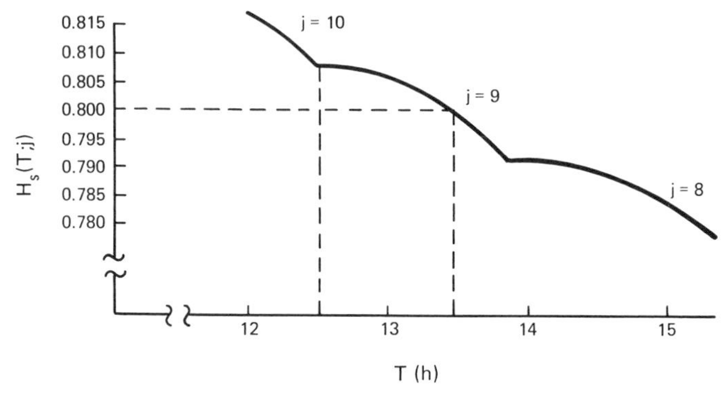

Figure 12.10. The effects on standby system reliability due to various periodic maintenance schedules.

Suppose the same reliability goal as described in Example 12.10 is given for the standby system. We then determine the smallest value of j for which

$$0.80 \leq \exp[-0.0177(125)](1+0.0177T)^{j}(1+0.0177\tau)$$

$$=0.10943(1+0.0177T)^{j}[1+0.0177(125-jT)]$$

$$=H_s(T;\ j),\ \text{say.}$$

For specified values of j, Figure 12.10 is a graph of T versus $H_s(T;\ j)$. From the graph we see that the required number of maintenance periods is nine (as compared to 19 for the parallel system). The values of T may range from 12.51 to 13.45 h (as compared to a range of from 6.26 to 6.41 h for the parallel system). Consequently, if periodic maintenance is performed, for example, every 13 h on the standby system, then with a probability of at least 0.95 system reliability at $t=125$ h will be at least 0.80. ■

EXERCISES

12.1. Consider a system consisting of three independently operating repairable components in series. There is a repair facility for each component which effects a repair independently of that of any other repair facility (see Figure 12.1). Moreover, while a failed component is undergoing repair (during which time the system is down), nonfailed components continue to operate. "Snapshot" observations on each component yielded the following results.

Component 1	10 "snapshots" taken with 9 operating states observed
Component 2	7 "snapshots" taken with 7 operating states observed
Component 3	12 "snapshots" taken with 11 operating states observed

Assume uniform prior pdf's on each component availability.

(a) Find the Bayes point estimate of system availability for a squared-error loss function.

(b) Determine the posterior cdf of system availability by using (i) Mellin transform techniques, and (ii) the integral expression in Exercise 11.8(a).

(c) Draw the graph of the posterior cdf obtained in (b). From the graph, find a symmetric 95% TBPI for system availability.

12.2. Consider a repairable component that has a constant failure rate λ and a constant repair rate μ. Let δ denote an estimate of component availability A based on "patch" data from k failure/repair cycles, where the observed total operating and repair times are x_+ and y_+, respectively. No "snapshots" are taken. Assume uniform prior pdf's on both λ and μ.

(a) Show that the Bayes point estimate of A for an absolute-error loss function, $L(A,\delta)=|A-\delta|$, is given by

$$\delta^* = \frac{x_+}{x_+ + y_+ \mathfrak{F}_{0.5}[2(k+1), 2(k+1)]}.$$

(b) The loss function given by $L(A,\delta)=(A-\delta)^2/A(1-A)$ is sometimes used when it is desirable to place more weight (importance) on an error made at, say, $A=0.95$ or 0.05 than at $A=0.50$. Determine the Bayes point estimate of A for this loss function. What is the limiting form of this estimate as k becomes large?

(c) Compare the Bayes point estimates obtained by evaluating the results in (a) and (b) with that found in Example 12.3 for a squared-error loss function.

12.3. Show that

$$E\left[A^i(1-A)^j|x_+,y_+\right]=\left(\frac{x_++\xi}{y_++\eta}\right)^j\frac{B(i+k+\beta,j+k+\alpha)}{B(k+\beta,k+\alpha)}\times$$

$${}_2F_1\left[i+j,j+k+\alpha;i+j+2k+\alpha+\beta;1-\frac{y_++\eta}{x_++\xi}\right],$$

where A has an Euler distribution given by (12.26). Hint: Make use of (12.21).

12.4. Suppose k failure/repair cycles are observed on a component that has a constant failure rate λ and a constant repair rate μ. The observed total operating and repair times are x_+ and y_+, respectively. Assume that λ and μ each have independent gamma prior pdf's given by

$$g_1(\lambda;\alpha,\xi)=[\Gamma(\alpha)]^{-1}\xi^\alpha\lambda^{\alpha-1}\exp(-\xi\lambda),\qquad \lambda>0$$

and

$$g_2(\mu;\beta,\eta)=[\Gamma(\beta)]^{-1}\eta^\beta\mu^{\beta-1}\exp(-\eta\mu),\qquad \mu>0.$$

(a) Show that the joint posterior pdf of λ and μ may be written as

$$g_{\lambda,\mu}(\lambda,\mu|x_+,y_+)=g_1(\lambda;k+\alpha,x_++\xi)g_2(\mu;k+\beta,y_++\eta).$$

(b) Let $\phi(\lambda,\mu)$ be a function of λ and μ. For hypothetical values a, b, x_a, y_b show that

$$E_g\left[\phi(\lambda,\mu)\lambda^a\exp(-x_a\lambda)\mu^b\exp(-y_b\mu)|x_+,y_+\right]$$

$$=\frac{\Gamma(a+k+\alpha)\Gamma(b+k+\beta)(x_++\xi)^{k+\alpha}(y_++\eta)^{k+\beta}}{\Gamma(k+\alpha)\Gamma(k+\beta)(x_a+x_++\xi)^{a+k+\alpha}(y_b+y_++\eta)^{b+k+\beta}}$$

$$\times E_{g^*}\left[\phi(\lambda,\mu)|x_a+x_+,y_b+y_+\right],$$

where E_g denotes expectation taken under the joint posterior pdf of λ and μ given by $g_{\lambda,\mu}(\lambda,\mu|x_+,y_+)=g_1(\lambda;k+\alpha,x_++\xi)g_2(\mu;k+\beta,y_++\eta)$, and E_{g^*} denotes expectation taken under the hypothetical joint posterior pdf of λ and μ given by $g^*_{\lambda,\mu}(\lambda,\mu|x_a+x_+,y_b+y_+)=g_1(\lambda;a+k+\alpha,x_a+x_++\xi)\times g_2(\mu;b+k+\beta,y_b+y_++\eta)$.

12.5. *Instantaneous component availability* at time t is defined as the probability that the component is in an operating state at time t. For a component with a constant failure rate λ and a constant repair rate μ, instantaneous component availability at time t is given by [see Section 4.8.2]

$$A(t)=\frac{\mu}{\lambda+\mu}+\frac{\lambda}{\lambda+\mu}\exp[-(\lambda+\mu)t],$$

provided the component was initially in an operating state. Suppose the component is observed for k failure/repair cycles, where x_+ and y_+ are the total operating and repair times, respectively. Assume gamma prior pdfs, $\mathcal{G}_1(\alpha,\xi)$ and $\mathcal{G}_1(\beta,\eta)$ on λ and μ, respectively. By using the results in Exercises 12.3 and 12.4(b), show that the Bayes point estimator of $A(t)$ for a squared-error loss function is given by

$$\begin{aligned}E[A(t)|x_+,y_+]\\ &=\left(\frac{k+\beta}{2k+\alpha+\beta}\right){}_2F_1\left(1,k+\alpha;2k+\alpha+\beta+1;1-\frac{y_++\eta}{x_++\xi}\right)\\ &\quad+\frac{(k+\alpha)(x_++\xi)^{k+\alpha}(y_++\eta)^{k+\beta}}{(2k+\alpha+\beta)(t+x_++\xi)^{k+\alpha-1}(t+y_++\eta)^{k+\beta+1}}\\ &\quad\times{}_2F_1\left(1,k+\alpha+1;2k+\alpha+\beta+1;1-\frac{t+y_++\eta}{t+x_++\xi}\right).\end{aligned}$$

12.6. *Mission oriented component availability* is defined as the probability that the component is in an operating state upon demand under steady-state conditions and remains operative for at least a mission interval of length t. For a component that has a constant failure rate λ and a constant repair rate μ, mission oriented component availability is given by [see Brender (1968a)]

$$A(t)=\frac{\mu}{\lambda+\mu}\exp(-\lambda t).$$

Suppose the component is observed for k failure/repair cycles, where x_+ and y_+ are the total operating and repair times, respectively. Assume gamma prior pdf's, $\mathcal{G}_1(\alpha,\xi)$ and $\mathcal{G}_1(\beta,\eta)$, on λ and μ, respectively.

(a) By using the results in Exercises 12.3 and 12.4(b), show that the mth moment of the posterior distribution of $A(t)$ is

$$E\{[A(t)]^m | x_+, y_+\} = \left(\frac{x_+ + \xi}{mt + x_+ + \xi}\right)^{k+\alpha} \frac{(k+\beta)_m}{(2k+\alpha+\beta)_m} \times$$

$$_2F_1\left[m, k+\alpha; m+2k+\alpha+\beta; 1 - \frac{y_+ + \eta}{mt + x_+ + \xi}\right].$$

(b) The exact posterior distribution of $A(t)$ would appear to be very difficult to determine. Instead, suppose this distribution is approximated by a beta distribution $\mathcal{B}_1(a, b)$ having the same first two moments. Show that

$$a = b\frac{E[A(t)|x_+, y_+]}{1 - E[A(t)|x_+, y_+]}$$

$$b = \{1 - E[A(t)|x_+, y_+]\} \times$$

$$\left(\frac{E[A(t)|x_+, y_+]\{1 - E[A(t)|x_+, y_+]\}}{\mathrm{Var}[A(t)|x_+, y_+]} - 1\right).$$

(c) (Continuation of Example 12.3) By using the result in (b), determine an approximation to the posterior cdf of $A(t)$, where the mission duration is of length $t = 75$ h. Find an approximate 90% LBPI for $A(75)$.

12.7. (Continuation of Example 12.4) By using the integral expression in Exercise 11.8(a), derive the posterior cdf of system availability.

12.8. Consider a series system consisting of two identical repairable components each with a constant failure rate λ and a constant repair rate μ. Suppose there is a single repair facility available to service the system. In addition, while the repair of a failed component is taking place, the other component continues to operate (although the system is down). In the event it too fails during the repair period, its repair is initiated immediately upon the completion of the repair of the preceding failed component. In this situation system availability is [see Sandler (1963) pp. 121–123]

$$A = \frac{\mu^2}{\mu^2 + 2\lambda\mu + 2\lambda^2}.$$

Suppose k failure/repair cycles are observed on the (common) component, where x_+ and y_+ are the total operating and repair times, respectively. Assume noninformative prior pdf's on both λ and μ.

(a) Show that the posterior cdf of A, say $G(A|x_+, y_+)$, is given by

$$G(A|x_+, y_+) = 1 - \mathcal{F}\left[(x_+/y_+)\left(-1+\sqrt{2/A-1}\right)/2; 2k, 2k\right].$$

Hint: Make use of the posterior pdf of the service factor $\rho = \lambda/\mu$ given by (12.44).

(b) Determine a symmetric $100(1-\gamma)\%$ TBPI for system availability.

(c) Suppose eight failure/repair cycles are observed on the (common) component that resulted in $y_+/x_+ = 0.105$. Find a 95% LBPI for system availability.

12.9. Consider a parallel system consisting of N identical repairable components each of which has a constant failure rate λ and a constant repair rate μ. There is a repair facility for each component that initiates a repair immediately upon occurrence of a failure (see Figure 12.5). An availability goal, which is part of a contract specification, requires that system availability be at least γ_1 and the probability that this be attained be at least γ_2. Suppose k failure/repair cycles are observed on the (common) component, where x_+ and y_+ are the total operating and repair times, respectively. Assume noninformative prior pdf's on both λ and μ.

(a) Show that the smallest value of N so that the availability goal is satisfied is given by

$$N = \left[\!\left[\frac{\ln(1-\gamma_1)}{-\ln\left[1+(x_+/y_+)\mathcal{F}_{1-\gamma_2}(2k,2k)\right]} \right]\!\right] + 1.$$

(b) Suppose that N is fixed. Determine an upper bound on the ratio of the observed average repair time to the observed average failure time.

(c) (Continuation of Example 12.6) Suppose an availability goal is such that system availability is to be at least 0.999 with probability at least 0.95. What is the minimum number of components required in the system to achieve this goal? Assuming the number of components in the system is fixed at two, approximately what should be the estimated repair rate (in terms of the estimated failure rate) to achieve the availability goal?

12.10. A parallel system consisting of two identical repairable components each with a constant failure rate λ and a constant repair rate μ has a single repair facility to service both components (see Figure 12.6). The system *mean time to first failure* (MTTFF) is given by [see Sandler (1963), p. 139]

$$\text{MTTFF} = \frac{3\lambda + \mu}{2\lambda^2}.$$

Suppose k failure/repair cycles are observed on the (common) component, where x_+ and y_+ are the total operating and repair times, respectively. Assume gamma prior pdf's, $\mathcal{G}_1(\alpha, \xi)$ and $\mathcal{G}_1(\beta, \eta)$ on λ and μ, respectively.

(a) By using Exercise 12.4(b), derive a point estimator of the system MTTFF for a squared-error loss function.

(b) (Continuation of Example 12.7) By evaluating the result in (a), find a point estimator of the MTTFF of the system that was considered in Example 12.7.

12.11. Consider an N-component standby redundant system with one repair facility. The components are identical with a constant failure rate λ and repairable with a constant repair rate μ. Assume perfect switching and, in addition, an off-line component cannot fail until it is switched to the on-line position. System availability is given by [see Sandler (1963), pp. 134–135]

$$A = 1 - \left[\sum_{j=0}^{N} \left(\frac{\mu}{\lambda} \right)^j \right]^{-1}.$$

Suppose k failure/repair cycles are observed on the (common) component, where x_+ and y_+ are the total operating and repair times, respectively. Gamma prior pdf's, $\mathcal{G}_1(\alpha, \xi)$ and $\mathcal{G}_1(\beta, \eta)$, are assigned to λ and μ, respectively.

(a) By first writing A as the quotient of two polynomials in the service factor $\rho = \lambda/\mu$, show that the posterior cdf of A, say $G(A|x_+, y_+)$, is given by

$$G(A|x_+, y_+) = 1 -$$

$$\mathcal{F}\left[(k+\beta)(x_+ + \xi)a/(k+\alpha)(y_+ + \eta); 2(k+\alpha), 2(k+\beta)\right],$$

where a is the unique positive root to the equation $(1+a+\cdots+a^{N-1})/(1+a+\cdots+a^N)=A$.

(b) Determine a symmetric $100(1-\gamma)\%$ TBPI for system availability.

(c) (Continuation of Example 12.8) Suppose an availability goal is such that system availability is required to be at last 0.999, and the probability that this be attained be at least 0.95. Based on the data in Example 12.8, determine the smallest value of N for which this availability goal is satisfied.

12.12. (Continuation of Example 12.9) Suppose periodic maintenance is performed every $T=24$ h on the two-component parallel system described in Example 12.9. Find a 95% LBPI for system reliability at $t=60$ h.

12.13. (Continuation of Example 12.11) Find the Bayes point estimate for system reliability at $t=25$ h (assuming a squared-error loss function) for the two-component standby redundant system for which periodic maintenance is performed every 6 hours and 20 minutes.

12.14. A repairable component with a constant failure rate λ is such that its repair time has a gamma distribution [see Section 12.4]. With regard to the repair of the component, three types of repairs are possible. Based on a previous history of repairs, the proportion of the time that each type of repair was made along with the number of exponential phases (each of average length $1/\mu$) that it goes through to effect the repair is given below.

Number of Exponential Phases (κ_i)	1	2	4
$p_i=\Pr(\kappa=\kappa_i)$	0.6	0.3	0.1

Suppose the component was observed through seven failure/repair cycles with the following results:

Failure Times (h)	Repair Times (h)
87.7	2.0
10.5	0.6
22.2	7.8
23.7	4.1
73.9	4.9
122.8	0.5
33.1	2.4

Assume a uniform prior pdf on λ. In addition, for given κ_i, $i=1,2,3$, a uniform prior pdf ($\beta_i=1$, $\eta_i=0$) is assigned to μ.

(a) Show that the (marginal) posterior pdf of κ is

$$p_1^* = \Pr(\kappa=1|\underset{\sim}{y}) = 0.5775,$$

$$p_2^* = \Pr(\kappa=2|\underset{\sim}{y}) = 0.4109,$$

$$p_3^* = \Pr(\kappa=4|\underset{\sim}{y}) = 0.0116.$$

(b) Determine the Bayes point estimate of component availability for a squared-error loss function.

(c) Show that the posterior cdf of component availability given by (12.75) reduces to

$$G(A|x_+, y_+) = \sum_{i=1}^{3} p_i^* \mathcal{F}[0.4771\kappa_i A/(7\kappa_i+1)(1-A); 2(7\kappa_i+1), 16].$$

From its graph, find a 95% LBPI for component availability.

REFERENCES

Barlow, R. E. and Proschan, F. (1975). *Statistical Theory of Reliability and Life-Testing*, Holt, Rinehart & Winston, New York.

Bazovsky, I. (1961). *Reliability Theory and Practice*, Prentice-Hall, Englewood Cliffs, NJ.

Brender, D. M. (1968a). The Prediction and Measurement of System Availability: A Bayesian Treatment, *IEEE Transactions on Reliability*, Vol. R-17, pp. 127–138.

Brender, D. M. (1968b). The Bayesian Assessment of System Availability: Advanced Applications and Techniques, *IEEE Transactions on Reliability*, Vol. R-17, pp. 138–147.

Butterworth, R. W. and Nikolaisen, T. (1973). Bounds on the Availability Function, *Naval Research Logistics Quarterly*, Vol. 20, pp. 289–296.

Dyer, D. (1982). The Convolution of Generalized-F Distributions, to appear in the *Journal of the American Statistical Association*.

Erdélyi, A. (1953). *Higher Transcendental Functions*, *Vol. I*, McGraw-Hill, New York.

Gaver, D. P. and Mazumdar, M. (1967). Statistical Estimation in a Problem of System Reliability, *Naval Research Logistics Quarterly*, Vol. 14, pp. 473–488.

Gaver, D. P. and Mazumdar, M. (1969). Some Bayes Estimates of Long-Run Availability in a Two-State System, *IEEE Transactions on Reliability*, Vol. R-18, pp. 184–189.

Gradshteyn, I. S. and Ryzhik, I. M. (1965). *Tables of Integrals*, *Series*, *and Products*, Academic, New York.

Karlin, S. and Taylor, H. M. (1975). *A First Course in Stochastic Processes*, (2nd ed.), Academic, New York.

Lie, C. H., Hwang, C. L. and Tillman, F. A. (1977). Availability of Maintained Systems: A State-of-the-Art Survey, *AIIE Transactions*, Vol. 9, pp. 247–259.

Luke, Y. L. (1977). *Algorithms for the Computation of Mathematical Functions*, Academic, New York.

Lwin, T. and Singh, N. (1974). Bayesian Analysis of the Gamma Distribution Model in Reliability Estimation, *IEEE Transactions on Reliability*, Vol. R-23, pp. 314–318.

Rohn, W. B. (1959). Reliability Prediction for Complex Systems, *Proceedings 5th National Symposium on Reliability and Quality Control*, pp. 381–388.

Sandler G. H. (1963). *System Reliability Engineering*, Prentice-Hall, Englewood Cliffs, NJ.

Shooman, M. L. (1968). *Probabilistic Reliability: An Engineering Approach*, McGraw-Hill, New York.

Smith, D. J. (1972). *Reliability Engineering*, Barnes & Noble, New York.

Springer, M. D. (1979). *The Algebra of Random Variables*, Wiley, New York.

Thompson, M. (1966). Lower Confidence Limits and a Test of Hypotheses for System Availability, *IEEE Transactions on Reliability*, Vol. R-15, pp. 32–36.

Thompson, W. E. and Palicio, P. A. (1975). Bayesian Confidence Limits for the Availability of Systems, *IEEE Transactions on Reliability*, Vol. R-24, pp. 118–120.

Thompson, W. E. and Springer, M. D. (1972). A Bayes Analysis of Availability for a System Consisting of Several Independent Subsystems, *IEEE Transactions on Reliability*, Vol. R-21, pp. 212–214.

Von Alven, W. H. (1964). *Reliability Engineering*, ARINC Research Corporation, Prentice-Hall, Englewood Cliffs, NJ.

Von Mises, R. (1964). *Mathematical Theory of Probability and Statistics*, Academic, New York.

CHAPTER 13

Empirical Bayes Reliability Estimation

This chapter departs markedly in philosophy from the remainder of the book. It has been included in order to address a fundamental question; namely, can the Bayesian procedure be used if the underlying prior distribution is unidentified? The theoretic structure of the methodology known as empirical Bayes (EB) decision theory is presented in Sections 13.1 and 13.2, whereas remaining sections contain the results for the binomial sampling model and the $\mathcal{E}(\lambda)$, $\mathcal{W}_1(\alpha, \beta)$, $\mathcal{G}(p, \theta)$, and $\mathcal{LN}(\xi, \sigma^2)$ failure time models. Sections 13.1 and 13.2 may be omitted by the reader who is interested only in reliability applications.

In previous chapters we considered a variety of Bayesian reliability estimation procedures. In all of these procedures an underlying prior distribution was assumed to exist and to have either a degree of belief or a frequency interpretation. Furthermore, the parametric form of the prior distribution was either assumed to be completely known and specified or known except for the values of certain parameters that had to be estimated from available data [see Sections 7.6 and 7.7]. It is sometimes the case that the prior distribution has a frequency interpretation and can be estimated through suitable observations. However, suppose that the distributional form of the prior remains unknown. In this case Bayes estimation methods cannot be employed apart from a hit-or-miss assumption about the unknown prior. Suppose now that such an assumption is made and further that the parameter to be estimated does indeed follow an unknown prior distribution having a frequency interpretation. The calculated Bayes estimates based on the assumed prior may or may not accurately approximate the true Bayes estimate that could be obtained if the true prior distribution were known. In such cases the accuracy of the approximating assumption is never really known. One can only demonstrate how well the assumed prior distribution behaves when the true distribution departs from the assumption. This procedure was used by Canavos (1975, 1977) and Higgins and Tsokos (1976, 1977).

Based on the above discussion it is clear that the assignment of a prior distribution contains an element of risk for the investigator. It is appropriate to ask if there is any alternative to the use of an assumed prior or of having to ignore altogether that the prior distribution exists. The answer is affirmative and several procedures that can be used to approximate the true, but unknown, Bayes estimators are described in this chapter. These procedures assume the existence of a prior distribution but require no explicit assumption regarding its distributional form, such as $\mathcal{G}_1(\alpha_0, \beta_0)$, $\mathcal{B}(x_0, n_0)$, or $\mathcal{U}(\alpha_0, \beta_0)$. In this sense the prior distribution remains completely unknown and unspecified. Past data is used to bypass the necessity for identification of the correct prior distribution. This is commonly referred to as the *empirical Bayes(EB) procedure* defined as follows:

Definition 13.1. *Empirical Bayes (EB) decision procedures* are a class of decision theoretic procedures that utilize past data as a means for bypassing the necessity of identifying a completely unknown and unspecified prior distribution having a frequency interpretation.

Although this definition of empirical Bayes procedures is the one adopted here, it is *not* a universal definition. Some other authors, such as Cox and Hinkley (1974), expand the definition to include those cases in which a prior distributional form is stated up to the values of the prior parameters that are then estimated by means of past data. They designate the procedures given in Sections 7.6 and 7.7 as empirical Bayes methods, and refer to the situation in which the prior distribution form is completely unspecified as *nonparametric empirical Bayes*. However, we prefer the more restrictive use of the term "empirical Bayes" because it indicates a sharper departure from pure Bayesian methods.

Looked at in a different way, there is another desirable reason for the use of EB procedures. The use of these procedures avoids the need to identify a prior distribution. The assignment of a prior distribution frequently represents a practical difficulty in the application of Bayesian methods and the psychological stigma of this often difficult task is thus avoided. It is mentioned here at the outset that EB procedures cannot always be employed, depending on the availability of certain data requirements to be discussed. When they can be used, however, it is often desirable to do so due to their greater dependency on empirical data and fewer assumptions than strict Bayesian methods.

There are several similarities between Bayes and EB procedures. However, in EB procedures, the postulated prior distribution must have a frequency interpretation and certain "past" data suitable for estimating this distribution are assumed to be available. Mathematical techniques akin to those used in the Bayesian procedure are used to determine the form of the required EB estimator for the parameter of interest. The EB estimator is expressed either as a function of the prior or marginal distribution, which is then empirically estimated by means of the past data.

There is an alternate interpretation of the EB approach. Such methods can be considered to be a means for pooling (or combining) either existing sets of reliability test data or reliability parameter estimates themselves, such as a set of failure rate estimates from various sources. Let us illustrate this with an example. Suppose that an analyst wishes to estimate the constant failure rate for a certain component. Further suppose that there exists a set of failure data, as well as several failure rate estimates provided by other sources, for this device. If the true failure rate underlying each estimate is the same apart from some degree of random variation between them, then these estimates may be pooled according to the EB method. The resultant EB estimate will have a smaller Bayes risk than the estimate based only upon the failure data above. In addition, the empirical prior distribution can be used to study the failure rate population itself without the necessity of identifying the form of the prior distribution.

Although early examples of an EB approach are given by Good (1953) and Von Mises (1942), the recent work in this field has been stimulated largely by the work of Robbins (1955), who is credited with the term "empirical Bayes." Since then the field has steadily grown both in theory as well as in a variety of areas of application. Notable among these developments have been the applications to the distributions commonly used in reliability analysis.

13.1. EMPIRICAL BAYES DECISION THEORY

The decision theoretic elements on which the EB procedures are based are the same as for the Bayes procedure discussed in Chapter 5.

Recall from Chapter 5, that the Bayes decision function $\delta_g(\underset{\sim}{x})$ is defined as the decision function which minimizes the *Bayes risk* represented as

$$r(g,\delta)=\int_{\mathscr{X}} f(\underset{\sim}{x})\left\{\int_{\Omega_{\underset{\sim}{\theta}}} L\{\underset{\sim}{\theta},\delta(\underset{\sim}{x})\}g(\underset{\sim}{\theta}|\underset{\sim}{x})\,d\underset{\sim}{\theta}\right\}d\underset{\sim}{x}, \tag{13.1}$$

provided that such a decision function exists. The minimum value of (13.1) over δ is known as the *minimum Bayes risk* and is given by

$$r(g)=r(g,\delta_g)=\min_{\delta\in D} r(g,\delta). \tag{13.2}$$

When g is known, $\delta_g(\underset{\sim}{x})$ can be determined; however, when g is unknown, the Bayes decision function cannot be obtained.

In the EB approach the complete determination and specification of the actual prior distribution is unnecessary. Instead, it is assumed that the decision problem has occurred repeatedly and independently with the same unknown prior distribution throughout the sequence of occurrences. Each occurrence is referred to as an "experiment." Thus there exists a sequence

$$(\underset{\sim}{X}_1,\underset{\sim}{\Theta}_1),(\underset{\sim}{X}_2,\underset{\sim}{\Theta}_2),\ldots,(\underset{\sim}{X}_N,\underset{\sim}{\Theta}_N),\ldots \tag{13.3}$$

of independent pairs of r.v.'s $(\underset{\sim}{X},\underset{\sim}{\Theta})$ such that at the time the present or Nth decision is to be made the sequence of observations $\underset{\sim}{x}_1,\underset{\sim}{x}_2,\ldots,\underset{\sim}{x}_{N-1}$ as well as $\underset{\sim}{x}_N$, is available to the decision maker. For convenience we will let $\underset{\sim}{x}=\underset{\sim}{x}_N$ and $\underset{\sim}{\theta}=\underset{\sim}{\theta}_N$. Of course, the sequence $\underset{\sim}{\theta}_1,\underset{\sim}{\theta}_2,\ldots,\underset{\sim}{\theta}_N$ remains unknown.

This sequence of additional information can be used to obtain a decision function of $\underset{\sim}{x}$ based upon $\underset{\sim}{x}_1,\underset{\sim}{x}_2,\ldots,\underset{\sim}{x}_N$ represented by

$$\delta_N(\underset{\sim}{x})=\delta_N(\underset{\sim}{x};\underset{\sim}{x}_1,\underset{\sim}{x}_2,\ldots,\underset{\sim}{x}_N). \tag{13.4}$$

Hence when $\underset{\sim}{x}$ is observed, action $\delta_N(\underset{\sim}{x})$ is taken and loss $L\{\underset{\sim}{\theta},\delta_N(\underset{\sim}{x})\}$ incurred. The sequence $\underset{\sim}{x}_1,\underset{\sim}{x}_2,\ldots,\underset{\sim}{x}_N$ is used to extract information about g in such a way that $\delta_N(\underset{\sim}{x})$ approximates the true (but unknown) Bayes decision function $\delta_g(\underset{\sim}{x})$.

Robbins (1964) introduces the concept of asymptotic optimality of an EB decision function. An EB decision function $\delta_N(\underset{\sim}{x})$ is said to be *asymptotically optimal* for estimating $\delta_g(\underset{\sim}{x})$ if

$$\lim_{N\to\infty} E_N r(g,\delta_N)=r(g), \tag{13.5}$$

for all g belonging to a certain class of distributions. In the expression E_N denotes a joint expectation over the r.v.'s $\underset{\sim}{x}_1,\ldots,\underset{\sim}{x}_N$, as $r(g,\delta_N)$ is regarded as an r.v. due to its functional dependency on the past data. $E_N r(g,\delta_N)$ is known as the *expected Bayes risk* of the EB rule $\delta_N(\underset{\sim}{x})$ or, simply, the EB *risk* of δ_N.

In many applications of the EB approach decisions are made sequentially as the quantity of past data increases. That is, decisions are required to be sequentially made about every past $\underset{\sim}{\theta}_j$, $j=1,2,\ldots,N$. Prior to the current observation $\underset{\sim}{x}$, $\underset{\sim}{x}_{N-1}$ was the "current observation," and prior to that, $\underset{\sim}{x}_{N-2}$,

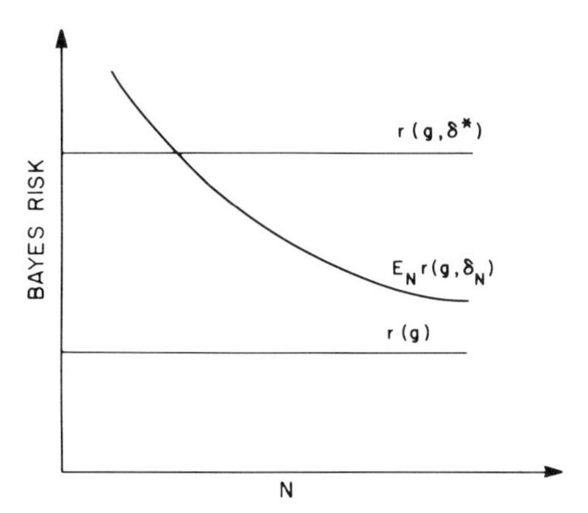

Figure 13.1. A typical relationship between the risks of classical, Bayes, and empirical Bayes decision functions.

and so on. In this case it is important to consider the rate of convergence of $E_N r(g, \delta_N)$ to $r(g)$. In most applications there will be a good nonBayes decision function δ^*, such as the ML or MVU estimator. Comparisons between $E_N r(g, \delta_N)$, $r(g, \delta^*)$, and $r(g)$ should then be made. Many EB investigations include such comparisons that are generally graphically represented. Typically these graphs appear as in Figure 13.1. It is noted that there is generally some value of N for which the EB risk becomes less than the Bayes risk of the classical rule. Numerous EB studies have observed this value to be as small as $N=2$. It is also noted that the rate of convergence of the EB risk to the minimum Bayes risk is generally quite rapid, the maximum gradient occurring for $N \leqslant 15$. This will be illustrated in later sections.

A rather large class of EB rules have been proposed that are *not* asymptotically optimal but which do possess desirable EB risk for small values of N. The Bayes risks may still be compared as in Figure 13.1, but with the understanding that no convergence is guaranteed. Such a class of EB rules will be presented in the next section.

13.2. EMPIRICAL BAYES ESTIMATION

The familar squared-error loss function is often considered when estimating a parameter θ. Recall that in such a case the decision function $\delta(\underset{\sim}{x})$ is referred to as an estimator. It was shown in Section 5.5.1 that the corresponding Bayes estimator $\delta_g(\underset{\sim}{x}) = \tilde{\underset{\sim}{\theta}}$ for $\underset{\sim}{\theta}$ is the mean of the posterior distribution given by

$$\tilde{\underset{\sim}{\theta}} = E(\underset{\sim}{\Theta}|\underset{\sim}{x}) = \int_{\Omega_{\underset{\sim}{\theta}}} \underset{\sim}{\theta} g(\underset{\sim}{\theta}|\underset{\sim}{x})\, d\underset{\sim}{x} = \frac{\int_{\Omega_{\underset{\sim}{\theta}}} \underset{\sim}{\theta} f(\underset{\sim}{x}|\underset{\sim}{\theta}) g(\underset{\sim}{\theta})\, d\underset{\sim}{\theta}}{\int_{\Omega_{\underset{\sim}{\theta}}} f(\underset{\sim}{x}|\underset{\sim}{\theta}) g(\underset{\sim}{\theta})\, d\underset{\sim}{\theta}}, \tag{13.6}$$

where

1. $g(\underset{\sim}{\theta}|\underset{\sim}{x})$ = posterior distribution of $\underset{\sim}{\Theta}$ given $\underset{\sim}{x}$,
2. $f(\underset{\sim}{x}|\underset{\sim}{\theta})$ = sampling distribution of $\underset{\sim}{X}$ given $\underset{\sim}{\theta}$,
3. $g(\underset{\sim}{\theta})$ = prior distribution of $\underset{\sim}{\Theta}$.

It was also shown in Section 5.5.2 that if there exists a sufficient statistic $\hat{\underset{\sim}{\theta}}$ for $\underset{\sim}{\theta}$ with pdf $f(\hat{\underset{\sim}{\theta}}|\underset{\sim}{\theta})$, then the Bayes estimator given by (13.6) can be written as

$$\tilde{\underset{\sim}{\theta}} = E(\underset{\sim}{\Theta}|\hat{\underset{\sim}{\theta}}) = \int_{\Omega_{\underset{\sim}{\theta}}} \underset{\sim}{\theta} g(\underset{\sim}{\theta}|\hat{\underset{\sim}{\theta}})\, d\underset{\sim}{\theta} = \frac{\int_{\Omega_{\underset{\sim}{\theta}}} \underset{\sim}{\theta} f(\hat{\underset{\sim}{\theta}}|\underset{\sim}{\theta}) g(\underset{\sim}{\theta})\, d\underset{\sim}{\theta}}{\int_{\Omega_{\underset{\sim}{\theta}}} f(\hat{\underset{\sim}{\theta}}|\underset{\sim}{\theta}) g(\underset{\sim}{\theta})\, d\underset{\sim}{\theta}}. \tag{13.7}$$

The EB risk of EB estimators generally falls somewhere between that of a sampling theory or classical (nonBayes) estimator and a Bayes estimator, as illustrated in Figure 13.1. Table 13.1 summarizes certain generic characteristics of all three estimation procedures. It is recognized that EB estimation methods are in some sense a hybrid combination of both classical methods and Bayesian methods, due to the fact that the parameter is treated as an r.v. having a prior distribution with a frequency interpretation.

Table 13.1. A Comparison of Certain Characteristics of Sampling Theory, Bayes, and Empirical Bayes Estimators

Characteristic	Sampling Theory	Bayes	Empirical Bayes
θ	Unknown constant	Unknown value of an r.v.	Unknown value of an r.v.
Prior distribution form	(Does not exist)	Known	Unknown
Sampling distribution	Known	Known	Known
Marginal distribution	(Does not exist)	Known	Unknown
Posterior distribution	(Does not exist)	Known	Unknown

Table 13.2. A Comparison of Certain Characteristics of Empirical Bayes Estimation Methods

Characteristic	Implicitly Estimated Prior	Explicitly Estimated Prior
Prior distribution	Implicit	Explicitly estimated
Sampling distribution	Given	Given
Marginal distribution	Explicitly estimated	Explicit
Posterior distribution	Implicit	Implicit[a]

[a]Explicitly computed when calculating interval estimates.

There are two main classes of EB estimators. The first class consists of those methods that attempt to approximate the Bayes estimator without explicitly estimating the unknown prior distribution. Historically this is the traditional EB procedure. Clemmer and Krutchkoff (1968), Lin (1972), Martz and Krutchkoff (1969), Miyawasa (1961), Nichols and Tsokos (1972), O'Bryan (1976, 1979), O'Bryan and Susarla (1977) and Robbins (1955, 1964) use this approach. The second class of EB estimators consists of those methods in which the unknown prior distribution is explicitly estimated. The so-called "smooth" EB methods developed by Bennett and Martz (1972), Lemon and Krutchkoff (1969), and Maritz (1966, 1970) are based on this approach. Table 13.2 summarizes the important distinctions between both classes of EB methods. The posterior distribution is easily obtained for the second class of EB estimators. Consequently, EB probability interval estimators can be obtained for this class, as well as further analysis performed based on the posterior distribution.

13.2.1 Point Estimation—Implicitly Estimated Prior Distribution

Rutherford and Krutchkoff (1969) develop a general approach used to construct EB estimators of the first type. They consider a family $\{f(x|\theta): \theta \in \Omega_\theta\}$, where X and Θ are univariate r.v.'s and

1. the r.v. X is discrete for all $\theta \in \Omega_\theta$, and
2. the probability mass function $f(x|\theta)$ satisfies

$$\frac{f(x+1|\theta)}{f(x|\theta)} = a(x) + b(x)\theta, \tag{13.8}$$

where $a(x)$ and $b(x)$ are any functions such that $b(x) \neq 0$.

Solving (13.8) for θ yields

$$\theta = \frac{f(x+1|\theta)}{b(x)f(x|\theta)} - \frac{a(x)}{b(x)}.$$

Therefore, according to (13.6),

$$E(\Theta|x) = \frac{1}{b(x)} \int_{\Omega_\theta} \frac{f(x+1|\theta)}{f(x|\theta)} g(\theta|x)\, d\theta - \frac{a(x)}{b(x)}$$

$$= \frac{1}{b(x)} \int_{\Omega_\theta} \frac{f(x+1|\theta)}{f(x|\theta)} \frac{f(x|\theta)}{f(x)} g(\theta)\, d\theta - \frac{a(x)}{b(x)}.$$

Letting

$$f(x) = \int_{\Omega_\theta} f(x|\theta) g(\theta)\, d\theta = \text{marginal distribution of } X,$$

then

$$E(\Theta|x) = \frac{f(x+1)}{b(x)f(x)} - \frac{a(x)}{b(x)}, \tag{13.9}$$

where

$$f(x+1) = \int_{\Omega_\theta} f(x+1|\theta) g(\theta)\, d\theta.$$

In (13.9) both $a(x)$ and $b(x)$ are known, whereas $f(x)$ and $f(x+1)$ are unknown values of marginal distributions that must be estimated. Both $f(x)$ and $f(x+1)$ can be estimated from the sequence of past data $x_1, x_2, \ldots, x_{N-1}$ as well as the current observation x. Denoting by $S_N(x)$ the total number of these N observations having the value x, of which there is at least one, it is well-known that

$$f_N(x) = \frac{S_N(x)}{N} \to f(x) \tag{13.10}$$

in probability as $N \to \infty$ [see Section 3.7]. That is, $f_N(x)$ is a *consistent estimator* for $f(x)$. In a similar way $f_N(x+1)$ consistently estimates $f(x+1)$. Thus the EB estimator for θ becomes

$$E_N(\Theta|x) = \frac{f_N(x+1)}{b(x)f_N(x)} - \frac{a(x)}{b(x)}, \tag{13.11}$$

and has the property that $E_N(\Theta|x) \to E(\Theta|x)$ in probability as $N \to \infty$. Under mild conditions on the moments of the prior distribution, Rutherford and Krutchkoff (1969) show that this implies that, in the limit as $N \to \infty$, the EB risk of the EB estimator approaches the minimum Bayes risk. Thus $E_N(\Theta|x)$ is asymptotically optimal for estimating $E(\Theta|x)$. The Poisson, negative binomial, and logarithmic distributions are examples of distributions that satisfy (13.8).

There is an important restriction on the practical applicability of the first class of EB estimators. Suppose there exists a sufficient statistic $\hat{\theta}$ for θ. The Bayes estimator is usually expressed as some function of the marginal distribution $f(\hat{\theta})$ given by

$$f(\hat{\theta}) = \int_{\Omega_\theta} f(\hat{\theta}|\theta) g(\theta)\, d\theta, \tag{13.12}$$

which implicitly involves the unknown prior distribution. The past data $\hat{\theta}_1, \hat{\theta}_2, \ldots, \hat{\theta}_N$ comprise an independent sample that is identically distributed according to $f(\hat{\theta})$, provided $f(\hat{\underset{\sim}{\theta}}|\underset{\sim}{\theta})$ is the same in each past experiment. This is not always the case. For example, suppose that the ith experiment in a series of N experiments consists of observing the number X_i of failures upon demand of a certain device in n_i trials. In this case the marginal density function of X_i depends upon n_i and, if n_i are not all equal for $i=1,2,\ldots,N$, then the past data $x_1, x_2, \ldots, x_N$ do *not* represent a sample from the same distribution. When using the first class of EB estimators, it is generally required that the past data be an i.i.d. sample from $f(\hat{\theta})$. Thus this class of estimators has restricted application, requiring such assumptions as a common sample size in each experiment, common nuisance parameter values, and so on. For this, and other reasons as well, this class of EB estimators is not as useful in reliability applications as the second class of estimators to be discussed.

13.2.2 Point Estimation—Explicitly Estimated Prior Distribution

Maritz (1966) observes that the rate of convergence of the EB risk to the minimum Bayes risk, although insured due to the asymptotic optimality of the EB estimator, can be quite slow. The reason for this is that, for small values of N, the empirically estimated marginal distribution $f_N(0), f_N(1), \ldots,$ can be irregular. For example, some values of $f_N(x)$ may be zero, whereas adjacent values exceed zero. This lack of "smoothness" of $f_N(x)$ yields a comparably nonsmooth EB estimator with accompanying large EB risk. There are several ways to "smooth" EB estimators and such estimators represent the second major class of EB estimators. Perhaps the most straightforward way of smoothing has already been considered in previous chapters. This occurs when the prior distribution is assumed to be a member of a certain parametric family of distributions. For example, when estimating a constant failure rate λ, it is frequently assumed that the prior distribution is a $\mathcal{G}_1(\alpha_0, \beta_0)$ distribution. This was considered in Chapter 8. If the $\mathcal{G}_1(\alpha_0, \beta_0)$ shape and scale parameters are unknown, then the past data can be used to estimate these parameters [see Section 7.7]. This is certainly a smoothing operation, as each past observation is used in computing the estimates; however, it is not considered to be an EB estimation procedure according to Definition 13.1 as the prior distribution is assumed to be a member of a specified parametric family.

There exists a variety of EB smoothing methods. In a series of articles Maritz (1966, 1967, 1968, 1970) proposes estimating the prior distribution function by a step function with steps taken at points determined either by ML or the method of moments [see Section 3.6]. Maritz (1970) also proposes smoothing EB estimators directly by fitting some type of reasonable curve through the observed points using a weighted least squares procedure.

Several other authors have addressed the problem of explicitly estimating the prior distribution. Blum and Susarla (1977), Choi and Bulgren (1968), Deely and Kruse (1968), and Tortella and O'Bryan (1979) estimate G by means of a step function. Gaffey (1959) provides some applications of estimation of the prior distribution. Meeden (1972) and Rolph (1968) treat the estimation of g in a Bayesian context, whereas Lenth (1976) considers the estimation of g in the framework of EB.

Lemon and Krutchkoff (1969) also propose a smoothing procedure which in one special case may be interpreted as the approximation of the prior distribution function by a step function having steps of equal height $1/N$ at each of the past classical estimates that are assumed to be based on a sufficient statistic. Suppose θ is univariate. If $\hat{\theta}_j$ denotes the classical estimate of θ_j in the jth experiment, $j=1,2,\ldots,N$, then according to (13.7) the EB estimator of θ_i in the ith experiment becomes

$$E_N(\Theta_i|\hat{\theta}_i) = \frac{\sum_{j=1}^{N} \hat{\theta}_j f(\hat{\theta}_i|\hat{\theta}_j)}{\sum_{j=1}^{N} f(\hat{\theta}_i|\hat{\theta}_j)}, \qquad i=1,2,\ldots,N, \tag{13.13}$$

where $f(\hat{\theta}_i|\hat{\theta}_j)$ is used here to represent either a pdf in the case of a continuous r.v. or a probability mass function in the case of a discrete r.v. The estimator proposed by Maritz (1967) is similar in spirit to (13.13) except that the values of $\hat{\theta}_i$ are chosen in a different way. Robbins (1951) also suggests replacing the prior distribution function $G(\theta)$ by its empirical distribution function for the compound decision problem. The proposed estimator is "smooth" in the sense that each past estimate $\hat{\theta}_i$ is explicitly present in (13.13). Lemon and Krutchkoff (1969) also suggest a second possible interaction of (13.13) with each classical estimate $\hat{\theta}_j$ replaced by the corresponding EB estimate $E_N(\Theta_j|\hat{\theta}_j)$ from the first iteration. The EB risk of the iterated estimator was found to be smaller than that of (13.13) for several sampling distributions that are investigated by Monte Carlo simulation by Bennett and Martz (1972) and Lemon and Krutchkoff (1969).

Bennett and Martz (1972) propose a further refinement to the smooth EB procedures. They suggest approximating the unknown prior pdf with a pdf estimator such as that suggested by Parzen (1962), rather than with a discrete function. This approach assigns mass to values of θ other than the empirical point estimates. The important gains are point estimates with smaller EB risk [Bennett and Martz (1972, 1973), Canavos (1971), Padgett and Robinson (1978)] and smoother interval estimates, (that is, interval estimates whose coverages are closer to a prespecified value). Such estimators will be referred to as *continuously smoothed EB (CSEB) estimators.*

The main disadvantage of the CSEB estimators is the computational complexity required for their use. Generally, a computer implementation is required, whereas the smooth EB estimator given in (13.13) can be generally computed on a handheld calculator.

Parzen (1962) presents a class of pdf estimators that have been extensively used in constructing CSEB estimators. He presents a consistent pdf estimator of the general form

$$\hat{g}_N(\theta) = \frac{1}{Nh(N)} \sum_{i=1}^{N} W\left[\frac{\theta - \hat{\theta}_i}{h(N)}\right], \tag{13.14}$$

where $W(\cdot)$ is the *window (kernel, weighting,* or *smoothing) function* satisfying certain boundedness and regularity conditions and $h(N)$ a constant chosen in such a way that $h(N) \to 0$ and $Nh(N) \to \infty$ in the limit as $N \to \infty$. These restrictions are placed on $W(\cdot)$ and $h(N)$ to insure the consistency of the estimator. Using this estimator, Bennett and Martz (1972) present the CSEB estimator for θ given by

$$E_N^{(1)}(\Theta|\hat{\theta}) = \frac{\int_{\Omega_\theta} \theta f(\hat{\theta}|\theta)\hat{g}_N(\theta)\,d\theta}{\int_{\Omega_\theta} f(\hat{\theta}|\theta)\hat{g}_N(\theta)\,d\theta}. \tag{13.15}$$

Unlike the estimator in (13.13) which gives weight only to the observed values $\hat{\theta}_1, \ldots, \hat{\theta}_N$, this estimator weights values of θ that are "close" to the observed values according to the form of the window $W(\cdot)$ used in (13.14).

Bennett and Martz (1972) use Monte Carlo simulation to show that the EB risk of (13.15) is significantly influenced by only the first two moments of the prior distribution. This observation suggested an additional approximation to the prior density function based on a sequence of linearly transformed observed values. In arriving at this sequence it was assumed that $E(\hat{\theta}|\theta) = f(k)\theta$, where $f(k)$ is any known but arbitrary function of k. It thus follows that

$$E(\hat{\theta}) = E_{\theta}\left[E_{\hat{\theta}|\theta}(\hat{\theta}|\Theta)\right] = f(k)E(\Theta) \tag{13.16}$$

and

$$\mathrm{Var}(\hat{\theta}) = f^2(k)\mathrm{Var}(\Theta) + E_{\theta}\left[\mathrm{Var}_{\hat{\theta}|\theta}(\hat{\theta}|\Theta)\right]. \tag{13.17}$$

Hence, by defining the linear transformation of $\hat{\theta}_i$,

$$\theta_i^* = \left[\frac{\mathrm{Var}(\Theta)}{\mathrm{Var}(\hat{\theta})}\right]^{1/2}\left[\hat{\theta}_i - E(\hat{\theta}_i)\right] + E(\Theta), \tag{13.18}$$

they obtain the desired result; namely,

$$E(\theta_i^*) = E(\Theta), \qquad \mathrm{Var}(\theta_i^*) = \mathrm{Var}(\Theta).$$

As the mean and variance of Θ are frequently unknown, consistent estimates for these quantities are required. Consistent estimates for these quantities are given by

$$\widehat{E(\Theta)} = \frac{1}{f(k)} \sum_{i=1}^{N} \frac{1}{N}\hat{\theta}_i = \frac{1}{f(k)}\bar{\theta} \tag{13.19}$$

and

$$\widehat{\mathrm{Var}(\Theta)} = \left\{s^2 - E_{\theta}\left[\mathrm{Var}_{\hat{\theta}|\theta}(\hat{\theta}|\Theta)\right]\right\}\frac{1}{f^2(k)}, \tag{13.20}$$

where $s^2 = \Sigma_{i=1}^{N}(\hat{\theta}_i - \bar{\theta})^2/N$. The estimate in (13.20) may be negative in which case a suitable adjustment must be made.

The sequence of transformed values θ_i^*, $i=1,2,\ldots,N$, may be used to obtain the alternative CSEB estimator for θ given by

$$E_N^{(2)}(\Theta|\hat{\theta}) = \frac{\int_{\Omega_\theta} \theta f(\hat{\theta}|\theta) g_N^*(\theta)\, d\theta}{\int_{\Omega_\theta} f(\hat{\theta}|\theta) g_N^*(\theta)\, d\theta}, \tag{13.21}$$

where

$$g_N^*(\theta) = \frac{1}{Nh(N)} \sum_{i=1}^{N} W\left[\frac{\theta - \theta_i^*}{h(N)}\right].$$

13.2.3 Interval Estimation

Now let us consider EB probability interval estimates of θ. In the case of a step function approximation to the unknown prior distribution (for example, the Maritz and Lemon and Krutchkoff procedures), the empirical posterior distribution is a discrete distribution with mass at those values of θ where the steps occur. Let $g_N(\tilde{\theta}_j|\hat{\theta})$ denote the empirical posterior probability mass function corresponding to the iterated Lemon and Krutchkoff procedure, where $\tilde{\theta}_j = E_N(\Theta_j|\hat{\theta}_j)$, the EB point estimate of θ according to (13.13). The subscript N denotes the fact that this is an empirically estimated probability distribution based on the N past estimates of θ. In the case of a noniterated estimator, $\tilde{\theta}_j$ would be replaced by $\hat{\theta}_j$, a classical estimate of θ_j; however, since the iterated estimator is generally superior, it will be used here. Thus

$$g_N(\tilde{\theta}_j|\hat{\theta}) = \Pr(\Theta = \tilde{\theta}_j|\hat{\theta}) = \frac{f(\hat{\theta}|\tilde{\theta}_j)}{\sum_{i=1}^{N} f(\hat{\theta}|\tilde{\theta}_i)}, \qquad j=1,2,\ldots,N. \tag{13.22}$$

A symmetric $100(1-\gamma)\%$ (or larger) smooth *two-sided EB probability interval* (TEBPI) estimate of θ_i in the ith experiment is given by

$$100(1-\gamma)\%\text{ TEBPI for }\theta_i\text{: } \left[\tilde{\theta}_{(j^-+1)};\ \tilde{\theta}_{(j^+)}\right], \qquad i=1,2,\ldots,N, \tag{13.23}$$

where $\tilde{\theta}_{(1)},\ldots,\tilde{\theta}_{(N)}$ denotes the ordered (smallest to largest) sequence of iterated Lemon and Krutchkoff EB estimates, j^+ is the smallest integer that satisfies the relation

$$\sum_{j=1}^{j^+} g_N\left(\tilde{\theta}_j|\hat{\theta}_i\right) \geq 1 - \frac{\gamma}{2}, \tag{13.24}$$

and j^- is the largest integer satisfying the relation

$$\sum_{j=1}^{j^-} g_N\left(\tilde{\theta}_j|\hat{\theta}_i\right) \leq \frac{\gamma}{2}. \tag{13.25}$$

If no such integer j^- exists between 1 and N, then j^- is set equal to zero and the lower limit of the estimator will thus be $\tilde{\theta}_{(1)}$. The actual empirical probability that Θ_i lies within the interval can be calculated by summing the appropriate posterior probabilities.

A corresponding $100(1-\gamma)\%$ (or larger) smooth *lower one-sided EB probability interval* (*LEBPI*) estimate of θ_i is given by

$$100(1-\gamma)\% \text{ LEBPI for } \theta_i\text{: } \left[\tilde{\theta}_{(j^-+1)};\ \infty\right], \qquad i=1,2,\ldots,N, \tag{13.26}$$

where now j^- is the largest integer satisfying the relation

$$\sum_{j=1}^{j^-} g_N\left(\tilde{\theta}_j|\hat{\theta}_i\right) \leq \gamma. \tag{13.27}$$

Similarly, a $100(1-\gamma)\%$ (or larger) smooth *upper one-sided EB probability interval* (*UEBPI*) estimate may be constructed.

The CSEB procedure may also be used to obtain interval estimates. The corresponding empirical posterior probability mass function or pdf, whichever the case may be, is required and is given by

$$g_N\left(\theta|\hat{\theta}\right) = \frac{f\left(\hat{\theta}|\theta\right)\hat{g}_N(\theta)}{\int_{\Omega_\theta} f\left(\hat{\theta}|\theta\right)\hat{g}_N(\theta)\,d\theta}, \tag{13.28}$$

where $\hat{g}_N(\theta)$ is defined in (13.14), and $f(\hat{\theta}|\theta)$ is the sampling distribution. The symmetric $100(1-\gamma)\%$ continuously smooth TEBPI estimate of θ_i is given by

$$100(1-\gamma)\% \text{ TEBPI for } \theta_i\text{: } [\theta_L;\ \theta_U], \qquad i=1,2,\ldots,N, \tag{13.29}$$

where θ_L and θ_U satisfy

$$\begin{aligned} &1. \quad \int_{-\infty}^{\theta_L} g_N(\theta|\hat{\theta}_i)\,d\theta = \frac{\gamma}{2}, \\ &2. \quad \int_{-\infty}^{\theta_U} g_N(\theta|\hat{\theta}_i)\,d\theta = 1 - \frac{\gamma}{2}. \end{aligned} \tag{13.30}$$

Thus $\Pr(\theta_L \leqslant \Theta_i \leqslant \theta_U | \hat{\theta}_i) = 1 - \gamma$.

A corresponding $100(1-\gamma)\%$ continuously smooth LEBPI estimate of θ_i is given by

$$100(1-\gamma)\% \text{ LEBPI for } \theta_i\text{: } [\theta_L; \infty], \qquad i = 1, 2, \ldots, N, \tag{13.31}$$

where θ_L now satisfies

$$\int_{-\infty}^{\theta_L} g_N(\theta|\hat{\theta}_i)\,d\theta = \gamma. \tag{13.32}$$

An upper one-sided estimate may be similarly constructed. Numerical integration will usually be required in order to solve (13.30) or (13.32), as illustrated in Section 13.3.2.

The EB estimators presented in this section are applied to some of the most commonly used reliability sampling models in the remaining sections of the chapter. Selected examples will serve to illustrate their use in practice.

13.3. ESTIMATION OF THE SURVIVAL PROBABILITY—THE BINOMIAL DISTRIBUTION

The binomial distribution was discussed in Chapter 7 from the Bayesian point of view. Recall that the reliability measure of interest is the probability of survival in a life test in which n units are tested for a fixed time t. The observations are assumed to be statistically independent, each having the same probability θ of surviving the test. We have changed notation slightly, whereas in Chapter 7 this probability was denoted by p, it is now denoted by θ.

Now consider a sequence of binomial sampling experiments in which the ith experiment results in x_i survivors in n_i trials. It is assumed that $N-1$ previous experiments have been conducted prior to the current or Nth experiment. For convenience let $x = x_N$, $n = n_N$, and $\theta = \theta_N$. It is further assumed that: (1) the x_i values conditioned on θ_i are independent, and (2) the survival probabilities $\theta_1, \ldots, \theta_N$ are independent, unknown, and unobservable realizations of an r.v. Θ according to a completely unknown and unspecified prior distribution $g(\theta)$. Further, let μ and σ^2 represent the prior mean and variance, respectively, and let $y = x/n$.

13.3.1 Point Estimation of θ

In addition to the smooth estimators given in (13.13) and (13.15), several other EB point estimators have been proposed for estimating θ. Christensen (1977), Copas (1972), George (1971), and Griffin and Krutchkoff (1971) all consider EB estimators for θ. Martz and Lian (1974) study the EB risk of nine EB estimators for θ. They observe that the second estimator presented by Copas (1972) was superior for large N and small n_i generally over the entire range of σ^2 for $\mu(1-\mu)<3/16$ ($\mu<0.25$ or $\mu>0.75$). For $\mu(1-\mu)$ extremely small (i.e. in the high reliability region), this estimator is nearly uniformly superior to all other estimators, provided that N is sufficiently large. As this situation is precisely that encountered in many reliability applications, this estimator is now presented.

The Copas (1972) estimator is given by

$$\hat{\theta}_C = (y + W_N \mu_{1N})/(1+W_N), \tag{13.33}$$

where $\mu_{1N} = \sum_{i=1}^{N} y_i/N$, $W_N = [\mu_{1N}(1-\mu_{1N}) - \sigma_{1N}^2]/n\sigma_{1N}^2$, $\sigma_{1N}^2 = \max[0, \{\sum_{i=1}^{N}(y_i - \mu_{1N})^2 - k\mu_{1N}(1-\mu_{1N})\}/(N-k)]$, and $k = \sum_{i=1}^{N} n_i^{-1}$. It is also noted that $\hat{\theta}_C$ is Bayes with respect to a certain class of $\mathscr{B}(x_0, n_0)$ prior distributions.

The Lemon and Krutchkoff (1969) iterated estimator of θ becomes

$$\hat{\theta}_L = \left[\sum_{i=1}^{N} \tilde{\theta}_i^{x+1}(1-\tilde{\theta}_i)^{n-x}\right] \Big/ \left[\sum_{i=1}^{N} \tilde{\theta}_i^{x}(1-\tilde{\theta}_i)^{n-x}\right], \tag{13.34}$$

where $\tilde{\theta}_i$ is calculated according to (13.13) as

$$\tilde{\theta}_i = \left[\sum_{j=1}^{N} y_j^{x_i+1}(1-y_j)^{n_i-x_i}\right] \Big/ \left[\sum_{j=1}^{N} y_j^{x_i}(1-y_j)^{n_i-x_i}\right], \qquad N \geqslant 2, \tag{13.35}$$

and $y_j = x_j/n_j$. The estimator in (13.35) is also considered by Lemon (1972).

Finally, Martz and Lian (1974) consider the use of (13.15) in which the smoothing functions in (13.14) are chosen to be triangular functions with total base width $2h(N)$ and height n_i. The resulting estimator that they refer to as the *smooth incomplete beta* estimator is given by

$$\hat{\theta}_S = \left(\frac{x+1}{n+3}\right) \frac{\sum_{i=1}^{N} H_i(y_i) Q_i(x, n; 2)}{\sum_{i=1}^{N} H_i(y_i) Q_i(x, n; 1)}, \tag{13.36}$$

where

1.
$$H_i(y_i) = \begin{cases} n_i, & y_i - h \geqslant 0, y_i + h \leqslant 1 \\ 2n_i h^2 / \{2h^2 - (y_i - h)^2\}, & y_i - h < 0 \\ 2n_i h^2 / \{2h^2 - (1 - y_i - h)^2\}, & y_i + h > 1, \end{cases}$$

2.
$$h = h(N) = \begin{cases} s_N N^{-1/5}, & s_N > 0 \\ 2N^{-1/5}/n, & s_N = 0, \end{cases}$$

3.
$$\begin{aligned} Q_i(x, n; a) = {} & 2(x+a) I(y_i; x+a+1, n-x+1) \\ & -(x+a) I(A_i; x+a+1, n-x+1) \\ & -(x+a) I(B_i; x+a+1, n-x+1) \\ & +(h+y_i)(n+a+1) I(B_i; x+a, n-x+1) \\ & -(h-y_i)(n+a+1) I(A_i; x+a, n-x+1) \\ & -2y_i(n+a+1) I(y_i; x+a, n-x+1), \end{aligned}$$

and where $A_i = \max(0, y_i - h)$, $B_i = \min(1, y_i + h)$, $s_N^2 = \Sigma_{i=1}^N n_i (y_i - \bar{y})^2 / M$, $\bar{y} = \Sigma_{i=1}^N n_i y_i / M$, $M = \Sigma_{i=1}^N n_i$, and $I(z; a, b)$ is the standard incomplete beta function defined in (7.26).

Unfortunately, the EB risk in using $\hat{\theta}_C$, $\hat{\theta}_L$, or $\hat{\theta}_S$ is unavailable in closed form. However, Martz and Lian (1974) calculate the sample EB risk for over 2000 parameter combinations using Monte Carlo simulation and use the results to fit regression equations for the ln EB risk of each of these estimators. The explanatory variables initially selected were n, N, σ^2, μ, and

Table 13.3. Coefficients of Prediction Equations for the ln EB Risk of Several Empirical Bayes Estimators for the Binomial Parameter θ

	Estimator		
Term	$\hat{\theta}_C$	$\hat{\theta}_L$	$\hat{\theta}_S$
Std. Dev.	0.2428	0.3347	0.2301
Constant	−0.37081	0.12146	−0.09872
$\ln \sigma^2$	1.15368	0.45478	1.03764
$1/\sigma^2$	−0.00096	*	0.00003
$\sqrt{n}$	*	−0.37875	*
$\ln n$	0.21507	*	−0.06130
$1/n$	*	*	−0.83751
$1/N\sqrt{n}$	2.15511	−1.60856	1.06678
$\ln(Nn)$	*	−0.33935	*
z	*	−0.00865	*
$\ln z$	0.96137	*	0.90947
nz	*	0.00800	*

$z=(\mu-\mu^2-\sigma^2)/(n\sigma^2)$, which is the ratio of the Bayes risk of the ML estimator y to the prior variance. Table 13.3 presents the coefficients of the final prediction equations as well as the standard deviations about the fitted equations. An asterisk * is used to indicate that a variable is not included in the models. These equations can be used to decide which EB estimator to use in practice. This will now be illustrated.

Example 13.1. [Martz and Lian (1974)] A large United States naval shipyard routinely assesses the reliability of submitted lots of vendor produced material. The following data consist of the number of survivors x_i of a specified materials properties test in samples of size 5 from five past lots of critical fasteners. The past data are (5, 4, 5, 5, 0) and in the current (sixth) lot, $x=5$. It is desired to estimate the probability of surviving the current test. Computing $\mu_{1N}=\hat{\mu}=0.80$, and $\sigma_{1N}^2=\hat{\sigma}^2=0.12667$, the estimates and their corresponding estimated EB risks from the equations in Table 13.3, are computed according to (13.33), (13.34), and (13.36) to be

$$\hat{\theta}_C=0.990 \qquad \text{(Estimated EB risk}=0.00683\text{)},$$

$$\hat{\theta}_L=0.975 \qquad \text{(Estimated EB risk}=0.05224\text{)},$$

$$\hat{\theta}_S=0.923 \qquad \text{(Estimated EB risk}=0.00668\text{)}.$$

■

13.3.2 Interval Estimation of θ

Suppose that an EB probability interval estimate of θ_i is desired. Both procedures discussed in Section 13.2.3 can be used. For the first procedure leading to (13.23) the empirical posterior probability mass function given by (13.22) here becomes

$$g_N(\tilde{\theta}_i|x;n) = \begin{cases} \left[\tilde{\theta}_i^x(1-\tilde{\theta}_i)^{n-x}\right] \Big/ \left[\sum_{j=1}^{N} \tilde{\theta}_j^x(1-\tilde{\theta}_j)^{n-x}\right], & i=1,2,\ldots,N \\ 0, & \text{otherwise.} \end{cases} \tag{13.37}$$

Example 13.2. For the data in Example 13.1 we have $\tilde{\theta}_1 = \tilde{\theta}_3 = \tilde{\theta}_4 = \tilde{\theta}_6 = 0.985$, $\tilde{\theta}_2 = 0.996$, and $\tilde{\theta}_5 = 0.800$ according to (13.35). Thus, for the current lot, the empirical posterior distribution is easily computed to be $g_6(0.800|5;5) = 0.065$, $g_6(0.985|5;5) = 4(0.185) = 0.740$, and $g_6(0.996|5;5) = 0.195$. ■

The symmetric $100(1-\gamma)\%$ (or larger) smooth TEBPI estimate of θ_i given by (13.23) is obtained by solving

$$\sum_{j=1}^{j^+} \left[\tilde{\theta}_j^{x_i}(1-\tilde{\theta}_j)^{n_i-x_i}\right] \Big/ \left[\sum_{j=1}^{N} \tilde{\theta}_j^{x_i}(1-\tilde{\theta}_j)^{n_i-x_i}\right] \geqslant 1-\frac{\gamma}{2}$$

and (13.38)

$$\sum_{j=1}^{j^-} \left[\tilde{\theta}_j^{x_i}(1-\tilde{\theta}_j)^{n_i-x_i}\right] \Big/ \left[\sum_{j=1}^{N} \tilde{\theta}_j^{x_i}(1-\tilde{\theta}_j)^{n_i-x_i}\right] \leqslant \frac{\gamma}{2}$$

for the integers j^- and j^+, and is given by $[\tilde{\theta}_{(j^-+1)}; \tilde{\theta}_{(j^+)}]$. Similarly, a $100(1-\gamma)\%$ (or larger) smooth LEBPI (UEBPI) estimate of θ is obtained by solving the second (first) equation in (13.38) for j^- (j^+) with $\gamma/2$ replaced by γ.

Example 13.3. Suppose that in Example 13.1 an 80% (or larger) symmetric smooth TEBPI estimate of θ_6 is desired. From the posterior distribution given in the previous example, the solution to (13.38) for $\gamma = 0.20$ is easily found to be $j^- = 1$ and $j^+ = 3$.

Note. The original set of six observations has been collapsed to three distinct values as ties are present. Thus the desired interval estimate is [0.985; 0.996] and $\Pr(0.985 \leqslant \Theta_6 \leqslant 0.996 | 5; 5) = 0.935$. ■

Now consider the CSEB procedure leading to (13.29). For the case of triangular smoothing functions discussed prior to (13.36), and the binomial sampling distribution, the corresponding empirical posterior cdf obtained by integrating (13.28) from 0 to θ becomes

$$G_N(\theta | x; n) = \begin{cases} 0, & \theta \leqslant \max\left\{0, \min_j \left(y_j - h\right)\right\}, \\ \dfrac{\sum_{j=1}^{N} H_j(y_j) b(\theta, A_j) R_j(\theta, x, n)}{\sum_{j=1}^{N} H_j(y_j) Q_j(x, n; 1)}, & \max\left\{0, \min_j \left(y_j - h\right)\right\} < \\ & \theta \leqslant \min\left\{1, \max_j \left(y_j + h\right)\right\}, \\ 1, & \theta > \min\left\{1, \max_j \left(y_j + h\right)\right\}, \end{cases} \tag{13.39}$$

provided that the denominator is not equal to zero, where

$$b(u, v) = \begin{cases} 1, & u > v \\ 0, & u \leqslant v, \end{cases}$$

and where we have defined

$$\begin{aligned} R_j(\theta, x, n) = {} & (n+2)(h - y_j)\left[I(C_j; x+1, n-x+1) - I(A_j; x+1, n-x+1)\right] \\ & + (x+1)\left[I(C_j; x+2, n-x+1) - I(A_j; x+2, n-x+1)\right] \\ & + b(\theta, y_j)\left\{(n+2)(h + y_j)\left[I(D_j; x+1, n-x+1)\right.\right. \\ & \left. - I(y_j; x+1, n-x+1)\right] - (x+1)\left[I(D_j; x+2, n-x+1)\right. \\ & \left.\left. - I(y_j; x+2, n-x+1)\right]\right\}. \end{aligned} \tag{13.40}$$

Also, in (13.40), we have defined $C_j = \min(\theta, y_j)$ and $D_j = \min(\theta, B_j)$. The remaining terms and functions are defined following (13.36). The empirical posterior cdf in (13.39) was first presented by Martz (1975b) and used to develop EB single-sample acceptance sampling plans.

Example 13.4. Let us reconsider Example 13.1. The posterior empirical cdf corresponding to the sixth lot is computed according to (13.39) and is plotted in Figure 13.2.

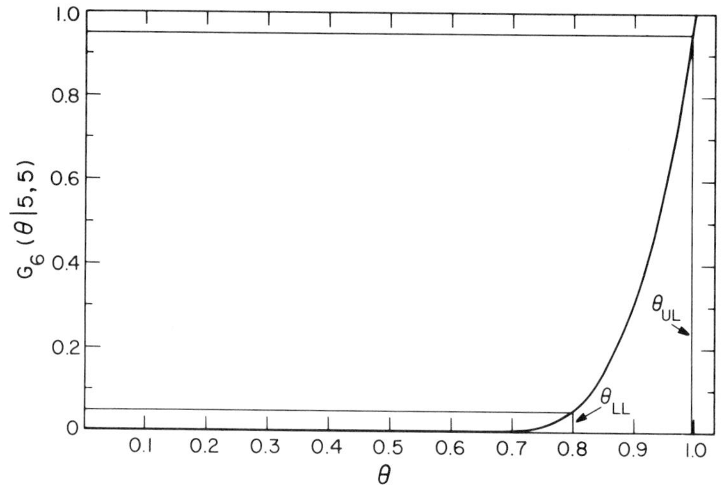

Figure 13.2. The empirical posterior cumulative distribution function in Example 13.4.

■

The symmetric $100(1-\gamma)\%$ continuously smooth TEBPI estimate of θ is obtained by numerically solving the pair of equations given by

$$G_N(\theta_L|x;n)=\frac{\gamma}{2}$$

and (13.41)

$$G_N(\theta_U|x;n)=1-\frac{\gamma}{2},$$

where G_N is given in (13.39), for θ_L and θ_U. A corresponding $100(1-\gamma)\%$ continuously smooth LEBPI (UEBPI) estimator of θ may be obtained by solving the first (second) equation in (13.41) with $\gamma/2$ replaced by γ. If the posterior cdf has been obtained and plotted, the desired interval may often be found directly from the plotted function.

Example 13.5. Let us again reconsider Examples 13.1 and 13.4 and compute a 90% continuously smooth TEBPI estimate of θ_6. Solving (13.41) with $\gamma = 0.10$ yields $\theta_L = 0.799$ and $\theta_U = 0.995$. The desired interval is thus (0.799, 0.995) and $\Pr(0.799 \leqslant \Theta_6 \leqslant 0.995 | 5; 5) = 0.90$. The interval is also indicated in Figure 13.2. ■

13.4. FAILURE RATE ESTIMATION—POISSON SAMPLING

Recall from Chapter 7 that the distribution of the number of failures S occurring in a fixed total test (or operating) time t, when testing with replacement, follows a Poisson distribution with parameter $\lambda \mathrm{t}$, where λ is the unknown failure rate parameter to be estimated. Further suppose that a sequence of N such experiments has been conducted in which s_i failures in total test time t_i has been observed in the ith experiment, $i = 1, 2, \ldots, N$. In the Nth or current experiment, it is assumed that the sequence of data given by $(s_1, \mathrm{t}_1), (s_2, \mathrm{t}_2), \ldots, (s_N, \mathrm{t}_N)$ is available for use in the estimation process. For convenience let $s = s_N$, $\mathrm{t} = \mathrm{t}_N$, and $\lambda = \lambda_N$. It is further assumed that (1) the s_i values, conditional on λ_i, are independent, and (2) the underlying failure rate values $\lambda_1, \lambda_2, \ldots, \lambda_N$ are independent, unknown, and unobservable realizations of an r.v. Λ according to a completely unknown and unspecified prior distribution $g(\lambda)$.

13.4.1 Point Estimation of λ

Three EB point estimators will be considered; namely, a consistent (implicitly estimated prior) estimator and two smooth (explicitly estimated prior) estimators.

Consistent Estimator. Suppose that we wish to estimate λ in the current experiment. In using this approach it is further required that $\mathrm{t}_i = \mathrm{t}$, $i = 1, 2, \ldots, N$, that is, the same total test time underlies each experiment. From the Poisson sampling model given in (7.4) we find that

$$\frac{f(s+1|\theta)}{f(s|\theta)} = \frac{\theta}{s+1},$$

which is of the form $a(s)+b(s)\theta$ in (13.8) where $a(s)=0$ and $b(s)=1/(s+1)$. Thus, according to (13.11), the EB estimator for λ becomes

$$E_N(\lambda|s)=\frac{(s+1)f_N(s+1)}{\mathrm{t}f_N(s)}=$$

$$\frac{(s+1)(\text{Total number of experiments in which } (s+1) \text{ failures were observed})}{\mathrm{t}\times(\text{Total number of experiments in which } s \text{ failures were observed})} \tag{13.42}$$

Example 13.6. Suppose that a 10^5 h life test has been routinely conducted for each of a sequence of eleven production lots of a certain high reliability device in which 0, 1, 0, 0, 1, 2, 0, 1, 0, 0, and 0 failures were respectively observed. The traditional consistent EB point estimate of λ_{11}, the failure rate in the current or 11th lot, is

$$E_{11}(\lambda|0)=\frac{(1)(3)}{(10^5\text{ h})(7)}\cong 4.3\times 10^{-6}\text{ f/h} \qquad \blacksquare$$

Smooth Estimators. The Lemon and Krutchkoff (1969) iterated estimator for λ is

$$\hat{\lambda}_L=\frac{\sum_{i=1}^{N}\tilde{\lambda}_i^{s+1}\exp(-\tilde{\lambda}_i\mathrm{t})}{\sum_{i=1}^{N}\tilde{\lambda}_i^{s}\exp(-\tilde{\lambda}_i\mathrm{t})}, \tag{13.43}$$

where $\tilde{\lambda}_i$, according to (13.13), is given by

$$\tilde{\lambda}_i=\frac{\sum_{j=1}^{N}\left(\frac{s_j}{\mathrm{t}_j}\right)^{s_i+1}\exp\left\{-\left(\frac{s_j}{\mathrm{t}_j}\right)\mathrm{t}_i\right\}}{\sum_{j=1}^{N}\left(\frac{s_j}{\mathrm{t}_j}\right)^{s_i}\exp\left\{-\left(\frac{s_j}{\mathrm{t}_j}\right)\mathrm{t}_i\right\}}, \qquad i=1,2,\ldots,N;\ N\geqslant 2, \tag{13.44}$$

since s_j/t_j is the MVU, as well as the ML, estimate of λ_j. It is observed that t_i is permitted to vary from experiment to experiment. Lemon (1972) considers a similar estimator for the case where $t_i = 1$, $i = 1, 2, \ldots, N$. Canavos (1973) likewise considers (13.44), as well as two corresponding estimators for the reliability function $R(t) = \exp(-\lambda t)$, $t > 0$. Using Monte Carlo simulation, Canavos concludes that for $N \geq 2$ (13.44) has almost uniformly smaller EB risk than any of the conventional (nonBayes) estimators.

Example 13.7. Let us consider Example 13.6 and find the smooth EB estimates of λ_{11} according to (13.44) and (13.43). Now, according to (13.44), $\tilde{\lambda}_{11} = 1.67 \times 10^{-6}$ f/h, and also $\tilde{\lambda}_1 = \tilde{\lambda}_3 = \tilde{\lambda}_4 = \tilde{\lambda}_7 = \tilde{\lambda}_9 = \tilde{\lambda}_{10} = 1.67 \times 10$ f/h. Similarly, $\tilde{\lambda}_2 = \tilde{\lambda}_5 = \tilde{\lambda}_8 = 1.20 \times 10^{-5}$ f/h, and $\tilde{\lambda}_6 = 1.33 \times 10^{-5}$ f/h. Using these EB estimates in (13.43), the iterated EB estimate of λ_{11} is easily computed to be $\hat{\lambda}_L = 3.42 \times 10^{-6}$ f/h. ∎

A CSEB point estimator of λ is considered by Bennett and Martz (1972) and Bennett (1977) for the special case in which $t_i = 1$, $i = 1, 2, \ldots, N$. The CSEB estimator is observed to possess a smaller EB risk than the estimator given in (13.43). The details of constructing such CSEB estimators may be found in the indicated references.

13.4.2 Interval Estimation of λ

Let us now consider a smooth EB interval estimate of λ_i. The empirical posterior distribution expressed in (13.22) for the case of a Poisson sampling distribution is given by

$$g_N(\tilde{\lambda}_i | s; t) = \begin{cases} \dfrac{\tilde{\lambda}_i^s \exp(-\tilde{\lambda}_i t)}{\sum_{j=1}^{N} \tilde{\lambda}_j^s \exp(-\tilde{\lambda}_j t)}, & i = 1, 2, \ldots, N \\ 0, & \text{otherwise,} \end{cases} \tag{13.45}$$

where $\tilde{\lambda}_i$ is given in (13.44). The symmetric $100(1-\gamma)\%$ (or larger) smooth TEBPI estimate of λ_i given by (13.23) is obtained by solving

$$\sum_{j=1}^{j^+} \left[\tilde{\lambda}_j^{s_i} \exp(-\tilde{\lambda}_j t_i)\right] \Big/ \left[\sum_{j=1}^{N} \tilde{\lambda}_j^{s_i} \exp(-\tilde{\lambda}_j t_i)\right] \geq 1 - \frac{\gamma}{2}$$

and (13.46)

$$\sum_{j=1}^{j^-} \left[\tilde{\lambda}_j^{s_i} \exp(-\tilde{\lambda}_j t_i)\right] \Big/ \left[\sum_{j=1}^{N} \tilde{\lambda}_j^{s_i} \exp(-\tilde{\lambda}_j t_i)\right] \leq \frac{\gamma}{2}$$

for the integers j^+ and j^-, from which the interval is then given by (13.23). Similarly, a $100(1-\gamma)\%$ (or larger) smooth LEBPI (UEBPI) estimate of θ may be obtained by solving the second (first) equation in (13.46) for j^- (j^+) with $\gamma/2$ replaced by γ.

Example 13.8. Again let us reconsider Examples 13.6 and 13.7. Suppose we desire a 95% (or larger) smooth UEBPI estimate of λ in the current production lot. The empirical posterior distribution in the current lot is computed according to (13.45) as

$$g_{11}(1.67\times10^{-6}\text{f/h}|0;10^5\text{h})=\frac{(7)\exp[-(1.67\times10^{-6})10^5]}{7.0915}=0.8353,$$

$$g_{11}(1.20\times10^{-5}\text{f/h}|0;10^5\text{h})=\frac{(3)\exp[-(1.20\times10^{-5})10^5]}{7.0915}=0.1274,$$

$$g_{11}(1.33\times10^{-5}\text{f/h}|0;10^5\text{h})=\frac{\exp[-(1.33\times10^{-5})10^5]}{7.0915}=0.0373.$$

Solving the first equation in (13.46) with $\gamma/2$ replaced by γ and for $\gamma=0.05$ yields $j^+=2$. Thus the desired interval estimate is $[0; 1.20\times10^{-5}\text{f/h}]$ and $\Pr(\Lambda_{11}\leqslant 1.20\times10^{-5}\text{f/h}|0; 10^5\text{ h})=0.9627$. ■

13.5. ESTIMATION IN THE WEIBULL DISTRIBUTION

The $\mathcal{W}_1(\lambda,\beta)$ distribution has been the subject of several EB investigations, most of which have considered smooth EB estimation methods. Couture and Martz (1972) present CSEB estimators for both the scale and shape parameters and use these estimators to develop a sample size reduction scheme. Bennett (1970) presents CSEB estimators for the unknown scale parameter with known shape parameter, for the unknown shape parameter with known scale parameter, and for both the unknown scale and unknown shape parameters. Couture (1970) also considers CSEB estimators for both the scale and shape parameters. Canavos (1971) considers both traditional and smooth EB estimators of the scale parameter with known and fixed shape parameter. Bennett and Martz (1973) consider estimating the scale parameter with known shape parameter. Both CSEB point and interval estimators for the reliability function and MTTF, when the shape parameter is known, are considered by Lian (1975) and Martz and Lian (1977). Bennett (1977) uses the $\mathcal{W}_1(\lambda, \beta)$ distribution as an example in an expository paper on EB methods. Since the $\mathcal{E}(\lambda)$ distribution is a special case of the $\mathcal{W}_1(\lambda, \beta)$ (when $\beta = 1$), these estimators can be used to estimate the constant failure rate of the $\mathcal{E}(\lambda)$ model based on gamma sampling. Section 13.5.1 considers this important special case.

Recall that the two-parameter $\mathcal{W}_1(\lambda, \beta)$ distribution has a pdf given by

$$f(t|\lambda, \beta) = \lambda\beta t^{\beta-1}\exp[-\lambda t^{\beta}], \qquad t, \lambda, \beta > 0, \tag{13.47}$$

where λ is the scale parameter and β is the shape parameter. We assume that a sequence of $N \geq 2$ Type II item-censored life test experiments has been conducted in which n_i items are placed on life test which is terminated after s_i, $1 \leq s_i \leq n_i$, failures occur in the ith experiment, $i = 1, 2, \ldots, N$. The ordered failure times $t_{i1} \leq t_{i2} \leq \cdots \leq t_{is_i}$ are recorded in each experiment. In the Nth or current experiment it is assumed that the sequence of data given by $(s_1, n_1, \underset{\sim}{t}_1), (s_2, n_2, \underset{\sim}{t}_2), \ldots, (s_N, n_N, \underset{\sim}{t}_N)$ is available for use in the estimation process, where $\underset{\sim}{t}_i' = (t_{i1}, t_{i2}, \ldots, t_{is_i})$ is the vector of ordered failure times in the ith experiment. For convenience let $s = s_N$, $n = n_N$, $\underset{\sim}{t} = \underset{\sim}{t}_N$, $\lambda = \lambda_N$ and $\beta = \beta_N$. It is further assumed that: (1) the $\underset{\sim}{t}_i$ values, conditional on the unknown random parameter(s), are independent, and (2) the underlying values of the random parameter(s) in the ith experiment are independent, unknown, and unobservable relizations according to either a completely unknown and unspecified univariate or bivariate prior distribution, depending upon whether only Λ or both Λ and β are r.v.'s.

13.5.1 Estimation of the Scale Parameter with Known Shape Parameter

Suppose now that the shape parameter β is known in each experiment. As in Chapter 9, let us again consider the statistic

$$\mathcal{W} = \sum_{j=1}^{s} T_j^{\beta} + (n-s)T_s^{\beta}. \tag{13.48}$$

Recall from Chapter 9 that $\mathcal{W}$ is a sufficient statistic for estimating the scale parameter λ and, conditional on λ, has a $\mathcal{G}_1(s, \lambda)$ distribution. It is also easily shown that the ML estimator of λ, which is $\hat{\lambda} = s/\mathcal{W}$, has an $\mathcal{IG}(s, \lambda s)$ distribution.

Point Estimation. We will present three smooth EB point estimators of λ. Canavos (1971) also derives a consistent EB point estimator of $\theta = 1/\lambda$. However, Canavos concludes that the smooth estimators had a smaller EB risk and for this reason the consistent estimator will not be presented here.

According to (13.13) the Lemon and Krutchkoff (1969) iterated EB point estimator, $\hat{\lambda}_L$, for λ becomes

$$\hat{\lambda}_L = \frac{\sum_{i=1}^{N} \tilde{\lambda}_i^{s+1} \exp[-\tilde{\lambda}_i s/\hat{\lambda}]}{\sum_{i=1}^{N} \tilde{\lambda}_i^{s} \exp[-\tilde{\lambda}_i s/\hat{\lambda}]}, \qquad (13.49)$$

where $\tilde{\lambda}_i$, according to (13.13), is given by

$$\tilde{\lambda}_i = \frac{\sum_{j=1}^{N} \hat{\lambda}_j^{s_i+1} \exp[-\hat{\lambda}_j s_i/\hat{\lambda}_i]}{\sum_{j=1}^{N} \hat{\lambda}_j^{s_i} \exp[-\hat{\lambda}_j s_i/\hat{\lambda}_i]}, \qquad i=1,2,\ldots,N, \qquad (13.50)$$

and where

$$\hat{\lambda}_i = s_i/\mathscr{W}_i = s_i \bigg/ \left[\sum_{j=1}^{s_i} t_{ij}^{\beta_i} + (n_i - s_i) t_{is_i}^{\beta_i} \right].$$

Here t_{ij} is the jth recorded failure time and β_i is the known shape parameter value in the ith experiment.

Example 13.9. Example 4.13 presents data on the unscheduled interruptions of two 115 kv power transmission circuits emanating from a common New Mexico generating station. Suppose that we are interested in estimating the failure rates (outages/100 mile-years) for both circuits using EB methods. First let us consider the smooth EB estimates according to (13.49). Suppose that $\beta_i = 1$, $i=1,2$, that is, the failure times follow an $\mathscr{E}(\lambda_i)$ distribution. Also $s_1 = n_1 = 9$ and $s_2 = n_2 = 11$. For the first and second data sets given in Example 4.13, the ML estimates are $\hat{\lambda}_1 = 100(9)/812.75$ mile-years $= 1.11$ per 100 mile-years, and $\hat{\lambda}_2 = 2.89$ per 100 mile-years. The noniterated smooth EB estimate of the failure rate for the first circuit is

$$\tilde{\lambda}_1 = \frac{(1.11)^{10}\exp[-(1.11)(9)/(1.11)] + (2.89)^{10}\exp[-(2.89)(9)/(1.11)]}{(1.11)^{9}\exp[-(1.11)(9)/(1.11)] + (2.89)^{9}\exp[-(2.89)(9)/(1.11)]}$$

$$= 1.12 \text{ per 100 mile-years.}$$

Similarly, $\tilde{\lambda}_2 = 2.85$ per 100 mile-years. Upon substituting these values into (13.49) the desired iterated smooth EB estimate of the failure rate for the first circuit is computed to be $\hat{\lambda}_{L,1} = 1.01$ per 100 mile-years. Also, $\hat{\lambda}_{L,2} =$ 2.81 per 100 mile-years. ■

Now let us consider a CSEB estimator of the $\mathcal{W}_1(\lambda, \beta)$ scale parameter λ. Bennett and Martz (1973) propose an estimator of the parameter λ based on the use of the weighting function $W(y) = (1/2\pi)(\sin y/y)^2$ in (13.14). The proposed estimator is

$$\hat{\lambda}_B = \frac{\sum_{i=1}^{N} \left(\int_{\min(\hat{\lambda}_i)}^{\max(\hat{\lambda}_i)} [\lambda^{s+1} \exp(-\lambda s/\hat{\lambda})] \{\sin[(\lambda - \hat{\lambda}_i)/2h] / [(\lambda - \hat{\lambda}_i)/2h]\}^2 d\lambda \right)}{\sum_{i=1}^{N} \left(\int_{\min(\hat{\lambda}_i)}^{\max(\hat{\lambda}_i)} [\lambda^{s} \exp(-\lambda s/\hat{\lambda})] \{\sin[(\lambda - \hat{\lambda}_i)/2h] / [(\lambda - \hat{\lambda}_i)/2h]\}^2 d\lambda \right)}, \tag{13.51}$$

where

$$h = h(N) = N^{-1/5} \left[\sum_{i=1}^{N} \frac{(\hat{\lambda}_i - \bar{\lambda})^2}{N} \right]^{1/2}, \qquad \bar{\lambda} = \sum_{i=1}^{N} \frac{\hat{\lambda}_i}{N}.$$

Numerical integration must be used in computing (13.51) and Bennett and Martz accomplish this using the 11-point Gauss quadrature formula. Bennett and Martz also find that the use of a linearly transformed past data sequence [according to (13.18)] yields an estimator of λ with even smaller EB risk than (13.51). The details may be found in the Bennett and Martz paper. An estimator similar to (13.51) is also considered by Canavos (1971) and is found to possess the smallest EB risk of three EB estimators he considers.

The numerical integration used in computing (13.51) represents a practical restriction on the use of (13.51). To overcome this handicap, Couture and Martz (1972) propose a CSEB estimator which involves the use of cumulative $\mathcal{N}(0,1)$ probabilities. Using the well-known fact that the ML estimator $\hat{\lambda}$ has a conditional asymptotic $\mathcal{N}(\lambda, \lambda^2/s)$ distribution, and a $\mathcal{N}(0,1)$ weighting function in (13.14), Couture and Martz propose the estimator of the scale parameter λ given by

$$\hat{\lambda}_C = \frac{\sum_{i=1}^{N} \exp\left[-\frac{1}{2\sigma^2}(\gamma_i - \delta_i^2)\right]\left[\delta_i \Phi(\delta_i/\sigma) + \frac{\sigma}{\sqrt{2\pi}} \exp(-\delta_i^2/2\sigma^2)\right]}{\sum_{i=1}^{N} \exp\left[-\frac{1}{2\sigma^2}(\gamma_i - \delta_i^2)\right]\Phi(\delta_i/\sigma)}, \tag{13.52}$$

where

$$\sigma^2 = \frac{\sigma_1^2\sigma_2^2}{\sigma_1^2 + \sigma_2^2}, \qquad \sigma_1^2 = \frac{\hat{\lambda}^2}{s}, \qquad \sigma_2^2 = \frac{\bar{\lambda}^2}{sN^{2/5}},$$

$$\gamma_i = \frac{\sigma_1^2\hat{\lambda}_i^2 + \sigma_2^2\hat{\lambda}^2}{\sigma_1^2 + \sigma_2^2}, \qquad \delta_i = \frac{\sigma_1^2\hat{\lambda}_i + \sigma_2^2\hat{\lambda}}{\sigma_1^2 + \sigma_2^2},$$

and where $\Phi(z)$ denotes the $\mathfrak{N}(0,1)$ cdf defined in (4.26).

Example 13.10. [Couture and Martz (1972)] Suppose that we wish to estimate the constant failure rate in a sequence of production lots of a certain low reliability device. Further suppose that the current lot is the third in a sequence of similar lots for which the EB assumptions are reasonably satisfied. The following failure data have been obtained for complete life tests based on a sample of size 15 from each lot:

Production Lot No.	Failure Data (h)
1	1.5, 2.7, 2.7, 4.1, 4.6, 5.8, 8.3, 10.7, 11.9, 12.3, 14.1, 15.3, 15.7, 18.4, 23.9
2	0.1, 2.1, 2.5, 4.1, 4.3, 4.5, 4.7, 4.9, 6.8, 12.2, 14.4, 19.5, 22.7, 26.9, 30.2
3	0.2, 0.7, 0.9, 2.2, 3.4, 3.6, 4.4, 4.9, 5.6, 6.4, 7.0, 9.7, 11.4, 24.0, 33.1

We desire an EB point estimate of the failure rate λ_3, assuming an $\mathcal{E}(\lambda_i)$

failure time model for the above data. Now $\hat{\lambda}_1 = 0.0986$ f/h. A more convenient value for use in the calculations is $\hat{\lambda}_1 = 20(0.0986)$ f/20 h $= 1.9720$ f/20 h. For convenience the units will be dropped for the remaining calculations. Also, $\hat{\lambda}_2 = 1.8740$ and $\hat{\lambda}_3 = 2.4720$. Now $\bar{\lambda} = 2.1060$, $\sigma_1^2 = 0.4074$, $\sigma_2^2 = 0.1905$, and $\sigma^2 = 0.1298$. Also $\delta_1 = 2.1313$, $\delta_2 = 2.0645$, $\delta_3 = 2.4720$ and $\gamma_1 = 4.5968$, $\gamma_2 = 4.3400$, $\gamma_3 = 6.1001$. Substituting these values into (13.52) yields the CSEB estimate of λ as $\hat{\lambda}_C = 2.2454$ f/20h $= 0.1123$ f/h. This estimate is only slightly smaller than the ML estimate; however, this estimate is estimated to have an associated EB risk which is roughly 40 percent smaller than the ML estimate $\hat{\lambda}_3$[Couture and Martz (1972)]. ■

The important question of which EB estimator has the smallest EB risk is fundamental in deciding which EB estimator to use. It is true that all three EB estimators have significantly smaller EB risk than the ML estimator in those situations most likely to be encountered in practice. Let us illustrate this fact. Figure 13.3 gives a typical plot of the ratio R of the EB risk to the Bayes risk of the ML estimator as a function of the number of experiences N for the estimators given by (13.49), (13.51), and (13.52), based on the use of Monte Carlo simulation results presented by Bennett (1970, 1977). The curves are indexed on the ratio $Z = E^2(\Lambda)/[s \text{ Var }(\Lambda)]$, which is the ratio of the expected asymptotic variance of $\hat{\lambda}$ given λ to the prior variance, and s, the test truncation number. It is observed that as few as two test data sets

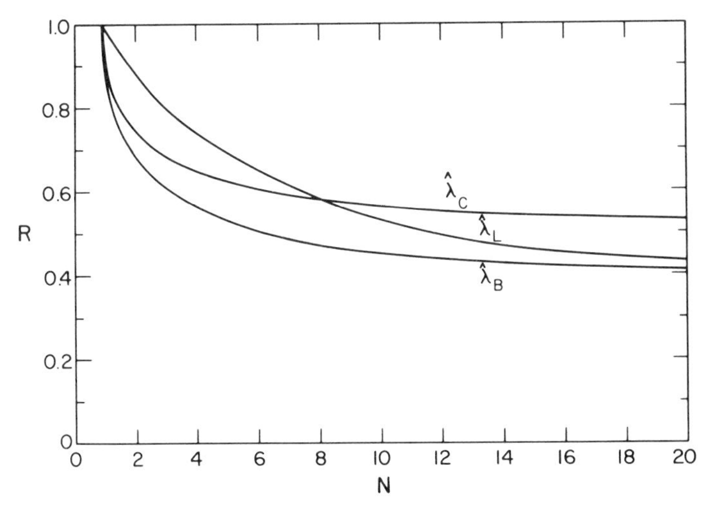

Figure 13.3. A comparison of the risks of three empirical Bayes estimators of the Weibull scale parameter for $Z = 2.0$ and $s = 10$.

(two experiences) yield significantly smaller EB risks than the Bayes risk of the ML estimator. All three estimators yield approximately 50% EB risk improvement over the ML estimator for $N=20$.

The Lemon and Krutchkoff estimator, $\hat{\lambda}_L$, improves in performance as N increases. The Couture estimator, $\hat{\lambda}_C$, approaches the Bennett estimator, $\hat{\lambda}_B$, in performance as s increases, that is, as the asymptotic sampling distribution of the ML estimator $\hat{\lambda}$ more closely approximates the exact $\mathcal{IG}(s, \lambda s)$ distribution. Numerous other Monte Carlo comparisons of the EB risk of competing EB estimators of the scale parameter $\theta = 1/\lambda$ may be found in Canavos (1971). Canavos concludes that the CSEB estimator generally possesses the smallest EB risk among the class consisting of a consistent, a smooth EB, and a CSEB estimator.

Interval Estimation. The empirical posterior distribution given by (13.22) for the case of an $\mathcal{IG}(s, \lambda s)$ sampling distribution on $\hat{\lambda}$ is given by

$$g_N(\tilde{\lambda}_i|\hat{\lambda}) = \begin{cases} \dfrac{\tilde{\lambda}_i^s \exp(-\tilde{\lambda}_i s/\hat{\lambda})}{\sum_{j=1}^{N} \tilde{\lambda}_j^s \exp(-\tilde{\lambda}_j s/\hat{\lambda})}, & i=1,2,\ldots,N \\ 0, & \text{otherwise,} \end{cases} \tag{13.53}$$

where $\tilde{\lambda}_i$ is given by (13.50). The symmetric $100(1-\gamma)\%$ (or larger) smooth TEBPI estimate of λ_i is obtained by solving

$$\frac{\sum_{j=1}^{j^+} \left[\tilde{\lambda}_j^{s_i} \exp(-\tilde{\lambda}_j s_i/\hat{\lambda}_i)\right]}{\sum_{j=1}^{N} \tilde{\lambda}_j^{s_i} \exp(-\tilde{\lambda}_j s_i/\hat{\lambda}_i)} \geqslant 1 - \frac{\gamma}{2}$$

and (13.54)

$$\frac{\sum_{j=1}^{j^-} \left[\tilde{\lambda}_j^{s_i} \exp(-\tilde{\lambda}_j s_i/\hat{\lambda}_i)\right]}{\sum_{j=1}^{N} \tilde{\lambda}_j^{s_i} \exp(-\tilde{\lambda}_j s_i/\hat{\lambda}_i)} \leqslant \frac{\gamma}{2}$$

for the integers j^+ and j^-. A $100(1-\gamma)\%$ (or larger) smooth LEBPI (UEBPI) estimate of λ_i is obtained by solving the second (first) equation in (13.54) for j^- (j^+) with $\gamma/2$ replaced by γ.

Example 13.11. A well-known light bulb manufacturer routinely conducts life tests on successive production lots of 60 W light bulbs. Samples of size 10 from successive lots yielded the following failure data:

Bulb Number	Failure Time (1000 h) Sample 1	Sample 2	Sample 3	Sample 4	Sample 5
1	0.80	0.60	5.05	3.54	0.15
2	4.82	0.74	0.89	1.35	3.54
3	5.99	1.27	0.71	5.05	4.41
4	1.99	1.35	2.28	0.42	0.50
5	2.10	1.69	1.43	7.01	0.17
6	0.33	1.60	0.77	1.64	1.74
7	0.89	0.77	0.08	2.28	1.55
8	1.20	0.08	4.08	3.22	0.66
9	1.83	0.58	3.03	0.37	1.47
10	0.58	1.06	0.21	3.79	1.27

Further suppose that the data are $\mathcal{E}(\lambda)$ distributed and that the EB assumptions are true. It is desired to construct both a point and a 90% (or larger) smooth UEBPI estimate on the failure rate in the fifth (current) production lot. For computational ease let us consider the noniterated procedure by replacing $\tilde{\lambda}_j$ with $\hat{\lambda}_j$ in (13.54). Now $\hat{\lambda}_1 = 0.49$, $\hat{\lambda}_2 = 1.03$, $\hat{\lambda}_3 = 0.54$, $\hat{\lambda}_4 = 0.35$, and $\hat{\lambda}_5 = 0.65$, in units of "failures per 1000 h" which will now be dropped for the remaining calculations. The desired point estimate according to (13.50) is $\tilde{\lambda}_5 = 0.60$. Also $g_5(0.35|\hat{\lambda}_5) = 0.068$, $g_5(0.49|\hat{\lambda}_5) = 0.228$, $g_5(0.54|\hat{\lambda}_5) = 0.280$, $g_5(0.65|\hat{\lambda}_5) = 0.329$, and $g_5(1.03|\hat{\lambda}_5) = 0.095$. Thus, for $\gamma = 0.10$, we find that $j^+ = 4$. The desired interval estimate is $[0; 0.65]$ and $\Pr(\Lambda_5 \leq 0.65|\hat{\lambda}_5) = 0.905$. ■

Martz and Lian (1977) consider a continuously smooth TEBPI estimate of the reliability function $r(t_0) = \exp[-\lambda t_0^\beta]$, which can also be used to construct a corresponding interval estimate of the scale parameter λ. However, this procedure requires the shape parameter to be known and have the same value in each experiment. The CSEB estimator proposed by Martz and Lian considers the use of triangular weighting functions in (13.14) each centered about the ML estimate $\hat{r}_i = \hat{r}_i(t_0) = \exp(-\hat{\lambda}_i t_0^\beta)$ with total base width $2h(N)$ and height C_i, where

$$C_i = \left(\ln \hat{r}_i^{-1}\right)^{-2} \hat{r}_i^{-2} s_i. \tag{13.55}$$

Since $C_i \cong \text{Var}^{-1}(\hat{r}_i|\lambda_i)$ it acts as a "weight" to account for varying sample size n_i and test truncation number s_i.

The symmetric $100(1-\gamma)\%$ continuously smooth TEBPI estimate of $r(t_0)$ is obtained by numerically solving the pair of equations for R_L and R_U given by

$$\frac{\sum_{i=1}^{N} \omega_i[(h-\hat{r}_i)J_0(e_{1i}, e_{2i}) + J_1(e_{1i}, e_{2i}) + (h+\hat{r}_i)J_0(e_{2i}, e_{3i}) + J_1(e_{2i}, e_{3i})]}{\sum_{i=1}^{N} \omega_i[(h-\hat{r}_i)J_0(d_{1i}, \hat{r}_i) + J_1(d_{1i}, \hat{r}_i) + (h+\hat{r}_i)J_0(\hat{r}_i, d_{2i}) + J_1(\hat{r}_i, d_{2i})]}$$

$$= \frac{\gamma}{2}, \tag{13.56}$$

where $e_{1i} = \max[0, \min_i(\hat{r}_i - h, R_L)]$, $e_{2i} = \min(\hat{r}_i, R_L)$, and $e_{3i} = \min[1, \max_i(\hat{r}_i + h, R_L)]$, $d_{1i} = \max(0, \hat{r}_i - h)$, $d_{2i} = \min(1, \hat{r}_i + h)$; and

$$\frac{\sum_{i=1}^{N} \omega_i[(h-\hat{r}_i)J_0(e'_{1i}, e'_{2i}) + J_1(e'_{1i}, e'_{2i}) + (h+\hat{r}_i)J_0(e'_{2i}, e'_{3i}) + J_1(e'_{2i}, e'_{3i})]}{\sum_{j=1}^{N} \omega_i[(h-\hat{r}_i)J_0(d_{1i}, \hat{r}_i) + J_1(d_{1i}, \hat{r}_i) + (h+\hat{r}_i)J_0(\hat{r}_i, d_{2i}) + J_1(\hat{r}_i, d_{2i})]}$$

$$= \frac{\gamma}{2}, \tag{13.57}$$

where $e'_{1i} = \max(\hat{r}_i - h, R_U)$, $e'_{2i} = \max(\hat{r}_i, R_U)$, and $e'_{3i} = \max[R_U, \min_i(1, \hat{r}_i + h)]$, and where we have defined

1. $$\omega_i = \begin{cases} C_i, & \hat{r}_i - h \geqslant 0 \quad \text{and} \quad \hat{r}_i + h \leqslant 1 \\ 2C_i h^2 / \left[2h^2 - (\hat{r}_i - h)^2\right], & \hat{r}_i - h < 0 \\ 2C_i h^2 / \left[2h^2 - (1 - \hat{r}_i - h)^2\right], & \hat{r}_i + h > 1, \end{cases}$$

2. $$h = h(N) = \begin{cases} SN^{-1/5}, & S > 0 \\ N^{-1/5}, & S = 0, \end{cases}$$

3. $$S^2 = \frac{\sum_{i=1}^{N} C_i(\hat{r}_i - \bar{r})^2}{M}, \qquad \bar{r} = \frac{\sum_{i=1}^{N} C_i \hat{r}_i}{M}, \qquad M = \sum_{i=1}^{N} C_i,$$

4. $J_q(x, y) = (\mathscr{W} t_0^{-\beta} + q + 1)^{-s-1}\{\Gamma^c[s+1, -(\mathscr{W} t_0^{-\beta} + q + 1)\ln y]$

$$- \Gamma^c[s+1, -(\mathscr{W} t_0^{-\beta} + q + 1)\ln x]\},$$

and where $\Gamma^c(a, z)$ is the complement of the incomplete gamma function defined in (9.24). The above equations may be efficiently solved on a computer. Corresponding one-sided intervals are obtained by appropriately solving either (13.56) or (13.57). Once R_L and R_U have been found, the desired symmetric $100(1-\gamma)\%$ continuously smooth TEBPI estimate for the scale parameter is easily computed to be (λ_L, λ_U), where $\lambda_L = -t_0^{-\beta}\ln R_U$ and $\lambda_U = -t_0^{-\beta}\ln R_L$, and where t_0 is the mission time for which R_U and R_L were computed.

Example 13.12. A sequence of five life test experiments have been conducted on a certain device in which the EB conditions are satisfied. In each experiment 20 devices were tested and the test was terminated at the time of the fifth failure. Further suppose that the $\mathscr{E}(\lambda)$ distribution is appropriate for describing the failure data. The total time on test in each experiment is $\mathscr{W}_1 = 25488$, $\mathscr{W}_2 = 5726$, $\mathscr{W}_3 = 69831$, $\mathscr{W}_4 = 38516$, and $\mathscr{W}_5 = 9257$ h, respectively. It is desired to find a symmetric 90% continuously smooth TEBPI interval estimate of the failure rate in the fifth experiment. It is convenient to consider $t_0 = 100$ h. The sequence of ML estimates of $r_i(100)$ become $\hat{r}_1 = 0.9806$, $\hat{r}_2 = 0.9164$, $\hat{r}_3 = 0.9929$, $\hat{r}_4 = 0.9871$, and $\hat{r}_5 = 0.9474$. By means of a computer the required interval estimate of $r_5(100)$ is calculated according to (13.56) and (13.57) to be (0.9188),0.9901). Thus $\lambda_L = -(100)^{-1}\ln(0.9901) = 9.9 \times 10^{-5}$ f/h, and $\lambda_U = -(100)^{-1}\ln(0.9188) = 8.5 \times 10^{-4}$ f/h and hence the desired 90% interval estimate of λ_5 is (9.9×10^{-5} f/h, 8.5×10^{-4} f/h). ■

Estimating the Failure Rate of an Exponential Distribution. The Couture and Martz (1972) estimator given in (13.52) can be simplified for use in estimating the failure rate λ of an $\mathscr{E}(\lambda)$ distribution. In addition, this estimator can be used for both item-censored or time-truncated life test data either with or without the replacement of failed items. Let $\hat{\lambda}_i = s_i / \mathrm{t}_i$ denote the ML estimate of λ_i in the ith experiment, where t_i is the observed total time on test defined in Section 4.4. Further define $\bar{\lambda} = \sum_{i=1}^{N} \hat{\lambda}_i / N$. The

CSEB point estimator of λ_j as presented by Martz (1975a) is

$$\tilde{\lambda}_j = \frac{\hat{\lambda}_j}{\theta_{jj}} \frac{\sum_{i=1}^{N} \phi(\psi_{ij})\left[\theta_{ij}\Phi(\theta_{ij}) + \phi(\theta_{ij})\right]}{\sum_{i=1}^{N} \phi(\psi_{ij})\Phi(\theta_{ij})}, \qquad j=1,2,\ldots,N, \quad (13.58)$$

where

$$\psi_{ij} = \nu_j(\rho_i - \rho_j), \qquad \theta_{ij} = \nu_j(1+\rho_i\rho_j), \qquad \nu_j^2 = s_j/(1+\rho_j^2),$$

$$\rho_j = N^{0.2}\hat{\lambda}_j/\bar{\lambda},$$

and where $\phi(\cdot)$ and $\Phi(\cdot)$ denote the $\mathfrak{N}(0,1)$ pdf and cdf defined in (4.24) and (4.26), respectively.

Ofttimes it is true that $\Phi(\theta_{ij}) \cong 1$ and $\phi(\theta_{ij}) \ll \theta_{ij}$ for every i, $j=1,2,\ldots,N$. In this case, (13.58) becomes

$$\tilde{\lambda}_j \cong \hat{\lambda}_j \sum_{i=1}^{N} \theta_{ij} c_{ij}/\theta_{jj}, \qquad j=1,2,\ldots,N, \quad (13.59)$$

where

$$c_{ij} = \phi(\psi_{ij}) \Big/ \sum_{i=1}^{N} \phi(\psi_{ij}).$$

Example 13.13. Let us reconsider the data in Example 4.13 and calculate the smooth EB point estimate of the failure rate for the first circuit using the above estimators. The following quantities are easily computed:

j	s_j	$\hat{\lambda}_j$	ρ_j	ν_j
1	9	0.0111	0.638	2.579
2	12	0.0289	1.660	1.788

	θ_{ij}		ψ_{ij}		$\phi(\psi_{ij})$	
i	$j=1$	$j=2$	$j=1$	$j=2$	$j=1$	$j=2$
1	3.558	3.682	0	−1.827	0.3989	0.0748
2	5.207	6.715	2.585	0	0.0139	0.3989

It is observed that $\Phi(\theta_{ij}) \cong 1$ and $\phi(\theta_{ij}) \ll \theta_{ij}$. Thus it is appropriate to consider the approximation given in (13.59). Substituting the above quantities into (13.59) yields

$$\tilde{\lambda}_1 = \frac{0.0111}{3.558} \frac{3.558(0.3989) + 5.207(0.0139)}{0.3989 + 0.0139} = 1.13 \times 10^{-2}$$

outages per mile-year = 1.13 outages per 100 mile-years. ■

13.5.2 Estimation of the Scale and Shape Parameters

Now let us consider the case where both the $\mathcal{W}_1(\lambda, \beta)$ scale parameter λ and shape parameter β are r.v.'s that take on values in each of a sequence of life test experiments according to a completely unknown and unspecified joint prior distribution $g(\lambda, \beta)$. Couture and Martz (1972) present CSEB estimators for both λ and β under the assumption that Λ and β have independent prior distributions, that is, $g(\lambda, \beta) = g(\lambda)g(\beta)$. However, the estimators were still observed to perform well in the case of prior dependency between Λ and β. Couture and Martz make use of the fact that the ML estimates $\hat{\lambda}$ and $\hat{\beta}$ [Mann, Schafer, and Singpurwalla (1974), p. 190] have a conditional asymptotic bivariate normal distribution with mean $(\lambda, \beta)'$ and covariance matrix

$$V = \begin{bmatrix} \dfrac{\lambda^2}{s}\left\{1 + \Delta^2\left(\dfrac{6}{\pi^2}\right)\right\} & \dfrac{-\lambda\beta\Delta}{s}\left(\dfrac{6}{\pi^2}\right) \\ \dfrac{-\lambda\beta\Delta}{s}\left(\dfrac{6}{\pi^2}\right) & \dfrac{\beta^2}{s}\left(\dfrac{6}{\pi^2}\right) \end{bmatrix}, \tag{13.60}$$

where $\Delta = 0.4227843 - \ln\lambda$. Using a product version of the pdf estimator in (13.14) with $\mathcal{N}(0, 1)$ weighting functions for λ and β, respectively, and replacing λ and β in (13.60) with ML estimates $\hat{\lambda}_N$ and $\hat{\beta}_N$, ultimately yields the CSEB estimators of λ and β given by

$$\tilde{\lambda} = \frac{\sum_{i=1}^{N} \sum_{j=1}^{N} (wx_i - vy_j)\exp(c_{ij}/2)}{\sum_{i=1}^{N} \sum_{j=1}^{N} \exp(c_{ij}/2)} \tag{13.61}$$

and

$$\tilde{\beta}=\frac{\sum_{i=1}^{N}\sum_{j=1}^{N}(uy_j-vx_i)\exp(c_{ij}/2)}{\sum_{i=1}^{N}\sum_{j=1}^{N}\exp(c_{ij}/2)}, \tag{13.62}$$

where

1. $\alpha=\pi^2/6+\hat{\Delta}^2,\ \hat{\Delta}=0.4227843-\ln\hat{\lambda}$,

2. $h_1=N^{-1/5}(\bar{\lambda}/\sqrt{s}),\ \bar{\lambda}=\sum_{j=1}^{N}\hat{\lambda}_j/N$,

3. $h_2=N^{-1/5}(\bar{\beta}/\sqrt{s}),\ \bar{\beta}=\sum_{j=1}^{N}\hat{\beta}_j/N$,

4. $\delta=(h_1^2 s+\hat{\lambda}^2)(\alpha h_2^2 s+\hat{\beta}^2)-h_1^2 h_2^2 s^2\hat{\Delta}^2$,

5. $w=h_1^2\hat{\lambda}^2(\alpha h_2^2 s+\hat{\beta}^2)/\delta$,

6. $v=h_1^2 h_2^2 s\hat{\Delta}\hat{\lambda}\hat{\beta}/\delta$,

7. $u=h_2^2\hat{\beta}^2(h_1^2 s+\hat{\lambda}^2)/\delta$,

8. $x_i=s/\hat{\lambda}+s\hat{\Delta}/\hat{\lambda}+\hat{\lambda}_i/h_1^2$,

9. $y_i=s\hat{\Delta}/\hat{\beta}+\alpha s/\hat{\beta}+\hat{\beta}_i/h_2^2$,

10. $c_{ij}=x_i^2 s-2x_i y_i v+y_j^2 u-\hat{\lambda}_i^2/h_1^2-\hat{\beta}_j^2/h_2^2-s(\alpha+2\hat{\Delta}+1)$.

The performance properties of (13.61) and (13.62), as well as a sample size determination scheme, are considered by Couture and Martz (1972).

13.5.3 Estimation of Reliability

Martz and Lian (1977) develop CSEB point and interval estimators of the $\mathcal{W}_1(\lambda,\beta)$ reliability function $r(t_0)=\exp(-\lambda t_0^{\beta})$ when β is known.

Point Estimation. Using triangular weighting functions in (13.14), as discussed prior to (13.55), yields the empirical prior pdf of $R = R(t_0)$ given by

$$\hat{g}_N(r) = \begin{cases} \dfrac{1}{Nh} \sum_{i=1}^{N} \omega_i \left(1 - \left|\dfrac{r - \hat{r}_i}{h}\right|\right), & \left|\dfrac{r - \hat{r}_i}{h}\right| \leqslant 1, \quad 0 \leqslant r \leqslant 1 \\ 0, & \text{otherwise,} \end{cases} \tag{13.63}$$

where $\hat{r}_i = \exp(-\hat{\lambda}_i t_0^{\beta})$ and where ω_i is defined in (1) following (13.57). The corresponding empirical posterior pdf becomes

$$g_N(r \mid \mathscr{W}) = \frac{f(\mathscr{W} \mid r)\hat{g}_N(r)}{\int_0^1 f(\mathscr{W} \mid r)\hat{g}_N(r)\, dr}, \qquad \max\left[0, \min_i(\hat{r}_i - h)\right] \leqslant r \leqslant \min\left[1, \max_i(\hat{r}_i + h)\right], \tag{13.64}$$

where $f(\mathscr{W} \mid r)$ is the sampling distribution of $\mathscr{W}$ given r. The denominator of (13.64) can be expressed as

$$\int_0^1 f(\mathscr{W} \mid r)\hat{g}_N(r)\, dr = \frac{t_0^{-s\beta}\mathscr{W}^{s-1}}{Nh^2(s-1)!} \sum_{i=1}^{N} \omega_i\big[(h - \hat{r}_i)J_0(d_{1i}, \hat{r}_i) + J_1(d_{1i}, \hat{r}_i) + (h + \hat{r})J_0(\hat{r}_i, d_{2i}) + J_1(\hat{r}_i, d_{2i})\big], \tag{13.65}$$

where h, ω_i, d_{1i}, d_{2i}, and $J_q(x, y)$ are defined in (13.57).

The CSEB point estimator of $r(t_0)$ in the Nth experiment is the mean of (13.64) which upon simplification becomes

$$\tilde{r}(t_0) = \frac{\sum_{i=1}^{N} \omega_i\big[(h - \hat{r}_i)J_1(d_{1i}, \hat{r}_i) + J_2(d_{1i}, \hat{r}_i) + (h + \hat{r}_i)J_1(\hat{r}_i, d_{2i}) + J_2(\hat{r}_i, d_{2i})\big]}{\sum_{i=1}^{N} \omega_i\big[(h - \hat{r}_i)J_0(d_{1i}, \hat{r}_i) + J_1(d_{1i}, \hat{r}_i) + (h + \hat{r}_i)J_0(\hat{r}_i, d_{2i}) + J_1(\hat{r}_i, d_{2i})\big]}. \tag{13.66}$$

Example 13.14. Let us reconsider Example 13.12 and compute a CSEB point estimate of the reliability of the device in the fifth experiment for $t_0 = 100$ h. Using the sequence of ML estimates given in Example 13.12, the desired estimate is found to be 0.9681, which is somewhat larger than the ML estimate 0.9474. The empirical prior and posterior pdf's for $t_0 = 100$ h are plotted in Figure 13.4. It is observed that the empirical posterior pdf responds well to the sampling value $\mathscr{W} = 9257$, and a peak is observed near $r = 0.95$. Since the ML estimate 0.9474 is somewhat less than the empirical prior mean 0.9895, we observe an increase of density below 0.96 in relation to the prior distribution. Recall from Example 13.12 that the symmetric 90% continuously smooth TEBPI estimate of $r_5(100)$ was found to be (0.9188,0.9901).

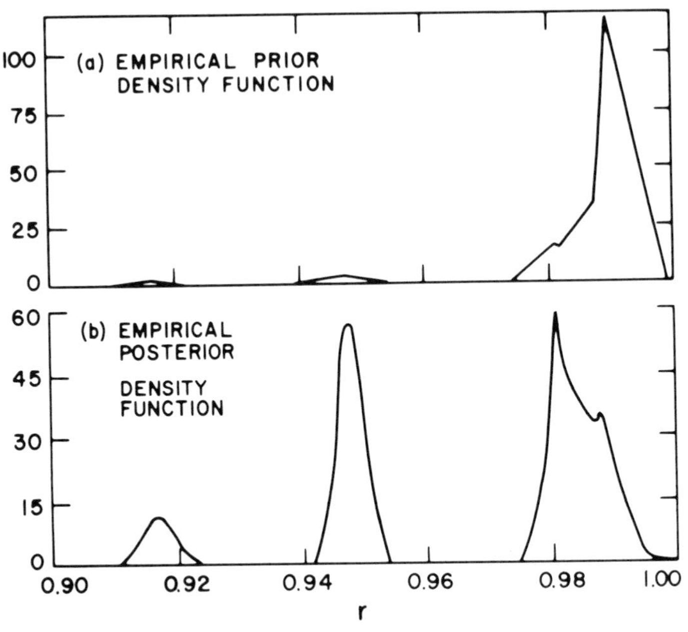

Figure 13.4. The empirical prior and posterior probability density function in Example 13.14 [Adapted, with permission, from Martz and Lian (1977)]. ■

Canavos (1973) finds, for the Poisson distribution, that the EB point estimate of reliability obtained by substituting the smooth EB estimate of the failure rate into the reliability function has a uniformly smaller EB risk than the estimator based on the expectation of the reliability function. If such is the case here, then the CSEB point estimator given by $\exp(-\hat{\lambda}_B t_0^{\beta})$, where $\hat{\lambda}_B$ is given by (13.51), should have good EB risk properties. However, there is no evidence at this time to support this conjecture. If both parameters are unknown values of r.v.'s then the corresponding CSEB point estimator given by $\exp(-\tilde{\lambda} t_0^{\tilde{\beta}})$, where $\tilde{\lambda}$ and $\tilde{\beta}$ are given by (13.61) and (13.62), respectively, would provide an EB method of estimating $r(t_0)$.

Interval Estimation. A procedure for determining a symmetric $100(1-\gamma)\%$ continuously smooth TEBPI estimate of $r(t_0)$ has been discussed in Section 13.5.1 and is given by (13.56) and (13.57). A $100(1-\gamma)\%$ continuously smooth LEBPI estimate may be obtained by solving (13.56) for R_L with $\gamma/2$ replaced by γ.

13.6. ESTIMATION IN THE GAMMA DISTRIBUTION

Canavos (1971) and Canavos and Tsokos (1971) derive EB point estimators for the scale parameter and reliability function of the $\mathcal{G}(p,\theta)$ distribution with known integer shape parameter. Recall that the two-parameter $\mathcal{G}(p,\theta)$ pdf is given by

$$f(t|\theta,p)=\frac{1}{\theta^p(p-1)!}t^{p-1}\exp(-t/\theta), \qquad 0<t<\infty,\ \theta>0,\ p=1,2,\ldots,$$

where the scale parameter θ is a value of an r.v. according to a completely unknown and unspecified prior distribution. It is desired to estimate θ when the shape parameter p is a known integer value. We further assume that a sequence of $N\geqslant 2$ complete life test experiments has been conducted in which n items are placed on life test in each experiment and the test is terminated only when all n items have failed. Let $\mathrm{t}_{i,n}=\Sigma_{j=1}^{n}t_{ij}$ represent the observed total time on test in the ith experiment, where t_{ij} is the jth observed ordered failure time in the ith experiment. Further let $\underset{\sim}{t}_i'=(t_{i1},t_{i2},\ldots,t_{in})$. In the Nth or current experiment it is assumed that the sequence of data given by $\mathrm{t}_{1,n},\mathrm{t}_{2,n},\ldots,\mathrm{t}_{N,n}$ is available for use in estimating θ_i, as $\mathcal{T}_{i,n}$ is a sufficient statistic for estimating θ_i [Canavos and Tsokos (1971)]. As before, let $\mathrm{t}=\mathrm{t}_{N,n}$ and $\theta=\theta_N$. In addition, the usual independence assumptions on $\mathcal{T}_{i,n}$ are assumed to hold.

13.6.1 Point Estimation of the Scale Parameter

As $\mathcal{T}_{i,n}$ is the sum of i.i.d. $\mathcal{G}(p,\theta)$ r.v.'s, the sampling distribution of $\mathcal{T}$ given θ (with p known) is given by

$$f(\mathrm{t}|\theta)=\frac{1}{(np-1)!\theta^{np}}\mathrm{t}^{np-1}\exp(-\mathrm{t}/\theta), \quad 0<\mathrm{t}<\infty, \qquad (13.67)$$

which is a $\mathcal{G}(np,\theta)$ distribution. Let us now consider consistent, smooth, and CSEB estimators of θ.

Consistent Estimator. Canavos and Tsokos (1971) derive the consistent EB estimator of θ given by

$$\tilde{\theta}_C=\frac{\mathrm{t}}{np}\frac{\sum_{j=1}^{N}\exp(-|\phi_j^*|)}{\sum_{j=1}^{N}\exp(-|\phi_j|)}, \qquad (13.68)$$

where

1. $\phi_j^*=(\mathrm{t}-\mathrm{t}_{j,n-1})/\delta, \quad j=1,2,\ldots,N,$

2. $\phi_j=(\mathrm{t}-\mathrm{t}_{j,n})/\delta, \quad j=1,2,\ldots,N,$

3. $\delta=N^{-1/5}\sqrt{\sum_{j=1}^{N}(\mathrm{t}_{j,n}-\bar{\mathrm{t}}_N)^2/N}, \qquad \bar{\mathrm{t}}_N=\sum_{j=1}^{N}\mathrm{t}_{j,n}/N,$

and where $\mathrm{t}_{i,n-1}=\sum_{j=1}^{n-1}t_{ij}$, the observed total time on test based only upon the first $(n-1)$ observed ordered failure times.

Smooth Estimators. The Lemon and Krutchkoff (1969) noniterated smooth EB estimator, $\tilde{\theta}_L$, of θ given in (13.13) here becomes

$$\tilde{\theta}_L=\frac{\sum_{j=1}^{N}\frac{1}{\mathrm{t}_{j,n}^{np-1}}\exp(-\mathrm{t}np/\mathrm{t}_{j,n})}{\sum_{j=1}^{N}\frac{1}{\mathrm{t}_{j,n}^{np}}\exp(-\mathrm{t}np/\mathrm{t}_{j,n})}, \quad np>1. \qquad (13.69)$$

Using exponential weighting functions in (13.14), Canavos (1971) obtains

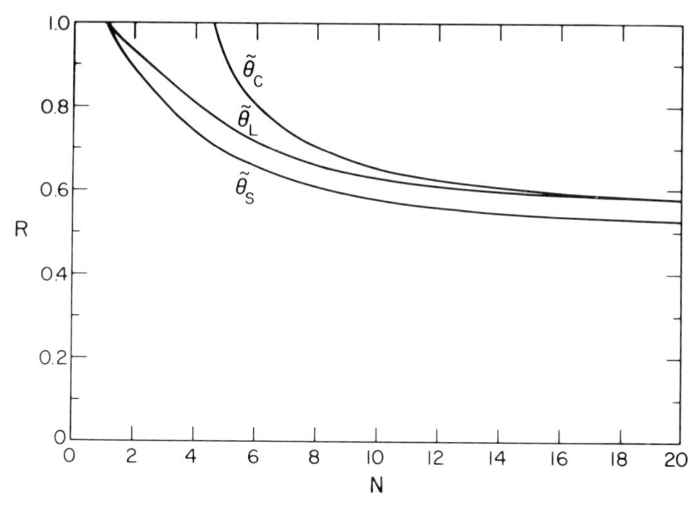

Figure 13.5. A comparison of the risks of three empirical Bayes estimators of the gamma scale parameter for $Z = 1.4$, $p = 2$, and $n = 3$.

the CSEB estimator of θ according to (13.15) as

$$\tilde{\theta}_S = \frac{\sum_{j=1}^{N} \left\{ \int_{\min(\hat{\theta}_j)}^{\max(\hat{\theta}_j)} \left[\frac{1}{\theta^{np-1}} \exp(-\mathrm{t}/\theta) \right] \left[\exp\left(-\left| \frac{\theta - \mathrm{t}_{j,n}/np}{\delta} \right| \right) \right] d\theta \right\}}{\sum_{j=1}^{N} \left\{ \int_{\min(\hat{\theta}_j)}^{\max(\hat{\theta}_j)} \left[\frac{1}{\theta^{np}} \exp(-\mathrm{t}/\theta) \right] \left[\exp\left(-\left| \frac{\theta - \mathrm{t}_{j,n}/np}{\delta} \right| \right) \right] d\theta \right\}}, \tag{13.70}$$

where $\hat{\theta}_j = \mathrm{t}_{j,n}/np$ and where δ is defined in 3 following (13.68). As before, numerical integration must be used to evaluate $\tilde{\theta}_s$ in practice. Figure 13.5 gives a typical plot of the ratio R of the estimated EB risk to the Bayes risk of the ML estimator for the EB estimators given in (13.68), (13.69), and (13.70) for the case where $p = 2$, $n = 3$, and $Z = E(\Theta^2)/[np\,\mathrm{Var}(\Theta)] = 1.4$, the ratio of the expected conditional variance of $\hat{\theta}$ given θ to the prior variance. These are the results of a Monte Carlo simulation carried out by Canavos (1971). Again, it is observed that $\tilde{\theta}_S$ has the smallest estimated EB risk.

Example 13.15. A sequence of three life testing experiments has been conducted on similar models of a certain mechanical component for which a $\mathcal{G}(2, \theta)$ failure model is deemed appropriate. In each experiment eight devices were tested to failure and the data (in h) are $t_{1,8}=200$, $t_{2,8}=165$, $t_{3,8}=230$, $t_{1,7}=185$, $t_{2,7}=135$, and $t_{3,7}=220$. Suppose that it is desired to estimate the value of the shape parameter θ in the third experiment.

The ML estimate is $\hat{\theta}_3=14.4$ h, the consistent EB estimate is $\tilde{\theta}_C=8.4$ h, and the smooth EB estimate is $\tilde{\theta}_L=13.0$ h. ■

13.6.2 Point Estimation of Reliability

Inasmuch as p is an integer, the reliability function is given by

$$r(t_0)=\left[\sum_{j=0}^{p-1}\frac{1}{j!}\left(\frac{t_0}{\theta}\right)^j\right]\exp(-t_0/\theta), \qquad t_0,\theta>0, \quad p=1,2,\ldots \tag{13.71}$$

It follows that an EB point estimator of $r=r(t_0)$ is given by

$$\tilde{r}(t_0)=\left[\sum_{j=0}^{p-1}\frac{1}{j!}\left(\frac{t_0}{\tilde{\theta}}\right)^j\right]\exp(-t_0/\tilde{\theta}), \tag{13.72}$$

where $\tilde{\theta}$ is given by either (13.68), (13.69), or (13.70). Canavos (1971) compares the EB risk in using (13.72) for all three EB estimators of θ using Monte Carlo simulation, and concludes that using $\tilde{\theta}_S$ in (13.72) generally yields the estimator having the smallest EB risk. However, for $N\geqslant 5$ all three estimators have smaller EB risks than the classical ML estimator.

13.7. ESTIMATION IN THE LOG NORMAL DISTRIBUTION

Padgett and Robinson (1978) derive both Lemon-type smooth and CSEB point estimators of the reliability function for a $\mathcal{LN}(\xi,\sigma^2)$ failure model for two cases: (1) ξ is an unknown value of an r.v. and σ^2 is known, and (2) ξ and σ are both unknown values of r.v.'s. Consider the $\mathcal{LN}(\xi,\sigma^2)$ pdf given by (4.31). Recall that the corresponding reliability function is given by

$$r(t_0)=1-\Phi\{[\ln(t_0)-\xi]/\sigma\}, \qquad t_0,\sigma>0,\ -\infty<\xi<\infty, \tag{13.73}$$

where Φ denotes the $\mathcal{N}(0,1)$ cdf. We further assume that a sequence of $N\geqslant 2$ life test experiments has been conducted in which n_i items are tested until failure in the ith experiment. Although the authors consider Type II censored testing in a supplemental report, only the complete testing case is considered here.

13.7.1 Point Estimation of Reliability With ξ A Random Variable and σ Known

Now suppose that σ is known and fixed and that ξ_i is a value of an r.v. according to a completely unknown and unspecified prior distribution $g(\xi)$. The ML estimate of ξ_i is $\hat{\xi}_i = \Sigma_{j=1}^{n_i} \ln t_{ij}/n_i$, where t_{ij} is the jth observed ordered failure time in the ith experiment. In the Nth or current experiment it is assumed that the sequence of data given by $(\hat{\xi}_1, n_1), (\hat{\xi}_2, n_2), \ldots, (\hat{\xi}_N, n_N)$ is available for estimating ξ_N. Further let $\hat{\xi} = \hat{\xi}_N$, $\xi = \xi_N$, and $n = n_N$. As before, the conditional independence of $\hat{\xi}_i$, $i = 1, \ldots, N$ is assumed. It follows that $\hat{\xi}_i$ given ξ_i has a $\mathfrak{N}(\xi_i, \sigma^2/n_i)$ distribution.

The noniterated smooth EB estimator of $r(t_0)$, according to (13.13), is given by

$$\tilde{r}_{L,1}(t_0) = \frac{\sum_{j=1}^{N} \Phi\{[\hat{\xi}_j - \ln(t_0)]/\sigma\} \exp\left[-n(\hat{\xi} - \hat{\xi}_j)^2/(2\sigma^2)\right]}{\sum_{j=1}^{N} \exp\left[-n(\hat{\xi} - \hat{\xi}_j)^2/(2\sigma^2)\right]}, \qquad t > 0. \quad (13.74)$$

Using $\mathfrak{N}(0,1)$ weighting functions in (13.14), the CSEB estimator of ξ, upon integration and simplification, becomes

$$\tilde{r}_{C,1}(t_0) = \frac{\sum_{j=1}^{N} \Phi\left[\frac{\hat{\psi}_j - \ln(t_0)}{\sigma\sqrt{1 + h^2/(\sigma^2 + nh^2)}}\right] \phi\left[(\hat{\xi} - \hat{\xi}_j)/\sigma_S\right]}{\sum_{j=1}^{N} \phi\left[(\hat{\xi} - \hat{\xi}_j)/\sigma_S\right]}, \qquad t > 0, \quad (13.75)$$

where

1. $$\hat{\psi}_j = \left(\hat{\xi} h^2 + \hat{\xi}_j \sigma^2/n\right)/\sigma_S^2,$$

2. $$\sigma_S^2 = \sigma^2/n + h^2,$$

3. $$h^2 = N^{-7/5} \sum_{j=1}^{N} \left(\sigma^2/n_j\right),$$

and where ϕ denotes the $\mathfrak{N}(0,1)$ pdf.

Padgett and Robinson (1978) conclude, based on Monte Carlo simulation, that $\tilde{r}_{L,1}(t_0)$ and $\tilde{r}_{C,1}(t_0)$ usually have a smaller EB risk than the ML estimate and that $\tilde{r}_{C,1}(t_0)$ generally has a smaller EB risk than $\tilde{r}_{L,1}(t_0)$.

13.7.2 Point Estimation of Reliability When Both ξ and σ Are Random Variables

Consider the case where both ξ_i and σ_i are unknown values of r.v.'s generated independently over i according to a completely unknown and unspecified bivariate prior distribution $g(\xi, \sigma)$. Then

$$\hat{\xi}_i = n_i^{-1} \sum_{j=1}^{n_i} \ln t_{ij} \qquad \text{and} \qquad \hat{\sigma}_i^2 = n_i^{-1} \sum_{j=1}^{n_i} \left(\ln t_{ij} - \hat{\xi}_i\right)^2$$

are the ML estimates of ξ_i and σ_i. In the Nth or current experiment it is assumed that the sequence of data given by $(\hat{\xi}_1, \hat{\sigma}_1, n_1)$, $(\hat{\xi}_2, \hat{\sigma}_2, n_2)$, $\ldots, (\hat{\xi}_N, \hat{\sigma}_N, n_N)$ is available for estimating both ξ_i and σ_i. Let $\hat{\xi} = \hat{\xi}_N$ and $\hat{\sigma} = \hat{\sigma}_N$. The usual EB independence assumptions are assumed to hold. Now, given ξ_i and σ_i, $\hat{\xi}_i$ has a $\mathcal{N}(\xi_i, \sigma_i^2/n_i)$ distribution, whereas $n_i\hat{\sigma}_i^2/\sigma_i^2$ has a $\chi^2(n_i - 1)$ distribution.

The noniterated smooth EB estimator of $r(t_0)$ in (13.73) is

$$\tilde{r}_{L,2}(t_0) = \frac{\sum_{j=1}^{N} \Phi\left\{\left[\hat{\xi}_j - \ln(t_0)\right]/\hat{\sigma}_j\right\} \hat{\sigma}_j^{-n} \exp\left\{-n\left[\left(\hat{\xi} - \hat{\xi}_j\right)^2 + \hat{\sigma}^2\right] \Big/ \left(2\hat{\sigma}_j^2\right)\right\}}{\sum_{j=1}^{N} \hat{\sigma}_j^{-n} \exp\left\{-n\left[\left(\hat{\xi} - \hat{\xi}_j\right)^2 + \hat{\sigma}^2\right] \Big/ \left(2\hat{\sigma}_j^2\right)\right\}}. \tag{13.76}$$

Now consider a CSEB estimator of $r(t_0)$. If ξ and σ are both unknown, the CSEB estimator of $r(t_0)$ is complicated and does not seem to perform any better than substituting the current estimate $\hat{\sigma}$ for σ in (13.75). If σ is approximately the same in each experiment, then the average estimate $\bar{\sigma}^2 = \sum_{j=1}^{N} \hat{\sigma}_j^2 / N$ could also be substituted for σ in (13.75). This estimate will be referred to as $\tilde{r}_{C,2}(t_0)$.

Example 13.16. A certain component having a $\mathcal{LN}(\xi, \sigma^2)$ failure time distribution is manufactured by two suppliers. The natural logarithms of a sample of size $n_1 = n_2 = 5$ from each supplier were found to be 2.72, 1.72, 2.85, 2.47, 1.67 from supplier A and 2.52, 2.01, 1.52, 2.21, 2.65 from supplier B, where the time units (1000 h) have been dropped for convenience. It is desired to estimate the reliability for $t_0 = 5.0$ of the component manufactured by supplier B. Although σ is unknown, let us assume that it is approximately the same for both experiments. Thus we will estimate $r(5)$ by means of (13.76) and (13.75) with σ replaced by $\bar{\sigma}$.

Now $\hat{\xi}_1 = 2.29$, $\hat{\xi}_2 = \hat{\xi} = 2.18$, $\hat{\sigma}_1 = 0.498$, $\hat{\sigma}_2 = \hat{\sigma} = 0.400$. Also $\Phi\{[\hat{\xi}_1 - \ln(t_0)]/\hat{\sigma}_1\} = 0.9147$, $\Phi\{[\hat{\xi}_2 - \ln(t_0)]/\hat{\sigma}_2\} = 0.9236$, $\exp\{-n[(\hat{\xi} - \hat{\xi}_1)^2 + \hat{\sigma}^2]/(2\hat{\sigma}_1^2)\} = 0.1764$, and $\exp\{-n[(\hat{\xi} - \hat{\xi}_2)^2 + \hat{\sigma}^2]/(2\hat{\sigma}_2^2)\} = 0.0821$. Upon substituting these values into (13.76), the desired smooth EB estimate of $r(5)$ becomes

$$\tilde{r}_{L,2}(5) = \frac{(0.9147)(0.498)^{-5}(0.1764) + (0.9236)(0.400)^{-5}(0.0821)}{(0.498)^{-5}(0.1764) + (0.400)^{-5}(0.0821)} = 0.92.$$

Also $\bar{\sigma} = 0.452$, $h^2 = 0.0310$, $\sigma_S^2 = 0.0719$, $\hat{\psi}_1 = 2.2413$, $\hat{\psi}_2 = 2.1788$, $(\hat{\xi} - \hat{\xi}_1)/\sigma_S = -0.4102$, $(\hat{\xi} - \hat{\xi}_2)/\sigma_S = 0$, and $\bar{\sigma}\sqrt{1 + h^2/(\bar{\sigma}^2 + nh^2)} = 0.4711$. Thus according to (13.75), we have that

$$\tilde{r}_{C,2}(5) = \frac{\Phi\left[\dfrac{2.2413 - \ln(5.0)}{0.4711}\right]\phi(-0.4102) + \Phi\left[\dfrac{2.1788 - \ln(5.0)}{0.4711}\right]\phi(0)}{\phi(-0.4102) + \phi(0)}$$

$$= \frac{(0.9099)(0.3668) + (0.8869)(0.3989)}{(0.3668) + (0.3989)} = 0.90. \qquad \blacksquare$$

One EB assumption that has been used throughout this chapter is the requirement that all past data are from the same underlying prior distribution. One may not be certain that this is always the case. Canavos (1973) studies the effect of changing priors for the Poisson parameter and reliability function and finds that the EB estimators continue to have significantly smaller EB risks than corresponding classical (nonBayes) estimators.

EXERCISES

13.1. Let $p(x|\theta)$ be the geometric distribution $p(x|\theta) = (1-\theta)\theta^x$, $x = 0, 1, \ldots$, $0 < \theta < 1$.

(a) Determine $a(x)$ and $b(x)$ in (13.8) for the geometric distribution.

(b) Find the posterior mean of Θ, given x, according to (13.9), and define $f(x)$ in terms of the appropriate integral.

13.2. Verify the mean and variance of θ_i^* defined in (13.18).

13.3. Suppose that a life test experiment is routinely conducted on successive production lots of a certain device in which the ith life test consists of observing the number of failures x_i in t hours with replacement of failed items. Further suppose that each item in the ith test fails independently with failure rate λ_i and it is desired to estimate the failure rate $\lambda = \lambda_n$ in the current production lot.

(a) Determine an EB point estimator of λ.

(b) Suppose that nine previous production lots have been tested in which 0, 1, 0, 0, 1, 2, 0, 1, 0, failures were observed in $t = 10^5$ h and zero failures were observed in the current life test of the same duration. What is the EB point estimate of λ?

13.4. Write out the prediction equations based on Table 13.3 for estimating the ln (Risk) of the three EB estimators $\hat{\theta}_C$, $\hat{\theta}_L$, and $\hat{\theta}_S$.

13.5. A large company that manufactures electronic components wishes to study the reliability of its product. The company has survival data from six past lots of electronic components. In samples of size 5 the numbers of survivors x_i in these samples are (4, 4, 5, 5, 0, 5). The number of survivors in the current (seventh) lot is $x_7 = x = 5$. It is desired to estimate the probability of surviving the current test.

(a) Compute $\hat{\mu}$ and $\hat{\sigma}^2$, the estimates of the prior mean and variance.

(b) Compute the estimates $\hat{\theta}_C$, $\hat{\theta}_L$, and $\hat{\theta}_S$, and their estimated EB risks. Use the results of Exercise 13.4 to compute the risks.

13.6. Using the data in Exercise 13.5, it is desired to compute EBPIs for the parameter θ in the current lot.

(a) Compute the values of $\tilde{\theta}_i$ according to (13.35).

(b) For the current lot compute the empirical posterior probability function according to (13.37).

(c) Compute an 80% TEBPI for θ_7 by using (13.38).

13.7. Consider the data in Exercise 13.5.

(a) Determine the empirical posterior cdf, corresponding to the seventh lot, according to (13.39).

(b) Graph the empirical posterior cdf in part (a).

(c) Solve (13.41) to obtain a 90% continuously smoothed TEBPI for θ_7.

13.8. Suppose a 1000 h life test has been conducted for each of a sequence of eight production lots of a certain electronic device in which 0, 1, 1, 0, 2, 0, 1, 0 failures were observed, respectively. Compute the EB point estimate of λ_8, the failure rate in the current lot, according to (13.42).

13.9. Consider the data in Exercise 13.8. Compute the smooth EB estimates of λ_8 according to (13.43) and (13.44).

13.10. Consider the data in Exercise 13.8.

(a) Compute the empirical posterior probability function in the current lot according to (13.45).

(b) Compute a 95% UEBPI for λ_8 according to the appropriate statement in (13.46).

13.11. In Example 13.9 verify the computation of $\tilde{\lambda}_2 = 2.85$ per 100 mile-years. Also, verify the values of the estimator in (13.49), that is, $\hat{\lambda}_{L,1} = 1.01$ per 100 mile-years and $\hat{\lambda}_{L,2} = 2.81$ per 100 mile-years.

13.12. Suppose in Example 13.9 that the time between outages has a $\mathscr{W}_1(\lambda_i, 2)$ distribution.

(a) Compute the ML estimates of λ_1 and λ_2; that is $\hat{\lambda}_1$ and $\hat{\lambda}_2$.

(b) Compute the noniterated smooth EB estimates of λ_1 and λ_2 according to (13.50).

13.13. Verify all the computations indicated in Example 13.10.

13.14. Consider the following failure times obtained in a sequence of three production lots of a certain component used in the manufacture of a given product:

Production Lot Number	Failure Times (100 h)
1	25.3, 6.3, 30.0, 10.3, 2.7, 3.7
2	0.8, 7.3, 8.0, 16.6, 1.3, 0.5
3	3.1, 3.4, 11.3, 10.2, 6.0, 6.3

Suppose the third lot is the current lot. A sample of size 6 was chosen from each lot for a complete life test. Assume the EB conditions are reasonably satisfied. We wish to find an EB estimate of the failure rate λ_3, assuming an $\mathscr{E}(\lambda_i)$ distribution for the above data.

(a) Compute the ML estimates $\hat{\lambda}_1$, $\hat{\lambda}_2$, $\hat{\lambda}_3$.

(b) Compute $\bar{\lambda}$, σ_1^2, σ_2^2, and σ^2, needed in (13.52).

(c) Compute δ_1, δ_2, δ_3 and γ_1, γ_2, γ_3, needed in (13.52).

(d) Use the values in (a), (b), and (c) to compute the EB estimate $\hat{\lambda}_C$ according to (13.52).

13.15. Suppose the data in Exercise 13.14 came from a $\mathcal{W}_1(\lambda_i, 2)$ distribution. Compute the EB estimate $\hat{\lambda}_C$ according to (13.52).

13.16. Consider the data in Exercise 13.14. Assume the data follow an $\mathcal{E}(\lambda_i)$ distribution. It is desired to construct a 95% UEBPI for the failure rate in the third (current) production lot.

(a) Compute $\tilde{\lambda}_1$, $\tilde{\lambda}_2$, $\tilde{\lambda}_3$, according to (13.50).

(b) Compute the empirical posterior probability function, according to (13.53), where $\hat{\lambda} = \hat{\lambda}_3$.

(c) Construct a 95% UEBPI for λ_3 according to (13.54).

13.17. A sequence of four life test experiments was conducted on a certain component in which the EB conditions were satisfied. In each experiment 10 components were tested and the test terminated at the fifth failure. Assume that an $\mathcal{E}(\lambda_i)$ distribution is appropriate for the failure data. The total times on test in the four experiments were $t_1 = 21545$, $t_2 = 4586$, $t_3 = 29648$, and $t_4 = 9485$ h, respectively.

(a) Carry the computations as far as possible, without a computer, to construct a continuously smoothed 95% TEBPI of the underlying failure rate in the fourth experiment.

(b) Use a computer, if available, to complete the necessary computations.

13.18. Use the data of Exercise 13.14 to calculate the smooth EB point estimate of the failure rate λ_3, according to (13.58).

(a) Summarize the values of s_j, $\hat{\lambda}_j$, ρ_j, ν_j.

(b) Summarize the values of θ_{ij}, ψ_{ij}, $\phi(\psi_{ij})$.

(c) Compute $\tilde{\lambda}_3$.

13.19. Use the sequence of ML estimators in Exercise 13.17 to compute a CSEB estimate of the reliability of the component in the fourth experiment for $t_0 = 500$ h.

(a) Compute a CSEB estimate of reliability according to (13.66) for $t_0 = 500$ h.

(b) Graph the empirical prior (13.63) and posterior (13.64) pdf's and discuss them.

13.20. Suppose that a sequence of three life test experiments were conducted on similar models of a mechanical device for which a $\mathcal{G}(2, \theta)$ failure model is appropriate. Six devices were tested in each experiment and the total test times are $t_{1,6} = 205$, $t_{2,6} = 180$, $t_{3,6} = 290$, $t_{1,5} = 160$, $t_{2,5} = 160$, $t_{3,5} = 230$ h. Compute an estimate of the shape parameter θ in the third experiment. In doing this summarize the computation of the quantities needed to compute the ML estimate $\hat{\theta}_3$; $\tilde{\theta}_C$ in (13.68); and $\tilde{\theta}_L$ in (13.69).

13.21. Consider a certain device that has a $\mathcal{LN}(\xi, \sigma^2)$ failure time distribution. Two companies manufacture this device. A sample of five devices were obtained from each company and the natural logarithms of the failure times were found to be:

Company 1:	1.86,	1.42,	2.38,	2.54,	1.60
Company 2:	2.48,	2.10,	1.64,	2.36,	2.82

It is desired to estimate the reliability at $t_0 = 10.0$ h. Assume σ is the same for both companies.

(a) Compute the reliability estimate according to (13.76). Summarize the computation of the quantities required to compute the reliability.

(b) Compute the reliability according to (13.75), with σ replaced with $\bar{\sigma}$. Summarize the computation of all the quantities required to compute this reliability.

REFERENCES

Bennett, G. K. (1970). *Smooth Empirical Bayes Estimation with Application to the Weibull Distribution*, Ph.D. Dissertation, Texas Tech University, Lubbock, TX.

Bennett, G. K. (1977). Basic Concepts of Empirical Bayes Methods with Some Results for the Weibull Distribution, *The Theory and Applications of Reliability*, *Vol. II*, Academic, New York, pp. 181–202.

Bennett, G. K. and Martz, H. F. (1972). A Continuous Empirical Bayes Smoothing Technique, *Biometrika*, Vol. 59, pp. 361–368.

Bennett, G. K. and Martz, H. F. (1973). An Empirical Bayes Estimator for the Scale Parameter of the Two-Parameter Weibull Distribution, *Naval Research Logistics Quarterly*, Vol. 20, pp. 387–393.

Blum, J. R. and Susarla, V. (1977). Estimation of a Mixing Distribution Function, *Annals of Probability*, Vol. 5, pp. 200–209.

Canavos, G. C. (1971). *A Bayesian Approach to Parameter and Reliability Estimation in Failure Distributions*, Ph.D. Dissertation, Virginia Polytechnic Institute and State University, Blacksburg, VA.

Canavos, G. C. (1973). An Empirical Bayes Approach for the Poisson Life Distribution, *IEEE Transactions on Reliability*, Vol. R-22, pp. 91–96.

Canavos, G. C. (1975). Bayesian Estimation: A Sensitivity Analysis, *Naval Research Logistics Quarterly*, Vol. 22, pp. 543–552.

Canavos, G. C. (1977). Robustness and the Prior Distribution in the Bayesian Model, *The Theory and Applications of Reliability*, *Vol II*, Academic, New York, pp. 173–180.

Canavos, G. C. and Tsokos, C. P. (1971). A Study of an Ordinary and Empirical Bayes Approach to Reliability Estimation in the Gamma Life Testing Model, *Proceedings 1971 Annual Symposium on Reliability*, pp. 343–349.

Choi, K. and Bulgren, W. G. (1968). An Estimation Procedure for Mixtures of Distributions, *Journal of the Royal Statistical Society, Series B*, Vol. 30, pp. 444–460.

Christensen, R. (1977). *A Ferguson-Antoniak Approach to the Empirical Bayes Estimation of a Binomial Parameter*, Technical Report No. 271, The University of Minnesota, School of Statistics, St. Paul, MN.

Clemmer, B. A. and Krutchkoff, R. G. (1968). The Use of Empirical Bayes Estimators in a Linear Regression Model, *Biometrika*, Vol. 55, pp. 525–534.

Copas, J. B. (1972). Empirical Bayes Methods and the Repeated Use of a Standard, *Biometrika*, Vol. 59, pp. 349–360.

Couture, D. J. (1970). *Some Practical Empirical Bayes Procedures for Use in Weibull Reliability Estimation*, Ph.D. Dissertation, Texas Tech University, Lubbock, TX.

Couture, D. J. and Martz H. F. (1972). Empirical Bayes Estimation in the Weibull Distribution, *IEEE Transactions on Reliability*, Vol. R-21, pp. 75–83.

Cox, D. R. and Hinkley, D. V. (1974). *Theoretical Statistics*, Chapman and Hall, London.

Deely, J. J. and Kruse, R. L. (1968). Construction of Sequences Estimating the Mixing Distribution, *Annals of Mathematical Statistics*, Vol. 39, pp. 268–288.

Gaffey, W. R. (1959). A Consistent Estimator of a Component of a Convolution, *Annals of Mathematical Statistics*, Vol. 30, pp. 198–205.

George, S. L. (1971). Evaluation of Empirical Bayes Estimators for Small Numbers of Past Samples, *Biometrika*, Vol. 58, p. 244.

Good, I. J. (1953). The Population Frequencies of Species and the Estimation of Population Parameters, *Biometrika*, Vol. 40, pp. 237–264.

Griffin, B. S. and Krutchkoff, R. G. (1971). Optimal Linear Estimators: An Empirical Bayes Version with Application to the Binomial Distribution, *Biometrika*, Vol. 58, pp. 195–201.

Higgins, J. J. and Tsokos, C. P. (1976). On the Behavior of Some Quantities Used in Bayesian Reliability Demonstration Tests, *IEEE Transactions on Reliability*, Vol. R-25, pp. 261–264.

Higgins, J. J. and Tsokos, C. P. (1977). Comparison of Bayesian Estimates of Failure Intensity for Fitted Priors of Life Data, *The Theory and Applications of Reliability, Vol. II*, Academic, New York, pp. 75–92.

Lemon, G. H. (1972). An Empirical Bayes Approach to Reliability, *IEEE Transactions on Reliability*, Vol. R-21, pp. 155–158.

Lemon, G. H. and Krutchkoff, R. G. (1969). An Empirical Bayes Smoothing Technique, *Biometrika*, Vol. 56, pp. 361–365.

Lenth, R. V. (1976). *Empirical Bayes Procedures Using Generalized Splines*, Technical Report No. 49, Department of Statistics, University of Iowa, Iowa City, IA.

Lian, M. G. (1975). *Bayes and Empirical Bayes Estimation of Reliability for the Weibull Model*, Ph.D. Dissertation, Texas Tech University, Lubbock, TX.

Lin, P. E. (1972). Rates of Convergence in Empirical Bayes Estimation Problems. Discrete Case, *Annals of the Institute of Statistical Mathematics*, Vol. 24, pp. 319–325.

Mann, N. R., Schafer, R. E., and Singpurwalla, N. D. (1974). *Methods for Statistical Analysis of Reliability and Life Data*, Wiley, New York.

Maritz, J. S. (1966). Smooth Empirical Bayes Estimation for One-Parameter Discrete Distributions, *Biometrika*, Vol. 53, pp. 417–429.

Maritz, J. S. (1967). Smooth Empirical Bayes Estimation for Continuous Distributions, *Biometrika*, Vol. 54, pp. 435–450.

Maritz, J. S. (1968). On the Smooth Empirical Bayes Approach to Testing of Hypotheses and the Compound Decision Problem, *Biometrika*, Vol. 55, pp. 83–100.

Maritz, J. S. (1970). *Empirical Bayes Methods*, Methuen, London.

Martz, H. F. (1975a). Pooling Life-Test Data by Means of the Empirical Bayes Method, *IEEE Transactions on Reliability*, Vol. R-24, pp. 27–30.

Martz, H. F. (1975b). Empirical Bayes Single Sampling Plans for Specified Posterior Consumer and Producer Risks, *Naval Research Logistics Quarterly*, Vol. 22, pp. 651–665.

Martz, H. F. and Krutchkoff, R. G. (1969). Empirical Bayes Estimators in a Multiple Linear Regression Model, *Biometrika*, Vol. 56, pp. 367–374.

Martz, H. F. and Lian, M. G. (1974). Empirical Bayes Estimation of the Binomial Parameter, *Biometrika*, Vol. 61, pp. 517–523.

Martz, H. F. and Lian, M. G. (1977). Bayes and Empirical Bayes Point and Interval Estimation of Reliability for the Weibull Model, *The Theory and Applications of Reliability, Vol. II*, Academic, New York, pp. 203–233.

Meeden, G. (1972). Bayes Estimation of the Mixing Distribution, the Discrete Case, *Annals of Mathematical Statistics*, Vol. 43, pp. 1993–1999.

Miyawasa, K. (1961). An Empirical Bayes Estimator of the Mean of a Normal Population, *Bulletin of the International Statistical Institute*, Vol. 39, pp. 181–188.

Nichols, W. G. and Tsokos, C. P. (1972). Empirical Bayes Point Estimation in a Family of Probability Distributions, *International Statistical Review*, Vol. 40, pp. 146–161.

O'Bryan, T. E. (1976). Some Empirical Bayes Results in the Case of Component Problems with Varying Sample Sizes for Discrete Exponential Families, *Annals of Statistics*, Vol. 4, pp. 1290–1293.

O'Bryan, T. E. (1979). Rates of Convergence in a Modified Empirical Bayes Estimation Problem Involving Poisson Distributions, *Communications in Statistics—Theory and Methods*, Vol. A8, pp. 167–174.

O'Bryan, T. E. and Susarla, V. (1977). Empirical Bayes Estimation with Non-Identical Components, Continuous Case. *The Australian Journal of Statistics*, Vol. 19, pp. 115–125.

Padgett, W. J. and Robinson, J. A. (1978). Empirical Bayes Estimators of Reliability for Lognormal Failure Model, *IEEE Transactions on Reliability*, Vol. R-27, pp. 332–336.

Parzen, E. (1962). On Estimation of a Probability Density Function and Mode, *Annals of Mathematical Statistics*, Vol. 33, pp. 1065–1076.

Robbins, H. (1951). Asymptotically Subminimax Solutions of Compound Statistical Decision Problems, *Proceedings Second Berkeley Symposium on Mathematical Statistics and Probability, Vol. 1*, University of California Press, pp. 131–148.

Robbins, H. (1955). An Empirical Bayes Approach to Statistics, *Proceedings Third Berkeley Symposium on Mathematical Statistics and Probability, Vol.* 1, University of California Press, pp. 157–163.

Robbins, H. (1964). An Empirical Bayes Approach to Statistical Decision Problems, *Annals of Mathematical Statistics*, Vol. 35, pp. 1–19.

Rolph, J. E. (1968). Bayesian Estimation of Mixing Distributions, *Annals of Mathematical Statistics*, Vol. 39, pp. 1289–1302.

Rutherford, J. R. and Krutchkoff, R. G. (1969). Some Empirical Bayes Techniques in Point Estimation, *Biometrika*, Vol. 5, pp. 133–137.

Tortella, M. and O'Bryan, T. (1979). Estimation of the Prior Distribution by Best Approximation in Uniformly Convex Function Spaces, *Bulletin of the Institute of Mathematics, Acedemia Sinica*, Vol. 7.

Von Mises, R. (1942). On the Correct Use of Bayes' Formula, *Annals of Mathematical Statistics*, Vol. 13, pp. 156–165.

APPENDIX A

Summary of Common Distributions

Table A1. Some Common Distributions

Distribution	Denotation	Probability Distribution
Binomial	—	$\binom{n}{x}p^x(1-p)^{n-x}, n=1,2,\ldots,$ $0<p<1, x=0,1,\ldots,n$
Poisson	—	$\lambda^x e^{-\lambda}/x!, x=0,1,\ldots,\lambda>0$
Negative Binomial	—	$\binom{r+x-1}{r-1}p^r(1-p)^x, x=0,1,\ldots,$ $r>0,\quad 0<p<1$
Normal	$\mathfrak{N}(\mu,\sigma^2)$	$\frac{1}{\sigma\sqrt{2\pi}}\exp\left[-\frac{1}{2\sigma^2}(x-\mu)^2\right],$ $-\infty<x<\infty,$ $-\infty<\mu<\infty,\quad \sigma>0$
Standard Normal	$\mathfrak{N}(0,1)$	$\frac{1}{\sqrt{2\pi}}\exp\left(-\frac{x^2}{2}\right),\quad -\infty<x<\infty$
t	$t(k)$	$\frac{[(k-1)/2]!}{\sqrt{n\pi}\,[(k-2)/2]!}(1+x^2/k)^{-(k+1)/2}$ $-\infty<x<\infty,\quad k=1,2,\ldots,$
Chi-Square	$\chi^2(n)$	$\frac{1}{\Gamma(n/2)2^{n/2}}x^{n/2-1}e^{-x/2},\quad x>0,$ $n=1,2,\ldots,$

Table A1. (*Continued*)

Distribution	Denotation	Probability Distribution
$\mathcal{F}$	$\mathcal{F}(n_1, n_2)$	$\dfrac{\Gamma[(n_1+n_2)/2]}{\Gamma(n_1/2)\Gamma(n_2/2)}(n_1/n_2)^{n_1/2} \times x^{n_1/2-1}(1+n_1 x/n_2)^{-(n_1+n_2)/2}$, $x>0, \quad n_1, n_2 = 1,2,\ldots,$
Two-Parameter Exponential	$\mathcal{E}(\lambda, \mu)$	$\lambda e^{-\lambda(x-\mu)}, \quad x>\mu \geqslant 0, \lambda>0$
Exponential	$\mathcal{E}(\lambda)$	$\lambda e^{-\lambda x}, \quad x>0, \lambda>0$
Exponential (Alternate Form)	$\mathcal{E}_1(\theta)$	$\dfrac{1}{\theta} e^{-x/\theta}, x>0, \theta>0$
Two-Parameter Exponential (Alternate Form)	$\mathcal{E}_1(\theta, \mu)$	$\dfrac{1}{\theta} e^{-(x-\mu)/\theta}, x>\mu \geqslant 0, \theta>0$
Weibull	$\mathcal{W}(\alpha, \beta, \theta)$	$\dfrac{\beta}{\alpha}\left(\dfrac{x-\theta}{\alpha}\right)^{\beta-1} \exp\left[-\left(\dfrac{x-\theta}{\alpha}\right)^{\beta}\right]$, $x>\theta \geqslant 0, \quad \alpha, \beta>0$
Two-Parameter Weibull	$\mathcal{W}(\alpha, \beta)$	$\dfrac{\beta}{\alpha}\left(\dfrac{x}{\alpha}\right)^{\beta-1} \exp\left[-\left(\dfrac{x}{\alpha}\right)^{\beta}\right], x>0,$ $\alpha, \beta>0$
Two-Parameter Weibull (Alternate Form 1)	$\mathcal{W}_1(\lambda, \beta)$	$\lambda \beta x^{\beta-1} \exp(-\lambda x^{\beta}), x>0, \quad \lambda, \beta>0$
Two-Parameter Weibull (Alternate Form 2)	$\mathcal{W}_2(\theta, \beta)$	$\dfrac{\beta}{\theta} x^{\beta-1} \exp(-x^{\beta}/\theta), x>0, \quad \theta, \beta>0$
Log Normal	$\mathcal{LN}(\xi, \sigma^2)$	$\dfrac{1}{x\sigma\sqrt{2\pi}} \exp\left[-\dfrac{1}{2\sigma^2}(\ln x-\xi)^2\right], x>0,$ $-\infty<\xi<\infty, \quad \sigma>0$
Uniform	$\mathcal{U}(\theta_1, \theta_2)$	$\dfrac{1}{\theta_2-\theta_1}, \theta_1<x<\theta_2$
Inverse Gaussian	$\mathcal{IN}(\mu, \lambda)$	$\left(\dfrac{\lambda}{2\pi x^3}\right)^{1/2} \exp\left[-\dfrac{\lambda}{2\mu^2 x}(x-\mu)^2\right]$, $x>0, \quad \mu, \lambda>0$
Inverse Gaussian (Alternate Form)	$\mathcal{IN}_1(\psi, \lambda)$	$\left(\dfrac{\lambda}{2\pi x^3}\right)^{1/2} \exp\left[-\dfrac{\lambda x}{2}\left(\psi-\dfrac{1}{x}\right)^2\right]$, $x>0, \quad \psi, \lambda>0$

Table A1. (*Continued*)

Distribution	Denotation	Probability Distribution
Gamma	$\mathcal{G}(\alpha,\beta)$	$\frac{1}{\beta^{\alpha}\Gamma(\alpha)}x^{\alpha-1}\exp\left(-\frac{x}{\beta}\right), x>0,$ $\alpha,\beta>0$
Gamma (Alternate Form)	$\mathcal{G}_1(\alpha,\theta)$	$\frac{\theta^{\alpha}}{\Gamma(\alpha)}x^{\alpha-1}e^{-\theta x}, x>0, \quad \alpha,\theta>0$
Inverted Gamma	$\mathcal{IG}(\alpha,\theta)$	$\frac{\theta^{\alpha}}{\Gamma(\alpha)}\left(\frac{1}{x}\right)^{\alpha+1}e^{-\theta/x}, x>0, \alpha,\theta>0$
Negative-Log Gamma	$\mathcal{NLG}(\alpha,\theta)$	$\frac{\theta^{\alpha}}{\Gamma(\alpha)}x^{\theta-1}(-\ln x)^{\alpha-1}, 0<x<1,$ $\alpha,\theta>0$
Negative-Log Gamma (Alternate Form)	$\mathcal{NLG}_1(\alpha,\beta)$	$\frac{(\beta+1)^{\alpha+1}}{\Gamma(\alpha+1)}x^{\beta}(-\ln x)^{\alpha}, 0<x<1,$ $\alpha,\beta>-1$
Beta	$\mathcal{B}(x_0,n_0)$	$\frac{1}{B(x_0,n_0-x_0)}x^{x_0-1}(1-x)^{n_0-x_0-1},$ $0<x<1, n_0>x_0>0$
Beta (Alternate Form)	$\mathcal{B}_1(p,q)$	$\frac{1}{B(p,q)}x^{p-1}(1-x)^{q-1}, 0<x<1,$ $p,q>0$

Table A2. The Mean and Variance of Some Common Distributions

Distribution[a]	Mean	Variance
Binomial	np	$np(1-p)$
Poisson	λ	λ
Negative Binomial	$r(1-p)/p$	$r(1-p)/p^2$
$\mathcal{N}(\mu,\sigma^2)$	μ	σ^2
$\mathcal{N}(0,1)$	0	1
$t(k)$	0	$k/(k-2), k>2$
$\chi^2(n)$	n	$2n$

Table A2. (*Continued*)

Distribution	Mean	Variance
$\mathcal{F}(n_1, n_2)$	$n_2/(n_2-2), n_2>2$	$\dfrac{2n_2^2(n_1+n_2-2)}{n_1(n_2-2)^2(n_2-4)}, \quad n_2>4$
$\mathcal{E}(\lambda, \mu)$	$\mu+1/\lambda$	$1/\lambda^2$
$\mathcal{E}(\lambda)$	$1/\lambda$	$1/\lambda^2$
$\mathcal{E}_1(\theta)$	θ	θ^2
$\mathcal{E}_1(\theta, \mu)$	$\mu+\theta$	θ^2
$\mathcal{W}(\alpha, \beta, \theta)$	$\theta+\alpha\Gamma\left(\dfrac{\beta+1}{\beta}\right)$	$\alpha^2\left[\Gamma\left(\dfrac{\beta+2}{\beta}\right)-\Gamma^2\left(\dfrac{\beta+1}{\beta}\right)\right]$
$\mathcal{W}(\alpha, \beta)$	$\alpha\Gamma(1+1/\beta)$	$\alpha^2[\Gamma(1+2/\beta)-\Gamma^2(1+1/\beta)]$
$\mathcal{W}_1(\lambda, \beta)$	$\lambda^{-1/\beta}\Gamma(1+1/\beta)$	$\lambda^{-2/\beta}[\Gamma(1+2/\beta)-\Gamma^2(1+1/\beta)]$
$\mathcal{W}_2(\theta, \beta)$	$\theta^{1/\beta}\Gamma(1+1/\beta)$	$\theta^{2/\beta}[\Gamma(1+2/\beta)-\Gamma^2(1+1/\beta)]$
$\mathcal{LN}(\xi, \sigma^2)$	$\exp(\xi+\sigma^2/2)$	$\exp(2\xi+2\sigma^2)-\exp(2\xi+\sigma^2)$
$\mathcal{U}(\theta_1, \theta_2)$	$(\theta_1+\theta_2)/2$	$(\theta_2-\theta_1)^2/12$
$\mathcal{IN}(\mu, \lambda)$	μ	μ^3/λ
$\mathcal{IN}_1(\psi, \lambda)$	$1/\psi$	$1/(\psi^3\lambda)$
$\mathcal{G}(\alpha, \beta)$	$\alpha\beta$	$\alpha\beta^2$
$\mathcal{G}_1(\alpha, \theta)$	α/θ	α/θ^2
$\mathcal{IG}(\alpha, \theta)$	$\dfrac{\theta}{\alpha-1}, \quad \alpha>1$	$\dfrac{\theta^2}{(\alpha-1)^2(\alpha-2)}, \quad \alpha>2$
$\mathcal{NLG}(\alpha, \theta)$	$(1+1/\theta)^{-\alpha}$	$(1+2/\theta)^{-\alpha}-(1+1/\theta)^{-2\alpha}$
$\mathcal{NLG}_1(\alpha, \beta)$	$\left(\dfrac{1+\beta}{2+\beta}\right)^{\alpha+1}$	$\left(\dfrac{1+\beta}{3+\beta}\right)^{\alpha+1}-\left(\dfrac{1+\beta}{2+\beta}\right)^{2\alpha+2}$
$\mathcal{B}(x_0, n_0)$	$\dfrac{x_0}{n_0}$	$\dfrac{x_0(n_0-x_0)}{n_0^2(n_0+1)}$
$\mathcal{B}_1(p, q)$	$\dfrac{p}{p+q}$	$\dfrac{pq}{(p+q)^2(p+q+1)}$

[a]See Table A1 for the associated name of the distribution

Table A3. Notation for Some Common Distributions

Distribution	Random Variable	100γth Percentile	Cumulative Distribution Function
$\mathcal{N}(0,1)$	Z	z_γ	$\Phi(x)$
$t(k)$	t	$t_\gamma(k)$	$t(x;k)$
$\chi^2(n)$	χ^2	$\chi^2_\gamma(n)$	$\chi^2(x;n)$
$\mathcal{F}(n_1,n_2)$	$\mathcal{F}$	$\mathcal{F}_\gamma(n_1,n_2)$	$\mathcal{F}(x;n_1,n_2)$

Examples: $\Pr(Z \leqslant z_\gamma) = \Phi(z_\gamma) = \gamma$

$\Pr[t \leqslant t_\gamma(k)] = t[t_\gamma(k);k] = \gamma$

$\Pr[\chi^2 \leqslant \chi^2_\gamma(n)] = \chi^2[\chi^2_\gamma(n);n] = \gamma$

$\Pr[\mathcal{F} \leqslant \mathcal{F}_\gamma(n_1,n_2)] = \mathcal{F}[\mathcal{F}_\gamma(n_1,n_2);n_1,n_2] = \gamma$

APPENDIX B

Statistical Tables

Table B1. Standard Normal Cumulative Distribution Function $\Phi(x)$

x	Φ(x)	x	Φ(x)	x	Φ(x)	x	Φ(x)	x	Φ(x)
-3.00	.0013	-2.50	.0062	-2.00	.0228	-1.50	.0668	-1.00	.1587
-2.99	.0014	-2.49	.0064	-1.99	.0233	-1.49	.0681	-.99	.1611
-2.98	.0014	-2.48	.0066	-1.98	.0239	-1.48	.0694	-.98	.1635
-2.97	.0015	-2.47	.0068	-1.97	.0244	-1.47	.0708	-.97	.1660
-2.96	.0015	-2.46	.0069	-1.96	.0250	-1.46	.0721	-.96	.1685
-2.95	.0016	-2.45	.0071	-1.95	.0256	-1.45	.0735	-.95	.1711
-2.94	.0016	-2.44	.0073	-1.94	.0262	-1.44	.0749	-.94	.1736
-2.93	.0017	-2.43	.0075	-1.93	.0268	-1.43	.0764	-.93	.1762
-2.92	.0018	-2.42	.0078	-1.92	.0274	-1.42	.0778	-.92	.1788
-2.91	.0018	-2.41	.0080	-1.91	.0281	-1.41	.0793	-.91	.1814
-2.90	.0019	-2.40	.0082	-1.90	.0287	-1.40	.0808	-.90	.1841
-2.89	.0019	-2.39	.0084	-1.89	.0294	-1.39	.0823	-.89	.1867
-2.88	.0020	-2.38	.0087	-1.88	.0301	-1.38	.0838	-.88	.1894
-2.87	.0021	-2.37	.0089	-1.87	.0307	-1.37	.0853	-.87	.1922
-2.86	.0021	-2.36	.0091	-1.86	.0314	-1.36	.0869	-.86	.1949
-2.85	.0022	-2.35	.0094	-1.85	.0322	-1.35	.0885	-.85	.1977
-2.84	.0023	-2.34	.0096	-1.84	.0329	-1.34	.0901	-.84	.2005
-2.83	.0023	-2.33	.0099	-1.83	.0336	-1.33	.0918	-.83	.2033
-2.82	.0024	-2.32	.0102	-1.82	.0344	-1.32	.0934	-.82	.2061
-2.81	.0025	-2.31	.0104	-1.81	.0351	-1.31	.0951	-.81	.2090
-2.80	.0026	-2.30	.0107	-1.80	.0359	-1.30	.0968	-.80	.2119
-2.79	.0026	-2.29	.0110	-1.79	.0367	-1.29	.0985	-.79	.2148
-2.78	.0027	-2.28	.0113	-1.78	.0375	-1.28	.1003	-.78	.2177
-2.77	.0028	-2.27	.0116	-1.77	.0384	-1.27	.1020	-.77	.2206
-2.76	.0029	-2.26	.0119	-1.76	.0392	-1.26	.1038	-.76	.2236
-2.75	.0030	-2.25	.0122	-1.75	.0401	-1.25	.1056	-.75	.2266
-2.74	.0031	-2.24	.0125	-1.74	.0409	-1.24	.1075	-.74	.2296
-2.73	.0032	-2.23	.0129	-1.73	.0418	-1.23	.1093	-.73	.2327
-2.72	.0033	-2.22	.0132	-1.72	.0427	-1.22	.1112	-.72	.2358
-2.71	.0034	-2.21	.0136	-1.71	.0436	-1.21	.1131	-.71	.2389
-2.70	.0035	-2.20	.0139	-1.70	.0446	-1.20	.1151	-.70	.2420
-2.69	.0036	-2.19	.0143	-1.69	.0455	-1.19	.1170	-.69	.2451
-2.68	.0037	-2.18	.0146	-1.68	.0465	-1.18	.1190	-.68	.2483
-2.67	.0038	-2.17	.0150	-1.67	.0475	-1.17	.1210	-.67	.2514
-2.66	.0039	-2.16	.0154	-1.66	.0485	-1.16	.1230	-.66	.2546
-2.65	.0040	-2.15	.0158	-1.65	.0495	-1.15	.1251	-.65	.2578
-2.64	.0041	-2.14	.0162	-1.64	.0505	-1.14	.1271	-.64	.2611
-2.63	.0043	-2.13	.0166	-1.63	.0516	-1.13	.1292	-.63	.2643
-2.62	.0044	-2.12	.0170	-1.62	.0526	-1.12	.1314	-.62	.2676
-2.61	.0045	-2.11	.0174	-1.61	.0537	-1.11	.1335	-.61	.2709
-2.60	.0047	-2.10	.0179	-1.60	.0548	-1.10	.1357	-.60	.2743
-2.59	.0048	-2.09	.0183	-1.59	.0559	-1.09	.1379	-.59	.2776
-2.58	.0049	-2.08	.0188	-1.58	.0571	-1.08	.1401	-.58	.2810
-2.57	.0051	-2.07	.0192	-1.57	.0582	-1.07	.1423	-.57	.2843
-2.56	.0052	-2.06	.0197	-1.56	.0594	-1.06	.1446	-.56	.2877
-2.55	.0054	-2.05	.0202	-1.55	.0606	-1.05	.1469	-.55	.2912
-2.54	.0055	-2.04	.0207	-1.54	.0618	-1.04	.1492	-.54	.2946
-2.53	.0057	-2.03	.0212	-1.53	.0630	-1.03	.1515	-.53	.2981
-2.52	.0059	-2.02	.0217	-1.52	.0643	-1.02	.1539	-.52	.3015
-2.51	.0060	-2.01	.0222	-1.51	.0655	-1.01	.1562	-.51	.3050

Table B1. (*Continued*)

x	Φ(x)	x	Φ(x)	x	Φ(x)	x	Φ(x)	x	Φ(x)
-.50	.3085	0.00	.5000	.50	.6915	1.00	.8413	1.50	.9332
-.49	.3121	.01	.5040	.51	.6950	1.01	.8438	1.51	.9345
-.48	.3156	.02	.5080	.52	.6985	1.02	.8461	1.52	.9357
-.47	.3192	.03	.5120	.53	.7019	1.03	.8485	1.53	.9370
-.46	.3228	.04	.5160	.54	.7054	1.04	.8508	1.54	.9382
-.45	.3264	.05	.5199	.55	.7088	1.05	.8531	1.55	.9394
-.44	.3300	.06	.5239	.56	.7123	1.06	.8554	1.56	.9406
-.43	.3336	.07	.5279	.57	.7157	1.07	.8577	1.57	.9418
-.42	.3372	.08	.5319	.58	.7190	1.08	.8599	1.58	.9429
-.41	.3409	.09	.5359	.59	.7224	1.09	.8621	1.59	.9441
-.40	.3446	.10	.5398	.60	.7257	1.10	.8643	1.60	.9452
-.39	.3483	.11	.5438	.61	.7291	1.11	.8665	1.61	.9463
-.38	.3520	.12	.5478	.62	.7324	1.12	.8686	1.62	.9474
-.37	.3557	.13	.5517	.63	.7357	1.13	.8708	1.63	.9484
-.36	.3594	.14	.5557	.64	.7389	1.14	.8729	1.64	.9495
-.35	.3632	.15	.5596	.65	.7422	1.15	.8749	1.65	.9505
-.34	.3669	.16	.5636	.66	.7454	1.16	.8770	1.66	.9515
-.33	.3707	.17	.5675	.67	.7486	1.17	.8790	1.67	.9525
-.32	.3745	.18	.5714	.68	.7517	1.18	.8810	1.68	.9535
-.31	.3783	.19	.5753	.69	.7549	1.19	.8830	1.69	.9545
-.30	.3821	.20	.5793	.70	.7580	1.20	.8849	1.70	.9554
-.29	.3859	.21	.5832	.71	.7611	1.21	.8869	1.71	.9564
-.28	.3897	.22	.5871	.72	.7642	1.22	.8888	1.72	.9573
-.27	.3936	.23	.5910	.73	.7673	1.23	.8907	1.73	.9582
-.26	.3974	.24	.5948	.74	.7704	1.24	.8925	1.74	.9591
-.25	.4013	.25	.5987	.75	.7734	1.25	.8944	1.75	.9599
-.24	.4052	.26	.6026	.76	.7764	1.26	.8962	1.76	.9608
-.23	.4090	.27	.6064	.77	.7794	1.27	.8980	1.77	.9616
-.22	.4129	.28	.6103	.78	.7823	1.28	.8997	1.78	.9625
-.21	.4168	.29	.6141	.79	.7852	1.29	.9015	1.79	.9633
-.20	.4207	.30	.6179	.80	.7881	1.30	.9032	1.80	.9641
-.19	.4247	.31	.6217	.81	.7910	1.31	.9049	1.81	.9649
-.18	.4286	.32	.6255	.82	.7939	1.32	.9066	1.82	.9656
-.17	.4325	.33	.6293	.83	.7967	1.33	.9082	1.83	.9664
-.16	.4364	.34	.6331	.84	.7995	1.34	.9099	1.84	.9671
-.15	.4404	.35	.6368	.85	.8023	1.35	.9115	1.85	.9678
-.14	.4443	.36	.6406	.86	.8051	1.36	.9131	1.86	.9686
-.13	.4483	.37	.6443	.87	.8078	1.37	.9147	1.87	.9693
-.12	.4522	.38	.6480	.88	.8106	1.38	.9162	1.88	.9699
-.11	.4562	.39	.6517	.89	.8133	1.39	.9177	1.89	.9706
-.10	.4602	.40	.6554	.90	.8159	1.40	.9192	1.90	.9713
-.09	.4641	.41	.6591	.91	.8186	1.41	.9207	1.91	.9719
-.08	.4681	.42	.6628	.92	.8212	1.42	.9222	1.92	.9726
-.07	.4721	.43	.6664	.93	.8238	1.43	.9236	1.93	.9732
-.06	.4761	.44	.6700	.94	.8264	1.44	.9251	1.94	.9738
-.05	.4801	.45	.6736	.95	.8289	1.45	.9265	1.95	.9744
-.04	.4840	.46	.6772	.96	.8315	1.46	.9279	1.96	.9750
-.03	.4880	.47	.6808	.97	.8340	1.47	.9292	1.97	.9756
-.02	.4920	.48	.6844	.98	.8365	1.48	.9306	1.98	.9761
-.01	.4960	.49	.6879	.99	.8389	1.49	.9319	1.99	.9767

Table B1. (*Continued*)

x	Φ(x)	x	Φ(x)	x	Φ(x)	x	Φ(x)	x	Φ(x)
2.00	.9772	2.20	.9861	2.40	.9918	2.60	.9953	2.80	.9974
2.01	.9778	2.21	.9864	2.41	.9920	2.61	.9955	2.81	.9975
2.02	.9783	2.22	.9868	2.42	.9922	2.62	.9956	2.82	.9976
2.03	.9788	2.23	.9871	2.43	.9925	2.63	.9957	2.83	.9977
2.04	.9793	2.24	.9875	2.44	.9927	2.64	.9959	2.84	.9977
2.05	.9798	2.25	.9878	2.45	.9929	2.65	.9960	2.85	.9978
2.06	.9803	2.26	.9881	2.46	.9931	2.66	.9961	2.86	.9979
2.07	.9808	2.27	.9884	2.47	.9932	2.67	.9962	2.87	.9979
2.08	.9812	2.28	.9887	2.48	.9934	2.68	.9963	2.88	.9980
2.09	.9817	2.29	.9890	2.49	.9936	2.69	.9964	2.89	.9981
2.10	.9821	2.30	.9893	2.50	.9938	2.70	.9965	2.90	.9981
2.11	.9826	2.31	.9896	2.51	.9940	2.71	.9966	2.91	.9982
2.12	.9830	2.32	.9898	2.52	.9941	2.72	.9967	2.92	.9982
2.13	.9834	2.33	.9901	2.53	.9943	2.73	.9968	2.93	.9983
2.14	.9838	2.34	.9904	2.54	.9945	2.74	.9969	2.94	.9984
2.15	.9842	2.35	.9906	2.55	.9946	2.75	.9970	2.95	.9984
2.16	.9846	2.36	.9909	2.56	.9948	2.76	.9971	2.96	.9985
2.17	.9850	2.37	.9911	2.57	.9949	2.77	.9972	2.97	.9985
2.18	.9854	2.38	.9913	2.58	.9951	2.78	.9973	2.98	.9986
2.19	.9857	2.39	.9916	2.59	.9952	2.79	.9974	2.99	.9986
								3.00	.9987

Example: $\Pr\{Z \leq 2.00\} = \Phi(2.00) = 0.9772$

$z_{0.90} = 1.282$, $z_{0.975} = 1.960$, $z_{0.995} = 2.575$

$z_{0.95} = 1.645$, $z_{0.99} = 2.327$

Table B2. Selected Percentiles $t_\gamma(k)$ of the $t(k)$ Distribution[a]

k	γ = .60	.70	.80	.90	.95	.975	.99	.995	.999
1	.325	.727	1.376	3.078	6.314	12.706	31.821	63.657	
2	.289	.617	1.061	1.886	2.920	4.303	6.965	9.925	22.327
3	.277	.584	.978	1.638	2.353	3.182	4.541	5.841	10.215
4	.271	.569	.941	1.533	2.132	2.776	3.747	4.604	7.173
5	.267	.559	.920	1.476	2.015	2.571	3.365	4.032	5.893
6	.265	.553	.906	1.440	1.943	2.447	3.143	3.707	5.208
7	.263	.549	.896	1.415	1.895	2.365	2.998	3.499	4.785
8	.262	.546	.889	1.397	1.860	2.306	2.896	3.355	4.501
9	.261	.543	.883	1.383	1.833	2.262	2.821	3.250	4.297
10	.260	.542	.879	1.372	1.812	2.228	2.764	3.169	4.144
11	.260	.540	.876	1.363	1.796	2.201	2.718	3.106	4.025
12	.259	.539	.873	1.356	1.782	2.179	2.681	3.055	3.930
13	.259	.538	.870	1.350	1.771	2.160	2.650	3.012	3.852
14	.258	.537	.868	1.345	1.761	2.145	2.625	2.977	3.787
15	.258	.536	.866	1.341	1.753	2.131	2.602	2.947	3.733
16	.258	.535	.865	1.337	1.746	2.120	2.584	2.921	3.686
17	.257	.534	.863	1.333	1.740	2.110	2.567	2.898	3.646
18	.257	.534	.862	1.330	1.734	2.101	2.552	2.878	3.610
19	.257	.533	.861	1.328	1.729	2.093	2.539	2.861	3.579
20	.257	.533	.860	1.325	1.725	2.086	2.528	2.845	3.552
21	.257	.532	.859	1.323	1.721	2.080	2.518	2.831	3.527
22	.256	.532	.858	1.321	1.717	2.074	2.508	2.819	3.505
23	.256	.532	.858	1.319	1.714	2.069	2.500	2.807	3.485
24	.256	.531	.857	1.318	1.711	2.064	2.492	2.797	3.467
25	.256	.531	.856	1.316	1.708	2.060	2.485	2.787	3.450
26	.256	.531	.856	1.315	1.706	2.056	2.479	2.779	3.435
27	.256	.531	.855	1.314	1.703	2.052	2.473	2.771	3.421
28	.256	.530	.855	1.313	1.701	2.048	2.467	2.763	3.408
29	.256	.530	.854	1.311	1.699	2.045	2.462	2.756	3.396
30	.256	.530	.854	1.310	1.697	2.042	2.457	2.750	3.385
40	.255	.529	.851	1.303	1.684	2.021	2.423	2.704	3.307
50	.255	.528	.849	1.299	1.676	2.009	2.403	2.678	3.261
60	.254	.527	.848	1.296	1.671	2.000	2.390	2.660	3.232
70	.254	.527	.847	1.294	1.667	1.994	2.381	2.648	3.211
80	.254	.527	.846	1.292	1.664	1.990	2.374	2.639	3.195
90	.254	.526	.846	1.291	1.662	1.987	2.369	2.632	3.183
100	.254	.526	.845	1.290	1.660	1.984	2.364	2.626	3.174
110	.254	.526	.845	1.289	1.659	1.982	2.361	2.621	3.165
120	.254	.526	.845	1.289	1.658	1.980	2.358	2.617	3.160
130	.254	.526	.844	1.288	1.657	1.978	2.355	2.614	3.154
140	.254	.526	.844	1.288	1.656	1.977	2.353	2.611	3.149
150	.254	.526	.844	1.287	1.655	1.976	2.351	2.609	3.145

Example: $\Pr\{t \leqslant t_{0.95}(10)\} = \Pr\{t \leqslant 1.812\} = 0.95$

[a] Table B2 is adapted from Table III of Fisher and Yates: *Statistical Tables for Biological, Agricultural and Medical Research*, published by Longman Group Ltd., London (1974), 6th Ed., (previously published by Oliver and Boyd Ltd., Edinburgh) and by permission of the authors and publishers.

Table B3. Selected Percentiles $\chi^2_\gamma(n)$ of the $\chi^2(n)$ Distribution[a]

n	γ = .005	.01	.025	.05	.10	.50	.90	.95	.975	.99	.995
1	.000	.000	.001	.004	.016	.455	2.706	3.841	5.024	6.635	7.879
2	.010	.020	.051	.103	.211	1.386	4.605	5.991	7.378	9.210	10.597
3	.072	.115	.216	.352	.584	2.366	6.251	7.815	9.348	11.345	12.838
4	.207	.297	.484	.711	1.064	3.357	7.779	9.488	11.143	13.277	14.860
5	.412	.554	.831	1.145	1.610	4.351	9.236	11.070	12.833	15.086	16.750
6	.676	.872	1.237	1.635	2.204	5.348	10.645	12.592	14.449	16.812	18.548
7	.989	1.239	1.690	2.167	2.833	6.346	12.017	14.067	16.013	18.475	20.278
8	1.344	1.646	2.180	2.733	3.490	7.344	13.362	15.507	17.535	20.090	21.955
9	1.735	2.088	2.700	3.325	4.168	8.343	14.684	16.919	19.023	21.666	23.589
10	2.156	2.558	3.247	3.940	4.865	9.342	15.987	18.307	20.483	23.209	25.188
11	2.603	3.053	3.816	4.575	5.578	10.341	17.275	19.675	21.920	24.725	26.757
12	3.074	3.571	4.404	5.226	6.304	11.340	18.549	21.026	23.337	26.217	28.300
13	3.565	4.107	5.009	5.892	7.042	12.340	19.812	22.362	24.736	27.688	29.819
14	4.075	4.660	5.629	6.571	7.790	13.339	21.064	23.685	26.119	29.141	31.319
15	4.601	5.229	6.262	7.261	8.547	14.339	22.307	24.996	27.488	30.578	32.801
16	5.142	5.812	6.908	7.962	9.312	15.338	23.542	26.296	28.845	32.000	34.267
17	5.697	6.408	7.564	8.672	10.085	16.338	24.769	27.587	30.191	33.409	35.718
18	6.265	7.015	8.231	9.390	10.865	17.338	25.989	28.869	31.526	34.805	37.156
19	6.844	7.633	8.907	10.117	11.651	18.338	27.204	30.144	32.852	36.191	38.582
20	7.434	8.260	9.591	10.851	12.443	19.337	28.412	31.410	34.170	37.566	39.997
21	8.034	8.897	10.283	11.591	13.240	20.337	29.615	32.671	35.479	38.932	41.401
22	8.643	9.542	10.982	12.338	14.041	21.337	30.813	33.924	36.781	40.289	42.796
23	9.260	10.196	11.689	13.091	14.848	22.337	32.007	35.172	38.076	41.638	44.181
24	9.886	10.856	12.401	13.848	15.659	23.337	33.196	36.415	39.364	42.980	45.559
25	10.520	11.524	13.120	14.611	16.473	24.337	34.382	37.652	40.646	44.314	46.928
26	11.160	12.198	13.844	15.379	17.292	25.336	35.563	38.885	41.923	45.642	48.290
27	11.808	12.879	14.573	16.151	18.114	26.336	36.741	40.113	43.195	46.963	49.645
28	12.461	13.565	15.308	16.928	18.939	27.336	37.916	41.337	44.461	48.278	50.993
29	13.121	14.256	16.047	17.708	19.768	28.336	39.087	42.557	45.722	49.588	52.336
30	13.787	14.953	16.791	18.493	20.599	29.336	40.256	43.773	46.979	50.892	53.672
40	20.707	22.164	24.433	26.509	29.051	39.335	51.805	55.758	59.342	63.691	66.766
50	27.991	29.707	32.357	34.764	37.689	49.335	63.167	67.505	71.420	76.154	79.490
60	35.534	37.485	40.482	43.188	46.459	59.335	74.397	79.082	83.298	88.379	91.952
70	43.275	45.442	48.758	51.739	55.329	69.334	85.527	90.531	95.023	100.425	104.215
80	51.172	53.540	57.153	60.391	64.278	79.334	96.578	101.879	106.629	112.329	116.321
90	59.196	61.754	65.647	69.126	73.291	89.334	107.565	113.145	118.136	124.116	128.299
100	67.328	70.065	74.222	77.929	82.358	99.334	118.498	124.342	129.561	135.807	140.169
110	75.550	78.458	82.867	86.792	91.471	109.334	129.385	135.480	140.917	147.414	151.948
120	83.852	86.923	91.573	95.705	100.624	119.334	140.233	146.567	152.211	158.950	163.648
130	92.222	95.451	100.331	104.662	109.811	129.334	151.045	157.610	163.453	170.423	175.278
140	100.655	104.034	109.137	113.659	119.029	139.334	161.827	168.613	174.648	181.840	186.847
150	109.142	112.668	117.985	122.692	128.275	149.334	172.581	179.581	185.800	193.208	198.360
160	117.679	121.346	126.870	131.756	137.546	159.334	183.311	190.516	196.915	204.530	209.824
170	126.261	130.064	135.790	140.849	146.839	169.334	194.017	201.423	207.995	215.812	221.242
180	134.884	138.820	144.741	149.969	156.153	179.334	204.704	212.304	219.044	227.056	232.620
190	143.545	147.610	153.721	159.113	165.485	189.334	215.371	223.160	230.064	238.266	243.959
200	152.241	156.432	162.728	168.279	174.835	199.334	226.021	233.994	241.058	249.445	255.264
225	174.116	178.609	185.348	191.281	198.278	224.334	252.578	260.992	268.438	277.269	283.390
250	196.161	200.939	208.098	214.392	221.806	249.334	279.050	287.882	295.689	304.940	311.346
275	218.349	223.399	230.957	237.595	245.406	274.334	305.451	314.678	322.829	332.480	339.158
300	240.663	245.972	253.912	260.878	269.068	299.334	331.789	341.395	349.874	359.906	366.844
325	263.087	268.645	275.951	284.232	292.785	324.334	358.072	368.042	376.836	387.234	394.421
350	285.608	291.406	300.064	307.648	316.550	349.334	384.306	394.626	403.723	414.474	421.900
375	308.216	314.245	323.243	331.119	340.359	374.334	410.477	421.154	430.544	441.635	449.293
400	330.903	337.155	346.482	354.641	364.207	399.334	436.649	447.632	457.305	468.724	476.606
450	376.483	383.163	393.119	401.817	412.007	449.334	488.849	500.456	510.670	522.717	531.026
500	422.303	429.388	439.936	449.147	459.926	499.333	540.930	553.127	563.852	576.493	585.207

Example: $\Pr\left\{\chi^2 \leqslant \chi^2_{0.95}(10)\right\} = \Pr\left\{\chi^2 \leqslant 18.307\right\} = 0.95$

[a]Adapted with permission of Biometrika Trustees, from E. S. Pearson and H. O. Hartley (1966). *Biometrika Tables for Statisticians, Vol. I*, Cambridge University Press, London.

Table B4. Selected Percentiles $\mathcal{F}_\gamma(n_1, n_2)$ of the $\mathcal{F}(n_1, n_2)$ Distribution[a]

$\gamma = 0.995$

n_2	n_1 = 1	2	3	4	5	6	7	8	9	10	12	15	20	25	30	40	60	120	∞
1	16211	20000	21615	22500	23056	23437	23715	23925	24091	24224	24426	24630	24836	24960	25044	25148	25253	25359	25464
2	198.5	199.0	199.2	199.2	199.3	199.3	199.4	199.4	199.4	199.4	199.4	199.4	199.4	199.5	199.5	199.5	199.5	199.5	199.5
3	55.55	49.80	47.47	46.19	45.39	44.84	44.43	44.13	43.88	43.69	43.39	43.08	42.78	42.59	42.47	42.31	42.15	41.99	41.83
4	31.33	26.28	24.26	23.15	22.46	21.97	21.62	21.35	21.14	20.97	20.70	20.44	20.17	20.00	19.89	19.75	19.61	19.47	19.32
5	22.78	18.31	16.53	15.56	14.94	14.51	14.20	13.96	13.77	13.62	13.38	13.15	12.90	12.76	12.66	12.53	12.40	12.27	12.14
6	18.63	14.54	12.92	12.03	11.46	11.07	10.79	10.57	10.39	10.25	10.03	9.81	9.59	9.45	9.36	9.24	9.12	9.00	8.88
7	16.24	12.40	10.88	10.05	9.52	9.16	8.89	8.68	8.51	8.38	8.18	7.97	7.75	7.62	7.53	7.42	7.31	7.19	7.08
8	14.69	11.04	9.60	8.81	8.30	7.95	7.69	7.50	7.34	7.21	7.01	6.81	6.61	6.48	6.40	6.29	6.18	6.06	5.95
9	13.61	10.11	8.72	7.96	7.47	7.13	6.88	6.69	6.54	6.42	6.23	6.03	5.83	5.71	5.62	5.52	5.41	5.30	5.19
10	12.83	9.43	8.08	7.34	6.87	6.54	6.30	6.12	5.97	5.85	5.66	5.47	5.27	5.15	5.07	4.97	4.86	4.75	4.64
11	12.23	8.91	7.60	6.88	6.42	6.10	5.86	5.68	5.54	5.42	5.24	5.05	4.86	4.74	4.65	4.55	4.45	4.34	4.23
12	11.75	8.51	7.23	6.52	6.07	5.76	5.52	5.35	5.20	5.09	4.91	4.72	4.53	4.41	4.33	4.23	4.12	4.01	3.90
13	11.37	8.19	6.93	6.23	5.79	5.48	5.25	5.08	4.94	4.82	4.64	4.46	4.27	4.15	4.07	3.97	3.87	3.76	3.65
14	11.06	7.92	6.68	6.00	5.56	5.26	5.03	4.86	4.72	4.60	4.43	4.25	4.06	3.94	3.86	3.76	3.66	3.55	3.44
15	10.80	7.70	6.48	5.80	5.37	5.07	4.85	4.67	4.54	4.42	4.25	4.07	3.88	3.77	3.69	3.58	3.48	3.37	3.26
16	10.58	7.51	6.30	5.64	5.21	4.91	4.69	4.52	4.38	4.27	4.10	3.92	3.73	3.62	3.54	3.44	3.33	3.22	3.11
17	10.38	7.35	6.16	5.50	5.07	4.78	4.56	4.39	4.25	4.14	3.97	3.79	3.61	3.49	3.41	3.31	3.21	3.10	2.98
18	10.22	7.21	6.03	5.37	4.96	4.66	4.44	4.28	4.14	4.03	3.86	3.68	3.50	3.38	3.30	3.20	3.10	2.99	2.87
19	10.07	7.09	5.92	5.27	4.85	4.56	4.34	4.18	4.04	3.93	3.76	3.59	3.40	3.29	3.21	3.11	3.00	2.89	2.78
20	9.94	6.99	5.82	5.17	4.76	4.47	4.26	4.09	3.96	3.85	3.68	3.50	3.32	3.20	3.12	3.02	2.92	2.81	2.69
21	9.83	6.89	5.73	5.09	4.68	4.39	4.18	4.01	3.88	3.77	3.60	3.43	3.24	3.13	3.05	2.95	2.84	2.73	2.61
22	9.73	6.81	5.65	5.02	4.61	4.32	4.11	3.94	3.81	3.70	3.54	3.36	3.18	3.06	2.98	2.88	2.77	2.66	2.55
23	9.63	6.73	5.58	4.95	4.54	4.26	4.05	3.88	3.75	3.64	3.47	3.30	3.12	3.00	2.92	2.82	2.71	2.60	2.48
24	9.55	6.66	5.52	4.89	4.49	4.20	3.99	3.83	3.69	3.59	3.42	3.25	3.06	2.95	2.87	2.77	2.66	2.55	2.43
25	9.48	6.60	5.46	4.84	4.43	4.15	3.94	3.78	3.64	3.54	3.37	3.20	3.01	2.90	2.82	2.72	2.61	2.50	2.38
26	9.41	6.54	5.41	4.79	4.38	4.10	3.89	3.73	3.60	3.49	3.33	3.15	2.97	2.85	2.77	2.67	2.56	2.45	2.33
27	9.34	6.49	5.36	4.74	4.34	4.06	3.85	3.69	3.56	3.45	3.28	3.11	2.93	2.81	2.73	2.63	2.52	2.41	2.29
28	9.28	6.44	5.32	4.70	4.30	4.02	3.81	3.65	3.52	3.41	3.25	3.07	2.89	2.77	2.69	2.59	2.48	2.37	2.25
29	9.23	6.40	5.28	4.66	4.26	3.98	3.77	3.61	3.48	3.38	3.21	3.04	2.86	2.74	2.66	2.56	2.45	2.33	2.21
30	9.18	6.35	5.24	4.62	4.23	3.95	3.74	3.58	3.45	3.34	3.18	3.01	2.82	2.71	2.63	2.52	2.42	2.30	2.18
40	8.83	6.07	4.98	4.37	3.99	3.71	3.51	3.35	3.22	3.12	2.95	2.78	2.60	2.48	2.40	2.30	2.18	2.06	1.93
60	8.49	5.79	4.73	4.14	3.76	3.49	3.29	3.13	3.01	2.90	2.74	2.57	2.39	2.27	2.19	2.08	1.96	1.83	1.69
120	8.18	5.54	4.50	3.92	3.55	3.28	3.09	2.93	2.81	2.71	2.54	2.37	2.19	2.07	1.98	1.87	1.75	1.61	1.43
∞	7.88	5.30	4.28	3.72	3.35	3.09	2.90	2.74	2.62	2.52	2.36	2.19	2.00	1.88	1.79	1.67	1.53	1.36	1.00

Example: $\Pr\{F \leq F_{0.995}(5,10)\} = \Pr\{F \leq 6.87\} = 0.995$

[a]Adapted with permission of Biometrika Tustees from E. S. Pearson and H. O. Hartley (1966). *Biometrika Tables for Statisticians, Vol. I*, Cambridge University Press, London.

Table B4. (*Continued*)

$\gamma = 0.99$

n_2 \ n_1	1	2	3	4	5	6	7	8	9	10	12	15	20	25	30	40	60	120	∞
1	4052	5000	5403	5625	5764	5859	5928	5981	6022	6056	6106	6157	6209	6240	6261	6287	6313	6339	6366
2	98.50	99.00	99.17	99.25	99.30	99.33	99.36	99.37	99.39	99.40	99.42	99.43	99.45	99.46	99.47	99.47	99.48	99.49	99.50
3	34.12	30.82	29.46	28.71	28.24	27.91	27.67	27.49	27.35	27.23	27.05	26.87	26.69	26.58	26.50	26.41	26.32	26.22	26.13
4	21.20	18.00	16.69	15.98	15.52	15.21	14.98	14.80	14.66	14.55	14.37	14.20	14.02	13.91	13.84	13.75	13.65	13.56	13.46
5	16.26	13.27	12.06	11.39	10.97	10.67	10.46	10.29	10.16	10.05	9.89	9.72	9.55	9.45	9.38	9.29	9.20	9.11	9.02
6	13.75	10.92	9.78	9.15	8.75	8.47	8.26	8.10	7.98	7.87	7.72	7.56	7.40	7.30	7.23	7.14	7.06	6.97	6.88
7	12.25	9.55	8.45	7.85	7.46	7.19	6.99	6.84	6.72	6.62	6.47	6.31	6.16	6.06	5.99	5.91	5.82	5.74	5.65
8	11.26	8.65	7.59	7.01	6.63	6.37	6.18	6.03	5.91	5.81	5.67	5.52	5.36	5.26	5.20	5.12	5.03	4.95	4.86
9	10.56	8.02	6.99	6.42	6.06	5.80	5.61	5.47	5.35	5.26	5.11	4.96	4.81	4.71	4.65	4.57	4.48	4.40	4.31
10	10.04	7.56	6.55	5.99	5.64	5.39	5.20	5.06	4.94	4.85	4.71	4.56	4.41	4.31	4.25	4.17	4.08	4.00	3.91
11	9.65	7.21	6.22	5.67	5.32	5.07	4.89	4.74	4.63	4.54	4.40	4.25	4.10	4.01	3.94	3.86	3.78	3.69	3.60
12	9.33	6.93	5.95	5.41	5.06	4.82	4.64	4.50	4.39	4.30	4.16	4.01	3.86	3.76	3.70	3.62	3.54	3.45	3.36
13	9.07	6.70	5.74	5.21	4.86	4.62	4.44	4.30	4.19	4.10	3.96	3.82	3.66	3.57	3.51	3.43	3.34	3.25	3.17
14	8.86	6.51	5.56	5.04	4.69	4.46	4.28	4.14	4.03	3.94	3.80	3.66	3.51	3.41	3.35	3.27	3.18	3.09	3.00
15	8.68	6.36	5.42	4.89	4.56	4.32	4.14	4.00	3.89	3.80	3.67	3.52	3.37	3.28	3.21	3.13	3.05	2.96	2.87
16	8.53	6.23	5.29	4.77	4.44	4.20	4.03	3.89	3.78	3.69	3.55	3.41	3.26	3.16	3.10	3.02	2.93	2.84	2.75
17	8.40	6.11	5.18	4.67	4.34	4.10	3.93	3.79	3.68	3.59	3.46	3.31	3.16	3.07	3.00	2.92	2.83	2.75	2.65
18	8.29	6.01	5.09	4.58	4.25	4.01	3.84	3.71	3.60	3.51	3.37	3.23	3.08	2.98	2.92	2.84	2.75	2.66	2.57
19	8.18	5.93	5.01	4.50	4.17	3.94	3.77	3.63	3.52	3.43	3.30	3.15	3.00	2.91	2.84	2.76	2.67	2.58	2.49
20	8.10	5.85	4.94	4.43	4.10	3.87	3.70	3.56	3.46	3.37	3.23	3.09	2.94	2.84	2.78	2.69	2.61	2.52	2.42
21	8.02	5.78	4.87	4.37	4.04	3.81	3.64	3.51	3.40	3.31	3.17	3.03	2.88	2.79	2.72	2.64	2.55	2.46	2.36
22	7.95	5.72	4.82	4.31	3.99	3.76	3.59	3.45	3.35	3.26	3.12	2.98	2.83	2.73	2.67	2.58	2.50	2.40	2.31
23	7.88	5.66	4.76	4.26	3.94	3.71	3.54	3.41	3.30	3.21	3.07	2.93	2.78	2.69	2.62	2.54	2.45	2.35	2.26
24	7.82	5.61	4.72	4.22	3.90	3.67	3.50	3.36	3.26	3.17	3.03	2.89	2.74	2.64	2.58	2.49	2.40	2.31	2.21
25	7.77	5.57	4.68	4.18	3.85	3.63	3.46	3.32	3.22	3.13	2.99	2.85	2.70	2.60	2.54	2.45	2.36	2.27	2.17
26	7.72	5.53	4.64	4.14	3.82	3.59	3.42	3.29	3.18	3.09	2.96	2.81	2.66	2.57	2.50	2.42	2.33	2.23	2.13
27	7.68	5.49	4.60	4.11	3.78	3.56	3.39	3.26	3.15	3.06	2.93	2.78	2.63	2.54	2.47	2.38	2.29	2.20	2.10
28	7.64	5.45	4.57	4.07	3.75	3.53	3.36	3.23	3.12	3.03	2.90	2.75	2.60	2.51	2.44	2.35	2.26	2.17	2.06
29	7.60	5.42	4.54	4.04	3.73	3.50	3.33	3.20	3.09	3.00	2.87	2.73	2.57	2.48	2.41	2.33	2.23	2.14	2.03
30	7.56	5.39	4.51	4.02	3.70	3.47	3.30	3.17	3.07	2.98	2.84	2.70	2.55	2.45	2.39	2.30	2.21	2.11	2.01
40	7.31	5.18	4.31	3.83	3.51	3.29	3.12	2.99	2.89	2.80	2.66	2.52	2.37	2.27	2.20	2.11	2.02	1.92	1.80
60	7.08	4.98	4.13	3.65	3.34	3.12	2.95	2.82	2.72	2.63	2.50	2.35	2.20	2.10	2.03	1.94	1.84	1.73	1.60
120	6.85	4.79	3.95	3.48	3.17	2.96	2.79	2.66	2.56	2.47	2.34	2.19	2.03	1.93	1.86	1.76	1.66	1.53	1.38
∞	6.63	4.61	3.78	3.32	3.02	2.80	2.64	2.51	2.41	2.32	2.18	2.04	1.88	1.77	1.70	1.59	1.47	1.32	1.00

$\gamma = 0.975$

n_2 \ n_1	1	2	3	4	5	6	7	8	9	10	12	15	20	25	30	40	60	120	∞
1	647.8	799.5	864.2	899.6	921.8	937.1	948.2	956.7	963.3	968.6	976.7	984.9	993.1	998.1	1001	1006	1010	1014	1018
2	38.51	39.00	39.17	39.25	39.30	39.33	39.36	39.37	39.39	39.40	39.41	39.43	39.45	39.46	39.46	39.47	39.48	39.49	39.50
3	17.44	16.04	15.44	15.10	14.88	14.73	14.62	14.54	14.47	14.42	14.34	14.25	14.17	14.12	14.08	14.04	13.99	13.95	13.90
4	12.22	10.65	9.98	9.60	9.36	9.20	9.07	8.98	8.90	8.84	8.75	8.66	8.56	8.50	8.46	8.41	8.36	8.31	8.26
5	10.01	8.43	7.76	7.39	7.15	6.98	6.85	6.76	6.68	6.62	6.52	6.43	6.33	6.27	6.23	6.18	6.12	6.07	6.02
6	8.81	7.26	6.60	6.23	5.99	5.82	5.70	5.60	5.52	5.46	5.37	5.27	5.17	5.11	5.07	5.01	4.96	4.90	4.85
7	8.07	6.54	5.89	5.52	5.29	5.12	4.99	4.90	4.82	4.76	4.67	4.57	4.47	4.40	4.36	4.31	4.25	4.20	4.14
8	7.57	6.06	5.42	5.05	4.82	4.65	4.53	4.43	4.36	4.30	4.20	4.10	4.00	3.94	3.89	3.84	3.78	3.73	3.67
9	7.21	5.71	5.08	4.72	4.48	4.32	4.20	4.10	4.03	3.96	3.87	3.77	3.67	3.60	3.56	3.51	3.45	3.39	3.33
10	6.94	5.46	4.83	4.47	4.24	4.07	3.95	3.85	3.78	3.72	3.62	3.52	3.42	3.35	3.31	3.26	3.20	3.14	3.08
11	6.72	5.26	4.63	4.28	4.04	3.88	3.76	3.66	3.59	3.53	3.43	3.33	3.23	3.16	3.12	3.06	3.00	2.94	2.88
12	6.55	5.10	4.47	4.12	3.89	3.73	3.61	3.51	3.44	3.37	3.28	3.18	3.07	3.01	2.96	2.91	2.85	2.79	2.72
13	6.41	4.97	4.35	4.00	3.77	3.60	3.48	3.39	3.31	3.25	3.15	3.05	2.95	2.88	2.84	2.78	2.72	2.66	2.60
14	6.30	4.86	4.24	3.89	3.66	3.50	3.38	3.29	3.21	3.15	3.05	2.95	2.84	2.78	2.73	2.67	2.61	2.55	2.49
15	6.20	4.77	4.15	3.80	3.58	3.41	3.29	3.20	3.12	3.06	2.96	2.86	2.76	2.69	2.64	2.59	2.52	2.46	2.40
16	6.12	4.69	4.08	3.73	3.50	3.34	3.22	3.12	3.05	2.99	2.89	2.79	2.68	2.61	2.57	2.51	2.45	2.38	2.32
17	6.04	4.62	4.01	3.66	3.44	3.28	3.16	3.06	2.98	2.92	2.82	2.72	2.62	2.55	2.50	2.44	2.38	2.32	2.25
18	5.98	4.56	3.95	3.61	3.38	3.22	3.10	3.01	2.93	2.87	2.77	2.67	2.56	2.49	2.44	2.38	2.32	2.26	2.19
19	5.92	4.51	3.90	3.56	3.33	3.17	3.05	2.96	2.88	2.82	2.72	2.62	2.51	2.44	2.39	2.33	2.27	2.20	2.13
20	5.87	4.46	3.86	3.51	3.29	3.13	3.01	2.91	2.84	2.77	2.68	2.57	2.46	2.40	2.35	2.29	2.22	2.16	2.09
21	5.83	4.42	3.82	3.48	3.25	3.09	2.97	2.87	2.80	2.73	2.64	2.53	2.42	2.36	2.31	2.25	2.18	2.11	2.04
22	5.79	4.38	3.78	3.44	3.22	3.05	2.93	2.84	2.76	2.70	2.60	2.50	2.39	2.32	2.27	2.21	2.14	2.08	2.00
23	5.75	4.35	3.75	3.41	3.18	3.02	2.90	2.81	2.73	2.67	2.57	2.47	2.36	2.29	2.24	2.18	2.11	2.04	1.97
24	5.72	4.32	3.72	3.38	3.15	2.99	2.87	2.78	2.70	2.64	2.54	2.44	2.33	2.26	2.21	2.15	2.08	2.01	1.94
25	5.69	4.29	3.69	3.35	3.13	2.97	2.85	2.75	2.68	2.61	2.51	2.41	2.30	2.23	2.18	2.12	2.05	1.98	1.91
26	5.66	4.27	3.67	3.33	3.10	2.94	2.82	2.73	2.65	2.59	2.49	2.39	2.28	2.21	2.16	2.09	2.03	1.95	1.88
27	5.63	4.24	3.65	3.31	3.08	2.92	2.80	2.71	2.63	2.57	2.47	2.36	2.25	2.18	2.13	2.07	2.00	1.93	1.85
28	5.61	4.22	3.63	3.29	3.06	2.90	2.78	2.69	2.61	2.55	2.45	2.34	2.23	2.16	2.11	2.05	1.98	1.91	1.83
29	5.59	4.20	3.61	3.27	3.04	2.88	2.76	2.67	2.59	2.53	2.43	2.32	2.21	2.14	2.09	2.03	1.96	1.89	1.81
30	5.57	4.18	3.59	3.25	3.03	2.87	2.75	2.65	2.57	2.51	2.41	2.31	2.20	2.12	2.07	2.01	1.94	1.87	1.79
40	5.42	4.05	3.46	3.13	2.90	2.74	2.62	2.53	2.45	2.39	2.29	2.18	2.07	1.99	1.94	1.88	1.80	1.72	1.64
60	5.29	3.93	3.34	3.01	2.79	2.63	2.51	2.41	2.33	2.27	2.17	2.06	1.94	1.87	1.82	1.74	1.67	1.58	1.48
120	5.15	3.80	3.23	2.89	2.67	2.52	2.39	2.30	2.22	2.16	2.05	1.94	1.82	1.75	1.69	1.61	1.53	1.43	1.31
∞	5.02	3.69	3.12	2.79	2.57	2.41	2.29	2.19	2.11	2.05	1.94	1.83	1.71	1.63	1.57	1.48	1.39	1.27	1.00

Table B4. (*Continued*)

$\gamma = 0.95$

n_2 \ n_1	1	2	3	4	5	6	7	8	9	10	12	15	20	25	30	40	60	120	∞
1	161.4	199.5	215.7	224.6	230.2	234.0	236.8	238.9	240.5	241.9	243.9	245.9	248.0	249.3	250.1	251.1	252.2	253.3	254.3
2	18.51	19.00	19.16	19.25	19.30	19.33	19.35	19.37	19.38	19.40	19.41	19.43	19.45	19.46	19.46	19.47	19.48	19.49	19.50
3	10.13	9.55	9.28	9.12	9.01	8.94	8.89	8.85	8.81	8.79	8.74	8.70	8.66	8.63	8.62	8.59	8.57	8.55	8.53
4	7.71	6.94	6.59	6.39	6.26	6.16	6.09	6.04	6.00	5.96	5.91	5.86	5.80	5.77	5.75	5.72	5.69	5.66	5.63
5	6.61	5.79	5.41	5.19	5.05	4.95	4.88	4.82	4.77	4.74	4.68	4.62	4.56	4.52	4.50	4.46	4.43	4.40	4.36
6	5.99	5.14	4.76	4.53	4.39	4.28	4.21	4.15	4.10	4.06	4.00	3.94	3.87	3.83	3.81	3.77	3.74	3.70	3.67
7	5.59	4.74	4.35	4.12	3.97	3.87	3.79	3.73	3.68	3.64	3.57	3.51	3.44	3.40	3.38	3.34	3.30	3.27	3.23
8	5.32	4.46	4.07	3.84	3.69	3.58	3.50	3.44	3.39	3.35	3.28	3.22	3.15	3.11	3.08	3.04	3.01	2.97	2.93
9	5.12	4.26	3.86	3.63	3.48	3.37	3.29	3.23	3.18	3.14	3.07	3.01	2.94	2.89	2.86	2.83	2.79	2.75	2.71
10	4.96	4.10	3.71	3.48	3.33	3.22	3.14	3.07	3.02	2.98	2.91	2.85	2.77	2.73	2.70	2.66	2.62	2.58	2.54
11	4.84	3.98	3.59	3.36	3.20	3.09	3.01	2.95	2.90	2.85	2.79	2.72	2.65	2.60	2.57	2.53	2.49	2.45	2.40
12	4.75	3.89	3.49	3.26	3.11	3.00	2.91	2.85	2.80	2.75	2.69	2.62	2.54	2.50	2.47	2.43	2.38	2.34	2.30
13	4.67	3.81	3.41	3.18	3.03	2.92	2.83	2.77	2.71	2.67	2.60	2.53	2.46	2.41	2.38	2.34	2.30	2.25	2.21
14	4.60	3.74	3.34	3.11	2.96	2.85	2.76	2.70	2.65	2.60	2.53	2.46	2.39	2.34	2.31	2.27	2.22	2.18	2.13
15	4.54	3.68	3.29	3.06	2.90	2.79	2.71	2.64	2.59	2.54	2.48	2.40	2.33	2.28	2.25	2.20	2.16	2.11	2.07
16	4.49	3.63	3.24	3.01	2.85	2.74	2.66	2.59	2.54	2.49	2.42	2.35	2.28	2.23	2.19	2.15	2.11	2.06	2.01
17	4.45	3.59	3.20	2.96	2.81	2.70	2.61	2.55	2.49	2.45	2.38	2.31	2.23	2.18	2.15	2.10	2.06	2.01	1.96
18	4.41	3.55	3.16	2.93	2.77	2.66	2.58	2.51	2.46	2.41	2.34	2.27	2.19	2.14	2.11	2.06	2.02	1.97	1.92
19	4.38	3.52	3.13	2.90	2.74	2.63	2.54	2.48	2.42	2.38	2.31	2.23	2.16	2.11	2.07	2.03	1.98	1.93	1.88
20	4.35	3.49	3.10	2.87	2.71	2.60	2.51	2.45	2.39	2.35	2.28	2.20	2.12	2.07	2.04	1.99	1.95	1.90	1.84
21	4.32	3.47	3.07	2.84	2.68	2.57	2.49	2.42	2.37	2.32	2.25	2.18	2.10	2.05	2.01	1.96	1.92	1.87	1.81
22	4.30	3.44	3.05	2.82	2.66	2.55	2.46	2.40	2.34	2.30	2.23	2.15	2.07	2.02	1.98	1.94	1.89	1.84	1.78
23	4.28	3.42	3.03	2.80	2.64	2.53	2.44	2.37	2.32	2.27	2.20	2.13	2.05	2.00	1.96	1.91	1.86	1.81	1.76
24	4.26	3.40	3.01	2.78	2.62	2.51	2.42	2.36	2.30	2.25	2.18	2.11	2.03	1.97	1.94	1.89	1.84	1.79	1.73
25	4.24	3.39	2.99	2.76	2.60	2.49	2.40	2.34	2.28	2.24	2.16	2.09	2.01	1.96	1.92	1.87	1.82	1.77	1.71
26	4.23	3.37	2.98	2.74	2.59	2.47	2.39	2.32	2.27	2.22	2.15	2.07	1.99	1.94	1.90	1.85	1.80	1.75	1.69
27	4.21	3.35	2.96	2.73	2.57	2.46	2.37	2.31	2.25	2.20	2.13	2.06	1.97	1.92	1.88	1.84	1.79	1.73	1.67
28	4.20	3.34	2.95	2.71	2.56	2.45	2.36	2.29	2.24	2.19	2.12	2.04	1.96	1.91	1.87	1.82	1.77	1.71	1.65
29	4.18	3.33	2.93	2.70	2.55	2.43	2.35	2.28	2.22	2.18	2.10	2.03	1.94	1.89	1.85	1.81	1.75	1.70	1.64
30	4.17	3.32	2.92	2.69	2.53	2.42	2.33	2.27	2.21	2.16	2.09	2.01	1.93	1.88	1.84	1.79	1.74	1.68	1.62
40	4.08	3.23	2.84	2.61	2.45	2.34	2.25	2.18	2.12	2.08	2.00	1.92	1.84	1.78	1.74	1.69	1.64	1.58	1.51
60	4.00	3.15	2.76	2.53	2.37	2.25	2.17	2.10	2.04	1.99	1.92	1.84	1.75	1.69	1.65	1.59	1.53	1.47	1.39
120	3.92	3.07	2.68	2.45	2.29	2.18	2.09	2.02	1.96	1.91	1.83	1.75	1.66	1.60	1.55	1.50	1.43	1.35	1.25
∞	3.84	3.00	2.60	2.37	2.21	2.10	2.01	1.94	1.88	1.83	1.75	1.67	1.57	1.51	1.46	1.39	1.32	1.22	1.00

$\gamma = 0.90$

n_2 \ n_1	1	2	3	4	5	6	7	8	9	10	12	15	20	25	30	40	60	120	∞
1	39.86	49.50	53.59	55.83	57.24	58.20	58.91	59.44	59.86	60.19	60.71	61.22	61.74	62.05	62.26	62.53	62.79	63.06	63.33
2	8.53	9.00	9.16	9.24	9.29	9.33	9.35	9.37	9.38	9.39	9.41	9.42	9.44	9.45	9.46	9.47	9.47	9.48	9.49
3	5.54	5.46	5.39	5.34	5.31	5.28	5.27	5.25	5.24	5.23	5.22	5.20	5.18	5.17	5.17	5.16	5.15	5.14	5.13
4	4.54	4.32	4.19	4.11	4.05	4.01	3.98	3.95	3.94	3.92	3.90	3.87	3.84	3.83	3.82	3.80	3.79	3.78	3.76
5	4.06	3.78	3.62	3.52	3.45	3.40	3.37	3.34	3.32	3.30	3.27	3.24	3.21	3.19	3.17	3.16	3.14	3.12	3.10
6	3.78	3.46	3.29	3.18	3.11	3.05	3.01	2.98	2.96	2.94	2.90	2.87	2.84	2.81	2.80	2.78	2.76	2.74	2.72
7	3.59	3.26	3.07	2.96	2.88	2.83	2.78	2.75	2.72	2.70	2.67	2.63	2.59	2.57	2.56	2.54	2.51	2.49	2.47
8	3.46	3.11	2.92	2.81	2.73	2.67	2.62	2.59	2.56	2.54	2.50	2.46	2.42	2.40	2.38	2.36	2.34	2.32	2.29
9	3.36	3.01	2.81	2.69	2.61	2.55	2.51	2.47	2.44	2.42	2.38	2.34	2.30	2.27	2.25	2.23	2.21	2.18	2.16
10	3.29	2.92	2.73	2.61	2.52	2.46	2.41	2.38	2.35	2.32	2.28	2.24	2.20	2.17	2.16	2.13	2.11	2.08	2.06
11	3.23	2.86	2.66	2.54	2.45	2.39	2.34	2.30	2.27	2.25	2.21	2.17	2.12	2.10	2.08	2.05	2.03	2.00	1.97
12	3.18	2.81	2.61	2.48	2.39	2.33	2.28	2.24	2.21	2.19	2.15	2.10	2.06	2.03	2.01	1.99	1.96	1.93	1.90
13	3.14	2.76	2.56	2.43	2.35	2.28	2.23	2.20	2.16	2.14	2.10	2.05	2.01	1.98	1.96	1.93	1.90	1.88	1.85
14	3.10	2.73	2.52	2.39	2.31	2.24	2.19	2.15	2.12	2.10	2.05	2.01	1.96	1.93	1.91	1.89	1.86	1.83	1.80
15	3.07	2.70	2.49	2.36	2.27	2.21	2.16	2.12	2.09	2.06	2.02	1.97	1.92	1.89	1.87	1.85	1.82	1.79	1.76
16	3.05	2.67	2.46	2.33	2.24	2.18	2.13	2.09	2.06	2.03	1.99	1.94	1.89	1.86	1.84	1.81	1.78	1.75	1.72
17	3.03	2.64	2.44	2.31	2.22	2.15	2.10	2.06	2.03	2.00	1.96	1.91	1.86	1.83	1.81	1.78	1.75	1.72	1.69
18	3.01	2.62	2.42	2.29	2.20	2.13	2.08	2.04	2.00	1.98	1.93	1.89	1.84	1.80	1.78	1.75	1.72	1.69	1.66
19	2.99	2.61	2.40	2.27	2.18	2.11	2.06	2.02	1.98	1.96	1.91	1.86	1.81	1.78	1.76	1.73	1.70	1.67	1.63
20	2.97	2.59	2.38	2.25	2.16	2.09	2.04	2.00	1.96	1.94	1.89	1.84	1.79	1.76	1.74	1.71	1.68	1.64	1.61
21	2.96	2.57	2.36	2.23	2.14	2.08	2.02	1.98	1.95	1.92	1.87	1.83	1.78	1.74	1.72	1.69	1.66	1.62	1.59
22	2.95	2.56	2.35	2.22	2.13	2.06	2.01	1.97	1.93	1.90	1.86	1.81	1.76	1.73	1.70	1.67	1.64	1.60	1.57
23	2.94	2.55	2.34	2.21	2.11	2.05	1.99	1.95	1.92	1.89	1.84	1.80	1.74	1.71	1.69	1.66	1.62	1.59	1.55
24	2.93	2.54	2.33	2.19	2.10	2.04	1.98	1.94	1.91	1.88	1.83	1.78	1.73	1.70	1.67	1.64	1.61	1.57	1.53
25	2.92	2.53	2.32	2.18	2.09	2.02	1.97	1.93	1.89	1.87	1.82	1.77	1.72	1.68	1.66	1.63	1.59	1.56	1.52
26	2.91	2.52	2.31	2.17	2.08	2.01	1.96	1.92	1.88	1.86	1.81	1.76	1.71	1.67	1.65	1.61	1.58	1.54	1.50
27	2.90	2.51	2.30	2.17	2.07	2.00	1.95	1.91	1.87	1.85	1.80	1.75	1.70	1.66	1.64	1.60	1.57	1.53	1.49
28	2.89	2.50	2.29	2.16	2.06	2.00	1.94	1.90	1.87	1.84	1.79	1.74	1.69	1.65	1.63	1.59	1.56	1.52	1.48
29	2.89	2.50	2.28	2.15	2.06	1.99	1.93	1.89	1.86	1.83	1.78	1.73	1.68	1.64	1.62	1.58	1.55	1.51	1.47
30	2.88	2.49	2.28	2.14	2.05	1.98	1.93	1.88	1.85	1.82	1.77	1.72	1.67	1.63	1.61	1.57	1.54	1.50	1.46
40	2.84	2.44	2.23	2.09	2.00	1.93	1.87	1.83	1.79	1.76	1.71	1.66	1.61	1.57	1.54	1.51	1.47	1.42	1.38
60	2.79	2.39	2.18	2.04	1.95	1.87	1.82	1.77	1.74	1.71	1.66	1.60	1.54	1.50	1.48	1.44	1.40	1.35	1.29
120	2.75	2.35	2.13	1.99	1.90	1.82	1.77	1.72	1.68	1.65	1.60	1.55	1.48	1.44	1.41	1.37	1.32	1.26	1.19
∞	2.71	2.30	2.08	1.94	1.85	1.77	1.72	1.67	1.63	1.60	1.55	1.49	1.42	1.38	1.34	1.30	1.24	1.17	1.00

Table B4. (*Continued*)

$\gamma = 0.80$

n_2	n_1 = 1	2	3	4	5	6	7	8	9	10	12	15	20	25	30	40	60	120	∞
1	9.47	12.00	13.06	13.64	14.01	14.26	14.44	14.58	14.68	14.77	14.90	15.04	15.17	15.25	15.31	15.37	15.44	15.51	15.58
2	3.56	4.00	4.16	4.24	4.28	4.32	4.34	4.36	4.37	4.38	4.40	4.42	4.43	4.44	4.45	4.46	4.46	4.47	4.48
3	2.68	2.89	2.94	2.96	2.97	2.97	2.97	2.98	2.98	2.98	2.98	2.98	2.98	2.98	2.98	2.98	2.98	2.98	2.98
4	2.35	2.47	2.48	2.48	2.48	2.47	2.47	2.47	2.46	2.46	2.46	2.45	2.44	2.44	2.44	2.44	2.43	2.43	2.43
5	2.18	2.26	2.25	2.24	2.23	2.22	2.21	2.20	2.20	2.19	2.18	2.18	2.17	2.16	2.16	2.15	2.15	2.14	2.13
6	2.07	2.13	2.11	2.09	2.08	2.06	2.05	2.04	2.03	2.03	2.02	2.01	2.00	1.99	1.98	1.98	1.97	1.96	1.95
7	2.00	2.04	2.02	1.99	1.97	1.96	1.94	1.93	1.93	1.92	1.91	1.89	1.88	1.87	1.86	1.86	1.85	1.84	1.83
8	1.95	1.98	1.95	1.92	1.90	1.88	1.87	1.86	1.85	1.84	1.83	1.81	1.80	1.79	1.78	1.77	1.76	1.75	1.74
9	1.91	1.93	1.90	1.87	1.85	1.83	1.81	1.80	1.79	1.78	1.76	1.75	1.73	1.72	1.71	1.70	1.69	1.68	1.67
10	1.88	1.90	1.86	1.83	1.80	1.78	1.77	1.75	1.74	1.73	1.72	1.70	1.68	1.67	1.66	1.65	1.64	1.63	1.62
11	1.86	1.87	1.83	1.80	1.77	1.75	1.73	1.72	1.70	1.69	1.68	1.66	1.64	1.63	1.62	1.61	1.60	1.59	1.57
12	1.84	1.85	1.80	1.77	1.74	1.72	1.70	1.69	1.67	1.66	1.65	1.63	1.61	1.60	1.59	1.58	1.56	1.55	1.54
13	1.82	1.83	1.78	1.75	1.72	1.69	1.68	1.66	1.65	1.64	1.62	1.60	1.58	1.57	1.56	1.55	1.53	1.52	1.51
14	1.81	1.81	1.76	1.73	1.70	1.67	1.65	1.64	1.63	1.62	1.60	1.58	1.56	1.54	1.53	1.52	1.51	1.49	1.48
15	1.80	1.80	1.75	1.71	1.69	1.66	1.64	1.62	1.61	1.60	1.58	1.56	1.54	1.52	1.51	1.50	1.49	1.47	1.46
16	1.79	1.78	1.74	1.70	1.67	1.64	1.62	1.61	1.59	1.58	1.56	1.54	1.52	1.50	1.49	1.48	1.47	1.45	1.43
17	1.78	1.77	1.72	1.68	1.65	1.63	1.61	1.59	1.58	1.57	1.55	1.53	1.50	1.49	1.48	1.46	1.45	1.43	1.42
18	1.77	1.76	1.71	1.67	1.64	1.62	1.60	1.58	1.56	1.55	1.53	1.51	1.49	1.47	1.46	1.45	1.43	1.42	1.40
19	1.76	1.75	1.70	1.66	1.63	1.61	1.58	1.57	1.55	1.54	1.52	1.50	1.48	1.46	1.45	1.44	1.42	1.40	1.39
20	1.76	1.75	1.70	1.65	1.62	1.60	1.58	1.56	1.54	1.53	1.51	1.49	1.47	1.45	1.44	1.42	1.41	1.39	1.37
21	1.75	1.74	1.69	1.65	1.61	1.59	1.57	1.55	1.53	1.52	1.50	1.48	1.46	1.44	1.43	1.41	1.40	1.38	1.36
22	1.75	1.73	1.69	1.64	1.61	1.58	1.56	1.54	1.53	1.51	1.49	1.47	1.45	1.43	1.42	1.40	1.39	1.37	1.35
23	1.74	1.73	1.68	1.63	1.60	1.57	1.55	1.53	1.52	1.51	1.49	1.46	1.44	1.42	1.41	1.39	1.38	1.36	1.34
24	1.74	1.72	1.67	1.63	1.59	1.57	1.55	1.53	1.51	1.50	1.48	1.46	1.43	1.41	1.40	1.39	1.37	1.35	1.33
25	1.73	1.72	1.66	1.62	1.59	1.56	1.54	1.52	1.51	1.49	1.47	1.45	1.42	1.41	1.39	1.38	1.36	1.34	1.32
26	1.73	1.71	1.66	1.62	1.59	1.56	1.53	1.52	1.50	1.49	1.47	1.44	1.42	1.40	1.39	1.37	1.35	1.33	1.31
27	1.73	1.71	1.66	1.61	1.58	1.55	1.53	1.51	1.49	1.48	1.46	1.44	1.41	1.39	1.38	1.36	1.35	1.33	1.30
28	1.72	1.71	1.65	1.61	1.57	1.55	1.52	1.51	1.49	1.48	1.46	1.43	1.41	1.39	1.37	1.36	1.34	1.32	1.30
29	1.72	1.70	1.65	1.60	1.57	1.54	1.52	1.50	1.49	1.47	1.45	1.43	1.40	1.38	1.37	1.35	1.33	1.31	1.29
30	1.72	1.70	1.64	1.60	1.57	1.54	1.52	1.50	1.48	1.47	1.45	1.42	1.39	1.38	1.36	1.35	1.33	1.31	1.28
40	1.70	1.68	1.62	1.57	1.54	1.51	1.49	1.47	1.45	1.44	1.41	1.39	1.36	1.34	1.33	1.31	1.29	1.26	1.24
60	1.68	1.65	1.60	1.55	1.51	1.48	1.46	1.44	1.42	1.41	1.38	1.35	1.32	1.30	1.29	1.27	1.24	1.22	1.18
120	1.66	1.63	1.57	1.52	1.48	1.45	1.43	1.41	1.39	1.37	1.35	1.32	1.29	1.27	1.25	1.23	1.20	1.17	1.12
∞	1.64	1.61	1.55	1.50	1.46	1.43	1.40	1.38	1.36	1.34	1.32	1.29	1.25	1.23	1.21	1.18	1.15	1.11	1.00

$\gamma = 0.50$

n_2 \ n_1	1	2	3	4	5	6	7	8	9	10	12	15	20	25	30	40	60	120	∞
1	1.00	1.50	1.71	1.82	1.89	1.94	1.98	2.00	2.03	2.04	2.07	2.09	2.12	2.13	2.15	2.16	2.17	2.18	2.20
2	.67	1.00	1.13	1.21	1.25	1.28	1.30	1.32	1.33	1.35	1.36	1.38	1.39	1.40	1.41	1.42	1.43	1.43	1.44
3	.59	.88	1.00	1.06	1.10	1.13	1.15	1.16	1.17	1.18	1.20	1.21	1.23	1.23	1.24	1.25	1.25	1.26	1.27
4	.55	.83	.94	1.00	1.04	1.06	1.08	1.09	1.10	1.11	1.13	1.14	1.15	1.16	1.16	1.17	1.18	1.18	1.19
5	.53	.80	.91	.96	1.00	1.02	1.04	1.05	1.06	1.07	1.09	1.10	1.11	1.12	1.12	1.13	1.14	1.14	1.15
6	.51	.78	.89	.94	.98	1.00	1.02	1.03	1.04	1.05	1.06	1.07	1.08	1.09	1.10	1.10	1.11	1.12	1.12
7	.51	.77	.87	.93	.96	.98	1.00	1.01	1.02	1.03	1.04	1.05	1.07	1.07	1.08	1.08	1.09	1.10	1.10
8	.50	.76	.86	.91	.95	.97	.99	1.00	1.01	1.02	1.03	1.04	1.05	1.06	1.07	1.07	1.08	1.08	1.09
9	.49	.75	.85	.91	.94	.96	.98	.99	1.00	1.01	1.02	1.03	1.04	1.05	1.05	1.06	1.07	1.07	1.08
10	.49	.74	.85	.90	.93	.95	.97	.98	.99	1.00	1.01	1.02	1.03	1.04	1.05	1.05	1.06	1.06	1.07
11	.49	.74	.84	.89	.93	.95	.96	.98	.99	.99	1.01	1.02	1.03	1.04	1.04	1.05	1.05	1.06	1.06
12	.48	.73	.84	.89	.92	.94	.96	.97	.98	.99	1.00	1.01	1.02	1.03	1.03	1.04	1.05	1.05	1.06
13	.48	.73	.83	.88	.92	.94	.96	.97	.98	.98	1.00	1.01	1.02	1.03	1.03	1.04	1.04	1.05	1.05
14	.48	.73	.83	.88	.91	.94	.95	.96	.97	.98	.99	1.00	1.01	1.02	1.03	1.03	1.04	1.04	1.05
15	.48	.73	.83	.88	.91	.93	.95	.96	.97	.98	.99	1.00	1.01	1.02	1.02	1.03	1.03	1.04	1.05
16	.48	.72	.82	.88	.91	.93	.95	.96	.97	.97	.99	1.00	1.01	1.02	1.02	1.03	1.03	1.04	1.04
17	.47	.72	.82	.87	.91	.93	.94	.96	.96	.97	.98	.99	1.01	1.01	1.02	1.02	1.03	1.03	1.04
18	.47	.72	.82	.87	.90	.93	.94	.95	.96	.97	.98	.99	1.00	1.01	1.02	1.02	1.03	1.03	1.04
19	.47	.72	.82	.87	.90	.92	.94	.95	.96	.97	.98	.99	1.00	1.01	1.01	1.02	1.02	1.03	1.04
20	.47	.72	.82	.87	.90	.92	.94	.95	.96	.97	.98	.99	1.00	1.01	1.01	1.02	1.02	1.03	1.03
21	.47	.72	.81	.87	.90	.92	.94	.95	.96	.96	.98	.99	1.00	1.01	1.01	1.02	1.02	1.03	1.03
22	.47	.72	.81	.87	.90	.92	.93	.95	.96	.96	.97	.99	1.00	1.00	1.01	1.01	1.02	1.03	1.03
23	.47	.71	.81	.86	.90	.92	.93	.95	.95	.96	.97	.98	1.00	1.00	1.01	1.01	1.02	1.02	1.03
24	.47	.71	.81	.86	.90	.92	.93	.94	.95	.96	.97	.98	.99	1.00	1.01	1.01	1.02	1.02	1.03
25	.47	.71	.81	.86	.89	.92	.93	.94	.95	.96	.97	.98	.99	1.00	1.00	1.01	1.02	1.02	1.03
26	.47	.71	.81	.86	.89	.91	.93	.94	.95	.96	.97	.98	.99	1.00	1.00	1.01	1.01	1.02	1.03
27	.47	.71	.81	.86	.89	.91	.93	.94	.95	.96	.97	.98	.99	1.00	1.00	1.01	1.01	1.02	1.03
28	.47	.71	.81	.86	.89	.91	.93	.94	.95	.96	.97	.98	.99	1.00	1.00	1.01	1.01	1.02	1.02
29	.47	.71	.81	.86	.89	.91	.93	.94	.95	.96	.97	.98	.99	1.00	1.00	1.01	1.01	1.02	1.02
30	.47	.71	.81	.86	.89	.91	.93	.94	.95	.96	.97	.98	.99	1.00	1.00	1.01	1.01	1.02	1.02
40	.46	.71	.80	.85	.89	.91	.92	.93	.94	.95	.96	.97	.98	.99	.99	1.00	1.01	1.01	1.02
60	.46	.70	.80	.85	.88	.90	.92	.93	.94	.94	.96	.97	.98	.98	.99	.99	1.00	1.01	1.01
120	.46	.70	.79	.84	.88	.90	.91	.92	.93	.94	.95	.96	.97	.98	.98	.99	.99	1.00	1.01
∞	.45	.69	.79	.84	.87	.89	.91	.92	.93	.93	.95	.96	.97	.97	.98	.98	.99	.99	1.00

Table B5. Critical Values $D_n^{(\gamma)}$ for the Kolmogorov Goodness-of-Fit Test[a]

	γ				
n	0.20	0.15	0.10	0.05	0.01
1	0.900	0.925	0.950	0.975	0.995
2	0.684	0.726	0.776	0.842	0.929
3	0.565	0.597	0.642	0.708	0.828
4	0.494	0.525	0.564	0.624	0.733
5	0.446	0.474	0.510	0.565	0.669
6	0.410	0.436	0.470	0.521	0.618
7	0.381	0.405	0.438	0.486	0.577
8	0.358	0.381	0.411	0.457	0.543
9	0.339	0.360	0.388	0.432	0.514
10	0.322	0.342	0.368	0.410	0.490
11	0.307	0.326	0.352	0.391	0.468
12	0.295	0.313	0.338	0.375	0.450
13	0.284	0.302	0.325	0.361	0.433
14	0.274	0.292	0.314	0.349	0.418
15	0.266	0.283	0.304	0.338	0.404
16	0.258	0.274	0.295	0.328	0.392
17	0.250	0.266	0.286	0.318	0.381
18	0.244	0.259	0.278	0.309	0.371
19	0.237	0.252	0.272	0.301	0.363
20	0.231	0.246	0.264	0.294	0.356
25	0.21	0.22	0.24	0.27	0.32
30	0.19	0.20	0.22	0.24	0.29
35	0.18	0.19	0.21	0.23	0.27
>35	$1.07/\sqrt{n}$	$1.14/\sqrt{n}$	$1.22/\sqrt{n}$	$1.36/\sqrt{n}$	$1.63/\sqrt{n}$

Example. $\Pr\{D_{\max} > D_{10}^{(0.05)}\} = \Pr\{D_{\max} > 0.410\} = 0.05$

[a]Adapted, with permission, from F. J. Massey (1951). The Kolmogorov-Smirnov Test for Goodness of Fit, *Journal of the American Statistical Association*, Vol. 46, p. 70.

Table B6. A Binomial Nomograph[a]

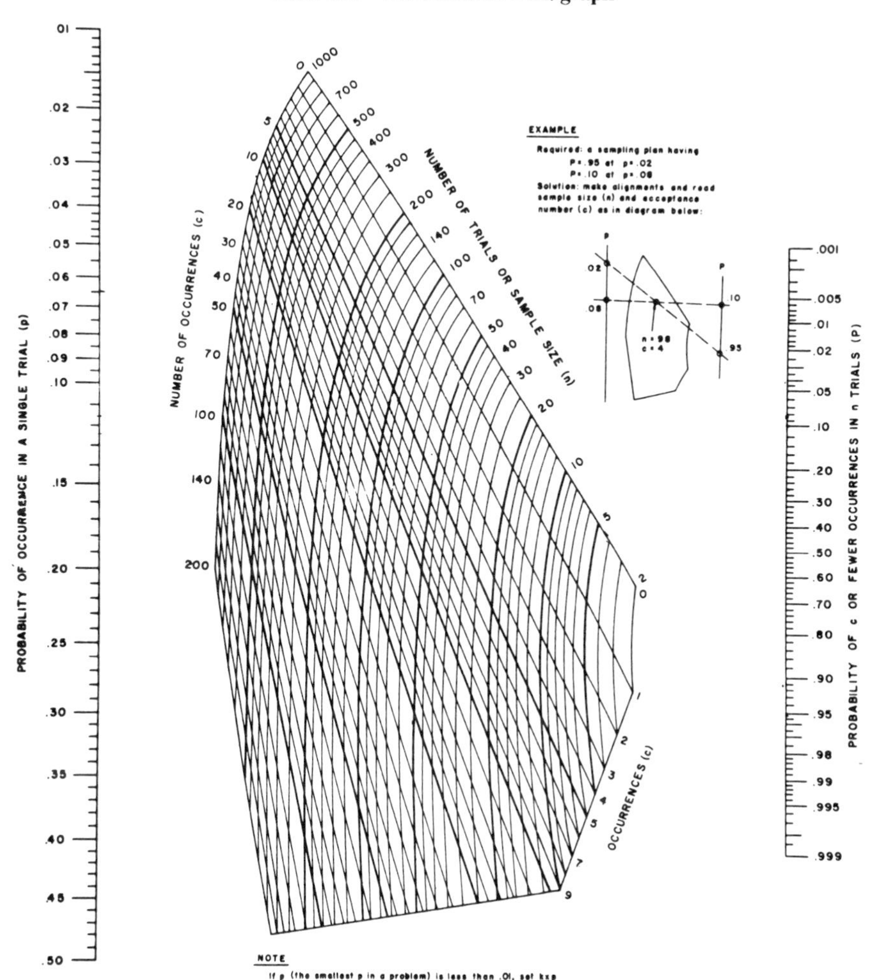

[a]Reproduced with permission, from H. R. Larson (1966). A Nomograph of the Cumulative Binomial Distribution, *Industrial Quality Control*, Vol. 23, p. 273.

Table B7. Values of ν for Selected Values of Prior Shape Parameter α_0 and Posterior Assurance $(1-\delta^*)$

α_0	$1-\delta^*$									
	0.99	.975	0.95	0.90	0.85	0.80	0.75	0.70	0.60	0.50
.0051	.085393	.003953	2.4167E-05	6.0128E-10	8.1604E-15	5.6127E-20	1.7920E-25	2.3889E-31	1.7839E-44	5.3167E-60
.0052	.089003	.004351	2.9327E-05	8.9465E-10	1.5062E-14	1.3021E-19	5.3026E-25	9.1692E-31	1.2245E-43	7.2575E-59
.0053	.092635	.004772	3.5329E-05	1.3113E-09	2.7166E-14	2.9262E-19	1.5061E-24	3.3452E-30	7.8154E-43	8.9762E-58
.0054	.096286	.005216	4.2267E-05	1.8951E-09	4.7937E-14	6.3818E-19	4.1157E-24	1.1633E-29	4.6574E-42	1.0115E-56
.0055	.099954	.005683	5.0240E-05	2.7023E-09	8.2862E-14	1.3529E-18	1.0843E-23	3.8662E-29	2.6010E-41	1.0436E-55
.0056	.103638	.006173	5.9349E-05	3.8048E-09	1.4046E-13	2.7922E-18	2.7595E-23	1.2310E-28	1.3660E-40	9.9073E-55
.0057	.107335	.006687	6.9701E-05	5.2932E-09	2.3372E-13	5.6181E-18	6.7963E-23	3.7633E-28	6.7689E-40	8.6910E-54
.0058	.111043	.007223	8.1407E-05	7.2805E-09	3.8215E-13	1.1035E-17	1.6226E-22	1.1070E-27	3.1739E-39	7.0739E-53
.0059	.114761	.007783	9.4580E-05	9.9063E-09	6.1451E-13	2.1163E-17	3.7615E-22	3.1394E-27	1.4123E-38	5.3627E-52
.0060	.118489	.008366	1.0934E-04	1.3342E-08	9.7262E-13	3.9791E-17	8.4788E-22	8.5993E-27	5.9795E-38	3.8000E-51
.0061	.122223	.008971	1.2580E-04	1.7794E-08	1.5164E-12	7.3216E-17	1.8609E-21	2.2789E-26	2.4146E-37	2.5253E-50
.0062	.125963	.009599	1.4409E-04	2.3512E-08	2.3307E-12	1.3209E-16	3.9821E-21	5.8525E-26	9.3212E-37	1.5787E-49
.0063	.129708	.010249	1.6432E-04	3.0794E-08	3.5336E-12	2.3389E-16	8.3180E-21	1.4586E-25	3.4473E-36	9.3116E-49
.0064	.133456	.010921	1.8663E-04	3.9994E-08	5.2882E-12	4.0683E-16	1.6979E-20	3.5332E-25	1.2230E-35	5.1959E-48
.0065	.137207	.011615	2.1115E-04	5.1525E-08	7.8164E-12	6.9566E-16	3.3907E-20	8.3284E-25	4.1788E-35	2.7500E-47
.0066	.140959	.012331	2.3799E-04	6.5874E-08	1.1417E-11	1.1704E-15	6.6307E-20	1.9128E-24	1.3747E-34	1.3838E-46
.0067	.144711	.013068	2.6729E-04	8.3604E-08	1.6490E-11	1.9388E-15	1.2710E-19	4.2856E-24	4.3647E-34	6.6352E-46
.0068	.148463	.013827	2.9918E-04	1.0537E-07	2.3560E-11	3.1642E-15	2.3900E-19	9.3765E-24	1.3394E-33	3.0382E-45
.0069	.152214	.014606	3.3378E-04	1.3190E-07	3.3315E-11	5.0915E-15	4.4128E-19	2.0055E-23	3.9791E-33	1.3312E-44
.0070	.155963	.015405	3.7122E-04	1.6407E-07	4.6644E-11	8.0821E-15	8.0061E-19	4.1972E-23	1.1459E-32	5.5915E-44
.0071	.159708	.016224	4.1163E-04	2.0283E-07	6.4692E-11	1.2663E-14	1.4284E-18	8.6034E-23	3.2029E-32	2.2556E-43
.0072	.163450	.017064	4.5513E-04	2.4967E-07	8.8911E-11	1.9596E-14	2.5077E-18	1.7287E-22	8.7007E-32	8.7531E-43
.0073	.167189	.017922	5.0185E-04	3.0463E-07	1.2114E-10	2.9962E-14	4.3353E-18	3.4078E-22	2.2997E-31	3.2729E-42
.0074	.170922	.018800	5.5191E-04	3.7026E-07	1.6367E-10	4.5290E-14	7.3848E-18	6.5956E-22	5.9210E-31	1.1809E-41
.0075	.174650	.019696	6.0543E-04	4.4770E-07	2.1937E-10	6.7710E-14	1.2402E-17	1.2543E-21	1.4865E-30	4.1177E-41
.0076	.178372	.020611	6.6253E-04	5.3864E-07	2.9177E-10	1.0016E-13	2.0545E-17	2.3452E-21	3.6425E-30	1.3894E-40
.0077	.182088	.021544	7.2332E-04	6.4495E-07	3.8520E-10	1.4667E-13	3.3593E-17	4.3144E-21	8.7203E-30	4.5421E-40
.0078	.185797	.022494	7.8793E-04	7.6869E-07	5.0494E-10	2.1268E-13	5.4238E-17	7.8139E-21	2.0415E-29	1.4405E-39
.0079	.189498	.023462	8.5645E-04	9.1210E-07	6.5738E-10	3.0552E-13	8.6517E-17	1.3941E-20	4.6775E-29	4.4368E-39
.0080	.193192	.024446	9.2900E-04	1.0777E-06	8.5022E-10	4.3492E-13	1.3640E-16	2.4514E-20	1.0497E-28	1.3287E-38
.0081	.196878	.025447	1.0057E-03	1.2680E-06	1.0927E-09	6.1376E-13	2.1265E-16	4.2511E-20	2.3092E-28	3.8726E-38
.0082	.200556	.026464	1.0866E-03	1.4861E-06	1.3957E-09	8.5890E-13	3.2795E-16	7.2737E-20	4.9832E-28	1.0997E-37
.0083	.204225	.027497	1.1719E-03	1.7351E-06	1.7723E-09	1.1922E-12	5.0051E-16	1.2285E-19	1.0556E-27	3.0451E-37
.0084	.207885	.028546	1.2615E-03	2.0183E-06	2.2377E-09	1.6421E-12	7.5622E-16	2.0493E-19	2.1966E-27	8.2300E-37
.0085	.211536	.029609	1.3557E-03	2.3394E-06	2.8099E-09	2.2447E-12	1.1315E-15	3.3774E-19	4.4925E-27	2.1729E-36
.0086	.215178	.030687	1.4545E-03	2.7023E-06	3.5097E-09	3.0463E-12	1.6773E-15	5.5020E-19	9.0367E-27	5.6091E-36
.0087	.218809	.031780	1.5580E-03	3.1112E-06	4.3616E-09	4.1052E-12	2.4640E-15	8.8632E-19	1.7888E-26	1.4167E-35
.0088	.222431	.032886	1.6663E-03	3.5705E-06	5.3935E-09	5.4947E-12	3.5881E-15	1.4124E-18	3.4863E-26	3.5034E-35
.0089	.226042	.034007	1.7794E-03	4.0850E-06	6.6378E-09	7.3066E-12	5.1811E-15	2.2273E-18	6.6936E-26	8.4896E-35
.0090	.229643	.035140	1.8975E-03	4.6597E-06	8.1316E-09	9.6547E-12	7.4206E-15	3.4769E-18	1.2667E-25	2.0172E-34
.0091	.233234	.036287	2.0205E-03	5.2998E-06	9.9173E-09	1.2680E-11	1.0544E-14	5.3748E-18	2.3636E-25	4.7026E-34
.0092	.236814	.037446	2.1487E-03	6.0111E-06	1.2043E-08	1.6554E-11	1.4869E-14	8.2304E-18	4.3511E-25	1.0763E-33
.0093	.240383	.038618	2.2819E-03	6.7994E-06	1.4563E-08	2.1488E-11	2.0814E-14	1.2488E-17	7.9054E-25	2.4200E-33
.0094	.243941	.039801	2.4203E-03	7.6709E-06	1.7540E-08	2.7739E-11	2.8927E-14	1.8781E-17	1.4182E-24	5.3490E-33
.0095	.247488	.040997	2.5640E-03	8.6322E-06	2.1043E-08	3.5616E-11	3.9925E-14	2.8004E-17	2.5130E-24	1.1623E-32
.0096	.251023	.042203	2.7129E-03	9.6901E-06	2.5150E-08	4.5492E-11	5.4737E-14	4.1411E-17	4.4003E-24	2.4857E-32
.0097	.254548	.043421	2.8672E-03	1.0852E-05	2.9948E-08	5.7815E-11	7.4556E-14	6.0742E-17	7.6166E-24	5.2330E-32
.0098	.258061	.044650	3.0269E-03	1.2125E-05	3.5535E-08	7.3118E-11	1.0091E-13	8.8406E-17	1.3037E-23	1.0851E-31
.0099	.261563	.045889	3.1920E-03	1.3517E-05	4.2018E-08	9.2034E-11	1.3576E-13	1.2770E-16	2.2074E-23	2.2170E-31
.0100	.265053	.047138	3.3626E-03	1.5036E-05	4.9518E-08	1.1531E-10	1.8155E-13	1.8310E-16	3.6983E-23	4.4655E-31

a_0	$1-\delta^*$ 0.99	.975	0.95	0.90	0.85	0.80	0.75	0.70	0.60	0.50
.0110	.299307	.060131	5.3753E-03	3.9217E-05	2.1715E-07	8.7746E-10	2.4840E-12	4.6906E-15	3.8473E-21	2.4370E-28
.0120	.332376	.073864	7.9546E-03	8.7195E-05	7.4443E-07	4.7615E-09	2.1980E-11	6.9998E-14	1.8458E-19	4.6531E-26
.0130	.364275	.088128	1.1094E-02	1.7147E-04	2.1116E-06	1.9921E-08	1.3908E-10	6.8928E-13	4.8825E-18	3.9608E-24
.0140	.395042	.102761	1.4771E-02	3.0621E-04	5.1616E-06	6.7942E-08	6.7622E-10	4.8962E-12	8.0900E-17	1.7871E-22
.0150	.424727	.117634	1.8950E-02	5.0624E-04	1.1200E-05	1.9678E-07	2.6631E-09	2.6782E-11	9.2198E-16	4.8526E-21
.0160	.453386	.132646	2.3591E-02	7.8611E-04	2.2064E-05	4.9992E-07	8.8373E-09	1.1847E-10	7.7525E-15	8.7220E-20
.0170	.481076	.147718	2.8649E-02	1.1594E-03	4.0136E-05	1.1344E-06	2.5469E-08	4.4001E-10	5.0747E-14	1.1161E-18
.0180	.507854	.162790	3.4083E-02	1.6380E-03	6.8320E-05	2.3540E-06	6.5261E-08	1.4126E-09	2.6963E-13	1.0760E-17
.0190	.533774	.177815	3.9849E-02	2.2341E-03	1.0998E-04	4.5240E-06	1.5147E-07	4.0115E-09	1.2017E-12	8.1726E-17
.0200	.558886	.192756	4.5910E-02	2.9497E-03	1.6882E-04	8.1451E-06	3.2319E-07	1.0264E-08	4.6127E-12	5.0687E-16
.0210	.583237	.207587	5.2228E-02	3.7969E-03	2.4880E-04	1.3867E-05	6.4162E-07	2.4014E-08	1.5578E-11	2.6423E-15
.0220	.606873	.222286	5.8770E-02	4.7779E-03	3.5401E-04	2.2496E-05	1.1969E-06	5.2011E-08	4.7104E-11	1.1855E-14
.0230	.629835	.236837	6.5506E-02	5.8950E-03	4.8855E-04	3.4993E-05	2.1151E-06	1.0534E-07	1.2937E-10	4.6686E-14
.0240	.652160	.251230	7.2409E-02	7.1492E-03	6.5644E-04	5.2468E-05	3.5646E-06	2.0116E-07	3.2666E-10	1.6401E-13
.0250	.673884	.265454	7.9454E-02	8.5399E-03	8.6152E-04	7.6168E-05	5.7623E-06	3.6481E-07	7.6597E-10	5.2112E-13
.0260	.695041	.279506	8.6620E-02	1.0066E-02	1.1074E-03	1.0746E-04	8.9775E-06	6.3201E-07	1.6821E-09	1.5150E-12
.0270	.715660	.293382	9.3888E-02	1.1724E-02	1.3974E-03	1.4779E-04	1.3536E-05	1.0513E-06	3.4851E-09	4.0699E-12
.0280	.735770	.307079	.101240	1.3511E-02	1.7345E-03	1.9870E-04	1.9820E-05	1.6865E-06	6.8551E-09	1.0189E-11
.0290	.755396	.320597	.108662	1.5423E-02	2.1213E-03	2.6177E-04	2.8269E-05	2.6187E-06	1.2870E-08	2.3944E-11
.0300	.774564	.333936	.116138	1.7457E-02	2.5601E-03	3.3860E-04	3.9380E-05	3.9490E-06	2.3170E-08	5.3155E-11
.0310	.793296	.347097	.123659	1.9606E-02	3.0528E-03	4.3080E-04	5.3699E-05	5.7995E-06	4.0162E-08	1.1209E-10
.0320	.811612	.360083	.131212	2.1869E-02	3.6011E-03	5.3996E-04	7.1824E-05	8.3156E-06	6.7267E-08	2.2562E-10
.0330	.829533	.372895	.138788	2.4236E-02	4.2060E-03	6.6763E-04	9.4393E-05	1.1666E-05	1.0920E-07	4.3531E-10
.0340	.847076	.385536	.146381	2.6706E-02	4.8685E-03	8.1529E-04	1.2208E-04	1.6045E-05	1.7231E-07	8.0806E-10
.0350	.864259	.398009	.153981	2.9274E-02	5.5892E-03	9.8439E-04	1.5560E-04	2.1670E-05	2.6490E-07	1.4480E-09
.0360	.881098	.410317	.161582	3.1930E-02	6.3685E-03	1.1763E-03	1.9567E-04	2.8783E-05	3.9764E-07	2.5120E-09
.0370	.897607	.422463	.169179	3.4675E-02	7.2064E-03	1.3922E-03	2.4304E-04	3.7651E-05	5.8397E-07	4.2302E-09
.0380	.913801	.434451	.176768	3.7502E-02	8.1028E-03	1.6333E-03	2.9848E-04	4.8562E-05	8.4047E-07	6.9311E-09
.0390	.929693	.446284	.184342	4.0407E-02	9.0573E-03	1.9007E-03	3.6273E-04	6.1826E-05	1.1873E-06	1.1073E-08
.0400	.945295	.457965	.191899	4.3384E-02	1.0069E-02	2.1953E-03	4.3656E-04	7.7771E-05	1.6486E-06	1.7282E-08
.0410	.960618	.469498	.199435	4.6430E-02	1.1138E-02	2.5181E-03	5.2073E-04	9.6744E-05	2.2529E-06	2.6393E-08
.0420	.975675	.480886	.206947	4.9541E-02	1.2264E-02	2.8697E-03	6.1595E-04	1.1910E-04	3.0333E-06	3.9504E-08
.0430	.990474	.492133	.214433	5.2713E-02	1.3444E-02	3.2598E-03	7.2296E-04	1.4523E-04	4.0280E-06	5.8032E-08
.0440	1.005026	.503242	.221890	5.5942E-02	1.4678E-02	3.6621E-03	8.4244E-04	1.7550E-04	5.2807E-06	8.3775E-08
.0450	1.019340	.514216	.229316	5.9223E-02	1.5966E-02	4.1039E-03	9.7508E-04	2.1032E-04	6.8404E-06	1.1899E-07
.0460	1.033425	.525058	.236710	6.2555E-02	1.7305E-02	4.5767E-03	1.1215E-03	2.5007E-04	8.7619E-06	1.6644E-07
.0470	1.047288	.535772	.244069	6.5933E-02	1.8695E-02	5.0897E-03	1.2823E-03	2.9517E-04	1.1106E-05	2.2953E-07
.0480	1.060938	.546361	.251394	6.9355E-02	2.0135E-02	5.6162E-03	1.4581E-03	3.4601E-04	1.3939E-05	3.1233E-07
.0490	1.074382	.556827	.258682	7.2817E-02	2.1622E-02	6.1833E-03	1.6494E-03	4.0301E-04	1.7334E-05	4.1971E-07
.0500	1.087627	.567174	.265932	7.6317E-02	2.3156E-02	6.7820E-03	1.8567E-03	4.6656E-04	2.1369E-05	5.5739E-07
.0510	1.100680	.577404	.273145	7.9852E-02	2.4735E-02	7.4124E-03	2.0805E-03	5.3706E-04	2.6130E-05	7.3206E-07
.0520	1.113547	.587520	.280319	8.3420E-02	2.6359E-02	8.0744E-03	2.3213E-03	6.1490E-04	3.1705E-05	9.5148E-07
.0530	1.126235	.597525	.287454	8.7018E-02	2.8025E-02	8.7679E-03	2.5793E-03	7.0047E-04	3.8192E-05	1.2245E-06
.0540	1.138747	.607422	.294549	9.0644E-02	2.9732E-02	9.4927E-03	2.8550E-03	7.9413E-04	4.5690E-05	1.5613E-06
.0550	1.151092	.617213	.301605	9.4295E-02	3.1480E-02	1.0249E-02	3.1488E-03	8.9624E-04	5.4308E-05	1.9733E-06
.0560	1.163273	.626900	.308620	9.7971E-02	3.3266E-02	1.1035E-02	3.4607E-03	1.0072E-03	6.4156E-05	2.4732E-06
.0570	1.175294	.636486	.315595	.101668	3.5089E-02	1.1853E-02	3.7913E-03	1.1272E-03	7.5350E-05	3.0754E-06
.0580	1.187163	.645973	.322530	.105386	3.6948E-02	1.2700E-02	4.1405E-03	1.2568E-03	8.8010E-05	3.7958E-06
.0590	1.198882	.655364	.329425	.109122	3.8843E-02	1.3577E-02	4.5087E-03	1.3961E-03	1.0226E-04	4.6517E-06
.0600	1.210456	.664660	.336278	.112876	4.0770E-02	1.4484E-02	4.8960E-03	1.5455E-03	1.1823E-04	5.6622E-06

Table B7. (*Continued*)

a_0	$1-\delta^*$: 0.99	.975	0.95	0.90	0.85	0.80	0.75	0.70	0.60	0.50
.0610	1.221890	.673865	.343092	.116644	4.2731E-02	1.5420E-02	5.3025E-03	1.7053E-03	1.3604E-04	6.8481E-06
.0620	1.233187	.682979	.349866	.120427	4.4722E-02	1.6384E-02	5.7282E-03	1.8757E-03	1.5584E-04	8.2320E-06
.0630	1.244351	.692005	.356599	.124222	4.6743E-02	1.7377E-02	6.1734E-03	2.0570E-03	1.7776E-04	9.8381E-06
.0640	1.255386	.700944	.363293	.128028	4.8794E-02	1.8397E-02	6.6379E-03	2.2494E-03	2.0193E-04	1.1692E-05
.0650	1.266295	.709800	.369947	.131845	5.0872E-02	1.9445E-02	7.1219E-03	2.4532E-03	2.2849E-04	1.3823E-05
.0660	1.277081	.718573	.376562	.135671	5.2976E-02	2.0519E-02	7.6253E-03	2.6684E-03	2.5759E-04	1.6260E-05
.0670	1.287749	.727265	.383137	.139504	5.5107E-02	2.1619E-02	8.1481E-03	2.8954E-03	2.8936E-04	1.9034E-05
.0680	1.298300	.735878	.389674	.143345	5.7262E-02	2.2745E-02	8.6902E-03	3.1343E-03	3.2396E-04	2.2179E-05
.0690	1.308739	.744414	.396172	.147192	5.9441E-02	2.3897E-02	9.2516E-03	3.3853E-03	3.6151E-04	2.5729E-05
.0700	1.319067	.752874	.402632	.151044	6.1643E-02	2.5073E-02	9.8322E-03	3.6484E-03	4.0216E-04	2.9723E-05
.0710	1.329288	.761260	.409054	.154900	6.3866E-02	2.6273E-02	1.0432E-02	3.9239E-03	4.4606E-04	3.4198E-05
.0720	1.339404	.769574	.415438	.158760	6.6111E-02	2.7496E-02	1.1050E-02	4.2118E-03	4.9334E-04	3.9194E-05
.0730	1.349417	.777816	.421785	.162623	6.8375E-02	2.8743E-02	1.1688E-02	4.5122E-03	5.4413E-04	4.4754E-05
.0740	1.359332	.785988	.428095	.166488	7.0659E-02	3.0013E-02	1.2344E-02	4.8253E-03	5.9859E-04	5.0920E-05
.0750	1.369148	.794093	.434369	.170354	7.2962E-02	3.1304E-02	1.3019E-02	5.1510E-03	6.5684E-04	5.7739E-05
.0760	1.378870	.802130	.440607	.174221	7.5282E-02	3.2617E-02	1.3712E-02	5.4896E-03	7.1901E-04	6.5255E-05
.0770	1.388499	.810102	.446809	.178088	7.7619E-02	3.3951E-02	1.4423E-02	5.8409E-03	7.8525E-04	7.3517E-05
.0780	1.398037	.818009	.452975	.181955	7.9972E-02	3.5306E-02	1.5152E-02	6.2051E-03	8.5566E-04	8.2575E-05
.0790	1.407486	.825854	.459106	.185821	8.2341E-02	3.6681E-02	1.5899E-02	6.5821E-03	9.3040E-04	9.2477E-05
.0800	1.416849	.833636	.465203	.189686	8.4725E-02	3.8075E-02	1.6664E-02	6.9721E-03	1.0096E-03	1.0328E-04
.0810	1.426127	.841358	.471266	.193549	8.7123E-02	3.9489E-02	1.7446E-02	7.3749E-03	1.0933E-03	1.1503E-04
.0820	1.435323	.849020	.477294	.197410	8.9534E-02	4.0921E-02	1.8245E-02	7.7906E-03	1.1817E-03	1.2778E-04
.0830	1.444437	.856624	.483289	.201268	9.1959E-02	4.2371E-02	1.9061E-02	8.2192E-03	1.2749E-03	1.4159E-04
.0840	1.453472	.864171	.489251	.205123	9.4396E-02	4.3839E-02	1.9894E-02	8.6607E-03	1.3729E-03	1.5651E-04
.0850	1.462429	.871661	.495181	.208974	9.6844E-02	4.5325E-02	2.0744E-02	9.1150E-03	1.4760E-03	1.7260E-04
.0860	1.471311	.879096	.501077	.212822	9.9304E-02	4.6827E-02	2.1610E-02	9.5820E-03	1.5842E-03	1.8991E-04
.0870	1.480119	.886477	.506942	.216666	.101775	4.8345E-02	2.2491E-02	1.0062E-02	1.6976E-03	2.0851E-04
.0880	1.488852	.893805	.512775	.220506	.104256	4.9880E-02	2.3389E-02	1.0554E-02	1.8163E-03	2.2844E-04
.0890	1.497515	.901080	.518577	.224341	.106746	5.1429E-02	2.4302E-02	1.1060E-02	1.9404E-03	2.4977E-04
.0900	1.506107	.908304	.524348	.228171	.109246	5.2994E-02	2.5230E-02	1.1577E-02	2.0700E-03	2.7256E-04
.0910	1.514632	.915477	.530088	.231996	.111755	5.4574E-02	2.6174E-02	1.2107E-02	2.2052E-03	2.9687E-04
.0920	1.523089	.922601	.535799	.235815	.114272	5.6168E-02	2.7133E-02	1.2650E-02	2.3460E-03	3.2274E-04
.0930	1.531480	.929676	.541479	.239629	.116797	5.7776E-02	2.8106E-02	1.3205E-02	2.4925E-03	3.5025E-04
.0940	1.539807	.936703	.547130	.243437	.119330	5.9397E-02	2.9093E-02	1.3772E-02	2.6449E-03	3.7945E-04
.0950	1.548070	.943684	.552752	.247239	.121870	6.1032E-02	3.0095E-02	1.4352E-02	2.8031E-03	4.1039E-04
.0960	1.556271	.950617	.558345	.251035	.124416	6.2679E-02	3.1110E-02	1.4943E-02	2.9672E-03	4.4315E-04
.0970	1.564412	.957506	.563909	.254825	.126969	6.4339E-02	3.2139E-02	1.5547E-02	3.1374E-03	4.7777E-04
.0980	1.572492	.964349	.569445	.258608	.129528	6.6011E-02	3.3182E-02	1.6162E-02	3.3136E-03	5.1431E-04
.0990	1.580514	.971149	.574954	.262385	.132093	6.7695E-02	3.4238E-02	1.6789E-02	3.4959E-03	5.5283E-04
.1000	1.588478	.977905	.580435	.266155	.134663	6.9390E-02	3.5306E-02	1.7428E-02	3.6845E-03	5.9339E-04
.1100	1.665165	1.043231	.633806	.303456	.160598	8.6902E-02	4.6657E-02	2.4432E-02	5.9177E-03	1.1232E-03
.1200	1.737082	1.104905	.684762	.339989	.186791	.105245	5.9065E-02	3.2473E-02	8.8009E-03	1.9143E-03
.1300	1.804937	1.163423	.733565	.375720	.213060	.124197	7.2338E-02	4.1427E-02	1.2338E-02	3.0102E-03
.1400	1.869291	1.219187	.780442	.410651	.239282	.143589	8.6317E-02	5.1175E-02	1.6514E-02	4.4431E-03
.1500	1.930593	1.272528	.825587	.444799	.265371	.163287	.100865	6.1608E-02	2.1299E-02	6.2348E-03
.1600	1.989215	1.323719	.869170	.478191	.291271	.183190	.115870	7.2629E-02	2.6657E-02	8.3974E-03
.1700	2.045460	1.372991	.911335	.510863	.316944	.203222	.131241	8.4151E-02	3.2549E-02	1.0935E-02
.1800	2.099584	1.420538	.952211	.542849	.342364	.223320	.146903	9.6101E-02	3.8932E-02	1.3844E-02
.1900	2.151805	1.466528	.991908	.574183	.367518	.243439	.162792	.108415	4.5766E-02	1.7119E-02
.2000	2.202305	1.511103	1.030526	.604902	.392398	.263544	.178859	.121038	5.3011E-02	2.0746E-02

a_0	$1 - \delta^*$									
	0.99	.975	0.95	0.90	0.85	0.80	0.75	0.70	0.60	0.50
.2100	2.251244	1.554388	1.068149	.635041	.417000	.283606	.195062	.133922	.060630	.024714
.2200	2.298758	1.596493	1.104856	.664630	.441326	.303605	.211366	.147027	.068589	.029007
.2300	2.344969	1.637511	1.140712	.693699	.465379	.323526	.227743	.160319	.076857	.033609
.2400	2.389980	1.677528	1.175781	.722278	.489164	.343355	.244169	.173768	.085405	.038503
.2500	2.433885	1.716618	1.210116	.750393	.512687	.363085	.260626	.187348	.094206	.043674
.2600	2.476767	1.754847	1.243767	.778068	.535954	.382709	.277098	.201039	.103236	.049105
.2700	2.518697	1.792275	1.276777	.805327	.558975	.402222	.293571	.214821	.112475	.054780
.2800	2.559743	1.828956	1.309187	.832190	.581755	.421622	.310034	.228578	.121902	.060686
.2900	2.599963	1.864938	1.341034	.858679	.604304	.440906	.326480	.242597	.131500	.066807
.3000	2.639409	1.900264	1.372350	.884811	.626628	.460074	.342899	.256565	.141253	.073131
.3100	2.678131	1.934974	1.403167	.910604	.648735	.479126	.359288	.270572	.151146	.079645
.3200	2.716171	1.969104	1.433511	.936074	.670632	.498062	.375640	.284609	.161168	.086336
.3300	2.753570	2.002688	1.463409	.961237	.692328	.516884	.391951	.298669	.171306	.093194
.3400	2.790363	2.035755	1.492884	.986106	.713829	.535593	.408220	.312745	.181549	.100208
.3500	2.826584	2.068333	1.521958	1.010695	.735141	.554190	.424442	.326831	.191888	.107369
.3600	2.862264	2.100447	1.550651	1.035016	.756272	.572677	.440616	.340922	.202315	.114668
.3700	2.897430	2.132122	1.578981	1.059081	.777228	.591057	.456740	.355013	.212822	.122096
.3800	2.932109	2.163379	1.606966	1.082900	.798014	.609331	.472814	.369102	.223401	.129645
.3900	2.966323	2.194237	1.634621	1.106484	.818637	.627502	.488836	.383184	.234047	.137308
.4000	3.000097	2.224716	1.661962	1.129843	.839102	.645571	.504806	.397257	.244752	.145078
.4100	3.033449	2.254833	1.689002	1.152985	.859414	.663541	.520724	.411319	.255513	.152949
.4200	3.066400	2.284604	1.715755	1.175918	.879579	.681415	.536589	.425367	.266324	.160915
.4300	3.098966	2.314045	1.742232	1.198652	.899601	.699194	.552402	.439400	.277181	.168970
.4400	3.131165	2.343168	1.768445	1.221193	.919485	.716881	.568163	.453416	.288079	.177110
.4500	3.163012	2.371988	1.794404	1.243549	.939235	.734478	.583871	.467414	.299016	.185328
.4600	3.194522	2.400517	1.820120	1.265726	.958856	.751987	.599529	.481393	.309987	.193622
.4700	3.225708	2.428767	1.845602	1.287731	.978352	.769411	.615135	.495351	.320990	.201987
.4800	3.256584	2.456747	1.870858	1.309570	.997726	.786751	.630690	.509288	.332022	.210419
.4900	3.287160	2.484470	1.895898	1.331248	1.016983	.804009	.646196	.523204	.343081	.218913
.5000	3.317448	2.511943	1.920729	1.352772	1.036125	.821187	.661652	.537097	.354163	.227468
.5100	3.347460	2.539177	1.945359	1.374145	1.055157	.838288	.677059	.550968	.365267	.236080
.5200	3.377204	2.566180	1.969794	1.395374	1.074082	.855313	.692419	.564815	.376391	.244745
.5300	3.406691	2.592959	1.994042	1.416462	1.092902	.872264	.707731	.578640	.387534	.253461
.5400	3.435930	2.619524	2.018108	1.437415	1.111621	.889143	.722997	.592441	.398692	.262226
.5500	3.464928	2.645880	2.041999	1.458236	1.130241	.905951	.738216	.606218	.409866	.271036
.5600	3.493695	2.672036	2.065720	1.478930	1.148766	.922690	.753391	.619971	.421052	.279891
.5700	3.522238	2.697997	2.089276	1.499500	1.167198	.939363	.768521	.633701	.432251	.288786
.5800	3.550563	2.723769	2.112674	1.519950	1.185539	.955969	.783608	.647407	.443461	.297721
.5900	3.578678	2.749360	2.135918	1.540283	1.203793	.972512	.798652	.661089	.454681	.306694
.6000	3.606589	2.774773	2.159012	1.560503	1.221960	.988992	.813654	.674748	.465909	.315702
.6100	3.634303	2.800016	2.181961	1.580614	1.240044	1.005411	.828614	.688383	.477146	.324744
.6200	3.661826	2.825091	2.204769	1.600617	1.258047	1.021770	.843533	.701995	.488389	.333819
.6300	3.689162	2.850006	2.227441	1.620517	1.275970	1.038070	.858413	.715583	.499638	.342925
.6400	3.716318	2.874764	2.249980	1.640316	1.293817	1.054314	.873253	.729148	.510893	.352060
.6500	3.743299	2.899370	2.272389	1.660016	1.311588	1.070502	.888055	.742690	.522152	.361224
.6600	3.770109	2.923827	2.294673	1.679620	1.329285	1.086635	.902818	.756210	.533416	.370415
.6700	3.796754	2.948141	2.316835	1.699131	1.346911	1.102714	.917545	.769706	.544683	.379631
.6800	3.823238	2.972314	2.338878	1.718552	1.364467	1.118742	.932235	.783181	.555953	.388872
.6900	3.849565	2.996352	2.360806	1.737883	1.381955	1.134718	.946889	.796633	.567226	.398137
.7000	3.875739	3.020256	2.382620	1.757129	1.399376	1.150644	.961507	.810064	.578500	.407424

Table B7. (*Continued*)

a_0	$1-\delta^*$									
	0.99	.975	0.95	0.90	0.85	0.80	0.75	0.70	0.60	0.50
.7100	3.901764	3.044032	2.404325	1.776289	1.416732	1.166522	.976091	.823473	.589776	.416733
.7200	3.927644	3.067681	2.425922	1.795368	1.434024	1.182351	.990641	.836860	.601053	.426062
.7300	3.953383	3.091207	2.447415	1.814366	1.451254	1.198133	1.005157	.850226	.612331	.435412
.7400	3.978985	3.114614	2.468806	1.833286	1.468424	1.213869	1.019640	.863571	.623609	.444780
.7500	4.004452	3.137904	2.490098	1.852130	1.485533	1.229561	1.034091	.876895	.634888	.454167
.7600	4.029787	3.161080	2.511292	1.870899	1.502585	1.245208	1.048511	.890199	.646166	.463571
.7700	4.054995	3.184144	2.532392	1.889594	1.519580	1.260811	1.062899	.903483	.657444	.472993
.7800	4.080078	3.207100	2.553399	1.908218	1.536519	1.276373	1.077256	.916747	.668721	.482430
.7900	4.105038	3.229949	2.574315	1.926773	1.553403	1.291892	1.091583	.929991	.679996	.491883
.8000	4.129879	3.252695	2.595144	1.945258	1.570234	1.307371	1.105881	.943215	.691271	.501351
.8100	4.154603	3.275339	2.615885	1.963677	1.587013	1.322810	1.120149	.956420	.702545	.510834
.8200	4.179214	3.297884	2.636542	1.982031	1.603741	1.338209	1.134388	.969607	.713816	.520330
.8300	4.203712	3.320331	2.657117	2.000321	1.620418	1.353570	1.148600	.982774	.725086	.529840
.8400	4.228102	3.342684	2.677610	2.018547	1.637046	1.368894	1.162784	.995923	.736355	.539362
.8500	4.252384	3.364944	2.698024	2.036713	1.653626	1.384180	1.176940	1.009054	.747621	.548897
.8600	4.276562	3.387113	2.718360	2.054818	1.670159	1.399429	1.191070	1.022166	.758884	.558444
.8700	4.300638	3.409192	2.738621	2.072864	1.686646	1.414643	1.205173	1.035261	.770146	.568003
.8800	4.324613	3.431185	2.758807	2.090852	1.703086	1.429822	1.219250	1.048338	.781405	.577572
.8900	4.348490	3.453092	2.778920	2.108784	1.719483	1.444966	1.233302	1.061398	.792661	.587152
.9000	4.372270	3.474915	2.798961	2.126660	1.735835	1.460076	1.247328	1.074441	.803915	.596743
.9100	4.395956	3.496656	2.818932	2.144482	1.752145	1.475153	1.261330	1.087467	.815166	.606344
.9200	4.419550	3.518317	2.838835	2.162250	1.768412	1.490198	1.275308	1.100476	.826415	.615954
.9300	4.443052	3.539898	2.858670	2.179966	1.784638	1.505210	1.289261	1.113468	.837660	.625574
.9400	4.466466	3.561402	2.878439	2.197630	1.800823	1.520190	1.303191	1.126445	.848902	.635202
.9500	4.489793	3.582830	2.898144	2.215243	1.816969	1.535140	1.317098	1.139405	.860142	.644839
.9600	4.513034	3.604184	2.917784	2.232808	1.833075	1.550058	1.330981	1.152350	.871378	.654485
.9700	4.536191	3.625464	2.937362	2.250323	1.849142	1.564947	1.344843	1.165278	.882611	.664139
.9800	4.559264	3.646673	2.956879	2.267790	1.865172	1.579806	1.358682	1.178192	.893841	.673801
.9900	4.582257	3.667811	2.976335	2.285211	1.881154	1.594636	1.372499	1.191090	.905067	.683470
1.0000	4.605170	3.688879	2.995732	2.302585	1.897120	1.609438	1.386294	1.203973	.916291	.693147
1.1000	4.830207	3.896009	3.186665	2.473953	2.054791	1.755974	1.523133	1.332016	1.028337	.790275
1.2000	5.048610	4.097368	3.372663	2.641460	2.209386	1.900089	1.658130	1.458746	1.140034	.887936
1.3000	5.261312	4.293777	3.554429	2.805648	2.361325	2.042103	1.791513	1.584312	1.251383	.985999
1.4000	5.469054	4.485882	3.732515	2.966944	2.510947	2.182274	1.923476	1.708839	1.362394	1.084372
1.5000	5.672434	4.674202	3.907365	3.125694	2.658524	2.320814	2.054173	1.832435	1.473083	1.182987
1.6000	5.871938	4.859161	4.079340	3.282184	2.804280	2.457898	2.183735	1.955190	1.583468	1.281796
1.7000	6.067967	5.041111	4.248742	3.436646	2.948404	2.593675	2.312273	2.077183	1.693566	1.380762
1.8000	6.260862	5.220348	4.415828	3.589282	3.091050	2.728265	2.439884	2.198480	1.803395	1.479857
1.9000	6.450908	5.397120	4.580807	3.740258	3.232358	2.861780	2.566647	2.319141	1.912973	1.579058
2.0000	6.638354	5.571643	4.743866	3.889720	3.372440	2.994308	2.692635	2.439216	2.022313	1.678347
2.1000	6.823411	5.744103	4.905160	4.037792	3.511401	3.125932	2.817907	2.558751	2.131432	1.777712
2.2000	7.006262	5.914661	5.064831	4.184584	3.649324	3.256717	2.942518	2.677785	2.240341	1.877141
2.3000	7.187075	6.083457	5.223002	4.330193	3.786287	3.386729	3.066517	2.796353	2.349054	1.976625
2.4000	7.365991	6.250616	5.379776	4.474698	3.922359	3.516020	3.189945	2.914488	2.457581	2.076157
2.5000	7.543136	6.416252	5.535249	4.618178	4.057600	3.644637	3.312840	3.032216	2.565934	2.175730
2.6000	7.718626	6.580461	5.689507	4.760700	4.192063	3.772629	3.435236	3.149562	2.674120	2.275339
2.7000	7.892568	6.743333	5.842625	4.902326	4.325799	3.900030	3.557164	3.266552	2.782151	2.374980
2.8000	8.065047	6.904947	5.994674	5.043103	4.458850	4.026877	3.678652	3.383203	2.890033	2.474650
2.9000	8.236147	7.065379	6.145709	5.183088	4.591255	4.153199	3.799724	3.499536	2.997774	2.574344
3.0000	8.405951	7.224687	6.295794	5.322320	4.723050	4.279030	3.920403	3.615568	3.105379	2.674061

a_0	$1 - \delta^*$									
	0.99	.975	0.95	0.90	0.85	0.80	0.75	0.70	0.60	0.50
3.1000	8.574519	7.382939	6.444977	5.460843	4.854271	4.404393	4.040710	3.731314	3.212857	2.773797
3.2000	8.741913	7.540188	6.593303	5.598691	4.984944	4.529312	4.160664	3.846789	3.320213	2.873551
3.3000	8.908206	7.696479	6.740820	5.735901	5.115098	4.653811	4.280279	3.962007	3.427453	2.973322
3.4000	9.073433	7.851864	6.887563	5.872497	5.244759	4.777909	4.399579	4.076978	3.534582	3.073107
3.5000	9.237654	8.006381	7.033568	6.008518	5.373948	4.901625	4.518574	4.191715	3.641604	3.172907
3.6000	9.400913	8.160075	7.178876	6.143985	5.502689	5.024977	4.637277	4.306229	3.748524	3.272717
3.7000	9.563252	8.312978	7.323508	6.278921	5.631001	5.147980	4.755703	4.420530	3.855346	3.372538
3.8000	9.724709	8.465125	7.467498	6.413355	5.758900	5.270650	4.873863	4.534621	3.962074	3.472370
3.9000	9.885325	8.616545	7.610876	6.547301	5.886407	5.393001	4.991767	4.648519	4.068712	3.572211
4.0000	10.045118	8.767274	7.753657	6.680783	6.013537	5.515046	5.109430	4.762229	4.175263	3.672061
4.1000	10.204140	8.917333	7.895874	6.813819	6.140305	5.636796	5.226851	4.875757	4.281728	3.771918
4.2000	10.362412	9.066751	8.037546	6.946426	6.266724	5.758266	5.344051	4.989110	4.388118	3.871783
4.3000	10.519965	9.215550	8.178691	7.078618	6.392811	5.879458	5.461033	5.102295	4.494426	3.971656
4.4000	10.676814	9.363754	8.319335	7.210415	6.518569	6.000392	5.577805	5.215319	4.600659	4.071533
4.5000	10.832998	9.511386	8.459489	7.341828	6.644020	6.121073	5.694376	5.328186	4.706820	4.171417
4.6000	10.988535	9.658461	8.599174	7.472874	6.769170	6.241509	5.810751	5.440905	4.812911	4.271305
4.7000	11.143445	9.805005	8.738404	7.603558	6.894031	6.361711	5.926936	5.553474	4.918937	4.371201
4.8000	11.297752	9.951026	8.877196	7.733902	7.018612	6.481687	6.042944	5.665904	5.024892	4.471100
4.9000	11.451474	10.096550	9.015564	7.863905	7.142921	6.601438	6.158772	5.778199	5.130786	4.571002
5.0000	11.604626	10.241591	9.153519	7.993589	7.266964	6.720979	6.274431	5.890363	5.236618	4.670909
5.5000	12.362488	10.960025	9.837569	8.637505	7.883548	7.315710	6.850347	6.449336	5.764916	5.170499
6.0000	13.108487	11.668333	10.513032	9.274674	8.494657	7.905993	7.422702	7.005551	6.291919	5.670162
6.5000	13.844121	12.367804	11.181017	9.905965	9.100988	8.492397	7.991953	7.559361	6.817786	6.169878
7.0000	14.570621	13.059480	11.842396	10.532072	9.703121	9.075384	8.558467	8.111049	7.342646	6.669638
7.5000	15.288963	13.744197	12.497896	11.153566	10.301504	9.655329	9.122543	8.660847	7.866611	7.169430
8.0000	15.999966	14.422678	13.148115	11.770915	10.896529	10.232544	9.684431	9.208947	8.389768	7.669251
8.5000	16.704332	15.095506	13.793557	12.384520	11.488514	10.807280	10.244338	9.755507	8.912195	8.169092
9.0000	17.402656	15.763192	14.434646	12.994712	12.077736	11.379774	10.802445	10.300677	9.433952	8.668950
9.5000	18.095438	16.426164	15.071766	13.601792	12.664427	11.950209	11.358903	10.844563	9.955096	9.168827
10.0000	18.783123	17.084810	15.705213	14.205991	13.248793	12.518752	11.913846	11.387274	10.475684	9.668716
11.0000	20.144684	18.390357	16.962222	15.406642	14.411228	13.650727	13.019633	12.469508	11.515331	10.668524
12.0000	21.489914	19.682034	18.207521	16.598123	15.566237	14.776661	14.120579	13.547985	12.553176	11.668363
13.0000	22.820845	20.961583	19.442570	17.781586	16.714734	15.897311	15.217283	14.623163	13.589441	12.668229
14.0000	24.139129	22.230398	20.668571	18.957964	17.857494	17.013283	16.310247	15.695437	14.624309	13.668115
15.0000	25.446105	23.489624	21.886488	20.128012	18.995128	18.125096	17.399879	16.765117	15.657928	14.668016
16.0000	26.742891	24.740213	23.097125	21.292374	20.128149	19.233157	18.486495	17.832455	16.690432	15.667930
17.0000	28.030459	25.982992	24.301184	22.451580	21.257004	20.337825	19.570390	18.897693	17.721915	16.667854
18.0000	29.309600	27.218645	25.499231	23.606091	22.382035	21.439396	20.651808	19.960991	18.752470	17.667791
19.0000	30.581062	28.447777	26.691771	24.756292	23.503585	22.538139	21.730954	21.022525	19.782180	18.667726
20.0000	31.845381	29.670858	27.879242	25.902540	24.621929	23.634281	22.808003	22.082438	20.811097	19.667673
21.0000	33.103128	30.888383	29.062021	27.045102	25.737296	24.727986	23.883126	23.140838	21.839302	20.667624
22.0000	34.354764	32.100744	30.240440	28.184271	26.849910	25.819462	24.956450	24.197847	22.866819	21.667583
23.0000	35.600707	33.308266	31.414812	29.320269	27.959956	26.908856	26.028098	25.253553	23.893720	22.667548
24.0000	36.841330	34.511301	32.585374	30.453305	29.067603	27.996296	27.098184	26.308045	24.920029	23.667503
25.0000	38.076956	35.710101	33.752414	31.583561	30.172998	29.081890	28.166803	27.361398	25.945792	24.667469
26.0000	39.307887	36.904948	34.916089	32.711215	31.276287	30.165788	29.234045	28.413679	26.971026	25.667437
27.0000	40.534390	38.096045	36.076610	33.836394	32.377578	31.248064	30.299991	29.464952	27.995799	26.667405
28.0000	41.756738	39.283586	37.234173	34.959258	33.476977	32.328819	31.364711	30.515272	29.020101	27.667379
29.0000	42.975095	40.467809	38.388918	36.079917	34.574607	33.408109	32.428274	31.564690	30.043968	28.667350
30.0000	44.189717	41.648842	39.540987	37.198505	35.670551	34.486044	33.490732	32.613239	31.067421	29.667329

APPENDIX C

Fitting Beta, Gamma, and Negative-Log Gamma Prior Distributions

Table C1. Values of n_0 and x_0 for Selected Values of R_1 (the Mean) and R_2 (the 95th Percentile)

R_1	R_2	n_o	x_o
.50000	.55000	269.72656	134.86328
.50000	.60000	66.77246	33.38623
.50000	.65000	29.18091	14.59045
.50000	.70000	16.00342	8.00171
.50000	.75000	9.87854	4.93927
.50000	.80000	6.52313	3.26157
.50000	.85000	4.46014	2.23007
.50000	.90000	3.06320	1.53160
.50000	.95000	2.00005	1.00002
.50000	.96000	1.80206	.90103
.50000	.97000	1.59969	.79985
.50000	.98000	1.38359	.69180
.50000	.99000	1.12762	.56381
.50000	.99500	.95406	.47703
.50000	.99550	.93241	.46620
.50000	.99600	.90942	.45471
.50000	.99650	.88463	.44231
.50000	.99700	.85783	.42892
.50000	.99750	.82817	.41409
.50000	.99800	.79460	.39730
.50000	.99850	.75531	.37766
.50000	.99900	.70629	.35315
.50000	.99950	.63610	.31805
.50000	.99960	.61636	.30818
.50000	.99970	.59280	.29640
.50000	.99980	.56248	.28124
.50000	.99990	.51746	.25873
.50000	.99991	.51126	.25563
.50000	.99992	.50449	.25225
.50000	.99993	.49706	.24853
.50000	.99994	.48866	.24433
.50000	.99995	.47922	.23961
.50000	.99996	.46806	.23403
.50000	.99997	.45452	.22726
.50000	.99998	.43669	.21834
.50000	.99999	.40922	.20461
.55000	.60000	264.69727	145.58350
.55000	.65000	64.91699	35.70435
.55000	.70000	28.06396	15.43518
.55000	.75000	15.19165	8.35541
.55000	.80000	9.21936	5.07065
.55000	.85000	5.94025	3.26714
.55000	.90000	3.89938	2.14466

Table C1. (*Continued*)

R_1	R_2	n_o	x_o
.55000	.95000	2.45247	1.34886
.55000	.96000	2.19383	1.20661
.55000	.97000	1.93329	1.06331
.55000	.98000	1.65939	.91267
.55000	.99000	1.34010	.73706
.55000	.99500	1.12724	.61998
.55000	.99550	1.10092	.60551
.55000	.99600	1.07307	.59019
.55000	.99650	1.04294	.57362
.55000	.99700	1.01051	.55578
.55000	.99750	.97466	.53606
.55000	.99800	.93422	.51382
.55000	.99850	.88692	.48780
.55000	.99900	.82817	.45549
.55000	.99950	.74425	.40934
.55000	.99960	.72088	.39649
.55000	.99970	.69284	.38106
.55000	.99980	.65689	.36129
.55000	.99990	.60358	.33197
.55000	.99991	.59624	.32793
.55000	.99992	.58823	.32352
.55000	.99993	.57945	.31870
.55000	.99994	.56963	.31330
.55000	.99995	.55847	.30716
.55000	.99996	.54531	.29992
.55000	.99997	.52938	.29116
.55000	.99998	.50831	.27957
.55000	.99999	.47617	.26189
.60000	.65000	254.24805	152.54883
.60000	.70000	61.68213	37.00928
.60000	.75000	26.31226	15.78735
.60000	.80000	13.99231	8.39539
.60000	.85000	8.28094	4.96857
.60000	.90000	5.11627	3.06976
.60000	.95000	3.06740	1.84044
.60000	.96000	2.71950	1.63170
.60000	.97000	2.37503	1.42502
.60000	.98000	2.01912	1.21147
.60000	.99000	1.61324	.96794
.60000	.99500	1.34754	.80853
.60000	.99550	1.31493	.78896
.60000	.99600	1.28040	.76824
.60000	.99650	1.24340	.74604
.60000	.99700	1.20335	.72201
.60000	.99750	1.15948	.69569
.60000	.99800	1.10989	.66593
.60000	.99850	1.05228	.63137

Table C1. (*Continued*)

R_1	R_2	n_o	x_o
.60000	.99900	.98076	.58846
.60000	.99950	.87938	.52763
.60000	.99960	.85115	.51069
.60000	.99970	.81749	.49049
.60000	.99980	.77438	.46463
.60000	.99990	.71049	.42629
.60000	.99991	.70171	.42103
.60000	.99992	.69218	.41531
.60000	.99993	.68169	.40901
.60000	.99994	.66996	.40197
.60000	.99995	.65660	.39396
.60000	.99996	.64096	.38458
.60000	.99997	.62199	.37319
.60000	.99998	.59700	.35820
.60000	.99999	.55876	.33525
.65000	.70000	238.33008	154.91455
.65000	.75000	57.03125	37.07031
.65000	.80000	23.89526	15.53192
.65000	.85000	12.38251	8.04863
.65000	.90000	7.01752	4.56139
.65000	.95000	3.94745	2.56584
.65000	.96000	3.45993	2.24895
.65000	.97000	2.98729	1.94174
.65000	.98000	2.50931	1.63105
.65000	.99000	1.97830	1.28590
.65000	.99500	1.63841	1.06497
.65000	.99550	1.59721	1.03819
.65000	.99600	1.55334	1.00967
.65000	.99650	1.50681	.97942
.65000	.99700	1.45664	.94682
.65000	.99750	1.40152	.91099
.65000	.99800	1.33972	.87082
.65000	.99850	1.26801	.82420
.65000	.99900	1.17931	.76655
.65000	.99950	1.05438	.68535
.65000	.99960	1.01967	.66278
.65000	.99970	.97847	.63601
.65000	.99980	.92573	.60173
.65000	.99990	.84801	.55120
.65000	.99991	.83733	.54426
.65000	.99992	.82579	.53676
.65000	.99993	.81310	.52852
.65000	.99994	.79880	.51922
.65000	.99995	.78268	.50874
.65000	.99996	.76370	.49641
.65000	.99997	.74072	.48147
.65000	.99998	.71058	.46188

Table C1. (*Continued*)

R_1	R_2	n_o	x_o
.65000	.99999	.66442	.43188
.70000	.75000	216.99219	151.89453
.70000	.80000	50.97656	35.68359
.70000	.85000	20.78857	14.55200
.70000	.90000	10.30121	7.21085
.70000	.95000	5.29633	3.70743
.70000	.96000	4.57077	3.19954
.70000	.97000	3.88718	2.72102
.70000	.98000	3.21503	2.25052
.70000	.99000	2.49138	1.74397
.70000	.99500	2.04163	1.42914
.70000	.99550	1.98746	1.39122
.70000	.99600	1.93062	1.35143
.70000	.99650	1.86996	1.30898
.70000	.99700	1.80492	1.26345
.70000	.99750	1.73378	1.21365
.70000	.99800	1.65424	1.15797
.70000	.99850	1.56231	1.09362
.70000	.99900	1.44939	1.01458
.70000	.99950	1.29128	.90389
.70000	.99960	1.24760	.87332
.70000	.99970	1.19572	.83700
.70000	.99980	1.12972	.79081
.70000	.99990	1.03283	.72298
.70000	.99991	1.01967	.71377
.70000	.99992	1.00517	.70362
.70000	.99993	.98934	.69254
.70000	.99994	.97170	.68019
.70000	.99995	.95158	.66610
.70000	.99996	.92821	.64975
.70000	.99997	.89970	.62979
.70000	.99998	.86250	.60375
.70000	.99999	.80566	.56396
.75000	.80000	190.08789	142.56592
.75000	.85000	43.43262	32.57446
.75000	.90000	16.90979	12.68234
.75000	.95000	7.56989	5.67741
.75000	.96000	6.39038	4.79279
.75000	.97000	5.31998	3.98998
.75000	.98000	4.30832	3.23124
.75000	.99000	3.26271	2.44703
.75000	.99500	2.63634	1.97725
.75000	.99550	2.56233	1.92175
.75000	.99600	2.48451	1.86338
.75000	.99650	2.40211	1.80159
.75000	.99700	2.31400	1.73550
.75000	.99750	2.21787	1.66340

Table C1. (*Continued*)

R_1	R_2	n_o	x_o
.75000	.99800	2.11143	1.58358
.75000	.99850	1.98860	1.49145
.75000	.99900	1.83907	1.37930
.75000	.99950	1.63097	1.22323
.75000	.99960	1.57394	1.18046
.75000	.99970	1.50642	1.12982
.75000	.99980	1.42078	1.06559
.75000	.99990	1.29566	.97175
.75000	.99991	1.27850	.95887
.75000	.99992	1.25999	.94500
.75000	.99993	1.23959	.92969
.75000	.99994	1.21689	.91267
.75000	.99995	1.19133	.89350
.75000	.99996	1.16119	.87090
.75000	.99997	1.12476	.84357
.75000	.99998	1.07727	.80795
.75000	.99999	1.00479	.75359
.80000	.85000	157.66602	126.13281
.80000	.90000	34.31396	27.45117
.80000	.95000	12.00867	9.60693
.80000	.96000	9.79004	7.83203
.80000	.97000	7.88727	6.30981
.80000	.98000	6.18896	4.95117
.80000	.99000	4.53568	3.62854
.80000	.99500	3.59421	2.87537
.80000	.99550	3.48549	2.78839
.80000	.99600	3.37181	2.69745
.80000	.99650	3.25165	2.60132
.80000	.99700	3.12424	2.49939
.80000	.99750	2.98615	2.38892
.80000	.99800	2.83318	2.26654
.80000	.99850	2.65884	2.12708
.80000	.99900	2.44827	1.95862
.80000	.99950	2.15874	1.72699
.80000	.99960	2.07996	1.66397
.80000	.99970	1.98708	1.58966
.80000	.99980	1.86958	1.49567
.80000	.99990	1.69907	1.35925
.80000	.99991	1.67618	1.34094
.80000	.99992	1.65100	1.32080
.80000	.99993	1.62354	1.29883
.80000	.99994	1.59283	1.27426
.80000	.99995	1.55811	1.24649
.80000	.99996	1.51768	1.21414
.80000	.99997	1.46866	1.17493
.80000	.99998	1.40495	1.12396
.80000	.99999	1.30825	1.04660

Table C1. (*Continued*)

R_1	R_2	n_o	x_o
.85000	.90000	119.45801	101.53931
.85000	.95000	23.22998	19.74548
.85000	.96000	17.73071	15.07111
.85000	.97000	13.47961	11.45767
.85000	.98000	10.02502	8.52127
.85000	.99000	6.97021	5.92468
.85000	.99500	5.36041	4.55635
.85000	.99550	5.18036	4.40331
.85000	.99600	4.99420	4.24507
.85000	.99650	4.79889	4.07906
.85000	.99700	4.59213	3.90331
.85000	.99750	4.37012	3.71460
.85000	.99800	4.12750	3.50838
.85000	.99850	3.85361	3.27557
.85000	.99900	3.52631	2.99736
.85000	.99950	3.08304	2.62058
.85000	.99960	2.96402	2.51942
.85000	.99970	2.82402	2.40042
.85000	.99980	2.64893	2.25159
.85000	.99990	2.39601	2.03661
.85000	.99991	2.36206	2.00775
.85000	.99992	2.32506	1.97630
.85000	.99993	2.28462	1.94193
.85000	.99994	2.23961	1.90367
.85000	.99995	2.18887	1.86054
.85000	.99996	2.12975	1.81028
.85000	.99997	2.05841	1.74965
.85000	.99998	1.96571	1.67086
.85000	.99999	1.82590	1.55202
.90000	.95000	74.81689	67.33521
.90000	.96000	47.74170	42.96753
.90000	.97000	31.51245	28.36121
.90000	.98000	20.84656	18.76190
.90000	.99000	13.04626	11.74164
.90000	.99500	9.50012	8.55011
.90000	.99550	9.12628	8.21365
.90000	.99600	8.74329	7.86896
.90000	.99650	8.34656	7.51190
.90000	.99700	7.93381	7.14043
.90000	.99750	7.49664	6.74698
.90000	.99800	7.02515	6.32263
.90000	.99850	6.50177	5.85159
.90000	.99900	5.88837	5.29953
.90000	.99950	5.07813	4.57031
.90000	.99960	4.86526	4.37874
.90000	.99970	4.61578	4.15421
.90000	.99980	4.30679	3.87611

Table C1. (*Continued*)

R_1	R_2	n_o	x_o
.90000	.99990	3.86810	3.48129
.90000	.99991	3.80936	3.42842
.90000	.99992	3.74603	3.37143
.90000	.99993	3.67661	3.30894
.90000	.99994	3.59955	3.23959
.90000	.99995	3.51295	3.16166
.90000	.99996	3.41263	3.07137
.90000	.99997	3.29170	2.96253
.90000	.99998	3.13568	2.82211
.90000	.99999	2.90146	2.61131
.95000	.96000	1175.19531	1116.43555
.95000	.97000	261.71875	248.63281
.95000	.98000	99.07227	94.11865
.95000	.99000	42.91992	40.77393
.95000	.99500	26.73340	25.39673
.95000	.99550	25.27771	24.01382
.95000	.99600	23.82813	22.63672
.95000	.99650	22.37549	21.25671
.95000	.99700	20.90759	19.86221
.95000	.99750	19.40613	18.43582
.95000	.99800	17.84668	16.95435
.95000	.99850	16.17889	15.36995
.95000	.99900	14.31122	13.59566
.95000	.99950	11.97815	11.37924
.95000	.99960	11.38763	10.81825
.95000	.99970	10.71014	10.17464
.95000	.99980	9.88617	9.39186
.95000	.99990	8.74634	8.30902
.95000	.99991	8.59680	8.16696
.95000	.99992	8.43506	8.01331
.95000	.99993	8.26035	7.84733
.95000	.99994	8.06732	7.66396
.95000	.99995	7.85065	7.45811
.95000	.99996	7.60193	7.22183
.95000	.99997	7.30362	6.93844
.95000	.99998	6.92291	6.57677
.95000	.99999	6.35910	6.04115
.96000	.97000	924.60938	887.62500
.96000	.98000	196.24023	188.39062
.96000	.99000	66.65039	63.98438
.96000	.99500	38.25684	36.72656
.96000	.99550	35.89478	34.45898
.96000	.99600	33.57544	32.23242
.96000	.99650	31.28662	30.03516
.96000	.99700	29.00391	27.84375
.96000	.99750	26.70898	25.64063
.96000	.99800	24.35608	23.38184

Table C1. (*Continued*)

R_1	R_2	n_o	x_o
.96000	.99850	21.88721	21.01172
.96000	.99900	19.17114	18.40430
.96000	.99950	15.85083	15.21680
.96000	.99960	15.02380	14.42285
.96000	.99970	14.08081	13.51758
.96000	.99980	12.94250	12.42480
.96000	.99990	11.38611	10.93066
.96000	.99991	11.18317	10.73584
.96000	.99992	10.96497	10.52637
.96000	.99993	10.72693	10.29785
.96000	.99994	10.46753	10.04883
.96000	.99995	10.17609	9.76904
.96000	.99996	9.84039	9.44678
.96000	.99997	9.44214	9.06445
.96000	.99998	8.93250	8.57520
.96000	.99999	8.18329	7.85596
.97000	.98000	666.79688	646.79297
.97000	.99000	126.29395	122.50513
.97000	.99500	62.45117	60.57764
.97000	.99550	57.83691	56.10181
.97000	.99600	53.40576	51.80359
.97000	.99650	49.13330	47.65930
.97000	.99700	44.97681	43.62750
.97000	.99750	40.88135	39.65491
.97000	.99800	36.79199	35.68823
.97000	.99850	32.60498	31.62683
.97000	.99900	28.13110	27.28717
.97000	.99950	22.83936	22.15417
.97000	.99960	21.55151	20.90497
.97000	.99970	20.09583	19.49295
.97000	.99980	18.36243	17.81155
.97000	.99990	16.02325	15.54256
.97000	.99991	15.72113	15.24950
.97000	.99992	15.39612	14.93423
.97000	.99993	15.04517	14.59381
.97000	.99994	14.66064	14.22083
.97000	.99995	14.23187	13.80492
.97000	.99996	13.74054	13.32832
.97000	.99997	13.15613	12.76144
.97000	.99998	12.41760	12.04507
.97000	.99999	11.33423	10.99420
.98000	.99000	398.53516	390.56445
.98000	.99500	134.91211	132.21387
.98000	.99550	121.45996	119.03076
.98000	.99600	109.13086	106.94824
.98000	.99650	97.76611	95.81079
.98000	.99700	87.19482	85.45093

Table C1. (*Continued*)

R_1	R_2	n_o	x_o
.98000	.99750	77.25830	75.71313
.98000	.99800	67.77344	66.41797
.98000	.99850	58.52051	57.35010
.98000	.99900	49.10889	48.12671
.98000	.99950	38.61694	37.84460
.98000	.99960	36.15723	35.43408
.98000	.99970	33.42896	32.76038
.98000	.99980	30.23987	29.63507
.98000	.99990	26.04370	25.52283
.98000	.99991	25.50659	24.99646
.98000	.99992	24.93896	24.44019
.98000	.99993	24.32251	23.83606
.98000	.99994	23.65112	23.17810
.98000	.99995	22.90344	22.44537
.98000	.99996	22.05505	21.61395
.98000	.99997	21.05103	20.63000
.98000	.99998	19.78760	19.39185
.98000	.99999	17.95959	17.60040
.99000	.99500	803.12500	795.09375
.99000	.99550	635.93750	629.57812
.99000	.99600	509.57031	504.47461
.99000	.99650	411.71875	407.60156
.99000	.99700	334.17969	330.83789
.99000	.99750	271.43555	268.72119
.99000	.99800	219.48242	217.28760
.99000	.99850	175.29297	173.54004
.99000	.99900	136.18164	134.81982
.99000	.99950	98.60840	97.62231
.99000	.99960	90.61279	89.70667
.99000	.99970	82.09229	81.27136
.99000	.99980	72.55859	71.83301
.99000	.99990	60.66895	60.06226
.99000	.99991	59.21021	58.61810
.99000	.99992	57.65991	57.08331
.99000	.99993	56.00586	55.44580
.99000	.99994	54.21143	53.66931
.99000	.99995	52.23389	51.71155
.99000	.99996	50.01831	49.51813
.99000	.99997	47.43652	46.96216
.99000	.99998	44.23218	43.78986
.99000	.99999	39.68506	39.28821
.99500	.99550	51512.50000	51254.93750
.99500	.99600	12256.25000	12194.96875
.99500	.99650	5150.78125	5125.02734
.99500	.99700	2716.79688	2703.21289
.99500	.99750	1612.50000	1604.43750
.99500	.99800	1022.65625	1017.54297

Table C1. (*Continued*)

R_1	R_2	n_o	x_o
.99500	.99850	670.41016	667.05811
.99500	.99900	440.13672	437.93604
.99500	.99950	272.94922	271.58447
.99500	.99960	243.04199	241.82678
.99500	.99970	212.98828	211.92334
.99500	.99980	181.56738	180.65955
.99500	.99990	145.36133	144.63452
.99500	.99991	141.13770	140.43201
.99500	.99992	136.69434	136.01086
.99500	.99993	131.99463	131.33466
.99500	.99994	126.97754	126.34265
.99500	.99995	121.53320	120.92554
.99500	.99996	115.50293	114.92542
.99500	.99997	108.58154	108.03864
.99500	.99998	100.18311	99.68219
.99500	.99999	88.57422	88.13135
.99550	.99600	46143.75000	45936.10312
.99550	.99650	10906.25000	10857.17187
.99550	.99700	4543.75000	4523.30312
.99550	.99750	2369.14063	2358.47949
.99550	.99800	1383.59375	1377.36758
.99550	.99850	856.25000	852.39687
.99550	.99900	538.18359	535.76177
.99550	.99950	322.11914	320.66960
.99550	.99960	284.86328	283.58140
.99550	.99970	247.92480	246.80914
.99550	.99980	209.81445	208.87029
.99550	.99990	166.55273	165.80325
.99550	.99991	161.54785	160.82089
.99550	.99992	156.29883	155.59548
.99550	.99993	150.75684	150.07843
.99550	.99994	144.87305	144.22112
.99550	.99995	138.47656	137.85342
.99550	.99996	131.42090	130.82950
.99550	.99997	123.36426	122.80912
.99550	.99998	113.59863	113.08744
.99550	.99999	100.19531	99.74443
.99600	.99650	40762.50000	40599.45000
.99600	.99700	9551.56250	9513.35625
.99600	.99750	3933.59375	3917.85937
.99600	.99800	2017.18750	2009.11875
.99600	.99850	1148.82812	1144.23281
.99600	.99900	680.95703	678.23320
.99600	.99950	389.30664	387.74941
.99600	.99960	341.35742	339.99199
.99600	.99970	294.58008	293.40176
.99600	.99980	247.07031	246.08203

Table C1. (*Continued*)

R_1	R_2	n_o	x_o
.99600	.99990	194.16504	193.38838
.99600	.99991	188.08594	187.33359
.99600	.99992	181.76270	181.03564
.99600	.99993	175.09766	174.39727
.99600	.99994	168.01758	167.34551
.99600	.99995	160.37598	159.73447
.99600	.99996	151.95312	151.34531
.99600	.99997	142.38281	141.81328
.99600	.99998	130.82275	130.29946
.99600	.99999	115.02686	114.56675
.99650	.99700	35381.25000	35257.41562
.99650	.99750	8192.18750	8163.51484
.99650	.99800	3317.18750	3305.57734
.99650	.99850	1657.81250	1652.01016
.99650	.99900	903.41797	900.25601
.99650	.99950	485.74219	484.04209
.99650	.99960	421.28906	419.81455
.99650	.99970	359.61914	358.36047
.99650	.99980	298.24219	297.19834
.99650	.99990	231.44531	230.63525
.99650	.99991	223.87695	223.09338
.99650	.99992	216.01563	215.25957
.99650	.99993	207.76367	207.03650
.99650	.99994	199.02344	198.32686
.99650	.99995	189.64844	188.98467
.99650	.99996	179.32129	178.69366
.99650	.99997	167.62695	167.04026
.99650	.99998	153.63770	153.09996
.99650	.99999	134.59473	134.12365
.99700	.99750	29981.25000	29891.30625
.99700	.99800	6825.00000	6804.52500
.99700	.99850	2691.40625	2683.33203
.99700	.99900	1285.54687	1281.69023
.99700	.99950	633.39844	631.49824
.99700	.99960	541.21094	539.58730
.99700	.99970	455.37109	454.00498
.99700	.99980	372.11914	371.00278
.99700	.99990	284.13086	283.27847
.99700	.99991	274.34082	273.51780
.99700	.99992	264.20898	263.41636
.99700	.99993	253.61328	252.85244
.99700	.99994	242.43164	241.70435
.99700	.99995	230.46875	229.77734
.99700	.99996	217.40723	216.75500
.99700	.99997	202.68555	202.07749
.99700	.99998	185.13184	184.57644
.99700	.99999	161.47461	160.99019

Table C1. (*Continued*)

R_1	R_2	n_o	x_o
.99750	.99800	24568.75000	24507.32812
.99750	.99850	5444.53125	5430.91992
.99750	.99900	2048.82813	2043.70605
.99750	.99950	881.44531	879.24170
.99750	.99960	737.30469	735.46143
.99750	.99970	607.71484	606.19556
.99750	.99980	486.62109	485.40454
.99750	.99990	363.47656	362.56787
.99750	.99991	350.14648	349.27112
.99750	.99992	336.32813	335.48730
.99750	.99993	322.02148	321.21643
.99750	.99994	306.98242	306.21497
.99750	.99995	290.96680	290.23938
.99750	.99996	273.58398	272.90002
.99750	.99997	254.15039	253.51501
.99750	.99998	231.15234	230.57446
.99750	.99999	200.51270	200.01141
.99800	.99850	19137.50000	19099.22500
.99800	.99900	4040.62500	4032.54375
.99800	.99950	1363.47656	1360.74961
.99800	.99960	1102.14844	1099.94414
.99800	.99970	879.88281	878.12305
.99800	.99980	683.39844	682.03164
.99800	.99990	494.53125	493.54219
.99800	.99991	474.70703	473.75762
.99800	.99992	454.44336	453.53447
.99800	.99993	433.49609	432.62910
.99800	.99994	411.66992	410.84658
.99800	.99995	388.57422	387.79707
.99800	.99996	363.76953	363.04199
.99800	.99997	336.27930	335.60674
.99800	.99998	304.10156	303.49336
.99800	.99999	261.81641	261.29277
.99850	.99900	13665.62500	13645.12656
.99850	.99950	2573.43750	2569.57734
.99850	.99960	1949.60937	1946.68496
.99850	.99970	1469.92187	1467.71699
.99850	.99980	1083.20312	1081.57832
.99850	.99990	744.72656	743.60947
.99850	.99991	711.03516	709.96860
.99850	.99992	676.85547	675.84019
.99850	.99993	641.99219	641.02920
.99850	.99994	606.05469	605.14561
.99850	.99995	568.55469	567.70186
.99850	.99996	528.71094	527.91787
.99850	.99997	485.15625	484.42852
.99850	.99998	435.00977	434.35725

Table C1. (*Continued*)

R_1	R_2	n_o	x_o
99850	.99999	370.41016	369.85454
.99900	.99950	8087.50000	8079.41250
.99900	.99960	5127.34375	5122.21641
.99900	.99970	3360.54687	3357.18633
.99900	.99980	2205.46875	2203.26328
.99900	.99990	1367.38281	1366.01543
.99900	.99991	1292.18750	1290.89531
.99900	.99992	1217.18750	1215.97031
.99900	.99993	1142.38281	1141.24043
.99900	.99994	1066.79687	1065.73008
.99900	.99995	989.45313	988.46367
.99900	.99996	909.17969	908.27051
.99900	.99997	823.63281	822.80918
.99900	.99998	727.83203	727.10420
.99900	.99999	608.39844	607.79004
.99950	.99970	27268.75000	27255.11562
.99950	.99980	10259.37500	10254.24531
.99950	.99990	4412.50000	4410.29375
.99950	.99991	4040.62500	4038.60469
.99950	.99992	3690.62500	3688.77969
.99950	.99993	3358.59375	3356.91445
.99950	.99994	3041.40625	3039.88555
.99950	.99995	2735.15625	2733.78867
.99950	.99996	2435.15625	2433.93867
.99950	.99997	2133.98437	2132.91738
.99950	.99998	1818.75000	1817.84062
.99950	.99999	1455.85937	1455.13145
.99960	.99970	95825.00000	95786.67000
.99960	.99980	20225.00000	20216.91000
.99960	.99990	6823.43750	6820.70813
.99960	.99991	6140.62500	6138.16875
.99960	.99992	5515.62500	5513.41875
.99960	.99993	4939.06250	4937.08687
.99960	.99994	4403.12500	4401.36375
.99960	.99995	3899.60937	3898.04953
.99960	.99996	3419.14062	3417.77297
.99960	.99997	2950.39062	2949.21047
.99960	.99998	2474.21875	2473.22906
.99960	.99999	1944.14062	1943.36297
.99970	.99980	68400.00000	68379.48000
.99970	.99990	12878.12500	12874.26156
.99970	.99991	11206.25000	11202.88812
.99970	.99992	9756.25000	9753.32312
.99970	.99993	8482.81250	8480.26766
.99970	.99994	7354.68750	7352.48109
.99970	.99995	6342.18750	6340.28484
.99970	.99996	5419.53125	5417.90539

Table C1. (*Continued*)

R_1	R_2	n_o	x_o
.99980	.99990	40462.50000	40454.40750
.99980	.99991	32025.00000	32018.59500
.99980	.99992	25650.00000	25644.87000
.99980	.99993	20718.75000	20714.60625
.99980	.99994	16812.50000	16809.13750
.99980	.99995	13650.00000	13647.27000
.99980	.99996	11032.81250	11030.60594
.99980	.99997	8807.81250	8806.05094
.99980	.99998	6839.06250	6837.69469
.99980	.99999	4949.21875	4948.22891
.99990	.99995	80925.00000	80916.90750
.99990	.99996	51306.25000	51301.11938
.99990	.99997	33625.00000	33621.63750
.99990	.99998	22065.62500	22063.41844
.99990	.99999	13678.12500	13676.75719
.99991	.99996	69400.00000	69393.75400
.99991	.99997	42931.25000	42927.38619
.99991	.99998	26975.00000	26972.57225
.99991	.99999	16137.50000	16136.04762
.99992	.99997	57587.50000	57582.89300
.99992	.99998	34125.00000	34122.27000
.99992	.99999	19500.00000	19498.44000
.99993	.99997	83100.00000	83094.18300
.99993	.99998	45262.50000	45259.33162
.99993	.99999	24325.00000	24323.29725
.99994	.99998	64400.00000	64396.13600
.99994	.99999	31715.62500	31713.72206
.99995	.99999	44137.50000	44135.29312
.99996	.99999	68250.00000	68247.27000

Table C2. Values of n_0 and x_0 for Selected Values of R_1 (the Mean) and R_3 (the 5th Percentile)

R_1	R_3	n_o	x_o
.55000	.50000	269.23828	148.08105
.60000	.50000	66.33301	39.79980
.60000	.55000	263.37891	158.02734
.65000	.50000	28.72314	18.67004
.65000	.55000	64.03809	41.62476
.65000	.60000	252.05078	163.83301
.70000	.50000	15.51819	10.86273
.70000	.55000	27.29492	19.10645
.70000	.60000	60.33936	42.23755
.70000	.65000	235.30273	164.71191
.75000	.50000	9.35516	7.01637
.75000	.55000	14.45313	10.83984
.75000	.60000	25.21973	18.91479
.75000	.65000	55.24902	41.43677
.75000	.70000	213.03711	159.77783
.80000	.50000	5.93109	4.74487
.80000	.55000	8.45490	6.76392
.80000	.60000	12.97607	10.38086
.80000	.65000	22.46094	17.96875
.80000	.70000	48.70605	38.96484
.80000	.75000	185.30273	148.24219
.85000	.50000	3.74603	3.18413
.85000	.55000	5.07660	4.31511
.85000	.60000	7.22046	6.13739
.85000	.65000	11.03210	9.37729
.85000	.70000	18.96362	16.11908
.85000	.75000	40.64941	34.55200
.85000	.80000	151.90430	129.11865
.90000	.50000	2.09808	1.88828
.90000	.55000	2.78778	2.50900
.90000	.60000	3.82843	3.44559
.90000	.65000	5.50232	4.95209
.90000	.70000	8.46252	7.61627
.90000	.75000	14.53857	13.08472
.90000	.80000	30.87158	27.78442
.90000	.85000	112.69531	101.42578
.95000	.50000	.01221	.01160
.95000	.55000	.32043	.30441
.95000	.60000	.77820	.73929
.95000	.65000	1.46484	1.39160
.95000	.70000	2.56500	2.43675
.95000	.75000	4.48914	4.26468
.95000	.80000	8.37097	7.95242
.95000	.85000	18.47534	17.55157
.95000	.90000	66.52832	63.20190

Table C2. (*Continued*)

R_1	R_3	n_o	x_o
.96000	.75000	2.66470	2.55812
.96000	.80000	5.63760	5.41210
.96000	.85000	12.66803	12.16131
.96000	.90000	40.07068	38.46785
.96000	.95000	1135.54687	1090.12500
.97000	.85000	7.53192	7.30596
.97000	.90000	23.68347	22.97297
.97000	.95000	240.57617	233.35889
.97000	.96000	883.59375	857.08594
.98000	.95000	82.54932	80.89833
.98000	.96000	173.38867	169.92090
.98000	.97000	624.21875	611.73437
.99000	.95000	22.50000	22.27500
.99000	.96000	44.10156	43.66055
.99000	.97000	98.20312	97.22109
.99000	.98000	351.46484	347.95020
.99500	.98000	88.71094	88.26738
.99500	.99000	707.61719	704.07910
.99550	.98000	72.65625	72.32930
.99550	.99000	545.50781	543.05303
.99550	.99500	50662.50000	50434.51875
.99600	.98000	56.05469	55.83047
.99600	.99000	422.26563	420.57656
.99600	.99500	11825.00000	11777.70000
.99600	.99550	45287.50000	45106.35000
.99650	.98000	35.46875	35.34461
.99650	.99000	325.39063	324.25176
.99650	.99500	4856.25000	4839.25312
.99650	.99550	10471.87500	10435.22344
.99650	.99600	39906.25000	39766.57812
.99700	.99000	246.28906	245.55020
.99700	.99500	2489.06250	2481.59531
.99700	.99550	4246.87500	4234.13437
.99700	.99600	9115.62500	9088.27812
.99700	.99650	34518.75000	34415.19375
.99750	.99000	177.92969	177.48486
.99750	.99500	1419.92188	1416.37207
.99750	.99550	2136.32812	2130.98730
.99750	.99600	3632.03125	3622.95117
.99750	.99650	7751.56250	7732.18359
.99750	.99700	29118.75000	29045.95312
.99800	.99000	111.91406	111.69023
.99800	.99500	846.87500	845.18125
.99800	.99550	1182.03125	1179.66719
.99800	.99600	1776.17187	1772.61953
.99800	.99650	3008.59375	3002.57656
.99800	.99700	6378.12500	6365.36875
.99800	.99750	23700.00000	23652.60000

Table C2. (*Continued*)

R_1	R_3	n_o	x_o
.99850	.99500	493.55469	492.81436
.99850	.99550	662.50000	661.50625
.99850	.99600	930.85938	929.46309
.99850	.99650	1401.95312	1399.85020
.99850	.99700	2369.53125	2365.97695
.99850	.99750	4987.50000	4980.01875
.99850	.99800	18262.50000	18235.10625
.99900	.99500	223.63281	223.40918
.99900	.99550	315.42969	315.11426
.99900	.99600	445.50781	445.06230
.99900	.99650	647.07031	646.42324
.99900	.99700	994.53125	993.53672
.99900	.99750	1696.09375	1694.39766
.99900	.99800	3557.03125	3553.47422
.99900	.99850	12771.87500	12759.10312
.99950	.99750	447.26563	447.04199
.99950	.99800	891.40625	890.96055
.99950	.99850	1990.62500	1989.62969
.99950	.99900	7118.75000	7115.19062
.99960	.99800	107.26562	107.22272
.99960	.99850	1336.71875	1336.18406
.99960	.99900	4243.75000	4242.05250
.99970	.99850	143.24219	143.19921
.99970	.99900	2471.87500	2471.13344
.99970	.99950	24968.75000	24961.25937
.99970	.99960	91425.00000	91397.57250
.99980	.99900	215.37109	215.32802
.99980	.99950	105.03906	105.01805
.99980	.99960	17803.12500	17799.56437
.99980	.99970	63912.50000	63899.71750
.99990	.99960	310.54687	310.51582
.99990	.99970	238.57422	238.55036
.99990	.99980	184.97070	184.95221
.99991	.99970	288.63281	288.60684
.99991	.99980	217.73437	217.71478
.99991	.99990	158.35937	158.34512
.99992	.99980	263.06641	263.04536
.99992	.99990	186.47461	186.45969
.99992	.99991	179.06250	179.04817
.99993	.99990	225.25391	225.23814
.99993	.99991	215.58594	215.57085
.99993	.99992	205.91797	205.90355
.99994	.99990	281.54297	281.52608
.99994	.99991	268.43750	268.42139
.99994	.99992	255.33203	255.31671
.99994	.99993	242.11914	242.10461
.99995	.99993	312.69531	312.67968
.99995	.99994	293.78906	293.77437

Table C3. Gamma Reference Scale Values b_0 for Selected Values of Shape Parameter α_0 and Prior Assurance p_0[a]

α_0	p_0		
	0.95	0.90	0.80
0.01	3.5227-155	4.4655-125	5.6607E-95
0.02	4.5019E-75	5.0687E-60	5.7068E-45
0.03	2.2797E-48	2.4673E-38	2.6702E-28
0.04	5.1503E-35	1.7282E-27	5.7988E-20
0.05	5.3157E-27	5.5739E-21	5.8446E-15
0.06	1.1726E-21	1.2199E-16	1.2691E-11
0.07	7.7079E-18	1.5394E-13	3.0746E-09
0.08	5.6375E-15	3.2656E-11	1.8916E-07
0.09	9.5383E-13	2.1098E-09	4.6669E-06
0.10	5.7917E-11	5.9307E-08	6.0730E-05
0.11	1.6690E-09	9.1013E-07	4.9629E-04
0.12	2.7502E-08	8.8704E-06	2.8611E-03
0.13	2.9479E-07	6.0971E-05	1.2611E-02
0.14	2.2538E-06	3.1852E-04	4.5014E-02
0.15	1.3150E-05	1.3360E-03	1.3573E-01
0.16	6.1598E-05	4.6882E-03	3.5682E-01
0.17	2.4079E-04	1.4204E-02	8.3786E-01
0.18	8.0955E-04	3.8074E-02	1.7907E+00
0.19	2.3972E-03	9.2062E-02	3.5355E+00
0.20	6.3725E-03	2.0392E-01	6.5255E+00
0.21	1.5443E-02	4.1900E-01	1.1368E+01
0.22	3.4551E-02	8.0682E-01	1.8841E+01
0.23	7.2112E-02	1.4684E+00	2.9900E+01
0.24	1.4163E-01	2.5435E+00	4.5682E+01
0.25	2.6366E-01	4.2186E+00	6.7501E+01
0.26	4.6815E-01	6.7327E+00	9.6835E+01
0.27	7.9697E-01	1.0384E+01	1.3531E+02
0.28	1.3067E+00	1.5534E+01	1.8469E+02
0.29	2.0715E+00	2.2611E+01	2.4685E+02
0.30	3.1857E+00	3.2110E+01	3.2372E+02
0.31	4.7668E+00	4.4597E+01	4.1734E+02
0.32	6.9577E+00	6.0702E+01	5.2976E+02
0.33	9.9288E+00	8.1121E+01	6.6303E+02
0.34	1.3880E+01	1.0661E+02	8.1924E+02
0.35	1.9041E+01	1.3798E+02	1.0004E+03
0.36	2.5675E+01	1.7610E+02	1.2086E+03
0.37	3.4074E+01	2.2186E+02	1.4457E+03
0.38	4.4566E+01	2.7622E+02	1.7135E+03
0.39	5.7506E+01	3.4015E+02	2.0140E+03
0.40	7.3283E+01	4.1465E+02	2.3489E+03
0.41	9.2314E+01	5.0074E+02	2.7197E+03
0.42	1.1504E+02	5.9945E+02	3.1280E+03
0.43	1.4194E+02	7.1180E+02	3.5752E+03
0.44	1.7351E+02	8.3884E+02	4.0627E+03
0.45	2.1026E+02	9.8160E+02	4.5917E+03
0.46	2.5272E+02	1.1411E+03	5.1633E+03
0.47	3.0146E+02	1.3183E+03	5.7786E+03
0.48	3.5703E+02	1.5142E+03	6.4385E+03
0.49	4.2003E+02	1.7298E+03	7.1438E+03
0.50	4.9103E+02	1.9661E+03	7.8954E+03
0.51	5.7065E+02	2.2238E+03	8.6938E+03
0.52	6.5947E+02	2.5039E+03	9.5397E+03
0.53	7.5810E+02	2.8073E+03	1.0434E+04
0.54	8.6715E+02	3.1347E+03	1.1376E+04
0.55	9.8723E+02	3.4869E+03	1.2367E+04
0.56	1.1189E+03	3.8647E+03	1.3407E+04
0.57	1.2628E+03	4.2688E+03	1.4496E+04
0.58	1.4196E+03	4.6997E+03	1.5635E+04
0.59	1.5897E+03	5.1582E+03	1.6823E+04
0.60	1.7737E+03	5.6448E+03	1.8060E+04

[a]Adapted with permission, from H. F. Martz, Jr. and R. A. Waller (1979). A Bayesian Zero-Failure (BAZE) Reliability Demonstration Testing Procedure, *Journal of Quality Technology*, Vol. 11, pp. 130–131.

Table C3. (*Continued*)

α_0	p_0		
	0.95	0.90	0.80
0.61	1.9723E+03	6.1602E+03	1.9348E+04
0.62	2.1859E+03	6.7047E+03	2.0685E+04
0.63	2.4152E+03	7.2789E+03	2.2071E+04
0.64	2.6605E+03	7.8833E+03	2.3506E+04
0.65	2.9225E+03	8.5182E+03	2.4991E+04
0.66	3.2016E+03	9.1840E+03	2.6525E+04
0.67	3.4982E+03	9.8811E+03	2.8108E+04
0.68	3.8130E+03	1.0610E+04	2.9739E+04
0.69	4.1462E+03	1.1370E+04	3.1418E+04
0.70	4.4983E+03	1.2163E+04	3.3145E+04
0.71	4.8698E+03	1.2988E+04	3.4921E+04
0.72	5.2610E+03	1.3846E+04	3.6743E+04
0.73	5.6722E+03	1.4737E+04	3.8612E+04
0.74	6.1040E+03	1.5660E+04	4.0528E+04
0.75	6.5565E+03	1.6616E+04	4.2491E+04
0.76	7.0302E+03	1.7606E+04	4.4499E+04
0.77	7.5253E+03	1.8629E+04	4.6552E+04
0.78	8.0421E+03	1.9685E+04	4.8651E+04
0.79	8.5809E+03	2.0775E+04	5.0794E+04
0.80	9.1420E+03	2.1898E+04	5.2982E+04
0.81	9.7256E+03	2.3054E+04	5.5213E+04
0.82	1.0332E+04	2.4244E+04	5.7488E+04
0.83	1.0961E+04	2.5467E+04	5.9806E+04
0.84	1.1614E+04	2.6723E+04	6.2166E+04
0.85	1.2289E+04	2.8013E+04	6.4569E+04
0.86	1.2989E+04	2.9336E+04	6.7013E+04
0.87	1.3712E+04	3.0692E+04	6.9498E+04
0.88	1.4459E+04	3.2082E+04	7.2024E+04
0.89	1.5229E+04	3.3504E+04	7.4590E+04
0.90	1.6024E+04	3.4959E+04	7.7197E+04
0.91	1.6843E+04	3.6448E+04	7.9842E+04
0.92	1.7687E+04	3.7968E+04	8.2527E+04
0.93	1.8555E+04	3.9522E+04	8.5251E+04
0.94	1.9447E+04	4.1107E+04	8.8012E+04
0.95	2.0364E+04	4.2725E+04	9.0812E+04
0.96	2.1305E+04	4.4375E+04	9.3649E+04
0.97	2.2271E+04	4.6057E+04	9.6522E+04
0.98	2.3262E+04	4.7771E+04	9.9432E+04
0.99	2.4278E+04	4.9516E+04	1.0238E+05
1.00	2.5318E+04	5.1293E+04	1.0536E+05
1.05	3.0891E+04	6.0643E+04	1.2079E+05
1.10	3.7082E+04	7.0752E+04	1.3705E+05
1.15	4.3888E+04	8.1594E+04	1.5410E+05
1.20	5.1301E+04	9.3145E+04	1.7190E+05
1.25	5.9311E+04	1.0538E+05	1.9039E+05
1.30	6.7908E+04	1.1827E+05	2.0956E+05
1.35	7.7079E+04	1.3179E+05	2.2935E+05
1.40	8.6810E+04	1.4593E+05	2.4974E+05
1.45	9.7087E+04	1.6064E+05	2.7069E+05
1.50	1.0790E+05	1.7592E+05	2.9219E+05
1.55	1.1923E+05	1.9174E+05	3.1419E+05
1.60	1.3106E+05	2.0808E+05	3.3669E+05
1.65	1.4338E+05	2.2492E+05	3.5965E+05
1.70	1.5618E+05	2.4225E+05	3.8306E+05
1.75	1.6944E+05	2.6003E+05	4.0689E+05
1.80	1.8314E+05	2.7827E+05	4.3113E+05
1.85	1.9729E+05	2.9693E+05	4.5575E+05
1.90	2.1186E+05	3.1601E+05	4.8075E+05
1.95	2.2683E+05	3.3549E+05	5.0611E+05
2.00	2.4221E+05	3.5536E+05	5.3181E+05

Table C3. (*Continued*)

α_0	p_0		
	0.95	0.90	0.80
2.05	2.5797E+05	3.7560E+05	5.5784E+05
2.10	2.7411E+05	3.9621E+05	5.8419E+05
2.15	2.9062E+05	4.1716E+05	6.1085E+05
2.20	3.0748E+05	4.3845E+05	6.3780E+05
2.25	3.2469E+05	4.6007E+05	6.6504E+05
2.30	3.4224E+05	4.8201E+05	6.9255E+05
2.35	3.6011E+05	5.0425E+05	7.2032E+05
2.40	3.7830E+05	5.2679E+05	7.4835E+05
2.45	3.9680E+05	5.4963E+05	7.7663E+05
2.50	4.1561E+05	5.7274E+05	8.0515E+05
2.55	4.3470E+05	5.9612E+05	8.3391E+05
2.60	4.5409E+05	6.1977E+05	8.6288E+05
2.65	4.7375E+05	6.4368E+05	8.9208E+05
2.70	4.9369E+05	6.6784E+05	9.2149E+05
2.75	5.1389E+05	6.9224E+05	9.5110E+05
2.80	5.3435E+05	7.1688E+05	9.8092E+05
2.85	5.5507E+05	7.4175E+05	1.0109E+06
2.90	5.7603E+05	7.6685E+05	1.0411E+06
2.95	5.9723E+05	7.9216E+05	1.0715E+06
3.00	6.1867E+05	8.1769E+05	1.1021E+06
3.05	6.4034E+05	8.4343E+05	1.1328E+06
3.10	6.6224E+05	8.6937E+05	1.1637E+06
3.15	6.8435E+05	8.9551E+05	1.1948E+06
3.20	7.0668E+05	9.2184E+05	1.2260E+06
3.25	7.2922E+05	9.4837E+05	1.2574E+06
3.30	7.5197E+05	9.7507E+05	1.2889E+06
3.35	7.7492E+05	1.0020E+06	1.3206E+06
3.40	7.9807E+05	1.0290E+06	1.3525E+06
3.45	8.2141E+05	1.0563E+06	1.3844E+06
3.50	8.4493E+05	1.0837E+06	1.4166E+06
3.55	8.6865E+05	1.1112E+06	1.4488E+06
3.60	8.9255E+05	1.1390E+06	1.4812E+06
3.65	9.1662E+05	1.1669E+06	1.5137E+06
3.70	9.4087E+05	1.1949E+06	1.5464E+06
3.75	9.6529E+05	1.2231E+06	1.5791E+06
3.80	9.8988E+05	1.2515E+06	1.6120E+06
3.85	1.0146E+06	1.2800E+06	1.6450E+06
3.90	1.0396E+06	1.3086E+06	1.6782E+06
3.95	1.0646E+06	1.3374E+06	1.7114E+06
4.00	1.0899E+06	1.3663E+06	1.7448E+06
4.05	1.1153E+06	1.3954E+06	1.7782E+06
4.10	1.1408E+06	1.4246E+06	1.8118E+06
4.15	1.1665E+06	1.4539E+06	1.8455E+06
4.20	1.1923E+06	1.4833E+06	1.8793E+06
4.25	1.2183E+06	1.5129E+06	1.9132E+06
4.30	1.2444E+06	1.5426E+06	1.9472E+06
4.35	1.2706E+06	1.5724E+06	1.9812E+06
4.40	1.2970E+06	1.6023E+06	2.0154E+06
4.45	1.3235E+06	1.6324E+06	2.0497E+06
4.50	1.3502E+06	1.6626E+06	2.0841E+06
4.55	1.3770E+06	1.6928E+06	2.1185E+06
4.60	1.4039E+06	1.7232E+06	2.1531E+06
4.65	1.4309E+06	1.7537E+06	2.1877E+06
4.70	1.4581E+06	1.7843E+06	2.2225E+06
4.75	1.4853E+06	1.8150E+06	2.2573E+06
4.80	1.5127E+06	1.8459E+06	2.2922E+06
4.85	1.5403E+06	1.8768E+06	2.3272E+06
4.90	1.5679E+06	1.9078E+06	2.3622E+06
4.95	1.5956E+06	1.9389E+06	2.3974E+06
5.00	1.6235E+06	1.9701E+06	2.4326E+06

Table C3. (*Continued*)

α_0	p_0		
	0.95	0.90	0.80
5.05	1.6515E+06	2.0015E+06	2.4679E+06
5.10	1.6795E+06	2.0329E+06	2.5033E+06
5.15	1.7077E+06	2.0644E+06	2.5387E+06
5.20	1.7360E+06	2.0960E+06	2.5742E+06
5.25	1.7644E+06	2.1277E+06	2.6098E+06
5.30	1.7929E+06	2.1594E+06	2.6455E+06
5.35	1.8215E+06	2.1913E+06	2.6813E+06
5.40	1.8502E+06	2.2232E+06	2.7171E+06
5.45	1.8790E+06	2.2553E+06	2.7529E+06
5.50	1.9079E+06	2.2874E+06	2.7889E+06
5.55	1.9369E+06	2.3196E+06	2.8249E+06
5.60	1.9659E+06	2.3519E+06	2.8610E+06
5.65	1.9951E+06	2.3843E+06	2.8971E+06
5.70	2.0244E+06	2.4167E+06	2.9333E+06
5.75	2.0538E+06	2.4492E+06	2.9696E+06
5.80	2.0832E+06	2.4818E+06	3.0060E+06
5.85	2.1128E+06	2.5145E+06	3.0423E+06
5.90	2.1424E+06	2.5473E+06	3.0788E+06
5.95	2.1721E+06	2.5801E+06	3.1153E+06
6.00	2.2019E+06	2.6130E+06	3.1519E+06
6.05	2.2318E+06	2.6460E+06	3.1885E+06
6.10	2.2617E+06	2.6790E+06	3.2252E+06
6.15	2.2918E+06	2.7122E+06	3.2620E+06
6.20	2.3219E+06	2.7454E+06	3.2988E+06
6.25	2.3521E+06	2.7786E+06	3.3356E+06
6.30	2.3824E+06	2.8119E+06	3.3726E+06
6.35	2.4128E+06	2.8453E+06	3.4095E+06
6.40	2.4433E+06	2.8788E+06	3.4465E+06
6.45	2.4738E+06	2.9123E+06	3.4836E+06
6.50	2.5044E+06	2.9459E+06	3.5208E+06
6.55	2.5351E+06	2.9796E+06	3.5579E+06
6.60	2.5658E+06	3.0133E+06	3.5952E+06
6.65	2.5966E+06	3.0471E+06	3.6324E+06
6.70	2.6275E+06	3.0810E+06	3.6698E+06
6.75	2.6585E+06	3.1149E+06	3.7072E+06
6.80	2.6895E+06	3.1488E+06	3.7446E+06
6.85	2.7206E+06	3.1829E+06	3.7821E+06
6.90	2.7518E+06	3.2170E+06	3.8196E+06
6.95	2.7831E+06	3.2511E+06	3.8572E+06
7.00	2.8144E+06	3.2853E+06	3.8948E+06
7.05	2.8457E+06	3.3196E+06	3.9324E+06
7.10	2.8772E+06	3.3539E+06	3.9701E+06
7.15	2.9087E+06	3.3883E+06	4.0079E+06
7.20	2.9403E+06	3.4227E+06	4.0457E+06
7.25	2.9719E+06	3.4572E+06	4.0835E+06
7.30	3.0036E+06	3.4918E+06	4.1214E+06
7.35	3.0354E+06	3.5264E+06	4.1593E+06
7.40	3.0672E+06	3.5610E+06	4.1973E+06
7.45	3.0991E+06	3.5957E+06	4.2353E+06
7.50	3.1311E+06	3.6305E+06	4.2734E+06
7.55	3.1631E+06	3.6653E+06	4.3115E+06
7.60	3.1952E+06	3.7001E+06	4.3496E+06
7.65	3.2273E+06	3.7351E+06	4.3878E+06
7.70	3.2595E+06	3.7700E+06	4.4260E+06
7.75	3.2917E+06	3.8050E+06	4.4643E+06
7.80	3.3240E+06	3.8401E+06	4.5026E+06
7.85	3.3564E+06	3.8752E+06	4.5409E+06
7.90	3.3888E+06	3.9104E+06	4.5793E+06
7.95	3.4213E+06	3.9456E+06	4.6177E+06
8.00	3.4538E+06	3.9808E+06	4.6561E+06

Table C3. (*Continued*)

α_0	p_0		
	0.95	0.90	0.80
8.05	3.4864E+06	4.0161E+06	4.6946E+06
8.10	3.5191E+06	4.0515E+06	4.7331E+06
8.15	3.5518E+06	4.0869E+06	4.7717E+06
8.20	3.5845E+06	4.1223E+06	4.8103E+06
8.25	3.6173E+06	4.1578E+06	4.8489E+06
8.30	3.6502E+06	4.1933E+06	4.8876E+06
8.35	3.6831E+06	4.2289E+06	4.9263E+06
8.40	3.7160E+06	4.2645E+06	4.9650E+06
8.45	3.7490E+06	4.3002E+06	5.0038E+06
8.50	3.7821E+06	4.3359E+06	5.0426E+06
8.55	3.8152E+06	4.3716E+06	5.0814E+06
8.60	3.8484E+06	4.4074E+06	5.1203E+06
8.65	3.8816E+06	4.4433E+06	5.1592E+06
8.70	3.9148E+06	4.4791E+06	5.1982E+06
8.75	3.9481E+06	4.5150E+06	5.2371E+06
8.80	3.9815E+06	4.5510E+06	5.2761E+06
8.85	4.0149E+06	4.5870E+06	5.3152E+06
8.90	4.0483E+06	4.6230E+06	5.3542E+06
8.95	4.0818E+06	4.6591E+06	5.3933E+06
9.00	4.1154E+06	4.6952E+06	5.4325E+06
9.05	4.1490E+06	4.7314E+06	5.4716E+06
9.10	4.1826E+06	4.7676E+06	5.5108E+06
9.15	4.2163E+06	4.8038E+06	5.5500E+06
9.20	4.2500E+06	4.8401E+06	5.5893E+06
9.25	4.2838E+06	4.8764E+06	5.6286E+06
9.30	4.3176E+06	4.9127E+06	5.6679E+06
9.35	4.3514E+06	4.9491E+06	5.7072E+06
9.40	4.3853E+06	4.9856E+06	5.7466E+06
9.45	4.4193E+06	5.0220E+06	5.7860E+06
9.50	4.4533E+06	5.0585E+06	5.8255E+06
9.55	4.4873E+06	5.0950E+06	5.8649E+06
9.60	4.5214E+06	5.1316E+06	5.9044E+06
9.65	4.5555E+06	5.1682E+06	5.9439E+06
9.70	4.5896E+06	5.2048E+06	5.9835E+06
9.75	4.6238E+06	5.2415E+06	6.0230E+06
9.80	4.6580E+06	5.2782E+06	6.0626E+06
9.85	4.6923E+06	5.3150E+06	6.1023E+06
9.90	4.7266E+06	5.3518E+06	6.1419E+06
9.95	4.7610E+06	5.3886E+06	6.1816E+06
10.00	4.7954E+06	5.4254E+06	6.2213E+06

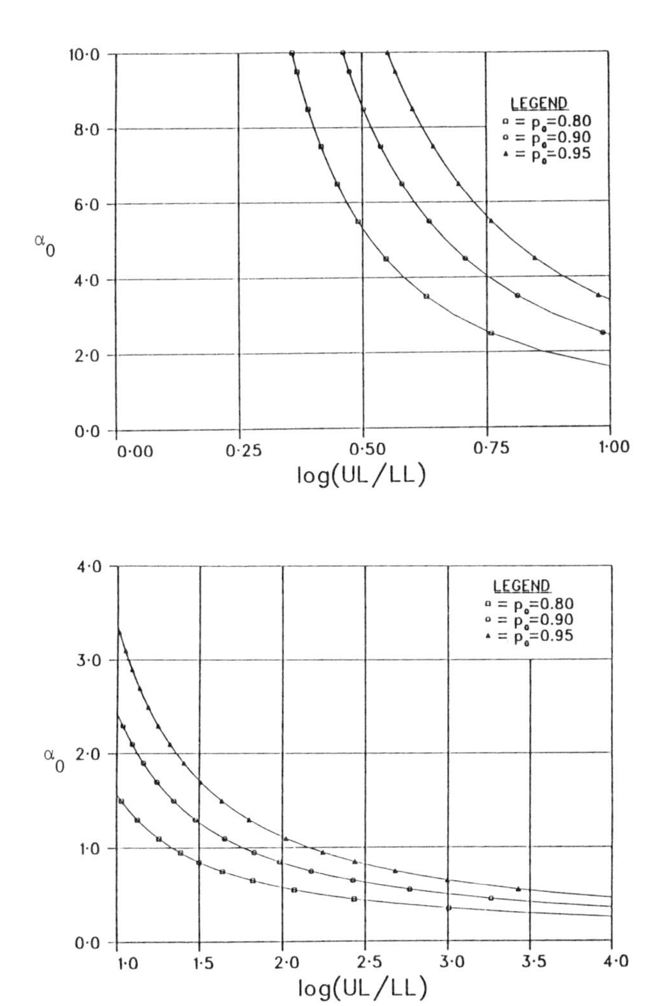

Figure C1. Gamma shape parameter α_0 as a function of $\log_{10}(UL/LL)$ [Reprinted, with permission, from H. F. Martz, Jr. and R. A. Waller (1979), A Bayesian Zero-Failure (BAZE) Reliability Demonstration Testing Procedure, *Journal of Quality Technology*, Vol. 11, pp. 130–131].

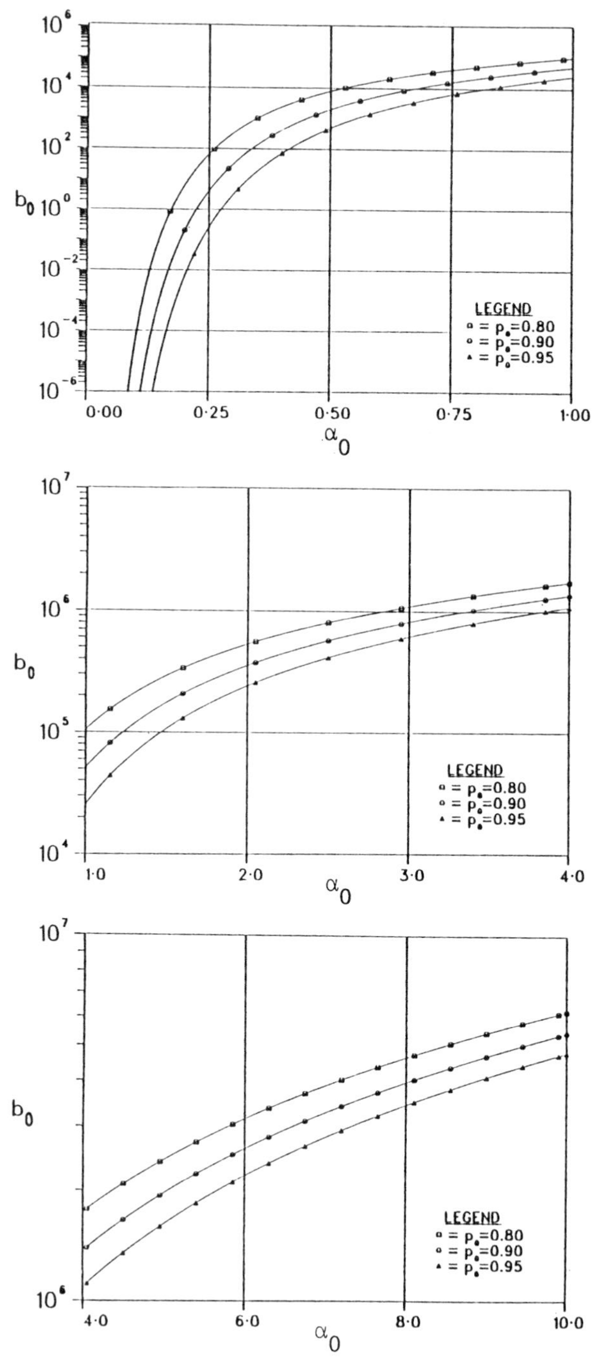

Figure C2. Gamma reference scale parameter b_0 as a function of the shape parameter α_0.

Table C4. Gamma Scale Parameter Values β_0 for Selected Values of p_0 and α_0 for $\lambda_0 = 1.0 \times 10^{-6}$[a]

λ_0=1.E-06

α_0/R_0	0.005	0.01	0.025	0.05	0.10	0.25	0.50
0.05	1.7941E+40	1.7110E+34	1.8812E+26	1.7941E+20	1.7110E+14	1.8812E+06	1.7941E+00
0.10	1.6861E+17	1.6466E+14	1.7266E+10	1.6861E+07	1.6466E+04	1.7266E+00	1.6852E-03
0.15	3.4743E+09	3.4198E+07	7.6043E+04	7.4851E+02	7.3676E+00	1.6382E-02	1.6039E-04
0.20	4.9039E+05	1.5325E+04	1.5692E+02	4.9039E+00	1.5324E-01	1.5684E-03	4.8201E-05
0.25	2.3705E+03	1.4815E+02	3.7928E+00	2.3705E-01	1.4815E-02	3.7848E-04	2.2897E-05
0.30	6.7096E+01	6.6568E+00	3.1390E-01	3.1143E-02	3.0890E-03	1.4493E-04	1.3674E-05
0.35	5.2164E+00	7.1991E-01	5.2517E-02	7.2474E-03	9.9958E-04	7.2231E-05	9.3136E-06
0.40	7.6286E-01	1.3485E-01	1.3646E-02	2.4117E-03	4.2574E-04	4.2436E-05	6.8928E-06
0.45	1.7005E-01	3.6443E-02	4.7561E-03	1.0187E-03	2.1779E-04	2.7822E-05	5.3958E-06
0.50	5.0929E-02	1.2732E-02	2.0365E-03	5.0863E-04	1.2666E-04	1.9698E-05	4.3962E-06
0.55	1.8910E-02	5.3621E-03	1.0129E-03	2.8678E-04	8.0862E-05	1.4752E-05	3.6895E-06
0.60	8.2511E-03	2.5985E-03	5.6379E-04	1.7715E-04	5.5370E-05	1.1524E-05	3.1675E-06
0.65	4.0765E-03	1.4030E-03	3.4217E-04	1.1740E-04	4.0014E-05	9.3027E-06	2.7684E-06
0.70	2.2208E-03	8.2466E-04	2.2231E-04	8.2215E-05	3.0170E-05	7.7062E-06	2.4544E-06
0.75	1.3083E-03	5.1887E-04	1.5252E-04	6.0182E-05	2.3535E-05	6.5186E-06	2.2018E-06
0.80	8.2145E-04	3.4506E-04	1.0939E-04	4.5667E-05	1.8874E-05	5.6093E-06	1.9946E-06
0.85	5.4355E-04	2.4018E-04	8.1371E-05	3.5698E-05	1.5487E-05	4.8962E-06	1.8218E-06
0.90	3.7575E-04	1.7367E-04	6.2405E-05	2.8605E-05	1.2954E-05	4.3254E-06	1.6758E-06
0.95	2.6953E-04	1.2967E-04	4.9107E-05	2.3405E-05	1.1012E-05	3.8605E-06	1.5508E-06
1.00	1.9950E-04	9.9499E-05	3.9498E-05	1.9496E-05	9.4912E-06	3.4761E-06	1.4427E-06
1.10	1.1807E-04	6.2654E-05	2.6967E-05	1.4134E-05	7.2964E-06	2.8808E-06	1.2654E-06
1.20	7.5829E-05	4.2357E-05	1.9493E-05	1.0736E-05	5.8174E-06	2.4448E-06	1.1262E-06
1.30	5.1865E-05	3.0250E-05	1.4726E-05	8.4552E-06	4.7720E-06	2.1140E-06	1.0142E-06
1.40	3.7280E-05	2.2558E-05	1.1519E-05	6.8528E-06	4.0042E-06	1.8557E-06	9.2219E-07
1.50	2.7886E-05	1.7417E-05	9.2680E-06	5.6843E-06	3.4225E-06	1.6494E-06	8.4532E-07
1.60	2.1548E-05	1.3834E-05	7.6302E-06	4.8058E-06	2.9701E-06	1.4814E-06	7.8016E-07
1.70	1.7104E-05	1.1250E-05	6.4030E-06	4.1280E-06	2.6105E-06	1.3422E-06	7.2424E-07
1.80	1.3885E-05	9.3305E-06	5.4602E-06	3.5937E-06	2.3195E-06	1.2252E-06	6.7574E-07
1.90	1.1489E-05	7.8687E-06	4.7202E-06	3.1644E-06	2.0801E-06	1.1258E-06	6.3329E-07
2.00	9.6623E-06	6.7315E-06	4.1287E-06	2.8140E-06	1.8804E-06	1.0403E-06	5.9582E-07

α_0/R_0	0.75	0.90	0.95	0.975	0.99	0.995
0.05	5.3858E-04	1.3103E-05	3.7604E-06	1.7631E-06	9.1943E-07	6.4741E-07
0.10	2.8324E-05	3.7572E-06	1.7228E-06	1.0226E-06	6.2953E-07	4.7743E-07
0.15	9.9143E-06	2.2482E-06	1.2113E-06	7.8584E-07	5.1798E-07	4.0590E-07
0.20	5.5910E-06	1.6532E-06	9.7038E-07	6.6177E-07	4.5407E-07	3.6302E-07
0.25	3.8369E-06	1.3326E-06	8.2637E-07	5.8254E-07	4.1087E-07	3.3313E-07
0.30	2.9163E-06	1.1302E-06	7.2868E-07	5.2624E-07	3.7887E-07	3.1050E-07
0.35	2.3560E-06	9.8942E-07	6.5705E-07	4.8348E-07	3.5378E-07	2.9245E-07
0.40	1.9810E-06	8.8508E-07	6.0170E-07	4.4950E-07	3.3332E-07	2.7750E-07
0.45	1.7127E-06	8.0415E-07	5.5729E-07	4.2159E-07	3.1615E-07	2.6482E-07
0.50	1.5114E-06	7.3922E-07	5.2064E-07	3.9810E-07	3.0144E-07	2.5383E-07
0.55	1.3546E-06	6.8576E-07	4.8972E-07	3.7795E-07	2.8861E-07	2.4416E-07
0.60	1.2290E-06	6.4082E-07	4.6318E-07	3.6039E-07	2.7727E-07	2.3554E-07
0.65	1.1261E-06	6.0240E-07	4.4007E-07	3.4490E-07	2.6714E-07	2.2780E-07
0.70	1.0400E-06	5.6911E-07	4.1971E-07	3.3110E-07	2.5802E-07	2.2076E-07
0.75	9.6703E-07	5.3992E-07	4.0159E-07	3.1868E-07	2.4972E-07	2.1434E-07
0.80	9.0426E-07	5.1407E-07	3.8534E-07	3.0744E-07	2.4214E-07	2.0843E-07
0.85	8.4966E-07	4.9099E-07	3.7064E-07	2.9718E-07	2.3516E-07	2.0297E-07
0.90	8.0171E-07	4.7022E-07	3.5728E-07	2.8778E-07	2.2871E-07	1.9790E-07
0.95	7.5925E-07	4.5142E-07	3.4505E-07	2.7911E-07	2.2273E-07	1.9317E-07
1.00	7.2135E-07	4.3429E-07	3.3381E-07	2.7109E-07	2.1715E-07	1.8874E-07
1.10	6.5654E-07	4.0421E-07	3.1381E-07	2.5667E-07	2.0703E-07	1.8067E-07
1.20	6.0309E-07	3.7858E-07	2.9650E-07	2.4406E-07	1.9807E-07	1.7348E-07
1.30	5.5819E-07	3.5642E-07	2.8134E-07	2.3290E-07	1.9007E-07	1.6701E-07
1.40	5.1989E-07	3.3705E-07	2.6792E-07	2.2292E-07	1.8285E-07	1.6114E-07
1.50	4.8681E-07	3.1993E-07	2.5593E-07	2.1394E-07	1.7629E-07	1.5579E-07
1.60	4.5793E-07	3.0468E-07	2.4514E-07	2.0580E-07	1.7030E-07	1.5087E-07
1.70	4.3247E-07	2.9098E-07	2.3536E-07	1.9837E-07	1.6480E-07	1.4634E-07
1.80	4.0986E-07	2.7861E-07	2.2646E-07	1.9156E-07	1.5972E-07	1.4214E-07
1.90	3.8961E-07	2.6736E-07	2.1830E-07	1.8528E-07	1.5502E-07	1.3823E-07
2.00	3.7138E-07	2.5709E-07	2.1080E-07	1.7948E-07	1.5064E-07	1.3459E-07

[a] Reprinted with permission of the Society for Industrial and Applied Mathematics, from *Nuclear Systems Reliability Engineering and Risk Assessment*, J. B. Fussell and G. R. Burdick, Ed.

Table C4 gives β_0 values that satisfy (6.13) for $p_0 = 0.005$, 0.01, 0.025, 0.05, 0.10, 0.25, 0.50, 0.75, 0.90, 0.95, 0.975, 0.99, 0.995, a selected set of α_0 values, and $\lambda_0 = 1.0 \times 10^{-6}$. We use the fact that β_0 is a scale parameter to obtain values of β_0 that correspond to failure rate percentiles different from $\lambda_0 = 1.0 \times 10^{-6}$. For given values of p_0 and α_0 in (6.13), the ratio λ_0/β_0 is constant. Therefore, in Table C4 for α_0 and p_0, multiplication of the β_0 value by $(\lambda_s/1.0 \times 10^{-6})$ yields the β_0 value corresponding to a p_0th percentile, λ_s.

Example. Let $p_0 = 0.05$ and $\alpha_0 = 0.25$. Then $\beta_0 = 2.3705 \times 10^{-1}$ for $\lambda_0 = 1.0 \times 10^{-6}$. Thus, for $\lambda_s = 1.0 \times 10^{-9}$, $\beta_0 = (1.0 \times 10^{-9}/1.0 \times 10^{-6})(2.3705 \times 10^{-1}) = 2.3705 \times 10^{-4}$. That is, the fifth percentile of a $\mathcal{G}(\alpha_0, \beta_0)$ distribution with $\alpha_0 = 0.25$ and $\beta_0 = 2.3705 \times 10^{-1}$ is 1.0×10^{-6} while the fifth percentile of a $\mathcal{G}(\alpha_0, \beta_0)$ distribution with $\alpha_0 = 0.25$ and $\beta_0 = 2.3705 \times 10^{-4}$ is 1.0×10^{-9}. ■

In Figures C3–C15 we have graphed α_0 and β_0 for selected values of λ_0. The results are derived from Table C4 as illustrated by the above example. To provide better resolution in the graphs, a logarithmic scale is used for β_0. The notation in Table C4 and Figures C3–C15 is defined as follows: $nEm = n \times 10^m$.

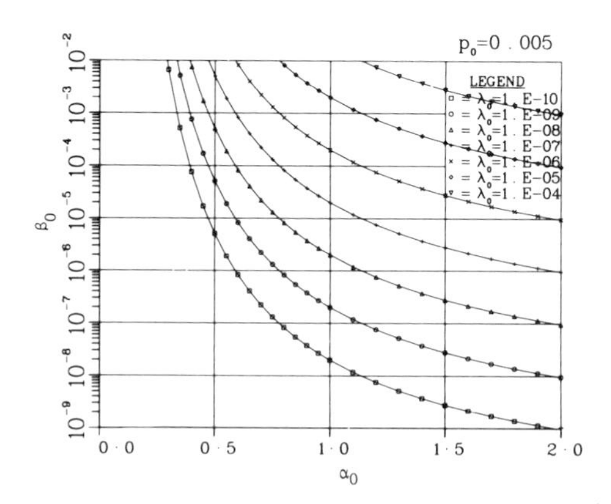

Figure C3. A graph of Table C4 for $p_0 = .005$ and a selected set of λ_0 values.

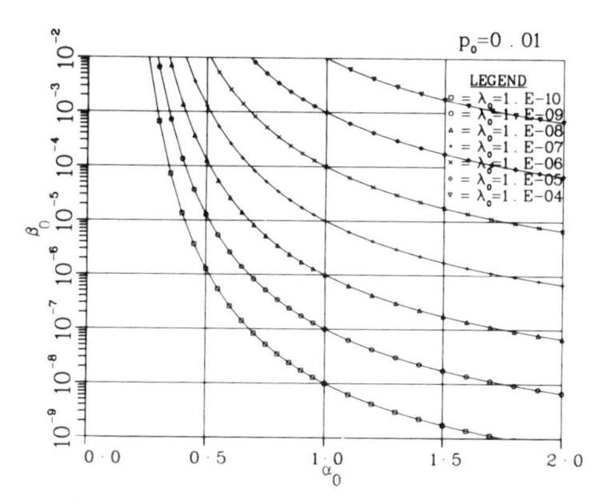

Figure C4. A graph of Table C4 for $p_0 = 0.01$ and a selected set of λ_0 values.

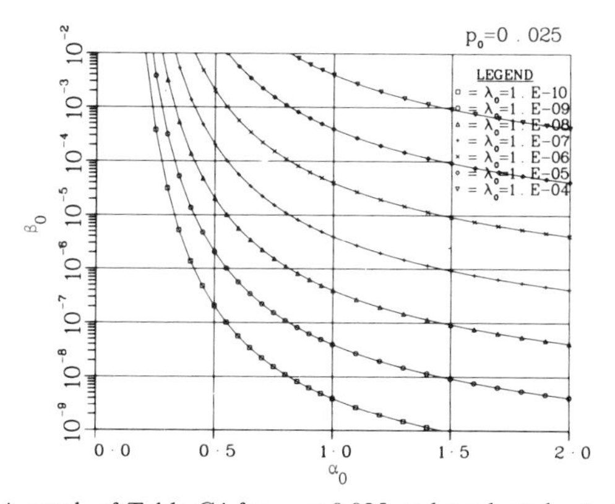

Figure C5. A graph of Table C4 for $p_0 = 0.025$ and a selected set of λ_0 values.

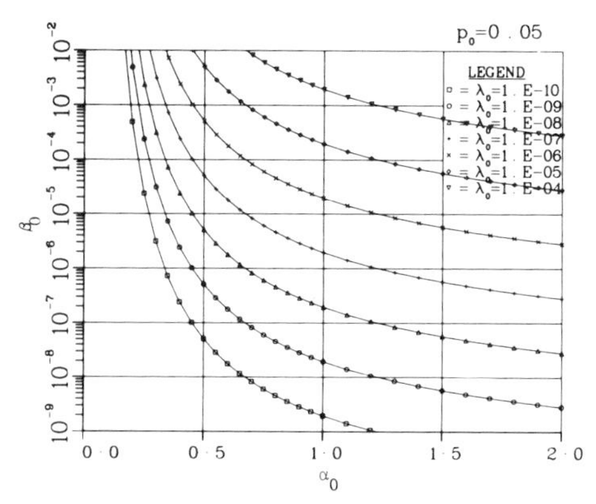

Figure C6. A graph of Table C4 for $p_0 = 0.05$ and a selected set of λ_0 values [Reprinted with permission of Society for Industrial and Applied Mathematics from *Nuclear Systems Reliability Engineering and Risk Assessment*, J. B. Fussell and G. R. Burdick, Eds. ©1977 Society for Industrial and Applied Mathematics, Philadelphia. All rights reserved.].

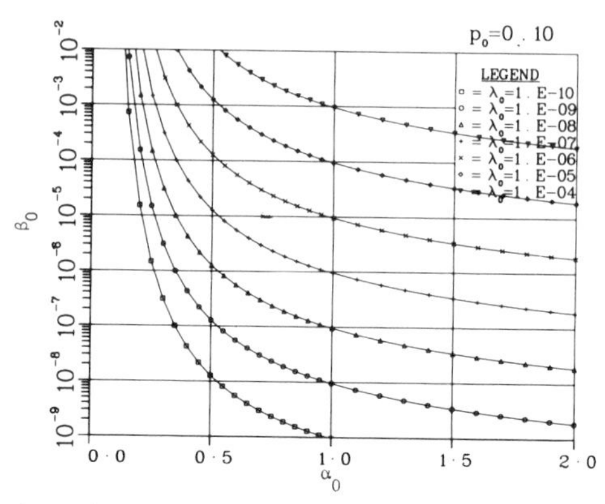

Figure C7. A graph of Table C4 for $p_0 = 0.10$ and a selected set of λ_0 values.

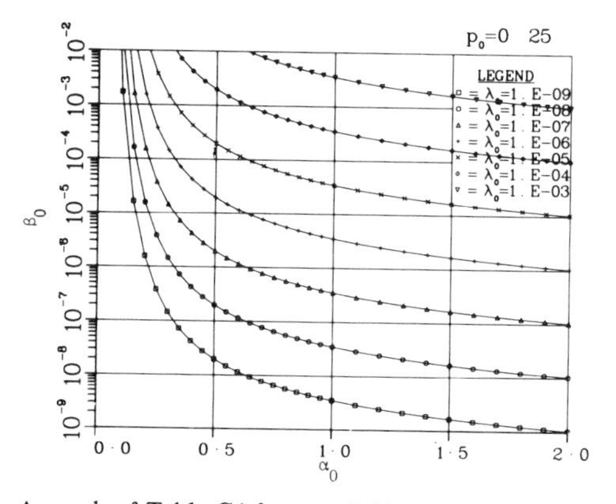

Figure C8. A graph of Table C4 for $p_0 = 0.25$ and a selected set of λ_0 values.

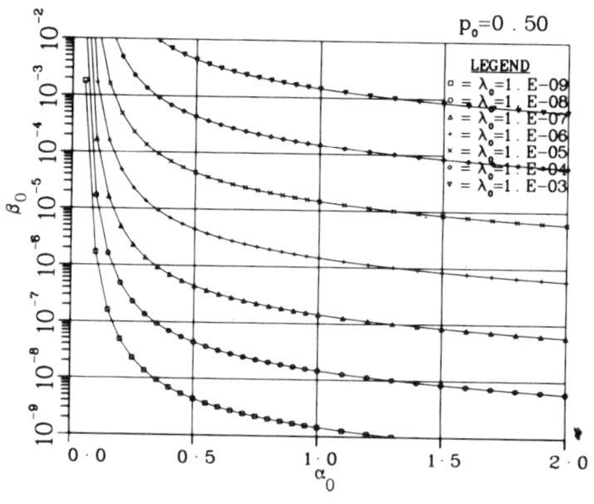

Figure C9. A graph of Table C4 for $p_0 = 0.50$ and a selected set of λ_0 values [Reprinted with permission of Society for Industrial and Applied Mathematics from *Nuclear Systems Reliability Engineering and Risk Assessment*, J. B. Fussell and G. R. Burdick, Eds. ©1977 Society for Industrial and Applied Mathematics, Philadelphia. All rights reserved.].

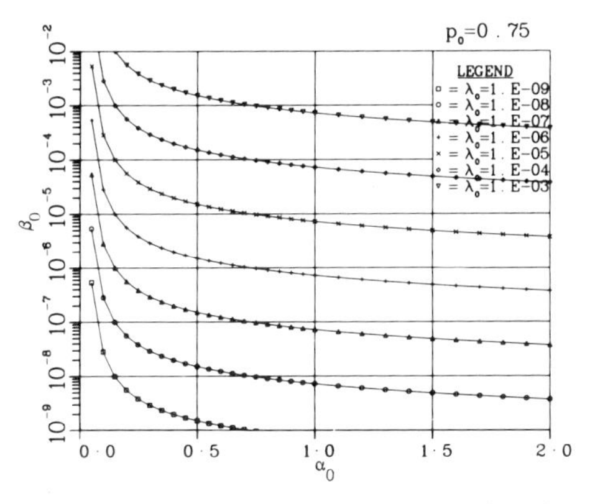

Figure C10. A graph of Table C4 for $p_0 = 0.75$ and a selected set of λ_0 values.

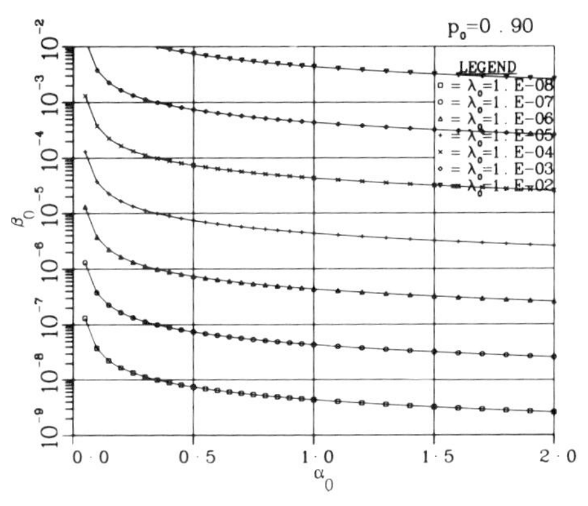

Figure C11. A graph of Table C4 for $p_0 = 0.90$ and a selected set of λ_0 values.

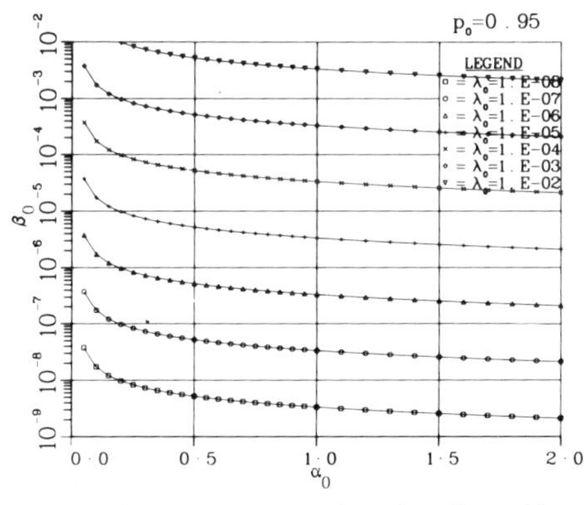

Figure C12. A graph of Table C4 for $p_0 = 0.95$ and a selected set of λ_0 values [Reprinted with permission of Society for Industrial and Applied Mathematics from *Nuclear Systems Reliability Engineering and Risk Assessment*, J. B. Fussell and G. R. Burdick, Eds. ©1977 Society for Industrial and Applied Mathematics, Philadelphia. All rights reserved.].

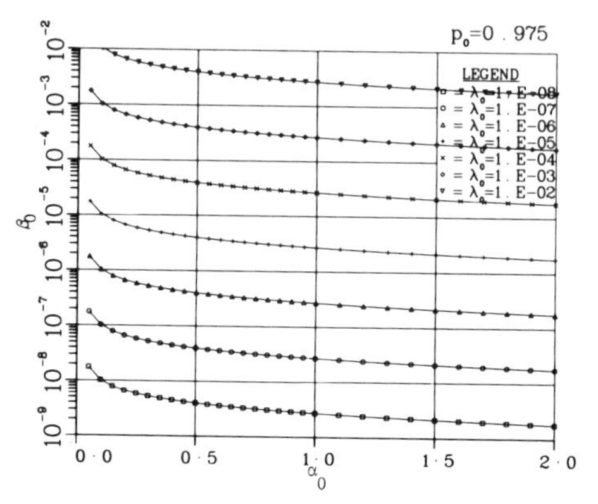

Figure C13. A graph of Table C4 for $p_0 = 0.975$ and a selected set of λ_0 values.

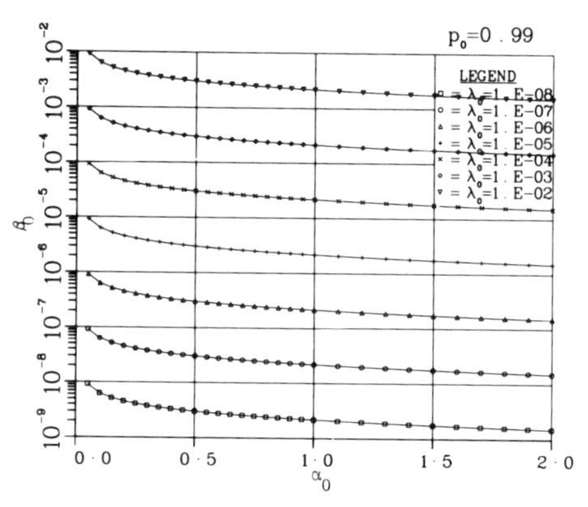

Figure C14. A graph of Table C4 for $p_0 = 0.99$ and a selected set of λ_0 values.

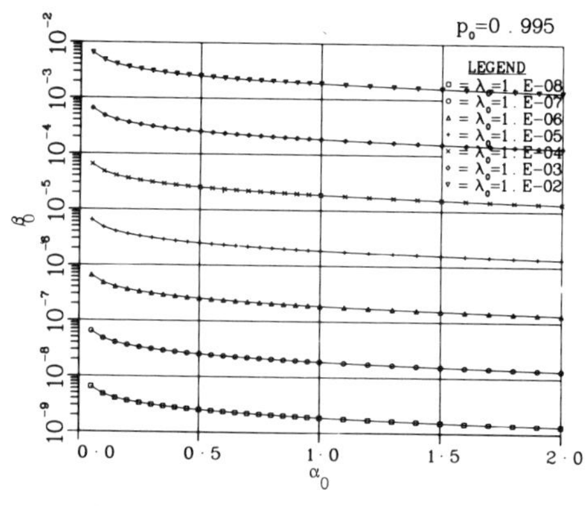

Figure C15. A graph of Table C4 for $p_0 = 0.995$ and a selected set of λ_0 values.

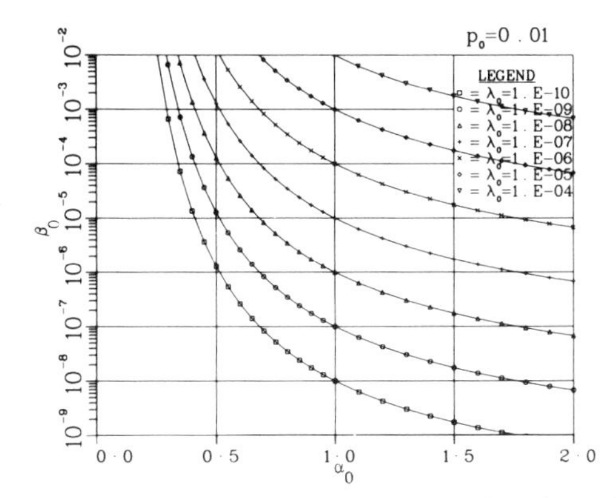

Figure C4. A graph of Table C4 for $p_0 = 0.01$ and a selected set of λ_0 values.

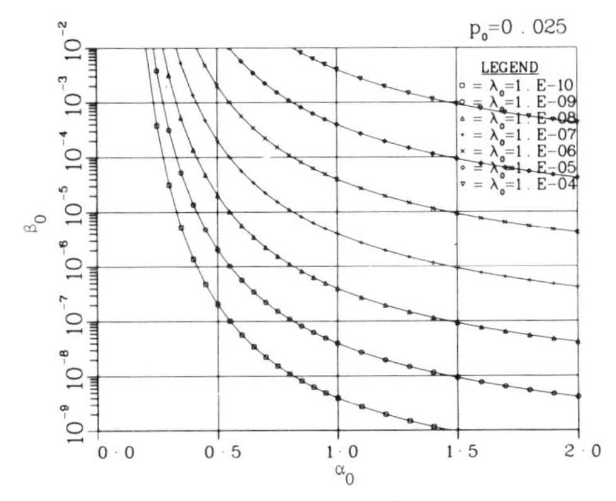

Figure C5. A graph of Table C4 for $p_0 = 0.025$ and a selected set of λ_0 values.

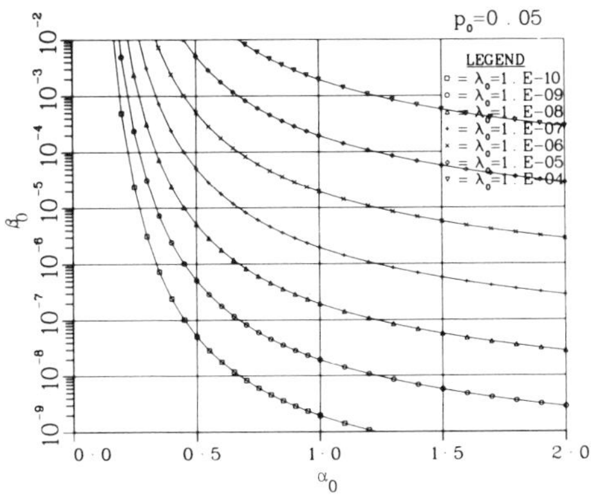

Figure C6. A graph of Table C4 for $p_0 = 0.05$ and a selected set of λ_0 values [Reprinted with permission of Society for Industrial and Applied Mathematics from *Nuclear Systems Reliability Engineering and Risk Assessment*, J. B. Fussell and G. R. Burdick, Eds. ©1977 Society for Industrial and Applied Mathematics, Philadelphia. All rights reserved.].

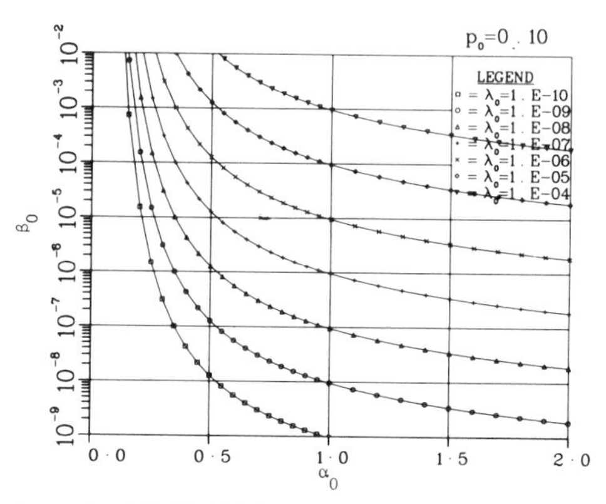

Figure C7. A graph of Table C4 for $p_0 = 0.10$ and a selected set of λ_0 values.

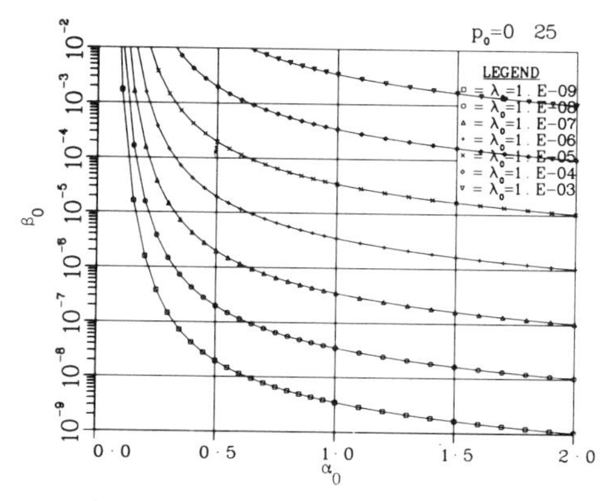

Figure C8. A graph of Table C4 for $p_0 = 0.25$ and a selected set of λ_0 values.

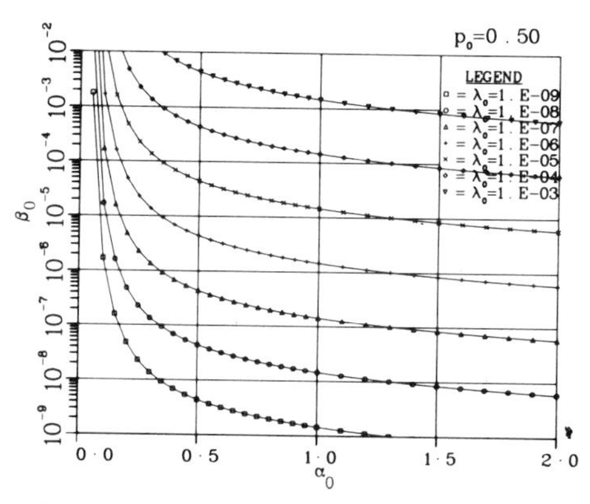

Figure C9. A graph of Table C4 for $p_0 = 0.50$ and a selected set of λ_0 values [Reprinted with permission of Society for Industrial and Applied Mathematics from *Nuclear Systems Reliability Engineering and Risk Assessment*, J. B. Fussell and G. R. Burdick, Eds. ©1977 Society for Industrial and Applied Mathematics, Philadelphia. All rights reserved.].

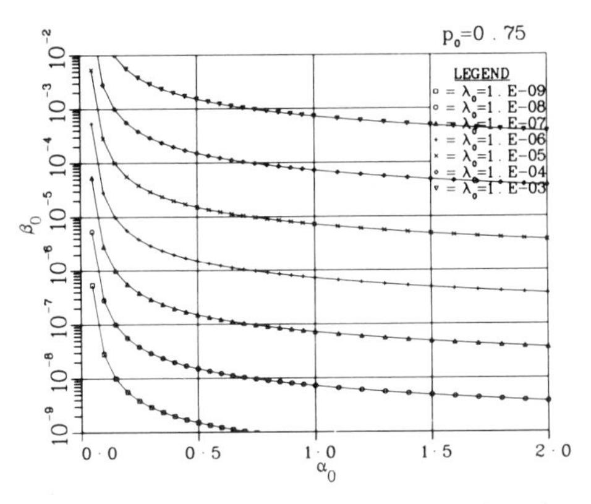

Figure C10. A graph of Table C4 for $p_0 = 0.75$ and a selected set of λ_0 values.

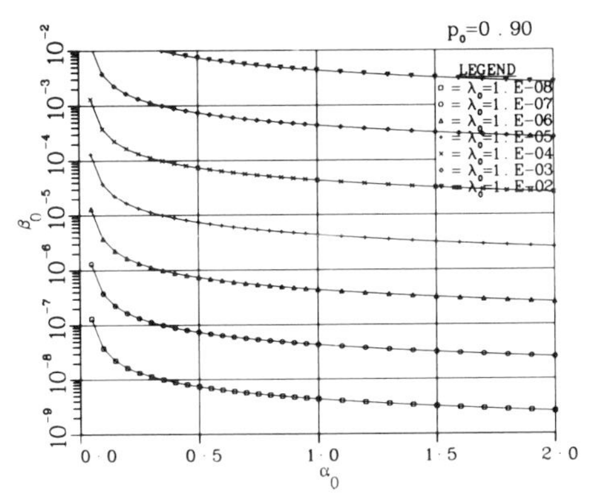

Figure C11. A graph of Table C4 for $p_0 = 0.90$ and a selected set of λ_0 values.

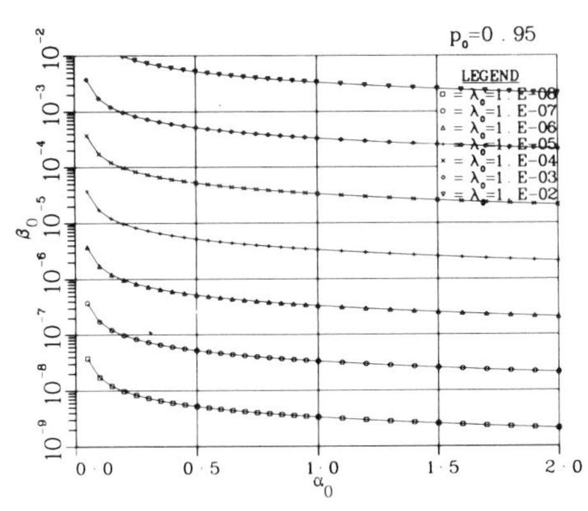

Figure C12. A graph of Table C4 for $p_0 = 0.95$ and a selected set of λ_0 values [Reprinted with permission of Society for Industrial and Applied Mathematics from *Nuclear Systems Reliability Engineering and Risk Assessment*, J. B. Fussell and G. R. Burdick, Eds. ©1977 Society for Industrial and Applied Mathematics, Philadelphia. All rights reserved.].

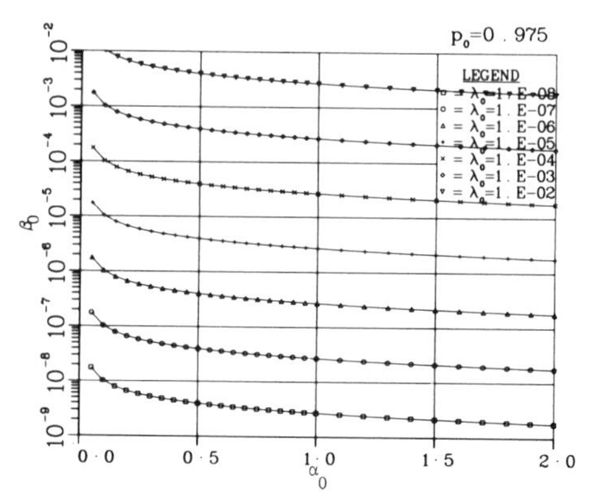

Figure C13. A graph of Table C4 for $p_0 = 0.975$ and a selected set of λ_0 values.

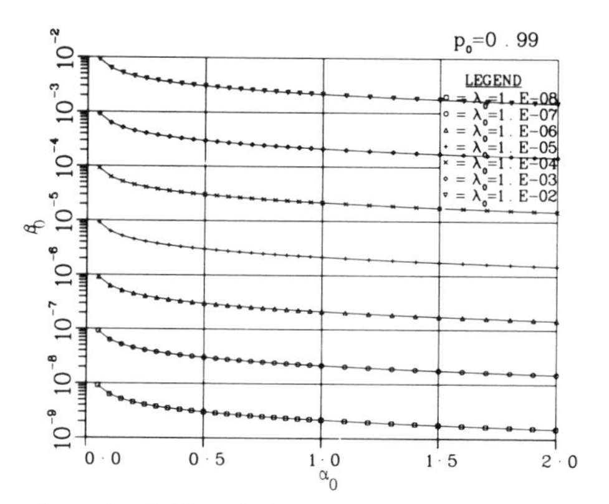

Figure C14. A graph of Table C4 for $p_0 = 0.99$ and a selected set of λ_0 values.

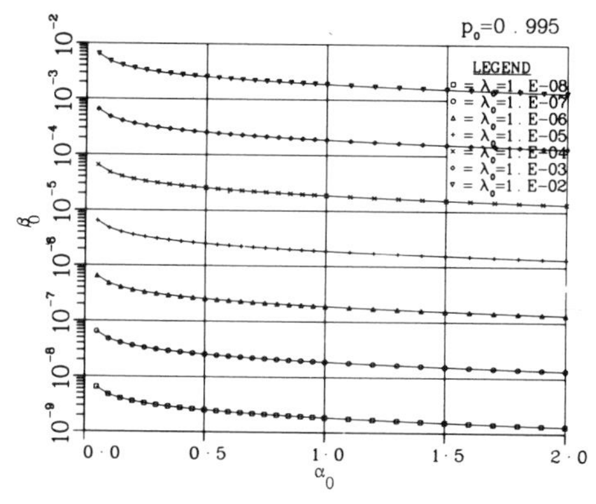

Figure C15. A graph of Table C4 for $p_0 = 0.995$ and a selected set of λ_0 values.

Table C5. A Table of γ_0 Values for $p_0 = 0.005$ and a Selected Set of Values for α_0 and R_0

$p_0 = 0.005$

α_0/R_0	0.50	0.60	0.70	0.80	0.90	0.95	0.99
0.05	4.4875E-01	3.3071E-01	2.3091E-01	1.4446E-01	6.8211E-02	3.3208E-02	6.5066E-03
0.10	3.3093E-01	2.4388E-01	1.7029E-01	1.0653E-01	5.0302E-02	2.4489E-02	4.7983E-03
0.15	2.8135E-01	2.0735E-01	1.4478E-01	9.0575E-02	4.2766E-02	2.0820E-02	4.0795E-03
0.20	2.5162E-01	1.8544E-01	1.2948E-01	8.1005E-02	3.8248E-02	1.8620E-02	3.6484E-03
0.25	2.3091E-01	1.7017E-01	1.1882E-01	7.4336E-02	3.5099E-02	1.7087E-02	3.3481E-03
0.30	2.1522E-01	1.5861E-01	1.1075E-01	6.9287E-02	3.2715E-02	1.5927E-02	3.1207E-03
0.35	2.0271E-01	1.4939E-01	1.0431E-01	6.5257E-02	3.0812E-02	1.5000E-02	2.9392E-03
0.40	1.9235E-01	1.4176E-01	9.8979E-02	6.1923E-02	2.9238E-02	1.4234E-02	2.7890E-03
0.45	1.8356E-01	1.3527E-01	9.4453E-02	5.9092E-02	2.7901E-02	1.3583E-02	2.6615E-03
0.50	1.7594E-01	1.2966E-01	9.0533E-02	5.6639E-02	2.6743E-02	1.3020E-02	2.5510E-03
0.55	1.6924E-01	1.2472E-01	8.7084E-02	5.4482E-02	2.5724E-02	1.2524E-02	2.4539E-03
0.60	1.6327E-01	1.2032E-01	8.4013E-02	5.2560E-02	2.4817E-02	1.2082E-02	2.3673E-03
0.65	1.5790E-01	1.1636E-01	8.1249E-02	5.0831E-02	2.4001E-02	1.1684E-02	2.2894E-03
0.70	1.5302E-01	1.1277E-01	7.8741E-02	4.9262E-02	2.3260E-02	1.1324E-02	2.2188E-03
0.75	1.4857E-01	1.0949E-01	7.6449E-02	4.7828E-02	2.2583E-02	1.0994E-02	2.1542E-03
0.80	1.4447E-01	1.0647E-01	7.4342E-02	4.6510E-02	2.1960E-02	1.0691E-02	2.0948E-03
0.85	1.4069E-01	1.0368E-01	7.2394E-02	4.5291E-02	2.1385E-02	1.0411E-02	2.0399E-03
0.90	1.3717E-01	1.0109E-01	7.0584E-02	4.4159E-02	2.0850E-02	1.0151E-02	1.9889E-03
0.95	1.3389E-01	9.8674E-02	6.8897E-02	4.3104E-02	2.0352E-02	9.9081E-03	1.9414E-03
1.00	1.3082E-01	9.6413E-02	6.7319E-02	4.2116E-02	1.9886E-02	9.6811E-03	1.8969E-03
1.10	1.2523E-01	9.2290E-02	6.4440E-02	4.0315E-02	1.9035E-02	9.2671E-03	1.8158E-03
1.20	1.2024E-01	8.8616E-02	6.1874E-02	3.8710E-02	1.8277E-02	8.8981E-03	1.7435E-03
1.30	1.1576E-01	8.5311E-02	5.9567E-02	3.7266E-02	1.7596E-02	8.5662E-03	1.6785E-03
1.40	1.1169E-01	8.2314E-02	5.7474E-02	3.5957E-02	1.6978E-02	8.2654E-03	1.6195E-03
1.50	1.0798E-01	7.9579E-02	5.5565E-02	3.4763E-02	1.6414E-02	7.9908E-03	1.5657E-03
1.60	1.0458E-01	7.7069E-02	5.3812E-02	3.3666E-02	1.5896E-02	7.7387E-03	1.5163E-03
1.70	1.0143E-01	7.4754E-02	5.2196E-02	3.2655E-02	1.5418E-02	7.5062E-03	1.4708E-03
1.80	9.8524E-02	7.2609E-02	5.0698E-02	3.1718E-02	1.4976E-02	7.2908E-03	1.4286E-03
1.90	9.5816E-02	7.0613E-02	4.9304E-02	3.0846E-02	1.4564E-02	7.0904E-03	1.3893E-03
2.00	9.3289E-02	6.8751E-02	4.8004E-02	3.0032E-02	1.4180E-02	6.9034E-03	1.3526E-03

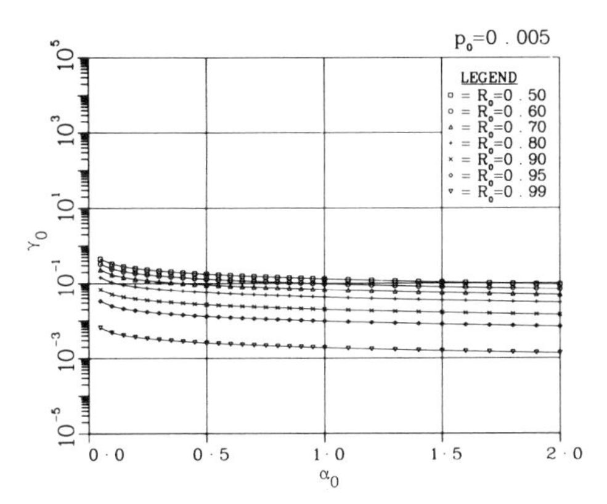

Figure C16. A graph of Table C5.

Table C6. A Table of γ_0 Values for $p_0 = 0.01$ and a Selected Set of Values for α_0 and R_0

$p_0 = 0.01$

α_0/R_0	0.50	0.60	0.70	0.80	0.90	0.95	0.99
0.05	6.3730E-01	4.6967E-01	3.2794E-01	2.0517E-01	9.6872E-02	4.7161E-02	9.2406E-03
0.10	4.3636E-01	3.2158E-01	2.2454E-01	1.4048E-01	6.6328E-02	3.2291E-02	6.3270E-03
0.15	3.5903E-01	2.6460E-01	1.8475E-01	1.1558E-01	5.4574E-02	2.6569E-02	5.2058E-03
0.20	3.1474E-01	2.3195E-01	1.6196E-01	1.0132E-01	4.7841E-02	2.3291E-02	4.5636E-03
0.25	2.8479E-01	2.0988E-01	1.4655E-01	9.1682E-02	4.3289E-02	2.1075E-02	4.1293E-03
0.30	2.6261E-01	1.9354E-01	1.3513E-01	8.4543E-02	3.9918E-02	1.9434E-02	3.8078E-03
0.35	2.4522E-01	1.8072E-01	1.2619E-01	7.8945E-02	3.7275E-02	1.8147E-02	3.5556E-03
0.40	2.3104E-01	1.7027E-01	1.1889E-01	7.4379E-02	3.5119E-02	1.7097E-02	3.3500E-03
0.45	2.1914E-01	1.6150E-01	1.1276E-01	7.0548E-02	3.3310E-02	1.6217E-02	3.1775E-03
0.50	2.0894E-01	1.5398E-01	1.0751E-01	6.7264E-02	3.1760E-02	1.5462E-02	3.0295E-03
0.55	2.0005E-01	1.4743E-01	1.0294E-01	6.4401E-02	3.0408E-02	1.4804E-02	2.9006E-03
0.60	1.9219E-01	1.4164E-01	9.8895E-02	6.1871E-02	2.9213E-02	1.4222E-02	2.7867E-03
0.65	1.8517E-01	1.3646E-01	9.5284E-02	5.9611E-02	2.8146E-02	1.3703E-02	2.6849E-03
0.70	1.7884E-01	1.3180E-01	9.2028E-02	5.7574E-02	2.7185E-02	1.3234E-02	2.5931E-03
0.75	1.7309E-01	1.2756E-01	8.9070E-02	5.5724E-02	2.6311E-02	1.2809E-02	2.5098E-03
0.80	1.6784E-01	1.2369E-01	8.6364E-02	5.4031E-02	2.5512E-02	1.2420E-02	2.4336E-03
0.85	1.6300E-01	1.2013E-01	8.3876E-02	5.2475E-02	2.4777E-02	1.2062E-02	2.3635E-03
0.90	1.5853E-01	1.1683E-01	8.1577E-02	5.1036E-02	2.4097E-02	1.1731E-02	2.2987E-03
0.95	1.5438E-01	1.1377E-01	7.9441E-02	4.9700E-02	2.3467E-02	1.1424E-02	2.2385E-03
1.00	1.5051E-01	1.1092E-01	7.7451E-02	4.8455E-02	2.2879E-02	1.1138E-02	2.1824E-03
1.10	1.4350E-01	1.0576E-01	7.3843E-02	4.6198E-02	2.1813E-02	1.0619E-02	2.0807E-03
1.20	1.3729E-01	1.0118E-01	7.0648E-02	4.4199E-02	2.0869E-02	1.0160E-02	1.9907E-03
1.30	1.3174E-01	9.7091E-02	6.7792E-02	4.2412E-02	2.0026E-02	9.7491E-03	1.9102E-03
1.40	1.2674E-01	9.3403E-02	6.5217E-02	4.0801E-02	1.9265E-02	9.3788E-03	1.8377E-03
1.50	1.2220E-01	9.0054E-02	6.2879E-02	3.9338E-02	1.8574E-02	9.0426E-03	1.7718E-03
1.60	1.1804E-01	8.6994E-02	6.0742E-02	3.8002E-02	1.7943E-02	8.7353E-03	1.7116E-03
1.70	1.1423E-01	8.4184E-02	5.8780E-02	3.6774E-02	1.7363E-02	8.4531E-03	1.6563E-03
1.80	1.1071E-01	8.1590E-02	5.6969E-02	3.5641E-02	1.6828E-02	8.1927E-03	1.6053E-03
1.90	1.0745E-01	7.9187E-02	5.5291E-02	3.4591E-02	1.6333E-02	7.9513E-03	1.5580E-03
2.00	1.0442E-01	7.6951E-02	5.3729E-02	3.3614E-02	1.5871E-02	7.7268E-03	1.5140E-03

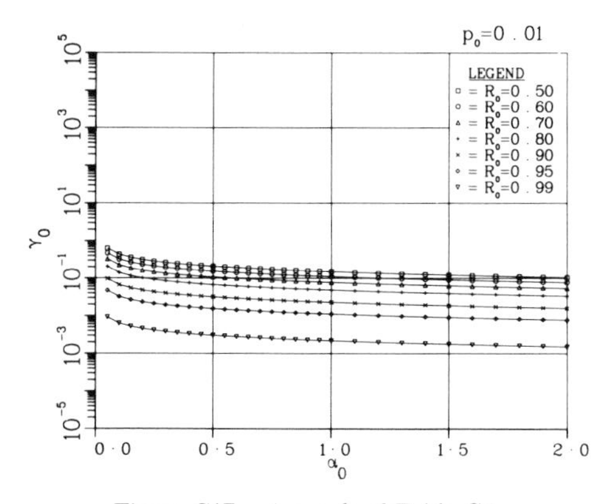

Figure C17. A graph of Table C6.

Table C7. A Table of γ_0 Values for $p_0 = 0.025$ and a Selected Set of Values for α_0 and R_0

$p_0 = 0.025$

α_0/R_0	0.50	0.60	0.70	0.80	0.90	0.95	0.99
0.05	1.2221E+00	9.0065E-01	6.2886E-01	3.9343E-01	1.8576E-01	9.0437E-02	1.7720E-02
0.10	7.0881E-01	5.2237E-01	3.6473E-01	2.2819E-01	1.0774E-01	5.2452E-02	1.0277E-02
0.15	5.4470E-01	4.0143E-01	2.8029E-01	1.7535E-01	8.2796E-02	4.0308E-02	7.8979E-03
0.20	4.5870E-01	3.3805E-01	2.3604E-01	1.4767E-01	6.9724E-02	3.3944E-02	6.6510E-03
0.25	4.0379E-01	2.9758E-01	2.0778E-01	1.2999E-01	6.1377E-02	2.9880E-02	5.8547E-03
0.30	3.6476E-01	2.6882E-01	1.8770E-01	1.1743E-01	5.5445E-02	2.6993E-02	5.2889E-03
0.35	3.3512E-01	2.4697E-01	1.7245E-01	1.0789E-01	5.0940E-02	2.4799E-02	4.8591E-03
0.40	3.1157E-01	2.2961E-01	1.6032E-01	1.0030E-01	4.7359E-02	2.3056E-02	4.5176E-03
0.45	2.9222E-01	2.1536E-01	1.5037E-01	9.4074E-02	4.4419E-02	2.1625E-02	4.2371E-03
0.50	2.7594E-01	2.0336E-01	1.4199E-01	8.8833E-02	4.1944E-02	2.0420E-02	4.0010E-03
0.55	2.6197E-01	1.9306E-01	1.3480E-01	8.4336E-02	3.9821E-02	1.9386E-02	3.7985E-03
0.60	2.4980E-01	1.8410E-01	1.2854E-01	8.0419E-02	3.7971E-02	1.8486E-02	3.6220E-03
0.65	2.3907E-01	1.7619E-01	1.2302E-01	7.6963E-02	3.6339E-02	1.7691E-02	3.4664E-03
0.70	2.2950E-01	1.6913E-01	1.1809E-01	7.3882E-02	3.4885E-02	1.6983E-02	3.3276E-03
0.75	2.2089E-01	1.6279E-01	1.1367E-01	7.1112E-02	3.3577E-02	1.6346E-02	3.2029E-03
0.80	2.1310E-01	1.5705E-01	1.0966E-01	6.8603E-02	3.2392E-02	1.5769E-02	3.0898E-03
0.85	2.0599E-01	1.5181E-01	1.0600E-01	6.6314E-02	3.1311E-02	1.5243E-02	2.9868E-03
0.90	1.9947E-01	1.4700E-01	1.0264E-01	6.4216E-02	3.0320E-02	1.4761E-02	2.8923E-03
0.95	1.9346E-01	1.4258E-01	9.9551E-02	6.2281E-02	2.9407E-02	1.4316E-02	2.8051E-03
1.00	1.8790E-01	1.3848E-01	9.6689E-02	6.0491E-02	2.8562E-02	1.3905E-02	2.7245E-03
1.10	1.7791E-01	1.3112E-01	9.1549E-02	5.7275E-02	2.7043E-02	1.3166E-02	2.5796E-03
1.20	1.6917E-01	1.2467E-01	8.7050E-02	5.4460E-02	2.5714E-02	1.2519E-02	2.4529E-03
1.30	1.6143E-01	1.1897E-01	8.3068E-02	5.1969E-02	2.4538E-02	1.1946E-02	2.3407E-03
1.40	1.5452E-01	1.1387E-01	7.9511E-02	4.9744E-02	2.3487E-02	1.1434E-02	2.2404E-03
1.50	1.4829E-01	1.0929E-01	7.6307E-02	4.7739E-02	2.2541E-02	1.0974E-02	2.1502E-03
1.60	1.4265E-01	1.0513E-01	7.3403E-02	4.5922E-02	2.1683E-02	1.0556E-02	2.0683E-03
1.70	1.3750E-01	1.0133E-01	7.0753E-02	4.4265E-02	2.0900E-02	1.0175E-02	1.9937E-03
1.80	1.3278E-01	9.7853E-02	6.8324E-02	4.2745E-02	2.0183E-02	9.8256E-03	1.9252E-03
1.90	1.2843E-01	9.4648E-02	6.6086E-02	4.1345E-02	1.9522E-02	9.5038E-03	1.8622E-03
2.00	1.2441E-01	9.1683E-02	6.4016E-02	4.0050E-02	1.8910E-02	9.2061E-03	1.8038E-03

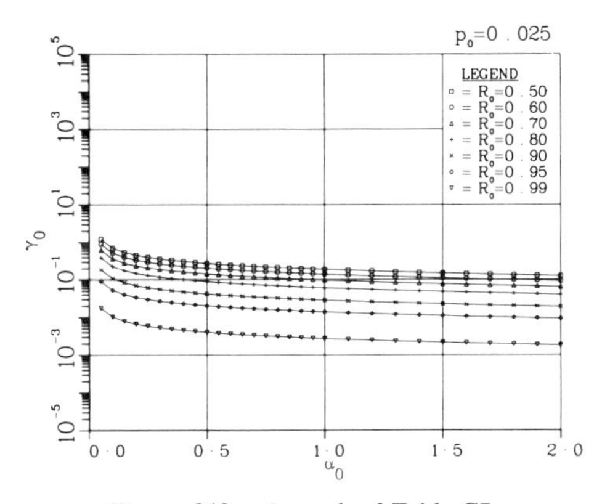

Figure C18. A graph of Table C7.

Table C8. A Table of γ_0 Values for $p_0 = 0.05$ and a Selected Set of Values for α_0 and R_0[a]

$p_0 = 0.05$

α_0/R_0	0.50	0.60	0.70	0.80	0.90	0.95	0.99
0.05	2.6065E+00	1.9209E+00	1.3412E+00	8.3910E-01	3.9619E-01	1.9288E-01	3.7793E-02
0.10	1.1942E+00	8.8007E-01	6.1450E-01	3.8444E-01	1.8152E-01	8.8370E-02	1.7315E-02
0.15	8.3958E-01	6.1874E-01	4.3203E-01	2.7028E-01	1.2762E-01	6.2129E-02	1.2174E-02
0.20	6.7261E-01	4.9569E-01	3.4611E-01	2.1653E-01	1.0224E-01	4.9774E-02	9.7526E-03
0.25	5.7279E-01	4.2213E-01	2.9474E-01	1.8440E-01	8.7066E-02	4.2387E-02	8.3053E-03
0.30	5.0508E-01	3.7223E-01	2.5990E-01	1.6260E-01	7.6774E-02	3.7376E-02	7.3235E-03
0.35	4.5543E-01	3.3564E-01	2.3435E-01	1.4662E-01	6.9227E-02	3.3702E-02	6.6036E-03
0.40	4.1707E-01	3.0736E-01	2.1461E-01	1.3427E-01	6.3395E-02	3.0863E-02	6.0473E-03
0.45	3.8628E-01	2.8468E-01	1.9877E-01	1.2436E-01	5.8716E-02	2.8585E-02	5.6009E-03
0.50	3.6088E-01	2.6595E-01	1.8570E-01	1.1618E-01	5.4854E-02	2.6705E-02	5.2326E-03
0.55	3.3945E-01	2.5016E-01	1.7467E-01	1.0928E-01	5.1597E-02	2.5119E-02	4.9218E-03
0.60	3.2105E-01	2.3660E-01	1.6520E-01	1.0335E-01	4.8800E-02	2.3758E-02	4.6551E-03
0.65	3.0503E-01	2.2480E-01	1.5696E-01	9.8198E-02	4.6366E-02	2.2572E-02	4.4228E-03
0.70	2.9092E-01	2.1440E-01	1.4970E-01	9.3655E-02	4.4220E-02	2.1528E-02	4.2182E-03
0.75	2.7836E-01	2.0514E-01	1.4324E-01	8.9612E-02	4.2312E-02	2.0599E-02	4.0361E-03
0.80	2.6709E-01	1.9684E-01	1.3744E-01	8.5985E-02	4.0599E-02	1.9765E-02	3.8727E-03
0.85	2.5691E-01	1.8933E-01	1.3220E-01	8.2706E-02	3.9051E-02	1.9011E-02	3.7251E-03
0.90	2.4764E-01	1.8251E-01	1.2743E-01	7.9724E-02	3.7643E-02	1.8326E-02	3.5907E-03
0.95	2.3917E-01	1.7626E-01	1.2307E-01	7.6995E-02	3.6354E-02	1.7699E-02	3.4679E-03
1.00	2.3138E-01	1.7052E-01	1.1906E-01	7.4487E-02	3.5170E-02	1.7122E-02	3.3549E-03
1.10	2.1751E-01	1.6030E-01	1.1193E-01	7.0024E-02	3.3063E-02	1.6096E-02	3.1539E-03
1.20	2.0552E-01	1.5146E-01	1.0575E-01	6.6162E-02	3.1240E-02	1.5209E-02	2.9799E-03
1.30	1.9501E-01	1.4372E-01	1.0035E-01	6.2779E-02	2.9642E-02	1.4431E-02	2.8276E-03
1.40	1.8571E-01	1.3686E-01	9.5559E-02	5.9784E-02	2.8228E-02	1.3742E-02	2.6926E-03
1.50	1.7740E-01	1.3073E-01	9.1283E-02	5.7108E-02	2.6965E-02	1.3127E-02	2.5722E-03
1.60	1.6992E-01	1.2522E-01	8.7434E-02	5.4701E-02	2.5828E-02	1.2574E-02	2.4637E-03
1.70	1.6314E-01	1.2023E-01	8.3948E-02	5.2520E-02	2.4798E-02	1.2073E-02	2.3655E-03
1.80	1.5697E-01	1.1568E-01	8.0772E-02	5.0533E-02	2.3860E-02	1.1616E-02	2.2760E-03
1.90	1.5132E-01	1.1151E-01	7.7863E-02	4.8713E-02	2.3000E-02	1.1197E-02	2.1940E-03
2.00	1.4611E-01	1.0768E-01	7.5187E-02	4.7038E-02	2.2210E-02	1.0813E-02	2.1186E-03

[a] Reprinted with permission of Society for Industrial and Applied Mathematics, from *Nuclear Systems Reliability Engineering and Risk Assessment*, J. B. Fussell and G. R. Burdick, Ed.

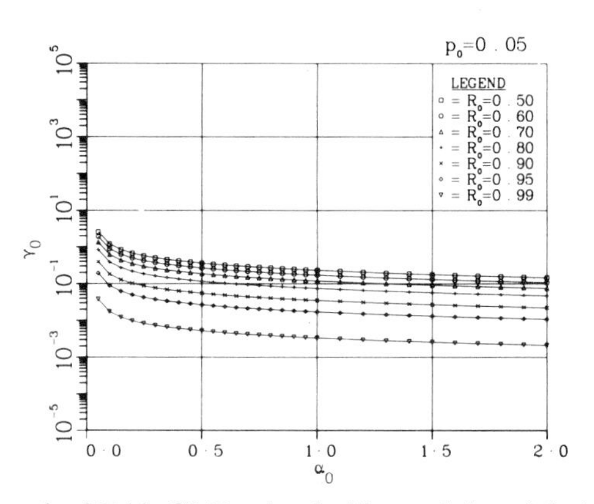

Figure C19. A graph of Table C8 [Reprinted with permission of Society for Industrial and Applied Mathematics from *Nuclear Systems Reliability Engineering and Risk Assessment*, J. B. Fussell and G. R. Burdick, Eds. ©1977 Society for Industrial and Applied Mathematics, Philadelphia. All rights reserved.].

Table C9. A Table of γ_0 Values for $p_0 = 0.10$ and a Selected Set of Values for α_0 and R_0

$p_0 = 0.10$

α_0/R_0	0.50	0.60	0.70	0.80	0.90	0.95	0.99
0.05	9.0825E+00	6.6935E+00	4.6736E+00	2.9239E+00	1.3806E+00	6.7211E-01	1.3169E-01
0.10	2.6043E+00	1.9193E+00	1.3401E+00	8.3840E-01	3.9586E-01	1.9272E-01	3.7761E-02
0.15	1.5583E+00	1.1484E+00	8.0188E-01	5.0167E-01	2.3687E-01	1.1532E-01	2.2595E-02
0.20	1.1459E+00	8.4448E-01	5.8964E-01	3.6889E-01	1.7418E-01	8.4796E-02	1.6615E-02
0.25	9.2371E-01	6.8074E-01	4.7532E-01	2.9737E-01	1.4041E-01	6.8355E-02	1.3393E-02
0.30	7.8338E-01	5.7733E-01	4.0311E-01	2.5219E-01	1.1908E-01	5.7971E-02	1.1359E-02
0.35	6.8581E-01	5.0542E-01	3.5290E-01	2.2078E-01	1.0425E-01	5.0751E-02	9.9440E-03
0.40	6.1349E-01	4.5212E-01	3.1569E-01	1.9750E-01	9.3252E-02	4.5399E-02	8.8953E-03
0.45	5.5739E-01	4.1078E-01	2.8682E-01	1.7944E-01	8.4726E-02	4.1247E-02	8.0820E-03
0.50	5.1239E-01	3.7761E-01	2.6366E-01	1.6495E-01	7.7885E-02	3.7917E-02	7.4294E-03
0.55	4.7533E-01	3.5030E-01	2.4459E-01	1.5302E-01	7.2252E-02	3.5175E-02	6.8921E-03
0.60	4.4418E-01	3.2735E-01	2.2856E-01	1.4299E-01	6.7517E-02	3.2870E-02	6.4404E-03
0.65	4.1755E-01	3.0772E-01	2.1486E-01	1.3442E-01	6.3470E-02	3.0899E-02	6.0544E-03
0.70	3.9448E-01	2.9072E-01	2.0299E-01	1.2699E-01	5.9962E-02	2.9192E-02	5.7197E-03
0.75	3.7424E-01	2.7580E-01	1.9258E-01	1.2048E-01	5.6886E-02	2.7694E-02	5.4264E-03
0.80	3.5633E-01	2.6260E-01	1.8336E-01	1.1471E-01	5.4163E-02	2.6368E-02	5.1666E-03
0.85	3.4033E-01	2.5081E-01	1.7512E-01	1.0956E-01	5.1731E-02	2.5184E-02	4.9346E-03
0.90	3.2593E-01	2.4020E-01	1.6772E-01	1.0493E-01	4.9543E-02	2.4119E-02	4.7259E-03
0.95	3.1290E-01	2.3060E-01	1.6101E-01	1.0073E-01	4.7562E-02	2.3155E-02	4.5369E-03
1.00	3.0103E-01	2.2185E-01	1.5490E-01	9.6910E-02	4.5757E-02	2.2276E-02	4.3648E-03
1.10	2.8018E-01	2.0648E-01	1.4417E-01	9.0197E-02	4.2588E-02	2.0733E-02	4.0625E-03
1.20	2.6241E-01	1.9339E-01	1.3503E-01	8.4477E-02	3.9887E-02	1.9419E-02	3.8048E-03
1.30	2.4705E-01	1.8207E-01	1.2713E-01	7.9534E-02	3.7553E-02	1.8282E-02	3.5822E-03
1.40	2.3362E-01	1.7217E-01	1.2022E-01	7.5210E-02	3.5511E-02	1.7288E-02	3.3874E-03
1.50	2.2176E-01	1.6343E-01	1.1411E-01	7.1390E-02	3.3708E-02	1.6410E-02	3.2154E-03
1.60	2.1118E-01	1.5564E-01	1.0867E-01	6.7986E-02	3.2101E-02	1.5628E-02	3.0621E-03
1.70	2.0169E-01	1.4864E-01	1.0379E-01	6.4931E-02	3.0658E-02	1.4925E-02	2.9245E-03
1.80	1.9312E-01	1.4232E-01	9.9372E-02	6.2169E-02	2.9354E-02	1.4291E-02	2.8001E-03
1.90	1.8532E-01	1.3657E-01	9.5361E-02	5.9660E-02	2.8169E-02	1.3714E-02	2.6871E-03
2.00	1.7820E-01	1.3133E-01	9.1697E-02	5.7368E-02	2.7087E-02	1.3187E-02	2.5838E-03

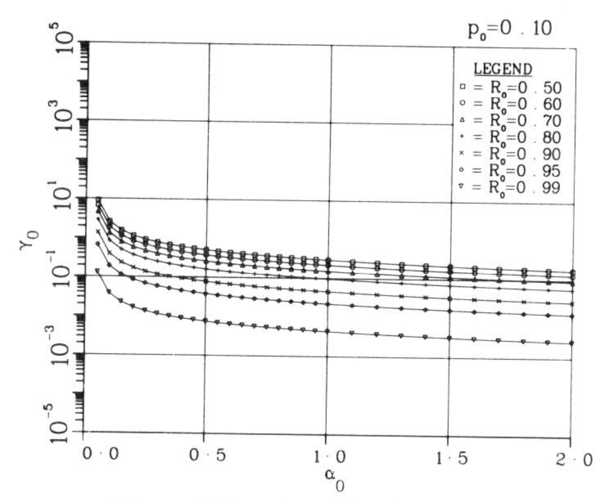

Figure C20. A graph of Table C9.

Table C10. A Table of γ_0 Values for $p_0 = 0.25$ and a Selected Set of Values for α_0 and R_0

$p_0 = 0.25$

α_0/R_0	0.7000	0.8000	0.9000	0.9500	0.9900	0.9990	0.9999
0.05	1.9210E+02	1.2018E+02	5.6745E+01	2.7626E+01	5.4129E+00	5.3885E-01	5.3861E-02
0.10	1.0102E+01	6.3202E+00	2.9842E+00	1.4528E+00	2.8466E-01	2.8338E-02	2.8325E-03
0.15	3.5362E+00	2.2123E+00	1.0446E+00	5.0854E-01	9.9642E-02	9.9192E-03	9.9148E-04
0.20	1.9942E+00	1.2476E+00	5.8907E-01	2.8678E-01	5.6191E-02	5.5938E-03	5.5913E-04
0.25	1.3685E+00	8.5618E-01	4.0426E-01	1.9681E-01	3.8562E-02	3.8388E-03	3.8371E-04
0.30	1.0402E+00	6.5076E-01	3.0726E-01	1.4959E-01	2.9310E-02	2.9178E-03	2.9165E-04
0.35	8.4034E-01	5.2573E-01	2.4823E-01	1.2085E-01	2.3679E-02	2.3572E-03	2.3562E-04
0.40	7.0656E-01	4.4204E-01	2.0871E-01	1.0161E-01	1.9909E-02	1.9819E-03	1.9811E-04
0.45	6.1088E-01	3.8218E-01	1.8045E-01	8.7850E-02	1.7213E-02	1.7136E-03	1.7128E-04
0.50	5.3907E-01	3.3725E-01	1.5924E-01	7.7523E-02	1.5190E-02	1.5121E-03	1.5114E-04
0.55	4.8316E-01	3.0227E-01	1.4272E-01	6.9483E-02	1.3614E-02	1.3553E-03	1.3547E-04
0.60	4.3836E-01	2.7425E-01	1.2949E-01	6.3041E-02	1.2352E-02	1.2296E-03	1.2291E-04
0.65	4.0164E-01	2.5127E-01	1.1864E-01	5.7759E-02	1.1317E-02	1.1266E-03	1.1261E-04
0.70	3.7095E-01	2.3208E-01	1.0958E-01	5.3347E-02	1.0453E-02	1.0406E-03	1.0401E-04
0.75	3.4492E-01	2.1579E-01	1.0189E-01	4.9602E-02	9.7190E-03	9.6752E-04	9.6708E-05
0.80	3.2253E-01	2.0178E-01	9.5273E-02	4.6382E-02	9.0881E-03	9.0471E-04	9.0430E-05
0.85	3.0305E-01	1.8960E-01	8.9521E-02	4.3582E-02	8.5394E-03	8.5009E-04	8.4970E-05
0.90	2.8595E-01	1.7890E-01	8.4469E-02	4.1123E-02	8.0575E-03	8.0211E-04	8.0175E-05
0.95	2.7080E-01	1.6942E-01	7.9994E-02	3.8944E-02	7.6307E-03	7.5963E-04	7.5928E-05
1.00	2.5729E-01	1.6096E-01	7.6002E-02	3.7000E-02	7.2498E-03	7.2171E-04	7.2138E-05
1.10	2.3417E-01	1.4650E-01	6.9174E-02	3.3676E-02	6.5985E-03	6.5687E-04	6.5657E-05
1.20	2.1511E-01	1.3458E-01	6.3542E-02	3.0934E-02	6.0612E-03	6.0339E-04	6.0312E-05
1.30	1.9909E-01	1.2456E-01	5.8811E-02	2.8631E-02	5.6100E-03	5.5847E-04	5.5822E-05
1.40	1.8543E-01	1.1601E-01	5.4776E-02	2.6667E-02	5.2251E-03	5.2015E-04	5.1992E-05
1.50	1.7363E-01	1.0863E-01	5.1291E-02	2.4970E-02	4.8926E-03	4.8706E-04	4.8684E-05
1.60	1.6333E-01	1.0218E-01	4.8248E-02	2.3489E-02	4.6024E-03	4.5816E-04	4.5795E-05
1.70	1.5425E-01	9.6504E-02	4.5566E-02	2.2183E-02	4.3465E-03	4.3269E-04	4.3250E-05
1.80	1.4619E-01	9.1457E-02	4.3183E-02	2.1023E-02	4.1192E-03	4.1006E-04	4.0988E-05
1.90	1.3897E-01	8.6940E-02	4.1050E-02	1.9985E-02	3.9157E-03	3.8981E-04	3.8963E-05
2.00	1.3246E-01	8.2872E-02	3.9129E-02	1.9049E-02	3.7325E-03	3.7157E-04	3.7140E-05

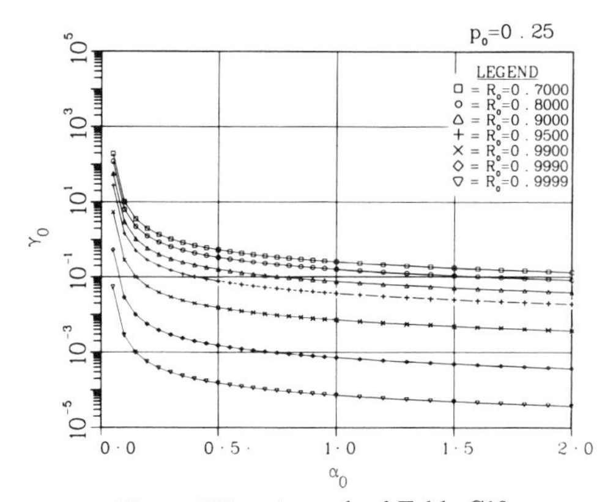

Figure C21. A graph of Table C10.

Table C11. A Table of γ_0 Values for $p_0 = 0.50$ and a Selected Set of Values for γ_0 and R_0[a]

$p_0 = 0.50$

α_0/R_0	0.7000	0.8000	0.9000	0.9500	0.9900	0.9990	0.9999
0.05	6.3990E+05	4.0034E+05	1.8903E+05	9.2024E+04	1.8031E+04	1.7950E+03	1.7942E+02
0.10	6.0108E+02	3.7605E+02	1.7756E+02	8.6441E+01	1.6937E+01	1.6861E+00	1.6853E-01
0.15	5.7207E+01	3.5790E+01	1.6899E+01	8.2269E+00	1.6120E+00	1.6047E-01	1.6040E-02
0.20	1.7192E+01	1.0756E+01	5.0785E+00	2.4724E+00	4.8444E-01	4.8225E-02	4.8204E-03
0.25	8.1668E+00	5.1093E+00	2.4124E+00	1.1745E+00	2.3012E-01	2.2908E-02	2.2898E-03
0.30	4.8772E+00	3.0513E+00	1.4407E+00	7.0139E-01	1.3743E-01	1.3681E-02	1.3675E-03
0.35	3.3219E+00	2.0783E+00	9.8129E-01	4.7773E-01	9.3605E-02	9.3183E-03	9.3141E-04
0.40	2.4585E+00	1.5381E+00	7.2623E-01	3.5356E-01	6.9275E-02	6.8963E-03	6.8932E-04
0.45	1.9246E+00	1.2040E+00	5.6851E-01	2.7677E-01	5.4230E-02	5.3985E-03	5.3961E-04
0.50	1.5680E+00	9.8099E-01	4.6319E-01	2.2550E-01	4.4183E-02	4.3984E-03	4.3964E-04
0.55	1.3160E+00	8.2330E-01	3.8873E-01	1.8925E-01	3.7081E-02	3.6914E-03	3.6897E-04
0.60	1.1298E+00	7.0682E-01	3.3373E-01	1.6247E-01	3.1835E-02	3.1691E-03	3.1677E-04
0.65	9.8741E-01	6.1774E-01	2.9168E-01	1.4200E-01	2.7823E-02	2.7698E-03	2.7685E-04
0.70	8.7544E-01	5.4769E-01	2.5860E-01	1.2590E-01	2.4668E-02	2.4557E-03	2.4546E-04
0.75	7.8534E-01	4.9132E-01	2.3199E-01	1.1294E-01	2.2129E-02	2.2029E-03	2.2019E-04
0.80	7.1143E-01	4.4508E-01	2.1015E-01	1.0231E-01	2.0046E-02	1.9956E-03	1.9947E-04
0.85	6.4980E-01	4.0653E-01	1.9195E-01	9.3448E-02	1.8310E-02	1.8227E-03	1.8219E-04
0.90	5.9770E-01	3.7394E-01	1.7656E-01	8.5955E-02	1.6842E-02	1.6766E-03	1.6758E-04
0.95	5.5312E-01	3.4605E-01	1.6339E-01	7.9544E-02	1.5586E-02	1.5515E-03	1.5509E-04
1.00	5.1457E-01	3.2193E-01	1.5200E-01	7.4001E-02	1.4500E-02	1.4434E-03	1.4428E-04
1.10	4.5133E-01	2.8236E-01	1.3332E-01	6.4906E-02	1.2718E-02	1.2660E-03	1.2654E-04
1.20	4.0169E-01	2.5131E-01	1.1866E-01	5.7767E-02	1.1319E-02	1.1268E-03	1.1263E-04
1.30	3.6174E-01	2.2631E-01	1.0686E-01	5.2022E-02	1.0193E-02	1.0147E-03	1.0143E-04
1.40	3.2892E-01	2.0578E-01	9.7163E-02	4.7302E-02	9.2684E-03	9.2265E-04	9.2224E-05
1.50	3.0150E-01	1.8863E-01	8.9063E-02	4.3359E-02	8.4957E-03	8.4574E-04	8.4536E-05
1.60	2.7826E-01	1.7409E-01	8.2198E-02	4.0017E-02	7.8408E-03	7.8055E-04	7.8019E-05
1.70	2.5832E-01	1.6161E-01	7.6306E-02	3.7149E-02	7.2788E-03	7.2460E-04	7.2427E-05
1.80	2.4102E-01	1.5079E-01	7.1196E-02	3.4661E-02	6.7914E-03	6.7608E-04	6.7577E-05
1.90	2.2588E-01	1.4131E-01	6.6724E-02	3.2483E-02	6.3648E-03	6.3361E-04	6.3332E-05
2.00	2.1252E-01	1.3295E-01	6.2776E-02	3.0562E-02	5.9882E-03	5.9612E-04	5.9585E-05

[a] Reprinted with permission of Society for Industrial and Applied Mathematics, from *Nuclear Systems Reliability Engineering and Risk Assessment*, J. B. Fussell and G. R. Burdick, Ed.

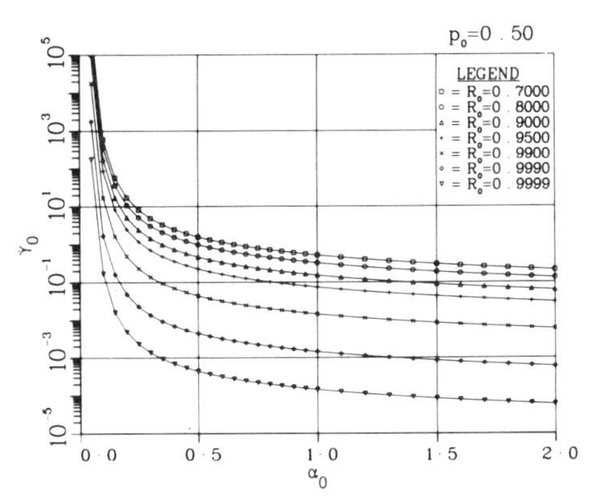

Figure C22. A graph of Table C11 [Reprinted with permission of Society for Industrial and Applied Mathematics from *Nuclear Systems Reliability Engineering and Risk Assessment*, J. B. Fussell and G. R. Burdick, Eds. ©1977 Society for Industrial and Applied Mathematics, Philadelphia. All rights reserved.].

Table C12. A Table of γ_0 Values for $p_0 = 0.75$ and a Selected Set of Values for α_0 and R_0

$p_0 = 0.75$

α_0/R_0	0.7000	0.8000	0.9000	0.9500	0.9900	0.9990	0.9999
0.05	6.7099E+11	4.1979E+11	1.9821E+11	9.6495E+10	1.8907E+10	1.8822E+09	1.8813E+08
0.10	6.1584E+05	3.8528E+05	1.8192E+05	8.8563E+04	1.7353E+04	1.7275E+03	1.7267E+02
0.15	5.8431E+03	3.6556E+03	1.7260E+03	8.4030E+02	1.6465E+02	1.6390E+01	1.6383E+00
0.20	5.5941E+02	3.4998E+02	1.6525E+02	8.0448E+01	1.5763E+01	1.5692E+00	1.5685E-01
0.25	1.3499E+02	8.4454E+01	3.9876E+01	1.9413E+01	3.8038E+00	3.7867E-01	3.7849E-02
0.30	5.1693E+01	3.2341E+01	1.5270E+01	7.4340E+00	1.4566E+00	1.4500E-01	1.4494E-02
0.35	2.5763E+01	1.6118E+01	7.6103E+00	3.7050E+00	7.2595E-01	7.2267E-02	7.2235E-03
0.40	1.5136E+01	9.4693E+00	4.4711E+00	2.1767E+00	4.2650E-01	4.2457E-02	4.2438E-03
0.45	9.9233E+00	6.2082E+00	2.9313E+00	1.4271E+00	2.7962E-01	2.7836E-02	2.7823E-03
0.50	7.0259E+00	4.3956E+00	2.0754E+00	1.0104E+00	1.9798E-01	1.9708E-02	1.9699E-03
0.55	5.2616E+00	3.2918E+00	1.5543E+00	7.5667E-01	1.4826E-01	1.4759E-02	1.4753E-03
0.60	4.1105E+00	2.5716E+00	1.2142E+00	5.9113E-01	1.1583E-01	1.1530E-02	1.1525E-03
0.65	3.3180E+00	2.0758E+00	9.8013E-01	4.7717E-01	9.3495E-02	9.3073E-03	9.3031E-04
0.70	2.7486E+00	1.7196E+00	8.1193E-01	3.9528E-01	7.7450E-02	7.7101E-03	7.7066E-04
0.75	2.3250E+00	1.4546E+00	6.8680E-01	3.3436E-01	6.5514E-02	6.5218E-03	6.5189E-04
0.80	2.0007E+00	1.2517E+00	5.9100E-01	2.8772E-01	5.6375E-02	5.6121E-03	5.6096E-04
0.85	1.7464E+00	1.0926E+00	5.1587E-01	2.5114E-01	4.9208E-02	4.8986E-03	4.8964E-04
0.90	1.5428E+00	9.6519E-01	4.5573E-01	2.2187E-01	4.3472E-02	4.3276E-03	4.3256E-04
0.95	1.3769E+00	8.6145E-01	4.0675E-01	1.9802E-01	3.8799E-02	3.8624E-03	3.8607E-04
1.00	1.2398E+00	7.7566E-01	3.6624E-01	1.7830E-01	3.4936E-02	3.4778E-03	3.4762E-04
1.10	1.0275E+00	6.4284E-01	3.0353E-01	1.4777E-01	2.8953E-02	2.8823E-03	2.8810E-04
1.20	8.7199E-01	5.4554E-01	2.5758E-01	1.2540E-01	2.4571E-02	2.4460E-03	2.4449E-04
1.30	7.5400E-01	4.7172E-01	2.2273E-01	1.0843E-01	2.1246E-02	2.1150E-03	2.1141E-04
1.40	6.6190E-01	4.1410E-01	1.9552E-01	9.5187E-02	1.8651E-02	1.8567E-03	1.8558E-04
1.50	5.8831E-01	3.6806E-01	1.7379E-01	8.4605E-02	1.6577E-02	1.6503E-03	1.6495E-04
1.60	5.2837E-01	3.3056E-01	1.5608E-01	7.5984E-02	1.4888E-02	1.4821E-03	1.4814E-04
1.70	4.7872E-01	2.9950E-01	1.4141E-01	6.8844E-02	1.3489E-02	1.3428E-03	1.3422E-04
1.80	4.3701E-01	2.7340E-01	1.2909E-01	6.2846E-02	1.2314E-02	1.2258E-03	1.2253E-04
1.90	4.0154E-01	2.5121E-01	1.1861E-01	5.7745E-02	1.1314E-02	1.1263E-03	1.1258E-04
2.00	3.7104E-01	2.3213E-01	1.0960E-01	5.3359E-02	1.0455E-02	1.0408E-03	1.0403E-04

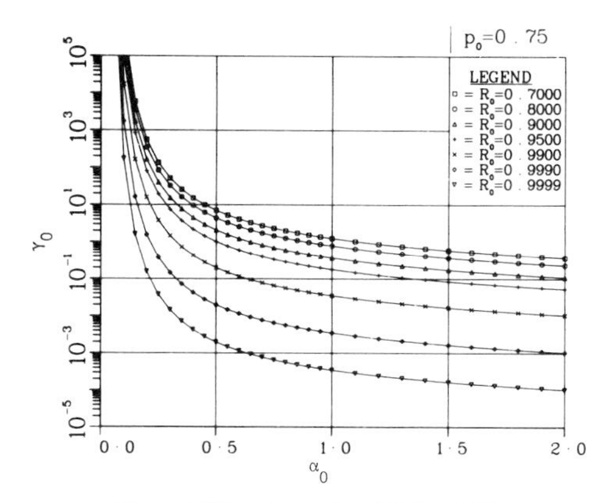

Figure C23. A graph of Table C12.

Table C13. A Table of γ_0 Values for $p_0 = 0.90$ and a Selected Set of Values for α_0 and R_0

$p_0 = 0.90$

α_0/R_0	0.900000	0.950000	0.990000	0.999000	0.999900	0.999990	0.999999
0.05	1.8027E+19	8.7761E+18	1.7196E+18	1.7118E+17	1.7111E+16	1.7110E+15	1.7110E+14
0.10	1.7349E+09	8.4461E+08	1.6549E+08	1.6474E+07	1.6467E+06	1.6466E+05	1.6466E+04
0.15	7.7626E+05	3.7791E+05	7.4047E+04	7.3713E+03	7.3680E+02	7.3677E+01	7.3676E+00
0.20	1.6146E+04	7.8604E+03	1.5402E+03	1.5332E+02	1.5325E+01	1.5325E+00	1.5324E-01
0.25	1.5609E+03	7.5989E+02	1.4889E+02	1.4822E+01	1.4815E+00	1.4815E-01	1.4815E-02
0.30	3.2546E+02	1.5845E+02	3.1046E+01	3.0906E+00	3.0892E-01	3.0891E-02	3.0890E-03
0.35	1.0532E+02	5.1272E+01	1.0046E+01	1.0001E+00	9.9963E-02	9.9958E-03	9.9958E-04
0.40	4.4856E+01	2.1837E+01	4.2788E+00	4.2595E-01	4.2576E-02	4.2574E-03	4.2574E-04
0.45	2.2946E+01	1.1171E+01	2.1888E+00	2.1790E-01	2.1780E-02	2.1779E-03	2.1779E-04
0.50	1.3345E+01	6.4966E+00	1.2729E+00	1.2672E-01	1.2666E-02	1.2666E-03	1.2666E-04
0.55	8.5197E+00	4.1477E+00	8.1269E-01	8.0903E-02	8.0866E-03	8.0863E-04	8.0862E-05
0.60	5.8338E+00	2.8401E+00	5.5648E-01	5.5397E-02	5.5372E-03	5.5370E-04	5.5370E-05
0.65	4.2159E+00	2.0524E+00	4.0215E-01	4.0034E-02	4.0016E-03	4.0014E-04	4.0014E-05
0.70	3.1787E+00	1.5475E+00	3.0322E-01	3.0185E-02	3.0172E-03	3.0170E-04	3.0170E-05
0.75	2.4796E+00	1.2072E+00	2.3653E-01	2.3546E-02	2.3536E-03	2.3535E-04	2.3535E-05
0.80	1.9886E+00	9.6813E-01	1.8969E-01	1.8884E-02	1.8875E-03	1.8874E-04	1.8874E-05
0.85	1.6318E+00	7.9440E-01	1.5565E-01	1.5495E-02	1.5488E-03	1.5487E-04	1.5487E-05
0.90	1.3648E+00	6.6445E-01	1.3019E-01	1.2960E-02	1.2955E-03	1.2954E-04	1.2954E-05
0.95	1.1602E+00	5.6483E-01	1.1067E-01	1.1017E-02	1.1012E-03	1.1012E-04	1.1012E-05
1.00	1.0000E+00	4.8684E-01	9.5390E-02	9.4960E-03	9.4917E-04	9.4913E-05	9.4912E-06
1.10	7.6875E-01	3.7425E-01	7.3331E-02	7.3000E-03	7.2967E-04	7.2964E-05	7.2964E-06
1.20	6.1292E-01	2.9839E-01	5.8467E-02	5.8203E-03	5.8177E-04	5.8174E-05	5.8174E-06
1.30	5.0278E-01	2.4477E-01	4.7960E-02	4.7744E-03	4.7722E-04	4.7720E-05	4.7720E-06
1.40	4.2189E-01	2.0539E-01	4.0244E-02	4.0062E-03	4.0044E-04	4.0042E-05	4.0042E-06
1.50	3.6059E-01	1.7555E-01	3.4397E-02	3.4242E-03	3.4226E-04	3.4225E-05	3.4225E-06
1.60	3.1293E-01	1.5235E-01	2.9850E-02	2.9716E-03	2.9702E-04	2.9701E-05	2.9701E-06
1.70	2.7505E-01	1.3390E-01	2.6237E-02	2.6119E-03	2.6107E-04	2.6106E-05	2.6106E-06
1.80	2.4438E-01	1.1898E-01	2.3312E-02	2.3207E-03	2.3196E-04	2.3195E-05	2.3195E-06
1.90	2.1916E-01	1.0669E-01	2.0905E-02	2.0811E-03	2.0802E-04	2.0801E-05	2.0801E-06
2.00	1.9812E-01	9.6450E-02	1.8898E-02	1.8813E-03	1.8805E-04	1.8804E-05	1.8804E-06

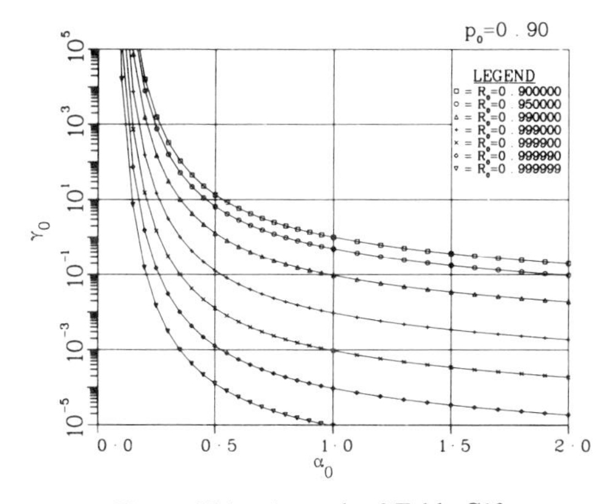

Figure C24. A graph of Table C13.

Table C14. A Table of γ_0 Values for $p_0 = 0.95$ and a Selected Set of Values for α_0 and R_0[a]

$p_0 = 0.95$

α_0/R_0	0.900000	0.950000	0.990000	0.999000	0.999900	0.999990	0.999999
0.05	1.8903E+25	9.2024E+24	1.8031E+24	1.7950E+23	1.7942E+22	1.7941E+21	1.7941E+20
0.10	1.7765E+12	8.6488E+11	1.6946E+11	1.6870E+10	1.6862E+09	1.6861E+08	1.6861E+07
0.15	7.8863E+07	3.8393E+07	7.5227E+06	7.4888E+05	7.4854E+04	7.4851E+03	7.4851E+02
0.20	5.1667E+05	2.5153E+05	4.9285E+04	4.9063E+03	4.9041E+02	4.9039E+01	4.9039E+00
0.25	2.4975E+04	1.2159E+04	2.3824E+03	2.3717E+02	2.3706E+01	2.3705E+00	2.3705E-01
0.30	3.2812E+03	1.5974E+03	3.1299E+02	3.1158E+01	3.1144E+00	3.1143E-01	3.1143E-02
0.35	7.6359E+02	3.7174E+02	7.2838E+01	7.2510E+00	7.2477E-01	7.2474E-02	7.2474E-03
0.40	2.5409E+02	1.2370E+02	2.4238E+01	2.4129E+00	2.4118E-01	2.4117E-02	2.4117E-03
0.45	1.0734E+02	5.2255E+01	1.0239E+01	1.0193E+00	1.0188E-01	1.0188E-02	1.0187E-03
0.50	5.3589E+01	2.6089E+01	5.1119E+00	5.0888E-01	5.0865E-02	5.0863E-03	5.0863E-04
0.55	3.0216E+01	1.4710E+01	2.8823E+00	2.8693E-01	2.8680E-02	2.8679E-03	2.8678E-04
0.60	1.8665E+01	9.0868E+00	1.7804E+00	1.7724E-01	1.7716E-02	1.7715E-03	1.7715E-04
0.65	1.2369E+01	6.0216E+00	1.1799E+00	1.1746E-01	1.1740E-02	1.1740E-03	1.1740E-04
0.70	8.6622E+00	4.2171E+00	8.2629E-01	8.2256E-02	8.2219E-03	8.2216E-04	8.2215E-05
0.75	6.3408E+00	3.0869E+00	6.0484E-01	6.0212E-02	6.0185E-03	6.0182E-04	6.0182E-05
0.80	4.8115E+00	2.3424E+00	4.5897E-01	4.5690E-02	4.5670E-03	4.5667E-04	4.5667E-05
0.85	3.7611E+00	1.8310E+00	3.5877E-01	3.5715E-02	3.5699E-03	3.5698E-04	3.5698E-05
0.90	3.0138E+00	1.4672E+00	2.8749E-01	2.8619E-02	2.8606E-03	2.8605E-04	2.8605E-05
0.95	2.4660E+00	1.2005E+00	2.3523E-01	2.3417E-02	2.3407E-03	2.3406E-04	2.3405E-05
1.00	2.0541E+00	1.0000E+00	1.9594E-01	1.9505E-02	1.9497E-03	1.9496E-04	1.9496E-05
1.10	1.4892E+00	7.2498E-01	1.4205E-01	1.4141E-02	1.4135E-03	1.4134E-04	1.4134E-05
1.20	1.1311E+00	5.5068E-01	1.0790E-01	1.0741E-02	1.0736E-03	1.0736E-04	1.0736E-05
1.30	8.9084E-01	4.3370E-01	8.4978E-02	8.4594E-03	8.4556E-04	8.4552E-05	8.4552E-06
1.40	7.2201E-01	3.5150E-01	6.8873E-02	6.8562E-03	6.8531E-04	6.8528E-05	6.8528E-06
1.50	5.9890E-01	2.9157E-01	5.7129E-02	5.6871E-03	5.6846E-04	5.6843E-05	5.6843E-06
1.60	5.0634E-01	2.4650E-01	4.8299E-02	4.8082E-03	4.8060E-04	4.8058E-05	4.8058E-06
1.70	4.3493E-01	2.1174E-01	4.1488E-02	4.1301E-03	4.1282E-04	4.1280E-05	4.1280E-06
1.80	3.7863E-01	1.8433E-01	3.6117E-02	3.5955E-03	3.5938E-04	3.5937E-05	3.5937E-06
1.90	3.3341E-01	1.6231E-01	3.1804E-02	3.1660E-03	3.1646E-04	3.1644E-05	3.1644E-06
2.00	2.9649E-01	1.4434E-01	2.8282E-02	2.8154E-03	2.8142E-04	2.8140E-05	2.8140E-06

[a]Reprinted with permission of Society for Industrial and Applied Mathematics, from *Nuclear Systems Reliability Engineering and Risk Assessment*, J. B. Fussell and G. R. Burdick, Ed.

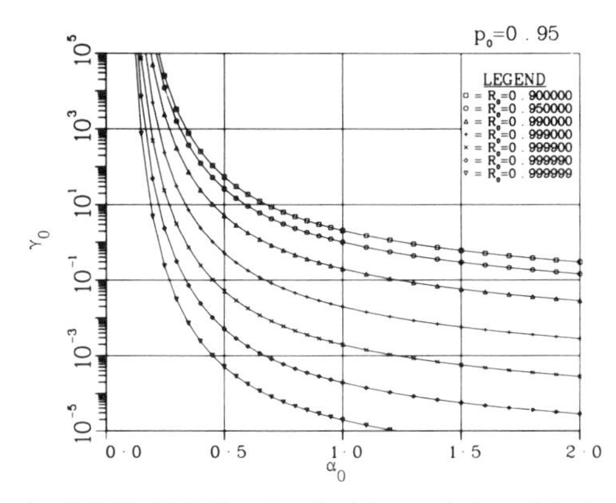

Figure C25. A graph of Table C14 [Reprinted with permission of Society for Industrial and Applied Mathematics from *Nuclear Systems Reliability Engineering and Risk Assessment*, J. B. Fussell and G. R. Burdick, Eds.].

Table C15. A Table of γ_0 Values for $p_0 = 0.975$ and a Selected Set of Values for α_0 and R_0

$p_0 = 0.975$

α_0/R_0	0.900000	0.950000	0.990000	0.999000	0.999900	0.999990	0.999999
0.05	1.9821E+31	9.6495E+30	1.8907E+30	1.8822E+29	1.8813E+28	1.8812E+27	1.8812E+26
0.10	1.8192E+15	8.8563E+14	1.7353E+14	1.7275E+13	1.7267E+12	1.7266E+11	1.7266E+10
0.15	8.0120E+09	3.9005E+09	7.6426E+08	7.6082E+07	7.6047E+06	7.6044E+05	7.6044E+04
0.20	1.6534E+07	8.0491E+06	1.5771E+06	1.5700E+05	1.5693E+04	1.5692E+03	1.5692E+02
0.25	3.9961E+05	1.9454E+05	3.8119E+04	3.7947E+03	3.7930E+02	3.7928E+01	3.7928E+00
0.30	3.3073E+04	1.6101E+04	3.1548E+03	3.1406E+02	3.1392E+01	3.1391E+00	3.1390E-01
0.35	5.5333E+03	2.6938E+03	5.2782E+02	5.2544E+01	5.2520E+00	5.2518E-01	5.2518E-02
0.40	1.4377E+03	6.9993E+02	1.3714E+02	1.3653E+01	1.3646E+00	1.3646E-01	1.3646E-02
0.45	5.0111E+02	2.4396E+02	4.7801E+01	4.7585E+00	4.7564E-01	4.7561E-02	4.7561E-03
0.50	2.1457E+02	1.0446E+02	2.0468E+01	2.0375E+00	2.0366E-01	2.0365E-02	2.0365E-03
0.55	1.0672E+02	5.1957E+01	1.0180E+01	1.0134E+00	1.0130E-01	1.0129E-02	1.0129E-03
0.60	5.9401E+01	2.8919E+01	5.6663E+00	5.6407E-01	5.6382E-02	5.6379E-03	5.6379E-04
0.65	3.6052E+01	1.7551E+01	3.4390E+00	3.4235E-01	3.4219E-02	3.4218E-03	3.4218E-04
0.70	2.3422E+01	1.1403E+01	2.2342E+00	2.2242E-01	2.2232E-02	2.2231E-03	2.2231E-04
0.75	1.6070E+01	7.8233E+00	1.5329E+00	1.5260E-01	1.5253E-02	1.5252E-03	1.5252E-04
0.80	1.1525E+01	5.6108E+00	1.0994E+00	1.0944E-01	1.0939E-02	1.0939E-03	1.0939E-04
0.85	8.5733E+00	4.1738E+00	8.1780E-01	8.1411E-02	8.1375E-03	8.1371E-04	8.13[illegible]1E-05
0.90	6.5750E+00	3.2010E+00	6.2719E-01	6.2436E-02	6.2408E-03	6.2406E-04	6.2405E-05
0.95	5.1739E+00	2.5189E+00	4.9354E-01	4.9132E-02	4.9109E-03	4.9107E-04	4.9107E-05
1.00	4.1615E+00	2.0260E+00	3.9697E-01	3.9518E-02	3.9500E-03	3.9498E-04	3.9498E-05
1.10	2.8412E+00	1.3832E+00	2.7103E-01	2.6980E-02	2.6968E-03	2.6967E-04	2.6967E-05
1.20	2.0538E+00	9.9985E-01	1.9591E-01	1.9503E-02	1.9494E-03	1.9493E-04	1.9493E-05
1.30	1.5515E+00	7.5533E-01	1.4800E-01	1.4733E-02	1.4727E-03	1.4726E-04	1.4726E-05
1.40	1.2137E+00	5.9087E-01	1.1577E-01	1.1525E-02	1.1520E-03	1.1520E-04	1.1519E-05
1.50	9.7649E-01	4.7539E-01	9.3147E-02	9.2727E-03	9.2685E-04	9.2681E-05	9.2680E-06
1.60	8.0392E-01	3.9138E-01	7.6686E-02	7.6340E-03	7.6306E-04	7.6302E-05	7.6302E-06
1.70	6.7462E-01	3.2843E-01	6.4352E-02	6.4062E-03	6.4033E-04	6.4030E-05	6.4030E-06
1.80	5.7529E-01	2.8007E-01	5.4876E-02	5.4629E-03	5.4604E-04	5.4602E-05	5.4602E-06
1.90	4.9732E-01	2.4211E-01	4.7440E-02	4.7226E-03	4.7204E-04	4.7202E-05	4.7202E-06
2.00	4.3500E-01	2.1177E-01	4.1494E-02	4.1307E-03	4.1289E-04	4.1287E-05	4.1287E-06

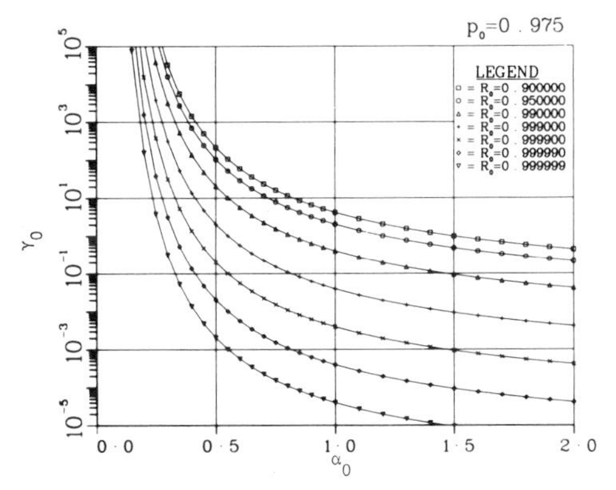

Figure C26. A graph of Table C15.

Table C16. A Table of γ_0 Values for $p_0 = 0.99$ and a Selected Set of Values for α_0 and R_0

$p_0 = 0.99$

α_0/R_0	0.900000	0.950000	0.990000	0.999000	0.999900	0.999990	0.999999
0.05	1.8027E+39	8.7761E+38	1.7196E+38	1.7118E+37	1.7111E+36	1.7110E+35	1.7110E+34
0.10	1.7349E+19	8.4461E+18	1.6549E+18	1.6474E+17	1.6467E+16	1.6466E+15	1.6466E+14
0.15	3.6031E+12	1.7541E+12	3.4370E+11	3.4215E+10	3.4199E+09	3.4198E+08	3.4198E+07
0.20	1.6146E+09	7.8605E+08	1.5402E+08	1.5332E+07	1.5325E+06	1.5325E+05	1.5325E+04
0.25	1.5610E+07	7.5993E+06	1.4890E+06	1.4823E+05	1.4816E+04	1.4816E+03	1.4815E+02
0.30	7.0136E+05	3.4145E+05	6.6903E+04	6.6601E+03	6.6571E+02	6.6568E+01	6.6568E+00
0.35	7.5851E+04	3.6927E+04	7.2354E+03	7.2027E+02	7.1995E+01	7.1992E+00	7.1991E-01
0.40	1.4208E+04	6.9171E+03	1.3553E+03	1.3492E+02	1.3486E+01	1.3486E+00	1.3485E-01
0.45	3.8397E+03	1.8693E+03	3.6627E+02	3.6462E+01	3.6445E+00	3.6444E-01	3.6443E-02
0.50	1.3414E+03	6.5305E+02	1.2796E+02	1.2738E+01	1.2732E+00	1.2732E-01	1.2732E-02
0.55	5.6495E+02	2.7504E+02	5.3891E+01	5.3648E+00	5.3624E-01	5.3621E-02	5.3621E-03
0.60	2.7378E+02	1.3329E+02	2.6116E+01	2.5998E+00	2.5986E-01	2.5985E-02	2.5985E-03
0.65	1.4782E+02	7.1962E+01	1.4100E+01	1.4037E+00	1.4030E-01	1.4030E-02	1.4030E-03
0.70	8.6886E+01	4.2299E+01	8.2881E+00	8.2507E-01	8.2470E-02	8.2466E-03	8.2466E-04
0.75	5.4669E+01	2.6615E+01	5.2149E+00	5.1913E-01	5.1890E-02	5.1888E-03	5.1887E-04
0.80	3.6355E+01	1.7699E+01	3.4679E+00	3.4523E-01	3.4507E-02	3.4506E-03	3.4506E-04
0.85	2.5306E+01	1.2320E+01	2.4139E+00	2.4030E-01	2.4019E-02	2.4018E-03	2.4018E-04
0.90	1.8298E+01	8.9080E+00	1.7454E+00	1.7375E-01	1.7368E-02	1.7367E-03	1.7367E-04
0.95	1.3662E+01	6.6512E+00	1.3032E+00	1.2973E-01	1.2968E-02	1.2967E-03	1.2967E-04
1.00	1.0483E+01	5.1036E+00	1.0000E+00	9.9549E-02	9.9504E-03	9.9500E-04	9.9499E-05
1.10	6.6012E+00	3.2137E+00	6.2969E-01	6.2685E-02	6.2657E-03	6.2654E-04	6.2654E-05
1.20	4.4628E+00	2.1726E+00	4.2570E-01	4.2378E-02	4.2359E-03	4.2357E-04	4.2357E-05
1.30	3.1871E+00	1.5516E+00	3.0402E-01	3.0265E-02	3.0251E-03	3.0250E-04	3.0250E-05
1.40	2.3767E+00	1.1571E+00	2.2671E-01	2.2569E-02	2.2559E-03	2.2558E-04	2.2558E-05
1.50	1.8350E+00	8.9336E-01	1.7504E-01	1.7425E-02	1.7418E-03	1.7417E-04	1.7417E-05
1.60	1.4576E+00	7.0961E-01	1.3904E-01	1.3841E-02	1.3835E-03	1.3834E-04	1.3834E-05
1.70	1.1853E+00	5.7705E-01	1.1307E-01	1.1256E-02	1.1251E-03	1.1250E-04	1.1250E-05
1.80	9.8307E-01	4.7859E-01	9.3775E-02	9.3352E-03	9.3310E-04	9.3305E-05	9.3305E-06
1.90	8.2906E-01	4.0361E-01	7.9084E-02	7.8727E-03	7.8691E-04	7.8688E-05	7.8688E-06
2.00	7.0924E-01	3.4528E-01	6.7654E-02	6.7349E-03	6.7319E-04	6.7316E-05	6.7315E-06

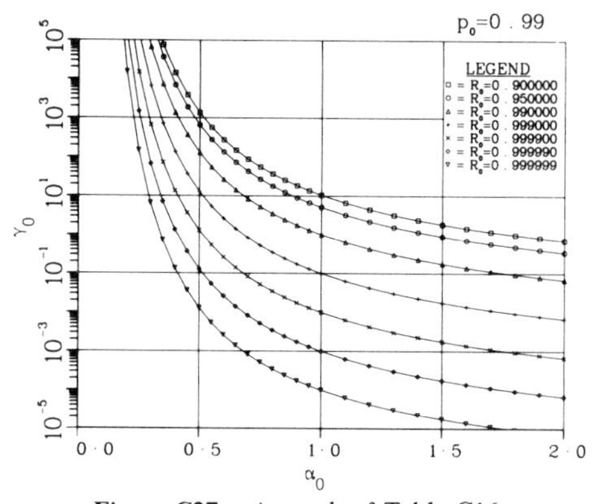

Figure C27. A graph of Table C16.

Table C17. A Table of γ_0 Values for $p_0 = 0.995$ and a Selected Set of Values for α_0 and R_0

$p_0 = 0.995$

α_0/R_0	0.900000	0.950000	0.990000	0.999000	0.999900	0.999990	0.999999
0.05	1.8903E+45	9.2024E+44	1.8031E+44	1.7950E+43	1.7942E+42	1.7941E+41	1.7941E+40
0.10	1.7765E+22	8.6488E+21	1.6946E+21	1.6870E+20	1.6862E+19	1.6861E+18	1.6861E+17
0.15	3.6605E+14	1.7821E+14	3.4917E+13	3.4760E+12	3.4744E+11	3.4743E+10	3.4743E+09
0.20	5.1667E+10	2.5153E+10	4.9285E+09	4.9063E+08	4.9041E+07	4.9039E+06	4.9039E+05
0.25	2.4975E+08	1.2159E+08	2.3824E+07	2.3717E+06	2.3706E+05	2.3705E+04	2.3705E+03
0.30	7.0693E+06	3.4416E+06	6.7434E+05	6.7130E+04	6.7100E+03	6.7097E+02	6.7096E+01
0.35	5.4960E+05	2.6756E+05	5.2426E+04	5.2190E+03	5.2166E+02	5.2164E+01	5.2164E+00
0.40	8.0375E+04	3.9129E+04	7.6670E+03	7.6324E+02	7.6289E+01	7.6286E+00	7.6286E-01
0.45	1.7917E+04	8.7225E+03	1.7091E+03	1.7014E+02	1.7006E+01	1.7005E+00	1.7005E-01
0.50	5.3659E+03	2.6123E+03	5.1185E+02	5.0954E+01	5.0931E+00	5.0929E-01	5.0929E-02
0.55	1.9924E+03	9.6997E+02	1.9006E+02	1.8920E+01	1.8911E+00	1.8910E-01	1.8910E-02
0.60	8.6934E+02	4.2323E+02	8.2927E+01	8.2553E+00	8.2515E-01	8.2512E-02	8.2511E-03
0.65	4.2951E+02	2.0910E+02	4.0971E+01	4.0786E+00	4.0767E-01	4.0766E-02	4.0765E-03
0.70	2.3399E+02	1.1391E+02	2.2320E+01	2.2219E+00	2.2209E-01	2.2208E-02	2.2208E-03
0.75	1.3785E+02	6.7110E+01	1.3149E+01	1.3090E+00	1.3084E-01	1.3084E-02	1.3083E-03
0.80	8.6549E+01	4.2135E+01	8.2559E+00	8.2186E-01	8.2149E-02	8.2146E-03	8.2145E-04
0.85	5.7268E+01	2.7880E+01	5.4628E+00	5.4382E-01	5.4357E-02	5.4355E-03	5.4355E-04
0.90	3.9590E+01	1.9274E+01	3.7765E+00	3.7594E-01	3.7577E-02	3.7576E-03	3.7575E-04
0.95	2.8397E+01	1.3825E+01	2.7088E+00	2.6966E-01	2.6954E-02	2.6953E-03	2.6953E-04
1.00	2.1019E+01	1.0233E+01	2.0050E+00	1.9960E-01	1.9951E-02	1.9950E-03	1.9950E-04
1.10	1.2440E+01	6.0564E+00	1.1867E+00	1.1813E-01	1.1808E-02	1.1807E-03	1.1807E-04
1.20	7.9894E+00	3.8895E+00	7.6211E-01	7.5867E-02	7.5833E-03	7.5830E-04	7.5829E-05
1.30	5.4646E+00	2.6603E+00	5.2126E-01	5.1891E-02	5.1868E-03	5.1866E-04	5.1865E-05
1.40	3.9278E+00	1.9122E+00	3.7468E-01	3.7299E-02	3.7282E-03	3.7280E-04	3.7280E-05
1.50	2.9380E+00	1.4303E+00	2.8026E-01	2.7899E-02	2.7887E-03	2.7886E-04	2.7886E-05
1.60	2.2703E+00	1.1052E+00	2.1656E-01	2.1558E-02	2.1549E-03	2.1548E-04	2.1548E-05
1.70	1.8020E+00	8.7730E-01	1.7190E-01	1.7112E-02	1.7104E-03	1.7104E-04	1.7104E-05
1.80	1.4629E+00	7.1221E-01	1.3955E-01	1.3892E-02	1.3886E-03	1.3885E-04	1.3885E-05
1.90	1.2105E+00	5.8931E-01	1.1547E-01	1.1495E-02	1.1490E-03	1.1489E-04	1.1489E-05
2.00	1.0180E+00	4.9561E-01	9.7110E-02	9.6672E-03	9.6628E-04	9.6624E-05	9.6623E-06

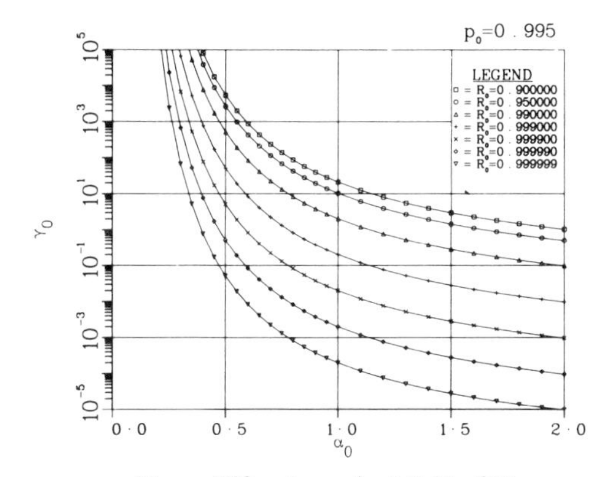

Figure C28. A graph of Table C17.

Author Index

Subject Index

Applied Probability and Statistics (Continued)

DANIEL and WOOD • Fitting Equations to Data: Computer Analysis of Multifactor Data, *Second Edition*
DAVID • Order Statistics, *Second Edition*
DEMING • Sample Design in Business Research
DODGE and ROMIG • Sampling Inspection Tables, *Second Edition*
DRAPER and SMITH • Applied Regression Analysis, *Second Edition*
DUNN • Basic Statistics: A Primer for the Biomedical Sciences, *Second Edition*
DUNN and CLARK • Applied Statistics: Analysis of Variance and Regression
ELANDT-JOHNSON • Probability Models and Statistical Methods in Genetics
ELANDT-JOHNSON and JOHNSON • Survival Models and Data Analysis
FLEISS • Statistical Methods for Rates and Proportions, *Second Edition*
FRANKEN, KÖNIG, ARNDT, and SCHMIDT • Queues and Point Processes
GALAMBOS • The Asymptotic Theory of Extreme Order Statistics
GIBBONS, OLKIN, and SOBEL • Selecting and Ordering Populations: A New Statistical Methodology
GNANADESIKAN • Methods for Statistical Data Analysis of Multivariate Observations
GOLDBERGER • Econometric Theory
GOLDSTEIN and DILLON • Discrete Discriminant Analysis
GROSS and CLARK • Survival Distributions: Reliability Applications in the Biomedical Sciences
GROSS and HARRIS • Fundamentals of Queueing Theory
GUPTA and PANCHAPAKESAN • Multiple Decision Procedures: Theory and Methodology of Selecting and Ranking Populations
GUTTMAN, WILKS, and HUNTER • Introductory Engineering Statistics, *Third Edition*
HAHN and SHAPIRO • Statistical Models in Engineering
HALD • Statistical Tables and Formulas
HALD • Statistical Theory with Engineering Applications
HAND • Discrimination and Classification
HARTIGAN • Clustering Algorithms
HILDEBRAND, LAING, and ROSENTHAL • Prediction Analysis of Cross Classifications
HOEL • Elementary Statistics, *Fourth Edition*
HOEL and JESSEN • Basic Statistics for Business and Economics, *Third Edition*
HOLLANDER and WOLFE • Nonparametric Statistical Methods
JAGERS • Branching Processes with Biological Applications
JESSEN • Statistical Survey Techniques
JOHNSON and KOTZ • Distributions in Statistics
Discrete Distributions
Continuous Univariate Distributions—1
Continuous Univariate Distributions—2
Continuous Multivariate Distributions
JOHNSON and KOTZ • Urn Models and Their Application: An Approach to Modern Discrete Probability Theory
JOHNSON and LEONE • Statistics and Experimental Design in Engineering and the Physical Sciences, Volumes I and II, *Second Edition*
JUDGE, HILL, GRIFFITHS, LÜTKEPOHL and LEE • Introduction to the Theory and Practice of Econometrics